CHILTON'S ROAD REPORT®

Daniel P. HERAUD

Chilton Book Company
Radnor, Pennsylvania

Factual information pertaining to the automobiles listed herein, specifically information under the headings "ENGINE," "TRANSMISSION," "PERFORMANCE," and "SPECIFICATIONS AND PRICES," has been supplied by the manufacturers of such automobiles and the author, publisher, and distributor of this book can accept no responsibility for the accuracy or completeness of such factual information. The commentary and recommendations contained herein are made to the best of the author's knowledge. Author, publisher and distributor accept no responsibility for any loss or damage caused by the use of any of the materials contained in this book, and they explicitly disclaim any and all guarantees.

5757, Cypihot, Saint-Laurent (Québec) Canada H4S 1X4

Published in Radnor, Pennsylvania 19089, by Chilton Book Company.

ISBN: 0-8019-8634-6
ISSN:1072-5377

Designed by Daniel P. Heraud
Printed and bound in Canada

1234567890ML987654 1234567890 4321098765
2678 MM14

SUMMARY

FOREWORD

Dear Readers:
I would like to thank you in my own name and for the Road Report team for having made this annual book a bestseller in North America. I appreciate your expressions of interest and want to thank all the readers who have sent to us, from every corner of the continent, their answers about the cars they drive, what they think of this book, and their completed car questionnaires. Thanks to you, I have understood that far from wanting to become the Barbara Cartland of the automobile, my only wish is to do all my work as a journalist and nothing else. This, while presenting you with the facts and more facts and always facts. I understand that my tastes and colors do not necessarily coincide with yours and that you are more than capable of getting an idea, beginning with the data I gather each year to allow you to make the choice in full possession of the facts. I work only for you, because it is you who pay me to make this the reference book that it has become. During the year, you have pointed out the small errors that have slipped into our data base and that was appreciated. We have, therefore, engaged the services of an associate who does nothing but crunch numbers to avoid errors. Numbers, always numbers. The automobile is not the land of dreamers, poets or sour-grape artists. With automobiles, any move begins with money, the price you have to pay for a car to buy or maintain it. For this purpose, Road Report innovates again by showing you the minimum price for which you can buy a given model. This gives you a feel for the maneuvering margin at the dealerships, without prejudice to anyone. The average price indicates the average value of the transactions effected in the country and the maximum that you should absolutely pay for your future car. The automobile is an exact science created by engineers and not by pamphleteers, and you have to have a solid comparison basis to be able to make judgment values, rather than relying on first impressions. There are, in all trades, amateurs and professionals, and I have selected to line up with the latter. In any case, the relationship that exists between you and the book is unique and that is what creates our main strength. That is how I know who you are and what you desire. Continue to be part of our team and we will continue to take into account your notes for everyone's benefit. Apart from figures, everyone dreams a little, and you'll find the stuff that dreams are made of in the World Round-up our first chapters. I plan a rendezvous for next year, same time, same bookstore, to share with you once again the pleasure of the automobile, and while waiting, beware of imitations...
Daniel Heraud

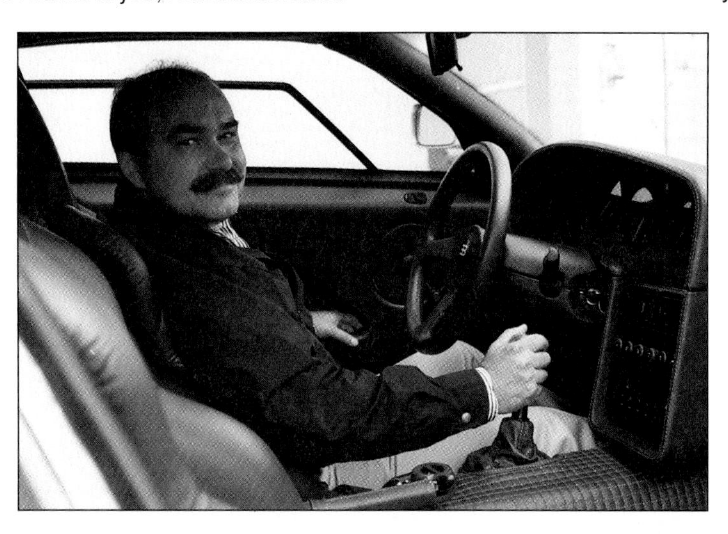

CREDITS

Cover page: PONTIAC Sunfire Speedster.
Credits for photos in these pages: Benoit Charette, Michel Condominas, Gérard Héraud, Daniel Heraud, Jean D'Hugues and the auto companies press services.

Chief Editor:	Daniel P. Heraud
Assistant:	Benoit Charette
Collaborators:	Michel Poirier-Defoy
	François Viau
Cover design:	Chilton Book Company
Design and make-up:	Daniel P. Heraud
Proofreader:	Françoise Moreau
Translation:	Alex Walordy
Color:	Scan Express
Printing:	Metropole Litho

INTERNATIONAL COLLABORATORS

United States:	Alex Walordy
France:	Michel Condominas
	Gérard Heraud
Great Britain:	Nick Bennett
Germany:	Helmut Herke

SPECIAL THANKS

To produce this edition of *Road Report*, we had to count on numerous individuals working in the automotive and publishing fields. To all of those whether close by or far away, who have participated in the creation of this book, a sincere and grateful thank you.

INDEX

() Canadian denominations

Model Identification

Flag of builder's country of origin

Manufacturer Identification

Historical facts, sales statistics, competitors, main improvements and equipment.

In "Data", the indices are explained as follows:
The **insurance rate** multiplied by the suggested retail price of a specific model gives the approximate total premium. It has been established for a person 35 years of age, married, living in a metropolitan area, without an accident in the last six years and no driver's license suspension. The base premium for $1 million of property and personal liability, $250,000 collision and $50 for multiple risks.
The **depreciation** risk is calculated based on our "Used car guide" for three years. If the vehicle exists for less than three years, the number of years follows in parenthesis.
The **cost per mile** is calculated by category. It is based on the use of the vehicle during a three-year period at the rate of 12,000 miles per year and includes the insurance rate, registration costs, interest rate, fuel cost, tire wear, maintenance and repairs, depreciation and parking costs.

Main available power train, engines and transmissions as well as their performance.

TOYOTA **Tercel-Paseo**

Growing pains...

Just like children, automobiles never cease to grow and the latest Tercel does not escape this phenomena which, one day or another, will force the builders to introduce a still smaller model to fill out the bottom end of the mini-compact range.

The Paseo coupe does not change for 1995, but the Tercel adopts a new body together with the more powerful engine already in the Paseo. Without wanting to pass judgment on the lines of the new arrival, it makes one think of the old Ford Escort rather than of a famous German. The Tercel is available in a 2- and 4-door, 3-volume sedan and a 3-volume, 2-door Paseo. The latter is sold at the level of a unique finish while the Tercel exists in versions S and DX.

STRONG POINTS

• Price/equipment: T: 90%; P: 80%
Even though they are always costlier than average, these cars justify their price by the excellent reputation for reliability and durability, more so than by the richness of their equipment, which is however in progress.
• Satisfaction: 85%
If the number of very satisfied owners is lower than for other models of the make, it does make for envy among the domestic builders.
• Safety: 80%
Two air bags and retractable belts compensate for the vulnerability of these small cars which however, have been made more rigid on the latest Tercels.
• Fuel consumption: 80%
Even though these models are not the most economical on the market, their fuel economy is good considering their more generous format and satisfactory performance.
• Quality/fit/finish: 70%
The interior has distinctly improved with plastics that have a richer appearance and are no longer malodorous. The assembly has remained strict and the finish carefully done; also the feel of a car that is economical at any cost, which characterized the preceding model, has disappeared.
• Driver's compartment: 70%
Even though the steering column is not adjustable, the driver readily gets comfortable. The dashboard is simple but well-organized and of good appearance. Instrumentation is legible and the controls conventional and well laid out. Visibility is better on the Tercel than on the Paseo whose tall tail and the presence of a spoiler reduce the view toward the rear.
• Suspension: 70%
The more consistent shock absorber control and longer wheel travel give these cars a driving comfort more characteristic of compacts than sub-compacts.
• Access: T: 70%; P: 50%
It has become easier to reach the rear Tercel 2-door bench seat than those of the Paseo which lacks height.
• Noise level: T: 60%; P: 50%
It has improved on the Tercel because even though the engine allows

DATA

Category: front-wheel drive sub-compact sedans and coupes.
Class: 3

HISTORIC

Introduced in: 1978
Modified in: 1982, 1987, 1991, 1995.
Made in: Takaoka, Japan.

RATINGS

Safety: 90 %
Satisfaction: 85 %
Depreciation: 53 %
Insurance: 7.0 % ($445)
Cost per mile: $0.30

NUMBER OF DEALERS

In U.S.A.: 1,233

SALES ON THE AMERICAN MARKET

Model	1992	1993	Result
Tercel	102,043	93,820	- 8.1 %
Paseo	22,838	24,466	+6.7 %

MAIN COMPETITORS

Tercel: DODGE Colt, HONDA Civic Hbk, GEO Metro (4dr.) HYUNDAI Accent, MAZDA 323 Hbk, NISSAN Sentra, VW Golf.
Paseo: FORD Escort GT, HONDA Civic SI & Del Sol, HYUNDAI Scoupe, MAZDA MX-3(1.6L) 323, SATURN SC1.

EQUIPMENT

TOYOTA Tercel TOYOTA Paseo	S	DX	 base
Automatic transmission:	-	O	O
Cruise control:	-	O	O
Power steering:	-	O	O
ABS brake:	O	O	O
Air conditioning:	O	O	O
Air bag (2):	S	S	S
Leather trim:	-	-	-
AM/FM/ Radio-cassette:	O	O	O
Power door locks:	-	-	-
Power windows:	-	-	-
Tilt steering column:	-	-	-
Dual power mirrors:	-	S	S
Intermittent wipers:	S	S	S
Light alloy wheels:	-	-	O
Sunroof:	-	-	O
Anti-theft system:	-	-	-

S : standard; O : optional; - : not available

COLORS AVAILABLE

Exterior: Tercel: White, Silver, Red, Green, Ruby.
Paseo: White, Black, Red, Green, Ruby.
Interior: Black, Gray.

MAINTENANCE

First revision: 4,000 miles
Frequency: 4,000 miles
Diagnostic plug: Yes

WHAT'S NEW IN 1995 ?

- Only two versions of the Tercel S and DX, the LS disappears.
- More powerful 1.5 L engine.

ENGINE						TRANSMISSION		PERFORMANCE										
Model/ versions *: standard	Type / timing valve / fuel system	Displacement cu/in	Power bhp @ rmn	Torque lb.ft @ rpm	Compres. ratio	Driving wheels / transmission	Final ratio	Acceler. 0-62 mph s	Stand. 1/4 mile s	Stand. 5/8 mile s	Braking 62-0 mph ft	Top speed mph	Lateral acceler. G	Noise level dBA	Fuel economy mpg city	hwy	Gasoline type / octane
Tercel	L4* 1.5 DOHC-16-EFI	91	93 @ 5400	100 @ 4400	9.4 :1	front - M4*	3.52	11.5	18.2	34.1	138	100	0.75	68	36.0	46.0	R 87
						front - M5*	3.72	11.0	18.0	33.7	144	103	0.75	68	35.0	45.0	R 87
						front - A3/A4	2.82	13.5	18.6	36.5	148	100	0.75	68	35.0	45.0	R 87
Paseo	L4* 1.5 SOHC-16-EFI	91	100 @ 6400	91 @ 3200	9.4 :1	front - M5*	3.94	10.5	17.3	33.0	144	106	0.77	68	36.0	44.0	R 87
						front - A4	2.82	10.8	17.7	33.5	151	103	0.77	68	31.0	43.0	R 87

362 Chilton's Road Report 1995

EVALUATION BASIS

CONCEPTION

Safety	Result of the American NHTSA tests
Aerodynamics	Based on the official coefficient of the car manufacturer
Interior space	Based on EPA methods
Trunk	According to EPA methods
Quality/fit/finish	According to 4 criteria (see table page 8)
DRIVING	
Driver's compartment	According to four criteria (see page 8)
Performance	Average of 6 accelerations from 0 to 62 mph
Handling:	Skid test on an asphalt circle of 250 ft. in diameter, given in G's
Steering	Based on five criteria (see page 8)
Braking	Average of six stops from 62 to 0 mph.

Note: The vehicles are listed under the manufacturer's name and not those who market them. Thus, the Chevy Tracker is described on the page of the Suzuki SideKick.
Always consult the index which will allow you to find the right page.

Tercel-Paseo **TOYOTA**

you to hear takeoffs and accelerations, the aerodynamic and rolling noises are less pronounced than on the preceding model.

• Steering: 60%
The standard manual is heavy at low speeds and lightens up as speed increases. Its high steering ratio makes it vague in the center. The standard power assist of the Paseo is more direct, precise and better modulated, in brief, more pleasant.

• Conveniences: 60%
Few, the storage areas include a glove box, door pockets and hollows on the dash of the deluxe versions.

• Seats: 60%
They offer good hold and lumbar support, thanks to improved shape and padding.

• Technical: 60%
The steel unitbody, of which some components are galvanized, uses a McPherson-type suspension up front and a torsional axle at the rear. The steering is manual and braking mixed on the standard and one is surprised not to find an ABS listed, even as an option. The simple lines show good aerodynamic efficiency with a Cd of 0.32 on a Tercel and 0.33 on the Paseo.

• Handling: T: 55%; P: 60%
The Tercel has become much more stable with less roll, more crisp and faster; it has almost become so much fun to drive that one wonders if the Paseo can add anything except for a sports feel.

• Insurance: 50%
The Tercel-Paseo do not cost any more to insure than any other sub-compact, which sets them off immediately as an interesting purchase for a young car owner, always penalized by the insurance companies.

• Performance: 50%
They are always more interesting with a stick shift than with an automatic, especially under load. A few extra horsepowers make the difference, but that still does not make them into dragsters. Accelerations are crisper with a manual transmission whose shifts are fast and precise than with the automatic which has satisfactory ratios and shifts.

WEAK POINTS

• Trunk: T: 30%; P: 20%
Roomy considering its size, it is possible to increase its content by lowering the rear bench seat. However, the tight Paseo opening makes it difficult to handle baggage.

• Interior space: T: 35%; P: 30%
Four good-sized adults will be at ease in the Tercel whose leg and headroom have been increased. This is not the case with the Paseo, limited to two adults and two children.

• Braking: 40%
Its modulation and fade resistance are adequate and efficiency is progressing in spite of the fact that the front wheels block rapidly, which lengthens the stopping distance.

• Depreciation: 45%
It keeps a good resale value and bargains are rare and costly.

CONCLUSION

• Overall rating: T: 61.0% P: 58.5%
More plush, the Tercel quietly mounts an attack on the Corolla because it is no longer a little car. However, its comfort and performances gain much together with its driving pleasure. 😐

OWNERS' SUGGESTIONS
(last year's model)
-Front seats that slide back further with better support
-Improved rearward Paseo visibility to better evaluate its length
-Larger wheels

SPECIFICATIONS & PRICES

Model	Version/Trim	Body/ Seats	Interior volume cu.ft.	Trunk volume cu ft.	Drag coef. Cd	Wheel-base in	Lgth x Wdth x Hght. in x in x in	Curb weight lb	Towing capacity std. lb	Susp. type ft/rr	Brake type ft/rr	Steering system	Turning diameter ft	Steer. turns nber	Fuel tank gal.	Standard tire size	Standard power train	PRICES US. $ 1994
TOYOTA		**General warranty: 3 years / 36,000 miles; power train: 5 years / 60,000 miles;corrosion:5 years / unlimited; without deductibles or transfert fees.**																
Tercel	S	2dr.sdn.4/5	80.4	9.3	0.32	93.7	162.2x65.4x53.1	1960	NR	i/si	dc/dr	r&p.	31.5	3.8	11.9	155/80R13	L4/1.5/M5	$8,698
Tercel	DX	2dr.sdn.4/5	80.4	9.3	0.32	93.7	162.2x65.4x53.1	1984	NR	i/si	dc/dr	r&p.	31.5	3.8	11.9	155/80R13	L4/1.5/M5	$10,148
Tercel	DX	4dr.sdn.4/5	81	9.3	0.32	93.7	162.2x65.4x53.1	2015	NR	i/si	dc/dr	r&p.	31.5	3.8	11.9	155/80R13	L4/1.5/M5	$10,248
Paseo	man.	2dr.cpe.2+2	77.2	7.7	0.33	93.7	163.2x65.2x50.2	2070	NR	i/si	dc/dr	pwr.r&p.	32.5	2.7	11.9	185/65R14	L4/1.5/M5	$12,838
Paseo	autom.	2dr.cpe.2+2	77.2	7.7	0.33	93.7	163.2x65.2x50.2	2160	NR	i/si	dc/dr	pwr.r&p.	32.5	2.7	11.9	185/65R14	L4/1.5/A4	$13,638

See page 393 for complete 1995 Price List

Chilton's Road Report 1995 363

Text of evaluation in twenty points, classified by decreasing order of qualities and defects with the conclusion and the average value which allows vehicles to be classified and compared.

The final score expressed:

☺ above 65%

😐 from 56% to 65%

☹ below 55%

Commentary from owners who responded to information requests, sent in preceding years.

Main technical specifications and price in preceding year.

AND EVALUATION OF VEHICLES

COMFORT

Access	According to four criteria (see table page 8)
Seats	According to five criteria (see table page 8)
Suspension	According to five criteria (see table page 8)
Noise level	An average of 3 measurements taken at 62 mph
Conveniences	According to five criteria (see table page 8)

BUDGET

Price/equipment	According to a scale between $6 and $30,000
Insurance	A percentage of the retail price excluding taxes
Fuel consumption	Measured during tests
Satisfaction	Derived from readers' reports
Depreciation	Average of three years based on our "Used Car Guide"

Road Report is written simultaneously for Quebec, Canada and the United States. It is therefore possible that some models shown will not exactly conform to those sold locally.

ABBREVIATIONS

TYPES OF ENGINES

H	Horizontal cylinders
L	In-line cylinders
V	V cylinders
R	Rotary engine

FUEL SUPPLY

EFI	Electronic fuel injection
EMPI	Electronic multiple point injection
MI	Mechanical injection
SFI	Sequential injection (multiple point)
TBI	Throttle body injection
EMI	Electronically controlled mechanical injection
RI	Rotary diesel injection
TR or TBO	Turbocharger
Ti	Turbocharger plus intercooler
S	Supercharger

GASOLINE

R	Regular grade
M	Medium grade
S	Superior grade

TRANSMISSIONS

M4;M5	Manual 4-speed; 5-speed
A3;A4	Automatic 3-speed; 4-speed
2WD	2-wheel drive
4WD	4-wheel drive

MODELS

cpe	coupe
sdn	sedan
wgn	station wagon
van	van
p-up	pickup
a-t	all terrain
2d	2-door
3d	3-door
4d	4-door

SIZES

cyl	cylinder
cu.ft	cubic feet
hp	horsepower (SAE)
in	inches
ft	feet
lbs/ft	torque: feet-pounds
L	volume in liter
kg	kilos
db	sound level in decibels
mph	miles per hour
mpg	miles per gallon

*	original equipment or standard
#	not available at print time
NA	not available
NR	not recommended (towing)
HD	heavy duty
Cd	air drag coefficient
dc	disc
dr	drum
ABS	anti-locking brake system
TCS	traction control system

si	semi-independent
r	rigid rear axle
r/p	rack and pinion
pwr	power assisted
bal.	recirculating ball steering
	sector and roller

1995 SCALE

Notes in %		**0**	10	20	30	40	**50**	60	70	80	90	**100**
DESIGN												
Safety:	NHTSA	structures		doors	environment			seat-belts	air bags			
Aerodynamics:	Cx	0.48	0.46	0.44	0.42	0.40	0.38	0.36	0.34	0.32	0.30	0.28
Interior space:	EPA cu.ft	57	64	70	78	85	88	92	99	106	113	120
Trunk:	EPA cu.ft	3.5	5.5	7	9	10.5	12.5	14	16	17.5	19	21
Quality/fit/finish			appearence			assembly			materials			finish
DRIVING												
Driver's compartment:			seat		visibility			commands		controls		
Performances:	0/62 in s.	16	15	14	13	12	11	10	9	8	7	5
Handling:	lat. G force	0.66	0.68	0.70	0.72	0.74	0.76	0.78	0.80	0.85	0.90	0.95
Steering:		ratio	precision		calibration			assist	return			
Breaks:	62 to 0 in ft.	215	200	180	165	150	130	125	120	110	105	100
COMFORT												
Access:				front		rear			trunk		Power train	
Seats:		seat		lateral support		back support			padding		trim	
Suspension:		damping		amplitude		roll			pitch		frequency	
Noise level:	dBA	73	72	71	70	69	68	67	66	65	64	63
Conveniences:		glove compartment			door pockets		console box			hollows		
BUDGET												
Price/equipment:	$ US	30,000	27,000	24,000	21,000	18,000	15,000	12,000	10,500	9,000	7,000	6,000
Insurance:% of retail price		16	14	12	10	9	8	7	6	5	4	3
Fuel consumption:	l./100	20	18	16	15	14	13	12	11	10	8	6
Depreciation: %		100	90	80	70	60	50	40	30	20	10	5
Satisfaction: %		5	10	20	30	40	50	60	70	80	90	100

GLOSSARY

ABS OR ALB: Initials for anti-locking brake system. This system detects and prevents wheel lockup during hard braking. Sensors at each disc or drum brake detect a sudden speed decrease indicating the beginning of lockup. Fluid pressure to that brake is reduced and reapplied at the rate of several pulsations per second till the wheel regains rotation speed. The sophistication of these systems varies between makes. Some are mechanical, others detect deceleration and lateral G-forces (centrifugal force in a turn) to prevent spinouts or large shifts in direction: always possible with more simplistic devices.
ADJUSTABLE STEERING COLUMN: Changes the angle of the steering wheel or its length, to accommodate the driver.
AIR DAM: A device to deflect air, usually under the front bumper or other locations. It can be fixed or movable.
ALL-WHEEL DRIVE and 4-WD: All wheels, front and rear, provide traction. A differential splits the torque between the front and rear axles.
ALTERNATOR: The belt-driven alternator supplies current for ignition, battery charge, AC and other accessories.
AUTOMATIC TRANSMISSION: Automatic shift change: offering 3, 4 or 5 ratios is also set up for holding the transmission in lower gears for torque or grade retard.
BAGGAGE COVER: Cloth or cover to conceal baggage in coupes, sedans or station wagons.
BODY BELT LINE: Artificial line dividing the sheet metal lower body from the glazed upper body.
BORE/STROKE: Dimensions that allow figuring the cylinder displacement. The bore is the diameter of the piston and the stroke is the distance it travels. Displacement formula is B/2 squared X Pi X S X number of cylinders = volume. Keep consistent measurements, inches or centimeters, to get cubic inches or cc volume. of the engine displacement
CAMBER: Lateral wheel angle, relative to a vertical plane, to compensate for tire roll in the turn.
CAMSHAFT: A steel part with lobes to operate intake and exhaust valves either directly or through rocker arms. It is gear, chain or belt-driven.
CARBURETOR: Meters fuel in proportion to the air needed by the engine and controls the air through one or more throttle blades.
CASTER: Forward or rearward inclination of the steering axis.
Cd: Coefficient of drag of the automobile body, usually measured in a wind tunnel. Multiply the Cd by the frontal area to get a comparison index from car to car. The lower the number, such as .029 , the higher the efficiency.
CLUTCH: Couples or disengages the engine and the transmission.Has one or more discs engaging by friction. Many clutches are used in automatic transmission. The clutch action can be timed for progressive engagement.
COMPRESSION RATIO: When the piston is at the bottom of its stroke, the cylinder and combustion chamber are filled with air. The ratio between that and the combustion chamber volume is the compression ratio. Higher compression ratio improves combustion efficiency. Supercharged or Turbo-charged engines may need a lower nominal compression ratio, while a Diesel needs 22:1 compression to initiate ignition.
COMPRESSION STROKE: After intake, the piston moves up to compress the airfuel mixture. More compression improves efficiency: the fuel octane limits the maximum compression that can be used.
CONNECTING ROD: The part which connects the piston to the crank. The rod conveys the alternating piston motion to rotary crank motion and vice versa.
CONTROLLED SHOCKS: The shock settings can be externally changed by an electronic control.
COUPE: 2-volume body with 2 main seats and 2 mini-seats, and a sports look.
CRANKSHAFT: The crankshaft turns on main bearings and has offset journals, just like a bicycle crank, to transform the alternating piston motion to a rotary motion.
CYLINDER HEAD: The part over the block which contains the combustion chambers and intake valves, as well as intake and exhaust ports. It carries the valves and valve train parts.
CYLINDERS: The barrel-shaped part of the block in which the piston forms a moving wall and which is closed off at the top by the combustion chamber. Cylinders can be laid out IN-LINE, in V6 or V8 or horizontally opposed (BOXER) as in VW Beetle or Subaru.
DIESEL: A denser fuel than gasoline. It requires much higher compression ratios to spontaneously ignite. ***Cetanes***, not octanes, rate Diesel fuel and is a measure of ignitability under compression, while octanes measure the resistance to ignition due to compression. A direct injection sprays into the cylinder. Glow plugs may be used but not a conventional ignition or fuel injection into the manifold.
DIFFERENTIAL: The gear train which allows the drive wheels to turn at different speeds up when going around a corner. You will find differentials in the front and rear drive trains in the 4WD, while transmitting torque to both wheels, as well as in the transfer case.
DISC BRAKE: The disc turns with the wheel and the caliper with pads applies friction for efficient stopping power. ***Ventilated disc:*** Fins between the 2 sides of the disc circulate air and cool the disc. Disc brakes work at very high temperatures and cooling is important.
DOHC: One cam operates the intakes, the other the exhausts, which allows timing refinements.
DRIVE RATIO: Determines the final drive ratio between the transmission output and the drive wheel rpm. Numerically higher ratios make the engine turn faster. More wear, but also more power.
DRUM BRAKE: A drum turns with the wheel. Shoes with friction-lining material rub against the drum and retard it during brake application.
ELECTRONIC IGNITION: Electronic ignition replaces points and may also control timing in conjunction with the computer.
ENGINE TORQUE: The turning effort delivered by the engine at any given rpm. Torque is pull on a moment (lever) arm, given in lbs/ft. or in m/kg. 1.0 lb/ft = 0.1382549 m/kg. Pulling on a wrench applies torque to the bolt.
FLOOR PAN: Bottom of the car, which also includes reinforcing and suspension mount members.
FUEL INJECTION: Atomizing the fuel under pressure, as it exits from a nozzle, so it can mix with air.
FULL-TIME 4WD: All 4 wheels always driving.
GAS SHOCK: May use gas pressure to prevent foaming up of the fluid or to produce additional lift. On some shocks, electronic controls adjust damping. In a suspension a spring temporarily stores, then releases energy and the shock absorber dissipates: damps out - energy and converts it to heat.
GRILL: Decorative air inlet to the radiator and engine.
HEAT EXCHANGER: A heat exchanger is used to transfer heat from engine oil or transmission oil or a power steering to the engine coolant or to ambiant air.
HYDRAULIC SHOCK: Uses internal oil forced through orifices by the motion of a piston in a cylinder to generate daming forces.
HYDROPNEUMATIC SUSPENSION: Uses gas pressure to maintain and adjust height, and fluid to control damping.
INDEPENDENT SUSPENSION: Each wheel moves independently of the other.
INDIRECT INJECTION: Used on Diesel engines with a pre-chamber to improve ignitability. ***Jets*** are calibrated orifices to meter fuel.
KNOCK SENSOR: Detects combustion noise and is used in conjunction with the computer to alter spark or fuel supply. See: ***Octane***.
LEAF SPRING SUSPENSION: A stack of flat leaf springs of decreasing lengths connects to the axle at a spring pad and pivots on the frame or body. Since the leaf gets longer when it deflects, a swing shackle compensates for the length change.
LIMITED SLIP DIFFERENTIAL: A device which limits the rpm difference between 2 wheels by engaging a clutch or other coupling.
LOCKING DIFFERENTIAL: It can lock the 2 output shafts to each other. ***Viscous coupling***: A clutch filled with high viscosity silicon to increase the lockup as the difference in rotation speed increases.
MACPHERSON: Independent suspension with a single lower control arm, connected to a housing with a shock absorber and a coil spring. The upper portion of the shock pivots

on the unit body. There are modified MacPhersons with upper and lower control arms, as on the Honda, which maintain better camber control when the car leans in a turn.

MANIFOLDS, INTAKE AND EXHAUST: Castings or tubular assemblies with the passages that connect each cylinder to the intake or exhaust side of the engine. Their shape can have a large affect on performance.

MANOMETER: One way to measure air or fluid pressure.

MANUAL TRANSMISSION: Provides a choice of gear ratios to match car speed with engine speed. The driver selects the gear ratio through cable or rod controls and couples the pair of gears needed to the input and output shafts, while disengaging the clutch to interrupt power flow.

MASTER CYLINDER: Part of the brake system, when the driver applies pedal pressure it provides the pumping action to deliver fluid to the individual wheel cylinders, drum or disc.

MECHANICAL INJECTION: A pump controls the fuel delivery. ***EFI***: Electronic fuel injection. ***MPI*** or ***EMPI***: Electronic multiple injection. SFI: sequential fuel injection, each cylinder fed in turn according to firing order.

MODULATION: How well the driver can control the brake or steering, so that the input force results in a proportioned output.

MULTIFUEL ENGINE: Engine capable of utilizing several types of fuel.

MULTIVALVE ENGINE: Engine with 3, 4 or more valves per cylinder.

OCTANE: The fuel's ability to resist detonation (knock) or preignition.

OIL PAN: The section under the engine which holds the oil supply and the oil pump.

OVERDRIVE: In the manual or automatic transmission, a final drive ratio that is higher than 1 to 1. The engine is turning slower than the transmission output shaft. Great economy but no acceleration.

OVERHANG: What extends behind the rear wheels or ahead of the front wheels.

OVERSTEER: The vehicle has a tendency to turn at a tighter radius than steering wheel input and at the limit the rear is kicking out. During oversteer, the driver applies reverse lock to the steering to keep from spinning out.

PANHARD ROD: It provides lateral control by connecting the body to the opposite side of the axle housing. Now the back of the car can't swing out on a turn.

PART-TIME 4WD: A differential or a slip coupling divides the power between the front and rear. Usualy includes a lockup feature.

PUSH ROD ENGINE V6/V8: A cam which runs inside the block operates the valves through lifters, pushrods and rocker arms, in succession.

QUARTER PANELS: Side body panels at the backlight, trunk and rear wheels.

RAILS: Lengthwise sections in the chassis or unit body, U, C or box-shaped. They are generally tied in by cross members.

RIGID SUSPENSION: The 2 wheels are connected by a rigid axle. On a rear-wheel drive car, the rigid axle includes the differential.

RACK-AND-PINION: Two hypoid gears at right angles to each other that transfer power from a drive shaft sitting lengthwise to the differential and the axle shafts of the rear wheels. It also provides a final drive gear ratio. The pinion is powered by the driveshaft and the ring gear drives the differential.

ROTARY ENGINES (WANKEL - MAZDA): A rotor controls intake and exhaust and moves eccentrically within the housing to complete intake compression and exhaust in one turn.

SEDAN: Body with 2 or 3 volumes and 2, 3, 4- or 5 doors offering 4 to 6 seats.

SERVO: Another word for assistance. The driver's input is multiplied by allowing it to control an outside power source.

SHOCK ABSORBER: Controls the amplitude (travel) and speed of suspension and wheel travel due to road undulations and also partially controls body movements such as roll, pitch, loading and unloading. Shock absorbers cannot substitute for larger springs or stabilizer bars.

SHOULDER BELT: Upper part of safety belt harness.

SIDE VALVE ENGINE: Used on antique, inline and V8 engines with side valves (valves next to the cylinder as in an old Ford).

SKIRT: Air deflector along the side of the car.

SOHC: Single overhead cam located in the cylinder head. It may operate the valves through cup followers or through rocker arms and controls both intakes and exhausts.

STABILIZER BAR: Long torsional bar connecting a pair of wheels and the body to limit roll during the turn. It does not support the car vertically, but stabilizes it by opposing roll.

SUPERCHARGER: A crank-powered belt or chain drives an external pump (supercharger), which feeds air to the engine to increase output, compared to a naturally-aspirated engine (the piston pumps its own air). There are Roots blowers (Ford Thunderbird SC) or spiral blowers (VW Corrado).

SYNCHRONIZATION: During the shift in a manual transmission, a small internal cone clutch equalizes gear and synchro sleeve speeds so engagement can be completed without grinding.

TIMING BELT DRIVE: A toothed timing belt that powers one or more cam shafts: flexible, quiet, doesn't require outside lubrication. When one breaks, you require a new engine.

TIMING CHAIN: A chain of either roller or toothed type that drives one or more cam shafts.

TIRE ASPECT RATIO: Speaks of the ratio of the height to the width of the tire cross section. Example: 185/60 R 14: the 60 says that 60 percent of 185 mm width is the height. Less than 60 percent (low profile) helps handling; larger than 65 percent gains comfort.

TOE-IN: A steering geometry designed to have the wheels tip toward each other.

TOE-OUT: The same, but with the wheels toeing away from each other. It is equally important at the front and as the rear.

TORQUE CONVERTER: A hydraulic coupling used with automatic transmissions to provide slip, as with a manual shift clutch, and to multiply torque.

TORQUE TUBE DRIVE: A tube or an arm extends forward from the drive wheel axle housing to near the center of the body to handle axle windup torque. An antique last seen on Opels, Vegas and early Fords, but very effective.

TORSION BAR SUSPENSION: A torsion bar is a straight coil spring. The bar works in twists: one side fixed in the frame and the other connected to the suspension arm. Most torsion bar suspensions offer initial ride height adjustment.

TRACTION CONTROL: Often combined with ABS to apply braking to the wheel that shows traction loss, which restores torque transfer to the opposite wheel.

TRANSFER CASE: Used on 4WD vehicles to shift into 2WD, N or 4WD in High or Low range.

TRANSMISSION HOUSING: The outer case of the transmission which contains the shafts, gears, clutches (automatic) and the transmission fluid.

TUNNEL: A long, tunnel-shaped section of the floor pan, in the middle of the car, makes room for the drive shaft and sometimes the exhaust.

TURBOCHARGER OR SUPERCHARGER HEAT EXCHANGER: Same concept. Either of the two devices adds heat. By cooling, the temperature of the air delivered to the engine is reduced and efficiency improves.

TURBOCHARGER: A form of supercharger which is driven by the engine's exhaust instead of by a crank-driven belt. It adds power, can add efficiency and is used not only in sports car performance but in big, Diesel trucks.

UNDERSTEER: At the limit, the vehicle is understeering when it tries to go out nose first; during understeering you need to increase the steering wheel angle.

UNIT BODY: The body, floor pan and various reinforcing sections combine into a single, structural shell as opposed to a separate frame and body connected by mounts.

VALVE: Works like a sink stopper, this mushroom-shaped part with a long stem is closed by a spring and opened by a cam shaft. Intake valves admit air, exhaust valves let out burned gases.

VENTURI: Senses air flow. The carburetor is designed to atomize the fuel, mix it with the air so the engine can burn it to make power. The number of throttle plates in the carburetor decides the number of barrels. A 2-barrel has 2 venturis, 2 throttles.

WASTE GATE: A valve controlled by intake manifold pressure which bypasses some of the exhaust to stay within pressure limits tolerated by the engine.

WHEEL RIM: Retains the tire.

TECHNICAL

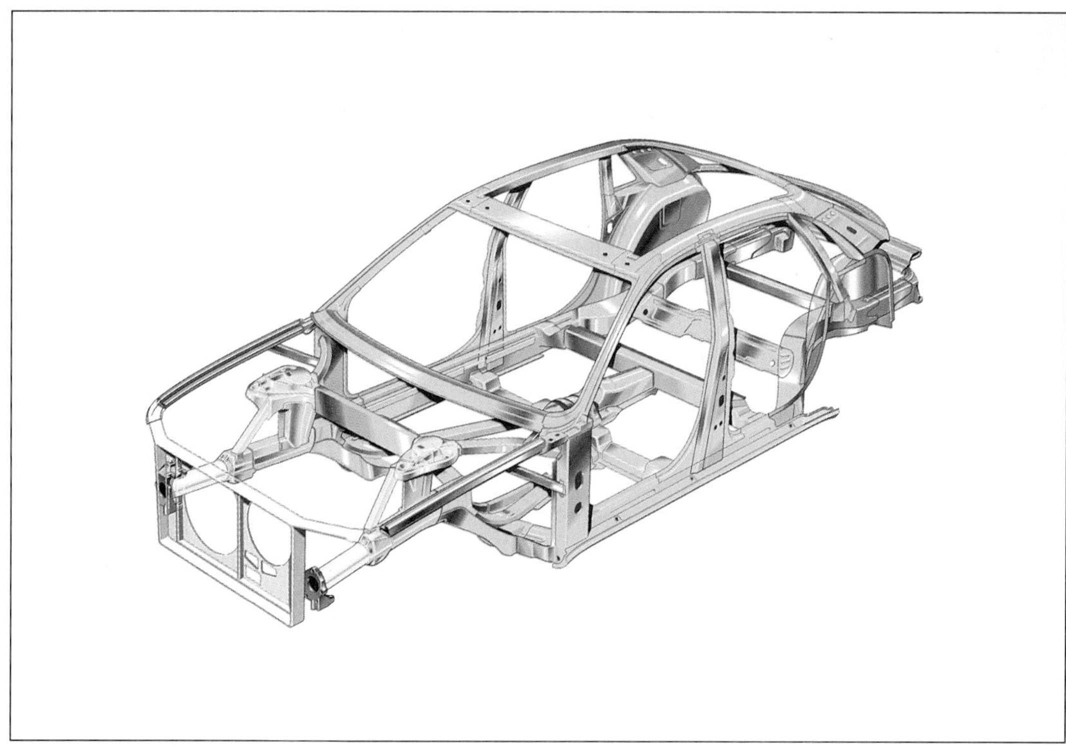

AUDI

The Audi A8 is the first production machine in the world whose body is built entirely of aluminum. This technique is revolutionary in more than one aspect and consists not only of building the main body panels in aluminum, which has been done on some exceptional models, but also load bearing structures (space frame below) which constitute the main car ribbing (French builder, Gregoire, did it in the sixties). Aluminum is a noble and ecological metal, lighter than steel, protectible from corrosion and almost infinitely recyclable. Audi's challenge was to insure optimal rigidity and occupant protection when one "knows" that aluminum is considered a soft metal.

TECHNICAL

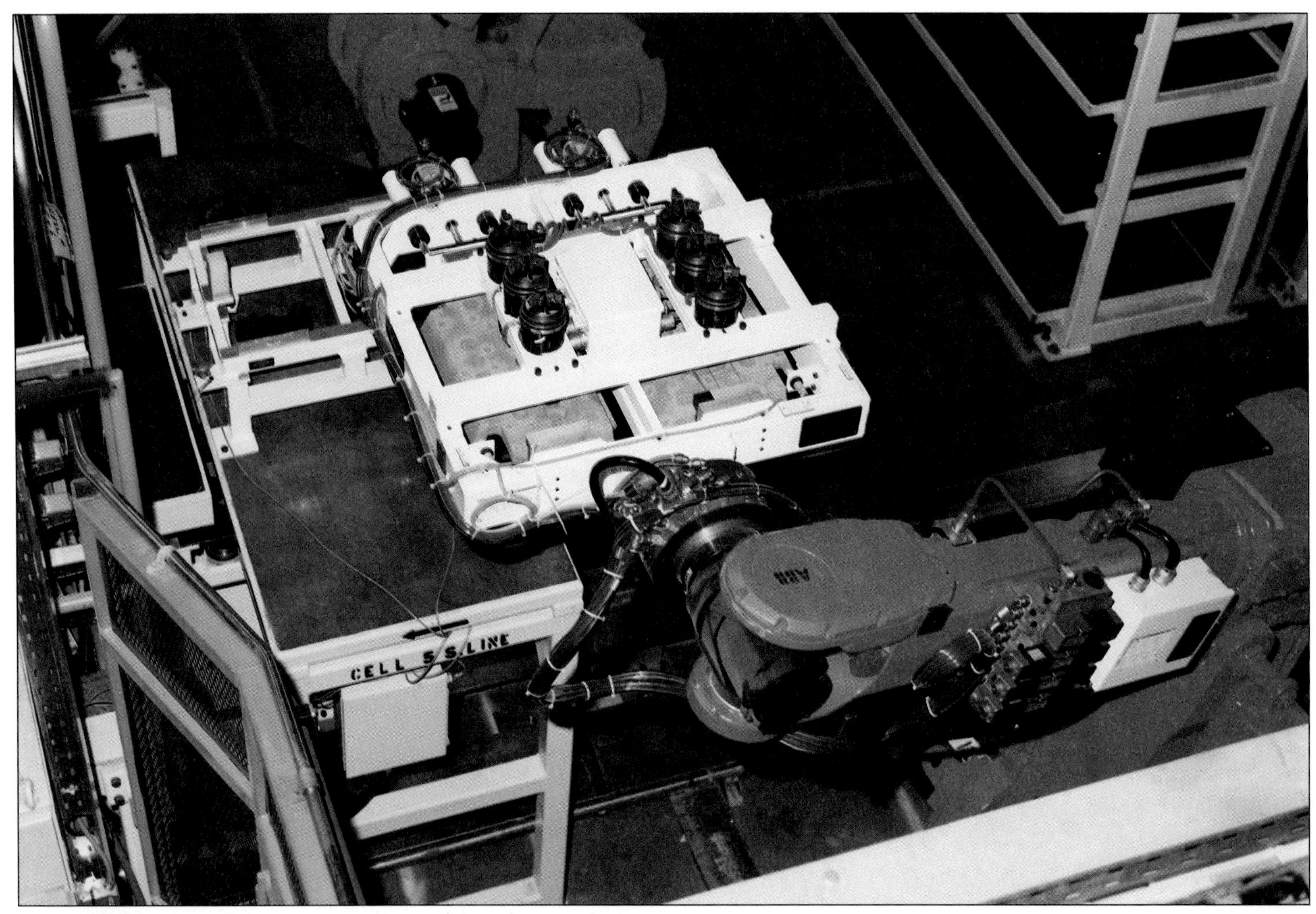

FORD

The search for saving energy and the vogue for recyling materials are in the process of giving aluminum a first-rate ranking in the automotive industry. Before bodies, it is engines that benefit from this material. Ford built in Windsor, Ontario, an ultra modern foundry employing 38 people and 80 robots. Based on Cosworth methods, it pours blocks for most of its V6 and V8 engines. These are assembled in Cleveland, Ohio, and Romeo, Michigan. The Windsor plant has an annual capacity of one million units. The work begins with two resin-bonded sand molds. During the aluminum pour, the mold is rocked to produce high quality parts free of gas bubbles. Ford succeeded in adapting to mass production a procedure to allow Cosworth to cast competition blocks.

TECHNICAL

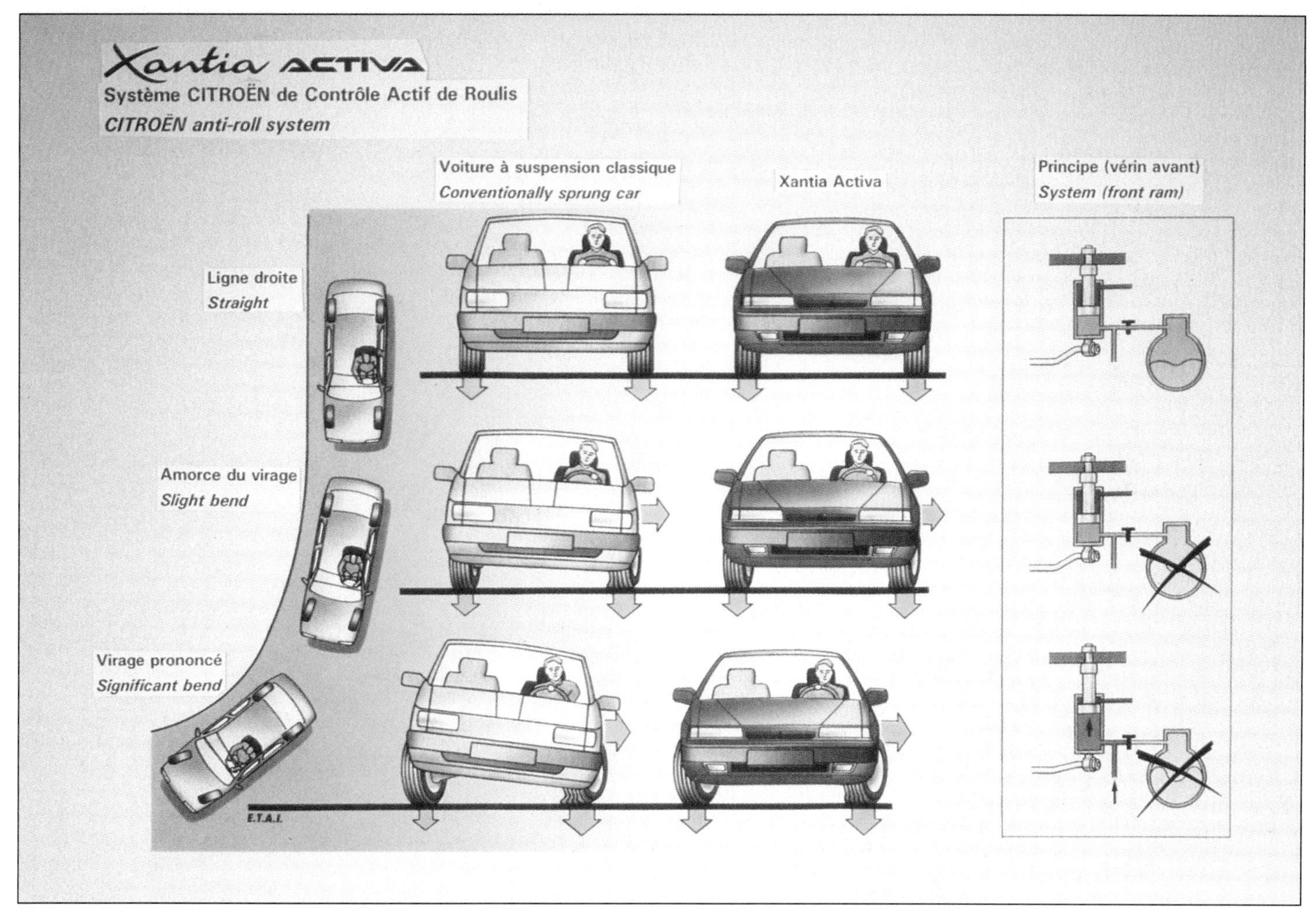

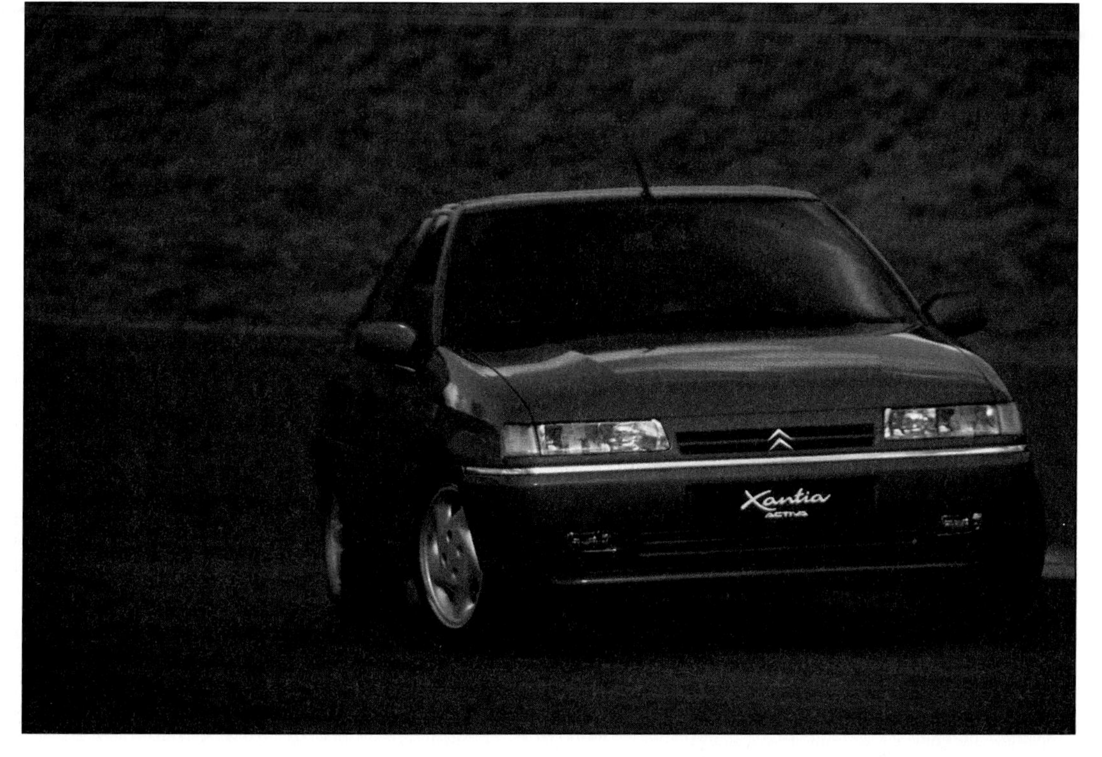

CITROËN

The French builder, wellknown for its original suspension designs, has shown a device, called the Activa, which allows Active Roll Control. It will be marketed, starting 1995, on one model of the range: the Xantia Activa. This rational driving assist associates electronics and hydraulics. When the car enters a curve, the electronics detect the body's angle of inclination and as soon as it reaches 0.30 degrees, two jacks act to reestablish the body equilibrium so that the occupants do not feel the roll. This technique sets a new standard in terms of road feel and perception.

TECHNICAL

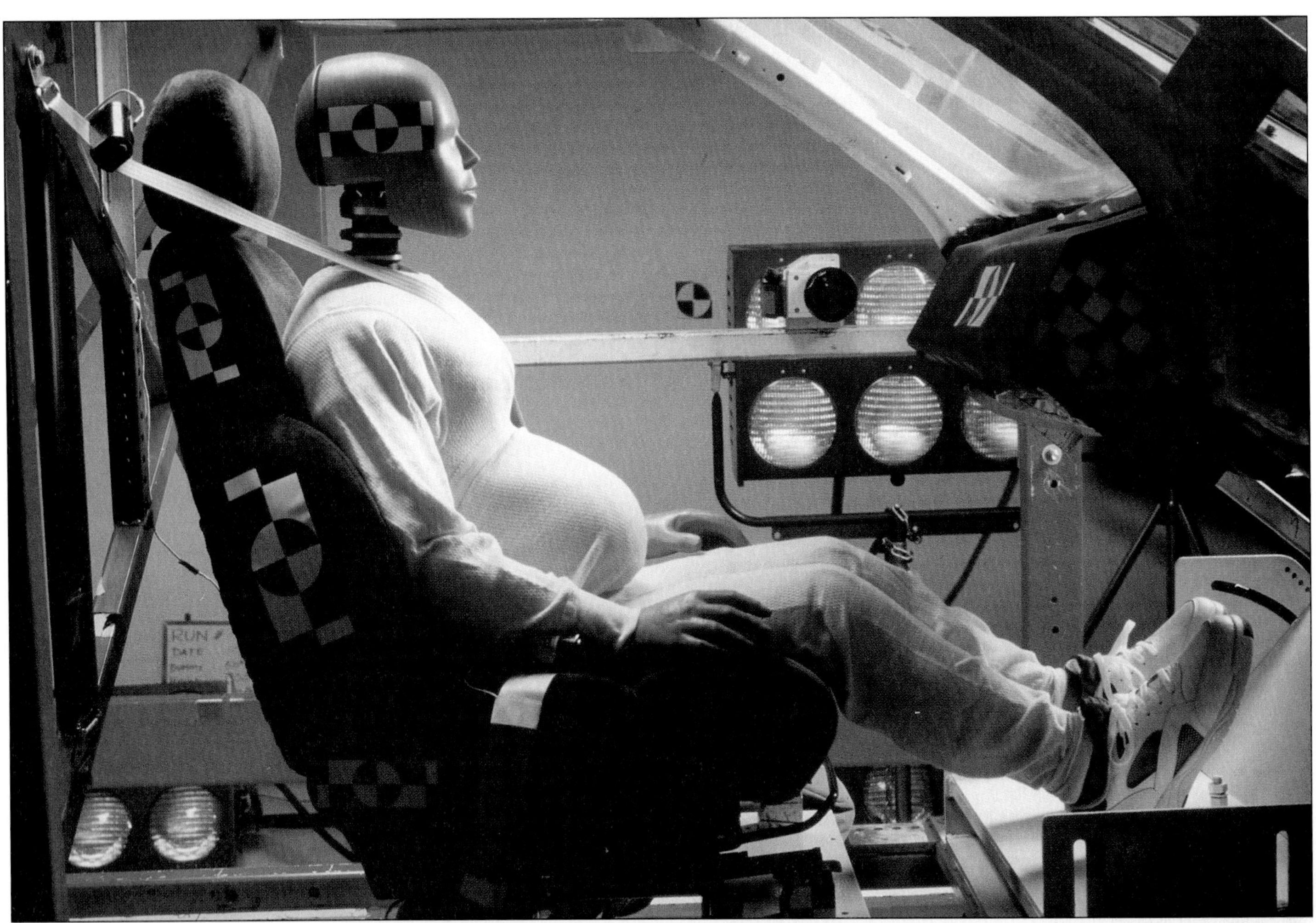

GM

General Motors has created the first Pregnant Dummy to study its reactions and those of the fetal implant during accident simulations. Developed jointly with the University of Michigan, this project will determine if seat belt wear is safe for pregnant women or if it will be necessary to devise a special device capable of protecting both the mother and the child.

The main work consists of actually calibrating the dummy, and the instrumentation registers the reaction of the "baby" to ensure that the data collected will allow finding new answers applicable to the majority of cases.

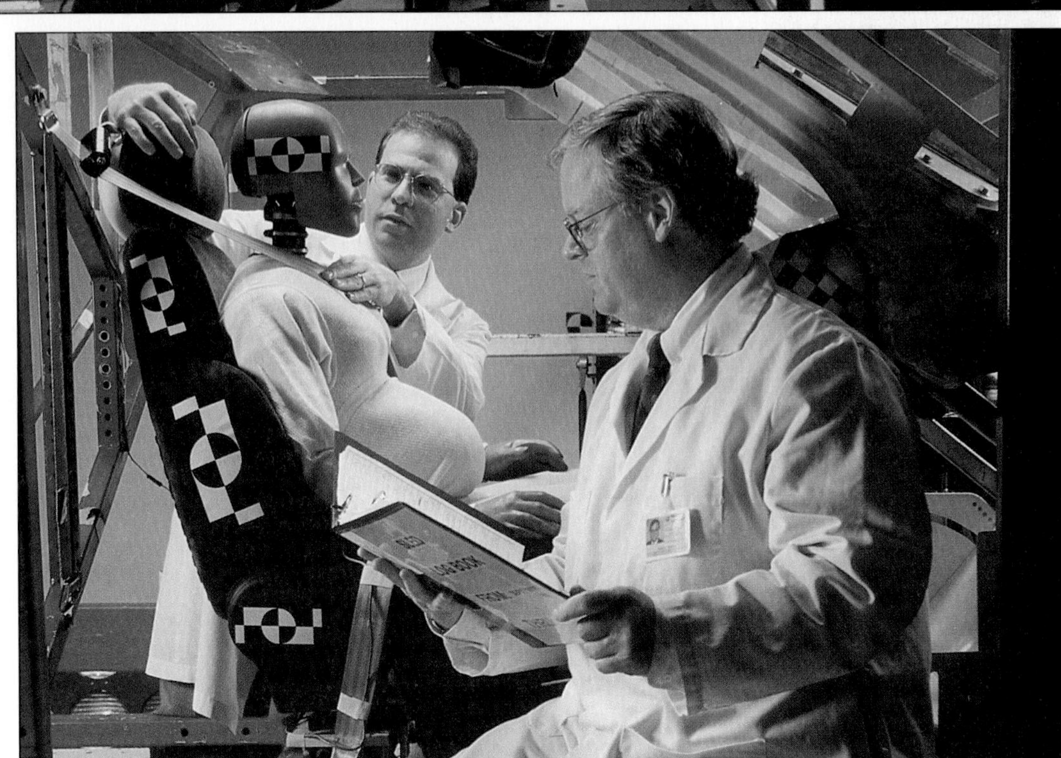

TECHNICAL

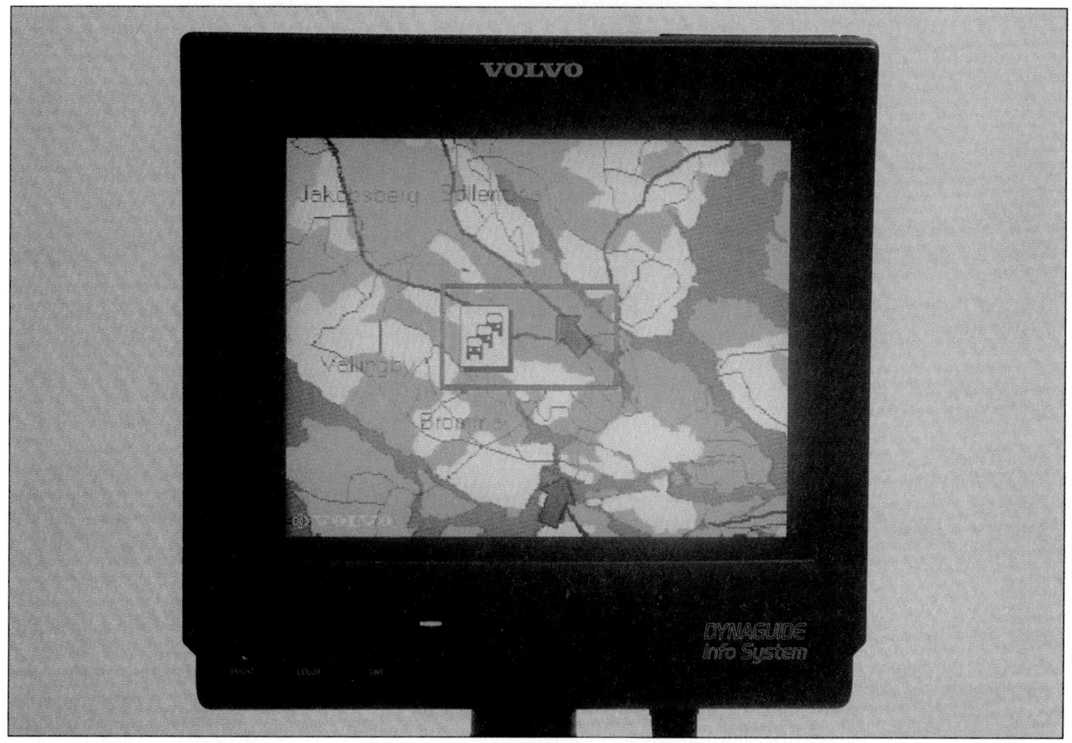

VOLVO

All builders are ready to market satellite navigation which will change the way we use our road system. Volvo's Dynaguide will allow combating bottlenecks by finding alternate routes. This reduces the driver stress and fuel consumption.

OLDSMOBILE

Oldsmobile's Navigational/Information System also uses the satellites of the "Global Positioning System" to facilitate navigation. By pointing out on a screen the spot to be found, the computer calculates vehicle position and the best road to follow for getting there. This system will also be able to provide a mass of other practical information.

TECHNICAL

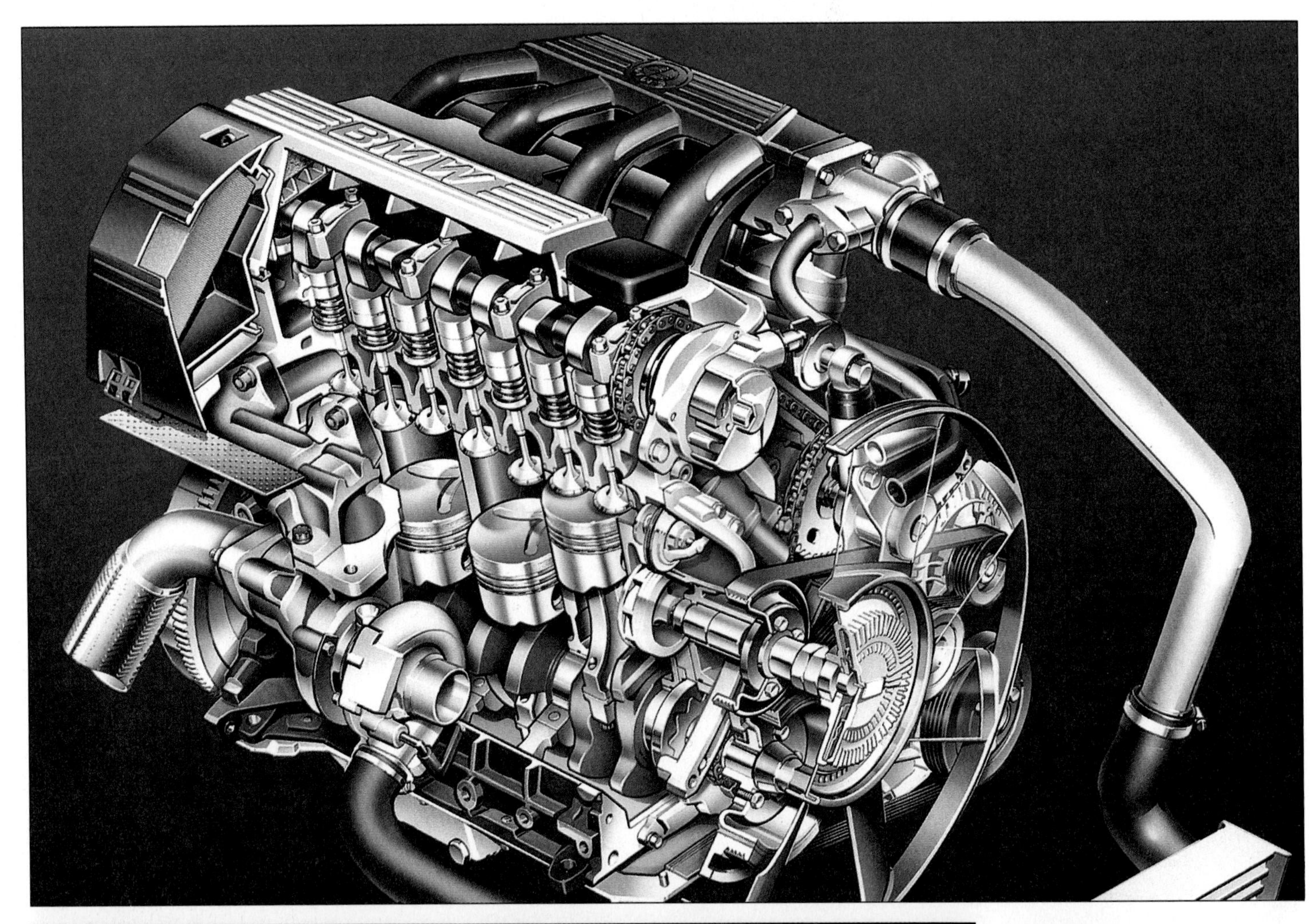

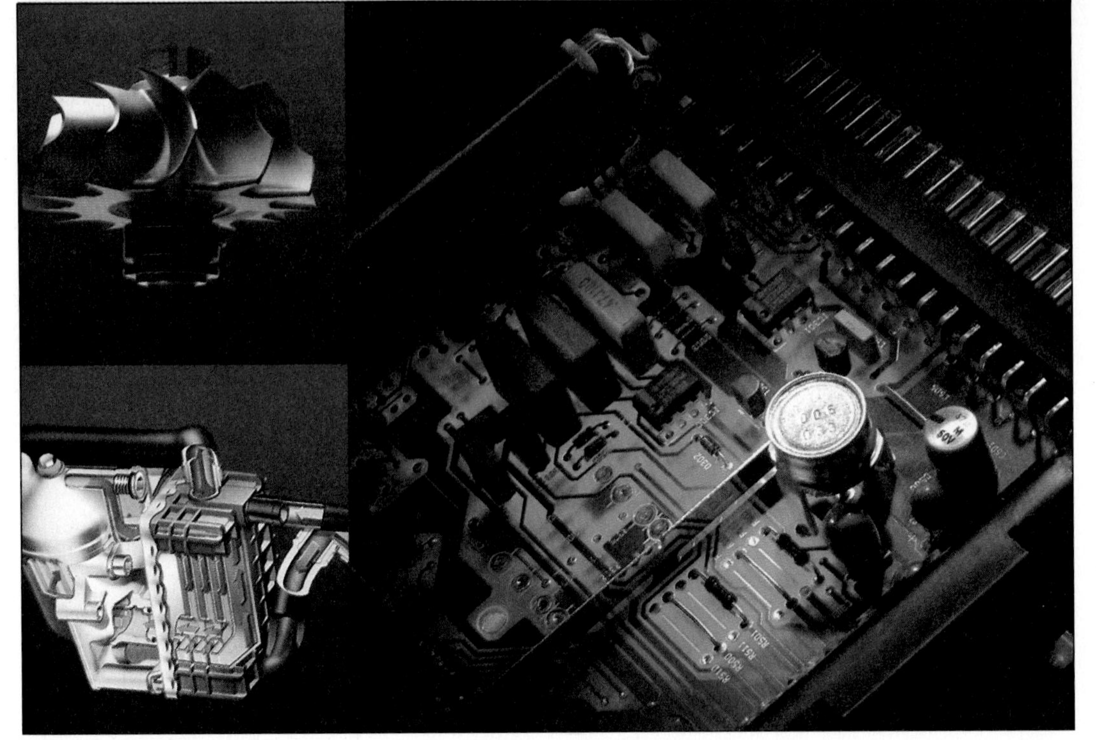

BMW

The Munich firm has just launched on the European market what some consider the best Diesel of the moment. It consists of a 1665 cc 4-cylinder which delivers 90 hp, thanks to a turbocharger. It owes its power to an indirect injection system into turbulence chambers modified by BMW. They have received a patent on a V-shaped cavity in the pistons, combined with a technique for cooling exhaust gas. Electronic controls allow maintaining a minimal pollution level by controlling the start and flow of the injection and the blower pressure. Finally, BMW developed a new injector which allows metering the fuel with greater precision.

STYLES

PONTIAC

At the last Detroit Show, Pontiac unveiled the Sunfire. Ultra simple, it only needed three months of development work, but is nonetheless very interesting and spectacular. The Pontiac stylists applied themselves at creating a model that comes closest to a production unit. The engine, placed forward between the drive wheels, is a turbocharged 2.4 L DOHC that produces 225 hp.

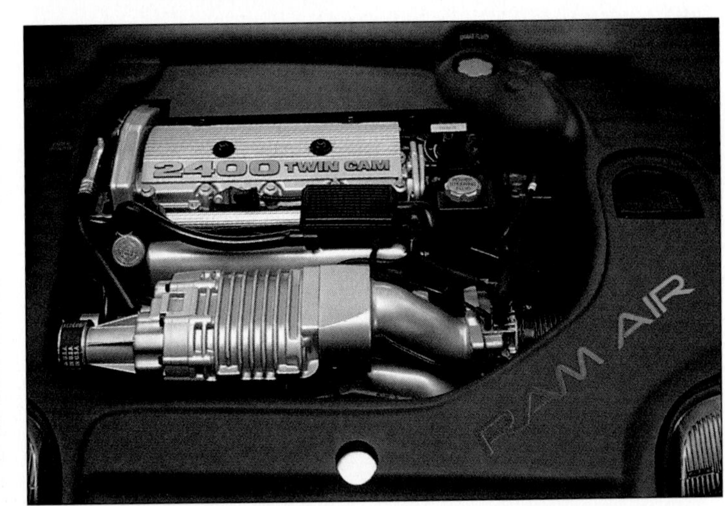

CADILLAC

This conservative little sedan, code named LSE, constitutes a test bed about which the GM luxury division would like to draw comment from the visiting show public. After the Cimmarron failure, Cadillac remains wary of small luxury cars, but the Japanese push in this area is so strong that GM owes it to itself to try again. The LSE (Luxury Sedan Eurostyle) is formatted like a Lumina and will accept five passengers. It has a rigid rear axle and a DOHC 3.0 L V6 with an output of 200 hp and as many lb/ft of torque. It is so "reasonable" in appearance that it could reach production in the near future and allow Cadillac to broadside the Lexus and Infiniti.

STYLES

FORD

This Lincoln Contempra, introduced at the Detroit Show, is a style study that led to the renewal of the Continental. Once again, the American designers hewed closely to reality to save time and money. The front lines are inspired by those initiated on the Mark VIII and the interior is not very futuristic; it differs from the production model just by the richness of the materials and equipment.

Below left the Ka, created by the Ford-of-Europe styling department, predicts the mini model which Ford will soon propose as the lower end of its range in the EEC countries. At the right, the profile is a sporty extrapolation of the Mondeo.

STYLES

FORD

Even though spectacular, this Ford Power Stroke which previews, they say, the next restyling of the F Series, misses good taste by its disquieting "look" and "smile". You do notice some imaginative detailing such as the integrated headlights, streamlined lighting and footsteps preceding the rear fender extension at the rear bumper. The lights which illuminate the loading area are interesting. This experimental vehicle receives a twin turbo V8 Diesel which brings up its performance to gasoline engine level.

STYLES

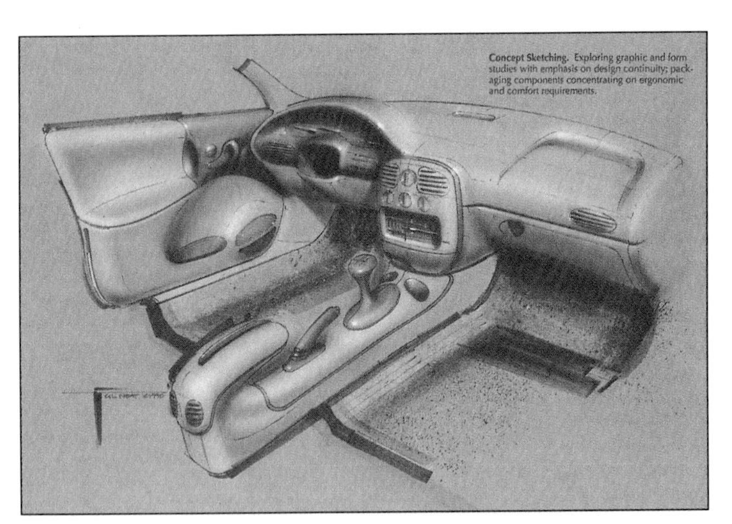

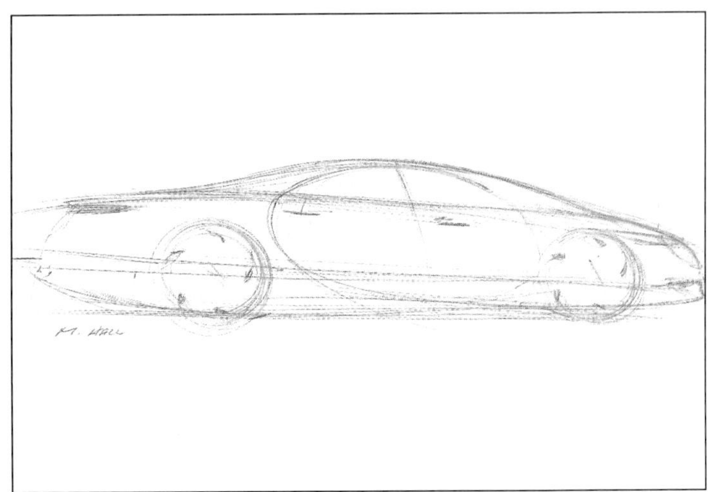

CHRYSLER

Above, you can admire the designs which led to the creation of the latest Chrysler Cirrus and Dodge Stratus. Below, the Aviat is a concept where the car borrows much from the airplane. At the bottom, the Expresso is an urban car, not necessarily miniaturized, which can serve as a taxi. New York City, invaded by Expressos, would be funnier than one filled with Chevrolet Caprices.

STYLES

CHRYSLER

For the needs of the television series "*Viper and the Defender*", the famous Dodge model (above), already very impressive, has received some cosmetic touches which make it frankly monstrous. Eagle pushed the Vision (below) to a paroxysm by creating the Aerie whose safety and performance are superior to those of the current model. Its 3.5 L V6 developes 275 hp. The hydraulic suspension is adjustable and the latest low profile Goodyears can continue to run after a flat. Inside, safety has been improved by a rollbar cage. All rear-view mirrors switch from day to night vision automatically. A satellite guidance system allows you to find a sky-rocket priced hotel or a restaurant anywhere in North America. Those who have problems with this system can always use the voice-activated, hand-free, cellular phone.

STYLES

TOYOTA

At the latest Tokyo Show, the Japanese builder presented the latest AXV-V. This simple, aerodynamic and realistic sedan could one day take its place among the Toyota offerings. Below, an all terrain leisure car,the RAV-4 similar to a Suzuki Sidekick, has nothing futuristic as it is already in production and will make its appearance in some markets in the course of the year.

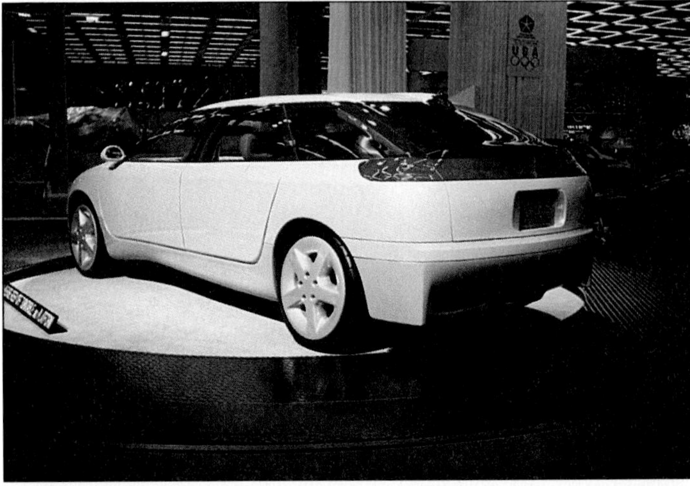

SUBARU

The Sagres concept car, unveiled at the latest Tokyo show, does not lack charm, in spite of its high belt line and forward cabin which result in unusual proportions. This style study, half sedan and half station wagon, receives a pancake six cylinder, an integral transmission and a continuously variable automatic known as the ECV-T.

STYLES

ISUZU

The XU-1 illustrates the futuristic vision of a sports utility machine. Based on the Trooper chassis, it is inspired (from afar) by the lines of a Stealth bomber, with four doors which open upward for ease of entry. The interior is very simple; it has four bucket seats installed along the large center console which extends from front to rear. Other interesting details: when the doors open, the lower portion of the body becomes a step and the spare tire is included in the housing integrated in the two halves of the tailgate. Two ideas which may yet see the light of day.... A CD-Rom system produces not only music but topographic and holographic markers on the video screen.

STYLES

RENAULT

The Zoom, already seen under the imprint of the Matra and Citroen, is an electrical city dweller with a fold up rear which allows it to park in a minimal space with its beetle-like, swing-up wing doors. They clear the body within its width for ultra-close parking. Placed up front, the 25 kw electric motor is fed by Nickel-Cad batteries placed under the seat. Weighing 1,760 lb (of which 770 is for the battery) the Zoom can reach 75 mph on the road and its cruise range is 95 miles in city driving or to 40 miles on the road at a speed of 30 mph. Will the manufacturer supply a ticket immunity certificate entitling the owner to obstruct traffic?

HEULIEZ

France Design has visualized this large, travel sedan-wagon, named Long Cours (Extended Voyage), based on a Renault Safrane.

RENAULT

No contest: the Argos is the most original concept of the year. Based on the Twingo platform and drive train, the body of this clean lined three seat convertible is inspired as much by a boat as by a plane. Composite panels are either satin or matte silver finished. The technical look is underlined by exposed bolts ribbing and the shape of the engine cover. The interior contrasts by the softness of its lines and materials. Almond green leather covers the seats and white silvered wood is used for the dash. Another innovation is a trunk lid that opens like that on a grand piano. The assembly only weighs 1650 lb.

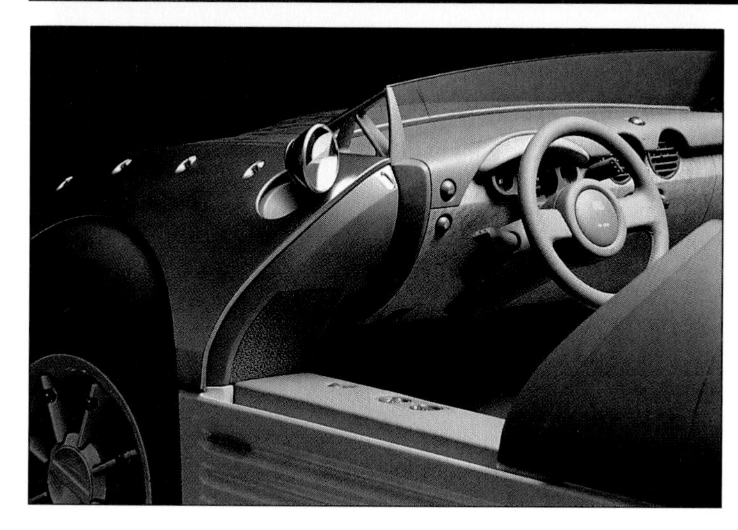

STYLES

SBARRO

This large, impressive coupe was one of the first projects of the students at the Espace Sbarro - a school where Franco Sbarro relays to young car fanatics his experience and know-how. Isatis is a high-tech coupe whose fiberglas-reinforced body connects to the chassis by hydraulic cylinders which stabilize it as a function of centrifugal force exerted on the four axles. The V12 BMW 850 engine produces 320 hp and the ZF transaxle is a 5-speed. The suspension is interactive and the assembly weighs just 2,755 lbs.

SBARRO

Urbi is not a car. It is a vehicle destined to circulate where cars are not admitted: on walks, in public parks and historical centers. It cannot be registered, has neither doors nor headlights, but can be rented by inserting a magnetic card. Ecological, it is powered by two 0.8 kw electric motors located within the rear wheels. Depending on terrain and the number of batteries, its cruise range is from 20 to 40 mph. A resin unit body and a minimal chassis allow its weight to stay in the 770 to 880 lb range.

STYLES

SBARRO

Created by the students of the Espace Sbarro, under the baton of the "maestro", the Oxalys is a light and simple 2+2 convertible which weighs just 1,875 lb. It is dedicated to the pleasure of sportscar driving without neglecting creature comforts. Based on a BMW M5 drive train, it delivers 305 hp and has a twin-walled fiberglas body. The Oxalys draws your attention by the space available at the front seats and the rear seats that have slide-away panels to form a file cabinet. Included is an experimental digital sound system and a front drive train inspired by those used in competition. As shown, the Oxalys could enter limited production and serves notice to budding amateur car manufacturers...

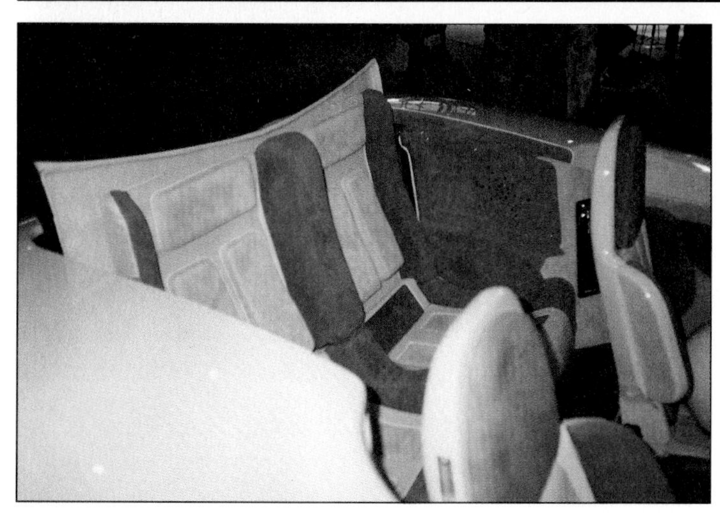

STYLES

BERTONE

The Karisma, presented at the March '94 Geneva Show, is based on the platform and drive train of a 911 Porsche. The idea was to build a 4-seater coupe based on a fabulous powertrain set at the rear. The interest in this enterprise was to provide ample seating and head-room in the rear seats. They also wanted more glazed area to avoid a blind-sided cubicle as is often the case with this type of vehicle. The Karisma commemorates 25 years of collaboration between Porsche and Bertone.

MERCEDES BENZ

The famous and serious German firm surprised the world by announcing in the course of an impromptu press conference its association with Swiss maker Hayek better known as the inventor and promoter of the Swatch watches. The "Swatcedes", really christened Micro Compact Car, will be offered in two forms: Eco-Sprinter and Eco-Speedster. The first puts the accent on ecological aspects while the second is a city machine which is young, maneuverable and can be transformed into a convertible in a few seconds. More "marketing" than technical innovations.

STYLES

ITAL DESIGN

The genial Giorgio Giugiaro is of uneven temper, as proof, this inventive Lucciola and this insipid Lexus Landau. The first is a young and versatile urban vehicle whose numerous tricks will not fail to be picked up elsewhere. It is compact, has a hybrid powertrain, two sunroofs, modular seats, a friendly demeanor - everything is there. For its second creation, you would say that Ital Design tried to insert a Lexus 300 interior into a little Fiat while grafting a V8 from an LS400. Giugiaro generally shows more elegant flamboyance. Here, except for a few interesting details such as the grill and transaxle, the total leaves you wondering.

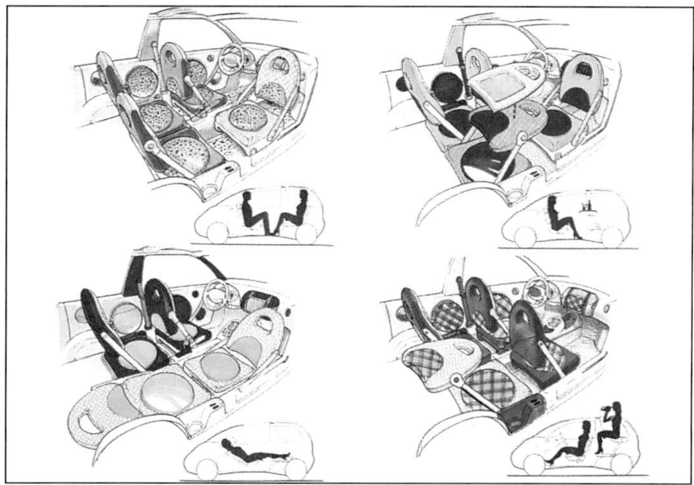

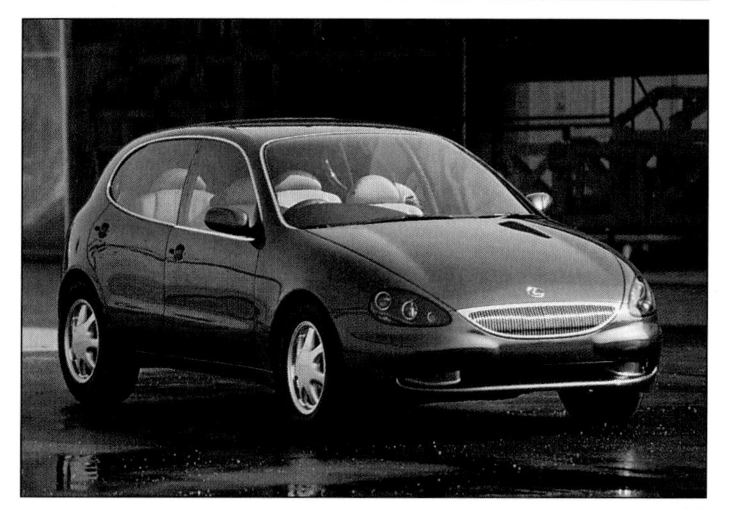

STYLES

VOLKSWAGEN

The Concept 1 revives the legendary Beetle style. Fittingly it was born in California, at the Volkswagen-of-America's Simi Valley Design Center. It is impossible not to recognize at first glance the inspirational source of this lovable design. However, only the lines are similar to the previous model, for here the engine is up front. The options are: a 2.0L Diesel, a 37 kW electric motor or a hybrid with a 3L Diesel and an 18 kW electric motor. The Diesel engine is coupled to an Ecomatic transmission (without a clutch pedal) to reduce fuel consumption and emissions. Just as with the original, all that remains is to sell 21 million copies.

1995 NEW RELEASES

The year 1995 promises to set new records in terms of models introduced in the North American market. You can count over thirty where for a normal year a dozen is a good average. Certainly, those vehicles are not all entirely new. You can notice that, as a result of difficult economic conditions, there is a tendancy to reuse the floor pan of the preceding model while redesigning the body and the interior. This saves nearly half the cost of a new design. Even though a true surge of new business is still not there, the most numerous introductions are those of luxury cars while the econoboxes account for only 1/6 of the new introductions.

CHRYSLER Cirrus

1995 NEW RELEASES

ASTON MARTIN

To accelerate its restart and to ensure the future of this famous British mark, Ford, the latest proprietor, has opted for the introduction of a product that is more "accessible". The new DB7 now fights the Porsche 911 Turbo and 928 and the Ferrari 355. Its 2+2 unit body is built of galvanized steel while the hood and rear deck, front fenders and bumpers are of composite materials.Its 3.2L in-line six cylinder engine develops 355 hp with the assistance of a turbocharger. Luxurious equipment includes leather, rare hardwoods, all the attributes of modern comfort, plus two air bags.

1995 NEW RELEASES

AUDI

The A8, the latest standard bearer of the Ingolstadt make, is a grand luxe sedan which will become available in North America during the year 1995. It will be preceded by the reputation of the most evolved car in the world, outfitted with the integral Quattro transmission, a V8 and an advanced automatic "Tiptronic". The all-aluminum body provides a major weight saving, excellent corrosion resistance and is fully recyclable. Its equipment, luxurious and complete, puts it on the same level as the big BMWs and Mercedes Benz.

1995 NEW RELEASES

BMW

At the latest Geneva Show, BMW revealed its most compact model. It is a 316, derived from a 3-Series platform, shortened 13 cm at the trunk level. Unlike other BMW models, it has a hatchback. In Europe, this car is outfitted with a 1.6L 2 valve, 4 cylinder which offers it 75 hp for a total weight of 2,675 lb. Power steering and anti-lock brakes are standard on this model which competes directly against the Golf GTI. BMW does not foresee for the moment importing the 316 into North America.

BMW

The 7-Series is entirely new for 1995, even though the outside look remains the same as on the previous model. It is "under the skin" that you will find the most important changes. Purists will recognize its larger grill integrated with the hood. The outlook is less massive and provoking than the S Series Mercedes. The new rear suspension provides a considerable road handling improvement and the performance of its V8 4.0L and V12 5L remains remarkable. Among the options is a satellite guidance system.

1995 NEW RELEASES

CHRYSLER

The new Chrysler Cirrus and Dodge Stratus will replace progressively the Acclaim and Spirit, inserting themselves between the Neon and the Intrepid-Concorde. Their lines are inspired by the "advanced-cab" concept, but receive a distinguishing grill. In the engine department you will find the base 2.0L 4-cylinder borrowed from the Neon and the new 2.5L V6. In the same class as the Honda Accord and Mazda 626, these cars will allow Chrysler exports to Europe a considerable chance of success.

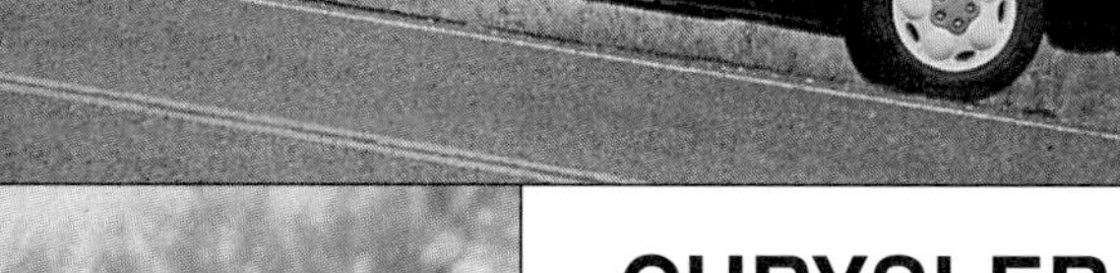

CHRYSLER

Other innovations expected from Chrysler this year are a 2-door version of the Neon as well as an extended cab for the Ram pickup. Also, the Eagle Talon has a new skin and is stockier. It borrows the 2.0L Neon engine for its base version while the TSi with an integral transmission retains the Mitsubishi Turbo 4-cylinder of the preceding model. Wheelbase and track have been increased to improve comfort and stability but the ABS brakes remain optional on all models. The Plymouth Laser disappears from the catalog.

1995 NEW RELEASES

FORD

The Ford Contour and Mercury Mystique replace the Tempo-Topaz which take a well-earned retirement. The latest arrivals illustrate the new world structure dear to the latest Ford president, Alex Trotman, since they are derived from the Mondeo launched in Europe last year. They are at the forefront of modern technology with their multi-valve 4-cylinder 2.0L (Zeta) or V6 2.5L; add fully independent suspension and disk brakes at all four wheels (V6). Available only as 4-door sedans, they directly confront Chrysler's Cirrus and Stratus from the preceding page.

1995 NEW RELEASES

FORD

The Explorer and the Lincoln Continental have taken a rejuvenation cure. The first has the front and rear ends redesigned in a more modern style and its contents have been brought up to the taste of the day by the addition of a second air bag, 4-wheel disc brakes with ABS and front shoulder harnesses whose heights are adjustable. There is a head rest at all seats.

The Continental is all-new, with lines inspired by the Mark VIII coupe from which it also borrows its drive train, including the famous 4.6L V8, the independent, semi-interactive suspension and ABS disk brakes at all four wheels. It will be offered with a unique finish and very complete equipment.

1995 NEW RELEASES

GM

Many improvements at GM, which grafts a mentality change to its deep restructuring. The Buick Riviera has progressed more in the direction of comfort, pleasure and general appearance, than purely on a technical plan, with an engine that dates over 35 years...

The Chevrolet Lumina has been freshened up and has a 2-door Monte Carlo version. They are pleasant to drive and efficient, but the platform and many other mechanical components remain as before.

GM

The Oldsmobile Aurora is the only authentic all-new car of the year at GM. It has been freshened up from known chassis and drive train components on up. This deluxe sedan, built in the Cadillac plant, competes in price with a dozen high-feathered medium class sedans, but delivers superior characteristics. It is well thought out, sharp in appearance, with a good finish, and ample equipment. It presents the new GM face which now offers to supply honest product at honest prices. At last!

1995 NEW RELEASES

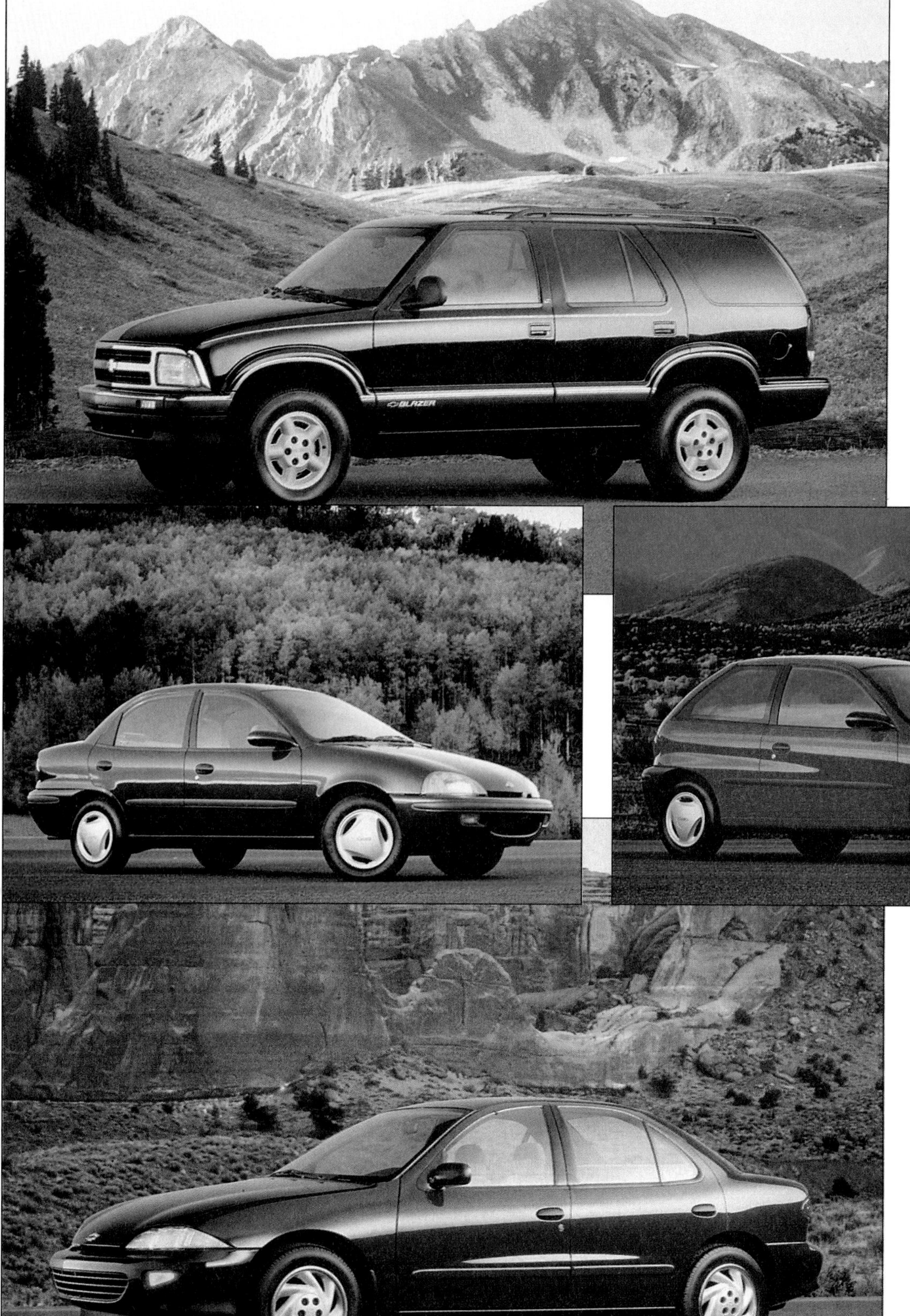

GM

The Blazer and the Jimmy have adopted mechanical and cosmetic changes brought last year to the utility S-10's. They have improved in chassis and body stiffness, have better roadholding, and cut better marks as all-terrain vehicles than the models they replaced. Their interior comfort and ergonomic inside layout have gained points. The Geo Metro also changed personality with a more streamlined style. A wider track improves in stability, but suspension and drive train are identical to preceding models.

GM

The Cavalier-Sunbird receive a body and interior redesign to improve an appearance that has not evolved much in the past 13 years. They will be offered in 1995 in 4-door sedan and 2-door coupe while the wagon and convertible (if such will be) will not join the range till next year. Their platform and base engine remain known, but the V6 will be replaced by the new 4-cylinder Quadfour engine.

1995 NEW RELEASES

HYUNDAI

Introduced at the last Frankfurt Show, the most recent version of the Sonata has seen its lines refined, its interior redesigned and safety improved by a more rigid body and two standard equipment air bags on the front seats of the LS. The drive train remains close to that of the model it replaces. The Accent (below) replaces the Excel as the most popular model in the Korean car builders range. Their design is 100% Hyundai and there will be a choice of 1.3 and 1.5L 4-cylinders with 12 valves.

SUBARU

As with many new 1995 models, the new Legacy is the ultimate evolution of the previous model. The drive train remains identical while the body has been recast into a more international style. The interior has received the same treatment and safety gains include two air bags that gain impact resistance and road handling. The Subaru puts back the emphasis on all-wheel drive which has made it famous and which differentiates it from its competitors, better entrenched in front-wheel drive.

1995 NEW RELEASES

MAZDA

The Millenia, the first of the 1995 Mazda introductions, fits between the 626 and 929 from which it will steal many customers. It is a stand out, noticeable for its high-class lines, the luxury of its equipment and especially by its V6 "Miller cycle" engine, powerful and economic. The 323 and Protegé have been updated in the form of a sporty looking coupe and a more conventional sedan. The drive train remains identical but the interior space is surprising.

KIA

This Korean builder has done more than generate talk about itself in the international arena. It has begun to invade the European market with its Sephia sedans and its sporty utility Sportage.

SEPHIA

SPORTAGE

1995 NEW RELEASES

NISSAN

The greatest innovation at Nissan this year is the new Maxima. Entirely redesigned, it is a superb car which distances itself from the previous styling with a stockier front end and a less elitist appearance. In the powertrain area, the Maxima includes a new all-aluminum 3.0L V6 which developes 190 hp and a multiple arm rear suspension. This keeps the rear wheels perpendicular to the ground for improved handling and side bite. Available in base versions at popular prices, the deluxe and sporty Maxima continues to offer a bit for all tastes.

FIAT

The latest Fiat coupe (above) has a very unorthodox appearance but does not lack performance. Its 4-cylinder 2.0L is placed crossways between the front drive wheels and delivers 190 hp. As a result, it does a 0-62 mph in 6.7 seconds and reaches a max of 148 mph hastily.

NISSAN

The latest 240SX coupe, pictured across, is available in one 2-door model with the 4-cylinder 2.4L.

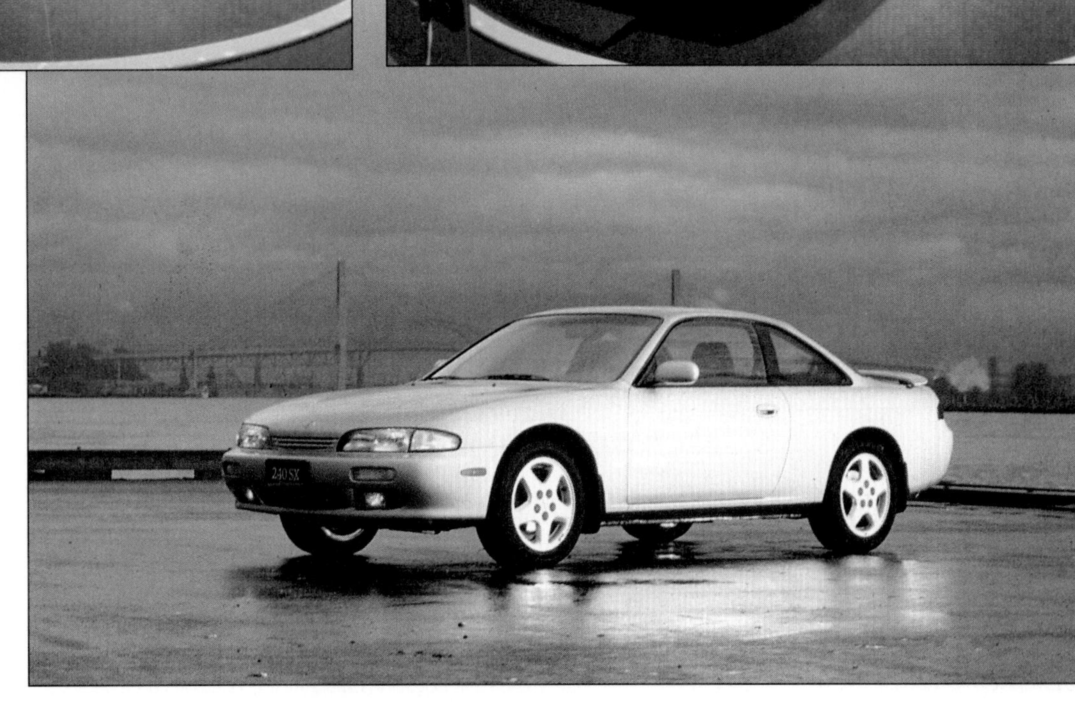

1995 NEW RELEASES

PEUGEOT

After 10 years of reflection, the Peugeot-Citroen group finally gave birth to its long expected mono-volume. At Peugeot, it is called 806 and at Citroen, Evasion. This same vehicle is delivered by Fiat and Lancia in Italy under the names of Ulysses and X. Their drive train and general design are identical but they differ in grills, front and rear headlights and interiors. Base power comes from a naturally-aspirated 4-cylinder 2.0L while the deluxe version uses a Turbo. Peugeot took advantage of the latest Geneva Show to also launch its 306 convertible.

RENAULT

The Laguna replaces the 21 with much elan. In addition to a superb line, it is both a performer and a comfortable machine. The interior is of international class and having tested it for a long time on European roads we can only regret that Renault no longer does business in North America. It will receive several 4-cylinder engines from 1.8 to 2.2L and a 3.0L V6. Its safety level is high and its components 90% recyclable. In the middle at the left: the convertible Renault 19, called Camargue.

1995 NEW RELEASES

SSANGYONG

This Korean builder sets its expansion sights on Europe but more particularly on North America, while its colleagues Hyundai and Kia do not want to see it grow too fast. Presented in Geneva, the latest version of its Korando Family (yes, that's the name!): the Musso...It will get you to the train on time. This is a 4-wheel drive, both sporty and utilitarian, in the class and format of an Isuzu Trooper and equipped with a 3.2L gasoline 6-cylinder Mercedes Benz or a 3.0L turbo Diesel.

SUZUKI

It is not known if you should seriously accept the Suzuki announcement made during the last Detroit Show, which claimed that it wanted to commercialize its new X90 leisure vehicle - sort of a sporty 4-wheel drive. The offspring of a Sidekick and a Swift, the X90 becomes a convertible when you conceal its roof panel in the trunk. Its standard equipment includes two air bags, ABS braking, an adjustable suspension as well as a 1.6L DOHC 100 horse engine.

1995 NEW RELEASES

LEXUS

The latest 400 needs not to prove anything in terms of efficiency and reliability. While the lines of its front end resemble those of the Mercedes a little less, it is now the rear that is inspired by the Stuttgart style. More rigid and more aerodynamic (Cd=0.28), it is also lighter by 220 lb of which only 6.6 were dieted from the body. It retains some of the attributes of the model it replaces such as a 4.0L V8 which gains 10 hp and also has a pneumatic self-leveling suspension that further enhances ride and handling.

TOYOTA

The Tercel whose retouches are more in depth than at the surface will have two bodies sharing a common 1.5L 16-valve engine that is already fitted to the Paseo.

The Avalon replaces the defunct Cressida at the top of the Toyota range. It stems from the platform and drive train of the V6 Camry and accommodates six passengers, thanks to the front seat arrangement and an automatic shift stick placed at the steering wheel. Designed and manufactured in North America, it classifies as a full-sized model by its interior and trunk volumes.

1995 NEW RELEASES

SAAB

The 9000CD has been touched up for 1995 in the grill and trunk areas and receives a new 24-valve V6 3.0L that developes 210 horsepower.

The convertible and coupe versions of the 900 join the sedan renewed last year. The coupe has optional naturally-aspirated 2.0L or 2.3L that make 133 or 150 hp or there is a 185 hp turbo and finally a 2.5L V6 that delivers 170 hp. The electrically-operated convertible will have the same powertrain choice, except for the base 2.0L.

VOLVO

Even though not evident at first glance, the 960 sedan and wagon have received a serious upgrade. Independent front and rear suspensions have been redesigned to improve handling and reduce the steering radius. The front end is redrawn in a less angular style and the body reinforced to offer superior stiffness. A drive train close to that of the older model includes a 2.9L 6-cylinder with an automatic transmission whose shift smoothness was improved. The rear axle of the sedan and wagon are now the same.

UP TO THE MINUTE

SPECIAL PARIS AUTO SHOW

STRAIGHT FROM THE PARIS AUTO SHOW

FROM OUR SPECIAL CORRESPONDENTS
Gérard HÉRAUD & Michel CONDOMINAS

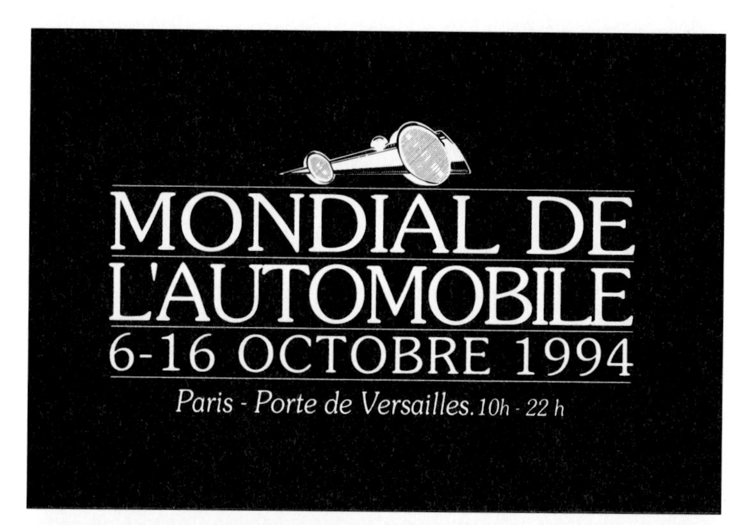

PARIS TOMORROW? MAYBE...

This synthesized image allows you to imagine what the French capital will be like in a few years, when futuristic vehicles will populate the streets with the century old mansions that are the charm of Paris. One can't help but note that the conceivers of this image have been strongly optimistic as to the density of circulation, when one knows of the congestion and pollution the inhabitants of that city are confronted with everyday. Renault affirms that urban vehicles will be minimalist and electric, just like public transport which explains the fluidity, clarity and cleanliness of the ambiance illustrated by this photo.

It is in Paris that the most important auto show of the European Community in 1994, the"Mondial of l'Automobile" was held.

SPECIAL PARIS AUTO SHOW

UP TO THE MINUTE

ALFA ROMÉO

Above are the latest Spider and Coupe GTV of the famous Milanese builder. Their very successful lines result from the collaboration between stylists from Alfa Romeo and from Pininfarina. Switched over to front wheel drive, these two sports cars are outfitted for export with a 192 hp 3.0L V6 that is already standard on the 164. Don't count on seeing these cars on our continent before the end of 1995.

ART CENTER

Below, this little roadster is based on the same minimalist philosophy as the famous Lotus VII. It was designed by the students at Art Center, an industrial design school based in Pasadena, California and in La Tour-du-Peilz in Switzerland. The automobile is one of the favorite subjects of that school's students. There they learn the aesthetics of the different, broad-based industrial products.

UP TO THE MINUTE

SPECIAL PARIS AUTO SHOW

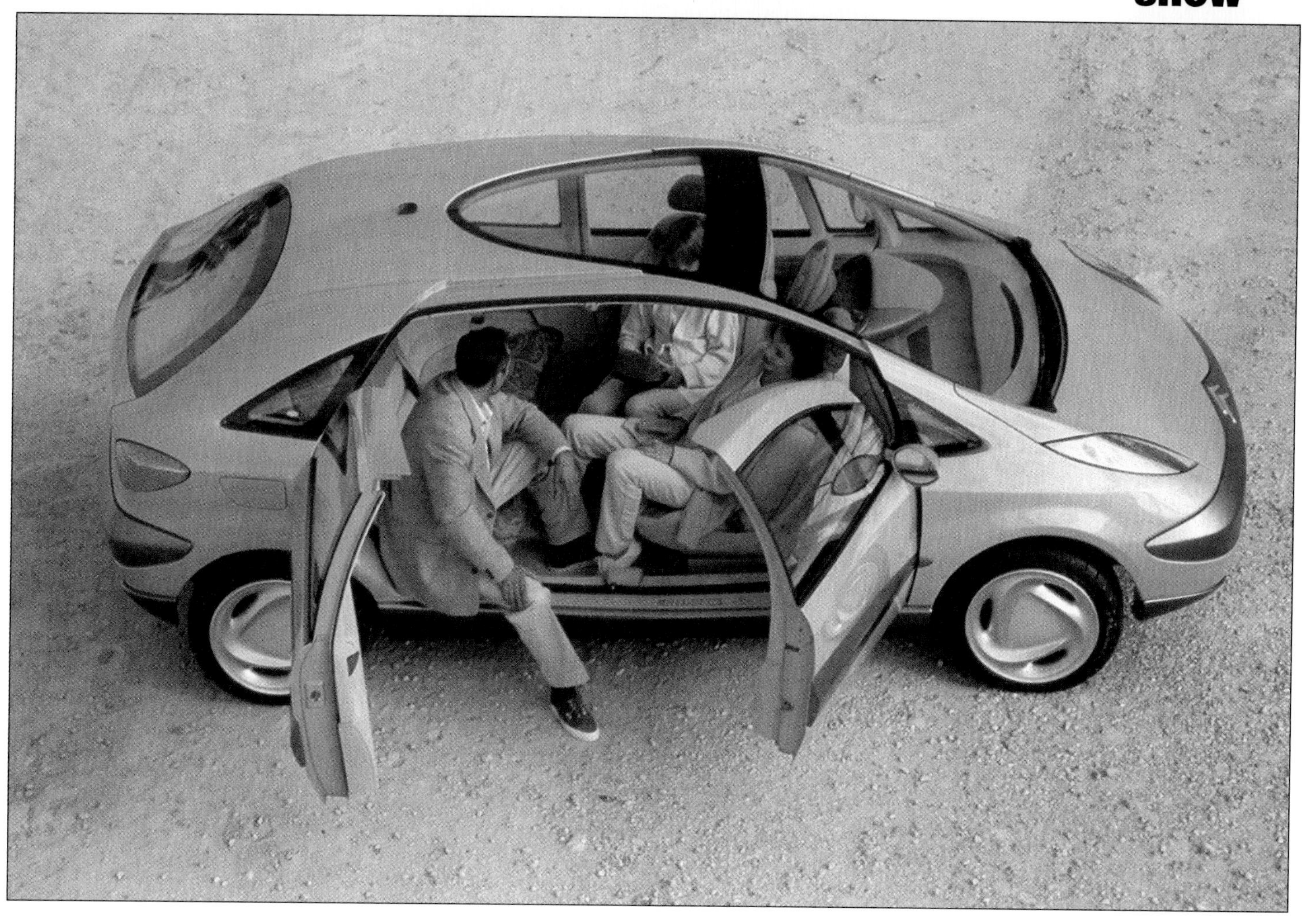

CITROËN

With the Xanae, the French builder wants to bring his contribution to the research into the new generation of vehicles that offer a strong convivial aspect. Mid-sedan and mid-monospace, Xanae offers a modular interior that can accept five people whose access is eased by doors without a center pillar and a major glass area that allows a permanent visual contact with the surroundings.

SPECIAL PARIS AUTO SHOW

UP TO THE MINUTE

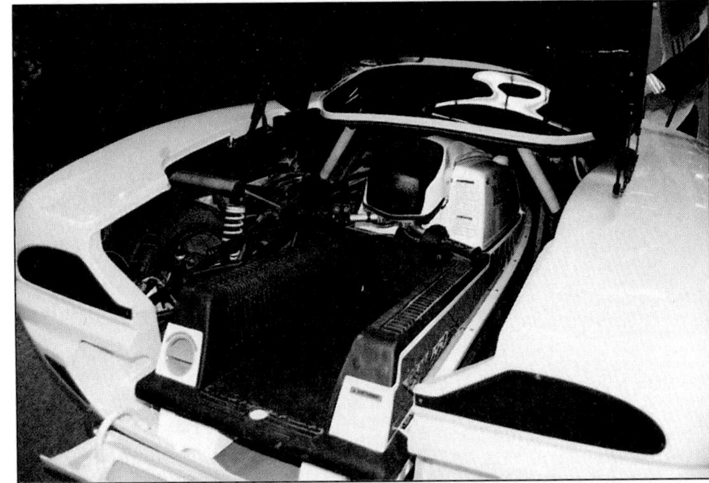

DESIGN PERFORMANCE

Above, the Barramunda is a prototype born from the passion of the automobile and the sea. The rear end is built to accept surf sailors or marine scooters.

FORD

At the right, the Galaxy is an experimental vehicle which Ford studied jointly with Volkswagen (see page 64) and which will be on the market in 1996. The engines are either a 2.0L 4 cylinder or a 2.8L V6 or 1.9L Turbo Diesel. Below, the Scorpio inaugurates an original style that starts out with polyellipsoidal headlights.

UP TO THE MINUTE

SPECIAL PARIS AUTO SHOW

FERRARI

First official launching for the new F355 (below) which replaces the 348 from which it is derived. Offered in a Berlinetta GTB with a Targa roof or in a 348 Spider, its 4.0L V8 now has five valves per cylinder, 40 in all and its power peaks at 380 hp. The lines and powertrain have been revised to improve its performance and handling.

The 512M is the latest evolution of the famous Testarossa. The front end has been spruced up, the headlights are no longer retractable and the grill has been rearranged around different air inlets and headlights. The 5.0L V12 now delivers 440 hp, allowing it to pass from 0 to 62.5 mph in 4.7 sec and attain a max speed of 195 miles per hour.

SPECIAL PARIS AUTO SHOW

UP TO THE MINUTE

LADA

If Russia moves forward as much on the political plan as on the economical one, the automobile side has not been forgotten, and it is in the process of shaking up several decades of lethargy. Following its privatization Autovaz/Lada has launched itself into the study of different projects. The first concerns coupes, sedans and wagons 110, 111 and 112 (above) to be launched in 1995, with modern lines, air bags, power from 1.5L engines with 8 or 16 valves and direct injection.

The Oka (opposite) is a small urban car proposed with either electric or gasoline with a 750cc 2-cylinder. The same engine powers the sports leisure minivehicle called Elfe shown below.

UP TO THE MINUTE

SPECIAL PARIS AUTO SHOW

RANGE ROVER

The legendary Range adopts a new image. Body lines have become more aerodynamic, the cab more spacious and ergonomic. Its engines are two V8s of 4.0L and 4.6L, with an automatic and a pneumatic suspension.

HONDA

As the Europeans are avid hatchback fans, Honda has just revealed its latest interpretation of the Civic. This sports sedan is equipped with a 1.5L VTEC and its luxury finish is decked out with leather and varnished wood.

MEGA

This small French-built sports utility, named Ranch, has a modular body that can be transformed into an open pickup or a closed 4-door sedan. Front wheel drives, these 4-cylinder Citroen engines have 1.1L and 1.3L.

SPECIAL PARIS AUTO SHOW

UP TO THE MINUTE

LAMBORGHINI

The SE is the latest evolution of the Diablo. This is a limited edition with just 150 copies to celebrate the 30th anniversary of the Santa Agata Bolognese firm. The body has been reworked at the front and rear bumpers, made of composite carbon fibers materiels. Its 5.7L V12 has seen its power climb to 492 hp which allows an exceptional 6.4 lbs/hp weight/ power ratio. This allows the SE Diablo to accelerate from 0 to 62.5 mph in 4 seconds and to attain a maximum speed of 206 mph. For this occasion, the interior is covered in leather and suede. The center tunnel is a carbon fiber lay up and the competition-type pedal is of light alloy.

UP TO THE MINUTE

SPECIAL PARIS AUTO SHOW

MERCEDES BENZ

To follow up on the SLK convertible unveiled last May at the most recent Turin Show, Bruno Sacco, the chief stylist of the Stuttgart car builder, created a coupe version "in tribute to metropolitan Paris"... It is a compact sports machine derived from the class C powertrain. This coupe has a two-part hard top capable of disappearing completely into the trunk in less than 30 seconds, thanks to hydroelectric controls.

Both modern and old-fashioned, the inside is displayed in a two tone mode, recalling the roadsters of the 30s'. This prototype is advance notice of a model which will be in the market beginning in 1996 and will oppose the one prepared by BMW of which the philosophy is similar.

UP TO THE MINUTE

MITSUBISHI

With its new Space Gear minivan, the Japanese builder has still not joined the front wheel drive ranks but has moved the engine forward of the driver's compartment. The lines are aerodynamic with a Cd of less than 0.36. The seats, some of which can be mounted on rails and can pivot, allow several original configurations. Different 4-cylinders of 2.0L and 2.4L displacements are available, together with all wheel drive.

PROTON

Below, the Proton 415 GLSi results from a joint venture between a Malaysian conglomerate and Mitsubishi, with the latter holding 30% of the capital. The Proton is a derivative of the Mirage known as the Colt in North America.

UP TO THE MINUTE

SPECIAL PARIS AUTO SHOW

OPEL

Presented last year at the Frankfurt Show in the form of a styling study by Franck Saucedo, this sports minicoupe named Tigra pleased the public so much that a European GM division has decided to put it into production. Based on a Corsa platform, a minicompact which is highly popular in Europe, it has good impact resistance, two high volume air bags and a standard ABS. Its aerodynamic coefficient of 0.31 is remarkable. Also, it is equipped with an air conditioner which includes a pollen filter. Its Ecotech 1.4 L 90 hp and 1.6 L 106 hp are clean, thanks to exhaust gas recirculation and an air pump injecting into the exhaust manifold.

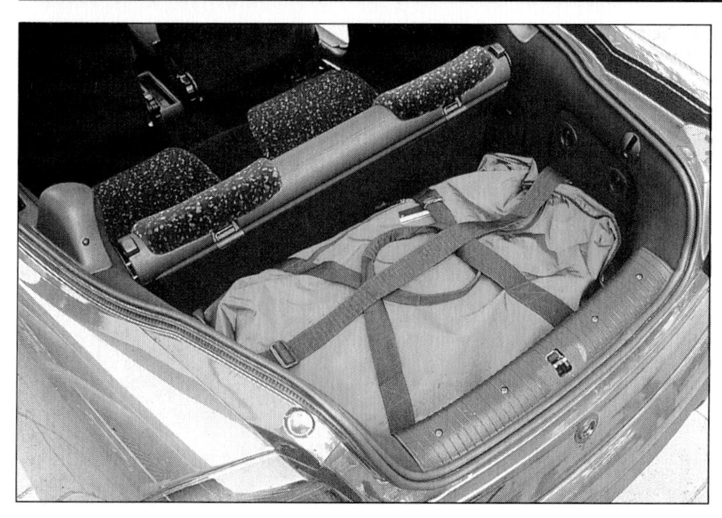

SPECIAL PARIS AUTO SHOW

UP TO THE MINUTE

PASSPORT

The firm Hobbycar, of French descent, (a detail worth mentioning) is well-known for its production of amphibious all-use vehicles. Here, it has given itself the task of creating "resolutely innovating" machines. The Passport finds itself halfway between a compact sedan deluxe and a monospace. It attracts attention by its four easy access sliding doors. A luxurious interior, all leather-covered, seats five in full comfort and is as richly equipped as a high end model. The format is very small, with the length comparable to that of a Toyota Tercel and the powertrain includes an all-wheel drive in a 2.0L engine of Opel descent which delivers 200 hp.

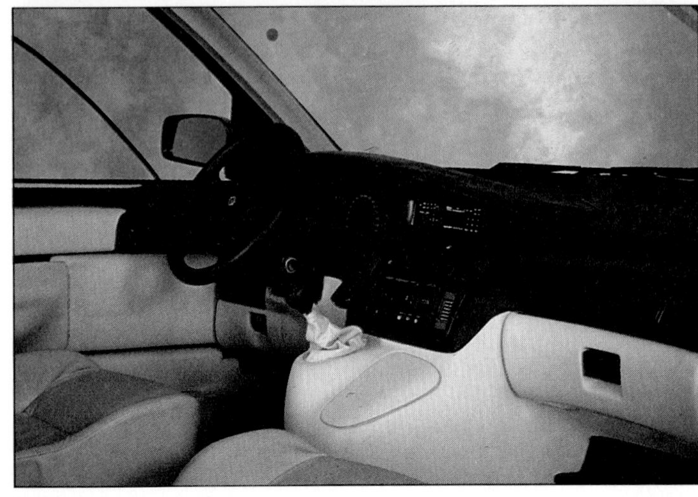

UP TO THE MINUTE

SPECIAL PARIS AUTO SHOW

PEUGEOT

This French builder has long leaned toward the use of electrical energy. Since this type of power is an excellent match to a suburban surrounding, Ion represents what this type of vehicle could be once a suitable means of storing electrical energy has been found. This prototype is tremendously innovative with door handles replaced by sensitive body zones, UV protection in all glazing, automatic temperature control, air filtering, not to speak of a screen at the center of the dash so children can use their video games... But not a word on the batteries, engine or cruise range. Could this be a negative ionic omen?

SPECIAL PARIS AUTO SHOW

UP TO THE MINUTE

PININFARINA

Can a car be at the same time roomy, practical, fun to drive and ecological? Ethos 3 is the Turinese stylist's response. He got prestigious partners like Hydroaluminum for the chassis, General Electric Plastics for body panels, Orbital for a two-cycle engine, BBS for the wheels and PPG for ecological paints. This little 4-door sedan is 20 inches shorter than a Geo Metro and can accommodate six people. Its 1.2 L three cylinder 95 hp engine allows it to reach a 112 mph speed. The most unusual is the steering and the main controls which can be moved laterally across the three front seats in search of a driver.

UP TO THE MINUTE

SPECIAL PARIS AUTO SHOW

RENAULT

Here, you can see two aspects of the creativity of France's prime car builder. Opposite is the Espace F1 minivan which will be driven as "Pace Car" on some Formula 1 circuits next year with Alain Prost at the wheel. It must be said that it is equipped with a famous V10 plus the transmission and rear axle of a Williams FW14. The power, unknown, allows going from 0 to 124 mph in 6.3 seconds and the engine, placed in the center of the interior, keeps musical company with three passengers. Thanks to a composite material body, total weight is only 2,425 pounds.

Below, Modus, a concept utility machine with a Diesel turbine engine, furnishes current to two electric motors.

SPECIAL PARIS AUTO SHOW

UP TO THE MINUTE

VOLKSWAGEN

Above, Sharan illustrates the next minivan in which VW shares paternity with Ford. It will arrive on the marketplace in 1995. Faithful to four hinged doors, it will accept six people on individual bucket seats. However, the production version will not be equipped with a 252 hp 2.8L turbo charged Porsche engine shown in this prototype. Instead, it will receive either a gasoline 115 hp 2.0L engine or 90 hp Diesel or a 174 hp V6.

Opposite, the latest Polo makes its entry under an appearance that resembles its big sister, the Golf. At the lower end of the European range, it receives 4 cylinder gasoline engines of 1.0, 1.3 or 1.6 L or a 1.9L Diesel.

UP TO THE MINUTE

SPECIAL PARIS AUTO SHOW

LANCIA

Little known in North America, this make is part of the Fiat syndicate. In effect, Lancias are fine machines both technically and aesthetically, which gives them an aristocratic touch. The model K, shown here, was revealed at the Paris Show and is at the top of the range. It is a deluxe sedan with a long wheelbase (106.2 inches) with three optional engines: two 5-cylinders of 2.0 L and 2.4 L which give respectively 145 and 175 hp and a V6 3.0L with 204 hp. It is a front-wheel drive with independent suspension and 4-wheel disc brakes and standard ABS. Very complete for Europe, its equipment includes two air bags, automatic air conditioning and a dashboard computer.

UP TO THE MINUTE

NISSAN

A curiosity, this Maxima, christened QX for the European market, comes with a Hercule Poirot mustache. It is identical to that marketed in North America.

INFINITI

The I30 which will be inserted between the G20 and J30, will be presented during the coming American shows. It is based on the platform and power train of the latest Maxima which is a front-wheel drive powered by a 3.0 L V6, coupled to a 4-speed automatic. As with the other models of the make, the I30 has a wide range of standard equipment, including two air bags and ABS.

UP TO THE MINUTE

HONDA

The Japanese car maker is on the eve of revealing this minivan, derived from the base Honda Accord and christened the Odyssey. Sold at between $20,000 and $25,000 American dollars, it will include for this considerable price much equipment as standard. Its 2.2 L. Accord engine will develop 140 hp.

CHRYSLER

Contrary to its custom, Chrysler announced without fanfare the coming introduction of two new sports models, the Chrysler Sebring and the Dodge Avenger. They are identical, except for a few cosmetic details, and built on the Mitsubishi Galant front- wheel drive platform by the Mitsubishi plant in Normal, Illinois.

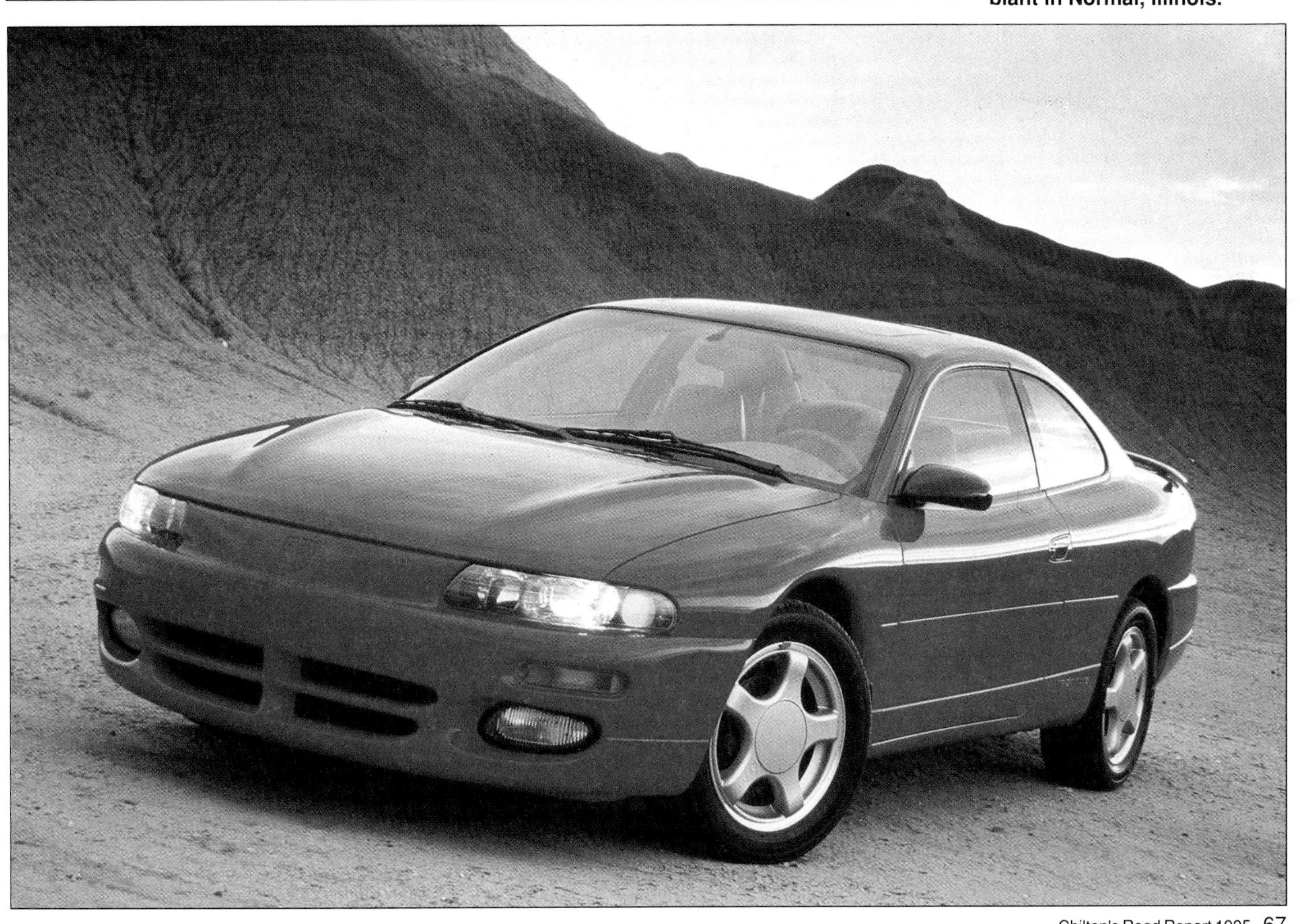

UP TO THE MINUTE

LADA

A second Russian revolution, the Lada improves its Niva (right) by giving it a fuel injection engine, a tailgate that comes down to the floor level, a complete dashboard and seats inspired by those of the Samara.

NEWSTREET

Intended for young French men, 14 to 18 years old, the Newstreet built by boat designer Jeanneau is a 2-wheel scooter with a 49 cc engine whose stand-out feature is that it is sold with a driving course, approved by the French government's accident prevention department.

TOYOTA

This convertible Celica will be sold in the USA during the 1995 model year.

JEANNEAU Newstreet

JEANNEAU Newstreet

TOYOTA Celica convertible

TOYOTA Celica convertible

★REPORT★

ANALYSIS

PONTIAC Sunfire Speedster

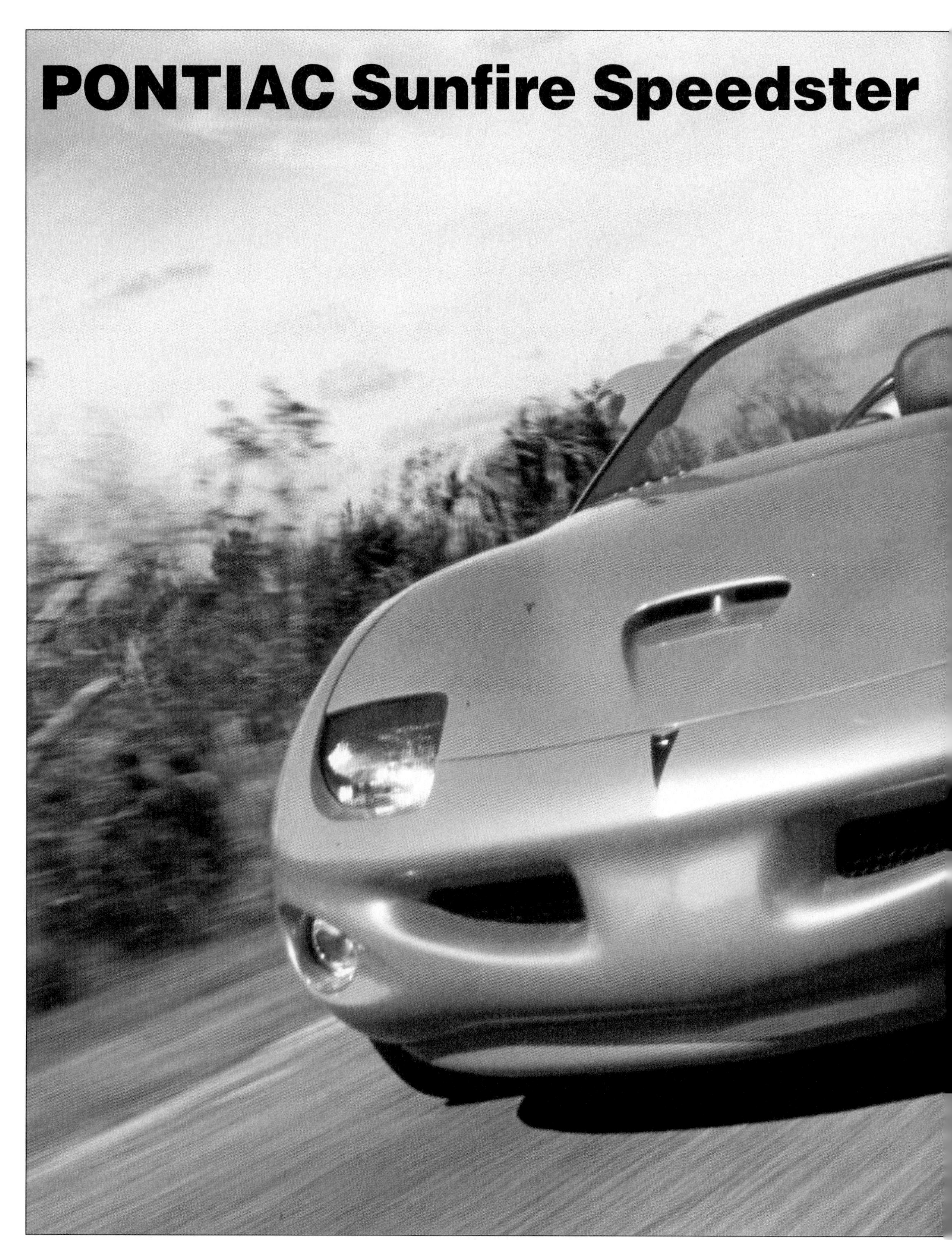

Could one dream of a better cover subject?

At the latest Detroit show Pontiac presented a styling study of its "Speedster" roadster. It previewed the August introduction of the Sunfire model range, designed to replace the Sunbird. This styling exercise was intended to excite people's curiosity and to collect their comments, an excellent means of market research to fine-tune Sunbird styling and engineering features. The Speedster Sunfire did not miss its goal and at Road Reports we have decided to illustrate our cover with it to underline the work of the GM stylists.

"I was in a large group of journalists crowded around the Pontiac stand, waiting for the unveiling of the styling exercise to give a revealing glimpse of the lines for the next replacement of the Sunbird.When the cover was pulled I knew that this Sunfire would crown our 1995 Road Report cover. Very American by its definition, it recalled the Firebirds of years past under a more realistic format. While not a crowning glory of aesthetics, it still had the immediate merit of being popular as shown by the interest of my colleagues and their complimentary comments. Before the turntables starting moving, I was able to glance into the interior and take some photos of this prototype. The first contact had taken place." The first feeling inspired by the spectacle of the Speedster Sunfire is to remember something known while bringing in something new. It is true that the line reminds me of other vehicles, better or less known, with typically North American character such as side exhaust exits, wheels sculptured to emphasize track width and the same for the hood air inlets that recall the Trans Am of the seventies. On the prototype, the lighting system is interesting by the way it combines high intensity high beams with fog lights. Blinker lights are installed in the outside rear-view mirrors They are mounted higher than usual because of the windshield slope which must appear as low as possible. One can see that the windshield is formed by the extension of a V which initiates above the grill and integrates into the RamAir. "A low windshield is the main visual characteristic allowing a roadster to be immediately identified," says Terry Henline, Pontiac's styling chief. "It is the best way to make the vehicle look lower, as though it was in some way part of the road."Another characteristic of this type of vehicle comes from the tonneau cover which creates the illusion of a 2-seater and can be set up in different ways to suit the personality of its owner. It is solid, fixed between the front seat backs and the bridge between the trunk lid and the backlight. It can be removed, to make room for seats transforming the car into a 2+2 convertible. In general, this spot will serve as a storage area for anything that does not fit into the trunk.

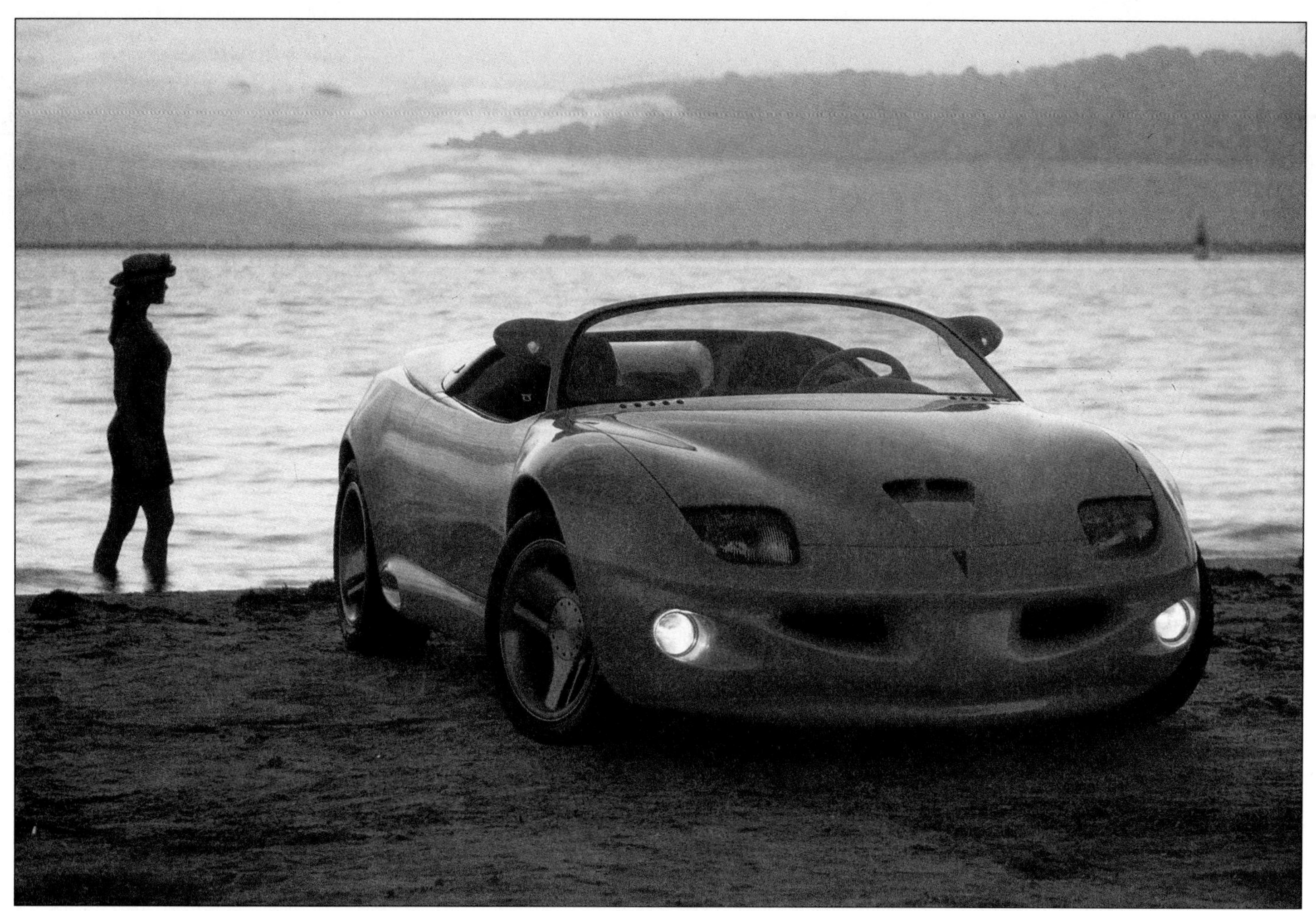

Above, you can see the Sunfire Speedster photographed on a California beach with everything that speaks of beauty and a good time.

Sunfire Mango color has been used for the exterior and for the inside, where it is part of the decor. No accent interrupts, except for the Pontiac and Sunfire logos. "If this vehicle went into production as such, some body sections would receive chrome accents to gain more punch for the total," says Jack Folden.The Sunfire Speedster is not just an inanimate mock-up. Under the hood, the Pontiac engineers installed an East-West 4-cylinder with a 2.4 L double overhead cam, 16 valves equipped with an Eaton supercharger and an air-to-water heat exchanger, as well as a RamAir. All of this hardware allows it to produce 225 hp. This engine comes through with a balance shaft. Thanks to it, you have a 4-cylinder with the feel of a V6. The hood has been designed to facilitate air circulation around the engine assembly. The layout of all control points has been the object of special attention. For instance, to facilitate maintenance, the handles on all dipsticks and the covers of all reservoirs have been painted in the same color as the car:Sunfire

The dashboard strangely resembles that of the existing Camaro and Firebird: chance or conversion ease?

Mango....
The automatic transmission has an electronic control which ties its operation with that of the engine and allows it to optimize up and down shifts. The front suspension unit has a high pressure shock while at the rear you find a tubular axle with helical springs. Steering assist varies as a function of speed and there are disc brakes and ABS at all four wheels. Technically, it is a study more oriented toward marketing than styling or engineering, but you can't market without the last two.

"A connection was needed with the production model..."

John C. Middlebrook, Pontiac's general manager.

"It is twice in four years that Pontiac has used the Sunfire name, which is becoming a refrain to create cars for the young or for the young at heart. This concept study blends Pontiac's standard values with the creation of prototypes that catch the public eye."

It only took three months to come up with this spider roadster beginning with the clay mock-up. This is a record in itself. Its aggressive style would bridge the first concept Sunfire (1990) and the Pontiac of the future. The idea is very close to reality so that the whole world would recognize similarities with the production model. This has helped us excite public curiosity at different shows where the Sunfire has been presented and at clinics where we have received many comments from persons participating in the testing. The polyester resin and fiberglas body shell was painted in an orange type "Sunfire Mango" which has surprised more than one and that was what we needed in that we know the public at which this roadster is aimed likes to be noticed. That is why we have given to this Sunfire a ripped and muscle look and have put the accent on the drive train which is also aggressive, with its DOHC engine and the RamAir feed to produce 225 hp.We wanted from the start to associate a reminiscence of the sixties with wide tracks, a massive look and a notion of a master driver of the nineties, by creating an interior which is organic and functional. The dashboard has supple low forms to give the impression of space and ownership of the road. You would be surprised to see how much you get the impression of being seated high in relation to the road when the dashboard is lowered. We have blended the body color with Graphite leather to confirm the usefulness of the body. The two principal stylists at Pontiac, Jack Folden and Tim Greig began work on this project in June 1994 while continuing to participate in the development of standard models by which these lines were inspired.At Pontiac, conceptual projects always end up in the standard car. Thus the Transport shown in 1988 projected the minivan which was marketed four years later. The Banshee shown in 1988 was the prelude to the Firebird changes of 1992. It is not yet known if the 1991 "ProtoSport4" and the "Salsa" of 1992 will have a production follow-up but one thing is certain, these cars are not created just to produce dreams or to decorate our shows, but rather to prepare the future of our new models.»

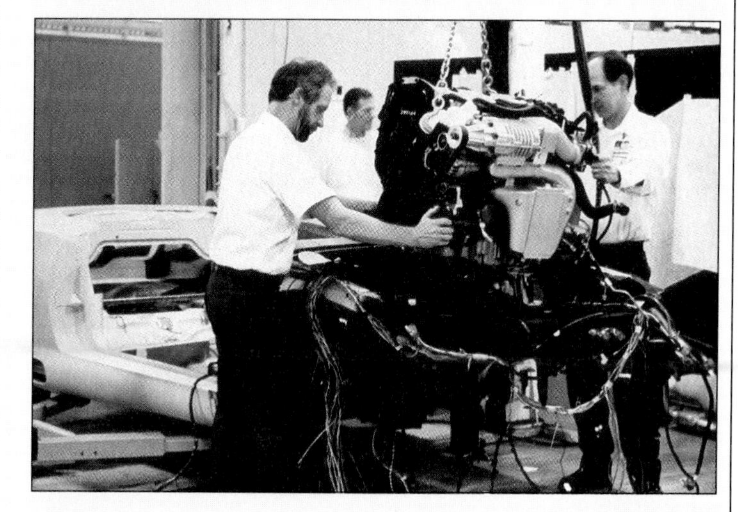

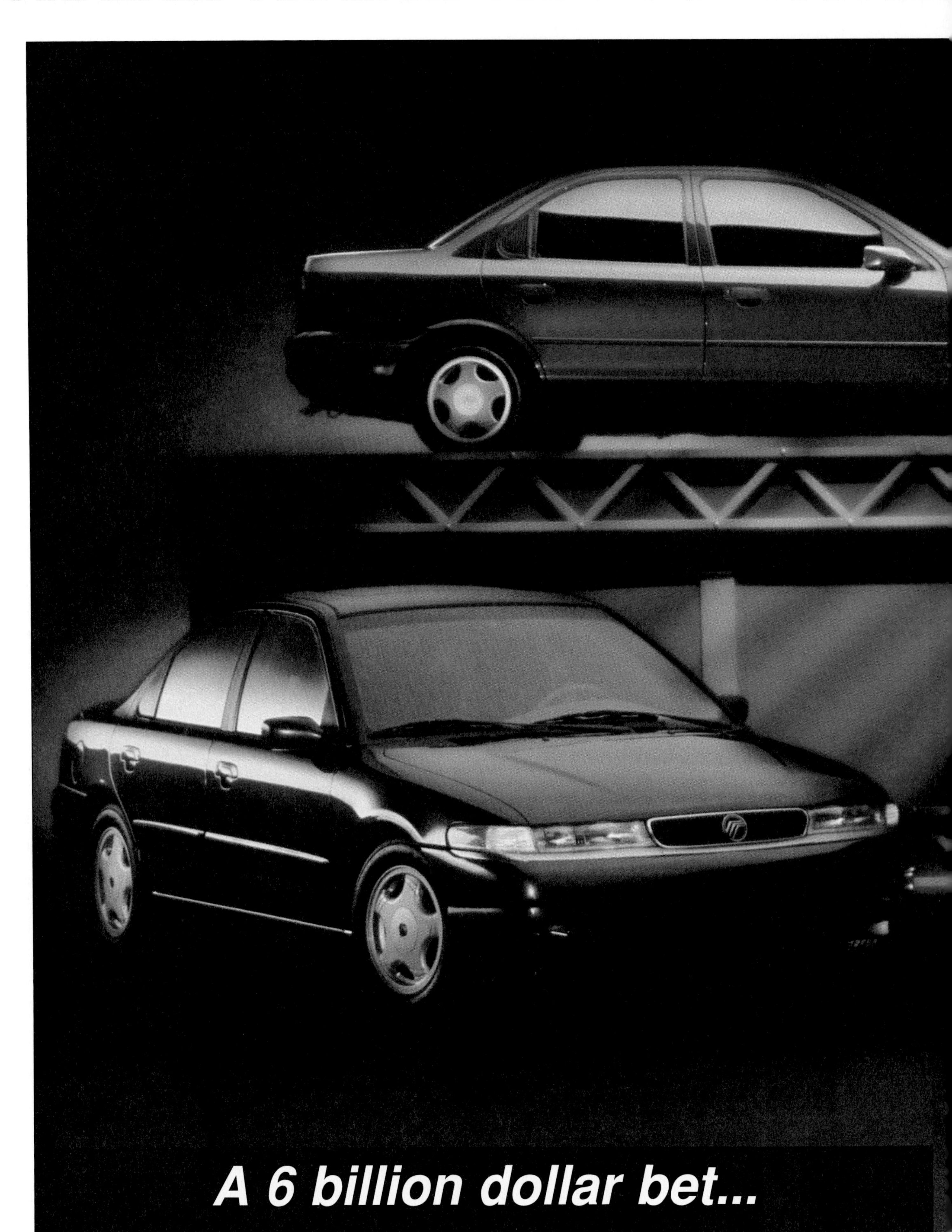

A 6 billion dollar bet...

FORD Contour

The front suspension below is based on a McPherson design, with an A-frame lower control arm, and a stabilizer bar attached directly to the suspension unit. The geometry includes a negative offset to provide better brake stability.

The 2.5 L. Duratec V6 with DOHC is a new aluminum engine, light and compact. It is manufactured at the Cleveland engine plant in Ohio but its wire harness, intake and exhaust manifolds were developed by Ford of Europe. It produces 170 hp and 165 lb/ft

PERFORMANCES OF THE FORD-MERCURY Contour-Mystique FACING THE COMPETITION

	ENGINE					TRANSMISSION		PERFORMANCE									
Model/ versions *: standard	Type / timing valve / fuel system	Displacement cu/in	Power bhp @ rmn	Torque lb.ft @ rpm	Compres. ratio	Driving wheels / transmission	Final ratio	Acceler. 0-62 mph s	Stand. 1/4 mile s	Stand. 5/8 mile s	Braking 62-0 mph ft	Top speed mph	Lateral acceler. G	Noise level dBA	Fuel economy mpg city	hwy	Gasoline type / octane
• FORD-MERCURY Contour-Mystique																	
1)	L4* 2.0 DOHC-16-MPSFI	-	125 @ 5500	130 @ 4000	9.6 :1	front- M5	3.82	10.2	16.8	31.8	144	103	0.76	67	28.0	35.5	R 87
						front-A4	3.92	11.4	17.4	32.4	148	100	0.76	68	26.0	33.0	R 87
2)	V6 2.5 DOHC-24-MPSFI	155	170 @ 6200	165 @ 4200	9.7 :1	front-M5	4.06	8.0	16.2	29.0	154	112	0.78	67	26.0	31.0	R 87
						front-A4	3.77	9.3	16.7	29.8	144	109	0.78	68	24.3	29.0	R 87
1) base 2) * Contour SE option Contour LX & Mystique.																	
• CHRYSLER Cirrus																	
1)	V6* 2.5-SOHC-24-MPFI	152	164 @ 5900	163 @ 4350	9.4 :1	front-A4*	3.90	10.5	15.75	31.5	144	115	0.88	67	21.0	31.0	R 87
• GM L&N-Series																	
1)	L4* 2.2 OHV-8-MFI	133	120 @ 5200	130 @ 4000	9.0 :1	front - M5*	3.83	10.0	16.8	31.0	138	100	0.77	68	25.0	34.0	R 87
						front-A3-A4	3.18	11.7	17.7	31.8	144	97	0.77	68	25.0	31.0	R 87
2)	L4* 2.3 DOHC-16-MFI	138	150 @ 6000	145 @ 4800	9.5 :1	front - M5*	3.94	9.5	16.6	30.6	154	112	0.75	67	23.0	35.0	R 87
						front - A3/A4	2.93	10.4							22.0	32.0	R 87
3)	V6* 3.1 OHV-12-MFI	191	155 @ 5200	185 @ 4000	9.6 :1	front-A3-A4	2.97	9.5	17.7	30.4	150	109	0.80	67	21.0	29.0	R 87
1) *Corsica/Beretta 2)* Grand Am & Achieva SC 3) * Beretta Z26 & Gran Sport, optional on other models																	
• HONDA Accord																	
EX-LX	L4* 2.2 SOHC-16-MFI	132	130 @ 5300	139 @ 4200	8.8 :1	front - M5*	4.062	9.0	16.8	29.2	128	131	0.78	65	30.0	41.0	R 87
						front - A4	4.133	10.2	17.6	30.8	135	124	0.78	65	27.0	39.0	R 87
EX-R	L4* 2.2 SOHC-16-MFI	132	145 @ 5600	147 @ 4500	8.8 :1	front - M5	4.062	8.5	15.8	28.0	121	136	0.80	64	30.0	41.0	R 87
						front - A4*	4.133	9.5e	16.5	31.0	125	131	0.80	64	27.0	39.0	R 87
• HYUNDAI Sonata																	
base	L4* 2.0 DOHC-16-MFI	122	137 @ 5800	129 @ 4000	9.0 :1	front - M5*	4.322	11.0	17.4	31.2	170	118	0.78	68	26.0	38.0	R 87
						front - A4	4.007	12.8	19.0	33.5	141	115	0.78	68	26.0	37.0	R 87
GLS V6	V6* 3.0 SOHC-12-MFI	181	142 @ 5000	168 @ 2500	8.9 :1	front - A4*	3.958	10.0	17.5	30.7	151	124	0.78	67	21.0	28.0	R 87
• MAZDA 626-Cronos																	
DX-LX	L4*2.0 DOHC-16-MFI	121	118 @ 5500	127 @ 4500	9.0 :1	front - M5*	4.11	10.2	16.8	30.2	138	112	0.78	65	30.0	43.0	R 87
						front - A4	3.77	11.3	17.6	30.8	144	109	0.78	65	26.0	39.0	R 87
ES	V6*2.5 DOHC-24-MPFI	152	164 @ 5600	160 @ 4800	9.2 :1	front - M5*	4.11	8.0	16.0	29.5	125	124	0.80	64	24.0	33.0	M 89
• MITSUBISHI Galant																	
S/ES/LS	L4* 2.4 SOHC-16-MPFI	143	141 @ 5500	148 @ 3000	9.5 :1	front - M5*	4.32	8.8	16.6	30.2	141	115	0.79	66	28.0	37.0	R 87
						front - A4	4.35	9.7	17.4	31.0	148	112	0.79	65	26.0	35.0	R 87
LS V6	V6* 2.5 SOHC-24-MPFI	152	155 @ 5500	161 @ 4400	10.0 :1	front - A4	3.91	9.0	16.5	30.8	138	124	0.79	65	21.0	28.0	R 87
• NISSAN Altima																	
base	L4* 2.4 DOHC-16-SFI	146	150 @ 5600	154 @ 4400	9.2 :1	front - M5*	3.650	8.9	16.6	29.7	128	124	0.82	65	29.0	39.0	R 87
						front - A4	3.619	9.5	17.0	30.8	131	118	0.82	65	25.0	37.0	R 87
SE	L4* 2.4 DOHC-16-ISFI	146	150 @ 5600	154 @ 4400	9.2 :1	front - M5	3.895	8.5	16.0	29.2	131	118	0.85	65	29.0	39.0	R 87
• SUBARU Legacy																	
4x2	H4* 2.2 SOHC-16-MPSFI	135	135 @ 5400	140 @ 4400	9.5 :1	front - M5*	3.45	11.0	17.2	32.3	144	112	0.76	67	28.0	37.0	R 87
						front - A4	3.45	12.2	18.7	34.8	148	109	0.76	67	26.0	38.0	R 87
4x4	H4* 2.2 SOHC-16-MPFI	135	135 @ 5400	140 @ 4400	9.5 :1	front - M5*	3.90	11.0	17.2	32.3	144	112	0.76	67	25.0	35.0	R 87
						front - A4	3.90	12.2	18.7	34.8	148	109	0.76	67	25.0	35.0	R 87
• TOYOTA Camry																	
base	L4*2.2 DOHC-16-MPFI	132	125 @ 5400	145 @ 4400	9.5 :1	front - M5	3.94	11.0	18.2	32.3	135	109	0.78	67	27.0	38.0	R 87
						front - A4	3.73	11.8	18.6	32.8	144	106	0.78	67	25.0	35.0	R 87
V6	V6 3.0 DOHC-24MPFI	183	188 @ 5200	203 @ 4400	9.6 :1	front - A4	3.42	9.5	17.6	29.8	151	118	0.80	66	22.0	31.0	M 89
• VOLKSWAGEN Passat																	
GLS diesel	L4* 1.9 SOHC-16-EFI	116	75 @ 4400	100 @ 2400	22.5 :1	front - M5*	3.94	17.8	23.1	37.6	131	103	0.80	68	39.0	49.0	D
GLX VR6	V6* 2.8 DOHC-12-SFI	170	172 @ 5800	177 @ 4200	10.0 :1	front - M5*	3.39	8.5	15.8	29.0	135	140	0.83	66	22.0	32.0	R 87
						front - A4	3.70	10.3	16.7	30.0	138	137	0.83	66	22.0	32.0	M 89

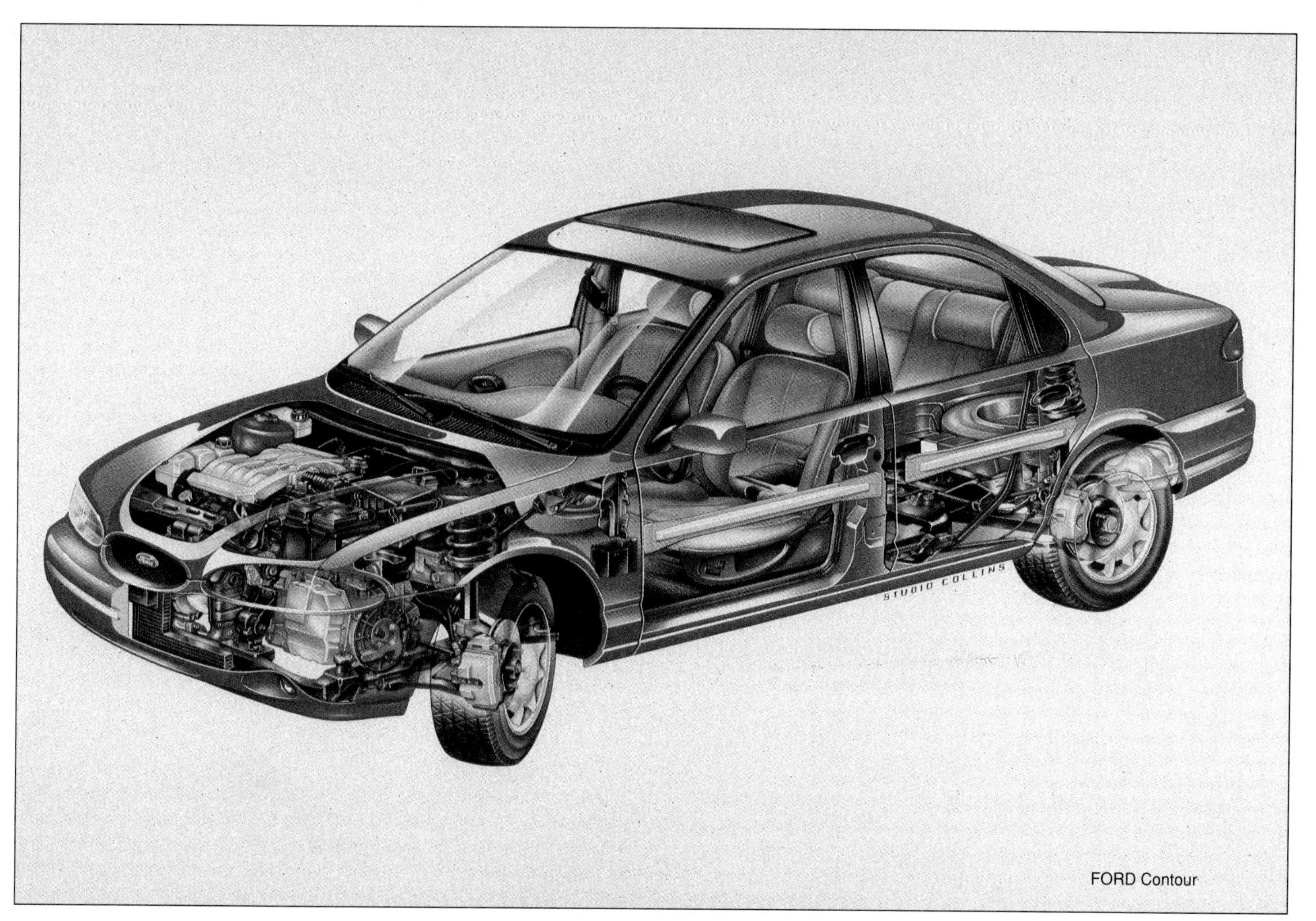

FORD Contour

SPECIFICATIONS & PRICES FOR THE FORD-MERCURY Contour-Mystique FACING THE COMPETITION

Model	Version/Trim	Body/ Seats	Interior volume cu.ft.	Trunk volume cu ft.	Drag coef. Cd	Wheel-base in	Lgth x Wdthx Hght. inx inx in	Curb weight lb	Towing capacity std. lb	Susp. type ft/rr	Brake type ft/rr	Steering system	Turning diameter ft	Steer. turns nber	Fuel. tank gal.	Standard tire size	Standard powertrain	PRICES US. $ 1994
FORD		**Total warranty & antipollution: 3 years / 36,000 miles;corrosion perforation: 5 years / unlimited.**																
Contour	GL	4dr.sdn. 5	89.4	14	0.31	106.5	183.9x69.0x54.4	2769	1000	i/i	dc/dr	pwr.r/p.	36.4	2.78	14.5	185/70R14	L4/2.0/M5	**$13,855**
Contour	LX	4dr.sdn. 5	89.4	14	0.31	106.5	183.9x69.0x54.4	2778	1000	i/i	dc/dr	pwr.r/p.	36.4	2.78	14.5	185/70R14	L4/2.0/M5	**$15,230**
Contour	SE	4dr.sdn. 5	89.4	14	0.31	106.5	183.9x69.0x54.4	2833	1000	i/i	dc/dc	pwr.r/p.	36.4	2.78	14.5	205/60R15	V6/2.5/M5	**$15,231**
MERCURY		**Total warranty & antipollution: 3 years / 36,000 miles;corrosion perforation: 5 years / unlimited.**																
Mystique	GS	4dr.sdn. 5	89.4	14	0.31	106.5	183.3x68.9x54.0	2824	1000	i/i	dc/dr	pwr.r/p.	36.4	2.78	14.5	185/70R14	L4/2.0/M5	**$13,855**
Mystique	LS	4dr.sdn. 5	89.4	14	0.31	106.5	183.3x68.9x54.0	2824	1000	i/i	dc/dr	pwr.r/p.	36.4	2.78	14.5	205/60R15	L4/2.0/M5	**$15,231**
CHRYSLER		**General Warranty: 3 years / 36,000 miles; surface rust: 1 year / 12,000 miles; perforation: 7 years / 100,000 miles; road assistance: 3 years / 36,000 miles.**																
Cirrus	LX	4 dr. sdn. 5	96	15.7	0.31	107.9	187.0x71.0x54.1	3146	1000	i/i	dc/dr/ABS	pwr.r/p.	37.1	3.09	15.8	195/65R15	V6/2.5/A4	**$17,970**
Cirrus	LXi	4 dr. sdn. 5	96	15.7	0.31	107.9	187.0x71.0x54.1	3146	1000	i/i	dc/dr/ABS	pwr.r/p.	37.1	3.09	15.8	195/65R15	V6/2.5/A4	**$19,900**
BUICK		**General warranty: 3 years / 36,000 miles; antipollution: 5 years / 50,000 miles; perforation corrosion: 6 years / 100,000 miles.**																
Skylark	Custom	4 dr.sdn.5	89.5	13.3	0.32	103.4	189.2x68.7x53.5	2941	1000	i/si	dc/dr/ABS	pwr.r/p.	35.3	2.33	15.2	195/70R14	L4/2.3/A3	**$13,734**
Skylark	Limited	4 dr.sdn.5	89.5	13.3	0.32	103.4	189.2x68.7x53.5	2941	1000	i/si	dc/dr/ABS	pwr.r/p.	35.3	2.33	15.2	195/70R14	L4/2.3/A3	**$16,334**
Skylark	Gran Sport	4 dr.sdn.5	89.5	13.3	0.32	103.4	189.2x68.7x53.5	3058	1000	i/si	dc/dr/ABS	pwr.r/p.	35.3	2.33	15.2	205/55R16	V6/3.1/A4	**$18,434**
CHEVROLET		**General warranty: 3 years / 36,000 miles; antipollution: 5 years / 50,000 miles; perforation corrosion: 6 years / 100,000 miles.**																
Corsica	base	4 dr.sdn.5	91.6	379	0.36	103.4	183.4x68.5x54.2	2745	1000	i/i	dc/dr/ABS	pwr.r/p.	35.3	2.33	15.2	195/70R14	L4/2.2/M5	**$13,315**
OLDSMOBILE		**General warranty: 3 years / 36,000 miles; antipollution: 5 years / 50,000 miles; perforation corrosion: 6 years / 100,000 miles.**																
Achieva	S	4 dr.sdn.5	89.8	14	0.33	103.4	189.2x67.5x52.2	2780	1000	i/si	dc/dr/ABS	pwr.r/p.	35.3	2.88	15.2	205/55R16	L4/2.3/M5	**$14,310**
Achieva	SL	4 dr.sdn.5	89.8	14	0.33	103.4	189.2x67.5x52.2	2877	1000	i/si	dc/dr/ABS	pwr.r/p.	35.3	2.88	15.2	205/55R16	L4/2.3/M5	**$17,710**
PONTIAC		**General warranty: 3 years / 36,000 miles; antipollution: 5 years / 50,000 miles; perforation corrosion: 6 years / 100,000 miles.**																
Grand Am	SE	4 dr.sdn.5	90.7	13.3	0.34	103.4	186.9x67.5x52.2	2881	1000	i/si	dc/dr/ABS	pwr.r/p.	35.3	2.5	15.2	195/70R14	L4/2.3/M5	**$12,614**
Grand Am	GT	4 dr.sdn.5	90.7	13.3	0.34	103.4	186.9x67.5x52.2	2940	NR	i/si	dc/dr/ABS	pwr.r/p.	35.3	2.5	15.2	205/55R16	L4/2.3/M5	**$15,114**
HONDA		**General warranty: 3 years / 36,000 miles; Power train: 5 years / 60,000 miles.**																
Accord	DX	4dr.sdn.5	94	13	0.33	106.8	184.0x70.0x55.1	2811	2000	i/i	dc/dr	pwr.r/p.	36.0	3.11	17.0	185/70R14	L4/2.2/M5	**$14,330**
Accord	LX	4dr.sdn.5	94	13	0.33	106.8	184.0x70.0x55.1	2877	2000	i/i	dc/dr	pwr.r/p.	36.0	3.11	17.0	185/70R14	L4/2.2/M5	**$17,230**
Accord	EX	4dr.sdn.5	94	13	0.33	106.8	184.0x70.0x55.1	3009	2000	i/i	dc/dc/ABS	pwr.r/p.	36.0	3.11	17.0	195/60R15	L4/2.2/M5	**$19,750**
HYUNDAI		**General warranty: 3 years / 36,000 miles; Power train: 5 years / 60,000 miles;perforation corrosion: 5 years / unlimited.**																
Sonata	GL	4dr.sdn.5	101	13.2	0.32	106.3	185.0x69.7x55.3	2769	1000	i/i	dc/dr	pwr.r/p.	34.6	3.1	17.2	195/70R14	L4/2.0/M5	**$12,799**
Sonata	GL V6	4dr.sdn.5	101	13.2	0.32	106.3	185.0x69.7x55.3	2921	1000	i/i	dc/dc	pwr.r/p.	34.6	3.1	17.2	195/70R14	V6/3.0/A4	**$14,369**
Sonata	GLS	4dr.sdn.5	101	13.2	0.32	106.3	185.0x69.7x55.3	2820	1000	i/i	dc/dr	pwr.r/p.	34.6	3.1	17.2	195/70R14	L4/2.0/A4	**$14,199**
Sonata	GLS V6	4dr.sdn.5	101	13.2	0.32	106.3	185.0x69.7x55.3	2972	1000	i/i	dc/dc	pwr.r/p.	34.6	3.1	17.2	195/70R14	V6/3.0/A4	**$15,769**
MAZDA		**General warranty: 3 years / 50,000 miles; Power train: 5 years / 60,000 miles; corrosion perforation: 5 years / unlimited.**																
626 Cronos	DX	4dr.sdn.5	97	13.8	0.32	102.8	184.4x68.9x55.1	2743	1000	i/i	dc/dr	pwr.r/p.	34.8	2.9	15.8	195/65R14	L4/2.0/M5	**$14,255**
626 Cronos	LX	4dr.sdn.5	97	13.8	0.32	102.8	184.4x68.9x55.1	2743	1000	i/i	dc/dr	pwr.r/p.	34.8	2.9	15.8	195/65R14	L4/2.0/M5	**$16,540**
626 Cronos	ES	4dr.sdn.5	97	13.8	0.32	102.8	184.4x68.9x55.1	2906	1000	i/i	dc/dc/ABS	pwr.r/p.	34.8	2.9	15.8	205/55VR15	V6/2.5/M5	**$21,545**
NISSAN		**General warranty: 3 years / 50,000 miles; Power train: 6 years / 60,000 miles; perforation corrosion & antipollution: 6 years / unlimited.**																
Altima	XE	4dr.sdn.5	93	14	0.35	103.1	180.5x67.1x55.9	2828	1000	i/i	dc/dr	pwr.r/p.	37.4	2.8	15.8	205/60R15	L4/2.4/M5	**$13,999**
Altima	GXE	4dr.sdn.5	93	14	0.35	103.1	180.5x67.1x55.9	2939	1000	i/i	dc/dr	pwr.r/p.	37.4	2.8	15.8	205/60R15	L4/2.4/M5	**$15,154**
Altima	SE	4dr.sdn.5	93	14	0.34	103.1	180.5x67.1x55.9	3027	1000	i/i	dc/dc/ABS	pwr.r/p.	37.4	2.8	15.8	205/60R15	L4/2.4/M5	**$18,179**
Altima	GLE	4dr.sdn.5	93	14	0.34	103.1	180.5x67.1x55.9	3093	1000	i/i	dc/dc/ABS	pwr.r/p.	37.4	2.8	15.8	205/60R15	L4/2.4/A4	**$19,179**
SUBARU		**Warranty: general: 3 years / 36,000 miles; Power train: 5 years / 60,000 miles; corrosion & antipollution: 5 years / unlimited.**																
Legacy	base 4x2	4dr.sdn.4/5	-	12.6	-	103.5	180.9x67.5x55.3	2570	2000	i/i	dc/dc	pwr.r/p.	34.8	3.2	15.8	185/70HR14	H4/2.2/M5	**$13,999**
TOYOTA		**General warranty: 3 years / 36,000 miles; Power train 5 years / 60,000 miles;corrosion:5 years / unlimited; without deductibles or transfert fees.**																
Camry	Dx	4dr.sdn..5	97.3	14.8	0.32	103.1	187.8x69.7x55.1	2932	2000	i/i	dc/dr	pwr.r/p.	35.4	3.06	18.5	195/70R14	L4/2.2/M5	**$16,428**
Camry	LE	4dr.sdn.5	97.3	14.8	0.32	103.1	187.8x69.7x55.1	3086	2000	i/i	dc/dr	pwr.r/p.	35.4	3.06	18.5	195/70R14	L4/2.2/A4	**$19,558**
Camry	SE	4dr.sdn.5	97.3	14.8	0.33	103.1	187.8x69.7x55.1	3208	2000	i/i	dc/dc/ABS	pwr.r/p.	36.7	2.98	18.5	205/60VR15	V6/3.0/A4	**$22,908**
VOLKSWAGEN		**General warranty: 3 years / 36,000 miles; Power train: 5 years / 60,000 miles; antipollution: 5 years / 50,000 miles; corrosion perforation: 6 years.**																
Passat	GLX VR6	4dr.sdn.5	95.5	33.5	0.32	103.3	184.4x67.7x56.3	3190	2000	i/si	dc/dc/ABS	pwr.r/p.	34.1	3.08	18.5	215/50R15	V6/2.8/M5	**$23,075**

Five years ago, Ford Motor Company made a vitally important strategic decision. It began development of a global vehicle capable of being manufactured and sold on both sides of the Atlantic as well as in Asian countries. It was necessary for the world car to include the best in technology and styling at introduction time.

Objectives

The decision was based on the fact that the tastes and needs of different peoples, to whom these vehicles were going to be offered, have come much closer during the last decades. The laws which control safety and emission are in the process of becoming close, if not identical. Ford's analysis showed that the vehicles conceived on one or the other side of the Atlantic where sufficiently similar, that the studies could be combined. This eliminated a number of duplications and substantially reduced development costs and operations and these savings were translated into better vehicle value. Finally, this orientation will offer the possibility of improving transoceanic communications by joining all their computers to share Ford's talents and resources, wherever they are in the world. It was finally decided that the decision center should be at one location, and Ford of Europe was delegated because of this organization's great experience with compact cars.

CONCEPTION

Body

When the stylists began to work on the Contour/Mystique they had the mandate to give it not only a new appearance but also to anticipate styling development of future years. The design needed to build in the high quality that Ford wanted to put in their manufacture and to have the least outside ornamentation.

Chassis

At the time of defining the structure of these two new models, the engineers took particular care to create a cage a "Safety Cell", that protects the interior. High Tensile steel beams built into the doors to protect against lateral impact will form part of the solid beltway around the unit body. After thousands of hours in a wind tunnel, the surface of the body was sufficiently cleaned up of all protrusions that could induce drag. The final 0.31Cd was made possible by cementing to the metal all fixed glass, perfectly integrating the headlights to the grill and designing the outside rear-view mirrors. Also, an air dam was integrated into the molded front bumpers. Even though they share the same platform, the Contour/Mystique were designed differently to join specifically targeted buying segments. Where the Contour shows a more sporty appearance, the Mystique wants to be more luxurious and refined. This way, they are identical except in grills, bumpers, headlights, directionals, hood and trunk lids, taillights, wheel treatment and interiors, which are different for each one of them.

traction control, which works at all speeds.

Engines

The new models are equipped with brand new engines. The 4-cylinder Zetec engine is a 2.0 L. DOHC with 16-valves that develops 125 hp and produces 130 lb/ft of torque. Built entirely out of aluminum, including the valve cover, it is fed by a sequential electronic fuel injection controlled by Ford's EEC-IV. The V6, called Duratech, is a 2.5 L. 60 degrees compact and light power plant, it has two overhead cams per bank of cylinders and 4-valves per cylinder. It produces 170 hp and a torque of 165 lb/ft. It is derive from the modular V8 installed in the Mark 8 coupe. Just like the Zetec, the Duratech uses a ribbed aluminum oil pan that ties into the engine block and reduces the noise propagation.

Transmission

The manual 5-speed MTX 75 is conventional, and the automatic 4-speed CD4E is remarkable in its compact format with an integral electronic computer and high internal efficiency. The Ford Mondeo, Probe, the Mazda 626 and the MX 6 are already equipped with it.

Conclusion

By investing 7 billion American dollars in the development of these vehicles, Ford has taken its most important bet in recent history. Lets just hope they win. See the test, performances and characteristics on pages 156 and 157.

The Contour/Mystique suspension is independent at all four wheels. Up front, there are McPherson struts and lower A-frames. The stabilizer bar is separated from the system to which it is attached by two struts connected to the suspension unit. The A-frame, stabilizer bar and rack-and-pinion are mounted on a sub-frame which allows reduction of road noise and vibrations into the body. At the rear, the "Quadralink" linkage uses the McPherson and four links per wheel, of which the two main ones are horizontal. The other elements work in a vertical direction to maintain the wheel in position under acceleration and braking. The sports suspension of the Contour SE differs by its higher

shock control and larger stabilizer bars.

Steering

It uses a pump derived from those which equip the Lincoln Mark VIII, compact, easy to drive and silent. Braking is mixed on the 4-cylinder engines and with 4-wheel discs on the V6. The ABS is offered as an option, same as the

OLDSMOBILE Aurora
A profound change of culture ?

Rack-and-pinion power steering varies with speed, using a magnetic control.

The front suspension, derived from a McPherson, includes numerous rubber bushings which isolate the front suspension from wheel noises and vibrations.

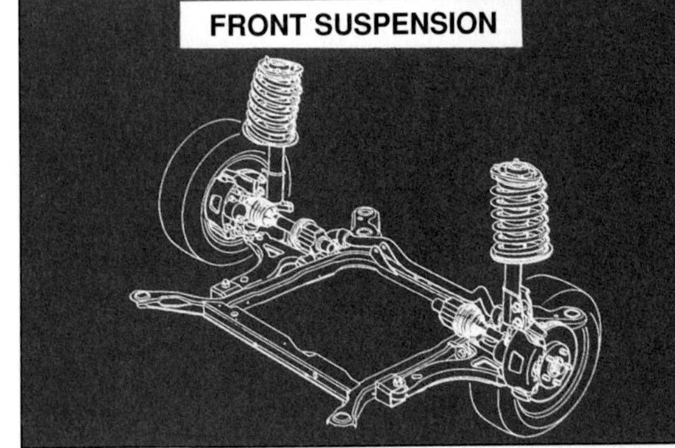

The hydraulic rack-and-pinion power assist varies as a function of speed, thanks to a magnetic device which controls a pump valve and changes the fluid pressure.

PERFORMANCES OF THE *OLDSMOBILE Aurora* FACING THE COMPETITION

	ENGINE					TRANSMISSION		PERFORMANCE									
Model/ versions *: standard	Type / timing valve / fuel system	Displacement cu/in	Power bhp @ rmn	Torque lb.ft @ rpm	Compres. ratio	Driving wheels / transmission	Final ratio	Acceler. 0-62 mph s	Stand. 1/4 mile s	Stand. 5/8 mile s	Braking 62-0 mph ft	Top speed mph	Lateral acceler. G	Noise level dBA	Fuel economy mpg city	Fuel economy hwy	Gasoline type / octane
OLDSMOBILE Aurora																	
Aurora	V8* 4.0 DOHC-32-SFI	244	250 @ 5600	260 @ 4400	10.3 :1	front - A4	3.48	9.5	16.7	30.6	157	134	0.80	65	19.0	27.0	M 89
ACURA																	
Legend	V6* 3.2 SOHC-24-MPFI	196	200 @ 5500	210 @ 4500	9.6 :1	front - A4*	4.37	9.0	16.7	30.2	144	124	0.80	67	22.0	32.0	S 91
Legend	V6* 3.2 SOHC-24-MPFI	196	230 @ 6200	206 @ 5000	9.6 :1	front - M6	4.49	7.8	15.6	28.2	138	131	0.85	66	21.0	34.0	S 91
AUDI																	
A6	V6*2.8 SOHC-12-EFI	169	172 @ 5500	184 @ 3000	10.3 :1	front-A4*	4.00	11.0	17.2	30.0	148	130*	0.78	65	22.0	31.0	S 91
S6	L5* 2.2 T SOHC-10-EFI	136	227 @ 5900	258 @ 1950	9.3 :1	four-M5*	4.11	7.0	14.7	27.8	115	130*	0.85	66	21.0	30.0	S 91
BMW																	
525i/iA	L6* 2.5 DOHC-24-EFI	152	189 @ 5900	184 @ 4200	10.5 :1	rear - M5*	3.23	9.0	16.8	28.8	111	128*	0.80	67	23.0	36.0	M 89
						rear - A4	4.10	9.8	17.4	29.5	115	128*	0.80	67	22.0	33.0	M 89
540i	V8* 4.0 DOHC-32-EFI	243	282 @ 5800	295 @ 4500	10.0 :1	rear - M6	2.93	7.3	15.5	28.4	125	128*	0.85	66	19.0	30.0	S 91
CADILLAC																	
Concours	V8* 4.6 DOHC-32-EFI	279	275 @ 5600	300 @ 4000	10.3 :1	front - A4*	3.11	8.0	16.4	29.0	154	124	0.80	66	16.0	25.0	S 91
De Ville	V8* 4.6 DOHC-32-EFI	279	300 @ 6000	295 @ 4400	10.3 :1	front - A4*	3.71	7.7	16.0	28.6	157	137	0.80	66	16.0	25.0	S 91
INFINITI																	
J30	V6* 3.0 OHC-24-MPSFI	181	210 @ 6400	193 @ 4800	10.5 :1	rear - A4*	3.917	8.7	16.3	29.5	138	131	0.78	65	22.0	30.0	M 89
J30t	V6* 3.0 OHC-24-MPSFI	181	210 @ 6400	193 @ 4800	10.5 :1	rear - A4*	3.917	8.7	16.3	29.5	138	131	0.80	65	22.0	30.0	M 89
Q45	V8* 4.5 DOHC-32-MPSFI	274	278 @ 6000	292 @ 4000	10.2 :1	rear - A4*	3.538	7.5	15.0	25.5	128	149	0.79	65	20.0	29.0	M 89
LEXUS																	
GS300	L6* 3.0 DOHC-24-MPFI	183	220 @ 5800	210 @ 4800	10.0 :1	rear - A4	4.083	9.7	17.7	31.7	125	136	0.80	65	21.0	30.0	M 89
LS 400	V8* 4.0 DOHC-32-EFI	242	260 @ 5300	270 @ 4500	10.4 :1	rear - A4*	3.615	8.5	16.5	29.0	121	136	0.78	65	19.0	28.0	S 91
LINCOLN																	
Continental	V6* 3.8 ACC-12-MPSFI	232	160 @ 4400	225 @ 3000	9.0 :1	front - A4*	3.37	10.0	16.7	30.8	154	115	0.75	66	21.0	33.0	R87
MAZDA																	
Millenia	V6*2.5 DOHC-24-MPFI	152	170 @ 5800	160 @ 4800	9.2: 1	front - A4	4.176	9.3	16.7	29.8	148	125	0.80	65	23.0	33.0	S 91
Millenia S	V6*2.3 DOHC-24-MPFI	138	210 2 5300	210 @ 3500	8.0: 1	front - A4	3.805	8.2	16.1	28.7	135	136	0.80	67	23.0	33.0	S 91
MERCEDES BENZ																	
C280	L6* 2.8 DOHC-24-EFI	171	158 @ 5800	162 @ 4600	10.0 :1	rear- A4	2.87	9.0	16.4	29.3	131	142	0.80	64	23.0	32.0	M 89
E420	V8*4.2 DOHC-32-MEFI	256	275 @ 5700	400 @ 3900	11.0 :1	rear - A4*	2.65	7.5	15.4	27.8	135	149	0.83	66	21.0	32.0	S 91
MITSUBISHI																	
Diamante ES	V6*3.0 SOHC-12-MPFI	181	175 @ 5500	185 @ 3000	10.0 :1	front - A4*	3.958	10.6	17.8	30.5	144	118	0.72	65	21.0	29.0	R 87
Diamante LS	V6*3.0 DOHC-24-MPFI	181	202 @ 6000	201 @ 3000	10.0 :1	front - A4*	3.958	9.2	17.0	29.8	151	131	0.74	65	21.0	28.0	M 89
SAAB 9000																	
base	L4* 2.3 DOHC-16-EFI	139	150 @ 5500	156 @ 3800	10.1:1	front-M5*	4.45	8.5	16.0	31.6	121	125	0.78	66	22.0	35.0	R 87
						front-A4	4.28	9.8	16.9	32.0	127	119	0.78	66	20.0	35.0	R 87
Turbo	L4* 2.3T DOHC-16-EFI	139	200 @ 5000	243 @ 2000	8.5:1	front-M5*	4.45	7.8	15.0	28.4	118	137	0.82	67	23.0	36.0	R 87
				221 @ 2000	8.5:1	front-A4	4.28	8.3	15.9	29.1	125	125	0.80	67	20.0	33.0	R 87
option	V6 3.0 OHV-16-EFI	181	210 @ 6200	200 @ 3300	10.8:1	front-A4	NA										
VOLVO 960																	
960	L6* 3.0 DOHC-24-EFI	178	181 @ 5200	199 @ 4100	10.7 :1	rear - A4*	3.73	9.0	17.2	30.0	121	125	0.78	66	20.0	32.0	R 87

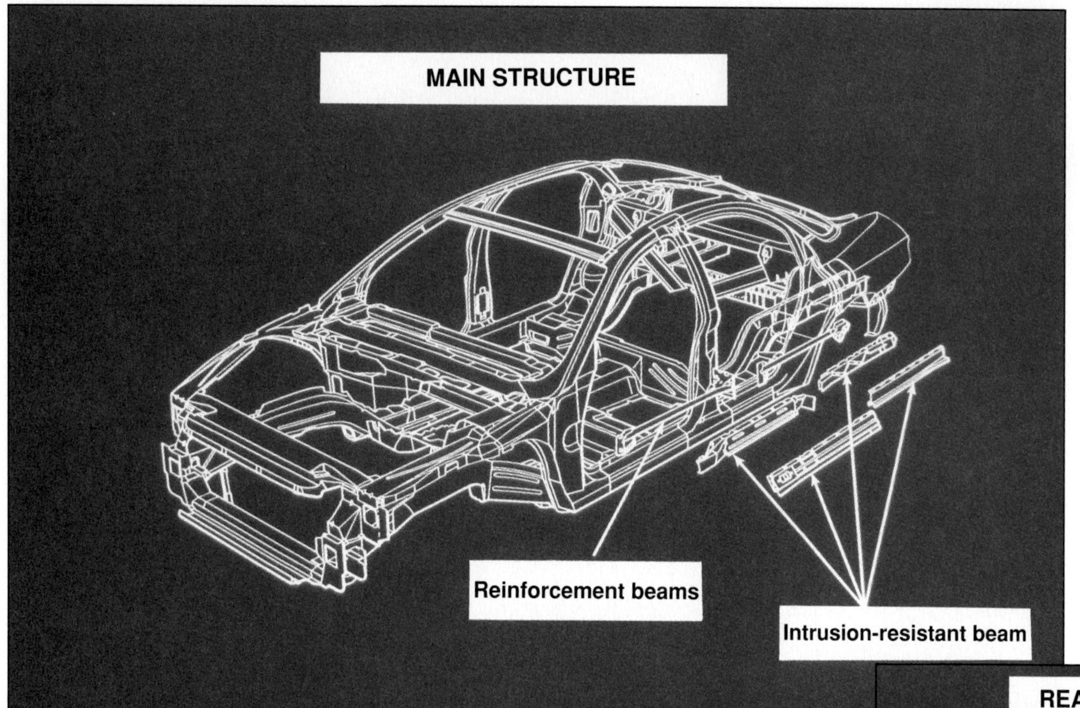

The rear suspension has trailing arms and satisfies both comfort and handling. It includes pneumatic suspension units which provide a leveling control.

The Aurora's unit body is extremely rigid, thanks to numerous reinforcements disposed around it to effectively protect the occupants against collisions and to allow improved handling and better noise and vibration isolation.

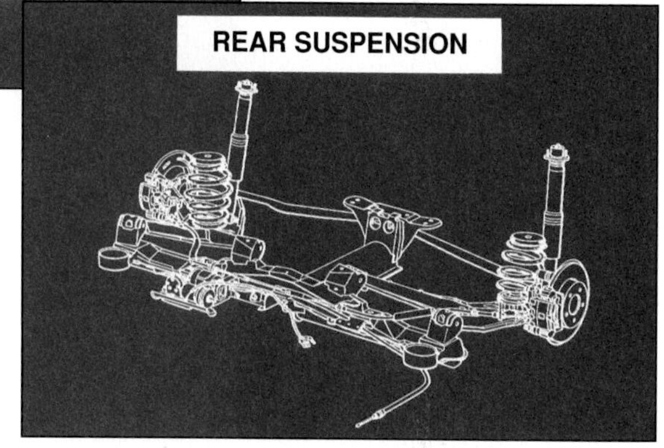

SPECIFICATIONS & PRICES FOR THE *OLDSMOBILE Aurora* FACING THE COMPETITION

Model	Version/Trim	Body/ Seats	Interior volume cu.ft.	Trunk volume cu ft.	Drag coef. Cd	Wheel-base in	Lgth x Wdthx Hght. inx inx in	Curb weight lb	Towing capacity std. lb	Susp. type ft/rr	Brake type ft/rr	Steering system	Turning diameter ft	Steer. turns nber	Fuel. tank gal.	Standard tire size	Standard powertrain	PRICES US. $ 1994
OLDSMOBILE		**General warranty: 3 years / 36,000 miles; antipollution: 5 years / 50,000 miles; perforation corrosion: 6 years / 100,000 miles.**																
Aurora	base	4dr. sdn. 5	102	16	-	113.8	205.4x74.4x55.4	3966	2000	i/i	dc/dc/ABS	pwr.r/p.	41	3.1	20	235/60R16	V8/4.0/A4	**$31,370**
ACURA		**General warranty: 4 years / 50,000 miles: Power train: 5 years / 60,000 miles; surface corrosion: 5 years/ unlimited.**																
Legend	LS	4dr.sdn. 5	93.4	14.8	0.34	114.6	194.9x71.3x55.1	-	2000	i/i	dc/dc/ABS	pwr.r/p.	34.8	3.64	18.0	205/60R15	V6/3.2/A4	**$38,600**
Legend	GS	4dr.sdn. 5	93.4	14.8	0.34	114.6	194.9x71.3x55.1	-	2000	i/i	dc/dc/ABS	pwr.r/p.	36.7	3.64	18.0	215/55VR16	V6/3.2/A4	**$40,700**
AUDI A6 & S6		**General warranty: 4 years / 60,000 miles; antipollution: 6 years / unlimited; perforation corrosion: 10 years.**																
A6	CS	4dr.sdn.5	96	18	0.29	105.8	192.6x70.0x56.3	3384	-	i/i	dc/ABS	pwr.r/p.	34.8	3.1	21.1	195/65HR15	V6/2.8/A4	**$40,570**
S6	S6	4dr.sdn.5	96	18	0.29	106.0	192.6x71.0x56.5	3781	NR	i/i	dc/ABS	pwr.r/p.	34.8	3.1	21.1	225/50ZR16	L5T/2.2/M5	**$49,070**
BMW		**Warranty: 4 years/ 50,000 miles; corrosion: 6 years / unlimited.**																
525	i/iA	4dr.sdn. 5	91	16.2	0.33	108.7	185.8x68.9x55.6	3483	1000	i/i	dc/dc/ABS	pwr.bal.	36.1	3.5	21.4	205/65HR15	L6/2.5/M5	**$38,425**
540	i/iA	4dr.sdn. 5	91	16.2	0.33	108.7	185.8x68.9x55.6	3803	1000	i/i	dc/dc/ABS	pwr.bal.	36.1	3.5	21.4	225/60HR15	V8/4.0/A5	**$47,500**
CADILLAC		**General warranty: 4 years / 50,000 miles; antipollution: 5 years / 50,000 miles; perforation corrosion: 6 years / 100,000 miles.**																
Seville	SLS	4 dr.sdn.5	105	14.4	0.33	111.0	204.0x74.2x54.5	3891	1000	i/i	dc/dc/ABS	pwr.r/p.	41.7	2.65	20	225/60R16	V8/4.6/A4	**$41,430**
Seville	STS	4 dr.sdn.5	105	14.4	0.33	111.0	204.0x74.2x54.5	3891	1000	i/i	dc/dc/ABS	pwr.r/p.	41.7	2.65	20	225/60ZR16	V8/4.6/A4	**$45,330**
INFINITI		**General warranty: 4 years / 60,000 miles; Power train & antipollution: 6 years / 60,000 miles; corrosion perforation: 7 years / unlimited .**																
J30	base	4dr.sdn. 4	86.5	10.1	0.35	108.7	191.3x69.7x54.7	3527	2000	i/i	dc/dc/ABS	pwr.r/p.	36.1	2.93	19	215/60R15	V6/3.0/A4	**$36,950**
J30	t	4dr.sdn. 4	86.5	10.1	0.34	108.7	191.3x69.7x54.7	3578	2000	i/i	dc/dc/ABS	pwr.r/p.	36.1	2.93	19	215/60R15	V6/3.0/A4	**$39,250**
Q45	base	4dr.sdn. 5	96	14.5	0.30	113.2	199.8x71.9x56.3	4039	2000	i/i	dc/dc/ABS	pwr.r/p.	37.4	2.6	22.5	215/65VR15	V8/4.5/A4	**$49,950**
Q45t	base	4dr.sdn. 5	96	14.5	0.30	113.2	199.8x71.9x56.3	4083	2000	i/i	dc/dc/ABS	pwr.r/p.	37.4	2.6	22.5	215/65VR15	V8/4.5/A4	**$54,950**
LEXUS		**General warranty: 4 years / 50,000 miles; Power train: 6 years / 70,000 miles; corrosion perforation: 6 years / unlimited & road assistance.**																
GS300		4dr.sdn. 4/5	104	13	0.31	109.4	194.9x70.7x55.1	3660	2000	i/i	dc/dcABS	pwr.r/p.	36.1	3.23	21.1	215/60VR16	L6/3.0/A4	**$41,100**
LS400		4dr.sdn.5	97	14	0.28	112.2	196.7x72.0x55.9	3693	2000	i/i	dc/dcABS	pwr.r/p.	34.8	3.46	22.4	225/60VR16	V8/4.0/A4	**$51,200**
LINCOLN		**General warranty: 3 years / 50,000 miles; corrosion perforation: 6 years / 100,000 miles..**																
Continental	Executive	4dr.sdn.6	104	19	0.34	109.0	205.1x72.3x55.4	3576	1000	i/i	dc/dcABS	pwr.r/p.	38.4	2.84	18.4	205/70R15	V6/3.8/A4	**$34,750**
Continental	Signature	4dr.sdn.6	104	19	0.34	109.0	205.1x72.3x55.4	3622	1000	i/i	dc/dcABS	pwr.r/p.	38.4	2.84	18.4	205/70R15	V6/3.8/A4	**$36,050**
MAZDA		**General warranty: 3 years / 50,000 miles; Power train: 5 years/ 60,000 miles; corrosion perforation: 5 years / unlimited**																
Millenia	fabric	4dr.sdn. 5	94	13.3	0.29	108.3	189.8x69.7x54.9	3216	NR	i/i	dc/dcABS	pwr.r/p.	37.4	2.9	18	205/65R15	V6/2.5/A4	**$25,995**
Millenia	leather	4dr.sdn. 5	94	13.3	0.29	108.3	189.8x69.7x54.9	3232	NR	i/i	dc/dcABS	pwr.r/p.	37.4	2.9	18	205/65R15	V6/2.5/A4	**$28,300**
Millenia	S	4dr.sdn. 5	94	13.3	0.29	108.3	189.8x69.7x54.9	3391	NR	i/i	dc/dcABS	pwr.r/p.	37.4	2.9	18	215/55R16	V6/2.3/A4	**$31,400**
MERCEDES-BENZ		**Total warranty: 4 years / 50,000 miles.**																
C 280		4dr.sdn.5	84.8	15.3	0.32	105.9	177.4x67.7x56.1	3291	1000	i/i	dc/dcABS	pwr.bal.	35.1	3.5	16.4	195/65R15	L6/2.8/A4	**$34,900**
E420	gas	4dr.sdn.5	94	14.6	0.31	110.2	187.2x68.5x56.3	3748	2000	i/i	dc/dcABS	pwr.bal.	37.1	3.1	18.5	195/65R15	V8/4.2/A4	**$42,500**
MITSUBISHI		**General warranty: 3 years / 36,000 miles; Power Train 6 years / 60,000 miles; corrosion perforation: 7 years / 100,000 miles.**																
Diamante	LS	4dr.sdn. 4/5	94	13.6	0.30	107.1	190.2x69.9x52.6	3604	1000	i/i	dc/dcABS	pwr.r/p.	36.7	3.15	19	205/65R15	V6/3.0/A4	**$32,825**
SAAB		**General warranty: 3 years / 36,000 miles; Power Train 6 years / 75,000 miles; corrosion perforation: 6 years / 100,000 miles.**																
9000	CS	5dr sdn.5	103	18	0.36	105.2	187.4x69.5x55.9	3108	2000	i/r	dc/dc/ABS	pwr.r/p.	35.8	3.0	17.4	195/65TR15	L4/2.3/M5	**$27,745**
9000	CSE	5dr sdn.5	103	24	0.34	105.2	183.5x69.5x55.9	3219	2000	i/r	dc/dc/ABS	pwr.r/p.	35.8	3.0	17.4	205/60ZR15	L4T/2.3/M5	**$33,045**
9000	CDE	4dr sdn.5	103	24	0.34	105.2	188.2x69.5x55.9	3175	NR	i/r	dc/dc/ABS	pwr.r/p.	35.8	3.0	17.4	195/65VR15	L4T/2.3/M5	**$32,685**
VOLVO		**General warranty: 4 years / 50,000 miles; corrosion: 8 years / unlimited; antipollution: 5 years / 50,000 miles.**																
964	SE	4dr.sdn.5	94	16.6	0.35	109.0	191.8x68.9x56.6	3521	2000	i/i	dc/dcABS	pwr.r/p.	31.8	3.5	21.1	205/55VR16	L6/3.0/A4	**$33,950**

The industry has not yet recovered. One year after the unveiling of the Aurora, the whole world is still asking if this car is a traffic accident, a happenstance, or if it announces a profound cultural change of which GM has great need. It seems that Oldsmobile inherited this model by chance, as it could have, with equal luck, been sold by Buick or even Chevrolet.

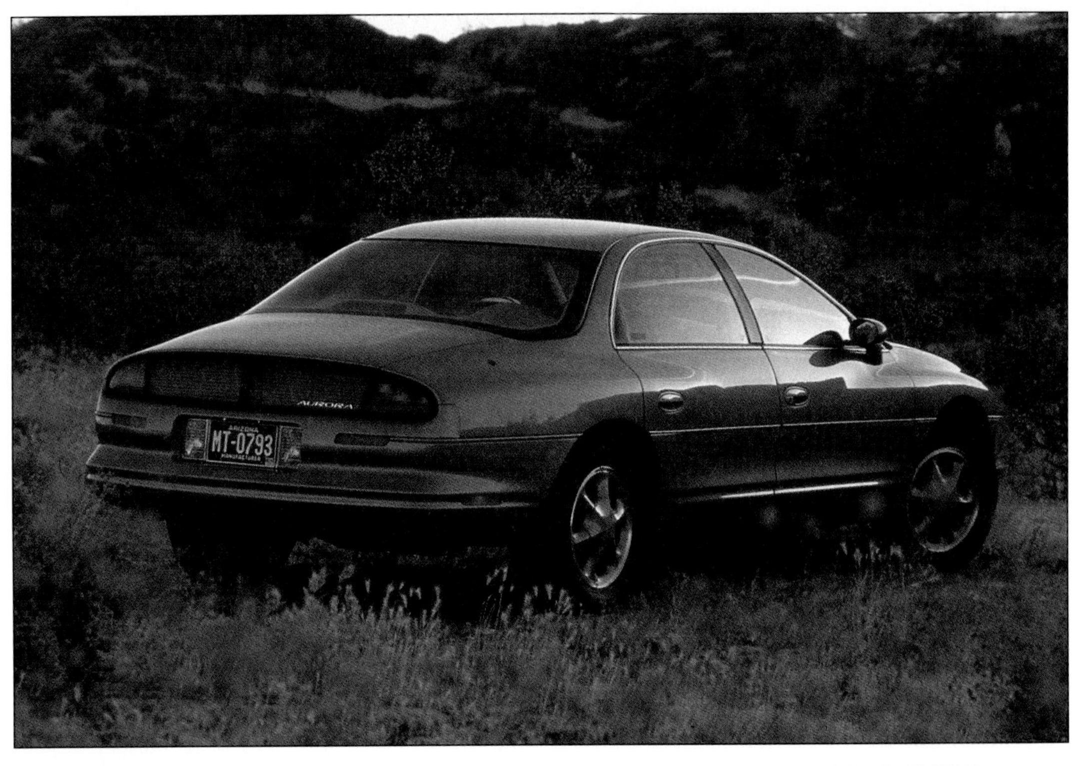

SITUATION

Intention

In the middle of the eighties, market studies made by GM and Oldsmobile revealed that one of the segments which was due to expand the most (75%) during the last decade of the twentieth century, was to buyers making around $50,000 a year. It was also the sector which would know the highest gain.

Objectives

The main objectives are finally simple and clear: a world-class car needed to be built, as reliable and durable as the competition, with a distinct appearance, a prestigious make capable of responding to the needs of a very demanding clientele and finally to emphasize the needs of the driver, occupants' safety, an exceptional price, value ratio and an excellent resale value. During the stage of defining the new vehicle, twenty public evaluation clinics allowed 4,000 people to give their input on this project. The main idea about the Aurora was to offer a target buyer group the car they wanted.

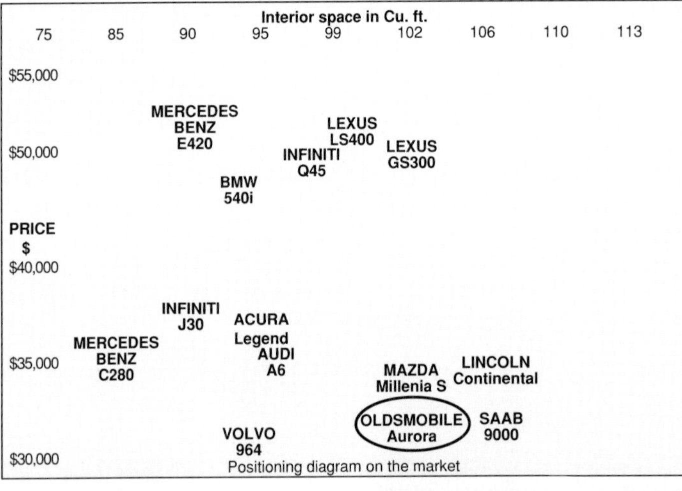

Positioning diagram on the market

CONCEPTION

Body

Cadillac's power train and manufacturing groups in Flint, Michigan were put in charge of the developing of this new model.

Chassis

From the outset, the idea was to build a quality vehicle in the same category where makes like Mercedes and Lexus operate for a long time. This led the Aurora designers to target their characteristics. For instance, the natural chassis frequency of those builders was 20 Hz (Hertz or cycle per second) and the Aurora target was set at 25 Hz to stiffen the car. The real father of the Aurora is Roger Mash, an excellent hands-on engineer responsible for all aspects of finalizing

this car which is now built in the Orion plant in Pontiac, Michigan. Those who designed the Aurora gave it this form of tube-type aircraft body to offer the impression of a massively robust but also very streamlined machine, able to accomplish its function. It's amusing to know that one of the cars studied for these qualities was the Porsche 928. That's why the chromed sections were reduced to a minimum and traditional elements, such as an eggcrate grill, were eliminated. Inside, they wanted the pilot's ergonomics to be perfect, so they treated the dashboard as a real information center. The seats had to be in the competition's class in terms of functionality and appearance, crucial in this segment. Even the equipment level, which had to include the usual service and gadgets, is interesting. For instance, the automatic rear-view interior mirror that switches from day to night was retained, together with the center armrest with a box and 2 cup carriers. As with other cars of this class, the outside rear-view mirrors are heated, air conditioning functions in 2 zones and real wood inserts complete beautiful leather work of good quality.

Engine

The Aurora is equipped with a 4.0 L. V8 derived from the Northstar. Its displacement was set to provide the best ratio between power, torque and economy. It produces 250 hp and a 260 ft/lb torque and its average fuel consumption is around 21 mpg, that of many of its 6-cylinder competitors.

Transmission

The Aurora is a front wheel drive whose automatic 4-speed is electronically controlled together with engine management.

Suspension

It is independent at all 4 wheels, based on the McPherson at the front and trailing arms at the back with self-leveling shocks. This type of suspension allows control of brake dive as well as lift on acceleration, and helps the car maintain a stable attitude in panic stops.

Steering

The hydraulic assist of the rack-and-pinion steering varies as a function of speed, thanks to a magnetic device which controls the pump valve, raising or lowering the pressure of the fluid in the system.

Brakes

The Aurora has 4 discs, controlled by an antiblocking device which also acts as a traction control when the drive wheel traction becomes border line.

DISTINCTIVE FEATURES

What is really distinctive in the Aurora is its price, very competitive for a car of its caliber equipped with a V8 engine. Most of its competitors have an interior volume which is more modest, an inline 6 cylinder or a V6, and are often more costly than the Aurora. The direct competitors range from the Acura Legend to the Chrysler LHS and includes the BMW 540, the Mercedes Benz C Class and the Lexus GS 300. When one compares the Aurora to products that are technically equivalent, such as the Audi A8, the Lexus LS 400, the Infinity Q45, and the Mercedes 400E, its price is practically half of that of its rivals.

CONCLUSION

If quality and reliability hold the rendezvous with the Aurora, we can affirm that something really changed at GM and hope that the new politics will apply to all models of the first world car builder. (See the test, performances and characteristics on pages 190-191.)

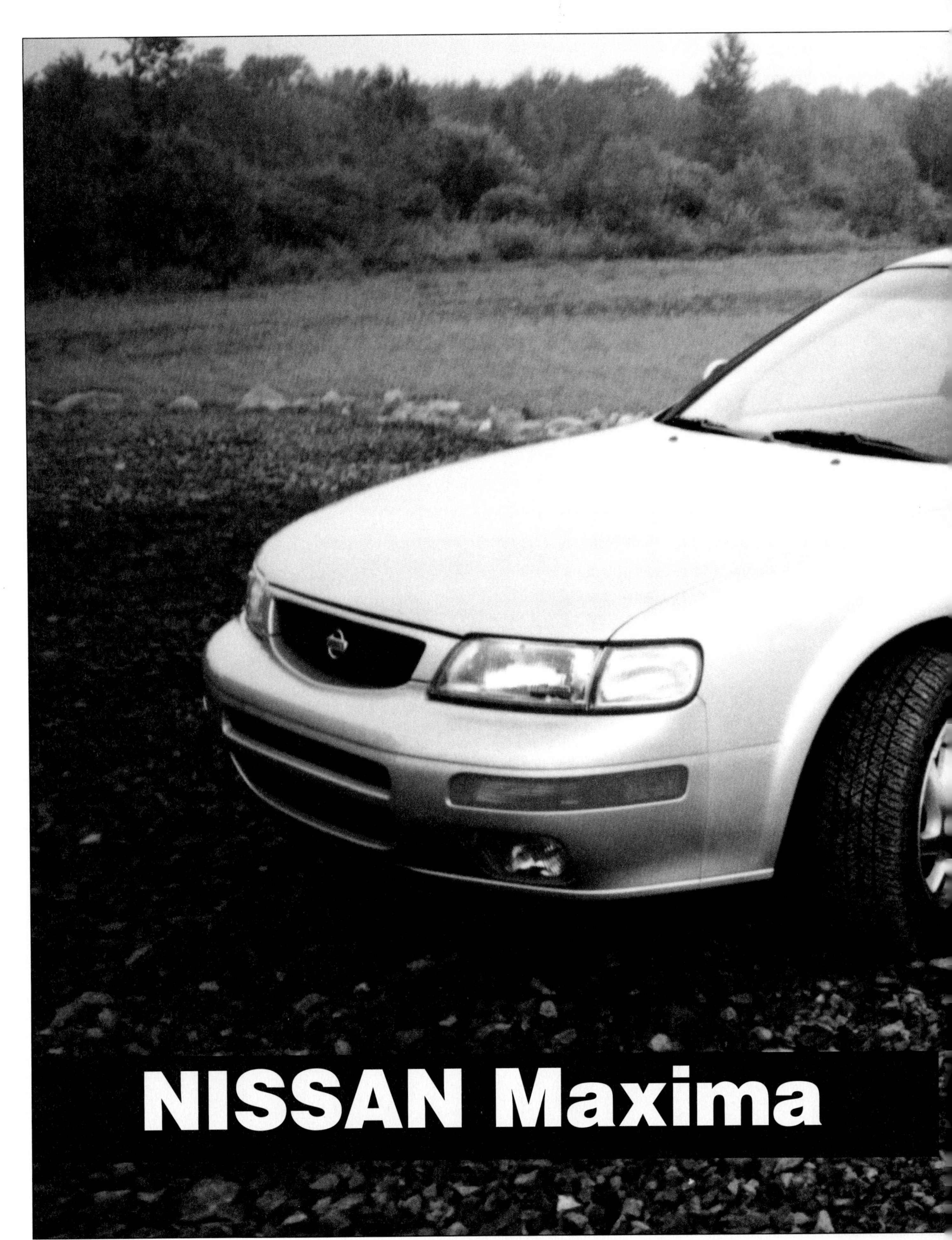

NISSAN Maxima

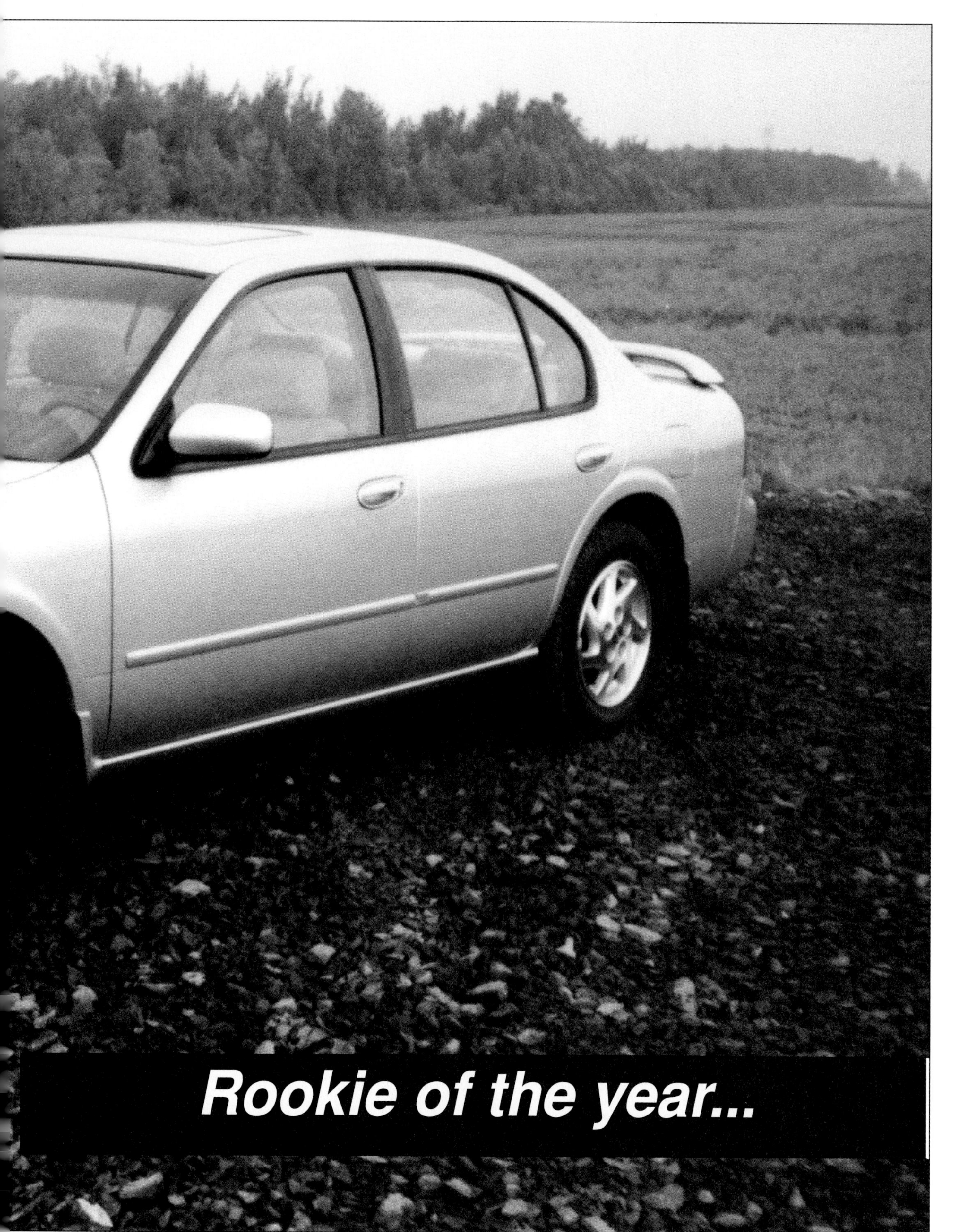

Rookie of the year...

One of the secrets of the Maxima engine resides in the cam drive. In an effort to obtain a compact structure and to reduce vibrations and noise, there are no more intermediate gears and both the primary and secondary chains are simple and reinforced, a design that also reduces engine height.

The Maxima engine is one of the best of its generation in this displacement. It develops 190 hp and a torque of 205 lb/ft. The fuel consumption has improved by 10%, the dimensions by 17% and the weight by 20%. It also includes 10% fewer pieces. Many of its components do not require any maintenance and it is practically recyclable.

PERFORMANCES OF THE *NISSAN Maxima* FACING THE COMPETITION

ENGINE						TRANSMISSION		PERFORMANCE									
Model/ versions *: standard	Type / timing valve / fuel system	Displacement cu/in	Power bhp @ rmn	Torque lb.ft @ rpm	Compres. ratio	Driving wheels / transmission	Final ratio	Acceler. 0-62 mph s	Stand. 1/4 mile s	Stand. 5/8 mile s	Braking 62-0 mph ft	Top speed mph	Lateral acceler. G	Noise level dBA	Fuel economy mpg city	hwy	Gasoline type / octane
NISSAN Maxima																	
base	V6* 3.0 DOHC-24-MPSFI	182	190 @ 5600	205 @ 4000	10.0 :1	front - M5*	3.823	7.0	14.8	27.6	138	138	0.83	66	27.0	35.0	S 91
						front - A4	3.619	7.7	15.4	28.5	147	131	0.83	66	25.0	35.0	S 91
ACURA Vigor																	
Vigor	L5 2.5 SOHC-20 MPFI	150	176 @ 6300	170 @ 3900	9.0 :1	front-M5	4.492	8.0	16.0	29.6	131	125	0.77	66	24.0	35.0	R 87
						front-A4	4.480	8.6	16.5	30.1	125	118	0.77	66	24.0	34.0	R 87
AUDI A4																	
V6* 2.8 SOHC-12-EFI		169	172 @ 5500	184 @ 3000	10.3 :1	front - A4*	4.00	9.5	17.7	30.5	137	131*	0.76	66	21.0	34.0	S 91
A4 Quattro	V6* 2.8 SOHC-12-EFI	169	172 @ 5500	184 @ 3000	10.3 :1	all - A4*	4.29	9.7	18.0	31.0	144	131*	0.76	66	22.0	31.0	S 91
A4 Sport	V6* 2.8 SOHC-12-EFI	169	172 @ 5500	184 @ 3000	10.3 :1	all - M5*	3.89	8.5	16.5	29.8	131	131*	0.79	66	22.0	31.0	S 91
HONDA Accord																	
EX-LX	L4* 2.2 SOHC-16-MFI	131	130 @ 5300	139 @ 4200	8.8 :1	front - M5*	4.062	9.0	16.8	29.2	128	131	0.78	65	30.0	41.0	R 87
						front - A4	4.133	10.2	17.6	30.8	134	125	0.78	65	27.0	39.0	R 87
EX-R	L4* 2.2 SOHC-16-MFI	131	145 @ 5600	147 @ 4500	8.8 :1	front - M5	4.062	8.5	15.8	28.0	121	138	0.80	64	29.0	40.0	R 87
						front - A4*	4.133	9.5e	16.5	31.0	125	131	0.80	64	27.0	39.0	R 87
LEXUS ES300																	
ES 300	V6*3.0 DOHC-24-MPFI	181	188 @ 5200	203 @ 4400	10.5:1	front - A4	3.72	9.2	16.2	29.3	125	131	0.80	64	22.0	31.0	R 87
MAZDA Millenia																	
Millenia	V6*2.5 DOHC-24-MPFI	152	170 @ 5800	160 @ 4800	9.2: 1	front - A4	4.176	9.3	16.7	29.8	148	125	0.80	65	23.0	33.0	S 91
Millenia S	V6*2.3 DOHC-24-MPFI	138	210 @ 5300	210 @ 3500	8.0: 1	front - A4	3.805	8.2	16.1	28.7	135	136	0.80	67	23.0	33.0	S 91
MITSUBISHI Diamante																	
ES	V6*3.0 SOHC-12-MPFI	181	175 @ 5500	185 @ 3000	10.0 :1	front - A4*	3.958	10.6	17.8	30.5	144	118	0.72	65	21.0	29.0	R 87
LS	V6*3.0 DOHC-24-MPFI	181	202 @ 6000	201 @ 3000	10.0 :1	front - A4*	3.958	9.2	17.0	29.8	151	131	0.74	65	21.0	28.0	M 89
SAAB 900																	
SE	V6 2.5 DOHC-24-EFI	152	170 @ 5900	167 @ 4200	10.8 :1	front - M5*	4.45	8.5	16.2	29.0	125	136	0.82	67	22.0	33.0	R 87
						front - A3	2.86	9.2	17.0	29.4	134	130	0.82	67	22.0	33.0	R
TOYOTA Avalon																	
base	V6 3.0 DOHC-24-MFI	183	192 @ 5200	210 @ 4400	10.5 :1	front - A4	3.625	8.7	16.8	29.0	125	125	0.80	64	25.0	35.0	M 89
TOYOTA Camry																	
base	L4 2.2 DOHC-16-MPFI	132	135 @ 5400	145 @ 4400	9.5:1	front- M5	3.94	11.0	18.2	32.3	128	109	0.78	65	27.0	38.0	R 87
						front-A4	3.94	11.8	18.6	32.8	135	106	0.78	65	25.0	35.0	R 87
V6	V6 3.0 DOHC-24-MPFI	152	188 @ 5200	203 @ 4400	9.6:1	front-A4	3.62	9.5	17.6	29.8	131	119	0.80	64	22.0	31.0	R 87
VOLVO 850																	
Turbo	L5T* 2.3 DOHC-20-MPFI	142	222 @ 5200	221 @2100	8.5 :1	front - M5	2.54	7.6	15.6	28.7	131	146	0.80	68	23.0	34.0	M 89
960	L6* 3.0 DOHC-24-EFI	178	181 @ 5200	199 @ 4100	10.7 :1	rear - A4*	3.73	9.0	17.2	30.0	121	125	0.78	66	20.0	32.0	R 87

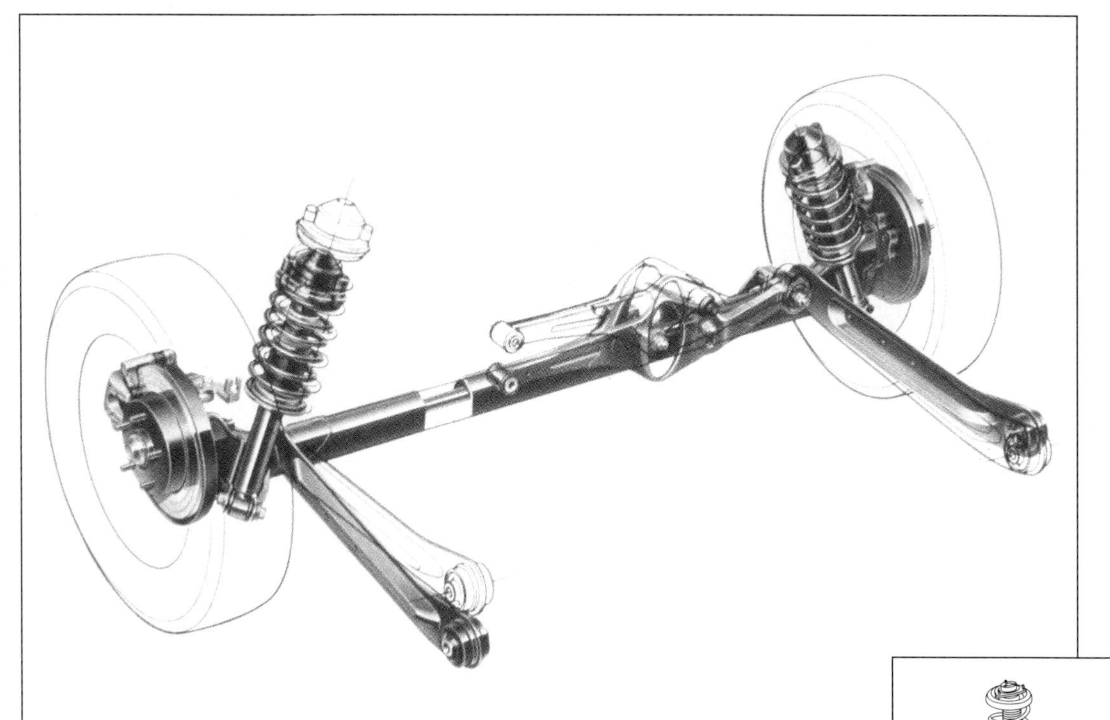

The front suspension (below) picks up the design of the J30 Infinity with a flexible mounting at the main cross member, to eliminate the vibrations and improve comfort. The antidive geometry and the antilift improve directional stability.

The rear suspension called Multilink (above) is based on the torsion axle design of Scott Russell, which is the one most suited for front wheel drives. The transverse arms have a better lateral road handling control. The shock absorbers also assist with axle control and eliminate camber changes at the wheels, which gains improved tire contact and cornering.

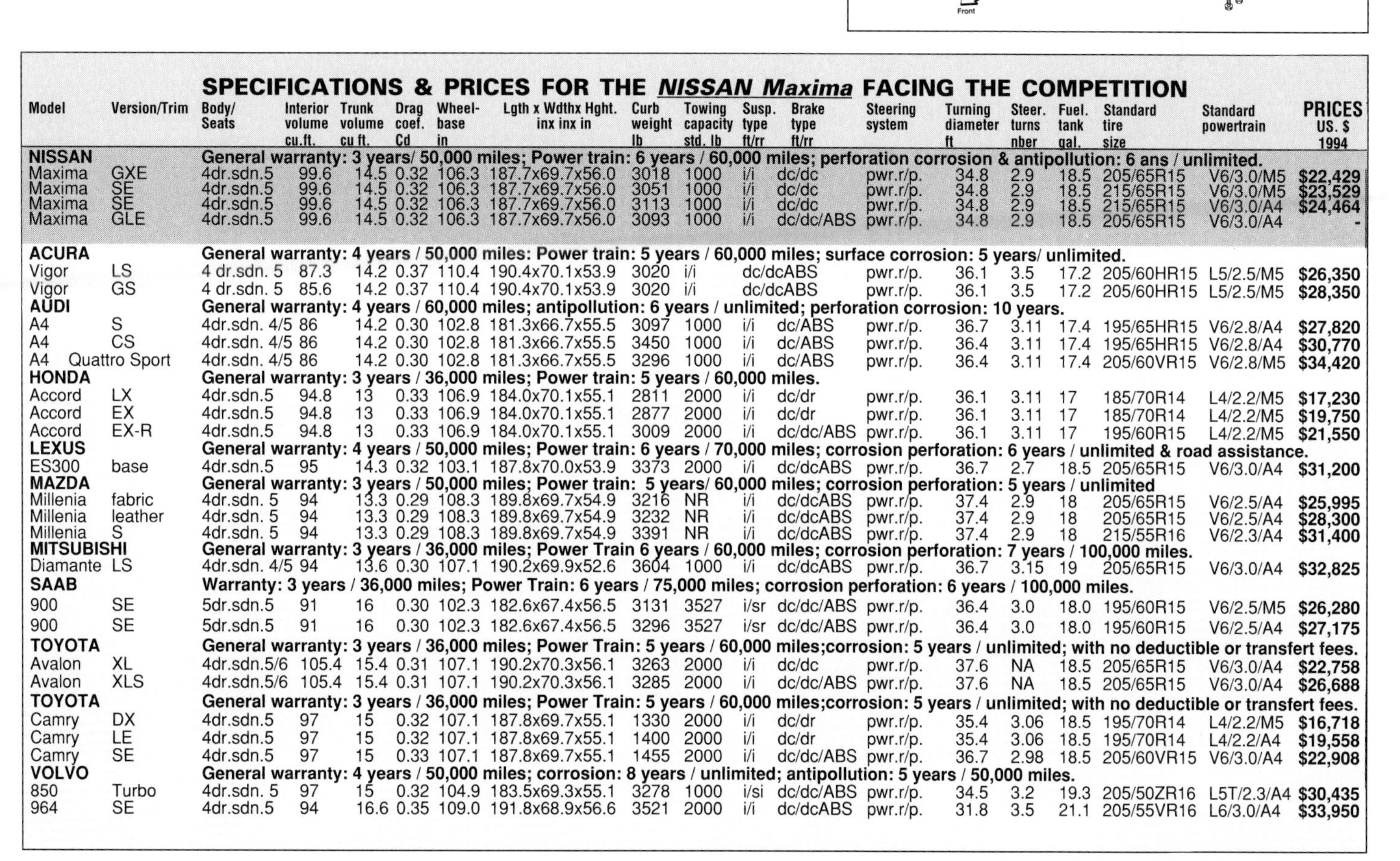

SPECIFICATIONS & PRICES FOR THE *NISSAN Maxima* FACING THE COMPETITION

Model	Version/Trim	Body/ Seats	Interior volume cu.ft.	Trunk volume cu ft.	Drag coef. Cd	Wheel-base in	Lgth x Wdthx Hght. inx inx in	Curb weight lb	Towing capacity std. lb	Susp. type ft/rr	Brake type ft/rr	Steering system	Turning diameter ft	Steer. turns nber	Fuel. tank gal.	Standard tire size	Standard powertrain	PRICES US. $ 1994
NISSAN		**General warranty: 3 years/ 50,000 miles; Power train: 6 years / 60,000 miles; perforation corrosion & antipollution: 6 ans / unlimited.**																
Maxima	GXE	4dr.sdn.5	99.6	14.5	0.32	106.3	187.7x69.7x56.0	3018	1000	i/i	dc/dc	pwr.r/p.	34.8	2.9	18.5	205/65R15	V6/3.0/M5	**$22,429**
Maxima	SE	4dr.sdn.5	99.6	14.5	0.32	106.3	187.7x69.7x56.0	3051	1000	i/i	dc/dc	pwr.r/p.	34.8	2.9	18.5	215/65R15	V6/3.0/M5	**$23,529**
Maxima	SE	4dr.sdn.5	99.6	14.5	0.32	106.3	187.7x69.7x56.0	3113	1000	i/i	dc/dc	pwr.r/p.	34.8	2.9	18.5	215/65R15	V6/3.0/A4	**$24,464**
Maxima	GLE	4dr.sdn.5	99.6	14.5	0.32	106.3	187.7x69.7x56.0	3093	1000	i/i	dc/dc/ABS	pwr.r/p.	34.8	2.9	18.5	205/65R15	V6/3.0/A4	-
ACURA		**General warranty: 4 years / 50,000 miles: Power train: 5 years / 60,000 miles; surface corrosion: 5 years/ unlimited.**																
Vigor	LS	4 dr.sdn. 5	87.3	14.2	0.37	110.4	190.4x70.1x53.9	3020		i/i	dc/dcABS	pwr.r/p.	36.1	3.5	17.2	205/60HR15	L5/2.5/M5	**$26,350**
Vigor	GS	4 dr.sdn. 5	85.6	14.2	0.37	110.4	190.4x70.1x53.9	3020		i/i	dc/dcABS	pwr.r/p.	36.1	3.5	17.2	205/60HR15	L5/2.5/M5	**$28,350**
AUDI		**General warranty: 4 years / 60,000 miles; antipollution: 6 years / unlimited; perforation corrosion: 10 years.**																
A4	S	4dr.sdn. 4/5	86	14.2	0.30	102.8	181.3x66.7x55.5	3097	1000	i/i	dc/ABS	pwr.r/p.	36.7	3.11	17.4	195/65HR15	V6/2.8/A4	**$27,820**
A4	CS	4dr.sdn. 4/5	86	14.2	0.30	102.8	181.3x66.7x55.5	3450	1000	i/i	dc/ABS	pwr.r/p.	36.4	3.11	17.4	195/65HR15	V6/2.8/A4	**$30,770**
A4	Quattro Sport	4dr.sdn. 4/5	86	14.2	0.30	102.8	181.3x66.7x55.5	3296	1000	i/i	dc/ABS	pwr.r/p.	36.4	3.11	17.4	205/60VR15	V6/2.8/M5	**$34,420**
HONDA		**General warranty: 3 years / 36,000 miles; Power train: 5 years / 60,000 miles.**																
Accord	LX	4dr.sdn.5	94.8	13	0.33	106.9	184.0x70.1x55.1	2811	2000	i/i	dc/dr	pwr.r/p.	36.1	3.11	17	185/70R14	L4/2.2/M5	**$17,230**
Accord	EX	4dr.sdn.5	94.8	13	0.33	106.9	184.0x70.1x55.1	2877	2000	i/i	dc/dr	pwr.r/p.	36.1	3.11	17	185/70R14	L4/2.2/M5	**$19,750**
Accord	EX-R	4dr.sdn.5	94.8	13	0.33	106.9	184.0x70.1x55.1	3009	2000	i/i	dc/dc/ABS	pwr.r/p.	36.1	3.11	17	195/60R15	L4/2.2/M5	**$21,550**
LEXUS		**General warranty: 4 years / 50,000 miles; Power train: 6 years / 70,000 miles; corrosion perforation: 6 years / unlimited & road assistance.**																
ES300	base	4dr.sdn.5	95	14.3	0.32	103.1	187.8x70.0x53.9	3373	2000	i/i	dc/dcABS	pwr.r/p.	36.7	2.7	18.5	205/65R15	V6/3.0/A4	**$31,200**
MAZDA		**General warranty: 3 years / 50,000 miles; Power train: 5 years/ 60,000 miles; corrosion perforation: 5 years / unlimited**																
Millenia	fabric	4dr.sdn. 5	94	13.3	0.29	108.3	189.8x69.7x54.9	3216	NR	i/i	dc/dcABS	pwr.r/p.	37.4	2.9	18	205/65R15	V6/2.5/A4	**$25,995**
Millenia	leather	4dr.sdn. 5	94	13.3	0.29	108.3	189.8x69.7x54.9	3232	NR	i/i	dc/dcABS	pwr.r/p.	37.4	2.9	18	205/65R15	V6/2.5/A4	**$28,300**
Millenia	S	4dr.sdn. 5	94	13.3	0.29	108.3	189.8x69.7x54.9	3391	NR	i/i	dc/dcABS	pwr.r/p.	37.4	2.9	18	215/55R16	V6/2.3/A4	**$31,400**
MITSUBISHI		**General warranty: 3 years / 36,000 miles; Power Train 6 years / 60,000 miles; corrosion perforation: 7 years / 100,000 miles.**																
Diamante	LS	4dr.sdn. 4/5	94	13.6	0.30	107.1	190.2x69.9x52.6	3604	1000	i/i	dc/dcABS	pwr.r/p.	36.7	3.15	19	205/65R15	V6/3.0/A4	**$32,825**
SAAB		**Warranty: 3 years / 36,000 miles; Power Train: 6 years / 75,000 miles; corrosion perforation: 6 years / 100,000 miles.**																
900	SE	5dr.sdn.5	91	16	0.30	102.3	182.6x67.4x56.5	3131	3527	i/sr	dc/dc/ABS	pwr.r/p.	36.4	3.0	18.0	195/60R15	V6/2.5/M5	**$26,280**
900	SE	5dr.sdn.5	91	16	0.30	102.3	182.6x67.4x56.5	3296	3527	i/sr	dc/dc/ABS	pwr.r/p.	36.4	3.0	18.0	195/60R15	V6/2.5/A4	**$27,175**
TOYOTA		**General warranty: 3 years / 36,000 miles; Power Train: 5 years / 60,000 miles;corrosion: 5 years / unlimited; with no deductible or transfert fees.**																
Avalon	XL	4dr.sdn.5/6	105.4	15.4	0.31	107.1	190.2x70.3x56.1	3263	2000	i/i	dc/dc	pwr.r/p.	37.6	NA	18.5	205/65R15	V6/3.0/A4	**$22,758**
Avalon	XLS	4dr.sdn.5/6	105.4	15.4	0.31	107.1	190.2x70.3x56.1	3285	2000	i/i	dc/dc/ABS	pwr.r/p.	37.6	NA	18.5	205/65R15	V6/3.0/A4	**$26,688**
TOYOTA		**General warranty: 3 years / 36,000 miles; Power Train: 5 years / 60,000 miles;corrosion: 5 years / unlimited; with no deductible or transfert fees.**																
Camry	DX	4dr.sdn.5	97	15	0.32	107.1	187.8x69.7x55.1	1330	2000	i/i	dc/dr	pwr.r/p.	35.4	3.06	18.5	195/70R14	L4/2.2/M5	**$16,718**
Camry	LE	4dr.sdn.5	97	15	0.32	107.1	187.8x69.7x55.1	1400	2000	i/i	dc/dr	pwr.r/p.	35.4	3.06	18.5	195/70R14	L4/2.2/A4	**$19,558**
Camry	SE	4dr.sdn.5	97	15	0.33	107.1	187.8x69.7x55.1	1455	2000	i/i	dc/dc/ABS	pwr.r/p.	36.7	2.98	18.5	205/60VR15	V6/3.0/A4	**$22,908**
VOLVO		**General warranty: 4 years / 50,000 miles; corrosion: 8 years / unlimited; antipollution: 5 years / 50,000 miles.**																
850	Turbo	4dr.sdn. 5	97	15	0.32	104.9	183.5x69.3x55.1	3278	1000	i/si	dc/dc/ABS	pwr.r/p.	34.5	3.2	19.3	205/50ZR16	L5T/2.3/A4	**$30,435**
964	SE	4dr.sdn.5	94	16.6	0.35	109.0	191.8x68.9x56.6	3521	2000	i/i	dc/dcABS	pwr.r/p.	31.8	3.5	21.1	205/55VR16	L6/3.0/A4	**$33,950**

By unveiling its first Maxima in the eighties, Nissan invented the low end of the luxury car segment. It was a rear wheel drive which converted to front wheel drive in 1984. Since then, this model has known a good commercial success and opened the way to everything which today constitutes the high end of the Japanese range. Now, it is in its 4th generation.

Situation

Intention

In renewing its flagship model, Nissan wanted to maintain the standards that made the renown of the Maxima. That is: improve the dynamic performances as well as the comfort and the practical side. It offered its clientele three differently oriented versions, one at a popular price, while still maintaining the same performance and safety.

Objectives

By renewing the Maxima, the Nissan managers decided to use innovative technology to create, as they say themselves, the next dimension in luxury and performance.

Conception

Body

Nissan's styling center in San Diego produced the first designs of the new body. Thanks to satellite connection, the designers at the Technical Center in Atsugi in Japan continued to progress with the forms of the project. The interior and the trunk are greater, thanks to lengthening the wheelbase. Even though Nissan doesn't publish its aerodynamic efficiency, it is excellent since the drag is now 0.30 as against the 0.32 previously used. There has been work done to preserve the "C" pillar that personalizes this model. The interior has been given more glass to improve visibility in all directions. It has simple soft forms and the layout of the main elements respects ergonomics rules.

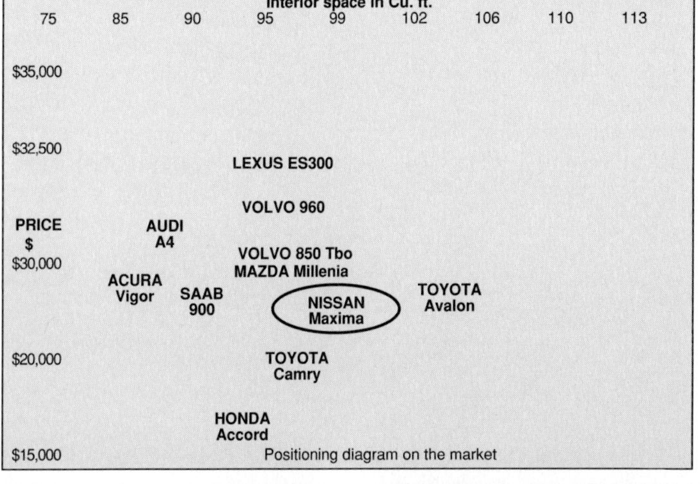

Positioning diagram on the market

Structure

The platform has been completely redefined to accept the new V6. The old Maxima, which was already know, as one of the most solid cars of its generation, had seen its stiffness improved by 10%. From the safety point of view, the front seat occupants are protected by air bags and retractable seat belts.

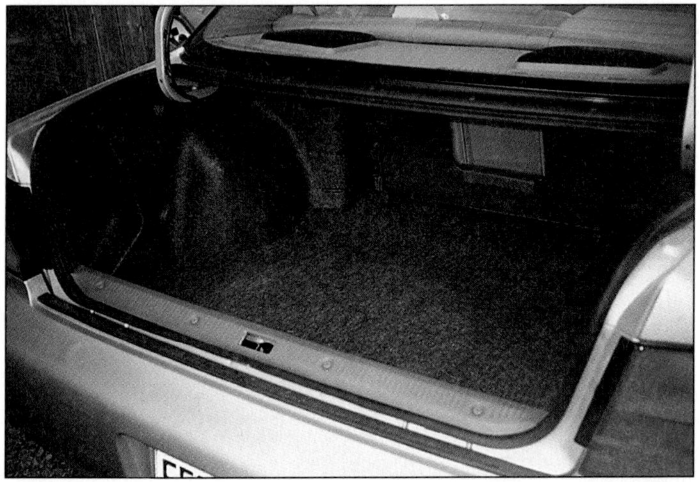

Suspension

This is the area in which Nissan worked hardest. To eliminate the defects of previously used systems (see page 94), the engineers developed and patented the Multilink axle whose main advantage is to maintain a constant camber angle to keep the tire patch contact at its maximum while the car is going through the turn in a roll attitude. The suspension allows ample comfort without interfering with handling. Also, the design of the new rear axle allowed more room for the rear seats and in the trunk. As to the front suspension, it is of a modified McPherson style.

Engine

After the structure and the rear suspension, the third major improvement involves the V6 engine. It is entirely new and one of the most performance-minded in the 3 liter displacement. It develops 190 hp and a 205 lb/ft torque; fuel consumption has improved 10%, weight by 20%, it has 10% fewer pieces and the dimensions are 17% more compact. Many of the components do not require any maintenance and are practically recyclable at 100%. To prevent vibrations, there are 7 crank counterweights, a flywheel is flexible and the engine is supported on the body by means of fluid filled and electronically controlled rubber mounts. For the interior, the Nissan engineers set to reducing all friction to improve fuel economy. The crank is micro polished and so are the camshafts to smooth their surface finish by 80%. The pistons have been redesigned and their skirts received a molybdenum disulfide coating. Also the rings are narrower. The cam drive has been redesigned (see page 90) to reduce the engine height and to eliminate the intermediate pinion. These measures combine friction reduction, high performance, and economy.

DISTINCTIVE FEATURES

The Nissan Maxima owes its noise and vibration isolation to a considerable amount of work completed to achieve optimum results. A foam has been applied at several strategic points of the body to damp out specific frequencies. Also triple seals have been used at the top of the doors to avoid wind whistles. Insulating sections have been added at the area where the shock absorbers are mounted on the body. The powertrain and front end are mounted on a floating sub-frame to filter out noise and vibrations from the road or the powertrain.

Marketing

The socio-demographic study demonstrates that the Maxima has an 80% male ownership of whom 80% are married with an average age of 50 and just under 25% are under 40. Most have a college education and an average income of $45,000. Nissan predicts that Maximas will be sold in the following proportions: 50% GXE, 25% GLE and 25% SE. Atsushi Fuji, the manager of this model at Nissan says at the Japanese builder continues its struggle against the dollar and the yen to better adapt to American marketing.

CONCLUSION

After long tests of the latest Maxima, one has to admit that without noisy pronouncements, it is a fearsome weapon in their arsenal. (See test performances and characteristics on pages 314-315).

	Trailing arms	Dual A-frame system	Torsion axle system
Rebounds in the same direction			
Roll in opposite direction			
	Impossible to maintain camber angle in the roll	Camber angle varies considerably in rebound	The torsion axle controls both roll and rebound

THE MULTILINK: HOW IT WORKS

Adopting the torsion bar axle presents several advantages for Nissan. This type of axle maintains a minimum camber as soon as the body begins to roll to one side or the other. This system allows effective control of body motions without needing to use spring-damper units and stiff stabilizer bars that detract from comfort. The torsion axle is a flex structure which consists of two lengthwise arms that determine the wheel position and the steel plate to connect the two wheels. This has a U-shape and is very rigid when the forces are applied horizontally or vertically but very flexible in torsion. Thus with this system it is possible to control toe in and camber on each of the wheels, which function independently from each other.

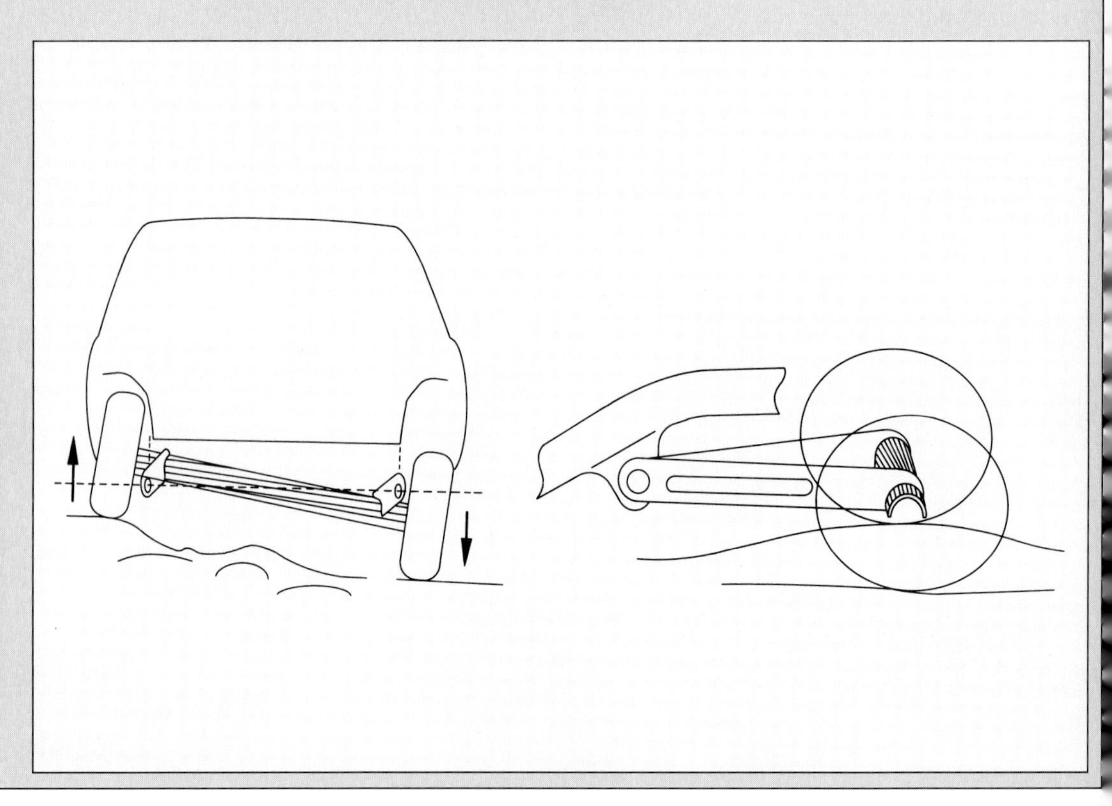

142 ROAD TESTS

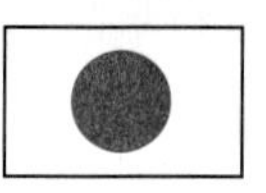

The most popular Acura...

Without contest, the Integra represents the highest sales numbers for Honda's luxury marquee. Not hard to understand after the collapse of the Legend... The Vigor never took off and the NSX flies very high over the market. Of the two models, the coupe produces the highest sales numbers.

Renewed last year, with the Honda Civic platform as a base, the Integras are sold as a 3-volume, 4-door sedan or as a 2-volume, 3-door coupe in RS, LS and GS-R trims. Two 4-cylinder 1.8 L engines with DOHC and 16 valves are offered. The one which develops 142 hp is standard on the RS and LS while the other with a VTEC intake system develops 170 HP and is installed on a GS-R. The standard transmission is a 5-speed manual and 4-speed automatics are optional.

STRONG POINTS

• Satisfaction: 90%
The Integras are among the most reliable, but their maintenance and replacements parts are costly.

• Safety: 90%
The structure has been substantially modified and the front seat occupants are protected by 2 air bags. This is standard on the LS and GS-R (the right air bag is an option on the RS).

• Technical: 80%
The steel Integra unit body has 4-wheel independent suspension and 4 wheel discs on the LS and GS-R, together with ABS. In spite of its sharp lines, the Integra rates a conservative aerodynamic coefficient, ranging between 0.32 and 0.34. The main Honda science includes sophisticated and brilliant powertrains. Paralleling the VTC system, the RS-R has a two-stage, intake manifold.

• Handling: 80%
It benefits from the increased body stiffness and a considerable longer wheelbase. The excellent suspension geometry allows it to remain neutral in a curve on dry pavement. The quality of the tires will make a difference in the rain, where the Integra is tougher to guide because its understeering and traction in cornering often lack as one driving wheel unloads.

• Fuel consumption 80%
In normal use, it corresponds to that of other engines of equivalent displacement. It increases rapidly with sportier driving.

• Steering: 80%
Even though its ratio is too high for cars with sporting tendency, it is precise, well-modulated and does not suffer from any torque steer during high acceleration.

• Driver's compartment: 80%
It is well-organized, with conventional controls, but its ergonomics leave much to be desired. The center console is far away and the radio controls are hard to reach. Visibility is good toward the front, but the "C" pillar obstructs towards 3/4 rear. On the coupe, the back light is narrow and encumbered by the spoiler. When it rains, the rear wiper doesn't clear much. One rapidly finds the best driving position thanks to combined controls of the seat and steering column. However, seating is very low and there is no adjustment for raising the front of the cushion.

DATA

Category: front wheel drive sedan and coupes.
Class : 3

HISTORIC

Introduced in: 1985 (Honda Quint) imported in 1987 with 1.6L.
Modified in: 1989 et 1994 renewed engine and body.
Made in: Suzuka, Japan.

RATINGS

Safety: 90 %
Satisfaction: 92 %
Depreciation: 50 %
Insurance: 8 % ($785)
Cost per mile: $0.43

NUMBER OF DEALERS

In U.S.A.: 289

SALES ON THE AMERICAN

Model	1992	1993	Result
Integra	55,174	58,757	+ 6.1 %

MAIN COMPETITORS

Sedan: HONDA Civic-Accord, HYUNDAI Elantra, MAZDA Protegé, TOYOTA Corolla, VOLKSWAGEN Golf-Jetta.
Coupe: EAGLE Talon, FORD Probe, HONDA Prelude, HYUNDAI Scoupe, MAZDA MX-6, NISSAN 240SX, TOYOTA Celica.

EQUIPMENT

ACURA Integra	RS	LS	GS-R
Automatic transmission:	O	O	O
Cruise control:	O	S	S
Power steering:	O	O	S
ABS brake:	O	S	S
Air conditioning:	S	S	S
Air bag:	O	O	S
Leather trim:	O	O	O
AM/FM/ Radio-cassette:	S	S	S
Power door locks:	O	S	S
Power windows:	O	S	S
Tilt steering column:	O	S	S
Dual power mirrors:	O	S	S
Intermittent wipers:	S	S	S
Light alloy wheels:	O	O	S
Sunroof:	O	O	O
Anti-theft system:	O	O	O

S : standard; O : optional; - : not available

COLORS AVAILABLE

Exterior: White, Metallic Gray, Red, Blue, Green, Black, Metallic Beige.
Interior: Black, Blue, Brown, Gray.

MAINTENANCE

First revision: 3,000 miles
Frequency: 6,000 miles
Diagnostic plug: No

WHAT'S NEW IN 1995 ?

- No information in time from the car maker.

Model/ versions *: standard	ENGINE Type / timing valve / fuel system	Displacement cu/in	Power bhp @ rmn	Torque lb.ft @ rpm	Compres. ratio	TRANSMISSION Driving wheels / transmission	Final ratio	PERFORMANCE Acceler. 0-62 mph s	Stand. 1/4 mile s	Stand. 5/8 mile s	Braking 62-0 mph ft	Top speed mph	Lateral acceler. G	Noise level dBA	Fuel economy mpg city	hwy	Gasoline type / octane
RS & LS	L4 * 1.8 DOHC-16-MPFI	112	142 @ 6300	127 @ 5200	9.2 :1	front - M5*	4.266	8.8	17.5	29.7	134	125	0.82	68	30.0	41.0	R 87
						front - A4	4.357	9.6	18.0	30.5	148	119	0.80	68	28.0	40.0	R 87
GS-R	L4* 1.8 DOHC-16-MPFI	110	170 @ 7600	128 @ 6200	10.0 :1	front - M5	4.400	7.5	16.6	29.0	131	131	0.85	67	30.0	40.0	M 89
						front - M4	4.357	8.6	17.5	29.8	138	125	0.85	67	29.0	39.0	M 89

• Quality/fit/finish: 70%
The assembly is of good quality, but the steel, like the paint, remains thin. Some materials, such as the plastic of the dashboard and accessories, are not as good as that of the competition.

• Seats: 70%
Better shaped and padded than before, the contours provide better lateral support. However, the seat cushion lacks length.

• Suspension: 70%
The damping is of good quality and allows effective control of body motions. It also levels out the road bumps. However, it is not soft and its reactions are often brutal at tar-strip crossings.

• Sound level: 70%
It is high because of drive train rolling and wind noises. The glass is thin and the opening sunroof adds to the total.

• Conveniences: 70%
A large glove box, console storage area and small door pockets are the main trinket collection centers.

• Braking: 60%
Easy to modulate in normal driving, it impresses more by its stability than by panic stop efficiency. Here, the distances are relatively long. When the ABS is installed, it causes an intermittent wheel lockup, more a nuisance than a danger.

• Access: 60%
Entry is without any problems up front but tall people will have a hard time reaching the rear seat bench of the coupe. Here, leg and headroom are tighter than on a sedan.

• Performance: 60%
The liveliness of the engines is the main attraction of these cars. If the punch of the GS-R is more evident, it is also live in the high rpm range, between 5,000 and 7,000 rpm (which are forbidden speeds). These multi-valves tend to have little mid-range torque. The manual transmission is effective, with well-staged ratios, but its synchronization shows signs of weakness. Just row the shift lever.

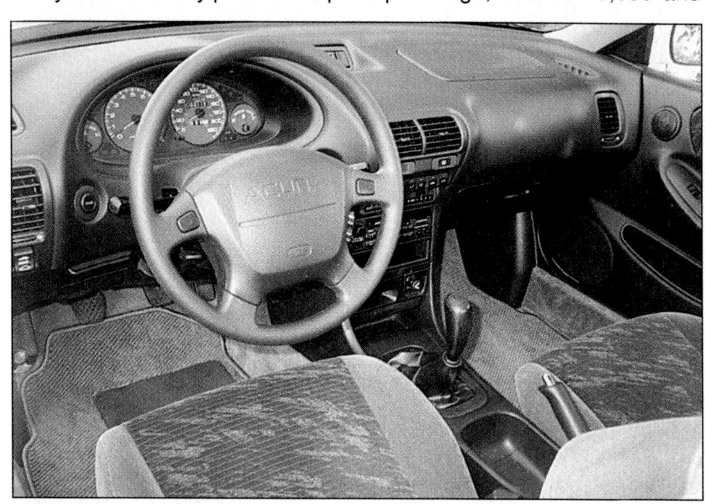

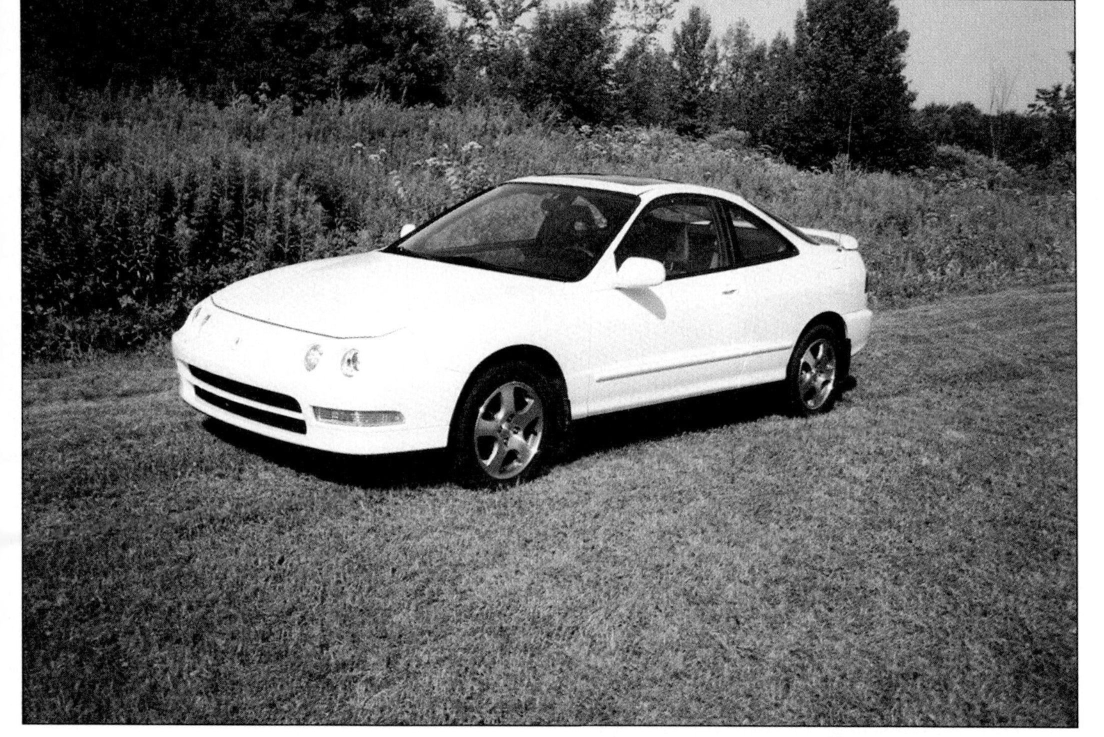

• Depreciation: 50%
An Integra loses approximately 50% of its value after three years of use. This can be considered normal. Better-equipped ones sell faster and for more.

WEAK POINTS

• Interior space: 40%
Two adults are comfortably seated up front, while the rear seats are intended more for children. Length may be sufficient but height under the roof is not.

• Trunk: 40%
It is not any greater in a sedan than a coupe but can be increased toward the interior by lowering the 40/60 bench seat.

• Price/equipment: 40%
Midway between the compacts and the sport machines, the Integras have increasingly more trouble justifying their status. One thing is certain, when they are affordable, they have little equipment and vice-versa.

• Insurance: 40%
Classed as luxury sports cars, they carry a high premium, which is even heftier for drivers under 25 years of age.

CONCLUSION

• Overall rating: 66.0%
Well-designed and offered at a reasonable price, the Integra has many competitors. None of them is direct, because it is in a class by itself. ☺

SPECIFICATIONS & PRICES

Model	Version/Trim	Body/ Seats	Interior volume cu.ft.	Trunk volume cu ft.	Drag coef. Cd	Wheel-base in	Lgth x Wdthx Hght. inx inx in	Curb weight lb	Towing capacity std. lb	Susp. type ft/rr	Brake type ft/rr	Steering system	Turning diameter ft	Steer. turns nber	Fuel. tank gal.	Standard tire size	Standard power train	PRICES US. $ 1994
ACURA		General warranty: 4 years / 50,000 miles: Power train: 5 years / 60,000 miles; surface corrosion: 5 years/ unlimited.																
Integra	RS	3 dr.cpe.4	76.8	13.4	0.32	101.2	172.4x67.3x52.6	2528	NR	i/i	dc/dc	pwr.r/p.	34.8	2.98	13.2	195/60R14	L4/1.8/M5	**$14,820**
Integra	RS	3 dr.cpe.4	76.8	13.4	0.32	101.2	172.4x67.3x52.6	2570	NR	i/i	dc/dc	pwr.r/p.	34.8	2.98	13.2	195/60R14	L4/1.8/A4	**$15,570**
Integra	LS	3 dr.cpe.4	76.8	13.4	0.32	101.2	172.4x67.3x52.6	2643	NR	i/i	dc/dc	pwr.r/p.	34.8	2.98	13.2	195/60R14	L4/1.8/M5	**$17,450**
Integra	LS	3 dr.cpe.4	76.8	13.4	0.32	101.2	172.4x67.3x52.6	2685	NR	i/i	dc/dc/ABS	pwr.r/p.	34.8	2.98	13.2	195/60R14	L4/1.8/A4	**$18,200**
Integra	GS-R	3 dr.cpe.4	76.8	13.4	0.32	101.2	172.4x67.3x52.6	2668	NR	i/i	dc/dc/ABS	pwr.r/p.	34.8	2.98	13.2	195/55R15	L4/1.8/M5	**$19,650**
Integra	RS	4dr.sdn.4	82.8	11	0.34	103.1	178.1x67.3x53.9	2617	NR	i/i	dc/dc	pwr.r/p.	35.4	2.98	13.2	195/60R14	L4/1.8/M5	**$15,580**
Integra	RS	4dr.sdn.4	82.8	11	0.34	103.1	178.1x67.3x53.9	2659	NR	i/i	dc/dc	pwr.r/p.	35.4	2.98	13.2	195/60R14	L4/1.8/A4	**$16,330**
Integra	LS	4dr.sdn.4	82.8	11	0.34	103.1	178.1x67.3x53.9	2698	NR	i/i	dc/dc/ABS	pwr.r/p.	35.4	2.98	13.2	195/60R14	L4/1.8/M5	**$17,450**
Integra	LS	4dr.sdn.4	82.8	11	0.34	103.1	178.1x67.3x53.9	2740	NR	i/i	dc/dc/ABS	pwr.r/p.	35.4	2.98	13.2	195/60R14	L4/1.8/A4	**$18,200**

See page 393 for complete 1995 Price List

History of a shipwreck...

While it was the reference in luxury car standards in the 80s', the Legend is now just a shadow of its former self. It suffers from the Honda syndrome: it was the fruit of bad decisions taken at the wrong time. No one's perfect.

The Legend is Acura's most prestigious model. While it is only equipped with a V6, it competes with the V8 Lexus and the V6 Infinity. It is marketed in the form of a 3-volume, 4-door or a 3-volume 2-door coupe based on the same platform but with a different wheelbase. It is equipped with a V6 3.2 L engine and either a standard 5-speed manual or an optional 4-speed automatic. The two bodies are offered in standard, L, or LS trims and differ only in their equipment level.

STRONG POINTS

• Satisfaction: 95%
This high rating is a measure of the quality and reliability. So why are sales dropping at an alarming rate?

• Safety: 90%
The bodies have been reinforced to improve their rigidity. They have 2 standard air bags at the front seat, except on the LS. However, it is strange that only the coupe is equipped with automatically tensioned belts.

• Technical: 80%
The steel unit body is more personalized than before, but aerodynamic efficiency is average. The 4-wheel independent suspension receives an improved geometry with better straight line stability, together with additional precision and agility in turns, the 4-wheel disc brakes with ABS are standard on all models.

• Quality/fit/finish: 80%
The assembly, finish and quality of the various materials are excellent. The leather trim is pleasant to touch, pleated as on the Italian cars. The wood appliqués are in good taste and give a warm and refined feel.

• Driver's compartment: 80%
The dashboard is ergonomical and the gauges and switches are well laid out. A comfortable driver's position is easy to find and visibility is

DATA

Category: front-wheel drive luxury coupes and sedans.
Class : 7

HISTORIC

Introduced in: 1986 (sedan), 1987 (coupe)
Modified in: 1987 2.7L engine, 1991 3.2L engine
Made in: Sayama, Japan.

RATINGS

Safety: 90 %
Satisfaction: 95 %
Depreciation: 54 %
Insurance: 3.3 % ($925)
Cost per mile: $0.94

NUMBER OF DEALERS

In U.S.A.: 289

SALES ON THE AMERICAN MARKET

Model	1992	1993	Result
Legend	49,226	38,866	-21.5 %

MAIN COMPETITORS

sedan: ALFA ROMEO 164, AUDI A6, BMW 5-Series, INFINITI Q45, LEXUS GS300 & LS400, MERCEDES-E Class, SAAB 9000, VOLVO 960.
coupe: DODGE Stealth R/T Turbo, LEXUS SC400, LINCOLN Mark VIII, NISSAN 300 ZX 2+2, PORSCHE 968, SUBARU SVX, TOYOTA Supra.

EQUIPMENT

Legend sedan	L	LS	GS		
Legend coupe				**L**	**LS**
Automatic transmission:	S	S	S	O	O
Cruise control:	S	S	S	S	S
Power steering:	S	S	S	S	S
ABS brake:	S	S	S	S	S(2)
Air conditioning:	S	S	S	S	S
Air bag (2):	S	S	S	S	S
Leather trim:	O	S	S	S	S
AM/FM/ Radio-cassette:	S	S	S	S	S
Power door locks:	S	S	S	S	S
Power windows:	S	S	S	S	S
Tilt steering column:	S	S	S	S	S
Dual power mirrors:	S	S	S	S	S
Intermittent wipers:	S	S	S	S	S
Light alloy wheels:	-	S	S	S	S
Sunroof:	S	S	S	S	S
Anti-theft system:	S	S	S	S	S

S : standard; O : optional; - : not available

COLORS AVAILABLE

Exterior: White, Black, Green, Red, Gray, Silver, Blue.
Interior: Black, Ivory, Mole, Charcoal.

MAINTENANCE

First revision: 3,000 miles
Frequency: 6,000 miles
Diagnostic plug: Yes

WHAT'S NEW IN 1995 ?

- Information not sent in time by the builder.

ENGINE						TRANSMISSION		PERFORMANCE										
Model/ versions *: standard	Type / timing valve / fuel system	Displacement cu/in	Power bhp @ rmn	Torque lb.ft @ rpm	Compres. ratio	Driving wheels / transmission	Final ratio	Acceler. 0-62 mph s	Stand. 1/4 mile s	Stand. 5/8 mile s	Braking 62-0 mph ft	Top speed mph	Lateral acceler. G	Noise level dBA	Fuel economy mpg city	Fuel economy hwy	Gasoline type / octane
1)	V6* 3.2 SOHC-24-MPFI	196	200 @ 5500	210 @ 4500	9.6 :1	front - A4*	4.37	9.0	16.7	30.2	144	124	0.80	67	22.0	32.0	S 91
2)	V6* 3.2 SOHC-24-MPFI	196	230 @ 6200	206 @ 5000	9.6 :1	front - M6	4.49	7.8	15.6	28.2	138	131	0.85	66	21.0	34.0	S 91
						front - A4	4.787	8.5	16.2	29.5	147	124	0.85	66	22.0	30.0	S 91

1) sedan L, LS 2) sedan GS, coupe

without reproach from any angle, thanks to thin roof pillars.

• Performance: 80%
In spite of good torque, the engine is not peaky and one has trouble believing its advertised 200 hp. The car works well with the automatic transmission, but only the stick shift coupe can pretend to sports car action (thanks to its 6-speed transmission). In spite of all technical refinements, this engine, which runs like a clock, offers little driving pleasure. It lacks low end torque and does not deliver power until the very high rpms. A little V8 could revive interest in this model.

• Steering: 80%
It still has too much ratio. However, its precision, modulation and progressivity are excellent.

• Access: 80%
The interior dimensions and the doors allow getting into the seats comfortably.

• Handling: 75%
It is generally neutral in most conditions, thanks to better control of the suspension and to good quality damping. Keep in mind that the weight of the car makes itself felt in tight turns, even on a coupe which rolls less and whose body motions are better under control.

• Conveniences: 75%
The storage consists of a large glove box, long door pockets and a storage area in the center console.

• Noise level: 70%
The soundproofing is well-done and the discretion of the power train maintains a low sound level, even on low roads in poor shape.

• Insurance: 70%
Taking into account its price, the Legend doesn't cost that much to insure.

• Seats: 70%
In spite of their flattering appearance, the front seats don't provide much lateral support or satisfactory lumbar support. While at the rear, the seat bench has been chopped away to give the impression of extra legroom.

• Suspension: 70%
Well-calibrated, with gas pressure dampers at the rear wheels, it is good at absorbing road defects and filtering out vibrations. However, its firmness dominates and in some instances becomes harsh. This takes away from a car of its class.

• Interior space: 60%
The interior is spacious, thanks to increased main dimensions. At the rear seats, the length is appreciable. Nevertheless, it would be difficult to include a third passenger on a long trip because of the seat and center console configuration.

• Trunk: 60%
As on some Japanese cars, the trunk volume is limited. This leads one to believe that the Japanese travel extra light.

• Braking: 50%
If it gives the impression of being powerful in normal use, it lacks effectiveness in panic stops. Here stopping distances are longer than average. However, in spite of the limited pedal stroke, it remains easy to control.

• Fuel consumption: 50%
It is about 19 mpg, about average in its category, considering the bulk and displacement of the engine.

WEAK POINTS

• Price/equipment: 10%
The equipment is complete, even on a base Legend which only lacks leather trim, a sliding roof and an air bag. They have more and more trouble justifying a price which approaches BMW or Mercedes figures. Yet, the latter have a better reputation for quality and resale value.

• Depreciation: 45%
Their resale value tumbles in a spectacular way, indicating that the Legend is not yet considered a secure value in this segment. Even prestigious makes suffer in this economic climate.

CONCLUSION

• Overall rating: 68.5%
The Legend remained frozen in place for several years. The market has moved and several competitors have entered the play. The aura which surrounded it when it first appeared has paled and is hardly remembered when prestigious Japanese cars are counted. However, it's still on the ball. ☺

SPECIFICATIONS & PRICES

Model	Version/Trim	Body/ Seats	Interior volume cu.ft.	Trunk volume cu ft.	Drag coef. Cd	Wheel-base in	Lgth x Wdthx Hght. inx inx in	Curb weight lb	Towing capacity std. lb	Susp. type ft/rr	Brake type ft/rr	Steering system	Turning diameter ft	Steer. turns nber	Fuel. tank gal.	Standard tire size	Standard power train	PRICES US. $ 1994
ACURA		**General warranty: 4 years / 50,000 miles: Power train: 5 years / 60,000 miles; surface corrosion: 5 years/ unlimited.**																
Legend	L	4dr.sdn. 5	93.4	14.8	0.34	114.6	194.9x71.3x55.1	3571	2000	i/i	dc/dc/ABS	pwr.r/p.	34.8	3.64	18.0	205/60R15	V6/3.2/A4	**$34,600**
Legend	LS	4dr.sdn. 5	93.4	14.8	0.34	114.6	194.9x71.3x55.1	-	2000	i/i	dc/dc/ABS	pwr.r/p.	34.8	3.64	18.0	205/60R15	V6/3.2/A4	**$38,600**
Legend	GS	4dr.sdn. 5	93.4	14.8	0.34	114.6	194.9x71.3x55.1	-	2000	i/i	dc/dc/ABS	pwr.r/p.	36.7	3.64	18.0	215/55VR16	V6/3.2/A4	**$40,700**
Legend	LS	2dr.cpe. 5	84.9	14.1	0.32	111.4	192.5x71.3x53.5	3549	NR	i/i	dc/dc/ABS	pwr.r/p.	34.1	3.64	18.0	215/55VR16	V6/3.2/M6	**$37,700**
Legend	LS	2dr.cpe. 5	84.9	14.1	0.32	111.4	192.5x71.3x53.5	-	NR	i/i	dc/dc/ABS	pwr.r/p.	36.4	3.64	18.0	215/55VR16	V6/3.2/M6	**$41,500**

See page 393 for complete 1995 Price List

Operation prestige...

Acura is the only Japanese builder which has had the audacity to market an exotic rival to the low-end Ferrari and Lamborghini. The experience wasn't conclusive because tradition and glory are not bought, even by investing heavily in competition.

Honda invested several years and much money in international competition. The NSX had as a mission to serve as a public symbol for this Honda high-tech image. The high performance exotic coupe is available as a 2-seater sedan with a central V6 and a manual 5-speed or a 4-speed automatic.

STRONG POINTS

• Technical: 100%

Fabricated almost entirely in aluminum (to lighten it), the unit body shaved almost 310 lbs, compared to steel. The suspension elements are forged of the same metal and only the springs, dampers and transmission half shafts are steel. Total weight approaches 2,975 pounds, of which 43% is at the front and 57% at the rear. This balance has been achieved by placing the engine in the center rear. To favor the high-speed stability, Honda engineers have not tried to improve on the aerodynamic drag and the Cd ranges between 0.31 and 0.32. The V6 engine block is that of the Legend, to which has been adopted a twin overhead cam 12-valve head on each bank. This is supplemented by a variable intake VTEC, which Honda protects with 350 patents. The technique, already applied to other models of the make, includes 3 cams and 3 rockers on the intake side. Two classical cams with rounded profiles operate the valves up to 5,000 rpm. Beyond that, a third cam, positioned between the first 2, picks up the action, operated by oil pressure. Its sharper profile allows the engine to breathe better at high rpm and gain extra power without sacrificing torque.

• Performance: 100%

With a weight-to-power ratio of 11lbs/hp and an output of 90 hp per liter, the NSX performances are not ordinary. It can accelerate from 0 to 62 mph in less than 5 seconds and attain a peak speed of155 mph.

DATA

Category: rear-wheel drive exotic coupes.
Class: GT

HISTORIC

Introduced in: 1989
Modified in: -
Made in: Tochigi, Japan.

RATINGS

Safety: 90 %
Satisfaction: 95 %
Depreciation: 49 %
Insurance: 3.0 % ($1,550)
Cost per mile: $1.15

NUMBER OF DEALERS

In U.S.A.: 289

SALES ON THE AMERICAN MARKET

Model	1992	1993	Result
NSX	1,154	652	-43.6 %

MAIN COMPETITORS

CHEVROLET Corvette ZR-1, DODGE Stealth R/T Turbo & Viper R/T10, FERRARI 355, NISSAN 300ZX Twin Turbo, PORSCHE 911.

EQUIPMENT

NSX	base
Automatic transmission:	O
Cruise control:	S
Power steering:	S
ABS brake:	S
Air conditioning:	S
Air bag (2):	S
Leather trim:	S
AM/FM/ Radio-cassette:	S
Power door locks:	S
Power windows:	S
Tilt steering column:	S
Dual power mirrors:	S
Intermittent wipers:	S
Light alloy wheels:	S
Sunroof:	-
Anti-theft system:	S

S : standard; O : optional; - : not available

COLORS AVAILABLE

Exterior: Black, White, Red, Metallic Gray, Green.
Interior: Leather: Black, Tan.

MAINTENANCE

First revision: 3,000 miles
Frequency: 6,000 miles
Diagnostic plug: Yes

WHAT'S NEW IN 1995 ?

- Information not sent in time by the builder.

ENGINE						TRANSMISSION		PERFORMANCE										
Model/ versions *: standard	Type / timing valve / fuel system	Displacement cu/in	Power bhp @ rmn	Torque lb.ft @ rpm	Compres. ratio	Driving wheels / transmission	Final ratio	Acceler. 0-62 mph s	Stand. 1/4 mile s	Stand. 5/8 mile s	Braking 62-0 mph ft	Top speed mph	Lateral acceler. G	Noise level dBA	Fuel economy mpg city	Fuel economy mpg hwy	Gasoline type / octane
NSX	V6 3.0 DOHC-24-MPFI	182	270 @ 7100	210 @ 5300	10.2 : 1	rear - M5*	4.062	5.8	13.9	24.5	125	161	0.90	68	22.0	31.0	S 91
			252 @ 6600	210 @ 5300	10.2 : 1	rear - A4	4.428	6.5	14.8	25.0	121	153	0.90	68	21.0	30.0	S 91

As on any Honda engine, the power is very high, always found above 4,200 rpm. The acceleration has nothing beastly about it and one is never glued to the seat as in some real exotics. The manual transmission shift lever is fast and precise to get the best of this sophisticated drive train.

• Satisfaction: 95%
The reliability causes no problem, which is undoubtedly Honda's greatest success. It demystifies the frivolous character of this type of car.

• Safety: 90%
The unit body of the NSX has good impact resistance and its occupants are well-protected by 2 air bags.

• Quality/fit/finish: 90%
The NSX is a product of high assembly quality and the finish and materials quality are above all suspicion.

• Handling: 90%
The extreme refinement of the suspension geometry provides constantly neutral handling and gets slightly understeering in sports driving conditions. On poor pavement it has a tendency to lighten considerably, but traction and resistance to traction loss in turns set new standards. Overall balance, a viscous coupling in the differential and tire quality all combine to give remarkable cornering.

• Driver's compartment: 80%
The driver's position is excellent, even though the seat design is simplistic for a car of this price. Visibility is good and blind spots have been reduced to a minimum: rare for this type of machine. Controls are conventional, except for switches installed in a sort of pod to control the head lights, windshield wiper and back light defrosting. Finally, the gauges are well-integrated into a very ergonomic assembly.

• Steering: 70%
It is precise at cruising speeds but becomes vague at high speeds, more because of the elasticity of the front structure than because of a conceptual defect. However, the large steering diameter and the steering gear ratio penalize maneuverability.

• Access: 70%
In spite of reduced roof height, one can slide fairly easily into the NSX seats. The doors are also ample.

• Seats: 70%
They provide good lateral and lumbar support but their design is not to the level of the rest of the vehicle.

• Insurance: 70%
While the premium rating is low, the high initial price should tell you to insure it only as needed.

• Braking: 60%
Decelerations are powerful, well-balanced and easy to control thanks to reasonable assembly weight, and the fade-free wear life of the pads is sufficient.

• Suspension: 60%
The sophisticated components provide ample wheel travel, while the shocks are firm and feed information about the road.

• Depreciation: 50%
The NSX has not yet reached a very high status so its resale value still fluctuates greatly.

• Fuel consumption: 50%
It remains at a reasonable level for this type of vehicle, considering the performance of which it is capable.

WEAK POINTS

• Price/equipment: 00%
Even though it is half an Italian of the same class, the price is overestimated and the equipment is ordinary.

• Interior space: 10%
Even though limited, you don't get a claustrophobic feel because the width is ample. It is the lack of height that will draw comments from big and tall.

• Trunk: 10%
Minuscule, it will accept only cloth bags and not even many of those.

• Noise level: 25%
Rolling noise often drowns out the song of the engine: melodious during strong rpm increases.

• Conveniences: 40%
Storage spaces are reduced to a minimum and consist of a glove box and a combination hiding place/arm rest in the center console.

CONCLUSION

• Overall rating: 61.5%
The NSX coupe has been designed knowing that it will cover more concrete on freeways and cities than in competition. That is why it is so civilized and also removed of the character that gives thrills and shutters in real exotics. In spite of all its qualities, it lacks the little spark that would damn. ☺

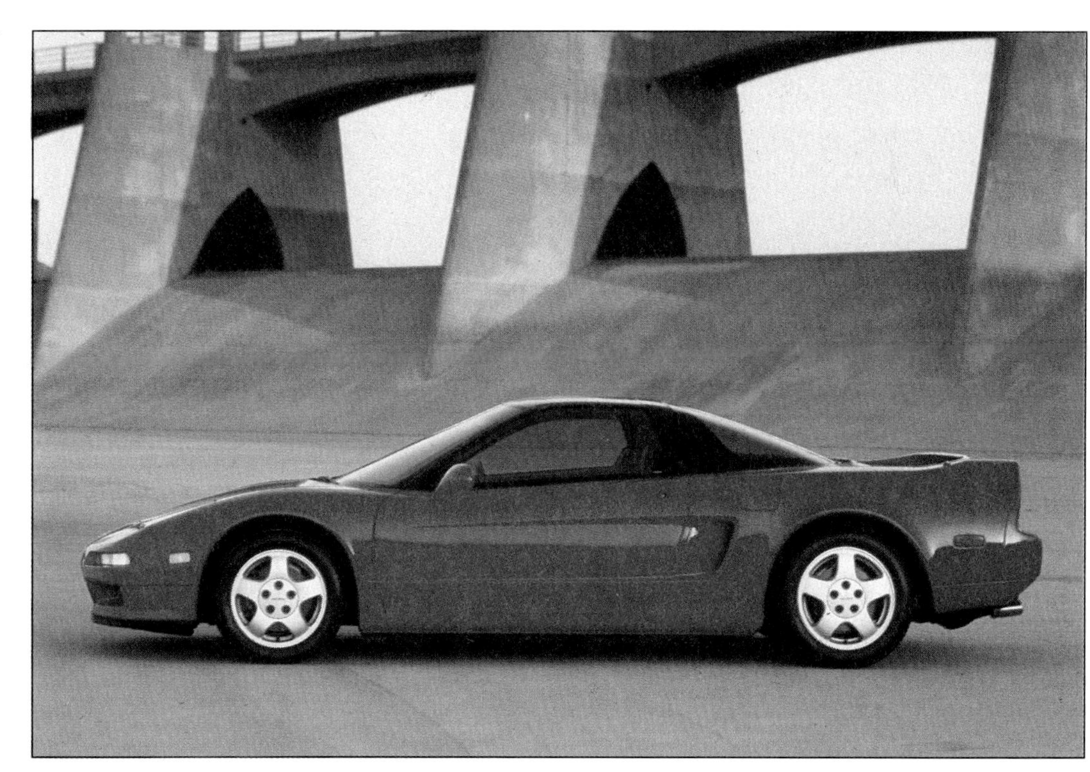

SPECIFICATIONS & PRICES

Model	Version/Trim	Body/ Seats	Interior volume cu.ft.	Trunk volume cu ft.	Drag coef. Cd	Wheel-base in	Lgth x Wdthx Hght. inx inx in	Curb weight lb	Towing capacity std. lb	Susp. type ft/rr	Brake type ft/rr	Steering system	Turning diameter ft	Steer. turns nber	Fuel. tank gal.	Standard tire size	Standard power train	PRICES US. $ 1994
ACURA		**General warranty: 4 years / 50,000 miles: Power train: 5 years / 60,000 miles; surface corrosion: 5 years/ unlimited.**																
NSX	base	2 dr.cpe. 2	49	5	0.32	2530	173.4x71.3x46.1	3020	NR	i/i	dc/dc/ABS	pwr.r/p.	38.1	3.24	18.5	215/45ZR16 245/40ZR17 (re.)	V6/3.0/M5	**$72,500**
NSX	base	2 dr.cpe. 2	49	5	0.32	2530	173.4x71.3x46.1	3108	NR	i/i	dc/dc/ABS	pwr.r/p.	38.1	3.57	18.5	205/45ZR16 245/40ZR17 (re.)	V6/3.0/A4	**$76,500**

See page 393 for complete 1995 Price List

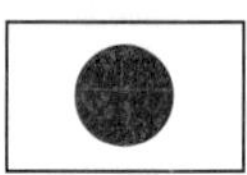

Everything to displease...

The Vigor constitutes, without a doubt, the best example of a car composed of good elements which at the bottom line has no personality and is not pleasant to drive or use. Ill- born, it was conceived to hastily plug the hole that existed between the Integra and the Legend.

To seduce the American market, Honda accumulated so many compromises that the Vigor has become a typical example of a colorless car with no odor or flavor. It will be renewed during 1995 and while waiting, the '94 continues to be offered as a 3-volume 4-door in two trim levels: LS and GS which share an identical power train.

STRONG POINTS

• Satisfaction: 90%
This constitutes the only but meager consolation for the risks of acquiring one.

• Safety: 80%
In spite of the good structural rigidity and the presence of front seat air bags, there is more to be done for the optimal protection of the occupants.

• Quality/fit/finish: 80%
Assembly and finish are carefully done and most materials have a good appearance. This is the case for leather or cloth seat trim, but the faux wood for the dashboard is not in the best taste.

• Driver's compartment: 80%
The driver will soon find the best position facing a classically organized dashboard, carefully done. Visibility is good, thanks to thin roof pillars and a major glazed area. The instruments fall under the eyes, which is not the case for the almost invisible dashboard switch locations.

• Insurance: 80%
The index and price combine to give an affordable premium for this type of car.

• Technical: 80%
The Vigor was created especially for the Japanese market, under the name of Inspire, to play the role of a deluxe Accord. Its platform was extended 85 mm and suspension calibrated differently to give it a straight-line stability worthy of its pretensions. In addition, this gained a little extra interior space. The steel unit body has fairly ordinary lines to remind you of the Accord and Legend. Its aerodynamic finesse must not have overly concerned the designers since it posts a mediocre 0.37, and its weight, which is particularly high, is over 3,000 pounds. The Vigor is radically different from the Accord by its 5-cylinder 2.5 L installed longitudinally over the front suspension, above the axes of the drive wheels. It connects to a 5-speed manual or a 4-speed automatic.

• Performance: 70%
The 5-speed provides easy takeoffs thanks to 2 fairly low gears and allows crisp takeoffs and lively accelerations until 60 mph. After that, this power fades out with the next gear ratios which are not as good at accommodating the weight. Vigors with an automatic transmission are penalized. The car name and performance have little in common.

• Fuel consumption: 70%
It is reasonable, taking into account the high weight of this car, and

DATA

Category: front-wheel drive exotic sedans.
Class: 7

HISTORIC

Introduced in: 1989 (Inspire in Japon); 1991 in New-York.
Modified in: -
Made in: Sayama, Japan.

RATINGS

Safety: 80 %
Satisfaction: 90 %
Depreciation: 45 %
Insurance: 4.2 % ($720)
Cost per mile: $0.62

NUMBER OF DEALERS

In U.S.A.: 289

SALES ON THE AMERICAN MARKET

Model	1992	1993	Result
Vigor	11,324	10,016	-11.6 %

MAIN COMPETITORS

AUDI A4, BMW 3 Series, INFINITI J30, LEXUS ES300, MAZDA Millenia, NISSAN Maxima, TOYOTA Avalon.

EQUIPMENT

ACURA Vigor	LS	GS
Automatic transmission:	O	O
Cruise control:	S	S
Power steering:	S	S
ABS brake:	S	S
Air conditioning:	S	S
Air bag (2):	S	S
Leather trim:	-	S
AM/FM/ Radio-cassette:	S	S
Power door locks:	-	S
Power windows:	S	S
Tilt steering column:	O	O
Dual power mirrors:	S	S
Intermittent wipers:	S	S
Light alloy wheels:	S	S
Sunroof:	-	S
Anti-theft system:	S	S

S : standard; O : optional; - : not available

COLORS AVAILABLE

Exterior: White, Black, Brown, Blue, Plum, Red.
Interior: Black, Cognac, Gray.

MAINTENANCE

First revision: 3,000 miles
Frequency: 6,000 miles
Diagnostic plug: Yes

WHAT'S NEW IN 1995 ?

- Information not sent in time by the builder.

Model/ versions *: standard	ENGINE Type / timing valve / fuel system	Displacement cu/in	Power bhp @ rmn	Torque lb.ft @ rpm	Compres. ratio	TRANSMISSION Driving wheels / transmission	Final ratio	PERFORMANCE Acceler. 0-62 mph s	Stand. 1/4 mile s	Stand. 5/8 mile s	Braking 62-0 mph ft	Top speed mph	Lateral acceler. G	Noise level dBA	Fuel economy mpg city	hwy	Gasoline type / octane
Vigor	L5 2.5 SOHC-20 MPFI	150	176 @ 6300	170 @ 3900	9.0 :1	front-M5	4.492	8.0	16.0	29.6	131	125	0.77	66	24.0	35.0	R 87
						front-A4	4.480	8.6	16.5	30.1	125	118	0.77	66	24.0	34.0	R 87

seldom goes under 25 mpg in normal driving.

• **Seats:** 70%
Whether leather or cloth-trimmed, they provide a firm lumbar support and an elusive lateral support. The seat cushions are too short, made to give the impression of space where none exists.

• **Noise level:** 70%
At cruise speed, the noise level is very low because the engine is very discreet. It doesn't sound off its presence, except under strong accelerations.

• **Depreciation:** 65%
It's relatively high because this model has not spread widely, and its competitors are better armed on a practical plane.

• **Braking:** 60%
It lacks bite, doesn't have miraculous stopping distances and is delicate to modulate. The pedal is particularly spongy.

• **Steering:** 60%
With a too-high ratio and a power assist that is either too high or too low, it is neither sufficiently direct nor precise to satisfy sports-type driving.

• **Trunk:** 60%
Deep enough, to accept a respectable baggage volume, it can be increased toward the interior by lowering the rear bench seat.

• **Handling:** 60%
It is balanced, thanks to ample tires and sports-like damping. This effectively control body motions, limits the roll and remains neutral when the pavement is favorable. As soon as it gets bad, the body rolls and pitches and handling rapidly turns to understeer. Front axle overload translates itself into a loss of traction in sharp turns, especially when it rains.

• **Access:** 60%
Moving back the center pillar reduced the rear door width, which now calls for contortions to reach the bench seat.

• **Conveniences:** 60%
The different storage areas are well-proportioned, except for the ridiculously small glove box.

• **Interior space:** 50%
It is sufficient at the front seats, but lacks at the rear seat because of tight dimensions and lack of legroom. Thick perforated soundproofing at the roof limits headroom. One of the problems is that the engine/transmission package takes up too much space in a platform whose wheelbase has not been sufficiently extended.

• **Suspension:** 50%
Everything works well on the freeway where it is comfortable, but the reactions are too harsh on bad pavement. Lack of wheel travel causes it to bottom out often and disagreeably.

WEAK POINTS

• **Price/equipment:** 35%
For only offering a 5-cylinder, the Vigor is too expensive in relation to its competitors equipped with a 3.0 L V6. The equipment is relatively complete, but at this price it is the least of things.

CONCLUSION

• **Overall rating:** 66.5%
While the Vigor pulls it off well on paper, in every day driving, it has everything to displease. Heavy, too tight in the back, equipped with an engine style long since passé, it is only at ease on American freeways at legal speeds. Too many dubious compromises cause this car to play on our data system without taking into account pleasurability, which is too personal a notion. ☺

SPECIFICATIONS & PRICES

Model	Version/Trim	Body/ Seats	Interior volume cu.ft.	Trunk volume cu ft.	Drag coef. Cd	Wheel-base in	Lgth x Wdthx Hght. inx inx in	Curb weight lb	Towing capacity std. lb	Susp. type ft/rr	Brake type ft/rr	Steering system	Turning diameter ft	Steer. turns nber	Fuel. tank gal.	Standard tire size	Standard power train	PRICES US. $ 1994
ACURA		General warranty: 4 years / 50,000 miles: Power train: 5 years / 60,000 miles; surface corrosion: 5 years/ unlimited.																
Vigor	LS	4 dr.sdn. 5	87.3	14.2	0.37	110.4	190.4x70.1x53.9	3020		i/i	dc/dcABS	pwr.r/p.	36.1	3.5	17.2	205/60HR15	L5/2.5/M5	**$26,350**
Vigor	GS	4 dr.sdn. 5	85.6	14.2	0.37	110.4	190.4x70.1x53.9	3020		i/i	dc/dcABS	pwr.r/p.	36.1	3.5	17.2	205/60HR15	L5/2.5/M5	**$28,350**

See page 393 for complete 1995 Price List

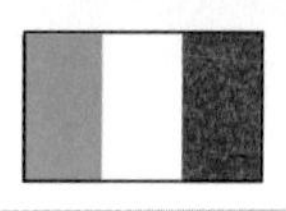

Antiquity...

The presentation of a successor to the Alfa Romeo Spider has tumbled the one known for the past thirty years to the rank of antiquity or collector's car. Even though it has been modernized several times, it is the last authentic witness to the great era of roadsters.

The North American organization of the Milanese car builder leaves something to be desired and is marginal to industry norms. It would seem usefull to expand the dealer network so the number of Italian cars would proliferate, but that is too logical. The convertible is in its last year and its replacement (which can be admired on page 50) was presented at the last Turin show. For 95, the convertible Spider range will be reduced to its simplest expression because it offers only one body, one engine and just two transmissions: a manual 5-speed and an optional 3-speed automatic. There are 2 trim levels: the standard Spider or the Veloce, on which the rigid roof is an option.

STRONG POINTS

• Insurance: 70%
Its favorable index and reasonable price give it an affordable premium. This may, according to the local conditions, be seasonal.

• Fuel consumption: 70%
It remains acceptable and averages around 25 to 28 mpg.

• Reliability: 70%
It has improved, but the rarity of dealers and spare parts makes long trips even more adventurous.

• Access: 70%
The doors are fairly long, the height of the roof is sufficient and causes little difficulty in climbing aboard.

• Quality/fit/finish: 70%
The assembly, the finish and the quality of the material as well as the components are typical of an Italian manufacturer that is, irregular.

• Driver's compartment: 70%
It spite of minor retouches, the dashboard is still rococo, the steering wheel offers a good grip, same as the gear shift, still placed at an unusual angle. Instrumentation is complete and legible, grouped around an elliptical housing built into the dashboard. The high dash hurts forward visibility. Also, laterally the convertible top generates a major blind spot, a problem resolved on the hard top by a greater gloss area. The clutch pedal is as firm as the shifter, and reverse is not always synchronized. To escape the macho driving style of the manual transmission, you can always opt for an automatic, which may not be a model of smoothness, but whose ratios are well-staged.

• Seats: 60%
Narrow but well-padded, they provide excellent lateral support as well as effective lumbar support. They give the combined impression of being one with the car.

• Performance: 60%
In spite of the age of its conception, the engine offers good efficiency with a wide torque range and power. Takeoffs and accelerations are heartening and provide sensations normally reserved for faster models. The automatic allows a more pleasant driving which will attract a larger feminine buyer group.

DATA

Category: rear-wheel drive sports coupes.
Class: 3S

HISTORIC

Introduced in: 1966
Modified in: 1971: 2.0L eng;1985 & 1991 aesthetic touch-ups.
Made in: Grugliasco, Torino, Italy.

RATINGS

Safety: 60 %
Satisfaction: 70 %
Depreciation: 60 %
Insurance: 5.0 % ($860)
Cost per mile: $0.43

NUMBER OF DEALERS

In U.S.A.: 120

SALES ON THE AMERICAN MARKET

Model	1992	1993	Result
Spider	1,102	605	-45.1 %

MAIN COMPETITORS

MAZDA Miata, FORD Mustang.

EQUIPMENT

ALFA ROMEO Spider	base	Veloce
Automatic transmission:	O	O
Cruise control:	-	O
Power steering:	-	-
ABS brake:	-	-
Air conditioning:	S	S
Air bag:	-	-
Leather trim:	S	S
AM/FM/ Radio-cassette:	S	S
Power door locks:	-	-
Power windows:	-	S
Tilt steering column:	-	O
Dual power mirrors:	S	S
Intermittent wipers:	S	S
Light alloy wheels:	O	S
Sunroof:	-	-
Anti-theft system:	-	-

S : standard; O : optional; - : not available

COLORS AVAILABLE

Exterior: Red, White, Black, Metallic Gray, Blue.
Interior: Beige, Black.

MAINTENANCE

First revision: 3,000 miles
Frequency: 7,500 miles
Diagnostic plug: No

WHAT'S NEW IN 1995 ?

- No major change.

Model/ versions *: standard	ENGINE Type / timing valve / fuel system	Displacement cu/in	Power bhp @ rmn	Torque lb.ft @ rpm	Compres. ratio	TRANSMISSION Driving wheels / transmission	Final ratio	PERFORMANCE Acceler. 0-62 mph s	Stand. 1/4 mile s	Stand. 5/8 mile s	Braking 62-0 mph ft	Top speed mph	Lateral acceler. G	Noise level dBA	Fuel economy mpg city	Fuel economy hwy	Gasoline type / octane
base	L4* 2.0 DOHC-8-MPFI	120	120 @ 5800	117 @ 2700	9.0 :1	rear - M5	4.10	9.5	17.0	29.6	151	115	0.77	74	26.0	39.0	R 87
						rear - A3	4.10	10.7	17.8	31.0	157	109	0.77	74	25.0	38.0	R 87

•Technical: 60%
In spite of the aggressivity of the lines, the Alfa Romeo convertible is not the champion of aerodynamic efficiency. Its Cd is only 0.40, a value equivalent to that of an old mini-van. Its steel unit body includes a front suspension with an A frame lower control arm and an upper arm with an angled strut. At the back a rigid rear axle with helical springs and a rear stabilizer bar complete the drive train. The engine is a four cylinder with a double overhead cam operating 8 valves. After years of waiting, Alfa has finally decided to offer an automatic transmission. However, it comes only with 3 speeds and a limited slip differential.

• Handling: 50%
Even though the body emits many cracking sounds, the rigidity is sufficient to provide acceptable road handling. Without being as refined as that of recent cars, it is more fun with an oversteer easily controlled with the steering wheel or the accelerator. The low center of gravity and the suspension firmness allow the Spider to corner flat. The limited slip differential and good tires allow it to benefit from good traction in a corner, but the rustic suspensions provoke much pitch. Also, the front end bottoms or rears up during brakings or accelerations.

• Safety: 50%
In this department the Spider doesn't shine because the rigidity of its body is more than dubious. The belts are ordinary, and you find no air bag, no roll cage and no door reinforcement.

• Steering: 50%
This is one of the last sector where roller systems are still used on a passenger car. Its outstanding feature is a vague center feel. However, it does have a good power assist and ratio.

WEAK POINTS

• Sound level: 00%
The numerous body noises, wind and tires make any conversation or listening to the radio unnecessary. Fortunately, the sounds are covered by the purring of the engine whose exhaust is generous.

• Interior space: 20%
There is little space between the transmission tunnel and the door, so the feet are thrust into a narrow tunnel of their own. However, there is sufficient space in length and height.

• Price/equipment: 30%
The Spider is costly for what it has to offer and its maintenance is reserved for fanatics who are as fortunate as they are good mechanics. Its equipment is not as complete as that of other more recent convertibles, another facet of its charming obsolescence..

• Braking: 40%
Penalized by overpowering assistance, a spongy pedal feel and the impossibility to modulate with precision. Rapid front wheel blockage occurs in panic stops.

• Trunk: 40%
It will actually accept your small suitcases or some soft bags but lacks height.

• Conveniences: 40%
They are limited to a good-sized glove box but one can use the space behind the seats to remedy this lack of conveniences.

• Suspension: 40%
It does not like rough pavement and reacts harshly by shaking up the occupants.

• Depreciation: 40%
It is strong because this model is designed for specific buyers, who are not numerous.

CONCLUSION

• Overall rating: 50.0%
Witness of an era, the Alfa Romeo Spider is long since a beautiful antiquity. Those who dream of it will be able to buy it brand new and make it last as long as possible. ☹

SPECIFICATIONS & PRICES

Model	Version/Trim	Body/ Seats	Interior volume cu.ft.	Trunk volume cu ft.	Drag coef. Cd	Wheel-base in	Lgth x Wdthx Hght. inx inx in	Curb weight lb	Towing capacity std. lb	Susp. type ft/rr	Brake type ft/rr	Steering system	Turning diameter ft	Steer. turns nber	Fuel. tank gal.	Standard tire size	Standard power train	PRICES US. $ 1994
ALFA ROMEO		**General warranty: 3 years / 36,000 miles; corrosion perforation: 6 years / 60,000 miles.**																
Spider	base	2dr.con. 2	37.3	10.3	0.40	88.6	167.6x63.9x49.7	2549	NR	i/r	dc/dc.	pwr.s&w.	35.4	3.2	12.2	195/60HR15	L4/2.0/M5	**$22,590**
Spider	Veloce	2dr.con. 2	37.3	10.3	0.40	88.6	167.6x63.9x49.7	2557	NR	i/r	dc/dc.	pwr.s&w.	35.4	3.2	12.2	195/60HR15	L4/2.0/M5	**$27,590**

See page 393 for complete 1995 Price List

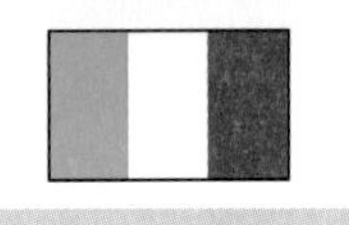

A costly passion...

Those who have Italian cars in their blood will appreciate the 164 which will acquaint them with the joy of nervous driving as practiced on the peninsula. The misfortune is that this demanding mistress consumes as many green-backs as adrenaline.

After Ferrari, the Alfa Romeos are the only representatives of the Italian industry in North America. This representation is very symbolic. It seems that the Europeans have more problems than the Japanese to establish themselves on the American market. The 164 is only sold as a 3-volume 4-door in the L and S trims, powered by the same V6 that was previously used on the Milano. That of the 164 S differs by the electronic computer control of a Bosch Motronic whose programming is different. The 5-speed manual is the standard transmission and only the L receives the 4-speed automatic as an option.

STRONG POINTS

• Performance: 85%
Takeoffs and accelerations are crisp, especially on the S version. This reaches 62.5 mph in less than 8 seconds, thanks to a favorable weight-to-power ratio. The 4-speed automatic of ZF provenance is smooth and better staged than the manual. First and second gears are short, and the next ones are too long, all in an effort to reduce fuel consumption on freeways.

• Seats: 80%
Rather firm, they provide better lateral support at the front than at the back. Here a fifth passenger will not feel comfortable in the center since the bench seat only offers 2 places outlined by formed padding.

• Suspension: 80%
Its damping is effective and wheel travel is sufficient. It effectively erases road defects and the handling is quite good.

• Technical: 80%
To reduce development costs of this model's framing, Alfa split the study cost of the 164 with Saab (9000), Fiat (Chroma) and Lancia (Thema). Each of these builders then designed their own body style and power train. The 3.0 L V6 aluminum which equips the 164 is a choice morsel. In the LS version it supplies 17 extra hp than on the L by modifying the software for fuel supply and ignition in the Bosch Motronic. The S, which is not offered with a manual transmission, has a higher final drive number in order to improve acceleration.

• Safety: 80%
The structure of the 164 has been reinforced to satisfy American safety standards. The driver is protected by an air bag and the passengers by 3-point seat belt and harness.

• Steering: 80%
Rapid, precise, with a good ratio, it allows good maneuverability but does not convey road feel.

• Access: 80%
It is as easy at the front seats as at the rear, thanks to a well-designed opening door angle.

• Insurance: 80%
For a car with an exotic character, its index is average. It is its high

DATA

Category: front-wheel drive luxury sedans.
Class: 7

HISTORIC

Introduced in: 1989
Modified in: -
Made in: Arese, Milano, Italy.

RATINGS

Safety: 90 %
Satisfaction: 75 %
Depreciation: 66 %
Insurance: 4.3 % ($995)
Cost per mile: $0.80

NUMBER OF DEALERS

In U.S.A.: 120

SALES ON THE AMERICAN MARKET

Model	1992	1993	Result
164	2,828	719	-74.6 %

MAIN COMPETITORS

ACURA Legend, AUDI A6, BMW 5-Series, INFINITI J30, LEXUS ES300, MAZDA Millenia, MERCEDES C Class, NISSAN Maxima, SAAB 9000, TOYOTA Avalon, VOLVO 960.

EQUIPMENT

ALFA ROMEO	164L	164S
Automatic transmission:	O	O
Cruise control:	S	S
Power steering:	S	S
ABS brake:	S	S
Air conditioning:	S	S
Air bag (left):	S	S
Leather trim:	O	S
AM/FM/ Radio-cassette:	S	S
Power door locks:	S	S
Power windows:	S	S
Tilt steering column:	S	S
Dual power mirrors:	S	S
Intermittent wipers:	S	S
Light alloy wheels:	S	S
Sunroof:	O	O
Anti-theft system:	O	S

S : standard; O : optional; - : not available

COLORS AVAILABLE

Exterior: Red, Black, Metallic Gray, White.
Interior: Beige, Black, Gray.

MAINTENANCE

First revision: 3,000 miles
Frequency: 7,500 miles
Diagnostic plug: No

WHAT'S NEW IN 1995 ?

- Not given in time by the car builder.

Model/ versions *: standard	ENGINE Type / timing valve / fuel system	Displacement cu/in	Power bhp @ rmn	Torque lb.ft @ rpm	Compres. ratio	TRANSMISSION Driving wheels / transmission	Final ratio	PERFORMANCE Acceler. 0-62 mph s	Stand. 1/4 mile s	Stand. 5/8 mile s	Braking 62-0 mph ft	Top speed mph	Lateral acceler. G	Noise level dBA	Fuel economy mpg city	hwy	Gasoline type / octane
164 L	V6* 3.0 SOHC-12-MPFI	181	183 @ 5800	185 @ 4400	9.5 :1	front - M5	3.11	8.7	16.4	29.2	121	137	0.76	67	21.0	32.0	M 89
						front - A4	4.28	9.5	17.3	29.8	125	131	0.76	66	19.0	29.0	M 89
164 S	V6* 3.0 SOHC-12-MPFI	181	200 @ 5800	189 @ 4400	10.0 :1	front - M5	3.41	7.5	15.6	27.0	118	144	0.78	67	21.0	32.0	M 89

initial cost which makes the premium climb.

• Trunk: 80%
Vast and with regular shape, but the high sill complicates loading.

• Interior space: 75%
The cabin can accept up to 5 people. While the length is satisfactory, the height under the ceiling is limiting.

• Driver's compartment: 75%
Driving position is ideal, thanks to multiple seat and steering column adjustments. The dashboard is barren, aesthetic and functional with a battery of switches aligned within hand's reach. Main controls are conveniently located and the instrumentation is simple and equally easy to reach.

• Quality/fit/finish: 75%
Interior presentation nears nudity. Assembly and finish are inspired by German techniques, a market of great importance to Alfa in Europe.

• Handling: 65%
It's during strong accelerations that one notes the limits of the suspension, more oriented towards comfort than sports driving. The front end finds itself hard put to make use of the available torque. On bumpy roads or in a tight turn, torque transmission to the road is not perfect. This results in disgraceful jumps and trajectories which need precision driver control, especially in the rain. On parkways, the stability is great both in a straight line and in wide curves. In a tight curve, the 164 is not vicious, but you have to get back on the gas gently since weight limits its agility.

• Reliability: 60%
Based on the experience of the Milano, which has the same drive train, maintenance and reliability are expensive when warranty ends. Service points are not numerous.

• Conveniences: 60%
The storage spaces are numerous, since in addition to the good-sized glove box, one discovers numerous door pockets and practical storage spaces.

• Braking: 60%
It is effective and fade-free and the ABS allows panic stops to be straight and progressive. The spongy pedal is sometimes difficult to modulate.

• Noise level: 50%
The soundproofing is so effective that you have to lower the windows to hear the Italian accents of the motor, of which the fans will never get tired. On passing road joints and tar strips, the tires put out a staccato noise which is tiring.

WEAK POINTS

• Price/equipment: 10%
Very high, it is one of the reasons for the limited distribution, since the Latin character of the 164 does not compensate for the extreme weakness of its network, its high parts prices, and the maintenance. The base version is better equipped, with climate control and automatic transmission. Light alloy wheels have replaced the plastic wheel covers.

• Fuel consumption: 30%
The 164 is naturally thirsty and the driving style that it presumes doesn't help matters.

• Depreciation: 35%
It will leave marks on the heart and in the wallet of the bold and foolhardy, who will regret having given in to a passionate impulse.

CONCLUSION

• Overall rating: 66.0%
The 164 is a tease of the worst kind because it is fascinating to drive in spite of some small defects. While it makes you suffer, on a budget level, some masochists will find it even more attractive. All tastes are in the nature. ☺

SPECIFICATIONS & PRICES

Model	Version/Trim	Body/ Seats	Interior volume cu.ft.	Trunk volume cu ft.	Drag coef. Cd	Wheel-base in	Lgth x Wdthx Hght. inx inx in	Curb weight lb	Towing capacity std. lb	Susp. type ft/rr	Brake type ft/rr	Steering system	Turning diameter ft	Steer. turns nber	Fuel. tank gal.	Standard tire size	Standard power train	PRICES US. $ 1994
ALFA ROMEO		**General warranty: 3 years / 36,000 miles; corrosion perforation: 6 years / 60,000 miles.**																
164	L	4dr.sdn. 4/5	103	15.9	0.30	104.7	179.3x69.3x55.1	3325	2000	i/i	dc/dc/ABS	pwr.r/p.	35.4	3.2	17.2	195/65VR15	V6/3.0/M5	**$34,890**
164	S	4dr.sdn. 4/5	103	15.9	0.30	104.7	179.3x69.3x55.1	3351	2000	i/i	dc/dc/ABS	pwr.r/p.	35.4	3.2	17.2	195/65VR15	V6/3.0/M5	**$37,690**

See page 393 for complete 1995 Price List

Arnold's toy...

Desert storm and Arnold Schwarzenegger have done the maximum to promote this extraordinary pickup nicknamed "Hummer". The least you can say is that it is as striking to meet it at the turn of a little road as on the freeway where it occupies much space.

The Hummer is the civilian version of a can-do vehicle with which the American army is equipped. It is offered in a 2-door pickup version or a 4-door cab with a short body or as a 5-door van with either soft or hard top. The 6.5 L V8 Diesel and an automatic 4-speed from GM power a transfer box with a high and low range.

STRONG POINTS

Technical:
The Hummer is based on a thick steel frame which carries at its extremities 2 cradles designed to receive suspensions, as well as front and rear differentials. The body is based on anodized aluminum panels which are assembled with 2,800 rivets, in addition to being cemented with Cybond. The front hood is made of a polyester resin reinforced with fiberglass and the doors are steel. Cataloged as a 3.5-ton truck, the Hummer does not have any bumpers. The suspension is independent and the braking has 4 wheel discs installed at the ends of the differentials to better protect them and to simplify pad changes. All wheels provide full-time drive, and the front and rear differentials are of the Thorsen type. The enormous wheels are shoed with Goodyear Wrangler tires. You can obtain, as an extra option, wheel assemblies solidly filled with rubber to allow this vehicle to roll even after a flat. As an option, in addition to a 6-ton capacity winch, the Hummer can be equipped with a system that allows tire pressure control while driving and maintains tire pressure after a flat.

Safety:
The extreme rigidity of the chassis and body,which carries an impressive roll bar cage, keeps the occupants safe from collisions, as only 3-point belts are offered, no air bag or ABS is available.

Interior space:
We did an exclusive test on this enormous size machine, last year. In spite of its unusual width, the interior of the station wagon contains only 4 seats, separated by a giant center console.

Storage:
In spite of the huge wheel wells, enormous amounts of baggage can be stored. However, the primitive 2-section hatch and the high ground clearance penalize access.

Quality/fit/finish:
Even though the construction is very robust and detailed, the appearance of the finish is utilitarian.

Driver's compartment:
The driver is enclosed in one of the four corners of the cab, defined by the width of the center console that conceals the drive train. Driving position is as good as the visibility, in spite of the thickness of the uprights and the height of the body belt, but the rear-view mirrors are big and well-located. The steering and main controls are well within hands' reach and the instrumentation is easy to read, even though the end dials are hidden by the driver's hands. The hand brake and the air

DATA

Category: 4 WD all-purpose vehicles.
Class: utility

HISTORIC

Introduced in: 1985
Modified in: 1992: civilian version.
Made in: South Bend, Indiana, U.S.A.

RATINGS

Safety: NA
Satisfaction: NA
Depreciation: NA
Insurance: 2.1 % ($720)
Cost per mile: NA

NUMBER OF DEALERS

In U.S.A.: 40

SALES ON THE AMERICAN MARKET

Model	1992	1993	Result
Hummer	NA	NA	NA

MAIN COMPETITORS

CHEVROLET-GMC Suburban-Tahoe, K pickups, DODGE Ram 350 4x4, FORD 350 4x4, LAND ROVER Defender 90.

EQUIPMENT

AM GENERAL Hummer	5dr. wgn.	2dr. p-u.	4dr. p-u.
Automatic transmission:	S	S	S
Cruise control:	O	O	O
Power steering:	O	O	O
ABS brake:	-	-	-
Air conditioning:	S	S	S
Air bag:	-	-	-
Leather trim:	-	-	-
AM/FM/ Radio-cassette:	O	O	O
Power door locks:	O	O	O
Power windows:	O	O	O
Tilt steering column:	-	-	-
Dual power mirrors:	O	O	O
Intermittent wipers:	S	S	S
Light alloy wheels:	O	O	O
Sunroof:	-		
Anti-theft system:	-		

S : standard; O : optional; - : not available

COLORS AVAILABLE

Exterior: Red, Black, White, Military Green, Green , Beige, Blue, Yellow.
Interior : Beige & Black, Beige & Gray.

MAINTENANCE

First revision: 4,000 miles
Frequency: 6,000 miles
Diagnostic plug: No

WHAT'S NEW IN 1995 ?

- **No major change.**

Model/ versions *: standard	ENGINE: Type / timing valve / fuel system	Displacement cu/in	Power bhp @ rmn	Torque lb.ft @ rpm	Compres. ratio	TRANSMISSION: Driving wheels / transmission	Final ratio	PERFORMANCE: Acceler. 0-62 mph s	Stand. 1/4 mile s	Stand. 5/8 mile s	Braking 62-0 mph ft	Top speed mph	Lateral acceler. G	Noise level dBA	Fuel economy mpg city	Fuel economy hwy	Gasoline type / octane
base	V8* 6.2 OHV-16-MI	379	150 @3600	250 @ 2000	21.3 :1	all- A3	2.73	20.0	22.0	42.0	197	78	0.67	74	17.0	22.5	D

conditioning controls are accessible because everything is ergonomic.

Performance:
If the Hummer beats any records it is that of slowness, because takeoffs and accelerations take an eternity and any passing on the road must be done with much caution (put your 50 on him and he'll move over). However, once moving on the freeway, it can maintain a 75 mph speed.

Handling:
The unusual size, the careful weight distribution of the components and a center of gravity lower than it appears, give the Hummer a surprising balance, because even when it rolls sideways on a 40% slope it will not flip. It crosses impressive obstacles, thanks to the diameter of its wheels, and the engine allows slow speed maneuvers while disposing of maximum torque. On the road, this monster vehicle handles like a sports car because its roll is limited and it reacts with much precision. The only driving problem comes from extreme width, which will call for you to measure some roads before venturing on them.

Steering:
Its easy steering makes it child play to drive. It is direct, precise, and rapid with a favorable ratio and the reversibility is totally convenient. Because of a sizable turning diameter, maneuverability is a bit of a problem.

Braking:
It is very effective, at least within the performance limits of this vehicle. Ordinary decelerations call for a light foot because the bite is brutal and the modulation is sensible. On a wet road an ABS would be very useful.

WEAK POINTS

Access:
In spite of the body height and the fact that the bottom is tub-shaped, it is not complicated to get on

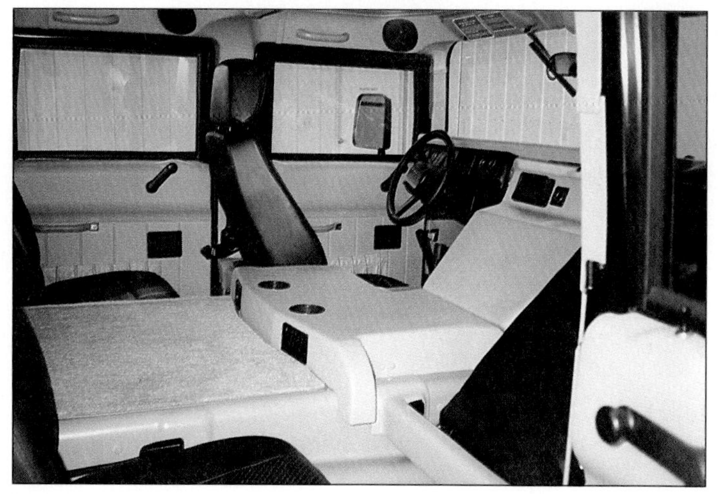

board, especially when dressed for the occasion.

Seats:
They are surprising, because they provide a fairly good lateral hold and good back support without hampering motions while getting in or out, and their padding is consistent.

Suspension:
Even though they are noisy in off road (based on obstacles to be crossed), it is not too harsh on the road and seems comfortable on the freeway, where you hardly feel the expansion joints.

Noise level:
In spite of major soundproofing, the environment is very musical, and at some speeds it is easier to communicate in sign language.

Conveniences:
If you consider the immense center console as storage space, you can install a complete office or a bed for living quarters.

Price/equipment:
The price of the Hummer is that of a highly specialized work tool, to which you still have to add some options to make its use fully comfortable.

Fuel consumption:
It holds at around 18 mpg but can go as low as 15 in difficult terrain.

CONCLUSION

The sight of a Hummer is not commonplace and triggers among those who see it pass some surprised reactions. With it you can go anywhere, it is only a question of imagination. Ask Arnold. ☺

SPECIFICATIONS & PRICES

Model	Version/Trim	Body Seats	Wheel-base in	Lgth x Wdthx Hght. inx inx in	Curb weight lb	Towing capacity std. lb	Susp. type ft/rr	Brake type ft/rr	Steering system	Turning diameter ft	Steer. turns nber	Fuel. tank gal.	Standard tire size	Standard power train	PRICES US. $ 1994
AM/GENERAL		**Warranty: 3 years / 36,000 miles.**													
Hummer	Recruit	4dr.p-u. 2	130.0	184.5x86.5x72.0	5800	9000	i/i	dc/dc	pwr.bal.	53.0	3.0	25.0	37x12.5R16.5	V8/6.2/A3	**$46,550**
		4 dr.sdn.4	130.0	184.5x86.5x72.0	6360	8600	i/i	dc/dc	pwr.bal.	53.0	3.0	25.0	37x12.5R16.5	V8/6.2/A3	**$49,950**
		4dr.wgn.4	130.0	184.5x86.5x72.0	6400	8400	i/i	dc/dc	pwr.bal.	53.0	3.0	25.0	37x12.5R16.5	V8/6.2/A3	**$52,950**

See page 393 for complete 1995 Price List

Reminiscences...

England, the land of tradition, is, along with Italy, the place where they fabricate the last of the sacred monsters of the car world. While the Rolls-Royce and the Bentley are among the best known, the Aston Martin evokes an era of victory in competition and also some James Bond films.

These coupes and convertibles of great luxury, manufactured by Aston Martin under the Virage and Volante names, built at the rate of 100 copies a year, are what is most chic and class, together with Rolls-Royce and Bentley, whose prices are comparable. Their drive train consists of a 5.3 L V8 which delivers 310 hp coupled to a manual or automatic transmission.

STRONG POINTS

Technical:
These exclusive cars are based on a rectangular steel tube chassis to which is attached a steel armature which supports the aluminum body parts. In spite of their impressive bulk, they offer acceptable aerodynamics with a Cd of 0.35 on the coupe and 0.39 on the convertible. The front suspension is based on 2 unequal upper and lower controls A arms, while the rear uses a De Dion axle whose support is at the back of the differential and connects to the unit body through tubular radius rods. A watts linkage insures full alignment of the assembly, which also receives a stabilizer bar.

Safety:
The car we tested had an air bag in the steering wheel and we were assured that the models sold in the USA will have a passenger one at the right. The structure of these rockets has been designed to satisfy impact-resistance standards.

Interior space:
The interior offers an unusual amount of space up front, thanks to the large-sized body. It does not go the same at the rear, where the space is reduced and only small bodies can travel there and then, not for long.

Trunk:
That of the Virage is greater than on a Volante which amputates it to make room for the soft top. Any excess baggage is easily placed in the back seats.

Quality/fit/finish:
For the crafts-type machine that it is, the fit on these models is impeccable. The coupe seems more homogeneous than the convertible, whose lack of rigidity becomes obvious on bad pavement. On the other hand, the top is remarkably well-done and very waterproof.

Driver's compartment:
Most controls and gauges are grouped around the steering wheel, which is like a twin brother to the one from an old Mustang. Its presence is explained by Aston Martin's need to offer an air bag without having to defray development costs. Regrettably, its massive hub hides the gauges and displays at the base of the instrument panel, while those disposed on the center console are out of sight. The front and rear defrosters also stem from large series models, and that can readily be noticed. The controls for the AC, located directly behind the transmission shift lever, are hard to reach.The seat is easily adjusted and its prominent side padding provides perfect support for hips and thighs,

DATA

Category: rear-wheel drive GT,coupes and convertibles.
Class: GT

HISTORIC

Introduced in: 1988: Virage; 1990: Volante.
Modified in:
Made in: Newport Pagnell, England.

RATINGS

Safety: 90 %
Satisfaction: NA %
Depreciation: NA %
Insurance: 3.0 % ($6,000)
Cost per mile: $3.25

NUMBER OF DEALERS

In U.S.A.: 42

SALES ON THE AMERICAN MARKET

Model	1992	1993	Result
Aston Martin	60	60	

MAIN COMPETITORS

BENTLEY Turbo R, BMW 850Ci, FERRARI 456GT, MERCEDES SC500 & SC600, PORSCHE 928GT.

EQUIPMENT

ASTON MARTIN	Virage coupe	Volante convertible
Automatic transmission:	O	O
Cruise control:	S	S
Power steering:	S	S
ABS brake:	S	S
Air conditioning:	S	S
Air bag (2):	S	S
Leather trim:	S	S
AM/FM/ Radio-cassette:	S	S
Power door locks:	S	S
Power windows:	S	S
Tilt steering column:	S	S
Dual power mirrors:	S	S
Intermittent wipers:	S	S
Light alloy wheels:	S	S
Sunroof:	-	
Anti-theft system:	-	

S : standard; O : optional; - : not available

COLORS AVAILABLE

Exterior: Black, Red, Metallic Gray, White, Champagne.
Interior: Black, Red, Blue.

MAINTENANCE

First revision: 1,500 miles
Frequency: 3,000 miles
Diagnostic plug: No

WHAT'S NEW IN 1995 ?

- No major change.

Model/ versions *: standard	ENGINE Type / timing valve / fuel system	Displacement cu/in	Power bhp @ rmn	Torque lb.ft @ rpm	Compres. ratio	TRANSMISSION Driving wheels / transmission	Final ratio	PERFORMANCE Acceler. 0-62 mph s	Stand. 1/4 mile s	Stand. 5/8 mile s	Braking 62-0 mph ft	Top speed mph	Lateral acceler. G	Noise level dBA	Fuel economy mpg city	hwy	Gasoline type / octane
Virage & Volante	V8* 5.3 DOHC-32-SFI	326	310 @ 6000	340 @ 3700	9.5 :1	rear - M5	3.54	6.6	15.2	27.6	125	155	0.84	66	10.0	17.0	S 91
						rear - A4	3.06	7.2	15.8	28.4	131	150	0.84	66	8.0	17.0	S 91

while the lumbar adjustment is pneumatic. Contrary to other exotic coupes and convertibles, visibility is very good on the Aston Martin's but better on the coupe than on the convertible whose top creates major blind spots towards 3/4 rear.

Performance:
Even though the accelerator is sensitive, one is surprised not to be glued to the seat during a strong acceleration. This 2-ton mass has a sizable inertia and, once launched, its power overflows. That has nothing to do with the 500 hp conversion which Aston Martin offers as a $50,000 option for owners in search of strong emotions.

Handling:
There is nothing there to scare yourself in the "normal" versions, except that these cars are much more at ease on parkways than on small twisty English roads, where the body takes up full width and the strong roll manifests itself in surprising ways. It is preferable to broach hairpin turns by slowing down sharply before entering them and to come out at full tilt. Otherwise, too fast an approach ends up in a sharp skid instead of a controlled drift. We must emphasis the cornering ability of the Goodyear Eagles which are able to transfer to the asphalt most of the available power. But on a bad road the directional stability is not as good because the rear axle hops around shamelessly.

Steering:
It is generally over-assisted, feels dead in the center, and then stiffens in an inconsiderate manner in a fast reverse lock so that the Virage is anything but agile, and one needs to learn how to measure out each trajectory change to avoid missteps. Four huge ventilated discs fed by individual systems ensure braking, on par with performance: powerful and balanced out by a brake antilock system. It will resist several panic stops without weakening too much and recovers rapidly.

WEAK POINTS

Access:
It is not difficult to slide into the rear seat, if you are small and nimble, while front seats are sumptuous, more in width than in length.

Seats:
Their shape and numerous electrical adjustments allow you to spend many hours in them without getting tired.

Suspension:
It handles better on the parkway than on bumpy roads where it forgets its phlegm and its good manners and shakes up the occupants.

Noise level:
In cruise, the exhaust produces just enough sound to add to the pleasure, but when you begin to play with the accelerator, comfort disappears.

Conveniences:
As bizarre as it may seem, one doesn't find a real glove box on these cars. It is replaced by a little housing in the center console and by small door pockets.

Price/equipment:
Anything an honest man could wish for is delivered in these cars for a price that matches.

Fuel consumption:
As a passing detail, the Aston Martin burns one gallon every 10 miles, and it must be worse in rapid driving. No wonder Monty ran out of gas.

CONCLUSION

In spite of constant updating and manufacturing care, these exceptional models are part of automotive history: heroic like the last cavalry charge at Baclava. Their technical advances seem primitive when compared to space age Mercedes SC or SL or even with standard BMWs. 😐

SPECIFICATIONS & PRICES

Model	Version/Trim	Body/ Seats	Interior volume cu.ft.	Trunk volume cu ft.	Drag coef. Cd	Wheel-base in	Lgth x Wdthx Hght. inx inx in	Curb weight lb	Towing capacity std. lb	Susp. type ft/rr	Brake type ft/rr	Steering system	Turning diameter ft	Steer. turns nber	Fuel. tank gal.	Standard tire size	Standard power train	PRICES US. $ 1994
ASTON MARTIN		**Warranty: 2 years / unlimited.**																
Virage		2dr.cpe. 2+2NA	12.4	0.35		102.8	186.8x73.1x53.5	4233	NR	i/i	dc/dc/ABS	pwr.r/p.	39.4	3.2	30	255/60ZR17	V8/5.3/M5	**$228,700**
Volante		2dr.cpe. 2+2NA	7.0	0.39		102.8	186.8x73.1x55.1	4409	NR	i/i	dc/dc/ABS	pwr.r/p.	39.4	3.2	25	255/60ZR17	V8/5.3/M5	**$255,848**

See page 393 for complete 1995 Price List

Confusion...

By calling their ancient 90 A4, Audi exhibits logic only for the European market, where these models are mostly equipped with 4 cylinders. In North America, these cars have 6 cylinders, just like the A6, the new name for the old Audi 100. What a jumble...

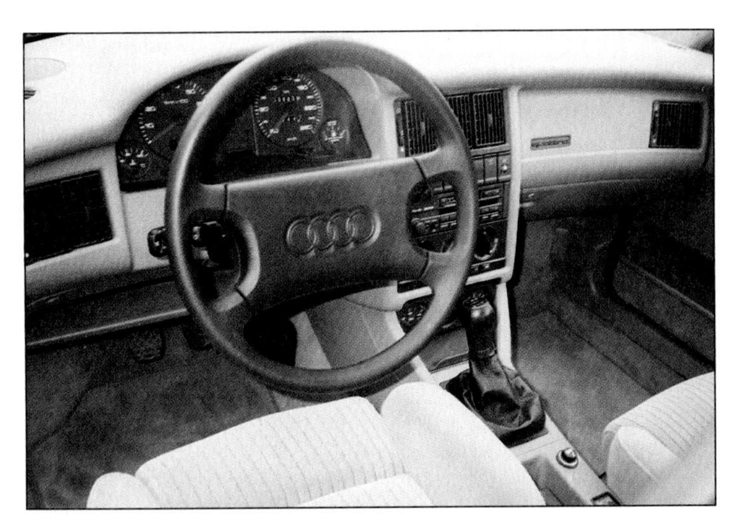

After a redesign a couple of years ago, with main dimensions extended, reinforced structure, suspensions redrawn and a 2.8 L V6 in the 100, here is the 90 changing names, henceforth to be known as the A4, like an immigrant coming to Ellis Island. This 4-door sedan will be offered with front-wheel drive or as a four-wheel drive Quattro, Quattro Sport and CS and as a convertible. The A4s come standard with a 4-speed automatic, except the Quattro Sport which benefits from a manual 5-speed.

STRONG POINTS

• Satisfaction: 90%
These models benefit from strict assembly control and a much above-average quality of materials. As to the finish, some detail immediately stand out as typically Audi.

• Technical: 90%
The massive body lines of the A4 do not alter its aerodynamic finesse and the drag coefficient ranges form 0.31 to 0.32. This steel unit body is electronically coated with zinc and gains the best corrosion protection. Four-wheel drive is used full-time with a central Torsen differential which automatically supplies 78% of the power to the front wheels, compensating for the difference in traction between the two drive trains. The 2.8L V6 is one of the most compact and lightest in its category and produces 172 hp.

• Safety: 90%
The structure of the little Audi allows a remarkable impact resistance and rigidity. A front seat belt retraction device and a steering column named Procon-Ten, as well as two standard front seat air bags and knee protectors, all contribute to occupant safety.

DATA

Category: front-wheel drive or AWD luxury sedans.
Class: 7

HISTORIC

Introduced in: 1987
Modified in: 1988: 20V engine; 1991: coupe.
Made in: Ingolstadt, Germany.

RATINGS

Safety: 90 %
Satisfaction: 88 %
Depreciation: 50 % (90)
Insurance: 5.1 % ($925) (90)
Cost per mile: $0.62

NUMBER OF DEALERS

In U.S.A.: 290

SALES ON THE AMERICAN MARKET

Model	1992	1993	Result
Audi 90	1,043	5,570	+ 534 %

MAIN COMPETITORS

ACURA Vigor, BMW 3-Series, INFINITI J30, LEXUS ES300, MAZDA Millenia, NISSAN Maxima, SAAB 900, TOYOTA Avalon, TOYOTA Avalon, VOLKSWAGEN Passat VR6, VOLVO 850.

EQUIPMENT

AUDI	90 S 2WD	90 S Quattro	90 CS Quattro
Automatic transmission:	S	S	M5
Cruise control:	S	S	S
Power steering:	O	O	O
ABS brake:	S	S	S
Air conditioning:	S	S	S
Air bag (2):	O	O	O
Leather trim:	O	O	O
AM/FM/ Radio-cassette:	S	S	S
Power door locks:	S	S	S
Power windows:	O	O	O
Tilt steering column:	S	S	S
Dual power mirrors:	S	S	S
Intermittent wipers:	S	S	S
Light alloy wheels:	O	O	S
Sunroof:	-	O	O
Anti-theft system:	-	-	O

S : standard; O : optional; - : not available

COLORS AVAILABLE

Exterior: Black, Silver White, Red, Mica, Titanium, Emerald.
Interior: Anthracite, Ecru, Platinum.

MAINTENANCE

First revision: 7,500 miles
Frequency: 7,500 miles
Diagnostic plug: Yes

WHAT'S NEW IN 1995 ?

- The new A4s, that are in reality new improved 90 versions, will be introduced in 1995.

Model/ versions *: standard	ENGINE Type / timing valve / fuel system	Displacement cu/in	Power bhp @ rmn	Torque lb.ft @ rpm	Compres. ratio	TRANSMISSION Driving wheels / transmission	Final ratio	PERFORMANCE Acceler. 0-62 mph s	Stand. 1/4 mile s	Stand. 5/8 mile s	Braking 62-0 mph ft	Top speed mph	Lateral acceler. G	Noise level dBA	Fuel economy mpg city	hwy	Gasoline type / octane
A4 S	V6* 2.8 SOHC-12-EFI	169	172 @ 5500	184 @ 3000	10.3 :1	front - A4*	4.00	9.5	17.7	30.5	137	131*	0.76	66	21.0	34.0	S 91
A4 Quattro	V6* 2.8 SOHC-12-EFI	169	172 @ 5500	184 @ 3000	10.3 :1	all - A4*	4.29	9.7	18.0	31.0	144	131*	0.76	66	22.0	31.0	S 91
A4 Sport	V6* 2.8 SOHC-12-EFI	169	172 @ 5500	184 @ 3000	10.3 :1	all - M5*	3.89	8.5	16.5	29.8	131	131*	0.79	66	22.0	31.0	S 91

*Electronically governed speed

• **Steering: 80%**
The steering is very successful and allows precise control and an above-average maneuverability, but its considerable assistance gives it a light feel.

• **Quality/fit/finish: 80%**
The materials, assembly and finish of the Audi are of great quality and distinguish themselves from BMW or Mercedes by a presentation vivid and elegant in colors and materials.

• **Access: 80%**
It is excellent since the doors open practically at 90 degrees and a cutout at the rear lowers the trunk sill.

• **Seats: 80%**
Even though definitely firm, they offer excellent lateral support, since they are as well-sculptured at the front as at the rear and allow long fatigue-free trips.

• **Conveniences: 80%**
They are generous, practical, amply-sized and in sufficient number. On the convertible the top comes down automatically in an impressive robotics cycle. Unfortunately, the plastic backlight does not have electric defrosting, which is unacceptable.

• **Driver's compartment: 80%**
The driver is gripped well by a seat with good contours but impractical height adjustment. He sees well thanks to third side windows and "open" headrests. Even though the steering column is fixed, the ergonomics are good and the main gauges are clear and well laid out. On the other hand, the levers controlling the headlights and the cruise control need to be gotten used to, as they are not laid out in conventional ways.

• **Suspension: 70%**
In spite of a certain typically German dryness, it is good at absorbing road defects, thanks to ample wheel travel and consistent shock control.

• **Performance: 70%**
The V6 engine is more at ease in the A4 than in the A6 whose weight is higher. With the weight-to-power ratio of 19.2 lbs/hp, the engine offers constant torque from 2000 rpm. While there is not much horsepower below this rpm, take-off and acceleration benefit, and in passing it loves revving past 4000 rpm. The excess weight of the convertible knocks out any sports-like velocity, which is too bad.

• **Noise level: 70%**
The combination of good soundproofing, body stiffness and quiet drive train lowers the sound level to that of the best limousines. Sports fans will complain of the excessively quiet exhaust.

• **Handling: 70%**
Except for a slight traction loss in take-offs, the road holding of this front-wheel drive is excellent and roll is well-controlled. That of the Quattro versions is more impressive, especially on wet roads. Because of its respectable weight, the A4 is more at ease on parkways than on small twisty roads. There it remains neutral and well-balanced.

• **Trunk: 60%**
Its capacity has been particularly improved by the extension of the platform, the redesign of the rear axle and the repositioning of the fuel tank. It will offer, from here on, a sizable volume that can be further increased by lowering the back of the two-part bench seat.

• **Fuel consumption: 50%**
The change of engine and displacement has not altered the fairly economical operation of the A4, whose appetite for fuel averages around 25 mpg.

• **Depreciation: 50%**
It seems stabilized at a very honest level (50%), which may be the sign of a return of confidence by the public in Audi products.

• **Braking: 50%**
It is one of the great qualities of the A4 because it has fearsome brakes, progressive and easy-to-modulate in spite of some pedal firmness. A standard ABS balances out panic stops and can be shut off by a console switch for driving on icy or snowy roads.

WEAK POINTS

• **Price/equipment: 30%**
Readjusted downward, the price of the A4 becomes more competitive. This way, the S front-wheel drive competes better against some Japanese and the Quattro option costs only $3,000 more.

• **Interior space: 40%**
Four people will be well at ease in the passenger compartment, but only a young child can take a place in the middle of the rear seat bench.

CONCLUSION

• **Overall rating: 65.5%**
More practical, better performing, more reliable and less costly, the Audi A4s have better chances of attracting attention. If the 4-wheel drive Quattro is not specifically needed, the front-wheel drive will do fine. ☺

SPECIFICATIONS & PRICES

Model	Version/Trim	Body/ Seats	Interior volume cu.ft.	Trunk volume cu ft.	Drag coef. Cd	Wheel-base in	Lgth x Wdthx Hght. inx inx in	Curb weight lb	Towing capacity std. lb	Susp. type ft/rr	Brake type ft/rr	Steering system	Turning diameter ft	Steer. turns nber	Fuel. tank gal.	Standard tire size	Standard powertrain	PRICES US. $ 1994
AUDI		**General warranty: 4 years / 60,000 miles; antipollution: 6 years / unlimited; perforation corrosion: 10 years.**																
A4	S	4dr.sdn. 4/5	86	14.2	0.30	102.8	181.3x66.7x55.5	3097	1000	i/i	dc/ABS	pwr.r/p.	36.7	3.11	17.4	195/65HR15	V6/2.8/A4	**$27,820**
A4	CS	4dr.sdn. 4/5	86	14.2	0.30	102.8	181.3x66.7x55.5	3450	1000	i/i	dc/ABS	pwr.r/p.	36.4	3.11	17.4	195/65HR15	V6/2.8/A4	**$30,770**
A4	Quattro Sport	4dr.sdn. 4/5	86	14.2	0.30	102.8	181.3x66.7x55.5	3296	1000	i/i	dc/ABS	pwr.r/p.	36.4	3.11	17.4	205/60VR15	V6/2.8/M5	**$34,420**
A4	Conv.	2dr.con.2+2	NA	8.1	0.30	100.6	181.3x66.7x54.3	3496	NR	i/i	dc/ABS	pwr.r/p.	36.4	3.11	17.4	195/65HR15	V6/2.8/A4	**$38,950**

See page 393 for complete 1995 Price List

Again in the saddle....

After 15 years of caution warnings for dubious reliability and sudden accelerations, the Audi has buckled down the buckle. After the product improvements and range restructuring, the German builder is again in business...

According to the latest Audi nomenclature, the 100 becomes the A6, which is logical because it is equipped with a 2.8L V6 which develops 172 hp, except for the sports S6 whose drive train is that of the old 200 Turbo, a turbocharged 5-cylinder which develops 23hp. It connects to a manual 5-speed. There is a choice between the sedan and the Avant station wagon whose power trains are identical. These different versions are not easy to identify except for a small emblem at the front of the vehicle. Together with the name change, they received cosmetic front and rear alterations including the headlights, grill, fenders, rim and taillights.

STRONG POINTS

• Safety: 90%
Audi has largely demonstrated the safety benefits of full-time 4-wheel drive by its wins in competition. Even though this option only attracts 23% of sales in the US, 25% in Canada and 6% in Europe, it grants Audi the active safety ribbon with cluster. However, passive safety has not been neglected since the structural rigidity has improved. The front seat belts and column retract under impact and two air bags are standard.

• Technical: 90 %
Oriented lengthwise, the V6 is both light and compact, it is distinguished by variable-length exhaust pipes controlled by a vacuum vane. A greater tube length favors low rpm range and a short one gains power above 4,000 rpm. The transmission on the A6 models is only available as a 4-speed automatic, while the electronic shift selection, derived from the Tiptronic and designed by Porsche, is based on driving style and road grade. Full-time 4-wheel drive includes three differentials: one at each end and the third of a Torsen type, which allows allocating torque to the end lacking traction. The front suspension is of the McPherson type and the rear uses radius rods and anti-rollbars.

• Satisfaction: 90%
After many lean years, Audi succeeded in returning to the winner's circle because the reliability of its vehicles is as good as those of very reputable Japanese models.

• Quality/fit/finish: 90%
The assembly is strict, the finish clinical and the quality of the materials exceptional. This earns high marks a make whose plastic materials look sharper than those of some Mercedes and BMWs.

• Trunk: 80%
It is huge and useful with regular forms; its low sill also makes it easy to load. In addition, you can make it larger by lowering the rear bench seat back.

• Steering: 80%
Your hands connect readily with the steering wheel but the power assist, which varies with speed, is much too high in all circumstances,

DATA

Category: front-wheel drive or AWD luxury sedans.
Class: 7

HISTORIC

Introduced in: 1982 (5000)
Modified in: 1988 & 1993
Made in: Ingolstadt, Germany.

RATINGS

Safety: 90%
Satisfaction: 88%
Depreciation: 53% (previous model)
Insurance: ($995) A6 4.1% ($1,135) S6
Cost per mile: $0.93

NUMBER OF DEALERS

In U.S.A.: 290

SALES ON THE AMERICAN MARKET

Model	1992	1993	Result
Audi 100	10,083	6,056	- 40.0 %

MAIN COMPETITORS

ACURA Legend, ALFA ROMEO 164, BMW 5-Series, INFINITI J30, LEXUS GS300, MAZDA Millenia, MERCEDES E Class, SAAB 9000, VOLVO 960

EQUIPMENT

AUDI	A6 S 2WD	A6S Quat.	A6CS Quat.	S6 Quat.
Automatic transmission:	S	S	S	-
Cruise control:	S	S	S	S
Power steering:	S	S	S	S
ABS brake:	S	S	S	S
Air conditioning:	S	S	S	S
Air bag (2):	S	S	S	S
Leather trim:	O	O	O	S
AM/FM/ Radio-cassette:	S	S	S	S
Power door locks:	S	S	S	S
Power windows:	S	S	S	S
Tilt steering column:	S	S	S	S
Dual power mirrors:	S	S	S	S
Intermittent wipers:	S	S	S	S
Light alloy wheels:	S	S	S	S
Sunroof:	S	S	S	S
Anti-theft system:	S	S	S	S

S : standard; O : optional; - : not available

COLORS AVAILABLE

Exterior: Black, Silver, White, Mica, Minerva, Ragusa, Titanium.
Interior: Charcoal Gray, Ecru, Platinum, Sapphire.

MAINTENANCE

First revision: 7,500 miles
Frequency: 7,500 miles
Diagnostic plug: Yes

WHAT'S NEW IN 1995 ?

- The Audi 100 becomes the A6 and the S4 becomes the S6.
- New style bumpers color-coordinated to the car.
- New headlight style.
- Cosmetic redesign of fenders, wheel rims, grill and taillights.
- Automatic air conditioning with filters for dust and pollen.

ENGINE					TRANSMISSION			PERFORMANCE										
Model/ versions *: standard	Type / timing valve / fuel system	Displacement cu/in	Power bhp @ rmn	Torque lb.ft @ rpm	Compres. ratio	Driving wheels / transmission	Final ratio	Acceler. 0-62 mph s	Stand. 1/4 mile s	Stand. 5/8 mile s	Braking 62-0 mph ft	Top speed mph	Lateral acceler. G	Noise level dBA	Fuel economy mpg city	hwy	Gasoline type / octane
A6	V6*2.8 SOHC-24-EFI	169	172 @ 5500	184 @ 3000	10.3 :1	front-A4*	4.00	11.0	17.2	30.0	148	130*	0.78	65	22.0	31.0	S 91
A6 Quattro	V6*2.8 SOHC-24-EFI	169	172 @ 5500	184 @ 3000	10.3 :1	four-A4*	3.97	12.0	18.0	30.0	160	130*	0.80	65	22.0	30.0	S 91
S6	L5* 2.2 T SOHC-20-EFI	136	227 @ 5900	258 @ 1950	9.3 :1	four-M5*	4.11	7.0	14.7	27.8	115	130*	0.85	66	21.0	30.0	S 91
A6 CS wgn	V6* 2.8 SOHC-24 EFI	169	172 @ 5500	184 @ 3000	10.3:1	front-A4*	3.97	11.5	18.0	30.0	138	130*	0.80	65	22.0	31.0	S 91

*Electronically governed speed.

very sensitive, calls for numerous corrections and takes getting used to.

• Access: 80%
It is eased by doors that open close to 90 degrees, which is rather rare.

• Seats: 80 %
Always firm, they support well at all levels, in spite of lacking a lumbar adjustment.

• Driver's compartment: 80%
Its ergonomics are near perfection, with many well laid out gauges and excellent visibility from almost any angle, even though the rear headrests close off some of the back light vision. The steering column and numerous seat adjustments allow the best driving position to be easily found. It will take the driver a little time to adapt to some of the controls. One of the switches allows switching off the ABS, which is very helpful when a road is icy or snowed in. Another, placed between the seats, allows locking the rear differential till 15 miles per hour is reached so as to maximize traction.

• Suspension: 80%
Its wheel travel and smoothness allow best absorption of road defects and substantially improve the ride and comfort of the occupants.

• Handling: 70%
Unlike older models, the suspension of the latest A6s is very soft and generates sizable roll and body motions. The S6 does not suffer from this inconvenience because its suspension is more rigid and its body rolls less. Their size and weight make these cars more at ease on freeways than on small twisty roads that show up their lack of agility.

• Braking: 70%
Its balance and fade-resistance are above criticism, and the efficiency is more evident on the S6 than on the A6.

• Noise level: 70%
The rigidity of the unit body, quality of the insulation and quiet power train keep the noise level at fairly low values.

AUDI S6

• Conveniences: 70%
Storage spaces are numerous and well laid out; it sets new standards.

• Interior space: 60%
The interior of the Audi A6 and S6 easily accepts four, but there is not enough room in all directions.

• Fuel consumption: 50%
In spite of a considerable weight, it rarely goes under 19 mpg in normal driving.

• Depreciation: 50%
One can consider the page turned because the residual value of the Audi is again in mid-range if not slightly above it.

WEAK POINTS

• Price/equipment: 00%
It is totally complete and you can find some accessories normally reserved for limousines such as the S-Series of the Mercedes or 7 of the BMW. This deluxe equipment explains the sticker shock, which can tally up to $40,000 before you can get into one of these cars.

• Performance: 40%
In spite of its smoothness, the engine is not as high torqued or powerful at low rpm as Audi pretends. Empty, the weight comes close to 3,500 pounds and the weight-to-power ratio nudges 22 lbs/hp which is very average. Therefore, you should not be surprised that take-offs and accelerations are soft. The S6 does distinctly better, which improves driving pleasure and gives it an exceptional character.

CONCLUSION

• Overall rating: 66.0%
With much patience and perseverance, Audi has finally returned to the lead in the quality race. Several positive signs seem to show that the public has more confidence in the Ingolstadt builder. ☺

SPECIFICATIONS & PRICES

Model	Version/Trim	Body/ Seats	Interior volume cu.ft.	Trunk volume cu ft.	Drag coef. Cd	Wheel-base in	Lgth x Wdthx Hght. inx inx in	Curb weight lb	Towing capacity std. lb	Susp. type ft/rr	Brake type ft/rr	Steering system	Turning diameter ft	Steer. turns nber	Fuel. tank gal.	Standard tire size	Standard powertrain	PRICES US. $ 1994
AUDI A6 & S6			General warranty: 4 years / 60,000 miles; antipollution: 6 years / unlimited; perforation corrosion: 10 years.															
A6	CS	4dr.sdn.5	96	18	0.29	105.8	192.6x70.0x56.3	3384	-	i/i	dc/ABS	pwr.r/p.	34.8	3.1	21.1	195/65HR15	V6/2.8/A4	**$40,570**
A6	S Quattro	4dr.sdn.5	96	18	0.29	106.0	192.6x70.0x56.6	3671	-	i/i	dc/ABS	pwr.r/p.	34.8	3.1	21.1	195/65HR15	V6/2.8/M5	**$43,020**
A6	CS Q wgn.	5dr.wgn.7	96	37.2	0.29	106.0	192.6x70.0x57.0	3891	-	i/i	dc/ABS	pwr.r/p.	34.8	3.1	21.1	195/65HR15	V6/2.8/A4	**$47,020**
S6	S6	4dr.sdn.5	96	18	0.29	106.0	192.6x71.0x56.5	3781	NR	i/i	dc/ABS	pwr.r/p.	34.8	3.1	21.1	225/50ZR16	L5T/2.2/M5	**$49,070**

See page 393 for complete 1995 Price List

Invitation to the club...

Driving a BMW remains for some a condition sine qua non: it's their ticket to the club, their status symbol and confirmation of a life style. BMW, which always wants to sell more cars, constantly widens the range toward the bottom, allowing the economically less advantaged to join the club...

The 3-Series has always been the most affordable ticket to the BMW club. Applicants have a choice between a coupe, a convertible,a 4-door sedan and they can be equipped with a 4-cylinder 1.8L or an in-line 6 2.0L and 2.5L. For 1995, the M3 sports version has departed and the new compact 318ti established on the same platform joins in, with a tail shortened by a foot.

STRONG POINTS

• Safety: **90%**
The front-seat occupants are protected by belts with automatic tensioners, plus two air bags. The interior structure is extremely rigid and includes preprogrammed collapsible crumple zones. On the convertible, a rollbar cage is concealed behind the headrests of the rear seats and appears only when the body takes a drastic inclination angle.

• Technical: **90%**
Even though they share a number of elements, these models offer much at the level of their structure so that their suspensions are matched to their separate temperaments. The steel unit bodies have 60% of their sheet metal galvanized and supplied with corrosion treatment. The standard lines of the marquee are combined with today's aerodynamics, so the drag coefficient is at 0.32. The 318 is a 1.8L 4-cylinder, while that of the 325 is an in-line 6-cylinder. Their cylinder heads include two overhead cams and four valves per cylinder. The standard transmission is a 5-speed manual or an automatic with electronic control which functions in standard or sport mode.

• Driver's compartment: **90%**
The interior of the latest 3-Series is a model of ergonomics with easy-

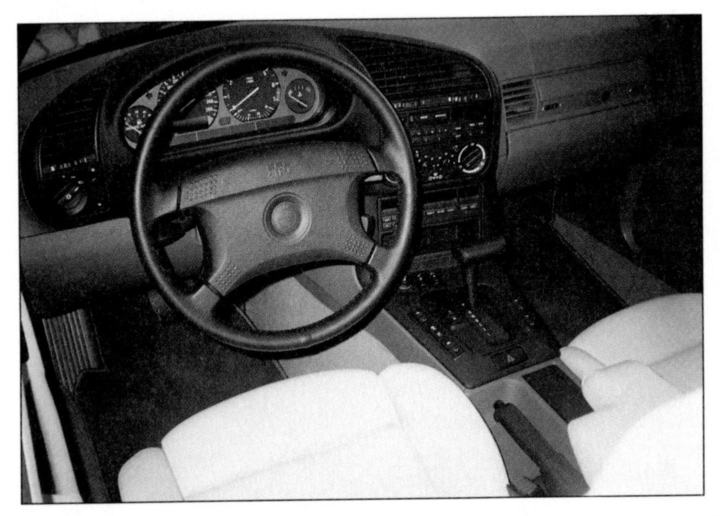

DATA

Category: rear-wheel drive luxury sedans.
Class: 7

HISTORIC

Introduced in: 1982 (320i 2dr.) 1991: latest model.
Modified in: 1992: 318; 1993: M3; 1994: cabriolet : 1995: compact
Made in: Dingolfing, Münich, Germany, Spartansburg, SC, USA

RATINGS

Safety: 90 %
Satisfaction: 80 %
Depreciation: 55 %
Insurance: 5.5 % ($860 to $1,140)
Cost per mile: $0.62

NUMBER OF DEALERS

In U.S.A.:

SALES ON THE AMERICAN MARKET

Model	1992	1993	Result
3-Series	37,964	45,590	+ 16.8%

MAIN COMPETITORS

ACURA Vigor, AUDI A4, INFINITI G-20 & J30, LEXUS ES300, MAZDA Millenia, NISSAN Maxima, SAAB 900, TOYOTA Camry XLE VOLVO 850.

EQUIPMENT

BMW 3-Series	318i	318is	325i	325iS
Automatic transmission:	O	O	O	O
Cruise control:	S	S	S	S
Power steering:	S	S	S	S
ABS brake:	S	S	S	S
Air conditioning:	S	S	S	S
Air bag (2):	S	S	S	S
Leather trim:	O	O	O	S
AM/FM/ Radio-cassette:	S	S	S	S
Power door locks:	S	-	S	S
Power windows:	S	S	S	S
Tilt steering column:	S	S	S	S
Dual power mirrors:	O	O	O	O
Intermittent wipers:	S	S	S	S
Light alloy wheels:	O	S	S	S
Sunroof:	O	O	S	S
Anti-theft system:	S	S	S	S

S : standard; O : optional; - : not available

COLORS AVAILABLE

Exterior: White, Black, Blue, Red, Gray, Silver, Purple, Green.
Interior: Fabric: Charcoal Gray, Silver Gray, Turquoise.
Leather: Black, Silver Gray, Beige, Yellow.

MAINTENANCE

First revision: 15,000 miles
Frequency: 15,000 miles
Diagnostic plug: Yes

WHAT'S NEW IN 1995 ?

- **New hatchback version available by Spring of '95.**
- **Front redesigned on the 318i convertible.**
- **Version M3 discontinued from the catalog.**

Model/ versions *: standard	ENGINE Type / timing valve / fuel system	Displacement cu/in	Power bhp @ rmn	Torque lb.ft @ rpm	Compres. ratio	TRANSMISSION Driving wheels / transmission	Final ratio	PERFORMANCE Acceler. 0-62 mph s	Stand. 1/4 mile s	Stand. 5/8 mile s	Braking 62-0 mph ft	Top speed mph	Lateral acceler. G	Noise level dBA	Fuel economy mpg city	hwy	Gasoline type / octane
316 ti	L4* 1.6 SOHC-16-EFI	97	102 @ 5500	108 @ 3900	9.7 :1	rear - M5*	3.38	12.3	-	33.6	-	115	0.80	-	30.0	50.0	M 89
						rear - A4	4.45	13.8	-	35.1	-	112	0.80	-	27.0	47.0	M 89
318i/is	L4* 1.8 DOHC-16-EFI	121	138 @ 6000	129 @ 4500	10.0 :1	rear - M5*	3.45	10.2	17.0	31.6	121	125	0.80	68	26.0	39.0	S 91
325i/is	L6* 2.5 DOHC-24-EFI	182	189 @ 5900	181 @ 4200	10.5 :1	rear - M5*	3.15	8.0	15.3	28.0	128	131	0.80	66	23.0	36.0	S 91
						rear - A4	3.91	8.8	16.0	28.4	131	125	0.80	66	24.0	36.0	S 91

BMW 316 ti

to-read analog instruments and console-mounted switches angled toward the driver and within hand's reach. The general presentation is less austere than before, with more rounded shapes. The seat provides a more effective driving position and the field of vision includes a minimum of blind spots.

• Quality/fit/finish: 80%
It is disappointing to note that the small BMWs are not as well finished as the others. To produce large quantities and to remain very profitable, the only solution is to cut corners. However, to reassure you, the assembly and finish remain strict, but the fabrics and plastics are less rich than on some Japanese models, and some of the finishing is worthy of Volkswagen Mexico.

• Satisfaction: 80%
Owners remain a little surprised to note that even BMWs, those cars of great reputation, can, from time to time, know annoying little problems, just as any other ordinary car.

• Seats: 80%
Well-shaped, they provide better lateral than lumbar support. Lumbar area is not adjustable and the leather trim is not the best choice because it is too slippery.

• Handling: 70%
The 318 and 325 are fun to drive because their reactions are cleaner than those of the previous models, especially on wet roads. Less oversteering than before, they handle in any radius curve and their straight line stability is excellent, even in a crosswind.

• Performance: 70%
If you except the 318i from which this point of view is very ordinary, those of the 320 and 325 are good but not outstanding. Many other popular models obtain equivalent results without making noise. Driving pleasure comes mostly from the power and torque of the in-line 6-cylinder, available at all rpm.

• Steering: 70%
Firm but precise, it shows flagrant lack of speed and takes 3.4 turns of the steering wheel lock-to-lock, which sometimes penalizes precision.

• Access: 70%
It is harder to get into the backs of the coupes and convertibles because of the low roof height.

• Suspension: 70%
Without being cushioned, it is not uncomfortable. When it reacts to road defects, it does so with a certain smoothness, thanks to good damping and ample tires and wheels.

• Fuel consumption: 60%
It is more realistic on the 318, but remains at about 21 mpg on the 325: not particularly economical for a car of this size.

• Braking: 55%
Even though progressive and easy to modulate, it offers nothing exceptional for cars of this caliber. Stopping distances are average, but perfectly rectilinear, thanks to the ABS.

• Noise level: 50%
It remains at a reasonable level, but even there, no feats. We have a great average of the vehicles of this format. The fans of a good horse race will favor the 6-cylinders, distinctly more melodious than the four, which speaks with the accent of the common man. Untermensch?

• Conveniences: 50%
Storage spaces do not abound, and the glove box, door pockets and center console hollow are minimalist. As to the air conditioning, it is powerful and fitted with a micro filter to keep out dust and pollen from the interior.

WEAK POINTS

• Price/equipment: 30%
For those who only want a BMW, the standard equipment will be sufficient for the price, while those in quest of a little more comfort will have to play with the option list.

• Interior: 30%
The volume of the interior will only allow seating four people with sufficient leg and headroom in the rear seats. However, the coupe and convertible are limited by roof height.

• Trunk: 40%
It is not the most generous when compared to others, but its form is regular and a cutout eases access.

• Depreciation: 45%
Difficult times have flexed somewhat the resale value of these cars, whose prices hold up better in normal times.

CONCLUSION

• Overall rating: 64.2%
Unless you want to belong to the club at any cost, it is time to say that there are a number of other models which offer as much for a more reasonable budget, but again, they do not carry the Munich builder's emblem. 😐

SPECIFICATIONS & PRICES

Model	Version/Trim	Body/ Seats	Interior volume cu.ft.	Trunk volume cu ft.	Drag coef. Cd	Wheel-base in	Lgth x Wdthx Hght. inx inx in	Curb weight lb	Towing capacity std. lb	Susp. type ft/rr	Brake type ft/rr	Steering system	Turning diameter ft	Steer. turns nber	Fuel. tank gal.	Standard tire size	Standard powertrain	PRICES US. $ 1994
BMW		**Warranty: 4 years/ 50,000 miles; corrosion: 6 years / unlimited; antipollution: 8 years / 80,000 miles.**																
316	ti	2dr.cpe.4	-	13.7	0.36	106.3	165.7x67.3x54.8	2623	NR	i/i	dc/dr/ABS	pwr.r/p.	34.1	-	13.7	185/65HR15	L4/1.6/M5	-
318	i	2dr.con.4	76	14.3	0.36	106.3	174.5x67.3x53.1	3119	NR	i/i	dc/dc/ABS	pwr.r/p.	34.1	3.4	17.2	205/60R15	L4/1.8/M5	-
318	is	2dr.cpe.4	82	14.3	0.31	106.3	174.5x67.3x53.8	2932	NR	i/i	dc/dc/ABS	pwr.r/p.	34.1	3.4	17.2	205/60R15	L4/1.8/M5	**$25,800**
318	i	4dr.sdn.4	86	15.4	0.31	106.3	174.5x66.9x54.8	2932	NR	i/i	dc/dc/ABS	pwr.r/p.	34.1	3.4	17.2	205/60R15	L4/1.8/M5	**$24,675**
320	i	4dr.sdn.4	86	15.4	0.31	106.3	174.5x66.9x54.8	3031	NR	i/i	dc/dc/ABS	pwr.r/p.	34.1	3.4	17.2	205/60R15	L6/2.0/M5	-
325	is	2dr.cpe.4	82	14.3	0.31	106.3	174.5x67.3x53.8	3020	NR	i/i	dc/dc/ABS	pwr.r/p.	34.1	3.4	17.2	205/60VR15	L6/2.5/M5	**$32,200**
325	i	4dr.sdn.4	86	15.4	0.31	106.3	174.5x66.9x54.8	3020	NR	i/i	dc/dc/ABS	pwr.r/p.	34.1	3.4	17.2	205/60VR15	L6/2.5/M5	**$30,850**
325	i	2dr.con.4	76	8.1	0.36	106.3	174.5x67.3x53.8	3351	NR	i/i	dc/dc/ABS	pwr.r/p.	34.1	3.4	17.2	205/60R15	L6/2.5/M5	**$38,800**

See page 393 for complete 1995 Price List

The belly button of the range...

For many, the 5-Series is the first real BMW, exclusive as much by its price as by its performances, more voluminous than the 3-Series but less than the 7 from which it takes some attributes. It will be replaced next year by a more inhabitable model.

The median point of the BMW range, the 5-Series is in its last production year under this form because it will be redesigned next year. We are dealing with a 4-door sedan offered in 525, 530 and 540 as well as a 530 touring station wagon. These are the solid burghers living within the town walls, as opposed to the rich 7s' from the castle and the aspiring 3s' on the wrong side of the moat.

STRONG POINTS

• Safety: 90%
A number of cross members have been incorporated to maximize torsional and beaming stiffness and to stiffen the bottom of the platform against side impacts, and, of course, two air bags protect the front-seat occupants.

• Technical: 90%
For a line which dates back several years and offers some similarity with that of the 7-Series, its aerodynamics are efficient since the Cd is around 0.32. The steel unit body of which most sheet metal is galvanized has an independent suspension, disc brakes with ABS at all 4 wheels and a Servotronic power assist that varies with speed. Three engines are available for these cars: the in-line 2.5 L 6-cylinder and two V8s of 3.0 and 4.0, respectively, with power ratings of 189, 215 and 282 hp. The sedans are equipped with 5- and 6-speed stick shifts or 4- or 5-speed automatics depending on the engine.

• Quality/fit/finish: 90%
If the interior presentation is less austere than in the past, it is partly because of the pleated leather trim of the 530i: more Latin than Germanic. As to their quality of assembly and finish, the materials are more in the class of the 7-Series than 3-Series.

• Interior space: 80%
The driver's position is ideal for savoring driving pleasure as its seat maintains perfectly. However, the ergonomics of the dashboard shows its age because some controls are hard to reach and the steering column does not have a telescopic adjustment. Visibility is good and the gauges, like the controls, are well laid out, but that of the cruise control is easy to confuse with the windshield wipers.

• Seats: 80%
They provide a lumbar and lateral support which is very comfortable, in spite of their hard padding and the particularly slippery leather trim.

• Steering: 80%
Subtly modulated, it sins by too high a ratio, which detracts from its precision and maneuverability, depending on the circumstances.

• Suspension: 80%
Even though it is not soft, it absorbs all the road defects well, thanks to its wheel travel and well-calibrated shock absorbers.

• Performance: 80%
The 5-Series does not linger on the road, but there is a world of difference between the three engines. The 525 combines an economical daily use with sufficient performance to circulate in North America.

DATA

Category: rear-wheel drive luxury sedans.
Class: 7

HISTORIC

Introduced in: 1972
Modified in: 1981 & 1988
Made in: Dingolfing, near Münich, Germany.

RATINGS

Safety: 100 %
Satisfaction: 80 %
Depreciation: NA %
Insurance: 3.1 % ($1,140 to $1,200)
Cost per mile: $0.94

NUMBER OF DEALERS

In U.S.A.: 363

SALES ON THE AMERICAN MARKET

Model	1992	1993	Result
5-Series	19,382	21,694	+ 10.7 %

MAIN COMPETITORS

ACURA Legend, ALFA ROMEO 164, AUDI A6, CADILLAC STS, INFINITI J30t, LEXUS GS300, MERCEDES E-Class, SAAB 9000, VOLVO 960.

EQUIPMENT

BMW 5 Series	525i sedan	525i Touring	530i	540
Automatic transmission:	O	S	O	No charge
Cruise control:	S	S	S	S
Power steering:	S	S	S	S
ABS brake:	S	S	S	S
Air conditioning:	S	S	S	S
Air bag (2):	S	S	S	S
Leather trim:	O	O	S	S
AM/FM/ Radio-cassette:	S	S	S	S
Power door locks:	S	S	S	S
Power windows:	S	S	S	S
Tilt steering column:	S	S	S	S
Dual power mirrors:	S	S	S	S
Intermittent wipers:	S	S	S	S
Light alloy wheels:	S	S	S	S
Sunroof:	S	S	S	S
Anti-theft system:	S	S	S	S

S : standard; O : optional; - : not available

COLORS AVAILABLE

Exterior: White, Red, Green, Black, Silver, Blue, Gray, Beige.
Interior: Black, Gray, Ultramarine, Parchment.

MAINTENANCE

First revision: 15,000 miles
Frequency: 15,000 miles
Diagnostic plug: Yes

WHAT'S NEW IN 1995 ?

- M5 version removed from the catalog.
- Manual shift transmission now offered in all sedans.
- Lower body shields now match the body color.
- New 7-Series steering wheel style.
- Double lock and protection against starting functions each time the central door lock system is activated.

Model/ versions *: standard	ENGINE Type / timing valve / fuel system	Displacement cu/in	Power bhp @ rmn	Torque lb.ft @ rpm	Compres. ratio	TRANSMISSION Driving wheels / transmission	Final ratio	PERFORMANCE Acceler. 0-62 mph s	Stand. 1/4 mile s	Stand. 5/8 mile s	Braking 62-0 mph ft	Top speed mph	Lateral acceler. G	Noise level dBA	Fuel economy mpg city	hwy	Gasoline type / octane
525i/iA	L6* 2.5 DOHC-24-EFI	152	189 @ 5900	184 @ 4200	10.5 :1	rear - M5*	3.23	9.0	16.8	28.8	111	128*	0.80	67	23.0	36.0	M 89
						rear - A4	4.10	9.8	17.4	29.5	115	128*	0.80	67	22.0	33.0	M 89
530i	V8* 3.0 DOHC-32-EFI	183	215 @ 5800	214 @ 4500	10.5 :1	rear - M5*	3.08	9.0	16.6	29.0	128	128*	0.84	67	20.0	33.0	S 91
540i	V8* 4.0 DOHC-32-EFI	243	282 @ 5800	295 @ 4500	10.0 :1	rear - M6	2.93	7.3	15.5	28.4	125	128*	0.85	66	19.0	30.0	S 91
						rear - A5	2.81	7.0	15.2	28.2	128	128*	0.85	66	19.0	30.0	S 91

*Electronically governed speed.

The V8 of the 530 and 540 brings more prestige than really usable power within legal limits. It is regrettable that the transmission ratios pull too far for the speed limits of our road network (more autobahn than interstate). They are easier to drive on dry pavement than on wet, where the tire selection will make all the difference.

• Satisfaction: 80%
It's not surprising that this index is this low because maintenance and repairs are often practiced at amazing prices, and the reliability of these cars does not stand up to all trials and tribulations.

• Equipment: 80%
It is very complete, even on the simplest models which include a standard travel computer.

• Handling: 75%
Stability and the ability to stay in a straight line at high speed are the best qualities of this car. It has become more difficult to kick out the rear in a sharp corner. Their neutral steer as well as traction in a corner under power are excellent. However, comfort takes precedence over performance, and they have lost some of the sports character of their predecessors without detracting from driving pleasure. Their weight is respectable but takes away much of the agility on a twisty road.

• Noise level: 70%
It is very low at cruise speed, but the unit body is well-insulated, and the engine only sounds off (agreeably) strong accelerations. The standard sunroof, which opens in two sections, allows the occupants of the front and rear seats to take in air and sun.

• Conveniences: 70%
Storage spaces include a large-capacity glove box, door pockets and a storage box/armrest combination located on the center console.

• Trunk: 70%
Inferior to that of some competitors which attain and pass the 14 cubic feet mark, it is easy to use, thanks to its regular forms. On a station wagon the storage can be increased by folding over the rear seat bench.

• Access: 60%
The doors and their opening angle are sufficiently large to allow seating without difficulty, unless your passengers are huge.

• Braking: 55%
It impresses by its bite and fade-free operation. The stability is very reassuring during panic stops, thanks to an advanced ABS system.

WEAK POINTS

• Price/equipment: 00%
In spite of their very complete equipment, the 5-Series is extremely expensive in a country where you cannot use their max speed and also taking into account the more plentiful interior space of some Japanese which also are more reliable and just as refined.

• Fuel consumption: 40%
These cars do not operate on holy water, and those who would like to find out how much fire the engine can produce will have to pay for the gas.

• Interior space: 40%
Accounted for to the nearest cubic inch, the interior volume will only accept four people in comfort. It is mostly the lack of length which limits rear legroom.

• Depreciation: 50%
As with most of the deluxe German cars, it holds its value favorably, but second owners must beware of the exorbitant spare parts and labor costs.

CONCLUSION

Overall rating: 68.0%
On the eve of being reconditioned, the 5-Series remains pertinent by its dynamic qualities, and it is mostly on the practical end that one awaits significant improvements. ☺

SPECIFICATIONS & PRICES

Model	Version/Trim	Body/ Seats	Interior volume cu.ft.	Trunk volume cu ft.	Drag coef. Cd	Wheel-base in	Lgth x Wdthx Hght. inx inx in	Curb weight lb	Towing capacity std. lb	Susp. type ft/rr	Brake type ft/rr	Steering system	Turning diameter ft	Steer. turns nber	Fuel. tank gal.	Standard tire size	Standard powertrain	PRICES US. $ 1994
BMW		**Warranty: 4 years/ 50,000 miles; corrosion: 6 years / unlimited.**																
525	i/iA	4dr.sdn. 5	91	16.2	0.33	108.7	185.8x68.9x55.6	3483	1000	i/i	dc/dc/ABS	pwr.bal.	36.1	3.5	21.4	205/65HR15	L6/2.5/M5	**$38,425**
530	i touring	4dr.wgn.5	89	31	0.33	108.7	185.8x68.9x55.8	3880	1000	i/i	dc/dc/ABS	pwr.bal.	36.1	3.5	21.4	225/60VR15	V8/3.0/A4	**$40,600**
530	i/iA	4dr.sdn. 5	91	16.2	0.32	108.7	185.8x68.9x55.6	3627	1000	i/i	dc/dc/ABS	pwr.bal.	36.1	3.5	21.4	225/60HR15	V8/3.0M5	**$41,500**
540	i/iA	4dr.sdn. 5	91	16.2	0.33	108.7	185.8x68.9x55.6	3803	1000	i/i	dc/dc/ABS	pwr.bal.	36.1	3.5	21.4	225/60HR15	V8/4.0/A5	**$47,500**

See page 393 for complete 1995 Price List

Grandeur and discretion...

Those not impassioned by automobiles will see absolutely no difference between the old and the new huge BMWs unless they look at them side by side and observe them attentively, simply because the main improvements are located "under the skin".

At the moment of renewing its prestige model, BMW has chosen to do so with discretion and efficiency, looking more to improve than to innovate. The 7-Series includes 4-door sedans, on 2 wheelbases: short for the 740i and long for the 740 and 750L.

STRONG POINTS

• Technical: 100%
Even though the new model strangely resembles the preceding one, its main dimensions have been increased although the weight has been maintained. To cover up its bulk, the Munich stylists have softened the angles, lengthened the lines and refined the front end to achieve a remarkable 0.30 drag coefficient. The unit body, which was already very rigid, was stiffened both for superior handling and impact protection. The suspension remains independent, with antidive and antilift geometry and the brakes are 4-wheel discs, assisted by ABS, which also serves as a traction device. The variable damping is electronically controlled, and an automatic height correction for the rear suspension is included. During 1995, the 750 will receive a height correction control in turns and will remain neutral in all circumstances. The engines are very near those of the first generations. This includes a 4.0 L V8 and a 5.4L V12 harnessed to a 5-speed automatic whose selector can work in several modes.

• Safety: 100%
In addition to all the active arsenal described above, these sedans have passive protection unusual for vehicles of our time: including a rollbar cage around the interior and air bags at the front seats.

• Quality/fit/finish: 90%
The interior is no longer typically German because the leather trim and wood appliques are in a character more Maserati than Mercedes.

DATA

Category: rear-wheel drive luxury sedans.
Class: 7

HISTORIC

Introduced in: 1986 (6 cyl.); 1987 (V12)
Modified in: 1989:735iL;1992: V8, 4.0L engine
Made in: Din golfing, near Münich, Germany.

RATINGS

Safety: 100 %
Satisfaction: 93 %
Depreciation: 50 %
Insurance: 3.2 % ($1,500 to $1,950)
Cost per mile: $1.60

NUMBER OF DEALERS

In U.S.A.: 363

SALES ON THE AMERICAN MARKET

Model	1992	1993	Result
7 Series	6,656	9,767	+ 31.9 %

MAIN COMPETITORS

INFINITI Q45, JAGUAR XJ6, LEXUS LS400, MERCEDES S-Class.

EQUIPMENT

BMW 7-Series	740i	740 iL	750 iL
Automatic transmission:	S	S	S
Cruise control:	S	S	S
Power steering:	S	S	S
ABS brake:	S	S	S
Air conditioning:	S	S	S
Air bag (2):	S	S	S
Leather trim:	S	S	S
AM/FM/ Radio-cassette:	S	S	S
Power door locks:	S	S	S
Power windows:	S	S	S
Tilt steering column:	S	S	S
Dual power mirrors:	S	S	S
Intermittent wipers:	S	S	S
Light alloy wheels:	S	S	S
Sunroof:	S	S	S
Anti-theft system:	S	S	S

S : standard; O : optional; - : not available

COLORS AVAILABLE

Exterior: White, Red, Green, Black, Silver, Blue.
Interior: Black, Gray, Beige, Red, Anthracite, Turquoise.

MAINTENANCE

First revision: 15,000 miles
Frequency: 15,000 miles
Diagnostic plug: Yes

WHAT'S NEW IN 1995 ?

- The car range is redesigned and there is a new 740i.
- Automatic transmission with 5-speed and an adaptive electronic control.
- Electric seats with multi-adjustments at the back (740iL and 750iL).

	ENGINE				TRANSMISSION			PERFORMANCE										
Model/ versions *: standard	Type / timing valve / fuel system	Displacement cu/in	Power bhp @ rmn	Torque lb.ft @ rpm	Compres. ratio	Driving wheels / transmission	Final ratio	Acceler. 0-62 mph s	Stand. 1/4 mile s	Stand. 5/8 mile s	Braking 62-0 mph ft	Top speed mph	Lateral acceler. G	Noise level dBA	Fuel economy mpg city	hwy	Gasoline type / octane
740i / iL	V8* 4.0 DOHC-32-MFI	243	282 @ 5800	295 @ 4500	10.0 :1	rear - A5*	3.15	7.2	15.9	27.8	125	131*	0.78	65	19.0	30.0	S 91
750iL	V12* 5.4 SOHC-24-MFI	304	322 @ 5000	361 @ 3900	10.0 :1	rear - A5*	3.15	6.4	15.2	28.4	131	131*	0.78	65	15.0	24.0	S 91

*Electronically governed speed.

• Noise level: 90%
The discretion of the motor, the extreme body rigidity and the effective soundproofing provide a cushy ambiance.

• Satisfaction: 90%
It is always surprising to find that in spite of the extreme complexity of this type of car, the builder manages to ensure an exceptional reliability level in the normal life span of the car.

• Driver's compartment: 90%
Thanks to multiple adjustments of the seat and steering column, the driver is royally seated and his environment very ergonomic. Visibility offers a minimum of dead spots, in spite of the narrowness of the backlight and the presence of headrests on the back seats. The instruments are simple and easy-to-read and the controls well laid out, except that of the cruise control which can be confused with the headlight switch. The very technical steering wheel regroups the controls, remote from the radio and cruise control. You have to get used to the little messages the car addresses to you and which parade across the little screen on the instrument panel.

• Seats: 80%
Their lateral support has been improved by accenting their contour and an adding an extra adjustment which articulates the front seat backs. While the front cushion is a little short up front, it is normal at the back, where the backrest is carefully sculptured.

• Suspension: 80%
Scientifically calculated, it magically swallows up most parasitic frequencies triggered by pavement defects.

• Steering: 80%
Power-assistance is reduced as speed increases and a short turning diameter gains surprising maneuverability.

• Performance: 80%
We had the opportunity to test one of the first North American versions of the 740i to rediscover the smooth power of the 282hp 4.0 L V8 which gains exceptional performance for this near 2-ton machine. It accelerates from 0 to 62 mph in a little over 9 seconds. The 5-speed automatic is not a stranger to this, since it has a shift mode allowing fast downshifts and corresponding accelerations. You can put it into "economy", "sport" or "winter" on demand.

• Interior space: 80%
While five people can fit, it is with four that optimal comfort is reached.

• Trunk: 80%
Connected with that of the interior, its volume allows much baggage, and long objects such as skis can poke through the center armrest thanks to a trap door.

• Handling: 70%
It will be difficult for common mortals to know the limits of this car because its dynamics are so refined, down to the smallest details. Everything combines to make the 7-Series stable in any road situation. What is surprising is the relative agility of this big elephant on a twisty road and even its ease in going through switchbacks. On the freeway, its ability to point straight down the road is superb and there, or in curves, sidewind doesn't affect it.

• Braking: 70%
Our panic stops were effective in 118 feet coming down from 62 mph: an excellent safety measure because the stability and fade-resistance of the pads remain intact after several tests.

• Access: 70%
It is just as easy to enter at the front as at the rear, considering the dimensions and the doors which open wide.

• Conveniences: 70%
They are numerous, practical and equally distributed between the front and rear of the interior.

WEAK POINTS

• Price/equiment: 00%
The one who invests in this vehicle is not going to ask if all the equipment and refinements are really necessary because at this price level, it is the prestige that rates.

• Fuel consumption: 10%
It is very reasonable considering the weight and potential of this 740i.

• Depreciation: 60%
The resale value of the BMWs has fewer bad surprises than that of some Japanese competitors.

CONCLUSION

• Overall rating: 69.0%
With its last 7-Series, BMW sets the high jump bar a little taller, with exceptional brio and discreet elegance, adjectives which fit these superb and refined cars like a glove. ☺

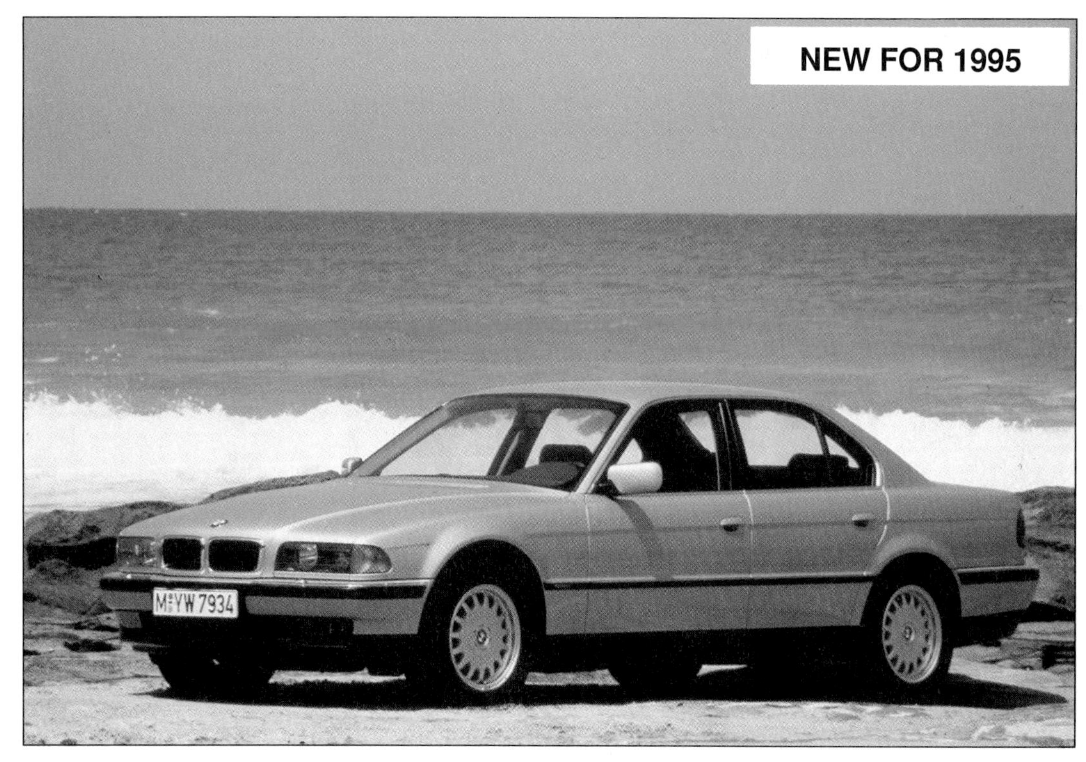

SPECIFICATIONS & PRICES

Model	Version/Trim	Body/ Seats	Interior volume cu.ft.	Trunk volume cu ft.	Drag coef. Cd	Wheel-base in	Lgth x Wdthx Hght. inx inx in	Curb weight lb	Towing capacity std. lb	Susp. type ft/rr	Brake type ft/rr	Steering system	Turning diameter ft	Steer. turns nber	Fuel. tank gal.	Standard tire size	Standard powertrain	PRICES US. $ 1994
BMW		**Warranty: 4 years/ 50,000 miles; corrosion: 6 years / unlimited; antipollution: 8 years / 80,000 miles.**																
740	i	4dr.sdn. 5	101	17.7	0.31	115.4	196.2x73.3x56.5	4145	2000	i/i	dc/dc/ABS	pwr.bal.	38.1	3.5	22.5	235/60HR16	V8/4.0/A4	**$55,950**
740	iL	4dr.sdn. 5	105	17.7	0.31	120.9	201.7x73.3x56.1	4200	2000	i/i	dc/dc/ABS	pwr.bal.	38.1	3.5	22.5	235/60HR16	V8/4.0/A4	**$59,950**
750	iL	4dr.sdn. 5	105	17.7	0.31	120.9	201.7x73.3x56.1	4497	2000	i/i	dc/dc/ABS	pwr.bal.	38.1	3.5	22.5	235/60HR16	V12/5.0/A4	**$83,950**

See page 393 for complete 1995 Price List

Test bench...

BMW and Mercedes are engaged in a fratricidal fight to see which of the two has the most advanced technology. This exercise is not as unnecessary as it seems because it triggers development of systems which tomorrow will equip less-pretentious models.

The BMW 8-Series gives the answer to Mercedes, Ferrari and Lamborghini on a different plane, because Germans are not fans of sporty 2-seaters but would rather see a 2+2 coupe. The common point between these vehicles with prestigious names is, of course, the V12 engines which they display with all their pride of technology. If the Italian designs remain fairly simple, the German ones reach increasing complexity and sophistication. The 850 is the most prestigious model in production by BMW. Mechanically the CSi derives from the 750iL of which it borrows the V12 that this year grows to 5.4 L. It exists, however, as a more popular Ci version powered by the 4.0 L V8 found on the 740i and the 540i which will be imported during 1995.

STRONG POINTS

• Technical: 100%

The steel unit body posts a remarkable Cd of just 0.29 in spite of cooling air openings and the width of the tires. There is an independent suspension and disc brakes at all 4 wheels. Up front it is inspired by the McPherson and at the rear five bars combine to attenuate lift at acceleration. Elastic mounts change the tow in and help set the car up into the turn. Electronic damping control allows programming the shocks to comfortable or sports. A third electronic control can lower the 6-inch ground clearance and allows a firmer response from the spring damper units. A traction control (ASC) tends to avoid any traction loss of the drive wheels, which must transmit up to 322 hp without tire slip, no matter what the pavement condition. It is integrated with the ABS and works on the intake, ignition and brakes to find maximum traction.

• Safety: 100%

The outside structure is extremely rigid with good impact resistance. The seats, which include seat belt take-up spools, have been the object of advanced research. There is also a kneecap protector and belts which adapt themselves to everyone's body shape.

• Quality/fit/finish: 90%

The presentation, assembly, materials quality and the care brought to finishing details are quasi perfect on the BMWs and far from Italian-production viewpoint.

• Satisfaction: 90 %

It is close to that of the 7-Series owners and proves that the various complex computers are amazingly reliable.

• Driver's comparment: 90%

Thanks to multiple adjustment combinations, the pilot soon finds an ideal position. Visibility is excellent, even laterally, thanks to the absence of a side pillar, and night vision benefits from the high-powered ellipsoidal headlights. The controls are easily handled, but one would like more gauges included in the dashboard which is a bit denuded.

• Performance: 90%

The V12 succeeds in powering the 850i from 0 to 62 mph in less than

DATA

Category: rear-wheel drive luxury sports coupes.
Class : GT

HISTORIC

Introduced in: 1990
Modified in: -
Made in: Din golfing, near Münich, Germany.

RATINGS

Safety:	100 %
Satisfaction:	91 %
Depreciation:	57.2 %
Insurance:	3.0 % ($2,320)
Cost per mile:	$1.70

NUMBER OF DEALERS

In U.S.A.: 363

SALES ON THE AMERICAN MARKET

Model	1992	1993	Result
8-Series	896	715	- 20.3 %

MAIN COMPETITORS

ACURA NS-X, LEXUS SC400, MERCEDES S 500 & S 600, JAGUAR XJR, PORSCHE 911 & 928.

EQUIPMENT

BMW 8-Series	840CiA	850 CSi
Automatic transmission:	S	S
Cruise control:	S	S
Power steering:	O	S
ABS brake:	S	S
Air conditioning:	S	S
Air bag (2):	S	S
Leather trim:	O	O
AM/FM/ Radio-cassette:	S	S
Power door locks:	S	S
Power windows:	S	S
Tilt steering column:	S	S
Dual power mirrors:	S	S
Intermittent wipers:	S	S
Light alloy wheels:	S	S
Sunroof:	S	S
Anti-theft system:	S	S

S : standard; O : optional; - : not available

COLORS AVAILABLE

Exterior: Black, White, Red, Blue, Gray, Green.
Interior: Black, Beige, Red, Gray.

MAINTENANCE

First revision:	15,000 miles
Frequency:	15,000 miles
Diagnostic plug:	Yes

WHAT'S NEW IN 1995 ?

- The V12 goes from 5.0 to 5.4 L on the 850.
- The manual 6-speed is not offered (850).

ENGINE					TRANSMISSION			PERFORMANCE										
Model/ versions *: standard	Type / timing valve / fuel system	Displacement cu/in	Power bhp @ rmn	Torque lb.ft @ rpm	Compres. ratio	Driving wheels / transmission	Final ratio	Acceler. 0-62 mph s	Stand. 1/4 mile s	Stand. 5/8 mile s	Braking 62-0 mph ft	Top speed mph	Lateral acceler. G	Noise level dBA	Fuel economy mpg city	Fuel economy mpg hwy	Gasoline type / octane
840CiA	V8* 4.0 DOHC-32-MFI	243	282 @ 5800	295 @ 4500	10.0:1	rear - A5	2.93	7.8	15.8	28.8	118	155*	0.83	66	18.0	29.0	S 91
850CSi	V12* 5.4 SOHC-24-MFI	304	322 @ 5000	361 @ 3900	10.0 :1	rear - A5	3.64	6.7	15.4	28.5	125	155*	0.85	66	14.0	25.0	S 91

*Electronically governed speed.

7 seconds in the comfort, luxury and safety that are out of the ordinary, especially considering that it weighs 4,123 pounds. Power and torque are remarkable, but standing starts are more effective than accelerations, which sometimes seem labored, keeping everything in proportion. To preserve precious driver license points of 850's buyers, BMW voluntarily limits top speed to 155 mph.

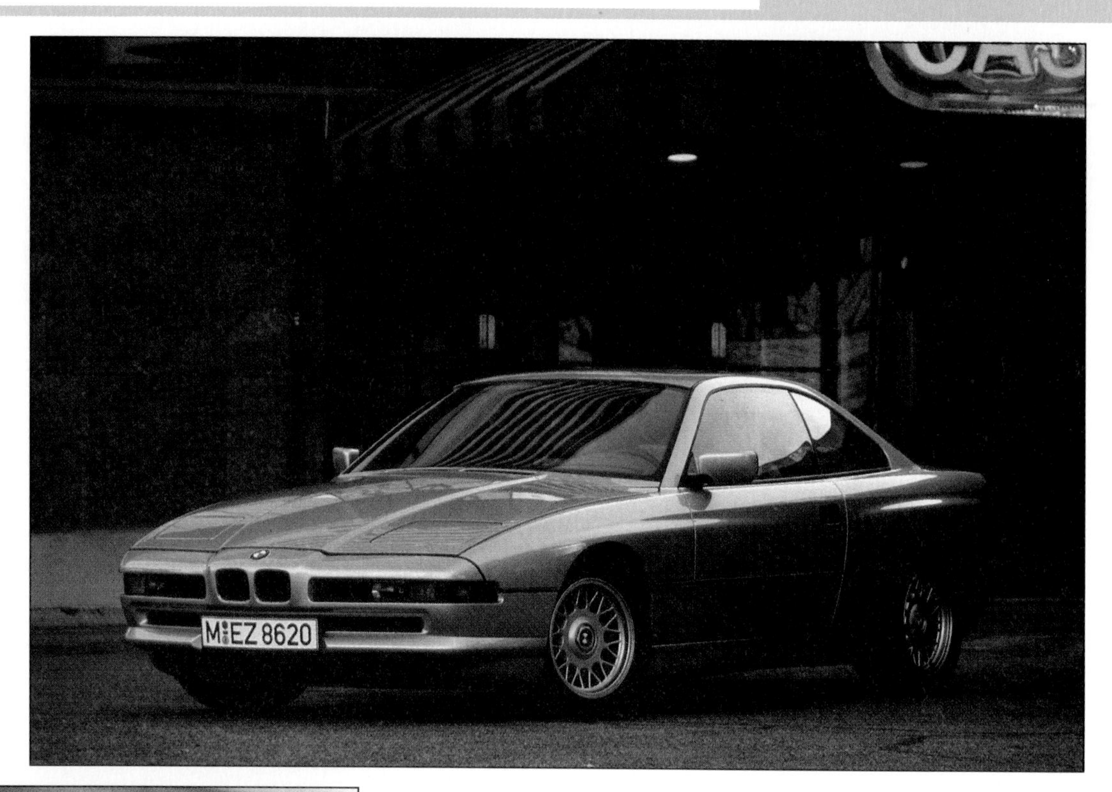

• Handling: 80%
On twisty roads, these coupes are penalized by their excess weight. They were created to gallop down German autobahns, which they swallow at a mad rate, rather than the American ones on which they die of boredom.

• Seats: 80%
They are not soft and they may not provide enough lateral hold, but their lumbar support is re-

markable.

• Suspension: 80%
Surprising for a car of this style, it papers over all road defects just like a sedan, without losing any of its aplomb.

• Steering: 80%
In addition to being well-modulated, it is fast and terribly precise because it allows positioning this pleasure tank with the diabolic precision of a missile.

• Access: 70%
The large doors ease entry into the front seat more so than at the back where headroom is more limited.

• Braking: 70%
It is up to the performance level, easy to modulate, very fade-free and of great precision. It is able to stop this moving mass from 62 mph in an average of 120 feet.

• Noise level: 50%
The soundproofing efficiency does not cease to surprise because it is hard to perceive the chant of the 12 cylinder, and it's only high-speed airstream and tires rolling on the asphalt that dominate.

• Conveniences: 50%
In spite of all sort of refinements, storage does not abound as much on the 8-Series as on the 7, especially around the center console.

WEAK POINTS

• Purchase price: 00%
You have to spend close to or more than $80,000 to profit from this high-tech rocket, which includes, as standard, everything that you can dream of: first aid kit, cellular phone, tool kit for do-it-yourself millionaires, and the only options are a manual 6-speed (no charge), light alloy rims and more sports like seats with velvet covering (no charge).

• Fuel consumption: 10%
As you can't make more than 275 miles on a full tank, the corner gas station becomes your buddy, unless you already own your own.

• Interior space: 35%
In spite of the imposing bulk of the body, the interior is only spacious up front because legroom is lacking at the back.

• Trunk: 45%
Its low capacity dissappoints because it corresponds to that of a Nissan 300ZX, but the cutout and low sill ease access.

• Depreciation: 40%
Losing 60% of $75,000 at the end of three years will not please anyone, but that is the price to pay for being seen in the company of this exotic beauty.

CONCLUSION

• Overall rating: 62.5%
The two models of the 8-Series are GT coupes which illustrate perfectly what BMW knows best in matters of terrestrial displacement. 😐

SPECIFICATIONS & PRICES

Model	Version/Trim	Body/ Seats	Interior volume cu.ft.	Trunk volume cu ft.	Drag coef. Cd	Wheel-base in	Lgth x Wdthx Hght. inx inx in	Curb weight lb	Towing capacity std. lb	Susp. type ft/rr	Brake type ft/rr	Steering system	Turning diameter ft	Steer. turns nber	Fuel. tank gal.	Standard tire size	Standard powertrain	PRICES US. $ 1994
BMW		Warranty: 4 years/ 50,000 miles; corrosion: 6 years / unlimited; antipollution: 8 years / 80,000 miles.																
840	CiA	3dr.cpe. 4	81	11.3	0.29	105.7	188.2x73.0x52.8	4123	NR	i/i	dc/dcABS	pwr.bal.	37.8	2.7	23.8	235/50ZR16	V8/4.0/A5	$68,100
850	CSi	3dr.cpe. 4	81	11.3	0.29	105.7	188.2x73.0x52.8	4123	NR	i/i	dc/dcABS	pwr.bal.	37.8	2.7	23.8	235/50ZR16	V12/5.4/A5	$85,500

See page 393 for complete 1995 Price List

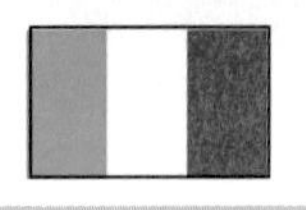

The grand sum of exotica...

Bugatti's name was not unknown to the public when they announced its resurrection four years ago, not in France, which was the cradle, but in Italy at Campogaliano in the triangle of Bologna, which is also the home of Lamborghini and Ferrari.

When he decided to launch himself into the manufacture of exotic cars, Romano Artioli knew very well what he was doing because he had already been a Ferrari distributor. He then put together a syndicate of companies famous for their high tech, such as French Aerospatiale, Elf Fuels, Messier Aviation, Michelin Tires, BBS Wheels, Schenk Chemical, etc. This syndicate purchased all that was left of Bugatti, which Aerospatiale had inherited. Beginning from that point the fast-car project production facilities were begun, and 3 years later came the Paris introduction of the EB110 and the building of a magnificent plant.

STRONG POINTS

Technical:
The start was to create a berlinetta so much at the state-of-the-art of advanced technology as to surpass the performances of its competitors, the Lamborghini and Ferrari. It is not by chance that the enterprise was built at Campogaliano, a suburb of Bologna. It is in this region that one finds the best craftsmen for everything that concerns the automobile stylists, engineers, engine block foundries, suppliers of special accessories, saddlers. Everything is there within hands reach. The EB 110 is a unit body built of Nomex honeycomb sandwiched between carbon fiber reinforced outer surfaces. This procedure stems directly from racing. What differentiates it from its competition is the permanent 4-wheel drive transfer case whose central differential is epicycloidal, allow the torque to be split at the rates of 27% for the front axle and 73% for the rear axle. The engine is a 3.5 L V12 with 60 valves, which should, at the end of development, deliver 600 hp, but actually tops out at 560, thanks to 4 turbo chargers. The suspension is biased at the front as at the rear by transverse 4-bar linkages with twin suspension units on the rear wheels. The four vented discs are controlled by a Bosch ABS. The assembly weighs 3,571 lbs of which 39% are at the front and 61 at the rear. The 18-inch BBS magnesium rims with Michelin MXX3 tires can run flat, which explains the absence of a spare tire.

Quality/fit/finish:
The criteria of assembly and detailing of this exceptional car are based on aeronautical practice.

Driver's compartment:
The very stripped down cockpit is without a doubt the most rational we have observed in this type of car. The ergonomics are perfect because the instruments are located in the pilot's vision field and all the controls fall under the finger tips. The visibility, normally critical, is as good toward the front as towards the sides and is satisfactory toward the rear, in spite of the narrowness of the back light. This, even though one is seated low and the body belt line is high. The bucket seats with electrical adjustments are of the type seen in race cars. They wrap around to support all body points efficiently and are equipped with a four-strap shoulder harness and seat belts with a central lock.

Performance:

DATA

Category: exotic AWD sport coupes.
Class: GT

HISTORIC

Introduced in: 1993
Modified in: -
Made in: Campogaliano, Italy

RATINGS

Safety: NA
Satisfaction: NA
Depreciation: NA
Insurance: NA
Cost per mile: NA

NUMBER OF DEALERS

In U.S.A.: NA

SALES ON THE AMRICAN MARKET

Model	1992	1993	Result
Bugatti EB110	NA	NA	

MAIN COMPETITORS

ASTON-MARTIN, Virage & Volante, BMW 850, FERRARI, 456, LAMBORGHINI Diablo

EQUIPMENT

BUGATTI- EB-110	GT
Automatic transmission:	-
Cruise control:	-
Power steering:	S
ABS brake:	S
Air conditioning:	S
Air bag (2):	S
Leather trim:	S
AM/FM/ Radio-cassette:	S
Power door locks:	S
Power windows:	S
Tilt steering column:	S
Dual power mirrors:	S
Intermittent wipers:	S
Light alloy wheels:	S
Sunroof:	-
Anti-theft system:	S

S : standard; O : optional; - : not available

COLORS AVAILABLE

Exterior: Silver, Black, Yellow, Blue & upon request.
Interior: Gray & upon request

MAINTENANCE

First revision: NA
Frequency: NA
Diagnostic plug: Yes

WHAT'S NEW IN 1995 ?

- Bugatti is in the process of completing its North American dealer network and will also market the Lotus which the firm bought from GM last year.

Model/ versions *: standard	ENGINE Type / timing valve / fuel system	Displacement cu/in	Power bhp @ rmn	Torque lb.ft @ rpm	TRANSMISSION Compres. ratio	Driving wheels / transmission	Final ratio	PERFORMANCE Acceler. 0-62 mph s	Stand. 1/4 mile s	Stand. 5/8 mile s	Braking 62-0 mph ft	Top speed mph	Lateral acceler. G	Noise level dBA	Fuel economy mpg city	hwy	Gasoline type / octane
EB-110	V12T* 3.5 DOHC-60-MFI	214	560 @ 8000	- @ 3750	7.5:1	all - M6	3.182	3.46	11.4	20.7	118	212	0.95	74	13.0	21.0	S 91

The day that we did a little driving on the first EB-110 GT, then called the 110 S Super Sport, we must have set the records for stumbling, lousiness, backfires and spitbacks because this engine, which had just been assembled, was not tuned and belched much unburned fuel. But that did not keep us from appreciating all the other parameters on this uncommon car. Besides, it was the same car which 3 weeks later set a new speed record on the Nardo circuit for a production car that officially reached 212 mph in the hands of factory driver Jean-Pierre Vitecocq. The factory indicates that the EB-110 GT does zero to 62 mph in 3.48 seconds, a time no production machine has been able to approach to within one second to date. One must say that the response time is a problem which must be resolved because the turbos don't come into action before 4,000 rpm, while the max. rpm is 8,000 which leaves too short a range to get the most out of the max power.

Handling:
The Bugatti surprises those who drive it for the first time by the ease with which it is handled. No need to be a champion to go fast with it because it holds the road all by itself and will forgive any errors of a novice. It proves its neutral handling at indecent speeds with a total absence of roll, and at the limits, it ends up with a little oversteer. The cornering traction is as phenomenal as in a straight line and one is often surprised that nothing much happens as the tach is happily climbing. Within all proportions it reminds of the Dodge Stealth by the fact that it can take tight turns at amazing speeds without having to fight with it to stay on the road or to have Nigel Mansel's talent.

Steering:
Superbly assisted, it is fast and very precise but does not feed back road conditions.

Braking:
At the speeds obtained, it has always shown itself balanced and easy to modulate with precision.

Access:
It is not as acrobatic as it looks and infinitely less so than that of some competitors.

Seats:
They are superb in lateral and lumbar support even though their padding is thin.

Suspension:
By its nature it is not uncomfortable, or perhaps you don't pay attention to that with this car.

Noise level:
Copious during standing starts and takeoffs, it still allows appreciation of the magnificent Nakamichi sound system at cruise speed (125 mph) or approximately 60% of its potential.

Conveniences:
Storage consists of a glove box of surprising capacity.

WEAK POINTS

Interior space:
Two occupants can be much at ease because the length and width are sufficient and only height is lacking.

Trunk:
It is nonexistent and you have to use a custom-made suitcase, which goes behind the seat backs and does not hold much.

Fuel consumption:
It is not that exorbitant because under 125 mph it holds at 14 miles per gallon.

CONCLUSION

Everything on this car is surprising but above all, its ease of driving and the civilized reactions, perhaps too civilized. It doesn't raise the adrenaline such as one feels in a Diablo or in a F512M. However, one thing is certain, the next productions from Ferrari and Lamborghini will have to take into account the EB-110, if only in ergonomic terms. ☺☺

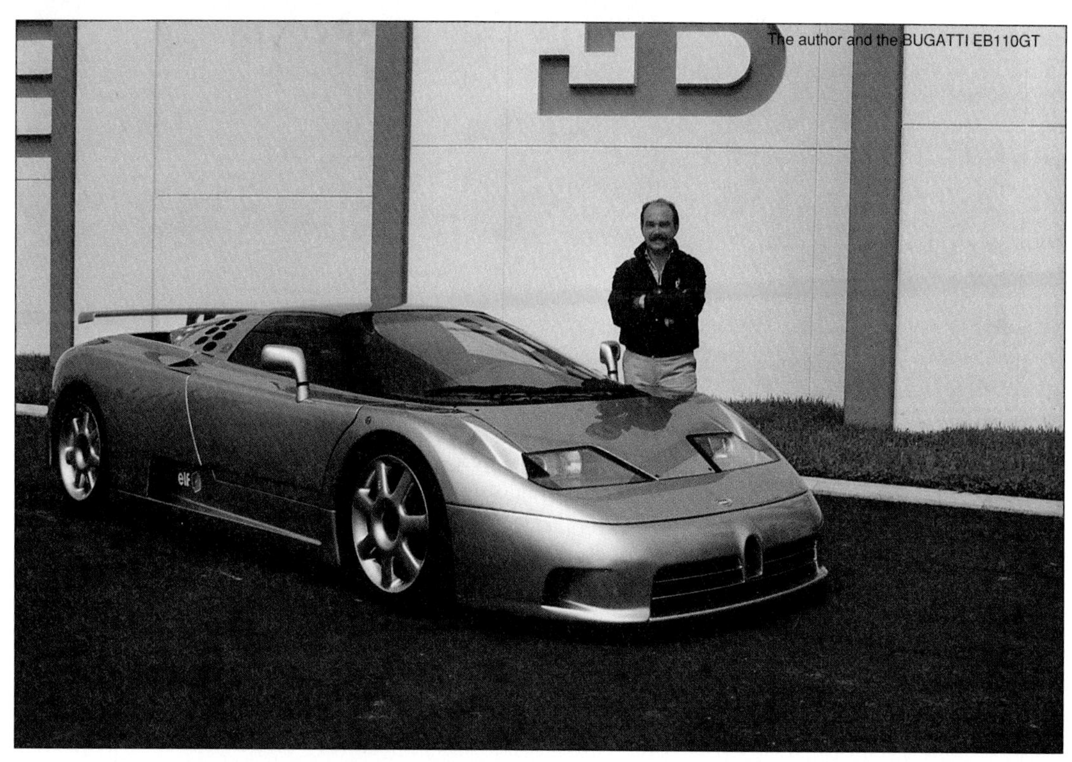

The author and the BUGATTI EB110GT

SPECIFICATIONS & PRICES

Model	Version/Trim	Body/ Seats	Interior volume cu.ft.	Trunk volume cu ft.	Drag coef. Cd	Wheel-base in	Lgth x Wdthx Hght. inx inx in	Curb weight lb	Towing capacity std. lb	Susp. type ft/rr	Brake type ft/rr	Steering system	Turning diameter ft	Steer. turns nber	Fuel. tank gal.	Standard tire size	Standard powertrain	PRICES US. $ 1994
BUGATTI																		
EB-110	GT	2dr.cpe. 2	390	2.5	0.29	100.4	173.2x77.2x44.3	3571	NR	i/i	dc/dc/ABS	pwr.r/p.	40	2.8	2x15.8	240/40R18 325/30R18 (re.)	V12T/3.5/M6	

See page 393 for complete 1995 Price List

Interim...

The Acclaim and Spirit take over, partially because the new Cirrus Stratus is not ready to replace them, but mostly because their difference in price will cause reflection by some buyers who are more sensitive to budget than to the attraction of a novelty.

The Acclaim/Spirit will still be sold in 1995. Identical except for a few cosmetic and equipment details, these two 4-doors will be offered in a single version equipped standard with a 2.5 L 4 cylinder with a 3-speed automatic. A V6 3.0 L is offered as an option with a 4-speed automatic.

STRONG POINTS

• Price/equipment: 80%
It is more favorable than that of its direct Japanese competitors, even with the V6 and automatic transmission. While equipment has been expanded, the options are limited to glass and electric door locks, disc brakes with ABS, and gold trim on the outside of the light-alloy rims.

• Satisfaction: 80%
These models have known few problems since most components have been long tested. Owners still complain of rapid wear of tires, shock absorbers and brake friction material.

• Safety: 65%
The safety of the structure is not perfect because its conception is dated. You don't find door beams to resist lateral impact, and only one air bag is offered standard on the driver's side.

• Interior space: 65%
It constitutes one of the advantages of this car. Five people can fit because leg and headroom are generous at the back.

• Trunk: 60%
Accessible, thanks to the center cutout, its capacity is only average because it lacks length. However, you can lower the back of the rear bench seat and use it for bulkier objects.

• Handling: 60%
It is never dangerous in normal use, but its simplicity calls for slowing down in tight turns and on slippery roads because of its excessive softness.

• Suspension: 60%
These cars are more at ease on the freeway than on little twisty ill-paved roads where the excessively soft suspension rolls and its limited wheel travel causes the rear to bottom out and react harshly.

• Technical: 60%
Descendants of the famous K cars, the Acclaim and Spirit don't shine by their very ordinary technical content. The steel unit body shows a very conservative appearance, the front suspension is of a McPherson type and at the rear there is a torsional axle. Braking is mixed, but 4 discs and ABS are offered as an option.

• Noise level: 60%
It is higher with a 4-cylinder, which is distinctly noisier than a V6. To the engine sounds are added wind and rolling noises which the lack of soundproofing let through.

• Quality/fit/finish: 60%
The inside appearance is somber and quiet, the assembly; just like the

DATA

Category: front-wheel drive compact sedans.
Class: 4

HISTORIC

Introduced in: 1988: Acclaim-Spirit; 1990: Le Baron.
Modified in: -
Made in: Toluca, Mexico.

RATINGS

Safety: 70 %
Satisfaction: 80 %
Depreciation: 65 %
Insurance: 5.0 % ($585)
Cost per mile: $0.40

NUMBER OF DEALERS

In U.S.A.: 1,822 ChryslerPlymouth, 1,887 Dodge.

SALES ON THE AMERICAN MARKET

Model	1992	1993	Result
Acclaim	73,220	78,374	+ 6.6 %
Spirit	65,547	87,421	+ 25.1 %

MAIN COMPETITORS

ACURA Integra sedan, CHEVROLET Corsica, FORD Contour, BUICK Skylarck, HONDA Accord, HYUNDAI Sonata, MAZDA 626, MERCURY Mystique, MITSUBISHI Galant, OLDSMOBILE Achieva, PONTIAC Grand Am, NISSAN Altima, SUBARU Legacy, TOYOTA Camry, VW Passat.

EQUIPMENT

DODGE Spirit	base
PLYMOUTH Acclaim	base
Automatic transmission:	S
Cruise control:	S
Power steering:	S
ABS brake:	O
Air conditioning:	S
Air bag (left):	S
Leather trim:	-
AM/FM/ Radio-cassette:	S
Power door locks:	O
Power windows:	O
Tilt steering column:	S
Dual power mirrors:	S
Intermittent wipers:	S
Light alloy wheels:	O
Sunroof:	-
Anti-theft system:	-

S : standard; O : optional; - : not available

COLORS AVAILABLE

Exterior: Black, Blue, Beige, Emerald, Red, Wildberry, White.
Interior: Blue, Champagne, Gray, Red.

MAINTENANCE

First revision: 5,000 miles
Frequency: 6 months
Diagnostic plug: Yes

WHAT'S NEW IN 1995 ?

- No major change.

ENGINE					TRANSMISSION			PERFORMANCE										
Model/ versions *: standard	Type / timing valve / fuel system	Displacement cu/in	Power bhp @ rmn	Torque lb.ft @ rpm	Compres. ratio	Driving wheels / transmission	Final ratio	Acceler. 0-62 mph s	Stand. 1/4 mile s	Stand. 5/8 mile s	Braking 62-0 mph ft	Top speed mph	Lateral acceler. G	Noise level dBA	Fuel economy mpg city	hwv	Gasoline type / octane
base	L4 2.5 SOHC-8-EFI	153	100 @ 4800	135 @ 2800	8.9 :1	front-A3	3.02	12.6	18.0	35.2	141	100	0.78	67	27.0	35.0	R 87
option	V6*3.0 OHV-12-SFI	181	142 @ 5000	171 @ 2400	8.9 :1	front-A3	2.28	11.0	17.0	33.8	134	115	0.80	66	25.0	35.0	R 87

finishing careful and the materials' quality is acceptable even though the plastics and fabrics reflect "utility".

• Driver's compartment: 55%
The dashboard is well organized and the instrumentation legible; also, the main controls are conventional. However, the best driving position is still to be discovered because the distance separating the steering wheel from the pedals is too long. Visibility is satisfactory from all angles, in spite of C pillar thickness.

• Access: 60%
Without a problem up front, it is less convenient at the back where the door opening angle is not sufficient.

• Fuel consumption: 50%
It is normal, considering the weight and displacement of these engines, but is sensitive to load and speed.

WEAK POINTS

• Seats: 30%
The seats are badly designed and do not provide any more lateral hold than lumbar support in the front or rear seats. At the rear, the bench seat is too short, its padding hard and without shape and there is no center armrest to help you get set.

• Braking: 40%
Its use is not pleasant because it is hard to modulate the spongy pedal. Panic stops are more hazardous than ordinary deceleration because the distances are long and the trajectory uncertain in the absence of 4-wheel disc brakes and ABS.

• Insurance: 40%
In relation to some of its competitors, the premium is inexplicably higher.

• Depreciation: 40%
Their resale value decreases as they approach the end and there are many of them on the used-car market.

PLYMOUTH Acclaim

DODGE Spirit

• Conveniences: 40%
Storage is limited to a minuscule glove box and to hollows in the center console, as well as in the tray built into the top of the dash, which do not effectively hold the small objects. One would like to find door pockets as well as storage within the back of the front seats.

• Performance: 45%
With the 4-cylinder engine and the automatic, the weight to power ratio does not favor standing starts or accelerations. You need a V6 for better mileage and pleasant driving.

• Steering: 45%
The steering diameter is too large to make these models maneuverable. The steering itself is well-modulated, but has a vague center feel.

CONCLUSION

• Overall rating: 54%
On the eve of their retirement, these domestic compacts constitute valid alternatives to the Japanese models which command high prices for their sophistication. While fairly ordinary and simplistic, the Chryslers contain few bad surprises. ☹

OWNERS' SUGGESTIONS

-Less belt noise.
-A suspension that doesn't bottom on every occasion. If bottoming does occur, soften it.
-Improving the pedal assembly, ventilation, defrosting and soundproofing of the base models.
-More storage.
-Improved quality of tires, shocks and brakes.

SPECIFICATIONS & PRICES

Model	Version/Trim	Body/ Seats	Interior volume cu.ft.	Trunk volume cu ft.	Drag coef. Cd	Wheel-base in	Lgth x Wdthx Hght. inx inx in	Curb weight lb	Towing capacity std. lb	Susp. type ft/rr	Brake type ft/rr	Steering system	Turning diameter ft	Steer. turns nber	Fuel. tank gal.	Standard tire size	Standard powertrain	PRICES US. $ 1994
DODGE		**General warranty: 3 years / 36,000 miles; surface corrosion: 1year / 12,000 miles; perforation: 7 years / 100,000 miles; road assistance: 3 years / 36,000 miles**																
Spirit	base	4dr.sdn.5	96	14	0.40	103.5	182.7x68.1x53.7	2864	2000	i/si	dc/dr	pwr.r/p.	37.1	2.6	16.1	185/70R14	L4/2.5/A3	**$12,470**
PLYMOUTH		**General warranty: 3 years / 36,000 miles; surface corrosion: 1year / 12,000 miles; perforation: 7 years / 100,000 miles; road assistance: 3 years / 36,000 miles**																
Acclaim	base	4dr.sdn.5	96	14	0.40	103.5	182.7x68.1x53.7	2861	2000	i/si	dc/dr	pwr.r/p.	37.1	2.6	16.1	185/70R14	L4/2.5/A3	**$12,470**

See page 393 for complete 1995 Price List

Le Baron convertible

Memories...

At a time when convertibles had disappeared from the catalogs of North American car builders, Chrysler relaunched them and the LeBaron J convertible was successful. It was another good idea from Lee A. Iaccoca.

The LeBaron convertible remains faithfully on duty while waiting for its replacement by a model descended from the latest Sebring and Avenger, it is offered in only one version with the appearance of the ancient GTC and comes standard with a 3.0L V6 and an electronically controlled 4-speed automatic. While its equipment has been enriched, there still remain a number of important options, such as the steering wheel, adjustable rear-view mirrors, cruise control and electric door locks, which should be included in the price.

STRONG POINTS

• Safety: 80 %
Stiffened up last year, the body shell satisfies impact requirements and the presence of two air bags ensures good front-occupant protection. No rollbar is available.

• Satisfaction: 70 %
The number of satisfied owners is dropping, rare in a model at the end of its career. Here all problems should have been solved. Owners still complain of the mediocrity of the shock absorbers, tires, front brake pads and electrical equipment.

• Technical: 65 %
Derived from the ancient K platform, the steel unit body has an independent McPherson front suspension and a semi-independent rear suspension with a torsional axle. Standard brakes are mixed disc drums; 4-wheel discs and ABS are options. The body lines are very efficient: for a convertible, a drag coefficient of 0.38 is remarkable.

• Quality/fit/finish: 60 %
The build is rugged and the finish superior to the previous ones, but the quality still leaves something to be desired. Even though powered, the top is not doubled or padded as the simple Golf.

• Access: 60 %
Even though the doors are as long as they are heavy, it is easier to reach the front seats than the back seat, where space is the smallest.

• Driver's compartment: 60 %
The most comfortable driving position is easy to find. Visibility suffers from low seating and from major blind spots due to the convertible top. The well-designed instrument panel offers sufficiently easy-to-read gauges and the main switches, placed at the edge of the dash, are within hand's reach. However, that of the cruise control, located at the bottom of the steering hub, is not easy to reach until its position has been memorized.

• Noise level: 60 %
The V6 engine is reasonably quiet, but the uninsulated top allows all exterior noises to come in, particularly rolling and wind noises.

DATA

Category: front-wheel drive mid-size convertibles.
Class: 4

HISTORIC

Introduced in: 1987
Modified in: 1992: V6 3.0L engine.
Made in: Newark, Delaware, U.S.A.

RATINGS

Safety: 75 %
Satisfaction: 70 %
Depreciation: 57 %
Insurance: 4.4 % ($725)
Cost per mile: $0.47

NUMBER OF DEALERS

In U.S.A.: 1,822 Chrysler-Plymouth.

SALES ON THE AMERICAN MARKET

Model	1992	1993	Result
Le Baron	42,946	34,519	- 19.7

MAIN COMPETITORS

FORD Mustang V6, OLDSMOBILE Cutlass Supreme, VW Golf

EQUIPMENT

Le Baron convertible	base
Automatic transmission:	S
Cruise control:	O
Power steering:	S
ABS brake:	O
Air conditioning:	S
Air bag (2):	S
Leather trim:	O
AM/FM/ Radio-cassette:	S
Power door locks:	O
Power windows:	S
Tilt steering column:	O
Dual power mirrors:	O
Intermittent wipers:	S
Light alloy wheels:	O
Sunroof:	S
Anti-theft system:	O

S : standard;-O : optional; - : not available

COLORS AVAILABLE

Exterior: Black, Aqua, Light Blue, Emerald, Red, White, Green, Blue.
Interior: Red, Blue, Champagne, Dark Gray.

MAINTENANCE

First revision: 5,000 miles
Frequency: 6 months
Diagnostic plug: Yes

WHAT'S NEW IN 1995 ?

- Remote door and trunk opening.
- Radio with laser CD.
- Change of emblems and light alloy wheel appearance.

	ENGINE					TRANSMISSION			PERFORMANCE									
Model/ versions *: standard	Type / timing valve / fuel system	Displacement cu/in	Power bhp @ rmn	Torque lb.ft @ rpm	Compres. ratio	Driving wheels / transmission	Final ratio	Acceler. 0-62 mph s	Stand. 1/4 mile s	Stand. 5/8 mile s	Braking 62-0 mph ft	Top speed mph	Lateral acceler. G	Noise level dBA	Fuel economy mpg city	Fuel economy mpg hwy	Gasoline type / octane
base	V6 3.0 SOHC-12-MPSFI	181	141 @ 5000	171 @ 2400	8.9 :1	front-A4*	3.62	11.0	17.5	31.0	161	109	0.78	68	24.0	35.0	R 87

Le Baron convertible

• Price/equipment: 60 %
The LeBaron convertible still reigns in a narrow niche where its price is affordable with what it offers. Its equipment is a little enriched, but it misses disc brakes with ABS.
• Insurance: 60 %
The premium on the LeBaron convertible is relatively reasonable when compared to a Miata.
• Fuel consumption: 60 %
It maintains itself within reasonable limits if you take into account the weight and engine displacement. However, do not force the radar roulette chances...
• Performance: 60 %
The V6 brings a certain driving pleasure, thanks to its broad torque range and operating quiet. Well-designed, transmission gear ratio spacing allows good performance.

• Handling: 60 %
Road handling is good in spite of the rustic suspension design. However, it is sensitive to tire quality. Soft springs and shock absorber inconsistency promote major body motion, and lack of wheel travel often causes front suspension bottoming.
• Suspension: 60 %
Its softness is a benefit only on freeways. Otherwise, inconsistent shock absorbers, indifferent tires and the lack of wheel travel generate hostile reactions on poorly maintained back roads.
• Seats: 50 %
Their very technical appearance gives the impression that they were scientifically studied to support the occupants like a glove. There is nothing to that, since the lateral and lumbar supports are purely symbolic.
• Depreciation: 50 %
The convertible loses a little less of its value and will bring better money when the sun shines.

WEAK POINTS

• Braking: 30 %
Average decelerations pose no problem except that the spongy pedal does not allow precision control. On the other hand, in a panic stop, the stopping distances are long even with the optional 4-wheel discs and ABS. The latter, at least, has the advantage of gaining the safer, straighter stopping travel path.
• Steering: 35 %
Light and vague in the center, its power assist is too high and the large turning diameter penalizes maneuverability.
• Interior space: 40 %
It is more spacious at the front than at the rear. Here, passenger space is limited by the convertible top.
• Trunk: 40 %
Again, the convertible top mortgages the space reserved for baggage. However, the rear bench seat can store what the trunk will not hold.
• Conveniences: 40 %
Storage space consists of a modest glove box, a hollow and a container built into the center console. The door pockets are not practical because they don't hold anything.

CONCLUSION

• Overall rating: 55.0 %
Designed to benefit from nice weather on a reasonable budget, the Le Baron convertible has no sporting pretensions since its driveability and performance could not be more ordinary. 😐

OWNERS' SUGGESTIONS

-Longer cruise range.
-Longer seat bottom.
-Disc brakes and ABS standard.
-More careful finish.
-Improved suspension.
-A more practical aspect.
-More sporting character.
-Improved service courtesy.
-Insulated convertible top.

SPECIFICATIONS & PRICES

Model	Version/Trim	Body/ Seats	Interior volume cu.ft.	Trunk volume cu ft.	Drag coef. Cd	Wheel-base in	Lgth x Wdthx Hght. inx inx in	Curb weight lb	Towing capacity std. lb	Susp. type ft/rr	Brake type ft/rr	Steering system	Turning diameter ft	Steer. turns nber	Fuel. tank gal.	Standard tire size	Standard powertrain	PRICES US. $ 1994
CHRYSLER		**General Warranty: 3 years / 36,000 miles; surface rust: 1 year / 12,000 miles; perforation: 7 years / 100,000 miles; road assistance: 3 years / 36,000 miles.**																
Le Baron	base	2dr. con. 4	84	10	0.38	100.5	184.8x69.2x52.4	1416	1000	i/si	dc/dr	pwr.r/p.	28.4	2.48	14	205/60R15	V6/3.0/A4	**$16,999**

See page 393 for complete 1995 Price List

CHRYSLER Cirrus
DODGE Stratus

Feet on the ground...

At Chrysler, one novelty does not wait for another. With the Cirrus and Stratus, the third American car builder is attempting to conquer a first-rank place in the segment of superior compacts, since these represent the current best-selling cars both in America and in the entire world.

The Chrysler Cirrus and the Dodge Stratus will progressively replace the LeBaron-Acclaim-Spirit. Entirely new, these cars borrow nothing from preceding models and constitute a redefinition of the compact car in the Chrysler range. The report that follows concerns only the Cirrus, the only one we have been able to test before going to press. The Stratus will arrive a little later, along with the model that will join the Plymouth range. The Cirrus is offered in two versions, LX and LXi, supplied with a new V6 and a 4-speed automatic.

STRONG POINTS

• Safety: **100 %**
As befits a 1995 car, the unit body of the JA platform is very rigid and contains many reinforcements. This allows it to resist impacts from all possible angles. Two air bags protect the front seat occupants and the seat belt heights are adjustable. Also, ABS brakes are standard on the Cirrus. An infant seat, integrated into the bench seat, is available as an option.

• Technical: **80 %**
Chrysler has not stinted on this point since the new model benefits from the best available. Its steel unit body receives many corrosion-proofing treatments and has good aerodynamic efficiency with a drag coefficient of 0.31. The 4-wheel independent suspension uses a modified McPherson suspension up front with long and short arms that incorporate rubber bushings. These isolate the front end from road vibrations without resorting to a separate cradle, as happens with some competitors. At the rear, one finds a system with unequal arms that induce a self-steering action. Stabilizer bars are used front and rear. Brakes are mixed, even though you would have expected to find discs on the luxury models. Power steering assist varies with speed and the original equipment tires are Michelin MX4. The Cirrus weighs 3,146 pounds with a front and rear weight distribution of 64-36 %.

• Steering: **80 %**
It is practically ideal, thanks to its speed-controlled power assist, steering ratio and precision. However, a larger-than-average turning radius penalizes maneuverability.

• Seats: **80 %**
Up front, they remind us of French cars that have always served as a reference in the matter of comfort, certainly more so than the German seating by which Japanese builders are too often inspired. They provide efficient support and holding, both lateral and lumbar. Without complicated adjustments, their padding is cushiony. At the rear the bench seat is a bit short: the back lacks contouring and is not fitted with a center arm rest, which is a mistake. Front seat occupants are better cared for than the rear ones: that, in our opinion, is nonsense.

• Suspension: **80 %**
Here too, a French inspiration makes itself positively felt. (Chrysler

DATA

Category: front-wheel drive compact sedans.
Class : 4

HISTORIC

Introduced in : 1994
Modified in : -
Made in: Sterling Heights, MI, U.S.A.

RATINGS

Safety: 90 % Stratus, 100 % Cirrus
Satisfaction: NA
Depreciation: NA
Insurance: 6.0 % ($ 585)
Cost per mile: $ 0.40

NUMBER OF DEALERS

In U.S.A.: 1,822 Chrysler-Plymouth, 1,887 Dodge .

SALES ON THE AMERICAN MARKET

Model	1992	1993	Result
Acclaim	73,220	78,374	+6.6 %
Spirit	65,547	87,421	+25.1 %

MAIN COMPETITORS

ACURA Integra 4dr., BUICK Skylark, CHEVROLET Corsica, DODGE Spirit, FORD Contour, HONDA Accord, HYUNDAI Sonata, MAZDA 626, MERCURY Mystique, MITSUBISHI Galant, NISSAN Altima, OLDSMOBILE Achieva, PLYMOUTH Acclaim, PONTIAC Grand Am, SUBARU Legacy, TOYOTA Camry, VW Passat.

EQUIPMENT

CHRYSLER Cirrus				LX	LXi
DODGE Stratus	-	-	-		
Automatic transmission:	O	O	O	S	S
Cruise control:	O	O	S	S	S
Power steering:	S	S	S	S	S
ABS brake:	O	O	S	S	S
Air conditioning:	O	O	S	S	S
Air bag:	S	S	S	S	S
Leather trim:	-	-	O	O	O
AM/FM/ Radio-cassette:	O	O	S	S	S
Power door locks:	-	O	S	S	S
Power windows:	-	O	S	S	S
Tilt steering column:	S	S	S	S	S
Dual power mirrors:	-	O	S	S	S
Intermittent wipers:	S	S	S	S	S
Light alloy wheels:	-	O	S	O	S
Sunroof:	-	-	-	-	-
Anti-theft system:	-	-	-	-	-

S : standard; O : optional; - : not available

COLORS AVAILABLE

Exterior: White, Black, Iris, Crimson, Green, Silver, Red, Blue.
Interior: Gray, Gray-green, Red.

MAINTENANCE

First revision: 5,000 miles
Frequency: 6 months
Diagnostic plug: Yes

WHAT'S NEW IN 1995 ?

-New models replacing the Acclaim-Spirit.

	ENGINE					TRANSMISSION		PERFORMANCE									
Model/ versions *: standard	Type / timing valve / fuel system	Displacement cu/in	Power bhp @ rmn	Torque lb.ft @ rpm	Compres. ratio	Driving wheels / transmission	Final ratio	Acceler. 0-62 mph s	Stand. 1/4 mile s	Stand. 5/8 mile s	Braking 62-0 mph ft	Top speed mph	Lateral acceler. G	Noise level dBA	Fuel economy mpg city	hwy	Gasoline type / octane
1)	L4* 2.0-DOHC-16-MPFI	121	132 @ 6000	129 @ 5000	9.8 :1	front- M5*	3.94	-	-	-	-	-	-	-	-	-	R 87
						front-A4	3.90	-	-	-	-	-	-	-	-	-	R 87
2)	L4* 2.4-DOHC-16-MPFI	148	140 @ 5200	160 @ 4000	9.4 :1	front- M5*	3.94	-	-	-	-	-	-	-	-	-	R 87
						front-A4	3.90	-	-	-	-	-	-	-	-	-	R 87
3)	V6* 2.5-SOHC-24-MPFI	152	164 @ 5900	163 @ 4350	9.4 :1	front-A4*	3.90	10.5	15.75	31.5	144	115	0.88	67	21.0	31.0	R 87

1) base Status 2) Option Status 3) base Cirrus, option Stratus

chief engineer, François Castaing is of French origin). In fact, Renault, Peugeot and Citroën are the only ones to bring suspensions to this level. Suppleness provides superior comfort with long wheel travel without a handling detriment. The vehicle rolls in the corner without altering the quality of its trajectory. The Chrysler JA does the same and its comfort marks a major point.

• **Conveniences: 80 %**
The storage areas are well laid out. The glove box is rather large, and so are the door pockets; however, their openings are limited. A large enclosed box in the center console and bellows-type soft bags in the back of the front seats, as well as holding traps at the roof, increase the storage space.

• **Driver's compartment: 80 %**
It is inspired more by the Neon than the LH models. Simple and curved, it contains a legible instrument block, but the center portion of the console is too low and not large enough to offer satisfactory ergonomics. The broad area that rejoins the bottom of the windshield provides too many reflections, to the point of obstructing forward vision. When the sun shines intensely, visibility is reduced by 3/4. Toward the rear, the height and narrowness of the backlight is also limited. Driver position is ideal: both fast and easy to find. It also allows long hours of driving without fatigue. The different controls are grouped on two practical stalks. Regarding visibility, headlights provide sufficient illumination at night, but the second speed of the windshield washer is too slow.

• **Quality/fit/finish: 75 %**
The general impression is better with leather than with fabric that looks more ordinary. The placement of false wood in the center console is not disturbing, but the quality of the dashboard plastics is the same as that of the Neon. Fit is relatively tight, and the main adjustments on the preproduction

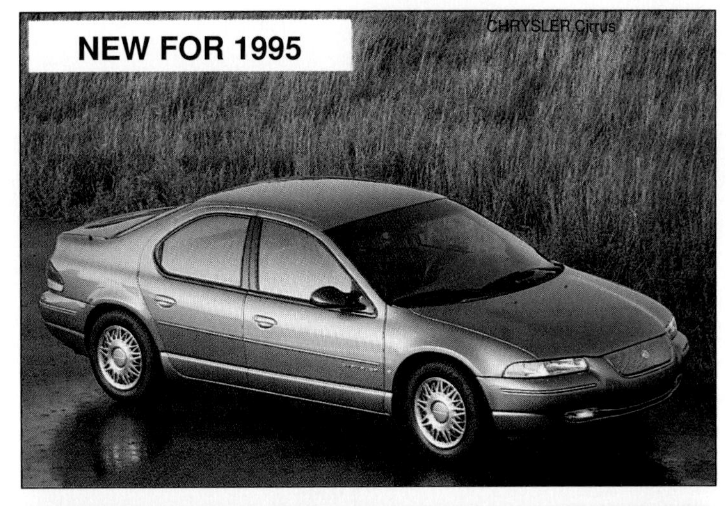

cars we looked at were good.

• **Handling: 75 %**
It is excellent and the car seems to grip the road on its own, in most circumstances. This gives the driver a strong safety feel. As we mentioned above, the roll is not annoying but in a tight turn the tires are pushed to provide best side bite. Stability, in a straight line or in wide curves, remains unperturbable regardless of the quality of the pavement or the crosswind.

• **Access: 72 %**
It is more convenient at the front, with the length of the doors and their opening angle, than at the rear.

• **Insurance: 70 %**
The premium on the Cirrus is in the middle of this popular category: at the same level as a Toyota Camry and a Honda Accord but less expensive than the Mazda 626.

• **Trunk: 70 %**
It is as wide as it is tall, and can be increased toward the interior by lowering the rear seat back. Thanks to new hinges, the trunk lid opens high and provides ample access.

• **Interior space: 65 %**
Interior dimensions are generous, especially at the rear where large-size people will find sufficient head and legroom.

• **Noise level: 60 %**
Mechanical and wind noises remain low while rolling noises dominate. Soundproofing is effective, thanks to the injection of plastic foam into the roof pillars.

• **Price/equipment: 60 %**
Definitely higher than the model it replaces, the price of the Cirrus is aligned just below that of the Japanese at which it aims, such as the Honda Accord that remains as the best reference in this category. This is partially justified by the rather ample standard equipment.

• **Performance: 55 %**
In spite of the good pedigree of this V6 engine of Mitsubishi origin, the pickup and acceleration are something of a letdown. While the weight-to-power ratio is favorable, the results are nothing earth-shaking, especially after those of the more modest Neon. Let's hope that the production models will show more punch, to bring driving pleasure to the level of a Mazda 626: hard to beat at this point.

• **Fuel consumption: 50 %**
It stays around 26 miles per gallon in normal driving, about the same as its competition, but offers lower performance.

WEAK POINTS

• **Braking: 45 %**
With standard ABS, the stopping distances we have established range between 144 and 167 ft. at 60 mph. This seems very long for this car, but it is easily explained by the absence of rear-wheel discs.

CONCLUSION

• **Overall rating: 64.0 %**
The Cirrus is a superb car that lacks in consistency because the excellence of the suspension and steering contrasts with mediocre performance and braking. Chrysler has gone a little too fast in defining the parameters and then detailing this model, which deserves better. Let's hope that the quality so dear to the president of Chrysler will become standard. ☺

SPECIFICATIONS & PRICES

Model	Version/Trim	Body/ Seats	Interior volume cu.ft.	Trunk volume cu ft.	Drag coef. Cd	Wheel-base in	Lgth x Wdthx Hght. inx inx in	Curb weight lb	Towing capacity std. lb	Susp. type ft/rr	Brake type ft/rr	Steering system	Turning diameter ft	Steer. turns nber	Fuel. tank gal.	Standard tire size	Standard powertrain	PRICES US. $ 1994
CHRYSLER		General Warranty: 3 years / 36,000 miles; surface rust: 1 year / 12,000 miles; perforation: 7 years / 1000,000 miles; road assistance: 3 years / 36,000 miles.																
Cirrus	LX	4 dr. sdn. 5	96	15.7	0.31	107.9	187.0x71.0x54.1	3146	1000	i/i	dc/dr/ABS	pwr.r/p.	37.1	3.09	15.8	195/65R15	V6/2.5/A4	**$17,970**
Cirrus	LXi	4 dr. sdn. 5	96	15.7	0.31	107.9	187.0x71.0x54.1	3146	1000	i/i	dc/dr/ABS	pwr.r/p.	37.1	3.09	15.8	195/65R15	V6/2.5/A4	**$19,900**

See page 393 for complete 1995 Price List

CHRYSLER Concorde, DODGE Intrepid, EAGLE Vision

Tidal wave...

The LH-Series cars find success similar to that which the Taurus/Sable knew in their days. Chrysler cannot produce enough of them to satisfy the demand, and it will be necessary to build other plants to profit from this manna while it is there.

The LH-Series cars occupy a particular market position. They are intermediate in their format and drive train but have a large format based on the volume of the trunk and interior and will do battle against several established competitors.

STRONG POINTS

• Safety: **100 %**
The unit body of the LH-Series offers good stiffness and many reinforcements have been integrated into the doors to resist lateral impact. Finally, two air bags protect the front seats. Too bad that the ABS is only installed as standard on the Concorde and Vision. For this reason, the Intrepid safety rating is only 90 %.

• Satisfaction: **90 %**
The owners consider themselves satisfied with their purchase, which is normal after a full year of sales.

• Technical: **80 %**
The idea of "advanced cabin" consists of pushing the wheels out to the four corners of the cab and not of the vehicle, as the builder would like you to believe. Thus, 75 % of the length of the vehicle is dedicated to the occupants and their baggage as opposed to 60 to 65 % average for the category. The frontal area is large, but this is offset by a low 0.31 drag coefficient. These front wheel drives are powered by 3.3L V6 engines that develop 161 hp and 3.5L DOHC engine, with 24 valves that deliver 214 hp. A 4-speed automatic with a recalibrated electronic shift control is the only transmission offered. Lacking a 4-wheel drive, an electronic traction control is coupled to the ABS. Four-wheel disc brakes are standard on the Concorde and Vision TSi and as an option elsewhere. The steel unit body includes a cradle for the drive train. An independent 4-wheel suspension with McPherson struts can become adjustable, thanks to an optional pneumatic device.

• Quality/fit/finish: **80 %**
These vehicles have very tight assembly tolerances, a finish of international caliber and materials with an excellent appearance. Some touchups were necessary, but in the main, these cars allow you to measure the progression in the Chrysler standards.

• Driver's compartment: **80 %**
The best driving position is hard to find, but visibility is good in spite of the pillar thickness. Among the complaints are an undersized rear-view mirror, the shape of the steering wheel and lack of lighting power from the Intrepid headlights, even after improvements. The dashboard is pleasant to look at and to use, and the headlight and AC controls have a German style. Only the parking brake seems archaic.

• Steering: **80 %**
It is smooth and precise with good power assist control. It would be appreciated if the steering had fewer turns, lock to lock (3.4 turns). Maneuverability is satisfactory and turning diameter is reasonable, but one has to beware of the extensive front overhang.

• Access: **80 %**
The large doors open wide and the trunk opening is cut out to assist baggage handling.

• Seats: **80 %**
Very successful, their lateral support is much more evident than their

DATA

Category: front-wheel drive full-size sedans.
Class : 6

HISTORIC

Introduced in : 1993
Modified in : 1994: 3.3L 161 hp.
Made in: Newark, Delaware, U.S.A. & Bramalea, Ontario, Canada.

RATINGS

Safety: 90 % (100 % with ABS std)
Satisfaction: 90 %
Depreciation: 42 % (2 years)
Insurance: 5.0 % ($ 585)
Cost per mile: $ 0.44

NUMBER OF DEALERS

In U.S.A.: 1,822 Chrysler-Plymouth 1,029 Jeep-Eagle

SALES ON THE AMERICAN MARKET

Model	1992	1993	Result
Concorde	-	56,029	-
Intrepid	-	89,127	-
Vision	-	24,429	-

MAIN COMPETITORS

Concorde: ACURA Legend, INFINITI J30, LEXUS ES300.
Intrepid: CHEVROLET Caprice, FORD Taurus-Crown Victoria, PONTIAC Bonneville & Grand Prix, TOYOTA Camry.
Vision: AUDI 100, BMW 525, SAAB 9000, VOLVO 960.

-

EQUIPMENT

CHRYSLER Concorde / DODGE Intrepid / EAGLE Vision	Intrepid base	Intrepid ES	Vision ESi	Vision TSi	Concorde base
Automatic transmission:	S	S	S	S	S
Cruise control:	O	S	S	S	S
Power steering:	S	S	S	S	S
ABS brake:	O	O	S	S	S
Air conditioning:	S	S	S	S	S
Air bag:	S	S	S	S	O
Leather trim:	-	-	-	O	O
AM/FM/ Radio-cassette:	S	S	S	S	S
Power door locks:	S	S	S	S	S
Power windows:	S	S	S	S	S
Tilt steering column:	S	S	S	S	S
Dual power mirrors:	S	S	S	S	S
Intermittent wipers:	S	S	S	S	S
Light alloy wheels:	-	O	S	S	O
Sunroof:	O	O	O	O	O
Anti-theft system:	-	O	O	O	O

S : standard; O : optional; - : not available

COLORS AVAILABLE

Exterior: Black,White, Blue, Driftwood, Emerald, Red ,Teal, Green, Gold.
Interior: Driftwood, Medium Quartz, Agate Blue, Spruce Blue.

MAINTENANCE

First revision: 5,000 miles
Frequency: 6 months
Diagnostic plug: Yes

WHAT'S NEW IN 1995 ?

-Revised: automatic transmission, emission system, sound system and remote door control.
-New color and new yellow color for the service points under the hood.

	ENGINE					TRANSMISSION		PERFORMANCE										
Model/ versions *: standard	Type / timing valve / fuel system	Displacement cu/in	Power bhp @ rmn	Torque lb.ft @ rpm	Compres. ratio	Driving wheels / transmission	Final ratio	Acceler. 0-62 mph s	Stand. 1/4 mile s	Stand. 5/8 mile s	Braking 62-0 mph ft	Top speed mph	Lateral acceler. G	Noise level dBA	Fuel economy mpg city	Fuel economy mpg hwy	Gasoline type / octane
base	V6* 3.3 OHV-12-SFI	201	161 @ 5300	181 @ 3200	8.9 :1	front - A4	3.66	11.0	17.6	30.5	151	112	0.78	66	23.0	35.5	R 87
option	V6 3.5 SOHC-24-SFI	215	214 @ 5850	221 @3100	9.6 :1	front - A4	3.66	8.8	16.4	29.7	144	124	0.80	65	21.5	30.5	M 89

lumbar one, adjustable only on the deluxe versions. The rear bench seat would gain from being longer, having a taller back and functional headrests.

• Conveniences: 70 %
A modest glove box is compensated for by a storage compartment in the console and ample pockets in the doors in the backs of the front seats.

• Insurance: 70 %
The insurance rate for these models is normal for this category.

• Performance: 70 %
The weight-to-power ratio for these two engines (20.5 lbs/hp for the 3.3 L and 16.5 lbs/hp for the 3.5 L) favors accelerations. By contrast, acceleration is softer with the 3.3 L when the vehicle is loaded or at altitude. The transmission provides well-staged ratios. However, it produces little engine braking during downhill driving. A manual gear box would allow slowdown without using the brakes.

• Suspension: 70 %
Effective wheel travel control provides a comfort similar to that of some high level Europeans.

• Handling: 60 %
It shows an excellent balance in most situations. Even major road defects do not alter the car's path. Cornering passing speed feels surprising self-assured because the car settles in smoothly and progressively. Both traction and motoring around corners with the inside wheel less loaded are excellent and you have to push hard to provoke understeering.

• Price/equipment: 60 %
Except for the base Intrepid, these cars offer complete equipment, and the price is attractive compared to some of the Japanese even in a lower class.

• Noise level: 60 %
The sound insulation is efficient, but rolling and wind noises disturb the interior quiet. The rolling noises were partially caused by inadequate seal in the hood.

• Interior space: 60 %
It is a main advantage of these cars that they can accept up to six people (base Intrepid with a full bench seat up front). Dimensions are generous and the front seats can be moved all the way back without limiting rear legroom.

• Trunk: 60 %
It is large and useful, with a flat floor, but the rear bench seat cannot be lowered to accommodate large objects.

• Depreciation: 60 %
After two years of availability, it seems more favorable than the average for other domestic cars in the same category.

• Fuel consumption: 50 %
It maintains an average of 23.5 miles per gallon: very reasonable for cars of this size.

WEAK POINTS

• Braking: 40 %
The best result was with the 4 discs and ABS on sports versions where distances and balance are satisfactory. Panic stops are less linear and surefooted in the base Intrepid. In all instances, the pedal is very spongy and does not allow precise control for simple slowdown or sudden stops.

CONCLUSION

• Overall rating: 70.0 %
By rapidly correcting problems as they presented themselves, Chrysler reinforced public confidence in the quality label that some began to doubt. ☺

EAGLE Vision TSi

SPECIFICATIONS & PRICES

Model	Version/Trim	Body/ Seats	Interior volume cu.ft.	Trunk volume cu ft.	Drag coef. Cd	Wheel-base in	Lgth x Wdthx Hght. inx inx in	Curb weight lb	Towing capacity std. lb	Susp. type ft/rr	Brake type ft/rr	Steering system	Turning diameter ft	Steer. turns nber	Fuel. tank gal.	Standard tire size	Standard powertrain	PRICES US. $ 1994
CHRYSLER		**General Warranty: 3 years / 36,000 miles; surface rust: 1 year / 12,000 miles; perforation: 7 years / 100,000 miles; road assistance: 3 years / 36,000 miles.**																
Concorde	base	4dr.sdn. 5	105	15	0.31	113.0	201.5x74.4x56.3	3375	2000	i/i	dc/dc/ABS	pwr.r/p.	37.7	3.38	18.0	205/70R15	V6/3.3/A4	**$19,457**
DODGE		**General Warranty: 3 years / 36,000 miles; surface rust: 1 year / 12,000 miles; perforation: 7 years / 100,000 miles; road assistance: 3 years / 36,000 miles.**																
Intredid	base	4dr.sdn. 5	105	15	0.31	113.0	201.7x74.4x56.3	3309	2000	i/i	dc/dr	pwr.r/p.	37.7	3.38	18.0	205/70R15	V6/3.3/A4	**$17,251**
Intrepid	GS	4dr.sdn. 5	105	15	0.31	113.0	201.7x74.4x56.3	3375	2000	i/i	dc/dc	pwr.r/p.	37.7	3.38	18.0	225/60R16	V6/3.3/A4	**$19,191**
EAGLE		**General Warranty: 3 years / 36,000 miles; surface rust: 1 year / 12,000 miles; perforation: 7 years / 100,000 miles; road assistance: 3 years / 36,000 miles.**																
Vision	ESi	4dr.sdn. 5	105	15	0.31	113.0	201.6x74.4x56.3	3408	2000	i/i	dc/dc	pwr.r/p.	37.7	3.38	18.0	205/70R15	V6/3.3/A4	**$19,308**
	TSi	4dr.sdn. 5	105	15	0.31	113.0	201.6x74.4x56.3	3507	2000	i/i	dc/dc/ABS	pwr.r/p.	37.7	3.38	18.0	225/60R16	V6/3.5/A4	**$22,773**

See page 393 for complete 1995 Price List

New Yorker & LHS

One stone, two good shots...

Originating from the LH cars, the New Yorker and LHS succeed in being able to address themselves to two diametrically opposite markets, while starting with the same bodies and drive trains. The New Yorker is as American as apple pie while the LHS invades the territory of the deluxe imports.

Identical in their format, these two models are the largest built by Chrysler. They reach two, highly different, niche markets. The New Yorker is a classic North American that battles the Chevrolet Caprice and Ford Crown Victoria while the LHS carves out its share from the legions of imports and medium-priced deluxe machines. In both instances, they are fighting strong opposition but have many valid points for being the car of choice.

STRONG POINTS

• Safety: 100 %
The stiffness of their unit body has been improved still further by reinforcements and door beams to improve their side impact resistance. Front occupants are protected by two air bags and by shoulder belts that have height adjustments.

• Technical: 80 %
The New Yorker and LHS are derived from the LH model from which they borrow the platform and drive train. They also share some body elements such as the hood, doors and front fenders of the steel unit body. While frontal area is large, the body is aerodynamically efficient, with a Cd of 0.31. Their style draws attention and borrows from classics such as the Jaguar and Bugatti of the past. The suspension is independent, the 4-wheel disc brakes with ABS are standard and the steering power assist is a function of speed. The only drive train available is a 3.5 L. V6 that develops 214 hp, and the electronically controlled 4-speed automatic has been revised.

• Driver's compartment: 80 %
Everything is organized as on the LH: there are enough easy-to-read gauges including a gear-shift indicator. The main controls are easy to reach and ergonomically laid out. However, the shape of the steering wheel is not practical as you cannot get your hands on the short thick

DATA

Category: front-wheel drive full-size sedans.
Class : 6

HISTORIC

Introduced in : 1993
Modified in : -
Made in: Bramalea, Ontario, Canada.

RATINGS

Safety: 100 %
Satisfaction: 82 %
Depreciation: 40 % (2 years)
Insurance: 3.2 % ($ 660)
Cost per mile: $ 0.45

NUMBER OF DEALERS

In U.S.A.: 1,822 Chrysler-Plymouth

SALES ON THE AMERICAN MARKET

Model	1992	1993	Result
N Y/LHS	-	37,390	-

MAIN COMPETITORS

New Yorker: BUICK Le Sabre, Park Avenue, Roadmaster, CHEVROLET Caprice, FORD Crown Victoria, MERCURY Grand Marquis, OLDSMOBILE 98 Regency, PONTIAC Bonneville.
LHS: BUICK Aurora & Riviera, CADILLAC De Ville & Seville, BMW 535, JAGUAR XJ6, LINCOLN Continental, MERCEDES BENZ E class, PONTIAC Bonneville SSEi, SAAB 9000, VOLVO 960.

EQUIPMENT

CHRYSLER	New Yorker	LHS
Automatic transmission:	S	S
Cruise control:	S	S
Power steering:	S	S
ABS brake:	S	S
Air conditioning:	S	S
Air bag:	S	S
Leather trim:	O	S
AM/FM/ Radio-cassette:	S	S
Power door locks:	S	S
Power windows:	S	S
Tilt steering column:	S	S
Dual power mirrors:	S	S
Intermittent wipers:	S	S
Light alloy wheels:	O	S
Sunroof:	O	S
Anti-theft system:	O	S

S : standard; O : optional; - : not available

COLORS AVAILABLE

Exterior: Black,White ,Red ,Driftwood,Teal,Green,Blue,Gold, Flaming red.
Interior: Medium Quartz, Driftwood, Agate Blue, Dark Gray, Spruce blue.

MAINTENANCE

First revision: 5,000 miles
Frequency: 6 months
Diagnostic plug: Yes

WHAT'S NEW IN 1995 ?

-Modifications: automatic transmission, emission controls, sound system, remote door opening.
-New interior color and all service points under the hood are marked in yellow.
- Assisted closing of the trunk.

Model/ versions *: standard	ENGINE Type / timing valve / fuel system	Displacement cu/in	Power bhp @ rmn	Torque lb.ft @ rpm	Compres. ratio	TRANSMISSION Driving wheels / transmission	Final ratio	PERFORMANCE Acceler. 0-62 mph s	Stand. 1/4 mile s	Stand. 5/8 mile s	Braking 62-0 mph ft	Top speed mph	Lateral acceler. G	Noise level dBA	Fuel economy mpg city	hwy	Gasoline type / octane
base	V6* 3.5 SOHC-24-MPFI	215	214 @ 5850	221 @ 3100	9.6 :1	front - A4*	3.66	8.6	16.2	29.5	125	124	0.80	66	21.5	30.5	M 89

CHRYSLER New Yorker

cross bars, and the cruise control switches are inconvenient. The headlights are powerful, but the windshield wipers leave a large dead spot at the upper right of the windshield.

• Trunk: 80 %
Its volume is average in this class and easily used, thanks to a flat floor and a tall and deep layout. A net retains small objects.

• Quality/fit/finish: 80 %
They are similar to that of their LH cousins. The care for aesthetics and the quality of the fits and materials are of international caliber.

• Satisfaction: 80 %
Buyers do not complain of major items, and most problems reported were tied into the recent introductions.

• Steering: 80 %
Except for the higher-than-average power assist, its control is ideal, the precision excellent and the maneuverability quite good for models of this size. While parking, take into account the extent of front overhang to avoid catching the bottom of the bumpers.

• Access: 80 %
It is just as easy at the front as at the rear, since the doors are sizable and open wide.

• Seats: 80 %
Well-padded and amply shaped, they provide more efficient support at the front than at the back. Yet, the rear seats include a large center armrest. The leather trims on the LHS are of a quality equivalent to some Japanese production.

• Insurance: 80 %
Its percentage rating is that of the luxury cars. Compared to the retail price, the premiums for the New Yorker and LHS are not very high.

• Interior space: 80 %
These models are as roomy as the Bonneville-98 Park Avenue and slightly less than the Crown Victoria-Grand Marquis. All dimensions are generous, especially at the rear where the headroom and leg space are worthy of the limousine.

• Performance: 70 %
By holding weight within reasonable limits, the power-to-weight ratio allows take-off and pick-up that are surprising for a V6 of this displacement, while holding good fuel consumption. The engine has good peak torque and a broad rpm range, but the transmission does not provide enough engine braking and is slow on down shifts.

• Suspension: 70 %
Of European inspiration, it remains firm on poor pavement but has enough wheel travel to absorb shocks without bottoming. On freeways, the comfort is royal.

• Conveniences: 70 %
The size of the glove box is reduced by the presence of a right-side air bag. However, the wide door pockets, the seat backs and the console container compensate for this inconvenience. At the rear, it would have been a good move to include a lock box into the retractable center armrest as on the Oldsmobile Aurora.

• Handling: 70 %
It is simply superb on the LHS, which results in a great combination for this type of vehicle. The attitude in corners is predictable and reassuring. You have to push hard before the neutral feel changes to the inevitable understeer and it is easy to control.

• Brakes: 55 %
Thanks to improvement in brake-pad material, the stopping distance with these cars was shorter than on the LH. Some slight beginning of ABS blocking can be detected during harsh stops, but this does not disturb the trajectory and balance, and fade resistance is good.

• Price/equipment: 50 %
Taking into account the comfort and performance, these cars are sold at a competitive price and their equipment is to the level called for by their status.

• Noise level: 50 %
Strangely, it is higher than on the LH, most likely because of resonance in the larger interior.

• Fuel consumption: 50 %
Another pleasant surprise, it is not exaggerated, considering the format and performance level of these cars.

• Depreciation: 50 %
The New Yorker and LHS hold a better resale value and are so much improved that they may well reach the head of their category.

WEAK POINTS

Anything that you can hold against these two models is strictly subjective. One might not like the grill style or the dashboard appearance, but outside of that they earn the title of best new entry for this year, just like last year.

CONCLUSION

• Overall rating: 69.0 %
Chrysler has found a way to match modern and efficient engineering with a North American format for the New Yorker and an international one for the LHS: a double hit. ☺

SPECIFICATIONS & PRICES

Model	Version/Trim	Body/ Seats	Interior volume cu.ft.	Trunk volume cu ft.	Drag coef. Cd	Wheel-base in	Lgth x Wdthx Hght. inx inx in	Curb weight lb	Towing capacity std. lb	Susp. type ft/rr	Brake type ft/rr	Steering system	Turning diameter ft	Steer. turns nber	Fuel. tank gal.	Standard tire size	Standard powertrain	PRICES US. $ 1994
CHRYSLER		General Warranty: 3 years / 36,000 miles; surface rust: 1 year / 12,000 miles; perforation: 7 years / 100,000 miles; road assistance: 3 years / 36,000 miles.																
New Yorker		4dr. sdn. 6	126	18	0.36	113.0	207.4x74.4x55.7	3591	2000	i/i	dc/dc/ABS	pwr.r/p.	37.7	3.38	18.0	225/60R16	V6/3.5/A4	**$25,386**
LHS		4dr. sdn. 5	126	18	0.36	113.0	207.4x74.4x55.7	3628	2000	i/i	dc/dc/ABS	pwr.r/p.	37.7	3.38	18.0	225/60R16	V6/3.5/A4	**$30,283**

See page 393 for complete 1995 Price List

DODGE Dakota

Monopoly...

For eight years the Dodge Dakota Pickup has been a lone rider - an intermediate pickup sandwiched between compacts and full-sized ones. Sales are satisfactory, and it is only the arrival of the T100 Toyota that may change something.

In the eight years of its existence, the Dakota still does not have any competitors, since the Toyota T100 does not dispose of a V8 and its price puts it out of reach of the mainstream public. Stocked, its carrying capacity can reach 1,650 pounds, and it can pull a trailer weighing up to 1.3 tons. It is offered with a pickup box and a standard or a stretch cab. The first offers a three-place bench seat, while the second can accept more baggage or two people in jump seats. The trims on the short cab are Sport base, Special SLT, while the Club Cab is only available in a SLT.

STRONG POINTS

• Technical: 70 %
The unit body is mounted on a ladder bar frame with five cross members, an all-steel construction. The base engine on a 4X2 is a 2.5 L 4-cylinder offered only on a manual 5-speed. The 3.9 L. Magnum V6 is standard on 4 WD and optional on 2 WD, while the 5.2 L. V8 Magnum is optional in both and is an advantage for those who must carry or tow high loads and need the low-end torque.

• Satisfaction: 70 %
Owners consider the Dakota strong and durable, but its appearance is aging rapidly.

• Price/equipment: 70 %
The price of the Dakota pickups seems affordable on the base version, but their equipment is sparse. The SLT versions are more costly, but not that fully equipped.

• Safety: 70 %
The structure is only average in impact resistance, but the driver has an air bag and the other occupants are relatively well-protected.

• Steering: 60 %
The use of a standard manual steering is particularly disagreeable. You are better off with a power steering that is fast and precise, has in a good ratio and proportioning. The large turning diameter does not help the maneuverability.

• Suspension: 60 %
While the softness of the base model is appreciable, the V6 and V8 are stiffer and provoke the large 4x4 tires unpredictable rebounds. In all instances, the rear goes through many excursions on bad pavement.

• Load capacity: 60 %
Interesting in the base model, it can reach 1.7 t as an option with V6 and V8: unusual in this size pickup. On the practical side, the bench seat on the Club Cabin can be moved out of the way to liberate additional, but hard to reach, cargo space.

• Insurance: 60 %
Insuring these utility machines is not cheap, especially the 4WD.

• Quality/fit/finish: 60 %
The general build is strong, but appearance and quality of materials as well as finishing care leave much to be desired.

DATA

Category: rear-wheel drive & 4x4 compact pickups.
Class : utility

HISTORIC

Introduced in : 1987
Modified in : 1991: Club Cab. 1992: V6 & V8 Magnum engines.
Made in: Dodge City, Warren, Michigan & Toledo, Ohio, U.S.A.

RATINGS

Safety: 80 %
Satisfaction: 68 %
Depreciation: 57 %
Insurance: 7.0 % ($ 519, 4x4: $ 585)
Cost per mile: $ 0.36

NUMBER OF DEALERS

In U.S.A.: 1,887 Dodge

SALES ON THE AMERICAN MARKET

Model	1992	1993	Result
Dakota	132,057	119,299	-9.7 %

MAIN COMPETITORS

FORD Ranger, CHEVROLET S-10, GMC Sonoma, ISUZU, MAZDA B, NISSAN Hardbody, TOYOTA compact & T100.

EQUIPMENT

DODGE Dakota	Standard cab Base	Standard cab Sport	Club Cab SLT	Club Cab SLT
Automatic transmission:	O	O	O	O
Cruise control:	-	O	O	S
Power steering:	O	S	S	S
ABS brake:	S	S	S	S
Air conditioning:	O	O	O	S
Air bag:	S	S	S	S
Leather trim:	-	-	-	-
AM/FM/ Radio-cassette:	O	S	S	S
Power door locks:	O	O	O	O
Power windows:	-	O	O	O
Tilt steering column:	O	O	O	S
Dual power mirrors:	O	O	S	S
Intermittent wipers:	S	S	S	S
Light alloy wheels:	-	S	O	O
Sunroof:	-	-	-	-
Anti-theft system:	-	-	-	-

S : standard; O : optional; - : not available

COLORS AVAILABLE

Exterior: Red, Blue, Turquoise, White, Beige, Black.
Interior: Gray, Red, Beige.

MAINTENANCE

First revision: 5,000 miles
Frequency: 6 months
Diagnostic plug: Yes

WHAT'S NEW IN 1995 ?

- **Natural gas version for the 6.2 L V8.**
- **Automatic transmission with a safety lock on the clutch and starter.**
- **Revised equipment for the sports version.**
- **Bench seat has a 60/40 split for improved comfort.**
- **Chassis-cab model deleted from the catalog.**

Model/ versions *: standard	ENGINE Type / timing valve / fuel system	Displacement cu/in	Power bhp @ rmn	Torque lb.ft @ rpm	Compres. ratio	TRANSMISSION Driving wheels / transmission	Final ratio	PERFORMANCE Acceler. 0-62 mph s	Stand. 1/4 mile s	Stand. 5/8 mile s	Braking 62-0 mph ft	Top speed mph	Lateral acceler. G	Noise level dBA	Fuel economy mpg city	Fuel economy hwy	Gasoline type / octane
1)	L4* 2.5 SOHC-8-EFI	153	99 @ 4500	132 @ 2800	9.0 :1	rear - M5*	3.55	14.2	20.5	39.7	154	97	0.73	68	27.0	35.5	R 87
2)	V6 3.9 OHV-12-MPFI	239	175 @ 4800	225 @ 3200	9.1 :1	rear - M5*	3.21	12.0	18.2	35.3	144	103	0.75	68	19.5	28.0	R 87
						rear - A4	3.55	13.3	19.4	37.4	151	100	0.75	68	19.5	27.5	R 87
3)	V6* 3.9 OHC-12-SFI	239	175 @ 4800	225 @ 3200	9.1 :1	four - M5*	3.55	13.0	18.7	36.9	151	103	0.75	68	18.0	27.0	R 87
						four - A4	3.55	13.8	20.5	37.8	154	100	0.75	68	19.0	26.5	R 87
4) 4x2	V8 5.2 OHV-16-MPFI	318	220 @ 4400	295 @ 3200	9.1 :1	rear - A4*	3.21	10.2	17.2	30.5	147	106	0.77	67	17.0	26.0	R 87
4x4	V8 5.2 OHV-16-MPFI	318	220 @ 4400	295 @ 3200	9.1 :1	rear - A4*	3.55	11.0	17.9	31.0	157	103	0.77	67	16.0	23.5	R 87

1) * 4x2 S & base 2) option 4x2 S & base 3) * 4x4 4) option

• Interior space: 60 %
The bench seat of the regular cab will accept a third occupant only temporarily because the center tunnel makes the middle seat uncomfortable, particularly with a manual transmission. At the rear of the club cabin, a second bench seat, facing the road, will accept two adults or three children although their legroom is more limited.

• Performance: 60 %
Those of the 2.5 L 4 cylinder are anemic even for light work, and a V6 is needed for more power under load. The optional V8 is an interesting and unique alternative in this category, and its torque and power are better suited to heavy work.

• Noise level: 50 %
The four cylinder is noisy during hard acceleration, while the V8 roars mostly when cold. Since the V8 is more discrete, wind and rolling noises are more noticeable.

• Handling: 50 %
Handling on the base versions is too soft. It speeds up the onset of oversteer and allows the rear end to walk around on bad pavement when riding empty. That of the V6, V8 and X4 is stiffer and rolls less, but takes into account the bounciness of all-terrain tires.

• Access: 50 %
Getting into the front seat poses no problem even in the Ford Explorer, but it's harder to get into the rear bench seat of the Club, especially for tall people.

• Driver's compartment: 50 %
The driver is more comfortable with the bucket seats of the stretch cab than with the bench seat of the standard cab. Unfortunately, the overly extended steering column imposes a tiring position. Visibility is satisfactory from all angles and the main controls are set up American style. Reduced to its simplest expression on the base models, the instrumentation is more elaborate and better laid out on the Sport SLT.

WEAK POINTS

• Conveniences: 30 %
The token storage consists of a glove box and a tray built into the dash.

• Fuel consumption: 30 %
It is never economical, especially with V6 and V8 engines, and while the natural gas version is an interesting approach, some owners regret that the Cummins Diesel is not offered in this vehicle.

• Braking: 40 %
It is not a Dakota strong point because stopping distances are long. The ABS acts only on the rear wheels and does not prevent the front wheels from blocking early.

• Seats: 40 %
The original bench seat is uncomfortable, lacking both lateral and lumbar supports. The original bucket seats offer a slight improvement but are not a cure.

• Depreciation: 40 %
It ranges in the middle of the category and depends mostly on usage, condition and mileage.

CONCLUSION

• Overall rating: 52.0%
Chrysler has had good reason to leave the beaten path and to create this intermediate pickup that sells well, has no competitors and offers the advantage to be powerful and compact. 😐

SPECIFICATIONS & PRICES

Model Version/Trim	Body/ Seats	Drag/ Coef	Max Payload lb	Wheel-base in	Lgth x Wdthx Hght. inx inx in	Curb weight lb	Towing capacity std. lb	Susp. type ft/rr	Brake type ft/rr	Steering system	Turning diameter ft	Steer. turns nber	Fuel. tank gal.	Standard tire size	Standard powertrain	PRICES US. $ 1994
DODGE	**General Warranty: 3 years / 36,000 miles; surface rust: 1 year / 12,000 miles; perforation: 7 years / 100,000 miles; road assistance: 3 years / 36,000 miles.**															
Dakota regular 4x2	2dr p-u.2	0.49	1250	112.0	189.0x69.3x64.2	3051	3064	i/r	dc/dr/ABS.re.r/p.		38.3	2.9	15	195/75R15	L4/2.5/M5	**$11,432**
Dakota long 4x2	2dr p-u.2	0.50	1800	123.9	207.5x69.3x65.0	3430	3155	i/r	dc/dr/ABS.re.pwr.r/p.		39.7	2.9	15	205/75R15	L4/2.5/M5	**$12,282**
Dakota ClubCab 4x2	2dr p-u.2+2	0.48	2000	131.0	208.0x69.3x65.6	3587	3507	i/r	dc/dr/ABS.re.pwr.r/p.		43.6	2.9	15	215/75R15	V6/3.9/M5	**$14,299**
Dakota regular 4x4	2dr p-u.2	0.56	1500	112.0	189.0x69.3x67.3	3688	3633	i/r	dc/dr/ABS.re.pwr.bal.		38.3	3.0	15	195/75R15	V6/3.9/M5	**$15,548**
Dakota long 4x4	2dr p-u.2	0.56	1500	123.9	207.5x69.3x67.3	3745	3752	i/r	dc/dr/ABS.re.pwr.bal.		42.0	3.0	15	205/75R15	V6/3.9/M5	**$15,723**
Dakota ClubCab 4x4	2dr p-u.2+2	0.52	1800	131.0	208.0x69.3x68.5	3955	3161	i/r	dc/dr/ABS.re.pwr.bal.		44.3	3.0	15	215/75R15	V6/3.9/M5	**$17,478**

See page 393 for complete 1995 Price List

Neon

Red alert...

Never in our era has the arrival of a new model psyched out the other car builders as much as the Neon. One should say it doesn't lack convincing points to set itself into the 90s', the age of the compacts...

After the Saturn, the Neon is the second answer of the North American industry to the Japanese industry in a popular car segment. The 4-door sedan is the base version, Highline and Sport have been joined by a 2-door coupe Highline or Sport. The two bodies have been provided with a 2.0L SOHC 132 hp engine, but the Sport coupe will have a 150 hp DOHC. The manual transmission is standard and an automatic with only three ratios is offered as an option, as is ABS braking.

STRONG POINTS

• Safety: 90%
The stiffness of the body, the lateral door reinforcements, two air bags to protect the front seats, shoulder belts with a height adjustment all combine to raise the safety level to new highs. Too bad the ABS is not offered as standard, which would constitute an interesting first.

• Technical: 85%
The Neon benefits from state-of-the-art technical advances and borrows nothing from the Shadow/Sundance that it replaces advantageously. Everything is modern, from the engine to the suspension design and the efficient and aerodynamic body. Snappy lines add much attraction to this model. The unit body is steel, the independent suspension is based on McPherson struts and braking is mixed.

• Fuel consumption: 85%
It is normal for an engine of this displacement and comparable to heavier models with 1.8 L power.

• Price/equipment: 80%
The base model, which carries the most popular price, doesn't have much to offer except as a valid alternative to some mini and subcompacts, less interesting from the point of view of format or safety. The price of better equipped versions falls just below that of more established and higher repute models.

• Satisfaction: 70%
The first owners are not unanimous about the manufacturing quality in spite of its popularity. Some start-up problems have not always been rapidly solved, and a few water leaks have remained a mystery.

• Driver's compartment: 70%
It is more spectacular than really practical since the center console is not as ergonomical as the builder would like you to believe. For one thing, the controls fall far from the pilot's hand. Those for the radio are too low and away from the AC. Main controls are well-located, and the instrumentation on the Highline and Sport includes many easily read gauges. Visibility is good from all angles, and driver's position is effective and easy to find.

• Handling: 70%
It provides surprising assurance in most situations. This is a 4-wheel independent suspension, designed with ample wheel travel. Gas shocks provide good damping action, and the combination offers a good compromise with well-controlled ride and roll.

DATA

Category: front-wheel drive compact coupes & sedans.
Class : 3S

HISTORIC

Introduced in : 1994
Modified in : -
Made in: Belvidere, Illinois, U.S.A.

RATINGS

Safety:	90 % (100 % with ABS)
Satisfaction:	NA
Depreciation:	NA
Insurance:	7.0 % ($ 519)
Cost per mile:	$ 0.30

NUMBER OF DEALERS

In U.S.A.: 1,822 Chrysler-Plymouth.

SALES ON THE AMERICAN MARKET

Model	1992	1993	Result	Market share
Neon	New model introduced in 1994.			

MAIN COMPETITORS

coupe : HONDA Civic SI & del sol, HYUNDAI Scoupe, MAZDA MX3, SATURN SC, TOYOTA Paseo
sedan : CHEVROLET Cavalier, FORD Escort, HONDA Civic 4dr., HYUNDAI Elantra, INFINITI G20, NISSAN Sentra, PONTIAC Sunfire, SATURN SL, SUBARU Impreza, TOYOTA Corolla, VW Golf.

EQUIPMENT

DODGE-PLYMOUTH Neon 2 dr. cpe 4 dr. sedan	 base	Hi-line Hi-line	Sport Sport
Automatic transmission:	O	O	O
Cruise control:	-	O	O
Power steering:	O	S	S
ABS brake:	O	O	S
Air conditioning:	O	O	O
Air bag:	S	S	S
Leather trim:	-	-	O (cpe)
AM/FM/ Radio-cassette:	O	O	O
Power door locks:	O	O	S
Power windows:	-	O	O
Tilt steering column:	-	O	S
Dual power mirrors:	-	S	S
Intermittent wipers:	-	S	S
Light alloy wheels:	-	O	S
Sunroof:	-	-	-
Anti-theft system:	-	-	-

S : standard; O : optional; - : not available

COLORS AVAILABLE

Exterior: White, Blue, Red, Black, Yellow.
Interior: Gray.

MAINTENANCE

First revision:	5,000 miles
Frequency:	6 months
Diagnostic plug:	Yes

WHAT'S NEW IN 1995 ?

- New Highline coupe with a 2.0 L. SOHC 132 hp or a Sport with a 2.0 L. DOHC with 150 hp and standard ABS.

Model/ versions *: standard	ENGINE Type / timing valve / fuel system	Displacement cu/in	Power bhp @ rmn	Torque lb.ft @ rpm	Compres. ratio	TRANSMISSION Driving wheels / transmission	Final ratio	PERFORMANCE Acceler. 0-62 mph s	Stand. 1/4 mile s	Stand. 5/8 mile s	Braking 62-0 mph ft	Top speed mph	Lateral acceler. G	Noise level dBA	Fuel economy mpg city	hwy	Gasoline type / octane
base	L4* SOHC-16-MPFI	122	132 @ 6000	129 @ 5000	9.6 :1	front-M5	3.55	9.5	16.6	30.4	148	115	0.80	68	30.0	40.0	R 87
						front-A3	2.98	10.6	17.3	31.0	157	112	0.80	69	27.0	37.5	R 87
option	L4 DOHC-16-MPFI	122	150 @ 6800	131 @ 5600	9.8 :1	front-M5	3.94	NA									
						front-A3	3.24	NA									

• Steering: 70%
It is agreeable, precise, with a convenient ratio and control when power-assisted. This is not the case with the manual steering of the less costly model.
• Seats: 70%
They offer better lateral and lumbar support at the front and rear seats. The split bench seat back, even in the deluxe versions, lacks shape, and the seat is too short.
• Performance: 65%
A higher-than-average engine displacement in a more compact body gains interesting performance and a favorable (17.4 lbs/hp) weight-to-power ratio. It is still more brilliant on a Sports coupe with DOHC, but we were unable to complete data testing on this.
• Access: 65%
It is more convenient at the front than at the back. Here the doors don't open far enough. Also, there is not enough headroom because of the body shape.
• Suspension: 65%
Its wheel travel is interesting for a model of this size. It reacts well to road irregularities, in spite of the firmness wanted for control.
• Quality/fit/finish: 65%
It is not startling, in a sense that the materials have a very ordinary appearance. Even in a deluxe version, the sound of the hood, trunk and doors tells us that insulating material was cheaply dispensed. Trunk trim is worthy of the most vulgar Korean.
• Interior space: 60%
Four persons will be at ease in this so-called forward cabin. Here the impression of spaciousness is more perceptible: high than low. There is a shortage of foot space for the rear occupants.
• Insurance: 60%
The premium for the newcomers is not especially inexpensive. This is explained by the novelty that often causes problems with securing parts and delays accident repairs.
• Trunk: 50%
Its average volume can be improved by lowering the back of the bench seat. It has a one-piece back on the base model and is divided 60/40 on other versions. The truck sill is low and the lid goes full height, but the opening doesn't accept bulky baggage.
• Depreciation: 50%
As with all new cars, we take an average, while waiting for market reaction to the first used Neons.

NEW FOR 1995

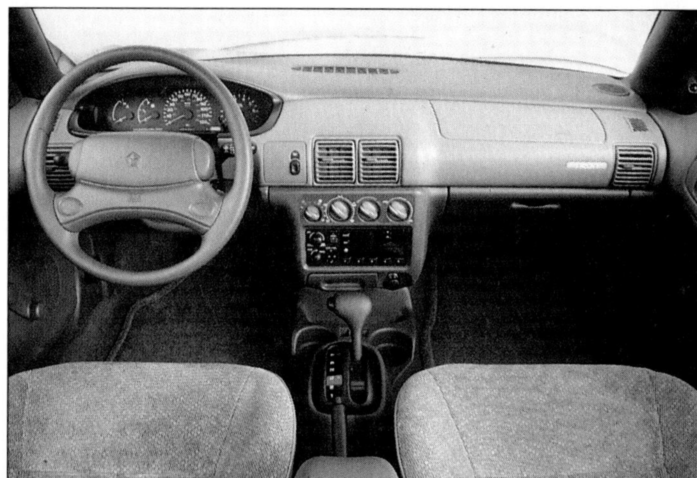

WEAK POINTS

• Braking: 30%
The stopping distances are too long for a car of this format. The ABS, which stabilizes the trajectory in panic stops, is only available as an option. In addition, the spongy pedal doesn't favor precise control.
• Noise level: 40%
Acceptable for the automatic transmission, it is higher for the manual. The automatic only uses 3 ratios and keeps the engine revving high while cruising the freeways.
• Conveniences: 45%
Hiding places do not abound on the base model. The glove box is minuscule, as are the hiding places on the higher level models. The door pockets are absent, but the front seats get full rights to 2 cup holders and a standard (no extra charge) small change container (for your factory refund??).

CONCLUSION

• Overall rating: 61.0%
As attractive technically as it is commercially, the Neon has known immediate success and plunged the market into a state of shock that will hold if the quality promised by Chrysler is really delivered. 😐

SPECIFICATIONS & PRICES

Model	Version/Trim	Body/ Seats	Interior volume cu.ft.	Trunk volume cu ft.	Drag coef. Cd	Wheel-base in	Lgth x Wdthx Hght. inx inx in	Curb weight lb	Towing capacity std. lb	Susp. type ft/rr	Brake type ft/rr	Steering system	Turning diameter ft	Steer. turns nber	Fuel. tank gal.	Standard tire size	Standard powertrain	PRICES US. $ 1994
DODGE-PLYMOUTH	**General Warranty: 3 years / 36,000 miles; surface rust: 1 year / 12,000 miles; perforation: 7 years / 100,000 miles; road assistance: 3 years / 36,000 miles.**																	
Neon	Highline	2 dr. cpe. 4	92	12	0.33	104.0	171.8x67.5x52.8	2383	1000	i/i	dc/dr	pwr.r/p.	35.4	3.2	11.2	185/70R13	L4/2.0/M5	**NA**
Neon	Sport	2 dr. cpe. 4	92	12	0.33	104.0	171.8x67.5x52.8	2471	1000	i/i	dc/dc//ABS	pwr.r/p.	35.4	2.8	11.2	185/65R14	L4/2.0/M5	**NA**
Neon	base	4 dr. sdn. 4	88.5	11.8	0.33	104.0	171.8x67.5x52.8	2339	1000	i/i	dc/dr	r/p.	35.4	3.9	11.2	165/80R13	L4/2.0/M5	**$8,975**
Neon	Highline	4 dr. sdn. 4	88.5	11.8	0.33	104.0	171.8x67.5x52.8	2405	1000	i/i	dc/dr	pwr.r/p.	35.4	3.2	11.2	185/70R13	L4/2.0/M5	**$10,690**
Neon	Sport	4 dr. sdn. 4	88.5	11.8	0.33	104.0	171.8x67.5x52.8	2447	1000	i/i	dc/drdABS	pwr.r/p.	35.4	3.2	11.2	185/65R14	L4/2.0/M5	**$12,215**

See page 393 for complete 1995 Price List

Competitive...

Chrysler must sincerely thanks its competitors for lacking judgment or audacity. The products they designed tried, in vain until now, to dethrone the Caravan-Voyager. The best of Chrysler is yet to come since the following generation arrives soon...

These mini-vans, already improved since 1993, have again been retouched, even in the smallest details for 1994. The Dodge and Plymouth ranges offer Caravan-Voyager (short base), SE and LE and the Grand Caravan-Grand Voyager (long) SE and LE. Chrysler proposes the unique Town and Country in an extended body. This year, 4-wheel drive is available as an option only on the extended bodies with a 3.3 L. V6. A new 3.8 L is available standard on the Town and Country and as an option on the long LE-SE with 4 WD.

STRONG POINTS

• Access: 100%
It is easy no matter which door you choose. At the rear, the hatchback is well-assisted and the side door always glides easily.

• Safety: 90%
Adding two air bags standard at the front seats installing reinforcements in the doors, are good points but ABS has fallen behind and only the LE and Town and Country have it as standard equipment. Again, let us point out the inadmissible options of headrests incorporated into the bench seats. It's regrettable because these safe and comfortable accessories are delivered standard on the models manufactured for Europe. Does Chrysler have two types of clients?

• Insurance: 80%
These vehicles are not expensive to insure compared to others, which helps the family budget.

• Interior space: 80%
The lengthened version, a true seven seater, is more spacious than the short one, which is ideal for five people. The dimensions are generous, but with four captain's seats, there is little room to circulate inside the van.

• Satisfaction: 80%
Owners still complain about the mediocre quality of some components but in terms of reliability and corrosion resistance, they feel unanimously that these vehicles are among the best in their category.

• Driver's compartment: 80%
The dashboard is ergonomic with accessible controls and easy-to-read gauges. We know, however, that the switches for the cruise control, located at the base of the steering wheel, are not easy to use. Driving position and visibility are ideal, the steering is well at hand and the shift lever can be operated with your finger tips, not the case for the hand brake, which remains archaic.

• Conveniences: 85%
The center console contains a practical storage space for emptying your pockets, while the drawer under the front seat compensates for the modest glove box. At the rear, there are well-designed side pockets, containers above the wheel wells and an ample number of cup holders.

DATA

Category: front-wheel drive or all-wheel drive compact vans.
Class : utility

HISTORIC

Introduced in : 1984: short; 1988: long.
Modified in : 1991: cosmetic touch-ups, AWD & ABS,1994: V6 3.8L.
Made in: Windsor, Ontario, Canada & St-Louis Missouri, U.S.A.

RATINGS

Safety: 90 %
Satisfaction: 80 %
Depreciation: 45 %
Insurance: 5.0 % ($519 , 4x4: $654)
Cost per mile: $ 0.42

NUMBER OF DEALERS

In U.S.A.: 1,822 Chrysler-Plymouth, 1,887 Dodge.

SALES ON THE AMERICAN MARKET

Model	1992	1993	Result
Caravan	244,147	262,838	+7.2 %
Voyager	201,016	211,813	+5.1 %

MAIN COMPETITORS

FORD Aerostar & Windstar, GM Astro & Lumina-Trans Sport-Silhouette, MAZDA MPV, MERCURY Villager, NISSAN Quest, TOYOTA Previa, VW Eurovan.

EQUIPMENT

Caravan/ Voyager Town & Country	base	SE	LE	 base
Automatic transmission:	S	S	S	S
Cruise control:	O	S	S	S
Power steering:	O	O	S	S
ABS brake:	-	-	S	S
Air conditioning:	O	O	S	S
Air bag:	S	S	S	S
Leather trim:	-	-	O	S
AM/FM/ Radio-cassette:	O	S	S	S
Power door locks:	O	O	S	S
Power windows:	O	O	O	S
Tilt steering column:	O	S	S	S
Dual power mirrors:	O	S	S	S
Intermittent wipers:	S	S	S	S
Light alloy wheels:	-	O	O	S
Sunroof:	-	-	-	-
Anti-theft system:	-	-	-	-

S : standard; O : optional; - : not available

COLORS AVAILABLE

Exterior: Blue, Driftwood, Green, Red, White, Teal.
Interior: Cloth: Red, Driftwood, Agate blue, Quartz.
Leather: Driftwood, Quartz.

MAINTENANCE

First revision: 5,000 miles
Frequency: 6 months
Diagnostic plug: Yes

WHAT'S NEW IN 1995 ?

-Revised: automatic transmission, remote door openings, ABS braking standard on the LE.
-Offered as an option is a 3.3 L V6 operating on natural gas.
-Manual shift was deleted from the catalog.

Model/ versions *: standard	ENGINE Type / timing valve / fuel system	Displacement cu/in	Power bhp @ rmn	Torque lb.ft @ rpm	Compres. ratio	TRANSMISSION Driving wheels / transmission	Final ratio	PERFORMANCE Acceler. 0-62 mph s	Stand. 1/4 mile s	Stand. 5/8 mile s	Braking 62-0 mph ft	Top speed mph	Lateral acceler. G	Noise level dBA	Fuel economy mpg city	hwy	Gasoline type / octane
1)	L4* 2.5 SOHC-8-IE	153	100 @ 4800	135 @ 2800	8.9 :1	front -A3*	3.22	14.5	19.6	38.6	141	94	0.71	68	24.5	31.0	R 87
2)	V6 3.0 SOHC-12-SFI	181	142 @ 5000	173 @ 2400	8.9 :1	front -A3*	3.02	11.8	18.5	35.4	154	106	0.73	68	22.0	30.5	R 87
3)	V6 3.0 SOHC-12-SFI	181	142 @ 5000	173 @ 2400	8.9 :1	front -A3*	3.02	12.5	18.8	35.8	151	103	0.73	68	22.0	30.5	R 87
4)	V6* 3.3 OHC-12-ISMP	200	162 @ 4800	194 @ 3600	8.9 :1	front -A4*	3.62	11.5	18.4	35.2	154	109	0.75	66	21.0	30.5	R 87
5)	V6* 3.3 OHC-12-ISMP	200	162 @ 4800	194 @ 3600	8.9 :1	four - A4*	3.62	12.3	18.6	35.5	151	112	0.75	66	21.0	30.5	R 87
6)	V6* 3.8 OHC-12-ISMP	231	162 @ 4400	213 @ 3600	9.0 :1	front -A4*	3.45	11.3	18.0	35.0	141	112	0.75	66	18.5	27.5	R 87
7)	V6* 3.8 OHC-12-ISMP	231	162 @ 4400	213 @ 3600	9.0 :1	four - A4*	3.45	11.6	18.4	35.5	148	109	0.77	66	18.0	27.0	R 87

1) base short 2) opt. base short 3) base. long 4) opt. long 5) std. long SE, LE & 4WD 6) std. Town & Country opt. long SE, LE & 7) 4WD.

DODGE Caravan SE

CHRYSLER Town & Country

• Price/equipment: 70%
Considering the price at which these vehicles are offered and the standard equipment, they continue to be an excellent buy.

• Technical: 70%
The steel unit body has a McPherson front suspension, while the rear has a rigid axle mounted on leaf springs. Standard braking is mixed, while only the LE and Town and Countries are available with ABS. Discs are not offered for the rear wheels. The base engine in a short body is a 2.5 L 4 cylinder. The 3.0 L. V6 with a 4-speed automatic is standard on the long body SE and LE with 2-wheel or 4-wheel drive. The latter has a transfer case and a viscous coupler that delivers no more than 10% of the power to the rear wheels under borderline traction conditions.

• Steering: 75%
Well-proportioned and with a good ratio, it allows precise and agreeable guidance. Unfortunately, the large turning diameter hurts maneuverability, especially on the long-bodied version.

• Suspension: 65%
The softness has improved, and the standard models do not roll as much as before. The rear suspension remains an outrage to modern technology because its simplicity generates disagreeable reactions on road joints and bad pavement.

• Quality/fit/finish: 65%
While the materials' quality, appearance and care in assembly progress constantly, the squeaks in the dash and the quality of the tires, dampers and brake pads all need improvement.

• Storage: 65%
You can increase the storage capacity by moving the last seat forward.The long bodied version offers ample space, but there is little room left in the short body when the last seats are occupied. No baggage cover is offered, even as an option.

• Fuel consumption: 60%
With a general average of 22.5 miles per gallon, it is reasonable. You can consider acquiring a natural gas version in areas where gasoline prices are too high.

• Depreciation: 55%
The used vans sell as fast as the new ones as long as they are fitted with a V6 engine, an automatic transmission and AC.

• Noise level: 50%
The V6 engines are discreet and the soundproofing covers up the rolling noise, but the dealerships are having problems muzzling the little noises in the dashboard.

• Handling: 50%
The sport suspension is better at absorbing road imperfections and is more effective at controlling body motions and the roll than the excessively soft base suspension.

WEAK POINTS

• Braking: 40%
Without ABS or standard rear discs, its effectiveness is average and the balance uncertain on wet roads because the front wheels lock early in panic stops.

• Performance: 45%
An unfavorable weight-to-power ratio (33 lbs/hp) results in mediocre performance with a base 4 cylinder. With a 3.0 L V6 (23.3 lbs /hp), accelerations are cleaner but accelerating up a grade is harder than with a 3.3 L or a 3.8 L. These torques are higher and allow pulling heavier trailers. If the 3-speed provides adequate engine braking, the 4-speed offers the disagreeable impression of being in free wheeling.

• Seats: 45%
The front seats are more comfortable than bench seats, while the latter are well-padded, they have little contouring and the backs are too low. They are heavy, hard to remove, their backs are not sufficiently inclined and they are not modular.

CONCLUSION

• Overall rating: 67.5%
While waiting the arrival of the next mini-van generation foreseen for the start of 1995, the existing models remain perfectly capable of monopolizing the sales of a segment that Chrysler invented 10 years ago. ☺

OWNERS' SUGGESTIONS

-Better mileage
-Standard ABS
-Superior finish, quality
-Firmer suspension
-Taller bench seats with headrests
-More practical parking brake
-Tires, brake linings, wiper blades and shock absorbers of better quality

SPECIFICATIONS & PRICES

Model	Version/Trim	Body/ Seats	Drag coef. Cd	Wheel-base in	Lgth x Wdthx Hght. inx inx in	Curb weight lb	Towing capacity std. lb	Susp. type ft/rr	Brake type ft/rr	Steering system	Turning diameter ft	Steer. turns nber	Fuel. tank gal.	Standard tire size	Standard powertrain	PRICES US. $ 1994
CHRYSLER	**General Warranty: 3 years / 36,000 miles; surface rust: 1 year / 12,000 miles; perforation: 7 years / 100,000 miles; road assistance: 3 years / 36,000 miles.**															
Town & Country	base	4dr van.7	0.39	119.3	192.8x72.0x68.8	3955	2000	i/r	dc/dr.ABS	pwr.r/p.	43.0	2.96	19.8	205/70R15	V6/3.8/A4	**$27,184**
Town & Country	4x4	4dr van.7	0.39	119.3	192.8x72.0x67.8	4270	2000	i/r	dc/dr.ABS	pwr.r/p.	43.0	2.96	18.0	205/70R15	V6/3.8/A4	**$29,280**
DODGE-PLYMOUTH	**General Warranty: 3 years / 36,000 miles; surface rust: 1 year / 12,000 miles; perforation: 7 years / 100,000 miles; road assistance: 3 years / 36,000 miles.**															
Caravan-Voyager	base	4dr van.5/7	0.39	112.3	178.1x72.0x66.0	3305	1000	i/r	dc/dr	pwr.r/p.	41.0	2.96	19.8	195/75R14	L4/2.5/A3	**$14,819**
Caravan -Voyager	SE	4dr van.5/7	0.39	112.3	178.1x72.0x66.0	3340	1500	i/r	dc/dr	pwr.r/p.	41.0	2.96	19.8	195/75R14	V6/3.0/A3	**$18,039**
Caravan-Voyager	LE	4dr van.5/7	0.39	112.3	178.1x72.0x66.0	3368	2000	i/r	dc/dr	pwr.r/p.	41.0	2.96	19.8	205/70R15	V6/3.0/A3	**$21,863**
Gd- Caravan-Voyager	SE	4dr van.7	0.39	119.3	192.8x72.0x67.8	3962	2000	i/r	dc/dr/ABS	pwr.r/p.	43.0	2.96	18.0	205/70R15	V6/3.3/A4	**$18,078**
Gd- Caravan-Voyager	LE	4dr van.7	0.39	119.3	192.8x72.0x66.6	2585	2000	i/r	dc/dr/ABS	pwr.r/p.	43.0	2.96	19.8	205/70R15	V6/3.3/A4	**$22,783**
Gd- Caravan-Voyager	LE 4x4	4dr van.7	0.39	119.3	192.8x72.0x67.8	3962	2000	i/r	dc/dr/ABS	pwr.r/p.	43.0	2.96	18.0	205/70R15	V6/3.3/A4	**$25,460**

See page 393 for complete 1995 Price List

No guts, no drive...

Chrysler created the Viper for several reasons: first, to respond to the public's love affair with the prototype second, to test a new work method and finally create a myth similar to the ZR1 Corvette.

Dodge made concrete the dreams of many cars fans to own relatively affordable fabulous cars: as only Americans can do. The Viper is a coupe, hard-liner with a savage outlook, monstrous proportions and limitless possibilities. It is also endowed with the latest in technical achievements. Costing the same as a ZR1 Corvette, it offers unique and exclusive facets that make its charm. Five thousand copies should be built, three thousand have been sold, and the coupe is awaited in 1995.

STRONG POINTS

Performance:
The pleasure begins when you start this enormous engine whose potential is unlimited when compared to the speed limits of the North American continent. The pickups and accelerations are exhilarating, and you need an iron hand to control with precision this particular ride. The numbers obtained are very exotic, and totally unimportant compared to the sensations you feel. Tremendous is just an understatement.

Handling:
Few pilots will know the limits of this rocket except on the track. The enormous tires ensure traction and side bite. The suspension doesn't know softness: it is strict and allows superb road holding in a wide curve or in a straight line. The Viper is not easy to fully master because the available power expresses itself better in a straight line than in tight turns. There, because of its weight and bulk, it lacks nimbleness, and you will need a little practice when applying full throttle on a wet road.

Technical:
The Viper is constructed on a laminated steel chassis stiffened by a tubular structure. The V10 engine is identical to that of the Ram trucks, except that the block is aluminum rather than cast iron. It has been reworked to get 400 hp and reach 6,000 rpm, rare for a production machine of this displacement. The only available transmission is a manual 6-speed. The suspension is independent and the disc brakes over-sized and vented at all four wheels, but ABS is still not available. The body panels are made of injection moldings, and the windshield frame is patented because it includes the top of the dash to gain stiffness.

Insurance:
The premium is commensurate with the gadget but costs $300 less than the ZR1, which must be a bargain of some sort.

Seats:
Even though Spartan, they provide effective support and holding power, and their padding is adequate. A more sophisticated adjustment system would help find a better driving position.

Depreciation:
Even in the recession an investor should be able to make a little money with a model this rare and unusual.

Quality/fit/finish:
Even though practically hand-built, the finish is more that of a kit car than a major production model. There is no doubt about the quality of the materials, but the interior is far from being overstyled.

Steering:
Hauling the Viper on the road calls for muscle. In spite of its precision

DATA

Category: rear-wheel drive exotic sports coupes.
Class : GT

HISTORIC

Introduced in : 1992
Modified in : -
Made in: New Mack, Detroit, Michigan, U.S.A.

RATINGS

Safety:	40 %
Satisfaction:	NA
Depreciation:	NA
Insurance:	3.0 % ($1,340)
Cost per mile:	$ 1.35

NUMBER OF DEALERS

In U.S.A.: 1,887 Dodge.

SALES ON THE AMERICAN MARKET

Model	1992	1993	Result
Viper	NA	1,396	-

MAIN COMPETITORS

CHEVROLET Corvette ZR-1.

EQUIPMENT

DODGE	Viper RT/10
Automatic transmission:	-
Cruise control:	NA
Power steering:	S
ABS brake:	-
Air conditioning:	O
Air bag:	-
Leather trim:	S
AM/FM/ Radio-cassette:	S
Power door locks:	-
Power windows:	-
Tilt steering column:	S
Dual power mirrors:	S
Intermittent wipers:	S
Light alloy wheels:	S
Sunroof:	S
Anti-theft system:	S

S : standard; O : optional; - : not available

COLORS AVAILABLE

Exterior: Emerald green, Yellow, Red, Black.
Interior: Black, Gray, Black & Tan.

MAINTENANCE

First revision:	5,000 miles
Frequency:	6 months
Diagnostic plug:	Yes

WHAT'S NEW IN 1995 ?

- No major changes.

Model/ versions *: standard	ENGINE Type / timing valve / fuel system	Displacement cu/in	Power bhp @ rmn	Torque lb.ft @ rpm	Compres. ratio	TRANSMISSION Driving wheels / transmission	Final ratio	PERFORMANCE Acceler. 0-62 mph s	Stand. 1/4 mile s	Stand. 5/8 mile s	Braking 62-0 mph ft	Top speed mph	Lateral acceler. G	Noise level dBA	Fuel economy mpg city	hwy	Gasoline type / octane
Viper	V10*8.0 OHV-20-MPFI	488	400 @ 4600	465 @ 3600	9.1 :1	rear - M6	3.07	4.8	13.2	21.8	128	165	0.97	75	18.0	29.0	M 89

and fast action, the power assist is not too strong and you need to get used to it.

Braking:
Effective and fade-free, its power assist is symbolic and it functions as an all-or-nothing proposition. You need to push firmly on the pedal to gain the slightest slowdown.

WEAK POINTS

Noise level:
The engine roars add to the driving excitement but are not the best setting for a sentimental conversation.

Price/equipment:
To be faithful to the spirit of the sport, the equipment is elemental and any notion of comfort is banished. As such, do not be surprised that the door handles or window cranks are among the missing. The AC serves mostly to protect the occupant from front-drive train heat.

Fuel consumption:
Since you can't imagine driving a car like this normally, you should expect neither city nor road mileage but full throttle track mpg (not listed on your sticker).

Conveniences:
This word doesn't mean anything on a car whose glove box is as roomy as the ashtray of a conventional machine.

Interior space:
Claustrophobics please abstain: you get the impression of being installed in the bottom of a bathtub while looking at the outside world through a slot, especially when the top is in place.

Trunk:
It offers just enough space to store the top, temporary side curtains and a couple of small handbags.

Access:
To get in and out of the Viper is not easy if the top is in place. Tall heavy-duty folks need practice to avoid muscle cramps, while smaller drivers must be careful not to sustain burns from the lateral mufflers.

Suspension:
In spite of its firmness, it is less uncomfortable than one would expect, but it rapidly dissuades one against long trips unless in the best of shape.

Safety:
The Viper has not yet been fitted with air bags, even though Chrysler does install them in other vehicles. While it has a roll-bar hoop and seat belts with 3 anchor points, it should be enriched with an ABS system as well as a roof with side windows to improve the visibility.

Driver's compartment:
To find the best driver's position is no easy thing. Because the seat is very low, its adjustments are limited. The pedal assembly is strongly off-centered and the center console is very tall. Steering and shifting will require football season strength. The main gauges are not distracting because most of the time they are hidden behind the driver's hands. In spite of Chrysler's promise to support the small owner's club, one must expect delays in securing specific replacement parts. Also keep a little extra spare money around for a small, ordinary car that allows getting to work and back.

CONCLUSION

Only violent love and passion can lead to acquiring a machine that denies all the creature comforts of a modern sports car. Here the budget, comfort and practical side all call for much self denial. ☺

SPECIFICATIONS & PRICES

Model	Version/Trim	Body/ Seats	Interior volume cu.ft.	Trunk volume cu ft.	Drag coef. Cd	Wheel-base in	Lgth x Wdthx Hght. inx inx in	Curb weight lb	Towing capacity std. lb	Susp. type ft/rr	Brake type ft/rr	Steering system	Turning diameter ft	Steer. turns nber	Fuel. tank gal.	Standard tire size	Standard powertrain	PRICES US. $ 1994
DODGE		**General Warranty: 3 years / 36,000 miles; surface rust: 1 year / 12,000 miles; perforation: 7 years / 100,000 miles; road assistance: 3 years / 36,000 miles.**																
Viper	RT/10	2 dr. con. 2	NA	13	0.55	96.2	175.1x75.7x44.0	3488	NR	i/i	dc/dc	pwr.r/p.	40.8	2.4	22	275/40ZR17 335/35ZR17	V10/8.0/M6	**$54,500**

See page 393 for complete 1995 Price List

DODGE Ram

New blood...

Chrysler needed a new truck for technical and aesthetic reasons as well as to regain the ground it had progressively lost. The arrival of the extended cab will help maintain demand after the little teething problems get taken care of.

The Ram pickups were entirely renovated last year in a style that leaves no one indifferent. Technically, everything is new, except the drive train inherited from the previous model, to which has been added a V10 to carry heavy loads. They are offered in 1500, 2500 and 3500 Series with a pickup box and standard or extended cabs. The latter, called Club Cab, is brand new for 1995. The base drive train uses a 3.9 L. V6 and the two new V8s are 5.2 L and 5.9 L and a 5-speed manual, an optional automatic, a Cummins Diesel and the famous 8.0 L. V10. An ABS brake system acting on all four wheels is offered as an option on all models.

STRONG POINTS

Technique:
The steel body is bolted to a ladder-style frame with 5 cross members and reinforcing gussets designed to gain maximum stiffness. The chassis is welded up front and riveted in the back. There is an independent front suspension and a rigid rear axle with leaf springs. The drag coefficient is more favorable than that of competitive vehicles, thanks to streamlined cab lines.

Suspension:
It is never brutal, has ample travel and even off-road it is practically impossible to bottom it out. Seldom is the traction lost on wavy pavement, thanks to effective damping.

Seats:
The Rams are, without a doubt, the most comfortable pickups on the market. The base bench seat includes headrests and slight hollows for lumbar support. Individual seats (2+1) provide better lateral support and their padding is softer. The rear bench seat in the Club Cab disappoints a little by its lack of contouring but still accepts 3-4 people, depending on size.

Noise level:
Soundproofing received particular care and is predictably more effective with gasoline engines than with the rather noisy turbo Diesel. Road noises dominate, but air whistles in the cab soon become annoying.

Conveniences:
While the base storage areas are limited to an average-size glove box and a tray formed over the dashboard, you can get, as an option, a roof console. Also available is a portable office built into the center of the seat. It will hold a computer or cellular phone or other office perks.

Interior space:
The standard cab accepts 3 people, and they will appreciate the ample roominess. The Club doubles the number of seats great for family use.

Safety:
It is at the same level as a car, thanks to excellent structural stiffness, doors that protect from lateral impact and an air bag at the center of the steering.

DATA

Category: rear-wheel drive or 4x4 full-size pickups.
Class : utility

HISTORIC

Introduced in : 1972: pickup; 1974: Ramcharger.
Modified in : 1994: complete renewal; 1995: Club Cab.
Made in: Warren, MI, U.S.A & Largo Alberto & Saltillo, Mexico (Club Cab).

RATINGS

Safety: 90 %
Satisfaction: 82 %
Depreciation: 55 %
Insurance: 6.5 % ($ 585 & $ 657 4x4)
Cost per mile: $ 0.55

NUMBER OF DEALERS

In U.S.A.: 1,887 Dodge

SALES ON THE AMERICAN MARKET

Model	1992	1993	Result
Ram	-	95,542	-

MAIN COMPETITORS

Pickups: CHEVROLET-GMC, FORD F-Series.

EQUIPMENT

DODGE Pickups Ram	LT	WS	ST	Laramie SLT
Automatic transmission:	O	O	O	S
Cruise control:	O	O	O	S
Power steering:	O	O	O	S
ABS brake:	O	O	O	O
Air conditioning:	O	O	O	S
Air bag:	S	S	S	S
Leather trim:	-	-	-	-
AM/FM/ Radio-cassette:	O	O	O	S
Power door locks:	-	-	-	S
Power windows:	-	-	-	S
Tilt steering column:	O	O	O	S
Dual power mirrors:	O	O	O	O
Intermittent wipers:	S	S	S	S
Light alloy wheels:	-	-	S	S
Sunroof:	-	-	-	-
Anti-theft system:	-	-	-	-

S : standard; O : optional; - : not available

COLORS AVAILABLE

Exterior: Red, Beige, Green , Pale & Dark Blue, Black, Gray, White.
Interior: Gray, Beige, Blue, Red.

MAINTENANCE

First revision: 5,000 miles
Frequency: 6 months
Diagnostic plug: Yes

WHAT'S NEW IN 1995 ?

-The extended Club Cab mounted on a longer chassis includes a 40/20/40 front bench seat, swing-out rear window and rear bench seat.
-Compressed natural gas is an option.
-Higher torque than the Cummins Diesel.

ENGINE						TRANSMISSION		PERFORMANCE									
Model/ versions *: standard	Type / timing valve / fuel system	Displacement cu/in	Power bhp @ rmn	Torque lb.ft @ rpm	Compres. ratio	Driving wheels / transmission	Final ratio	Acceler. 0-62 mph s	Stand. 1/4 mile s	Stand. 5/8 mile s	Braking 62-0 mph ft	Top speed mph	Lateral acceler. G	Noise level dBA	Fuel economy mpg city	Fuel economy mpg hwy	Gasoline type / octane
base 1500	V6* 3.9 OHV-12-MPFI	239	170 @ 4400	230 @ 3200	9.1 :1	rear - M5*	3.21	11.5	18.2	32.1	48	100	0.71	68	20.0	28.5	R 87
base 2500	V8* 5.2 OHV-16-MPFI	318	220 @ 4400	300 @ 3200	9.1 :1	re./4 - M5*	3.54	9.5	16.7	31.6	49	106	0.74	67	16.0	24.0	R 87
opt. GNC	V8 5.2 OHV-16-MPFI	318	200 @ 4400	250 @ 3600	9.1 :1	re./4 - M5*	3.54	NA									GNC
base 3500	V8 5.9 OHV16-MPFI	360	230 @ 4000	330 @ 3200	8.9 :1	re./4 - M5*	3.54	9.0	16.5	31.5	51	103	0.75	67	15.0	21.5	R 87
option	L6DT 5.9 OHV-12-MI	360	175 @ 2500	430 @ 1600	17.5 :1	re./4 - M5*	3.54	14.0	19.5	33.0	53	94	0.72	72	21.5	29.0	D
option	V10 8.0 OHV-20-SFI	488	300 @ 4000	450 @ 2400	8.6 :1	re./4 - M5*	3.54	NA									R 87

DODGE Ram

Trunk:
The space behind the front bench seat is useful and will accept 2 large suitcases and some cloth bags. In addition, Dodge offers as an option a clever system of modular containers and trays to adapt the volume to what is needed.
Quality/fit/finish:
The outside impression of sturdiness is undeniable and you only need to check out the size of the chassis to become convinced. The interior trim and quality of material are similar to that of cars but fastening of some parts needs improvement.
Driver's compartment:
The dashboard is well-organized and the controls ergonomically laid out. Except for the parking brake, which is too low, the main controls are within hands' reach and the gauges are well laid out and easy to read. The driver's position and his field of vision are satisfactory, in spite of the pillar width and large-sized side rear-view mirrors. On the four-wheel drive, the display that indicates traction mode is practically invisible, poorly placed at the bottom of the instrument panel, same as the cruise control switches, which are hard to tell apart. The effectiveness of the suspension, as well as the soundproofing on the luxury versions, create such a comfortable ambiance that one has a tendency to drive a little faster than the allowed limit.
Performance:
Each in its own way, the engines provide high torque, take offs and pick ups are strong, even with a diesel or the base V6. The V10 distinguishes itself from competitive V8s by its hard-to-equal power and smooth performance.
Handling:
It constitutes one of the most marked improvements in the new pickup generation. In spite of the classical solutions used, development refinements offer an excellent compromise. Even unloaded, on wet roads, the handling is civilized and safe.
Price/equipment:
They are competitive if you consider the traction and towing capabilities of these pickups. Limited to a minimum in a base version, the lists of accessories expand in the luxury versions and the number of options is generous. Regrettably, the ABS and the second air bag are not standard.

Satisfaction:
The first owners did not have special problems, except for the intolerably high sound level of the Cummins engine, which penalizes comfort.
Insurance:
The premium is reasonable considering the multifaceted character and price of these vehicles.

WEAK POINTS

Steering:
Its reduction is average, but the power assist comes in too slowly to maneuver, the precision is good, but a large steering diameter complicates maneuverability, especially in the long wheelbase Club Cab.
Braking:
The spongy pedal often comes on suddenly, which limits precise control and can produce surprises on wet roads when there is no ABS. When the ABS is installed (on 4 wheels) stability is insured, even on wavy pavement, but stopping distances are extended.
Access:
The step is higher on a 4X4, but Dodge engineers forgot to place a handle on "B" pillar.
Fuel consumption:
It is never economical with gasoline engines and only the diesel offers interesting efficiency, but you need to drive a lot to amortize the extra cost.

CONCLUSION

The Ram pickups have had immediate success with owners who appreciate their capacity, sturdiness, and original appearance. The addition of an extended cab will only increase their well-earned popularity.

SPECIFICATIONS & PRICES

Model	Version/Trim	Body/ Seats	Box Length ft	Max Payload lb	Drag coef. Cd	Wheel-base in	Lgth x Wdthx Hght. inx inx in	Curb weight lb	Towing capacity std. lb	Susp. type ft/rr	Brake type ft/rr	Steering system	Turning diameter ft	Steer. turns nber	Fuel. tank gal.	Standard tire size	Standard powertrain	PRICES US. $ 1994
DODGE	**General warranty: 3 years / 36,000 miles; surface rust: 1 year / 12,000 miles; perforation: 7 years / 100,000 miles; road assistance: 3 years / 36,000 miles.**																	
Pickups 2x4																		
1500	WS/LT regular	2dr.p-u.3	78	1979	0.44	118.7	204.1x78.4x72.0	4032	3600	i/r	dc/dr/ABS	re.pwr.bal.	40.6	3.2	26	225/75R16	V6/3.9/M5	**$13,833**
1500	LT long	2dr.p-u.3	96	1825	0.44	134.7	224.3x78.4x72.0	4160	4200	i/r	dc/dr/ABS	re.pwr.bal.	45.2	3.2	35	225/75R16	V6/3.9/M5	**$14,050**
2500	LT long	2dr.p-u.3	96	2899	0.44	134.7	224.3x78.4x72.4	4599	7400	i/r	dc/dr/ABS	re.pwr.bal.	45.5	3.9	35	225/75R16	V8/5.2/M5	**$15,110**
2500	LT Club Cab	2dr.p-u.6	96	3770	0.44	154.7	244.0x78.4x72.4	4881	7400	i/r	dc/dr/ABS	re.pwr.bal.	45.5	3.9	35	245/75R16	V8/5.9/M5	**$17,375**
3500	LT	2dr.p-u.3	96	5291	0.44	134.7	224.3x93.5x73.4	5209	11900	i/r	dc/dr/ABS	re.pwr.bal.	46.4	3.7	35	215/85R16	V8/5.9/M5	**$15,970**
Pickups 4x4																		
1500	LT regular	2dr.p-u.3	78	1861	0.48	118.7	204.1x78.4x75.9	4539	6700	r/r	dc/dr/ABS	re.pwr.bal.	40.6	3.0	26	225/75R16	V6/3.9/M5	**$16,389**
1500	LT long	2dr.p-u.3	96	1704	0.48	134.7	224.3x78.4x75.9	4694	7100	r/r	dc/dr/ABS	re.pwr.bal.	45.2	3.0	35	225/75R16	V6/3.9/M5	**$16,609**
2500	LT long	2dr.p-u.3	96	2595	0.48	134.7	224.3x78.4x75.9	4905	8800	r/r	dc/dr/ABS	re.pwr.bal.	45.3	3.7	35	225/75R16	V8/5.2/M5	**$17,691**
2500	LT Club Cab	2dr.p-u.6	96	3463	0.48	154.7	244.0x78.4x78.0	5187	8800	r/r	dc/dr/ABS	re.pwr.bal.	51.3	3.7	35	245/75R16	V8/5.9/M5	**$19,550**
3500	LT	2dr.p-u.3	96	4900	0.48	134.7	224.3x93.5x78.7	5624	11400	r/r	dc/dr/ABS	re.pwr.bal.	46.4	3.7	35	215/85R16	V8/5.9/M5	**$18,592**

See page 393 for complete 1995 Price List.

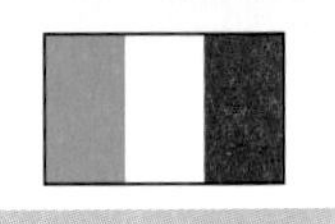

Plenitude...

By going back to the drawing board, Ferrari succeeded in correcting the imperfect copy that was the 348 and infusing a soul in a machine that lacked one. While it is the most modest of the cars with a prancing horse, the F355 has become as pleasing to handle as it is to contemplate.

The F355 replaces the 348, which was the most-sold of the Ferraris since ten of them exited from the Maranello workshops everyday as opposed to seven 512 M's and two 456 GT's. It will be offered this year only as a coupe whose drive train consists of a 3.5 L. V8 with 40 valves and a manual 6-speed transmission.

STRONG POINTS

Technical:
Pininfarina has skillfully retouched the lines of the 348 by erasing, here and there, some live corners and flags that decorated the lateral air inlets. Wheelbase length and width have been increased and the unit body structure, fitted with two cradles for the front and rear ends, has been stiffened. The floor includes two tunnels that create a ground effect that glues the car to the road, while dispensing with ungraceful extras. The hood and the brake calipers are aluminum while the door frames are made of composite materials. The suspension is electronically controlled to vary the damping level depending on the pavement quality. A longitudinally positioned engine has seen its displacement,the number of valves and the power increased to reach 3.5 L, 40 and 380 hp, forever placing the "small" Ferrari out of reach of less noble competition. Included is a 6-speed manual transmission, a power steering whose assist varies as a function of speed and an ABS system that can be shut off on demand. Finally, 40-Series tires are larger and the wheels are magnesium.

Performance:
It is still better, since the engine is worthy of the reputation of the famous Italian marque. More in power than in torque, it is well matched to the short transmission ratios. The later shifts better cold and allows propelling the assembly to a max of 186 mph and goes from 0 to 62 mph in 5.5 seconds.

Handling:
That is where the improvement is most significant, because the 348 was touchy to drive at high speed on a wet road, while the F355 has become more stable in all circumstances, thanks to the active suspension, larger tires and the ground effect.

Steering:
While Ferrari took a long time to convert to power steering, the delay was beneficial, and they have completely matched comfort and steering efficiency without altering the character of this rocket ship. However, the steering ratio and the turning diameter do not help the maneuverability.

Seats:
Their padding is not soft, but the driver now disposes of an electrically controlled side wing adjustment for an "a la carte" lateral support.

Driver's compartment:
The dashboard is well laid out but instrumentation is reduced to a minimum. The ergonomics could be better for a car capable of such

DATA

Category: exotic rear-wheel drive convertibles.
Class : GT

HISTORIC

Introduced in: 1989 (308) 1975 (328 coupe), 1977 (328 convertible.)
Modified in: 1982 V8 3.0L 32 valves engine, 3.2L in1985, 1989 (348)
Made in: 1994: F355 3.5/M6 engine.
Maranello near Modena in Italy.

RATINGS

Safety: 100 %
Satisfaction: 75 %
Depreciation: 35 %
Insurance: 3.7 % ($ 3,597)
Cost per mile: $1.65

NUMBER OF DEALERS

In U.S.A.: 36

SALES ON THE AMERICAN MARKET

Model	1992	1993	Result
Ferrari	First year on the market		

MAIN COMPETITORS

ACURA NSX, BMW 840i, MERCEDES SL500, PORSCHE 928.

EQUIPMENT

FERRARI F355	Berlinetta
Automatic transmission:	-
Cruise control:	S
Power steering:	S
ABS brake:	S
Air conditioning:	S
Air bag:	S
Leather trim:	S
AM/FM/ Radio-cassette:	S
Power door locks:	S
Power windows:	S
Tilt steering column:	S
Dual power mirrors:	S
Intermittent wipers:	S
Light alloy wheels:	S
Sunroof:	-
Anti-theft system:	-

S : standard; O : optional; - : not available

COLORS AVAILABLE

Exterior: Red, Yellow, White, Black,Metallic Gray.
Interior: Black, Tan, White.

MAINTENANCE

First revision: 300 miles
Frequency: 3,000 miles
Diagnostic plug: No

WHAT'S NEW IN 1995 ?

- **Body touched up.**
- **Larger displacement engine with five valves per cylinder and 380 hp.**
- **ABS can be turned off at will.**
- **Drivers seat width is adjustable (lateral support).**

Model/ versions *: standard	ENGINE Type / timing valve / fuel system	Displacement cu/in	Power bhp @ rmn	Torque lb.ft @ rpm	TRANSMISSION Compres. ratio	Driving wheels / transmission	Final ratio	PERFORMANCE Acceler. 0-62 mph s	Stand. 1/4 mile s	Stand. 5/8 mile s	Braking 62-0 mph ft	Top speed mph	Lateral acceler. G	Noise level dBA	Fuel economy mpg city	hwy	Gasoline type / octane
F355	V8* 3.5 QOHC-40 MPFI	213	380 @ 8250	268 @ 4000	10.4 :1	rear-M6*	3.56	5.5	13.8	24.7	135	183	0.95	77	10.0	14.5	S 91

performance, since driving position is oblique. Also 3/4 visibility suffers from a major blind spot. The tilt steering wheel selector is now less ornery.

Quality/fit/finish:
The assembly is strictly controlled but some parts and materials borrowed from large-scale production are very utilitarian and some trim details are out-of-place on the cars of this price.

Safety:
Finally, the F355 must be the first Ferrari provided with air bags, which, together with increased body stiffness, provide adequate occupant protection. It is strange that the high performance Italian vehicles lack in this prime area.

Satisfaction:
Ferrari owners complain less and less of the fragility and capriciousness of their acquisition since Ferrari has delegated the electronic management of its models to German Bosch computers. Maintenance and inspections are more frequent and directly connected to the reliability, sometimes problematical.

Depreciation:
The F355 will probably resell as well as the 348, but the number of buyers capable of facing up to this financial challenge is limited.

Braking:
Even though redoubtably efficient with deceleration and fade resistance, the addition of ABS has had the effect of lengthening the stopping distance to 131 feet from 62 mph, while the same stop without the ABS takes 115 feet (not recommended on a wet road).

Access:
It is not any easier to slide into an F355 than it was a 348, because the ceiling and floor are low: about sidewalk height.

Insurance:
Even though the percentage rates are among the lowest, once multiplied by the price, it results in a copious annual premium nearing $3,600.

Suspension:
Less firm than before, it is still no less bouncy on a bumpy road but becomes almost comfortable at high speeds on new asphalt.

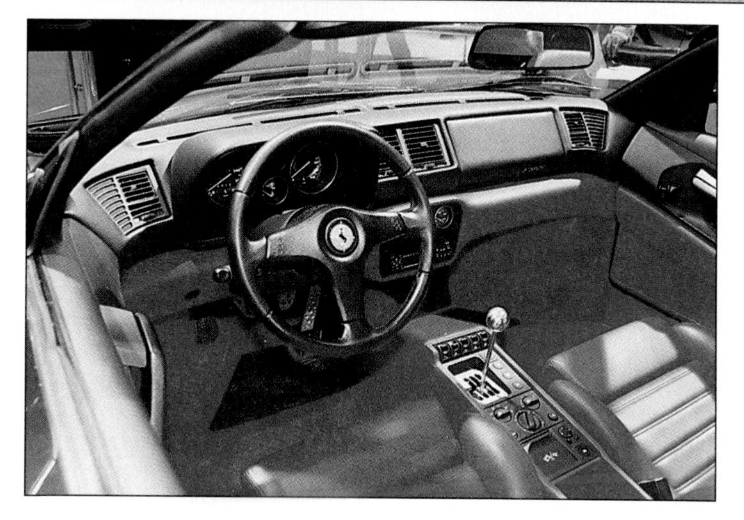

WEAK POINTS

Noise level:
Some will yell "insult" but if we consider that the lyrical exhaust volleys are charming at first audition,they become distinctly less melodious after 200 miles of freeway that constitutes its average cruise radius.

Fuel consumption:
The F355 swallows fuel as rapidly as the miles, which is not saying little, since it does an average of 12 miles per gallon.

Trunk:
The practical aspect is not the first criteria for selecting this type of car, and so the trunk volume regressed from 7 cubic feet on a 348 to 5.3 on an F355. There is ample room for a simple credit card.

Price/equipment:
The richer the equipment of these cars, the poorer their owners become. This paradox doesn't impede sales, which come close to 1,800 copies per year, and this should, without a doubt, increase the number of homeless.

Interior space:
The width and length are sufficient, but the low roof height will interfere with tall drivers.

Conveniences:
Storage is limited to a small glove box, a center console bin and a little space behind the seats.

CONCLUSION

Less ordinary than the 348, the F355 has regained rank, distanced itself from all other road machines running too closely and reconquered full respect.

SPECIFICATIONS & PRICES

Model	Version/Trim	Body/ Seats	Interior volume cu.ft.	Trunk volume cu ft.	Drag coef. Cd	Wheel-base in	Lgth x Wdthx Hght. inx inx in	Curb weight lb	Towing capacity std. lb	Susp. type ft/rr	Brake type ft/rr	Steering system	Turning diameter ft	Steer. turns nber	Fuel. tank gal.	Standard tire size	Standard powertrain	PRICES US. $ 1994
FERRARI		**Total warranty : 3 years / unlimited.**																
F355	Berlinetta	2dr. cpe. 2	NA	16	0.32	96.5	167.3x74.8x46.1	3186	NR	i/i	dc/dc/ABS	pwr.r/p.	12.0	3.25	88	fr.225/40ZR18 re.265/40ZR18	V8/3.5/M6	**$130,000**

See page 393 for complete 1995 Price List

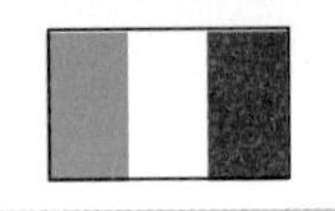

It revives a tradition...

By taking the place left by the ancient 512 in the Ferrari range, the 456 GT goes further than offering a traditional V12 four seater. Its lines recall that of the famous Daytona and gathers in all the traditional attributes of the famous Italian marque.

The 456 GT is a 2+2 sedan with an engine up front and 2 rear driving wheels. It replaces the old 512. It exists in only one version, and the factory makes 2 a day or 440 a year. Production is sold-out in advance to 1995, which means that North American buyers who would like to put their name at the bottom of the list will not be served before next year.

STRONG POINTS

Technical:
In spite of its classic structure, the 456 GT does not lack in letters patent of nobility because its conception is particularly sophisticated. For the body, Pininfarina has done beautiful work incorporating reminiscences of the famous Daytona at the back and the common family look with the F355 and the 512 M at the front. The aluminum alloy body is mounted on a tubular steel chassis to form a very rigid assembly. The doors and the trunk lid are also from an aluminum alloy, while the engine hood and the retractable headlights are based on a honey-comb panel covered with composite material.
The 456 has very effective aerodynamics, thanks among other things to a movable spoiler integrated into the rear bumper. Operated by an electric motor, the later changes the down-load as a function of speed. Independent 4-wheel suspension benefits from an electronic adjustment in three positions: hard, medium or soft, and a height control maintains a constant ground clearance. The power steering varies with speed and the four disc brakes have twin aluminum pads and ATE ABS system. The engine is an all aluminum V12 at 65° with 4 overhead cams, 48 valves and ignition as well as injection controlled by a Motronic 2.7 from Bosch.For better balance, the manual 6-speed transmission is integrated into the limited slip differential located between the rear wheels.

Performance:
With a power-to-weight ratio of 8.4 lbs/hp the only problem is to resist the temptation to experience the exceptional sensations that this extraordinary car can provide. Power and torque are bound at all rpm, and it is possible to re-accelerate in 5th gear from 30 mph.

Handling:
Typically Ferrari, the suspension geometries are very technical. Adding to them are shock absorber adjustments, made by the pilot, and the suspension that makes all the difference. In spite of its weight and bulk, the 456 is agile and precise in all circumstances and also very well balanced (56/44). The traction is excellent, even in the tightest turn, thanks to the limited slip and the tire adhesion.

Steering:
Ultra precise, and with a good ratio, it allows positioning the car to the nearest millimeter. A rarity, it offers exceptional sensitivity without transmitting front end reactions.

Braking:
An important element in a car capable of this high a performance level, it matches the output. The stops are easy to control and start, considering the weight and ABS.

DATA

Category: rear-wheel drive exotic coupes.
Class : GT

HISTORIC

Introduced in: 1993
Modified in: -
Made in: Grugliasco (body) & Maranello (engine), Italy.

RATINGS

Safety: 80 %
Satisfaction: NA
Depreciation: 35 %
Insurance: 3.2 % ($ 4,680)
Cost per mile: $ 2.00

NUMBER OF DEALERS

In U.S.A.: 36

SALES ON THE AMERICAN MARKET

Model	1992	1993	Result
Ferrari	500	550	+ 9.1%

MAIN COMPETITORS

ASTON MARTIN Virage 6, BMW 850CSi, MERCEDES BENZ 600SEC, PORSCHE 928 GTS.

EQUIPMENT

FERRARI	456 GT 2+2
Automatic transmission:	-
Cruise control:	S
Power steering:	S
ABS brake:	S
Air conditioning:	S
Air bag:	S
Leather trim:	S
AM/FM/ Radio-cassette:	S
Power door locks:	S
Power windows:	S
Tilt steering column:	S
Dual power mirrors:	S
Intermittent wipers:	S
Light alloy wheels:	S
Sunroof:	S
Anti-theft system:	O

S : standard; O : optional; - : not available

COLORS AVAILABLE

Exterior: Red, White, Black, Yellow, Blue.
Interior: Black,Blue, Beige, White, Red, Chestnut.

MAINTENANCE

First revision: 300 miles
Frequency: 3,000 miles
Diagnostic plug: No

WHAT'S NEW IN 1995 ?

- No major change.

Model/ versions *: standard	ENGINE Type / timing valve / fuel system	Displacement cu/in	Power bhp @ rmn	Torque lb.ft @ rpm	Compres. ratio	TRANSMISSION Driving wheels / transmission	Final ratio	PERFORMANCE Acceler. 0-62 mph s	Stand. 1/4 mile s	Stand. 5/8 mile s	Braking 62-0 mph ft	Top speed mph	Lateral acceler. G	Noise level dBA	Fuel economy mpg city	hwy	Gasoline type / octane
456 GT	V12* 5.5 DOHC-48-EFI	334	442 @ 6250	403 @ 4500	10.6 :1	rear-M6	3.63	5.8	13.6	23.5	137	186	0.92	68	8.0	22.0	S 91

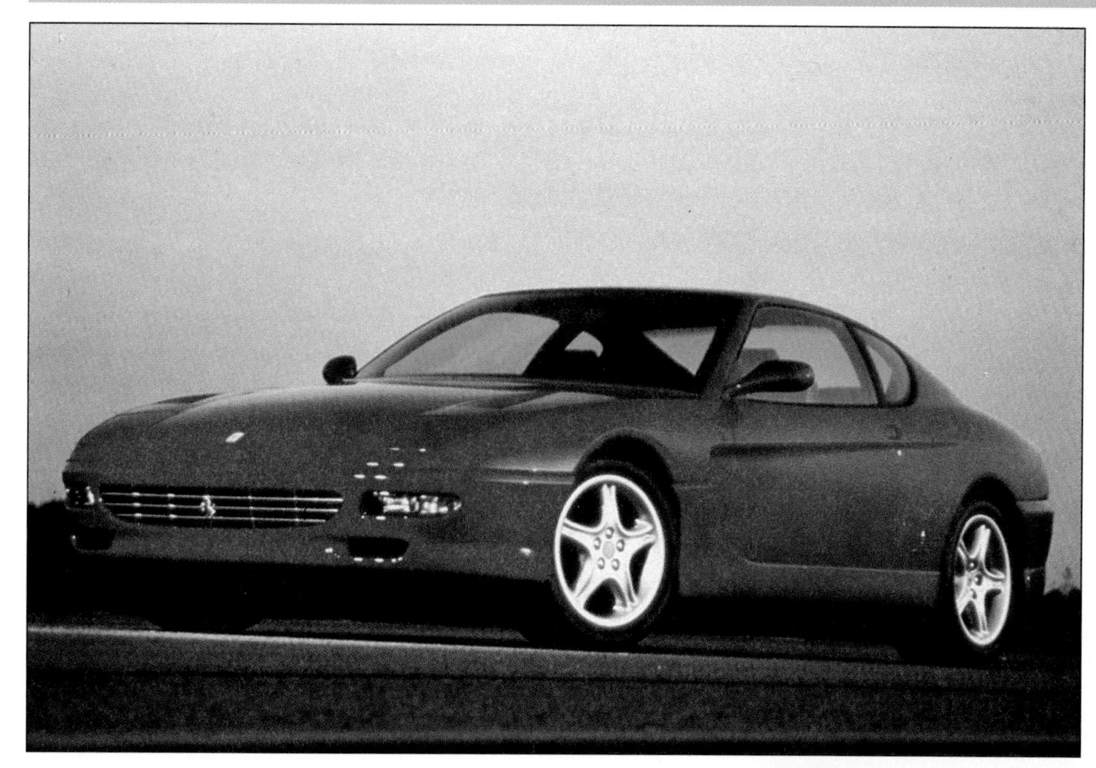

Safety:
The North American version will be equipped with 2 air bags that complement the excellent unit body rigidity and offer good occupant protection in the event of an impact.

Quality/fit/finish:
The 456 GT 2+2 breaks from a reputation of mediocrity which affected Ferrari models for the past 20 years.
An infinity of little details have been improved upon, and one no longer finds the inexpensive accessories borrowed from Fiat, since most switches and controls have been redefined.

Interior space:
While the available front space compares with that of a sedan, at the rear you will find a surprising length and height which delivers an very honorable comfort for this type of body.

Driver's compartment:
The driver's position is nearly perfect and the visibility excellent, except perhaps the "C" pillar which obstructs slightly. The steering wheel is well-at-hand and the post is adjustable in two axes. Also, the gear shift lever remains "imprisoned" in its aluminum guide grill. The main gauges, in the dashboard are easier to read than the ones on the center console. Here the controls for the radio, AC and some accessories are bizarrely installed behind the shifter and, therefore, poorly accessible to the pilot. The hand brake is located on the floor at the right of the driver in ergonomic fashion.

Access:
Thanks to long doors it is easy to enter or leave the 456, even from the rear seats. When one lowers the seat back from a front seat, the cushion is moved forward with an electric motor, as on a Lexus SC 400.

Seats:
Magnificently built and covered with Connolly leather the seats are equipped with headrests at all four places. They also provide good lateral support thanks to highly sculptured padding. However, at the rear, the center console, which separates the seats, is sometimes inconvenient.

Suspension:
In the "soft" mode, it provides an ultra-smooth ride comparable to that of a deluxe sedan. The intermediate position applies to small sine waved roads in good condition. While the hard mode is mandatory for very fast driving to gain the car optimum road holding, all comfort is then absent.

Noise level:
Even though engine noise is part of the pleasure, the soundproofing of the 456 GT is very effective at cruise speeds.

Conveniences:
The practical side of this 456 is a pleasant surprise because storage is practical and conveniently located.

WEAK POINTS

Price/equipment:
Even though the equipment is complete and luxurious, prices attain dizzy summit which reserves the 456 for the rare well-heeled buyers.

Fuel consumption:
Even though a side issue on a car of this class, its fuel appetite is voracious because you have to count on an average of 12 miles per gallon and 8 in the city, which gives this car a range of 280 miles, at best.

Trunk:
Three small custom-made suitcases will find space if you replace the spare with a refill can and use blowout-proof tires.

CONCLUSION

The 456 has joined a very select club of Grand Turismo coupes where reign the Mercedes, BMW, and Astin Martin.
Inaccessible to most, they will give a strong emotion to the lucky fan who will see one of them go by and recognizes it. ☺

SPECIFICATIONS & PRICES

Model	Version/Trim	Body/ Seats	Interior volume cu.ft.	Trunk volume cu ft.	Drag coef. Cd	Wheel-base in	Lgth x Wdthx Hght. inx inx in	Curb weight lb	Towing capacity std. lb	Susp. type ft/rr	Brake type ft/rr	Steering system	Turning diameter ft	Steer. turns nber	Fuel. tank gal.	Standard tire size	Standard powertrain	PRICES US. $ 1994
FERRARI			**Total warranty : 3 years / unlimited**															
456 GT	base	2dr. cpe.2+2		-	0.34	158.7	288.6x117.1x79.3	3725	NR	i/i	dc/dc/ABS	pwr.r/p.	NA	3.1	110	255/45ZR17 285/40ZR17 (re)	V12/5.5/M6	**$175,000**

See page 393 for complete 1995 Price List

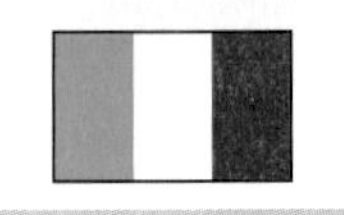

Sacred monster in decline...

The Testarossa has marked its era in a majestic fashion, as much at the style level as in performance; but ten years later, in spite of all the respect you can have for successful things, it is flaming out: great time for Ferrari to present the successor model.

The 512 M suffers as much from an economic context that is unfavorable to luxury cars as from the success of the Lamborghini Diablo which has caused Ferrari to limit its production. Brought up to the performance level of its competitor in 1992, it still does not have the 4-wheel drive which tends to make driving safer in the rain. From here on, it is waiting for the change of guard that should take place in 1996.

STRONG POINTS

Performance:
The 512 M is not stingy with emotions, thanks to the power and enormous noise of its fabulous engine. It is, however, more at ease at high rpm on a road than in urban traffic jams. Even though it is no longer exceptional to accelerate from 0 to 62 mph in 5 seconds or less, the maximum speed of this monster easily reaches 185 mph.

Handling:
Giant tires favor traction and stabilize straight-line road holding or in sweeping curves. The suspension seems better guided and more precise thanks to improvements made in the front end. Even though lighter, it still lacks in agility on a very sinuous stretch where its high torque becomes a handicap.

Technical:
The 512 M is a Testarossa on which Ferrari has corrected some defects. It has retained the tubular support structure, of which the rear part that supports the drive train disassembles to speed repairs. The body is aluminum alloy while the hood and doors are steel. There is a four-wheel independent suspension, and at the rear there are 2 shock absorbers per tire. The assembly has been lightened by 410 pounds, and the wheels have gone from 16 to 18 inches. Engine displacement is the same, but it has gained a new computer which allows a wider rpm range and pushes power beyond the 400 hp mark.

DATA

Category: exotic mid-engine rear-wheel drive coupes.
Class : GT

HISTORIC

Introduced in: 1984
Modified in: 1992 the 512 TR replaces the Testarossa;1995: 512 M
Made in: Grugliasco (body) & Maranello (engine), Italy.

RATINGS

Safety: 70 %
Satisfaction: 72 %
Depreciation: 45 %
Insurance: 3.5 % ($5,796)
Cost per mile: $ 2.20

NUMBER OF DEALERS

In U.S.A.: 36

SALES ON THE AMERICAN MARKET

Model	1992	1993	Result
Ferrari	500	550	+ 9.1%

MAIN COMPETITORS

BUGATTI EB-110, LAMBORGHINI Diablo, MERCEDES BENZ 600 SL, VECTOR W8.

EQUIPMENT

FERRARI	512 M
Automatic transmission:	-
Cruise control:	S
Power steering:	-
ABS brake:	S
Air conditioning:	S
Air bag:	S
Leather trim:	S
AM/FM/ Radio-cassette:	S
Power door locks:	S
Power windows:	S
Tilt steering column:	S
Dual power mirrors:	S
Intermittent wipers:	S
Light alloy wheels:	S
Sunroof:	-
Anti-theft system:	-

S : standard; O : optional; - : not available

COLORS AVAILABLE

Exterior: Red, White, Black, Yellow, Blue.
Interior: Black, Blue, Beige, White, Red, Chestnut.

MAINTENANCE

First revision: 300 miles
Frequency: 3,000 miles
Diagnostic plug: No

WHAT'S NEW IN 1995 ?

- No major change.

Model/ versions *: standard	ENGINE Type / timing valve / fuel system	Displacement cu/in	Power bhp @ rmn	Torque lb.ft @ rpm	Compres. ratio	TRANSMISSION Driving wheels / transmission	Final ratio	PERFORMANCE Acceler. 0-62 mph s	Stand. 1/4 mile s	Stand. 5/8 mile s	Braking 62-0 mph ft	Top speed mph	Lateral acceler. G	Noise level dBA	Fuel economy mpg city	Fuel economy mpg hwy	Gasoline type / octane
512 M	H12* 4.9 DOHC-48-MPFI	301	421 @ 6750	360 @ 5500	10.0 :1	rear - M5*	3.21	5.0	12.8	23.2	148	315	0.90	74	10.0	17.5	S 91

Steering:
Precise and tailored to high speeds, its power assist is cruelly missing at low rpm, and with tires of this size, parking is more effective than heading for the gym.
Braking:
Not equipped with ABS, it shows a remarkable effectiveness and stability, but its fade resistance is hampered by a high weight, and the pedal is hard and difficult to modulate.
Satisfaction:
It seems anachronistic to write that with time the reliability of the Ferrari is less capricious. One main improvement was a switch to German electronic controls and the robotization of some engine assembly.
Safety:
The 512 M should receive air bags to conform to North American standards. It is difficult to admit such a sophisticated car is, in some aspects, so far behind on this vital point.
Driver's compartment:
The bucket seats provide good lumbar and lateral support and their padding is satisfactory. Visibility remains mediocre from 3/4 and toward the rear where the back light is very narrow. Controls are shared between the dash and center console. The pedal assembly, offset toward the right because of the enormous wheel wells, causes the driver to sit crossways. The main controls are more legible than those located under the ventilation controls: which do little when you have your eyes glued to the road. The shift lever, guided by a chromed grill, allows you to listen to a particular metal sound with each shift. Also, the sequence is unconventional.
Depreciation:
Even when the times do not allow you to earn money by selling one of these cars used for more than you bought it new, the Ferrari's lose less than others, thanks to the cult of the make.
Seats:

Better designed and padded than on the ancient Testarossa, they provide more efficient lumbar and lateral support.
Access:
It is relatively easy to slide into the car since the doors open wide, but the roof height and the ground clearance penalize people taller than average.
Insurance:
The 512 M is the most expensive Ferrari to insure since the premium approaches $6,000.
Quality/fit/finish:
The Ferrari products have not yet reached a quality standard that can rival the products of German or Japanese industry. Here, most of the driving accessories come from Fiat, and many of the finish details look low-priced, unlike the F355 or the 456 GT whose manufacturing methods are more recent. Only the leather work does credit to the reputation of the Italian craftsmen.
Suspension:
Comfort does not form part of the standard equipment of this car sacrificed on the altar of performance. The suspension and the tires transmit and amplify all road defects, and romantic cruising calls for freeways rather than for narrow country back-roads.

WEAK POINTS

Price/equipment:
Some leading technical innovations inherited from the competition, limited production and finishing by individual craftsmen, explain the stratospheric prices requested for one of these cars. The equipment is, fortunately, complete, but some improvements are still necessary to make the 512 more agreeable and safer to use.
Fuel consumption:
Count on burning a gallon of fuel every 9 miles in "normal" driving and more if you drive with a Ferrari foot.
Noise level:
The sounds and vibration frequencies produced by the 512 M produce a mechanical melody of which the fans never tire.
Trunk:
There simply isn't any. To stack custom-made baggage, billed out as an option, you only dispose of the space once reserved for the spare tire (now replaced by a refill of self-sealing material). There is also little extra space that remains behind the seats. This will obligate the lady to travel light.
Interior space:
Seating width is sufficient, but length and height are lacking, with much of the space taken up by the drive train.
Conveniences:
They are rare because storage space remain inefficient. The ladies will at least find a huge courtesy mirror that allows them to verify their hairdo and makeup.

CONCLUSION

While waiting to be replaced, the 512 M justifies its title as the fastest production car, a title also bestowed on the Lamborghini Diablo, but few will be capable of splitting the honor shares. ☺

SPECIFICATIONS & PRICES

Model	Version/Trim	Body/ Seats	Interior volume cu.ft.	Trunk volume cu ft.	Drag coef. Cd	Wheel-base in	Lgth x Wdthx Hght. inx inx in	Curb weight lb	Towing capacity std. lb	Susp. type ft/rr	Brake type ft/rr	Steering system	Turning diameter ft	Steer. turns nber	Fuel. tank gal.	Standard tire size	Standard powertrain	PRICES US. $ 1994
FERRARI		Total warranty : 3 years / unlimited																
512	M	2dr. cpe. 2	77.7	12.0	0.32	155.6	273.4x120.6x69.0	3344	NR	i/i	vdc/vdc	pwr.r/p.	39.4	3.25	30.4	235/40ZR18 295/35ZR18 (re.)	H12/4.9/M5	$195,000

See page 393 for complete 1995 Price List

Reprieve...

Condemned to disappear as soon as the new Windstar showed the tip of its nose, the Aerostar sees its mandate extended for two good reasons: it is inexpensive to manufacture and sell and it finally became more reliable....

The Aerostar range has been singularly simplified. It is offered in 2-or 4-wheel drive, short and long body (on the same wheelbase) and a standard rear wheel ABS. There are 2 V6 engines, a 3.0 L on 2WD and 4.0 L standard on 4WD and optional on the extended body of the 2WD. Only the XLT trim level remains available.

STRONG POINTS

• Safety: 80%
In spite of the standard air bag mounted on the driver's side, children's seats integrated into the middle bench seat and safety beams in the doors, there is still much that remains to be done for optimum safety.

• Access: 80%
It is easy to reach the interior, no matter what door you select. At the rear, the hatchback is heavy and poorly assisted.

• Insurance: 75%
The Aerostar is in a good average but one must stress that the premium for the 4X4 is lower than that of a 2WD because it offers safer winter handling.

• Suspension: 70%
Very smooth on good pavement, it becomes frantic on a bad road, where the rear axle marches to its own drummer. In general, it offers good efficiency if you consider the design simplicity.

• Seats: 70%
While not really modular, they allow, as on most of the competition, several arrangements, one of which is with 2 or 4 seats and a bench-bed useful for long trips. The generous padding does not completely compensate for the lack of contour and support.

• Driver's compartment: 70%
The driver is comfortably installed in front of a dashboard of futuristic style and a well-designed ergonomy. The instrumentation is easy to read, the controls well-grouped and located and the visibility satisfactory, in spite of the complicated juncture between the front doors and the windshield.

• Satisfaction: 60%
It is improved, just like the reliability, but some owners still complain of the transmission, the front brakes, the electrical system and the rack-and-pinion.

• Interior space: 60%
Slightly limited in height, so it is not easy to circulate inside. The size of the seats and the drive train intrude on the space reserved for the legs of front-seat occupants.

• Trunk: 60%
On the short body the volume depends on the number of occupied seats, while on the extended body it is permanently more spacious.

• Quality/fit/finish: 60%
If the unit body and most components give a strong impression of sturdiness, the finish lacks refinement. Both plastics and fabrics have

DATA

Category: rear-wheel drive and AWD vans.
Class : utility

HISTORIC

Introduced in: 1985
Modified in: 1987: 3.0L engine, 1989: stretch version, 1990: V6 4.0L
Made in: St Louis, Missouri, U.S.A.

RATINGS

Safety: 100 %
Satisfaction: 60 %
Depreciation: 50 %
Insurance: 5.5 % ($585)
Cost per mile: $0.42

NUMBER OF DEALERS

In U.S.A.: 5,200 Ford-Lincoln-Mercury

SALES ON THE AMERICAN MARKET

Model	1992	1993	Result
Aerostar	166,951	191,148	+11.7 %

MAIN COMPETITORS

CHEVROLET Astro & Lumina, DODGE-PLYMOUTH Caravan-Voyager, FORD Windstar, PONTIAC Trans Sport, MAZDA MPV, MERCURY Villager, NISSAN Quest, TOYOTA Previa, VOLKSWAGEN Eurovan.

EQUIPMENT

FORD Aerostar XLT	Short	Long
Automatic transmission:	S	S
Cruise control:	O	S
Power steering:	S	S
ABS brake:	S	S
Air conditioning:	S	S
Air bag:	S	S
Leather trim:	-	-
AM/FM/ Radio-cassette:	O	O
Power door locks:	S	
Power windows:	O	O
Tilt steering column:	O	S
Dual power mirrors:	O	O
Intermittent wipers:	S	S
Light alloy wheels:	S	S
Sunroof:	O	O
Anti-theft system:	O	O

S : standard; O : optional; - : not available

COLORS AVAILABLE

Exterior: Mocha,Red, Green, Blue, Emerald, Platinum, Black, Silver, White.
Interior: Chrystal Blue, Red Ruby, Medium Gray, Mocha.

MAINTENANCE

First revision: 5,000 miles
Frequency: 6 months
Diagnostic plug: Yes

WHAT'S NEW IN 1995 ?

- Reinforcing beams in side doors.
- Only trim available: XLT in 2- or 4-wheel drive, standard or extended body with 4-speed automatic, AC and 7 seats standard.

Model/ versions *: standard	ENGINE Type / timing valve / fuel system	Displacement cu/in	Power bhp @ rmn	Torque lb.ft @ rpm	Compres. ratio	TRANSMISSION Driving wheels / transmission	Final ratio	PERFORMANCE Acceler. 0-62 mph s	Stand. 1/4 mile s	Stand. 5/8 mile s	Braking 62-0 mph ft	Top speed mph	Lateral acceler. G	Noise level dBA	Fuel economy mpg city	hwy	Gasoline type / octane
1)	V6* 3.0 OHV-12-EFI	182	135 @ 4600	160 @ 2800	9.2 :1	rear - A4	3.45	14.5	18.7	35.5	180	94	0.70	68	20.0	29.0	R 87
2)	V6* 4.0 OHV-12-EFI	245	155 @ 4000	230 @ 2400	9.0 :1	rear - A4	3.73	11.5	17.0	33.5	177	103	0.71	68	19.0	27.5	R 87
						all - A4	3.73	12.5	18.2	34.8	171	100	0.71	68	18.5	26.0	R 87

1) 2WD 2) * 4WD / option 2WD long.

the econo box look.

• Price/equipment: 55%
Taking into account the rustic design, the Aerostar is expensive after Ford raised its equipment level to make it more attractive.

• Steering: 55%
Fast and fairly precise, the power assist is also well-modulated, but the turning diameter affects maneuverability.

• Depreciation: 50%
While the reliability has improved and rejoins the group average, the extra traction of the 4X4 has gained additional converts since the two-wheel drive has borderline handling in winter driving.

• Technical: 50%
The Aerostar architecture is quite traditional: front engine, rear-wheel drive, a rigid axle and a longer wheelbase than that of its main competitors.
The suspension has a Twin I-beam at the front and helical springs at the rear. The curved front-end lines give it a Cd of 0.37.
Its overall height keeps the Aerostar out of many average-sized residential or underground garages.
Unlike front-wheel drive minivans, its traditional design with a rear-wheel drive, gives it better traction at full load.

• Noise level: 50%
Even though soundproofing is well-developed, the noise level is high because the engines lack quiet under load.

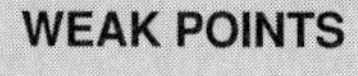

WEAK POINTS

• Braking: 30%
Its efficency and stability are not ideal in panic stops, where the front wheels block instantaneously (with a zero driver on glare ice) because ABS controls only the rear wheels.

• Handling: 40%
The softness of the front suspension and the presence of a rigid rear axle limit the handling. In winter use, the lack of traction in the rear-wheel drive justifies the more stable and efficient, but also costlier, 4-wheel drive.

• Performance: 40%
The 2 V6s lack low-end torque which penalizes acceleration, and the 4.0 L is essential if you intend to carry or tow high loads.

• Fuel consumption: 40%
It is high in all instances, and the power-to-weight ratio is not one of the most favorable.

• Conveniences: 40%
Practical storage is lacking because the door pockets and glove box are too small. By contrast, the sliding rear windows allow efficient, low-noise ventilation.

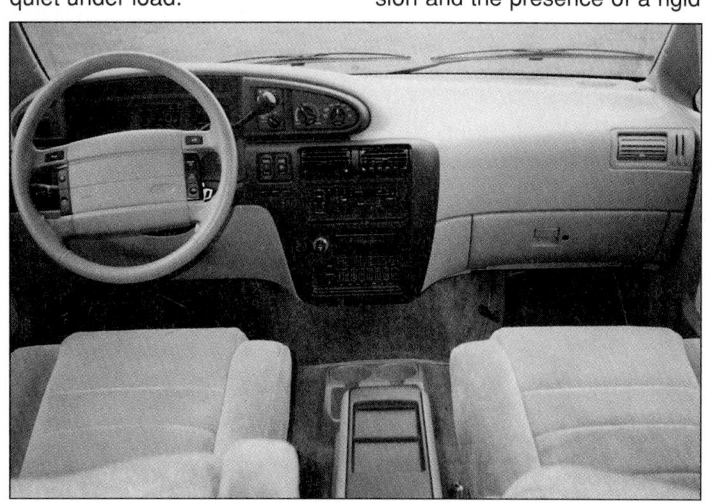

CONCLUSION

• Overall rating: 56.75%
The Aerostar has its enthusiastic fans who recognise its merits in terms of price and freeway comfort. For us, it already is part of history. 😐

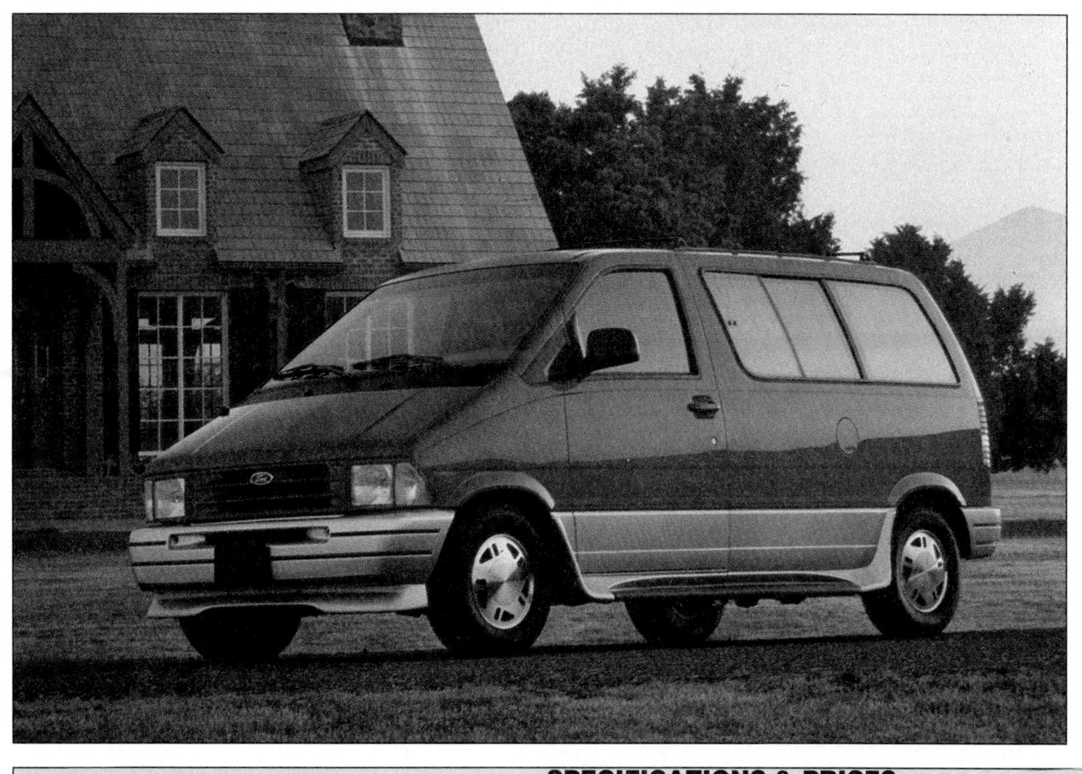

OWNERS' SUGGESTIONS
- Better road handling on slippery roads (4x2)
- More storage space
- Integral ABS
- More effective windshield wipers
- Bench seats with standard headrests
- Less cross wind sensitivity
- A base engine with more power and economy

SPECIFICATIONS & PRICES

Model	Version/Trim	Body Seats	Drag coef. Cd	Wheel-base in	Lgth x Wdthx Hght. inx inx in	Curb weight lb	Towing capacity std. lb	Susp. type ft/rr	Brake type ft/rr	Steering system	Turning diameter ft	Steer. turns nber	Fuel. tank gal.	Standard tire size	Standard powertrain	PRICES US. $ 1994
FORD		Total warranty & antipollution: 3 years / 36,000 miles;corrosion perforation: 5 years / unlimited.														
Aerostar	XLT short 4x2	4dr. van.5/7	0.37	167.7	174.9x71.7x72.2	3406	1951	i/r	dc/dr/ABS	pwr.r/p.	42.3	3.25	21	215/70R14	V6/3.0/A4	**$20,420**
Aerostar	XLT long 4x2	4dr. van.5/7	0.38	167.7	190.3x72.0x73.5	3648	1869	i/r	dc/dr/ABS	pwr.r/p.	42.3	3.25	21	215/70R14	V6/3.0/A4	**$20,900**
Aerostar	XLT long 4x4	4dr. van.5/7	0.38	167.7	190.3x72.0x73.5	3737	1649	i/r	dc/dr/ABS	pwr.r/p.	42.3	3.25	21	215/70R14	V6/4.0/A4	**$21,975**

See page 393 for complete 1995 Price List

Change just to change...

It is a temptation to which car builders succumb too often. The Festiva, which is the most practical work car in the North American market, took 4 years to build a name and a buyer group all its own. Here is Ford again stirring the pot.

The Aspire has replaced the Festiva as the low end of the range for Ford in North America. It has radically changed its philosophy with very rounded forms, a choice of two bodies, and a more luxurious safety. Inherited from the Festiva, the drive train remains a 1.3 L 4 cylinder, develops 63 hp and comes with either a manual 5-speed or an optional 3-speed automatic. The three-door sedan is offered in base or SE and the five-door is only available in base. Only the version equipped with the automatic transmission can receive power steering.

STRONG POINTS

• Safety: 80%

The unit body of the Aspire was designed to best resist collisions, considering its minimum format, and it has, as standard equipment, 2 air bags.

• Fuel consumption: 80%

Without setting any records on this score, the Aspire is not very thirsty and contents itself with between 40 and 53 miles per gallon. This is slightly more than the Festiva because of a weight increase.

• Price/equipment: 60%

Even though the price is higher than that of the old Festiva, the equipment list did not get that much longer. The only standard items are air bags and 2 non-adjustable rear-view mirrors. The windshield washer is not intermittent, and that of the backlight is optional, a practice that seems to be, to say the least, mean.

• Technical: 60%

The under side of the Aspire hides nothing revolutionary. The unit body is steel, the front suspension is a McPherson and at the rear it has a torsional axle and a semi-independent suspension. In spite of its ultra-profiled shape, the aerodynamic coefficient can't do better than 0.36. Brakes are mixed and the ABS is billed out as an option.

DATA

Category: front-wheel drive sub-compacts sedans.
Class : 2

HISTORIC

Introduced in: 1994
Modified in: -
Made in: Seoul, South Korea by Kia.

RATINGS

Safety: 90 %
Satisfaction: 82 % (Festiva)
Depreciation: 62 % (Festiva)
Insurance: 7 %($444)
Cost per mile: $0.30

NUMBER OF DEALERS

In U.S.A.: 5,200 Ford-Lincoln-Mercury

SALES ON THE AMERICAN MARKET

Model	1992	1993	Result
Festiva	33,828	31,918	- 5.7 %

MAIN COMPETITORS

GEO Metro, SUBARU Justy, SUZUKI Swift.

EQUIPMENT

FORD Aspire	base 3-5 dr.	SE 3 dr.	
Automatic transmission:	O	O	
Cruise control:	-	-	
Power steering:	S	-	on 5-dr. automatic
ABS brake:	O	O	
Air conditioning:	-	-	
Air bag:	S	S	
Leather trim:	-	-	
AM/FM/ Radio-cassette:	O	0	
Power door locks:	-	-	
Power windows:	-	-	
Tilt steering column:	-	-	
Dual power mirrors:	S	S	
Intermittent wipers:	-	-	
Light alloy wheels:	-	-	
Sunroof:	O	S	
Anti-theft system:	-	-	

S : standard; O : optional; - : not available

COLORS AVAILABLE

Exterior: Silver, Red, Blue, White, Green, Iris.
Interior: Opal Gray.

MAINTENANCE

First revision: 5,000 miles
Frequency: 6 months
Diagnostic plug: Yes

WHAT'S NEW IN 1995 ?

-New model introduced in 1994 to replace the Festiva and renamed to stress the change in philosophy.

Model/ versions *: standard	ENGINE Type / timing valve / fuel system	Displacement cu/in	Power bhp @ rmn	Torque lb.ft @ rpm	Compres. ratio	TRANSMISSION Driving wheels / transmission	Final ratio	PERFORMANCE Acceler. 0-62 mph s	Stand. 1/4 mile s	Stand. 5/8 mile s	Braking 62-0 mph ft	Top speed mph	Lateral acceler. G	Noise level dBA	Fuel economy mpg city	hwy	Gasoline type / octane
base	L4* 1.3 SOHC-8-EFI	81	63 @ 5000	74 @ 3000	9.7 :1	front - M5*	4.06	12.5	18.6	35.6	174	94	0.70	70	40.0	53.0	R 87
						front - A3	3.45	13.7	19.2	36.7	187	90	0.70	71	35.5	40.0	R 87

• Quality/fit/finish: 60%
Sheet metal fit is more rigid than on a Festiva and the assembly tolerances more precise. The quality of the plastics and of the cover cloths has improved and the interior is less utilitarian.

• Driver's compartment: 60%
The driver is well-seated and a substantial glazed area gives him excellent visibility. The dashboard can't be any simpler or more functional, and the main controls and switches are reduced to a strict minimum. However the instrumentation is more legible in the day than at night because of the colors of the fascia and numbers.

• Suspension: 60%
It is truly comfortable on the parkway and reacts less harshly to road defects than the preceding model but, to improve comfort and road handling, Ford should have adopted 14-inch wheels.

• Insurance: 60%
The Aspire premium remains a little higher than average in spite of the included safety items added to it, since cars of this format remain more vulnerable.

• Access: 60%
It is easier to reach the rear seats on a 3-door, than on the sedan where rear-doors are too narrow.

• Steering: 60%
The manual steering, which is standard on all models, is too vague in the center and has too many turns lock-to-lock. When it is power assisted, with the optional automatic, it becomes more precise and direct. In both instances, the short turning diameter and the small size make it highly maneuverable.

• Seats: 50%
One is better supported at the front than at the rear where the bench seat does not offer any contouring and the padding remains as thin as on the Festiva.

• Satisfaction: 50%
Since most of the main drive train components are known, we hope that the reliability doesn't offer any bad surprises. Our rating is average for any vehicle for its first year on the market.

• Noise level: 50%
The Aspire is no more restful than a Festiva, and the noise level is unusual for our era because of the lack of soundproofing material.

• Conveniences: 50%
The conveniences don't amount to much since the glove box and door panels are minuscule. The shoulder-belt height is not adjustable in the front seats, and the ashtray disappeared together with the cigarette lighter.

WEAK POINTS

• Braking: 20%
Its efficiency is very average because with ABS it requires up to 187 feet to stop from 62 mph. Luckily, the stability remained perfect.

• Interior space: 30%
In spite of its modest dimensions, four adults can conveniently take places and roof height will accept the taller body sizes.

• Performance: 30%
Even though the engine is peaky, it feels torqueless because the higher weight and long transmission ratios penalize takeoffs and accelerations.

• Handling: 30%
In spite of its small wheels and the height of its body, the Aspire is less understeering than the Festiva. Under some conditions, its self-assurance amazes because the roll is limited. It is less sensitive in the cross wind or to traction losses than before, but the side bite (lateral traction) on a wet road can still be taken to task.

• Trunk: 40%
Even though its volume is limited when a bench seat is used, folding down the back of the seat liberates a larger cargo space. On the LX and GL versions, the back is divided into 2 equal parts.

• Depreciation: 40%
Mini cars do not constitute the main market segment, which is undoubtedly why their resale value is weaker than that of subcompacts.

CONCLUSION

• Overall rating: 58.5%
Even though penalized by its format, the Aspire remains in the game, thanks to its safer appearance and large choice of trims and equipment. This should allow it to attract a larger buyer group. 😐

OWNERS' SUGGESTIONS

- More powerful engine
- More complete standard equipment
- Standard power steering and optional manual steering
- Larger wheels
- ABS standard
- Automatic transmission available on 3-doors
- More practical steering
- Better seat padding

SPECIFICATIONS & PRICES

Model	Version/Trim	Body/ Seats	Interior volume cu.ft.	Trunk volume cu ft.	Drag coef. Cd	Wheel-base in	Lgth x Wdthx Hght. inx inx in	Curb weight lb	Towing capacity std. lb	Susp. type ft/rr	Brake type ft/rr	Steering system	Turning diameter ft	Steer. turns nber	Fuel. tank gal.	Standard tire size	Standard powertrain	PRICES US. $ 1994
FORD		**Total warranty & antipollution: 3 years / 36,000 miles;corrosion perforation: 5 years / unlimited.**																
Aspire	base	3dr.sdn. 4	80.0	15	0.36	90.7	152.8x65.5x55.6	2004	NR	i/si	dc/dr	r/p.	29.5	4.08	10	165/70SR13	L4/1.3/M5	**$8,240**
Aspire	base	5dr.sdn. 4	82.3	17	0.36	93.9	155.9x65.5x55.6	2052	NR	i/si	dc/dr	pwr.r/p.	30.8	2.8	10	165/70SR13	L4/1.3/A4	**$8,855**
Aspire	SE	3dr.sdn. 4	80.0	15	0.36	90.7	152.8x65.5x55.6	2004	NR	i/si	dc/dr	r/p.	29.5	4.08	10	165/70SR13	L4/1.3/M5	**$8,895**

See page 393 for complete 1995 Price List

European grafitti...

Throwing more than six billion dollars into the battle, Ford has bet a large part of its world future. In Europe, the Mondeo has begun a satisfactory career and now its American clones will launch themselves into the battle of the compacts for the end of the century market.

The Ford Contour and Mercury Mystique replace the Tempo-Topaz that have taken a well-earned retirement. The two models differ in aesthetic details. The drive train is derived from the Mondeo, launched last year on the European market.

STRONG POINTS

• Safety: **90 %**
The structure is particularly well taken care of. It includes a very rigid cage to efficiently protect the occupants if there is a collision. Besides the two air bags, standard for the front seats, the shoulder belts have a height adjustment.

• Technical: **80%**
It is modern, of European inspiration, with independent McPherson suspension at the four corners. Rubber insulating bushings prevent vibrations that relayed by the two front-driving wheels reach the body and have a passive steering action at the rear, especially in tight corners. Self-stabilizing geometry includes front and rear stabilizer bars. Adjustable shock absorbers, offered on the Mondeo, are not available on the American models. A sport suspension with stiffer dampers and larger stabilizer bar is offered on the Contour SE. Power steering is standard and mixed disc and drum brakes are used on all models. The V6 receives rear discs and 4-wheel ABS is optional. Two new engines have been introduced: a 2.0L 4-cylinder that delivers 125 hp and a 2.5L V6 good for 170 hp. Manual transmission is standard and an electronically controlled 4-speed automatic is an extra-cost option. The body has good aerodynamics with a 0.31 Cd, but you can't say that its appearance, even though European, is of a stunning originality.

• Driver's compartment: **80 %**
It is well-organized, a mix of European and American styling. One immediately recognizes the Ford touches and some relationship with the Taurus instrument panel. The seat provides a comfortable driving position, sometimes impeded by the exaggerated cushion sizes. The main controls are not unusual and the gauges are easy to read.

• Fuel consumption: **75 %**
It is normal no matter which combination you chose. Taking into account the displacement and the reasonable weight of these compacts, you can count on an average of 28 to 30 miles per gallon.

• Seats: **75 %**
As mentioned above, their padding is generous, not common in this category. While the lumbar support is acceptable, the lateral support is not very effective, especially for smaller-sized people.

• Steering: **75 %**
It will please you from all points of view. Its power assist is well-controlled; it is precise and with the right ratio. The steering diameter is slightly higher than average.

DATA

Category: front-wheel drive compacts sedans.
Class : 4

HISTORIC

Introduced in: 1994
Modified in: -
Made in: Kansas City, MI, U.S.A.

RATINGS

Safety: 100 %
Satisfaction: 60 % (Tempo-Topaz)
Depreciation: 60 % (Tempo-Topaz)
Insurance: 8 % ($ 519)
Cost per mile: $ 0.41

NUMBER OF DEALERS

In U.S.A.: 5,200 Ford-Lincoln-Mercury

SALES ON THE AMERICAN MARKET

Model	1992	1993	Result
Tempo	207,173	217,644	+ 4.9 %
Topaz	80,754	77,267	- 4.4 %

MAIN COMPETITORS

ACURA Integra 4dr., BUICK Skylark, CHEVROLET Corsica, CHRYSLER Cirrus, DODGE Stratus, HONDA Accord, HYUNDAI Sonata, MAZDA 626, MITSUBISHI Galant, NISSAN Altima, OLDSMOBILE Achieva, PONTIAC Grand Am, SUBARU Legacy TOYOTA Camry 4 cyl., VW Passat.

EQUIPMENT

FORD Contour MERCURY Mystique	GL	LX GS	SE LS
Automatic transmission:	O	O	O
Cruise control:	O	O	O
Power steering:	S	S	S
ABS brake:	-	O	O
Air conditioning:	O	O	O
Air bag:	S	S	S
Leather trim:	-	-	O
AM/FM/ Radio-cassette:	O	O	O
Power door locks:	-	O	O
Power windows:	-	O	O
Tilt steering column:	S	S	S
Dual power mirrors:	S	S	S
Intermittent wipers:	S	S	S
Light alloy wheels:	O	O	S
Sunroof:	O	O	O
Anti-theft system:	O	O	O

S : standard; O : optional; - : not available

COLORS AVAILABLE

Exterior: Champagne, Red, Coral, Blue, Green, Gray, White.
Interior: Blue, Gray, Ruby, Pumice Stone.

MAINTENANCE

First revision: 5,000 miles
Frequency: 6 months
Diagnostic plug: Yes

WHAT'S NEW IN 1995 ?

-New models replace the Tempo-Topaz.

Model/ versions *: standard	ENGINE Type / timing valve / fuel system	Displacement cu/in	Power bhp @ rmn	Torque lb.ft @ rpm	Compres. ratio	TRANSMISSION Driving wheels / transmission	Final ratio	PERFORMANCE Acceler. 0-62 mph s	Stand. 1/4 mile s	Stand. 5/8 mile s	Braking 62-0 mph ft	Top speed mph	Lateral acceler. G	Noise level dBA	Fuel economy mpg city	hwy	Gasoline type / octane
1)	L4* 2.0 DOHC-16-MPSFI	-	125 @ 5500	130 @ 4000	9.6 :1	front- M5	3.82	10.2	16.8	31.8	144	103	0.76	67	28.0	35.5	R 87
						front-A4	3.92	11.4	17.4	32.4	148	100	0.76	68	26.0	33.0	R 87
2)	V6 2.5 DOHC-24-MPSFI	155	170 @ 6200	165 @ 4200	9.7 :1	front-M5	4.06	8.0	16.2	29.0	154	112	0.78	67	26.0	31.0	R 87
						front-A4	3.77	9.3	16.7	29.8	144	109	0.78	68	24.3	29.0	R 87

1) base 2) * Contour SE option Contour LX & Mystique.

• Conveniences: 72 %
The glove box, just like the door pockets and the console compartment, offers sufficient storage space.

• Handling: 70 %
It surprises by its poise, thanks to the sophistication of the design. It is particularly effective on the SE, whose body roll is well-controlled. Moderately understeering in a tight corner, the Contour-Mystique is remarkably stable in a straight line or in wide curves and readily forgives drivers' small errors.

• Suspension: 70 %
It offers an excellent compromise between efficiency and comfort because its travel is well-designed and unfavorable reactions are rare. There too, you feel the European influence in the fact that the car can offer a certain suppleness without negative road holding effects. The damping is very effective.

• Access: 70 %
It is easier to slide into the front seats than the rear ones where narrow doors with low opening angle and tighter legroom impede entry.

• Quality/fit/finish: 70 %
The quality of assembly, materials and finish is strict, but the presentation is without a flourish.

• Performance: 60 %
The cars are surprisingly live, especially with a V6 engine that produces driving sensations that remind you of the Taurus SHO. The takeoffs and accelerations are well-muscled for what is essentially a popular car. One could almost say that the Contour-Mystique have more punch than their great rival the Cirrus-Stratus that have not impressed anyone on this score.

• Price/equipment: 60 %
Since they are targeted between the compact classes, 3S and 4, the latest Ford models seem more affordable than some of their competitors. You have to peel out the lists of standard equipment and options to realize that everything must be paid for.

• Trunk: 60 %
Of good holding capacity, it can be increased in size by folding down the rear seat-back, which is easy, thanks to clever lock design.

• Noise level: 50 %
It is average for cars of this engine displacement, but the 4-cylinder is noisier in some rpm ranges during strong acceleration.

• Interior space: 50 %
It gives full satisfaction at the front seats but is a little less favorable at the back where length and height are not as generous as in the Cirrus-Stratus that mark a point on this score.

• Insurance: 50 %
The first numbers indicate a relatively high premium. This is often the case in the categories that meet with success, for example compacts during some years.

WEAK POINTS

• Braking: 40 %
While more stable with ABS, the stopping distances are that much longer despite being fitted with 4 discs. Precise pedal feel is hard to achieve, but brake fade resistance is above criticism.

CONCLUSION

• Overall rating: 66.5 %
Compared to the Tempo-Topaz they replace, the Contour-Mystique bring a world of change and modernism. Ford has gambled everything to place itself at the head of this category, but there is strong competition and the result of this battle is uncertain. ☺

SPECIFICATIONS & PRICES

Model	Version/Trim	Body/ Seats	Interior volume cu.ft.	Trunk volume cu ft.	Drag coef. Cd	Wheel-base in	Lgth x Wdthx Hght. inx inx in	Curb weight lb	Towing capacity std. lb	Susp. type ft/rr	Brake type ft/rr	Steering system	Turning diameter ft	Steer. turns nber	Fuel. tank gal.	Standard tire size	Standard powertrain	PRICES US. $ 1994
FORD		**Total warranty & antipollution: 3 years / 36,000 miles;corrosion perforation: 5 years / unlimited.**																
Contour	GL	4dr.sdn. 5	89.4	14	0.31	106.5	183.9x69.0x54.4	2769	1000	i/i	dc/dr	pwr.r/p.	36.4	2.78	14.5	185/70R14	L4/2.0/M5	-
Contour	LX	4dr.sdn. 5	89.4	14	0.31	106.5	183.9x69.0x54.4	2778	1000	i/i	dc/dr	pwr.r/p.	36.4	2.78	14.5	185/70R14	L4/2.0/M5	-
Contour	SE	4dr.sdn. 5	89.4	14	0.31	106.5	183.9x69.0x54.4	2833	1000	i/i	dc/dc	pwr.r/p.	36.4	2.78	14.5	205/60R15	V6/2.5/M5	-
MERCURY		**Total warranty & antipollution: 3 years / 36,000 miles;corrosion perforation: 5 years / unlimited.**																
Mystique	GS	4dr.sdn. 5	89.4	14	0.31	106.5	183.3x68.9x54.0	2824	1000	i/i	dc/dr	pwr.r/p.	36.4	2.78	14.5	185/70R14	L4/2.0/M5	-
Mystique	LS	4dr.sdn. 5	89.4	14	0.31	106.5	183.3x68.9x54.0	2824	1000	i/i	dc/dr	pwr.r/p.	36.4	2.78	14.5	205/60R15	L4/2.0/M5	-

See page 393 for complete 1995 Price List

Relics of the past...

These cars, with names that ring bells of the past, are part of the American heritage. They are seductive to anyone who has known the dinosaur era and American big-car superiority. Since that time many cars and opinions have changed.

The largest Ford and Mercury models are 3-volume sedans with 4 doors, equipped with a 4.6 L V8 engine with a 4-speed electronically controlled automatic. The Crown Victory is offered in S, base, LX and Touring; the Grand Marquis comes in GS or LS.

STRONG POINTS

• **Insurance:** 90%
These large cars do not cost much to insure, probably because they are less vulnerable than small ones. However, proportionately, the base-model premium is higher.

• **Satisfaction:** 85%
Numerous are the owners highly satisfied by these models. Some still complain of front brake problems, fuel leaks, and wind noise in the front door.

• **Safety:** 80%
An imposing structure, plus 2 air bags at the front seats, insure the status of a road tank to these rolling living rooms and the picture would be complete if the ABS was standard.

• **Trunk space:** 80%
It seems immense but its irregular shape and the spare-wheel placement do not allow full use of its capacity.

• **Access:** 80%
Well-dimensioned, the doors clear a satisfactory space, which would be better still if they could open more at the back. Up front, the hood opens wide allowing easy service and maintenance.

• **Suspension:** 80%
At cruising speed on a freeway, the comfort is imperial. The ride is velvety and body motion is limited to a minimum. On the other hand, bad pavement agitates the rear end and breaks the charm.

DATA

Category: rear-wheel drive full-size sedans.
Class : 6

HISTORIC

Introduced in: 1978
Modified in: 1988, 1991.
Made in: St Thomas, Ontario, Canada.

RATINGS

Safety: 80 %
Satisfaction: 84 %
Depreciation: 66 %
Insurance: 4.0 % ($585)
Cost per mile: $0.52

NUMBER OF DEALERS

In U.S.A.: 5,200Ford-Lincoln-Mercury

SALES ON THE AMERICAN MARKET

Model	1992	1993	Result
Crown Vic.	92,506	101,685	+ 9.1 %
Gd Marquis	94,607	85,195	-10.0 %

MAIN COMPETITORS

BUICK Le Sabre, BUICK Roadmaster, CHEVROLET Caprice, CHRYSLER New Yorker,OLDSMOBILE 88, PONTIAC Bonneville.

EQUIPMENT

Crown Victoria	**S**	**base**	**LX**
Grand Marquis	**-**	**GS**	**LS**
Automatic transmission:	S	S	S
Cruise control:	S	S	S
Power steering:	O	O	O
ABS brake:	S	S	S
Air conditioning:	S	S	S
Air bag:	O	O	O
Leather trim:	O	O	S
AM/FM/ Radio-cassette:	O	O	O
Power door locks:	O	O	O
Power windows:	S	S	S
Tilt steering column:	S	S	S
Dual power mirrors:	S	S	S
Intermittent wipers:	O	O	O
Light alloy wheels:	O	O	O
Sunroof:	O	O	O
Anti-theft system:			

S : standard; O : optional; - : not available

COLORS AVAILABLE

Exterior: Mocha, Blue, Black, Red, Willow Green, Opal, White, Silver, Champagne.
Interior: Garnet, Willow Green, Saddle Brown, Graphite, Portofino Blue.

MAINTENANCE

First revision: 5,000 miles
Frequency: 6 months
Diagnostic plug: Yes

WHAT'S NEW IN 1995 ?

- New front and rear appearance.
- New dashboard with false-wood applique.
- Changes in the location of controls.
- New seats and headrest.

Model/ versions *: standard	ENGINE Type / timing valve / fuel system	Displacement cu/in	Power bhp @ rmn	Torque lb.ft @ rpm	TRANSMISSION Compres. ratio	Driving wheels / transmission	Final ratio	PERFORMANCE Acceler. 0-62 mph s	Stand. 1/4 mile s	Stand. 5/8 mile s	Braking 62-0 mph ft	Top speed mph	Lateral acceler. G	Noise level dBA	Fuel economy mpg city	Fuel economy mpg hwy	Gasoline type / octane
base	V8* 4.6 SOHC-16-MPFI	281	190 @ 4250	260 @ 3250	9.0 :1	rear - A4*	2.73	11.0	18.2	31.5	157	112	0.78	64	21.0	30.5	R
	same with dual exhaust:		210 @ 4250	270 @ 3250	9.0 :1	rear - A4*	2.73	10.5	18.0	31.2	164	115	0.78	64	21.0	30.5	R

FORD Crown Victoria

MERCURY Grand Marquis

• **Noise level: 80%**
It remains very low on this type of car thanks to the discretion of the powertrain and the massive presence of insulating material.

• **Interior space: 70%**
The cars are among the largest on the market because they can accept 5 or 6 people according to the type of front seat. Unfortunately, front and rear legroom remain short, considering the length of these machines.

• **Technical: 60%**
The style of these vehicles has been tastefully brought up to the standards of the day. They have a favorable drag coefficient but a large frontal area. The ultra-stiff steel body is mounted on the perimeter chassis through insulating body mounts. The front suspension is independent with upper and lower control arms, and at the rear there is a rigid axle with helical springs, a 4-bar linkage and a stabilizer bar. There are 4-wheel disc brakes with optional ABS for all trims.

• **Handling: 60%**
It is clean in most conditions, but the weight, vehicle size, and low suspension rates, result in a sizable body roll which accents a final oversteer that appears late and is easier to control on the Touring whose heavy duty suspension is stiffer. On a road with potholes, the rigid rear axle makes itself known by the weavy feel as bumps and road joints make it dance. Here, Ford has not added an independent rear-wheel drive as on the T-Bird and Cougar.

• **Quality/fit/finish: 60%**
The assembly and finish quality is more obvious on the outside of the body than in the interior, where some adjustments remain vague and the plastic looks econo. There have been no owner complaints about rattles.

• **Seats: 60%**
They are deceiving because their shape does not provide enough support. Considering the age of the average buyer, Ford should take better care of this area with some high-tech computer designed real-people seats.

• **Driver's compartment: 50%**
In spite of all the available adjustments, the most comfortable driving position is hard to find in the bucket seat, which is nothing compared to searching for it in a full-length bench seat. Visibility is excellent, but the dashboard looks remain classical. Controls are laid out in the good old way, but the center console is not particularly ergonomic and remains out of reach. Instrumentation is very simplistic but easy to read.

• **Performance: 50%**
Taking into account the weight and bulk, the takeoffs and accelerations are quite good, as all drivers who have received tickets can duly note.

WEAK POINTS

• **Conveniences: 20%**
The storage in these immense vehicles is limited to a small glove box and 2 pockets in the front doors. Once again, the mountain gave birth to a mouse.

• **Braking : 30%**
In spite of the 4-wheel discs, fitted standard to these 2 cars, the stopping distances are too long and their modulation is delicate: too much too soon. The ABS, which is only offered as an option on all models, may prevent alternate wheel blocking.

• **Fuel consumption: 30%**
It easily stays around 20 miles per gallon in average cruising, but driving in pursuit-style noticeably increases this number.

• **Depreciation: 35%**
It remains higher than average for this type of automobile, depends on gas prices, and it is the best-equipped ones that resell fastest.

• **Steering: 40%**
It is fast, in spite of its high ratio, and relatively precise. But its high-speed power assist makes it touchy, and in a strong cross wind, its ability to stay in a straight line is reduced. Over all, bulk and large turning diameter make maneuvers a little harder.

• **Price/equipment 40%**
Well-equipped, these big Americans are not given away, and at that price the ABS should be delivered standard, which should be the least of things.

CONCLUSION

• **Overall rating: 59.0%**
The arrival of the latest New Yorker and Chrysler LHS has put a touch of the old on the large cars, not always logical in our era of efficiency and ecology.

OWNERS' SUGGESTIONS

-Less power assist
-More storage space
-Improved finish
-Improved sound insulation in base and S trim
-More legroom
-Eliminating automatic door locks

SPECIFICATIONS & PRICES

Model	Version/Trim	Body/ Seats	Interior volume cu.ft.	Trunk volume cu ft.	Drag coef. Cd	Wheel-base in	Lgth x Wdthx Hght. inx inx in	Curb weight lb	Towing capacity std. lb	Susp. type ft/rr	Brake type ft/rr	Steering system	Turning diameter ft	Steer. turns nber	Fuel. tank gal.	Standard tire size	Standard powertrain	PRICES US. $ 1994
FORD		**Total warranty & antipollution: 3 years / 36,000 miles;corrosion perforation: 5 years / unlimited.**																
Crown Victoria S		4dr.sdn.6	111.2	20.6	0.34	114.4	212.0x77.8x56.8	3761	2000	i/r	dc/dc	pwr.bal.	39.1	3.4	20.0	215/70R15	V8/4.6/A4	**$19,350**
Crown Victoria base		4dr.sdn.6	111.2	20.6	0.34	114.4	212.0x77.8x56.8	3792	2000	i/r	dc/dc	pwr.bal.	39.1	3.4	20.0	215/70R15	V8/4.6/A4	**$19,300**
Crown Victoria LX		4dr.sdn.6	111.2	20.6	0.34	114.4	212.0x77.8x56.8	3858	2000	i/r	dc/dc	pwr.bal.	39.1	3.4	20.0	215/70R15	V8/4.6/A4	**$20,715**
MERCURY		**Total warranty & antipollution: 3 years / 36,000 miles;corrosion perforation: 5 years / unlimited.**																
Gd Marquis GS		4dr.sdn.6	109.3	21.0	0.36	114.4	211.8x77.8x56.8	3761	2000	i/r	dc/dc	pwr.bal.	39.1	3.4	20.0	215/70R15	V8/4.6/A4	**$20,330**
Gd Marquis LS		4dr.sdn.6	109.3	21.0	0.36	114.4	211.8x77.8x56.8	3774	2000	i/r	dc/dc	pwr.bal.	39.1	3.4	20.0	215/70R15	V8/4.6/A4	**$21,500**

See page 393 for complete 1995 Price List

Escort-Tracer

Word of mouth doesn't work...

Paradoxically, as the owners' satisfaction improves, sales fall off. The builder should invest more to ensure an honorable career-ending for the Escorts, which are now so much better than the preceding ones.

Their format has changed from sub-compact to compact, and the Escorts now exist in four body styles: 3-,4- and 5-door sedans and a 5-door station wagon in an LX trim with a 1.9 L Ford engine.The GT trim is a 3-door with a 1.8 L Mazda engine. The Mercury Tracers are sold only in the U.S. in sedan or station wagon with a base 1.9 L engine or an LTS sedan with a 1.8L engine.

STRONG POINTS

• Safety: 90%
In 1995 all models received 2 front air bags as standard. The rigidity of the cab along with the well-calculated impact absorption zones in the front and rear allow good occupant protection, ABS braking is not offered.

• Technical: 80%
Derived from the Mazda 323-Protegé platform, the Escort borrows the steering and suspension from its Japanese cousins. However, the calibration of the 4-wheel suspension has been adapted to the American market. Disc and drum brakes are used on the LX and 4 discs on the GT.

• Fuel consumption: 80%
It remains very reasonable, even on a GT whose performance is higher.

• Satisfaction: 75%
Owners are complaining less and less about the Escorts, which have improved over the years, to the point where 87% of the owners would repeat the purchase today. Only the transmission and some dashboard elements still cause a few problems.

• Price/equipment: 75%
Compared with their most recent Japanese and American competitors, the basic price is attractive, but except for standard air bags there are many extra cost options.

DATA

Category: front-wheel drive compacts sedans and wagons.
Class : 3

HISTORIC

Introduced in: 1980, 1983 (GT) 1990
Modified in: 1985: 1.9L engine; 1990: 1.8L Mazda engine
Made in: Wayne, Michigan, U.S.A & Hermosillo, Mexico.

RATINGS

Safety: 100 %
Satisfaction: 78 %
Depreciation: 60 %
Insurance: 10 % ($514)
Cost per mile: $0.32

NUMBER OF DEALERS

In U.S.A.: 5,200 Ford-Lincoln-Mercury

SALES ON THE AMERICAN MARKET

Model	1992	1993	Result
Escort	236,622	269.034	+ 12.1 %
Tracer	43,127	45,754	+ 5.8 %

MAIN COMPETITORS

CHEVROLET Cavalier, DODGE-PLYMOUTH Neon, HONDA Civic 4dr., HYUNDAI Elantra, MAZDA Protegé, PONTIAC Sunbird, SATURN, SUBARU Impreza, TOYOTA Corolla, VOLKSWAGEN Golf-Jetta.

EQUIPMENT

FORD Escort	LX	GT
Automatic transmission:	O	O
Cruise control:	O	O
Power steering:	O	O
ABS brake:	-	-
Air conditioning:	O	O
Air bag:	S	S
Leather trim:	-	-
AM/FM/ Radio-cassette:	O	S
Power door locks:	O	O
Power windows:	O	O
Tilt steering column:	-	-
Dual power mirrors:	O	S
Intermittent wipers:	S	S
Light alloy wheels:	O	S
Sunroof:	O	O
Anti-theft system:	O	O

S : standard; O : optional; - : not available

COLORS AVAILABLE

Exterior: Green , Red , Coral, Black, White, Silver, Blue, Bronze, Iris.
Interior: Royal Blue, Mocha, Opal Gray.

MAINTENANCE

First revision: 5,000 miles
Frequency: 6 months
Diagnostic plug: Yes

WHAT'S NEW IN 1995 ?

- **Two standard air bags.**
- **Redesigned dashbord.**
- **No more leather steering wheel available on the GT.**
- **Integrated child seat available in option.**

Model/ versions *: standard	ENGINE Type / timing valve / fuel system	Displacement cu/in	Power bhp @ rmn	Torque lb.ft @ rpm	Compres. ratio	TRANSMISSION Driving wheels / transmission	Final ratio	PERFORMANCE Acceler. 0-62 mph s	Stand. 1/4 mile s	Stand. 5/8 mile s	Braking 62-0 mph ft	Top speed mph	Lateral acceler. G	Noise level dBA	Fuel economy mpg city	hwy	Gasoline type / octane
GT/LTS	L4* 1.8 DOHC-16-EFI	109	127 @ 6500	114 @ 4500	9.0 :1	front - M5*	4.10	9.0	16.7	30.3	141	122	0.82	68	30.0	39.0	R 87
						front - A4	3.74	10.2	17.7	31.7	148	118	0.82	68	28.0	37.5	R 87
LX/base	L4* 1.9 SOHC-8-SFI	114	88 @ 4400	108 @ 3800	9.0 :1	front - M5*	3.62	10.0	17.5	31.2	138	103	0.80	68	35.5	49.0	R 87
						front - A4	3.55	11.2	18.2	32.8	151	100	0.80	68	29.5	44.0	R 87

Escort-Tracer

MERCURY Tracer

FORD Escort LX

• Handling: 75%
The modern Escort suspension gives it a precision steering control but its tire quality can make a difference on a wet road. In curves it understeers but is easily controlled and predictable. In a straight line, when the pavement is imperfect, its pointing ability is darty even in the sports car version.

• Suspension: 65%
Its travel allows it to absorb road defects without too much damage to the occupants. On the sports versions it is very successful.

• Performance: 62%
The GT equipped with a 1.8 L engine, provides superior driving pleasure with more brilliant performance than those of standard versions. The favorable power-to-weight ratio is enhanced by a matching gear ratio spacing. Takeoffs and accelerations are sharp, without a fuel consumption penalty. The base engine works hard, even with an automatic but its power curve is less sharp.

• Driver's compartment: 60%
The driver area is open and the visibility excellent from all angles. Economy versions have less gauges than the sports version one but they are visible and easy to read. The main switches and controls are well within reach. Windshield wiper arms have a tendancy to lift from the windshield starting at 60 mph.

• Access : 60%
It is easier to enter the rear with a 4-door than with a 2-door the front seats of which do not offer enough space.

• Steering: 60%
Vague, with too high a steering ratio on the manual, it is fast and precise when power-assisted but takes away all road feel. Torque steer is not negligible and appears during hard acceleration.

• Storage: 60%
Storage is only practical on well-equipped versions where you find pockets built into the doors and on the back of the right front seat as well as in the center console.

• Trunk: 50%
It is equally ample on the hatchbacks and sedans. But it is the station wagon which offers the largest and most accessible space.

• Braking: 50%
Mediocre and unstable on the station wagon, they are effective, well-balanced and fade-free on the sports version in spite of a lack of initial bite.

• Interior space: 50%
Changes have made these models more spacious. Generous dimensions allow them to receive the tallest people, but with the GT the sunroof interfers with available height.

WEAK POINTS

• Quality/fit/finish: 40%
Assembly is better than in the past, but appearance and finish of some sections still look cheap as is the case whitch the plastic dashboard whose shinny surface attracts dust.

• Noise level: 40%
The soundproofing of all these models is insufficient and roaring noise dominates.

• Insurance: 40%
The Escorts are more costly to insure than most of their competitors and the rates for the LX are as high as for the sporty GT.

• Depreciation: 40%
Lower than those of the older models, it is still higher than that of the Japanese models in the same category.

• Seats: 48%
Better designed than on the older model, lateral and lumbar supports are still insufficient and the padding is hard which is unusual in an American car.

CONCLUSION

• Overall rating: 60.0%
In spite of definite qualities the public interest in the Escort-Tracer fades because their reputation is still based of the preceding model. 😐

OWNERS' SUGGESTIONS

- Better quality tires
- More efficient braking with ABS
- Better soundproofing
- More gauges (LX)
- More powerful and less noisy heating and defrosting devices
- Better overall quality
- More easy-to-adjust AC setting
- More waterproof body
- Better quality paint
- Truly adjustable steering wheel
- More effective windshield wipers

SPECIFICATIONS & PRICES

Model	Version/Trim	Body/ Seats	Interior volume cu.ft.	Trunk volume cu ft.	Drag coef. Cd	Wheel-base in	Lgth x Wdthx Hght. inx inx in	Curb weight lb	Towing capacity std. lb	Susp. type ft/rr	Brake type ft/rr	Steering system	Turning diameter ft	Steer. turns nber	Fuel. tank gal.	Standard tire size	Standard powertrain	PRICES US. $ 1994
FORD		**Total warranty and antipollution: 3 years / 36,000 miles;corrosion perforation: 5 years / unlimited.**																
Escort	LX	3dr. sdn. 4	15.6	17.3	0.34	2500	170.0x66.7x 52.5	2354	1 000	i/i	dc/dr	r&p.	31.5	4.3	11.9	175/65R14	L4/1.9/M5	**$9,890**
Escort	GT	3dr. sdn. 4	15.6	17.3	0.34	2500	170.0x66.7x 52.5	2458	1 000	i/i	dc/dc	pwr.r&p.	31.5	3.1	13.2	185/60R15	L4/1.8/M5	**$12,300**
Escort	LX	4 dr. sdn. 4	15.7	12.0	0.35	2500	170.9x66.7x 52.7	2383	1 000	i/i	dc/dr	r&p.	31.5	4.3	11.9	175/65R14	L4/1.9/M5	**$10,550**
Escort	LX	5 dr. sdn. 4	15.8	17.3	0.34	2500	170.0x66.7x 52.5	2403	1 000	i/i	dc/dr	r&p.	31.5	4.3	11.9	175/65R14	L4/1.9/M5	**$10,325**
Escort	LX	5 dr. wgn. 4	16.0	30.6	0.36	2500	171.3x66.7x 53.6	2451	1 000	i/i	dc/dr	r&p.	31.5	4.3	11.9	175/65R14	L4/1.9/M5	**$10,880**
MERCURY		**Total warranty and antipollution: 3 years / 36,000 miles;corrosion perforation: 5 years / unlimited.**																
Tracer	base	4 dr. sdn. 4	15.7	12.0	0.35	2500	170.9x66.7x 52.7	2418	1 000	i/i	dc/dr	r&p.	31.5	4.3	11.9	175/65R14	L4/1.9/M5	**$10,250**
Tracer	LTS	4 dr. sdn. 4	15.7	12.0	0.35	2500	170.9x66.7x 52.7	2463	1 000	i/i	dc/dr	r&p.	31.5	4.3	11.9	185/60R14	L4/1.8/M5	**$12,560**
Tracer	base	5 dr. wgn. 4	16.0	30.6	0.36	2500	171.3x66.7x 53.6	2498	1 000	i/i	dc/dr	r&p.	31.5	4.3	11.9	175/65R14	L4/1.9/M5	**$13,292**

See page 393 for complete 1995 Price List

Updating...

It was about time to update the Explorer as this sales champ of the sporty utilitarians was beginning to lag in suspension, steering and brakes. Now it's done and the look is more feminine.

With its appearance freshened up, the Explorer is offered in 2-and 4-doors and a four-wheel drive suspension on request. The only engine available is a 4.0 L V6 with the five-speed manual standard or a four-speed automatic optional (standard on the 4-door Limited).

The trims are XL, Sport and Expedition for the 2-door and XL, XLT, Eddie Bauer and Limited for the 4-door. The drive train remains similar to the preceding one. It is a 4.0 L V6 and a manual 5-speed base, or an optional 4-speed manual except on the Limited where the automatic is standard.

The Navajo, derived from the Explorer and sold by Mazda exclusively in the US, is not offered in 1995 since its last sales did not exceed 3,500 units.

STRONG POINTS

• Safety: 100%

Improved thanks to greater body rigidity, the presence of 2 front seat air bags, headrest on all seats, and 4-wheel ABS discs standard.

• Insurance: 80%

The premium on these vehicles differs little between 2 and 4 WD and it is reasonable.

• Technical: 70%

These all-terrain vehicles are based on a ladder type chassis with six cross members, derived from those of the Ranger, with a steel unit body retained on rubber body mounts. The overall streamlining is improved with the coefficient of drag dropping from 0.43 to 0.41 thanks to smoother lines in the front area. The front suspension changed the most. Ford replaced its twin I-Beam suspension by long and short A-Frames, typical at Ford. At the rear, the rigid axle connects through leaf springs.Braking now includes 4-wheel discs and a standard ABS.

DATA

Category: rear-wheel drive or AWD all terrain multipurpose.
Class : utility

HISTORIC

Introduced in : 1983: Bronco II
Modified in : 1991: Explorer & Mazda Navajo, renewed in 1995.
Made in: Louisville, Kentucky, U.S.A.

RATINGS

Safety: 100 %
Satisfaction: 60 %
Depreciation: 50 %
Insurance: 5.5 % ($585)
Cost per mile: $ 0.42

NUMBER OF DEALERS

In U.S.A: 5,200 Ford-Lincoln-Mercury

SALES ON THE AMERICAN MARKET

Model	1992	1993	Result
Explorer	306,681	302,201	-2.5 %

MAIN COMPETITORS

CHEVROLET Blazer, GMC Jimmy, JEEP Cherokee & Grand Cherokee, LAND-ROVER Discovery, MITSUBISHI Montero, NISSAN Pathfinder, TOYOTA Land Cruiser & 4Runner.

EQUIPMENT

FORD Explorer 4x4 2 dr.	XL	Sport	Expedition	
FORD Explorer 4x4 4 dr.	**XL**	**XLT**	**E.B.**	**Ltd**
Automatic transmission:	O	O	O	S
Cruise control:	-	S	S	S
Power steering:	S	S	S	S
ABS brake:	S	S	S	S
Air conditioning:	S	S	S	S
Air bag:	S	S	S	S
Leather trim:	-	-	-	S
AM/FM/ Radio-cassette:	O	O	O	S
Power door locks:	-	S	S	S
Power windows:	-	S	S	S
Tilt steering column:	-	S	S	S
Dual power mirrors:	O	S	S	S
Intermittent wipers:	S	S	S	S
Light alloy wheels:	O	O	O	S
Sunroof:	O	O	O	O
Anti-theft system:	O	O	O	O

S : standard; O : optional; -: not available

COLORS AVAILABLE

Exterior: Green, Emerald, Silver, Red, Blue, White, Black, Iris, Gray.
Interior: Crystal blue, Willow green, Saddle brown, Medium graphite .

MAINTENANCE

First revision: 5,000 miles
Frequency: 6 months/ 5,000 miles
Diagnostic plug: Yes

WHAT'S NEW IN 1995 ?

-New appearance.
-Fenders, grill and bumpers redesigned.
-New steel and light alloy wheels.
-New neon rear center stop lights.
-Two standard front air bags.
-Standard 4-wheel disc brakes.
-Height adjustable front seat belts.
-Headrests for the rear seats.
-Redesigned dashboard.

ENGINE						TRANSMISSION		PERFORMANCE										
Model/ versions *: standard	Type / timing valve / fuel system	Displacement cu/in	Power bhp @ rmn	Torque lb.ft @ rpm	Compres. ratio	Driving wheels / transmission	Final ratio	Acceler. 0-62 mph s	Stand. 1/4 mile s	Stand. 5/8 mile s	Braking 62-0 mph ft	Top speed mph	Lateral acceler. G	Noise level dBA	Fuel economy mpg city	hwy	Gasoline type / octane
4x4	V6* 4.0 OHV-12-MPFI	245	160 @ 4200	225 @ 2800	9.0 :1	rear - M5*	3.45	12.0	17.6	32.0	131.0	109	0.65	68	21.0	28.0	R 87
						rear - A4	3.45	12.5	18.0	32.5	137.7	106	0.65	68	18.5	26.0	R 87

Explorer

• **Satisfaction: 60%**
The rate remains relatively low which indicates what remains to be done for quality and reliability. Let's wish that the quality of the transmission, wiring harness, paint and some accessories has really improved.

• **Noise level: 60%**
A discrete drive train and efficient sound insulation make the luxury versions more comfortable but wind and rolling noises are noted occasionnally.

• **Cargo space: 60%**
Vast, because the spare tire is now under the rear floor pan. You can double the storage by folding down the rear seat.

• **Quality/fit/finish: 60%**
The general build and drive train are robust and the materials quality seems to have positively improved. The fit is a little tighter and the fabrics, leather and plastics have a richer appearance.

• **Steering: 60%**
Precise, even though its ratio is too high, it offers good maneuverability but its excessive assistance is too high and requires too much attention.

• **Driver's compartment: 60%**
A comfortable driving position is easier to find on the versions with multiple seat adjustments than on the base models.Visibility is satisfactory on all angles and the outside rear-view mirrors are amply sized.The dashboard has a less utilitarian appearance and the ergonomics are good since everything is within reach. On a center console, the radio controls overlooking those of the AC, are easy to use.

• **Interior space: 60%**
That of the 4-door cab is more generous thanks to its larger wheelbase and overall length. Four people can sit comfortably as the dimensions are well-sized, especially in height. A fifth passenger will only stand it for a brief trip.

• **Handling : 55%**
The new front suspension, mated to the rack-and-pinion steering has distinctly improved control in corners or in the straight, and the waddling of preceding models has virtually disappeared. However, one gains the impression that the structural rigidity could be better. Check the ground clearance, which is a little tight on the 4X2 versions with 15 inch tires.

• **Braking: 55%**
Its control is now more precise and its efficiency and balance have been improved by the use of a standard 4-disc and ABS. The life of the brake pads remains in need of improvement.

• **Suspension: 50%**
The front end is better at absorbing low amplitude disturbances. The rear suspension provides harsh notice of tar strips.

• **Access: 50%**
It is still hard to reach the seats in a 2-door because of the ground clearance and the narrowness of the doors or the lack of available space (2-dr).

• **Seats: 50%**
Better proportioned, formed and padded, they offer a superior support compared to the previous model. Up front adjustable headrests would be more comfortable.

• **Depreciation: 50%**
The 4-door Explorer is more popular than the 2-door and the 4X4 more than the 4X2.

WEAK POINTS

• **Fuel consumption: 30%**
Comparable to that of a large sedan on the freeway, it rises as quickly off road as in town.

• **Performance: 40%**
It remains average considering the displacement. The engine lacks low rpm torque and there is talk of a future supercharged model that produces 225 hp.

• **Price/equipment: 40%**
The average price of these vehicles comes to around $24,000 which requires a sizeable budget. The best purchase is in the sport (2-d), and XLT (4-d) versions whose equipment is complete.

CONCLUSION

• **Overall rating: 56.5%**
Largely ahead of sales in its category, the Explorer will keep its lead over its next nearest competitor, the Grand Cherokee. However in spite of its technical update, it cannot completely equal the Grand Cherokee. 😐

SPECIFICATIONS & PRICES

Model	Version/Trim	Body/ Seats	Interior volume cu.ft.	Trunk volume cu ft.	Drag coef. Cd	Wheel-base in	Lgth x Wdthx Hght. inx inx in	Curb weight lb	Towing capacity std. lb	Susp. type ft/rr	Brake type ft/rr	Steering system	Turning diameter ft	Steer. turns nber	Fuel. tank gal.	Standard tire size	Standard powertrain	PRICES US. $ 1994
FORD Explorer 4x4				**Total warranty and antipollution: 3 years / 36,000 miles;corrosion perforation: 5 years / unlimited.**														
Explorer	XL	3dr.wgn.4/5	-	120	0.41	101.6	178.5x70.2x67.0	3979	4800	i/r	dc/dc/ABS	pwr.r&p.	34.5	3.5	17.5	225/70R15	V6/4.0/M5	-
Explorer	Sport	3dr.wgn.4/5	-	120	0.41	101.6	178.5x70.2x67.0	4023	4800	i/r	dc/dc/ABS	pwr.r&p.	34.5	3.5	17.5	225/70R15	V6/4.0/M5	**$20,000**
Explorer	Expedition	3dr.wgn.4/5	-	120	0.41	101.6	178.5x70.2x67.0	4078	4800	i/r	dc/dc/ABS	pwr.r&p.	34.5	3.5	17.5	555/70R16	V6/4.0/M5	**NA**
Explorer	XL	5dr.wgn.4/5	-	141	0.41	111.4	188.5x70.2x67.0	4188	5000	i/r	dc/dc/ABS	pwr.r&p.	37.3	3.5	21.0	225/70R15	V6/4.0/M5	**$19,900**
Explorer	XLT	5dr.wgn.4/5	-	141	0.41	111.4	188.5x70.2x67.0	4233	5000	i/r	dc/dc/ABS	pwr.r&p.	37.3	3.5	21.0	225/70R15	V6/4.0/M5	**$22,410**
Explorer	E. Bauer	5dr.wgn.4/5	-	141	0.41	111.4	188.5x70.2x67.0	4288	5000	i/r	dc/dc/ABS	pwr.r&p.	37.3	3.5	21.0	255/70R16	V6/4.0/M5	**$26,205**
Explorer	Limited	5dr.wgn.4/5	-	141	0.41	111.4	188.5x70.2x67.0	4288	5000	i/r	dc/dc/ABS	pwr.r&p.	37.3	3.8	21.0	235/75R15	V6/4.0/M5	**$28,535**

See page 393 for complete 1995 Price List

Always in the lead...

A record year for Ford which has 6 vehicles in the top ten in sales on the North American continent. The F-Series trucks maintain the lead in their category with over 500,000 units sold.

In the F-Series, the choice is wide because you can count 3 types of cabin: regular, extended, and 4-door plus 2 types of beds: regular and step side. There are 2WD and 4WD drives in the ranges of F-150, 250, and 350. In terms of powertrain the 4.9 L6 is the base engine. And you have a choice of 3 gasoline V8's going from 5.0L to 7.5L or Diesel and turbo Diesel of 7.3 L. The manual 5-speed is standard on all pickups and 3- and 4-speed automatics are optional. The Bronco remains a 4X4 3-door wagon and comes with a choice of 3 engines, 2 transmissions and a transfer box.

STRONG POINTS

Technical:
These utility machines are based on a steel ladder frame with 8 cross members for the F-Series and 5 for the Bronco on which the steel body is set on rubber mounts. The front suspension is independent with twin I-Beams and a rigid axle with leaf springs at the back. The brakes are a combination disc and drum and the ABS acts only on the rear wheels since only the Bronco has integral ABS.

Quality/fit/finish:
Ford maintains itself at the head of the pickup domain thanks to a reputation for durability that its models have acquired over time. The finish is typically American and the base models are not as sharp as the XLT.

Safety:
The index of the Ford pickup improves thanks to a standard air bag mounted on the driver's side. Structural stiffness is in the upper average, as is occupant safety.

Interior space:
Thanks to generous cab dimensions these utility machines can accomodate 3 to 6 occupants depending on the number of doors and

DATA

Category: 4x2 & 4x4 pick-ups and utilities.
Class : utility

HISTORIC

Introduced in : 1953: F-Series; 1979: Bronco.
Modified in : 1983: V8 diesel, 1987 & 1992: appearence.
Made in: F-Series: Kansas-City, Missouri; Wayne, Michigan; Norfolk, Virginia; Twin Cities, Minnesota, U.S.A. & Oakville, Ontario, Canada. Bronco: Wayne, Michigan, U.S.A.

RATINGS

Safety: 75 %
Satisfaction: 68 %
Depreciation: 55 %
Insurance: 5.0 % (from $520 to $655)
Cost per mile: $ 0.48

NUMBER OF DEALERS

In U.S.A.: 5,200 Ford-Lincoln-Mercury

SALES ON THE AMERICAN MARKET

Model	1992	1993	Result
F-Series	488,539	565,082	+13.2 %
Bronco	24,752	29,729	+16.8 %

MAIN COMPETITORS

F-Series: CHEVROLET-GMC & DODGE Ram pick-ups.
Bronco: CHEVROLET-GMC Tahoe-Yukon, DODGE Ramcharger.

EQUIPMENT

F-Series	S	XL	XLT	Ed.Bauer
Bronco		XL	XLT	Ed.Bauer
Automatic transmission:	O	O	O	O
Cruise control:	O	O	S	S
Power steering:	S	S	S	S
ABS brake:	S	S	S	S
Air conditioning:	O	O	O	O
Air bag:	S	S	S	S
Leather trim:	-	-	-	-
AM/FM/ Radio-cassette:	O	O	O	O
Power door locks:	O	O	O	O
Power windows:	O	O	O	O
Tilt steering column:	O	O	O	O
Dual power mirrors:	O	O	O	O
Intermittent wipers:	S	S	S	S
Light alloy wheels:	O	O	O	O
Sunroof:	O	O	O	O
Anti-theft system:	O	O	O	O

S : standard; O : optional; -: not available

COLORS AVAILABLE

Exterior: Copper, Red, Black, White, Blue, Green, Bronze.
Interior: Royal blue, Ruby red, Opal gray, Medium mocha.

MAINTENANCE

First revision: 5,000 miles
Frequency: 6 months/ 5,000 miles
Diagnostic plug: Yes

WHAT'S NEW IN 1995 ?

-Headlights and rear bumpers redesigned.
-New directional alloy wheels.
-Door trim touch up.
-New axle ratio with 2.0 L engine.

Model/ versions *: standard	ENGINE Type / timing valve / fuel system	Displacement cu/in	Power bhp @ rmn	Torque lb.ft @ rpm	Compres. ratio	TRANSMISSION Driving wheels / transmission	Final ratio	PERFORMANCE Acceler. 0-62 mph s	Stand. 1/4 mile s	Stand. 5/8 mile s	Braking 62-0 mph ft	Top speed mph	Lateral acceler. G	Noise level dBA	Fuel economy mpg city	hwy	Gasoline type / octane
1)	L6*4.9 OHV-12-EFI	300	145 @ 3400	265 @ 2000	8.8 :1	rear - M5*	-	12.5	17.5	33.6	180	90	0.68	67	17.0	23.5	R 87
2)	V8 5.0 OHV-16-EFI	305	195 @ 4000	270 @ 3000	9.0 :1	rear - A4	-	12.5	17.7	34.4	171	100	0.70	66	16.6	26.0	R 87
3)	V8 5.8 OHV-16-EFI	352	210 @ 3800	325 @ 2800	8.8 :1	rear - A4	-	12.0	16.8	33.2	180	106	0.70	66	15.5	22.5	R 87
4)	V8TD 7.3 OHV-16-MI	445	210 @ 3000	425 @ 2000	17.5 :1	rear - A4	-	14.5	19.7	36.5	196	90	0.68	68	18.0	26.0	D
5)	V8 7.5 OHV-16-EFI	457	245 @ 4000	390 @ 2200	8.5 :1	rear - A4	-	12.5	17.0	34.0	196	103	0.68	69	14.5	22.0	R 87

1) base F-Series 2)base Bronco optional F-Series 3) optional F-Series & Bronco 4-5) optional F-Series.

F-Series & Bronco

FORD Bronco

FORD F-150

benches. The Bronco can handle 5-6 passengers, based on front seat layout.

Cargo capacity:
It can vary between 1,000 and 5,720 pounds between the F-150 and the F-350 depending on a drive train axle capacity and the number of drive wheels, single or dual.

Driver's compartment:
The dashboard is simplistic and the small number of gauges available is easy to check. The driver is better seated with individual buckets than on the standard bench which has poor support ability. The controls are conventional and visibility is satisfactory in all directions.

Performance:
The 5 L V8 is the minimum in power and torque for a utility machine of this format. For larger hauls, the gasoline 7.5 L V8 is the largest available. For better mileage and durability, the V8 Diesel and turbo Diesel are a better choice especially for long hauls.

Handling:
It is based on wheelbase length and on tire quality. It's surprising to note how some versions drive easily while offering sound handling with some oversteer at the limit,while others drive in sporty fashion and offer real enjoyment.

Braking:
Effective and easily controlled during moderate deceleration, it is less convincing in panic stops where the front wheels block rapidly. Also, depending on tire quality, traction is borderline. The Bronco is more stable with ABS on 4 wheels but stopping distances are too long.

Noise level:
It is generally low because gasoline V8's are more discrete but tire wind and suspension noise are very present.The Diesels are noiser in spite of the extra soundproofing both in starting and during fast acceleration.

Satisfaction:
Few problems that time hasn't already settled.

Depreciation:
Their resale value depends on use and maintenance.

WEAK POINTS

Steering:
Imprecise in the center and with too high a ratio, it produces some drift. There is too much power assist to inform the driver of the road. Also the maneuverability is not ideal due to the large turning diameter.

Access:
It is easier to enter the 4X2 versions than the 4X4 in which the ground clearance is higher.
And it is more complicated to get to the rear seats of the Supercab or Bronco because of the small available space than into an extended cab.

Seats:
The standard bench seat is the worst for this type of vehicle and it is definitely worth the extra cost for a better padded bucket seat.

Suspension:
That of two-wheel drive and long wheelbase vehicles is more comfortable than the 4WD which is jumpy and harsh on bad pavement.

• Storage:
Storage is only practical on the higher trim levels.

Price/equipment:
Depending on the usage, the price of this vehicle can reach unexpected heights because the list of options and option groups is infinite.

Insurance:
A little high for a car, it is normal for a utility machine of this type.

Fuel Consumption:
There is nothing economical about it with gasoline engines but the Diesel delivers better efficiency.

CONCLUSION

It seems that nothing can alter the steam roller success of these machines, not even the advent of a new competitor such as the Dodge Ram. ☺

SPECIFICATIONS & PRICES

Model	Version/Trim	Body/ Seats	Box Length ft	Max Payload lb	Drag coef. Cd	Wheel-base in	Lgth x Wdthx Hght. inx inx in	Curb weight lb	Towing capacity std. lb	Susp. type ft/rr	Brake type ft/rr	Steering system	Turning diameter ft	Steer. turns nber	Fuel. tank gal.	Standard tire size	Standard powertrain	PRICES US. $ 1994
FORD		Total warranty and antipollution: 3 years / 36,000 miles;corrosion perforation: 5 years / unlimited.																
F-150	4x2	Standard	S	2 dr. p-u.3		116.8	197.0x79.0x70.7	3896	1400	i/r	dc/dr/ABS	pwr.bal.	39.9	-	34.6	215/75R15	L6/4.9/M5	**$12,266**
F-150	4x2	Super Cab	XLT	2 dr. p-u.3		138.8	219.1x79.0x71.7	4187	1730	i/r	dc/dr/ABS	pwr.bal.	46.4	-	34.6	235/75R15	L6/4.9/M5	**$12,772**
F-250	4x2	Standard	XL	2 dr. p-u.3		133.0	213.3x79.0x73.4	4230	4156	i/r	dc/dr/ABS	pwr.bal.	45.8	-	34.6	235/85R16	L6/4.9/M5	**$15,002**
F-250	4x2	Super Cab	XLT	2 dr. p-u.3		155.0	235.3x79.0x71.7	4751	4156	i/r	dc/dr/ABS	pwr.bal.	52.6	-	34.6	215/85R16	V8/5.0/M5	**$16,587**
F-350	4x2	Standard	XL	2 dr. p-u.3		133.0	213.3x79.0x74.3	4881	5110	i/r	dc/dr/ABS	pwr.bal.	62.3	-	37.2	215/85R16	V8/5.0/M5	**$17,839**
F-350	4x2	Crew Cab	XL	4 dr. p-u.6		168.4	248.7x79.0x74.9	5388	4610	i/r	dc/dr/ABS	pwr.bal.	56.6	-	37.2	215/85R16	V8/5.0/M5	**$19,927**
F-150	4x4	Standard	S	2 dr. p-u.3		116.8	197.0x79.0x74.4	4041	2100	r/r	dc/dr/ABS	pwr.bal.	40.1	-	18.0	215/75R15	L6/4.9/M5	**$15,676**
F-150	4x4	Super Cab	XLT	2 dr. p-u.3		138.8	219.1x79.0x-	4422	2154	r/r	dc/dr/ABS	pwr.bal.	46.7	-	34.6	235/75R15	L6/4.9/M5	**$16,005**
F-250	4x4	Standard	XL	2 dr. p-u.3		133.0	213.3x79.0x74.4	4945	3920	r/r	dc/dr/ABS	pwr.bal.	46.3	-	34.6	235/85R16	V8/5.0/M5	**$18,996**
F-250	4x4	Super Cab	XLT	2 dr. p-u.3		155.0	235.3x79.0x-	5231	3920	r/r	dc/dr/ABS	pwr.bal.	53.0	-	34.6	215/85R16	V8/5.0/M5	**$20,552**
F-350	4x4	Standard	XL	2 dr. p-u.3		133.0	213.3x79.0x74.4	5046	4140	i/r	dc/dr/ABS	pwr.bal.	50.3	-	37.2	215/85R16	V8/5.0/M5	**$19,486**
F-350	4x4	Crew Cab	XL	4 dr. p-u.6		168.4	248.7x79.0x-	5624	4140	i/r	dc/dr/ABS	pwr.bal.	62.3	-	27.2	215/85R16	V8/5.0/M5	**$22,372**
Bronco	4x4	XL		2 dr. wgn.5104.7			183.6x79.1x74.4	4566	1049	i/r	dc/dr/ABS	pwr.bal.	36.6	3.3	32.0	235/75R15	V8/5.0/M5	**$21,515**
Bronco	4x4	XLT/Sport		2 dr. wgn.5104.7			183.6x79.1x74.4	4566	1049	i/r	dc/dr/ABS	pwr.bal.	36.6	3.3	32.0	235/75R15	V8/5.0/M5	**$23,755**
Bronco	4x4	Eddie Bauer		2 dr. wgn.5104.7			183.6x79.1x74.4	4566	1049	i/r	dc/dr/ABS	pwr.bal.	36.6	3.3	32.0	235/75R15	V8/5.8/A4	**$26,590**

See page 393 for complete 1995 Price List

Friskier than ever...

Ford has done well not to kill its little horse for the benefit of the Probe. By betting on two very different horses, it has assured for itself an enviable position in the race of sports coupes. The success of the Mustang coupes and convertibles has caused world-wide envy.

The Mustang is offered in base trim or GT in the form of a coupe or a convertible with an electric top. An optional rigid top is easily adapted to the convertible allowing 4-season use. The base engine is a 3.8L V6 while the GT and the Cobra have a 5.0 L V8' s whose power differs by 25 hp with a manual stick shift or an optional automatic.

STRONG POINTS

• Safety: **90%**

Body stiffness is more evident on the coupes than on the convertibles which do not benefit from roof rail stiffness, so body noises and vibrations on bad pavement are numerous.

• Technical: **72%**

Brought up to the taste of the day, the Mustang line is dynamic without being too aggressive and the streamlining targets a drag coefficient of 0.34. The main work consisted of redesigning the body starting with the platform of the preceding model. About 85% of the vehicle is totally new. Stiffening called for reinforcements and new pieces designed to improve the convertible.The front suspension is a McPherson type while at the rear you find the old rigid axle and its 4 bar linkage. Both models are equipped with stabilizer bars and the GT receives 4 rear shocks of which 2 handle spring wind-up. Fifteen inch wheels are used on a base body while the VT has 16 inch wheels with a Z designation. The brakes include 4 wheel discs on all models but ABS is only available as an option.

• Price/equipment: **70%**

The Mustang remains popular thanks to its price which stays around $16,000 and compares favorably to the stars of its category. The most evident progress concerns standard equipment which is more generous than the previous model.

DATA

Category: rear-wheel drive coupe & convertible sports cars.
Class : 3S

HISTORIC

Introduced in : 1964, 1976 (Mustang II), 1979, 1994 (actual model).
Modified in : 1982: V8 5.0L engine; 1983: 4 cylinders Turbo. 1983: convertible version; 1994: renewal.
Made in: Dearborn, Michigan, U.S.A.

RATINGS

Safety:	90 %
Satisfaction:	72 %
Depreciation:	60 %
Insurance:	9-11 % (from $660 to $860)
Cost per mile:	$ 0.47

NUMBER OF DEALERS

In U.S.A.: 5,200 Ford-Lincoln-Mercury

SALES ON THE AMERICAN MARKET

Model	1992	1993	Result
Mustang	86,036	98,648	+12.8 %

MAIN COMPETITORS

ACURA Integra, EAGLE Talon, CHEVROLET Camaro, FORD Probe, HONDA Prelude, MAZDA MX-6, NISSAN 240SX, PONTIAC Firebird, TOYOTA Celica, VW Corrado.

EQUIPMENT

FORD Mustang	LX	GT
Automatic transmission:	O	O
Cruise control:	O	O
Power steering:	S	S
ABS brake:	O	O
Air conditioning:	O	O
Air bag:	S	S
Leather trim:	-	-
AM/FM/ Radio-cassette:	O	O
Power door locks:	S(1)	S
Power windows:	S(1)	S
Tilt steering column:	S	S
Dual power mirrors:	S	S
Intermittent wipers:	S	S
Light alloy wheels:	O	S
Sunroof:	O (1)	O (1)
Anti-theft system:	O	S

(1) convertible

S : standard; O : optional; -: not available

COLORS AVAILABLE

Exterior: Red, Black, White, Blue, Green ,Yellow, Silver, Gray, Iris.
Interior: Bright red, Saddle brown, Opal gray, Black, White.

MAINTENANCE

First revision:	5,000 miles
Frequency:	6 months/ 5,000 miles
Diagnostic plug:	Yes

WHAT'S NEW IN 1995 ?

-New body colors : Brillant Red and Frosted Silver.
-Anti-theft standard on the GT version.

Model/ versions *: standard	ENGINE Type / timing valve / fuel system	Displacement cu/in	Power bhp @ rmn	Torque lb.ft @ rpm	Compres. ratio	TRANSMISSION Driving wheels / transmission	Final ratio	PERFORMANCE Acceler. 0-62 mph s	Stand. 1/4 mile s	Stand. 5/8 mile s	Braking 62-0 mph ft	Top speed mph	Lateral acceler. G	Noise level dBA	Fuel economy mpg city	hwy	Gasoline type / octane
LX	V6* 3.8 OHV-8-SFI	232	145 @ 4000	215 @ 2500	9.0 :1	rear-M5*	2.73	10.5	16.8	32.4	40	103	0.80	67	24.2	38.0	R 87
						rear-A4	2.73	11.6	17.5	33.1	42	100	0.80	67	23.5	37.5	R 87
GT	V8* 5.0 OHV-16-SFI	302	215 @ 4200	285 @ 3400	9.0 :1	rear-M5*	2.73	7.0	14.5	28.6	38	124	0.83	68	20.0	32.0	R 87
						rear-A4	2.73	8.0	15.0	29.5	40	121	0.83	68	21.0	31.5	R 87
Cobra	V8* 5.0 OHV-16-SFI	302	240 @ 4800	285 @ 4000	9.5 :1	rear-M5*	3.08	6.8	14.2	28.4	37	130	0.85	68	20.0	31.0	R 87

Mustang

• Driver's compartment: 70%
The most comfortable position is most easy to find thanks to standard electric seat controls. More technical than that of the base model, the GT seat offers good lateral and lumbar support.
Numerous instruments are well-grouped and easy to decipher. The steering wheel is well-designed, the shift lever and the hand brake are well-located.

• Performance: 70%
Between the highly muscled Cobra and the very placid but still pleasant to drive LX, the GT seems too much. Though the numbers obtained at its wheel are respectable, you will search in vain for a sporty character.

• Suspension: 70%
Driving comfort is much improved even on undulating highways and you have trouble recognizing the normal handling of a rigid axle. The response of the GT and the Cobra is firmer without being too uncomfortable.

• Steering: 70%
It is smooth and its power assist is better controlled and it is as rapid as it is precise, which enhances driving pleasure. However, the turning diameter a bit large penalizes maneuverability.

• Satisfaction: 70%
It already maintains a satisfactory level with the older model which augurs well for its replacement, many of whose components are well known.

• Access: 60%
Long heavy doors allow easy seating access up front and relatively easy access at the rear, depending on the persons' size.

• Quality/fit/finish: 60%
There still remains work to do on finish and appearance of some parts which bring back to memory the typical past. From this point of view, competitive pricing does not seem to be a valid excuse.

• Driving: 60%
The coupe is more stable than the convertible which jumps around on bad pavement because it lacks body stiffness. However, one has to work harder than before to make it break loose at the back on first quality tires. In the rain there are no special problems because power is not sufficiently abrupt to upset the balance. Also, drive wheel traction is much improved.

• Conveniences 60%
Practical hiding places are more numerous on a GT than on a base model, but front seat belt height is not adjustable. The soft convertible top is motorized while the hard top is easily handled by two people.

• Seats: 60%
Well-padded and comfortable but those of the GT will hold you better than the less contoured ones of the base model.

• Braking: 50%
Panic stop distance is above 150 feet for the base model and 130 on the GT which is more normal. Unfortunately, ABS is not available.

• Noise level: 50%
It is maintained at a comfortable level on both coupes and convertibles with well-tuned exhaust systems. The sporty sound does not become tiring during long runs.

WEAK POINTS

• Insurance: 28%
This type of toy is expensive to insure especially for young folks who will need to find out the size of the insurance premium before signing on the dotted line.

• Interior space: 30%
Limited by the intent of the vehicle styling, it is sufficient in the front in spite of the center console size and surprising at the rear where the height and length are more acceptable for children.

• Luggage space: 40%
In spite of the "Hatch Back" look of the coupe, the trunk is separated from the interior but can be increased in size by folding the two-part seat back. It is wide and deep but lacks height.and its narrow opening is not very practical. The one in the convertible holds even less.

• Fuel Consumption: 40%
More reasonable on a V6 than on a V8, it depends mostly on the pilot's mood and on the weight of his right foot.

• Depreciation: 40%
Let us bet that it will be lower than before, thanks to the improvement on manufacturing quality compared to previous models.

CONCLUSION

• Overall rating: 65.0%
The Mustang is again successful thanks to its provocative styling, competitive pricing and known drive train components that do not raise fears about reliability. It is a conservative combination for which Ford seems to hold the secret. ☺

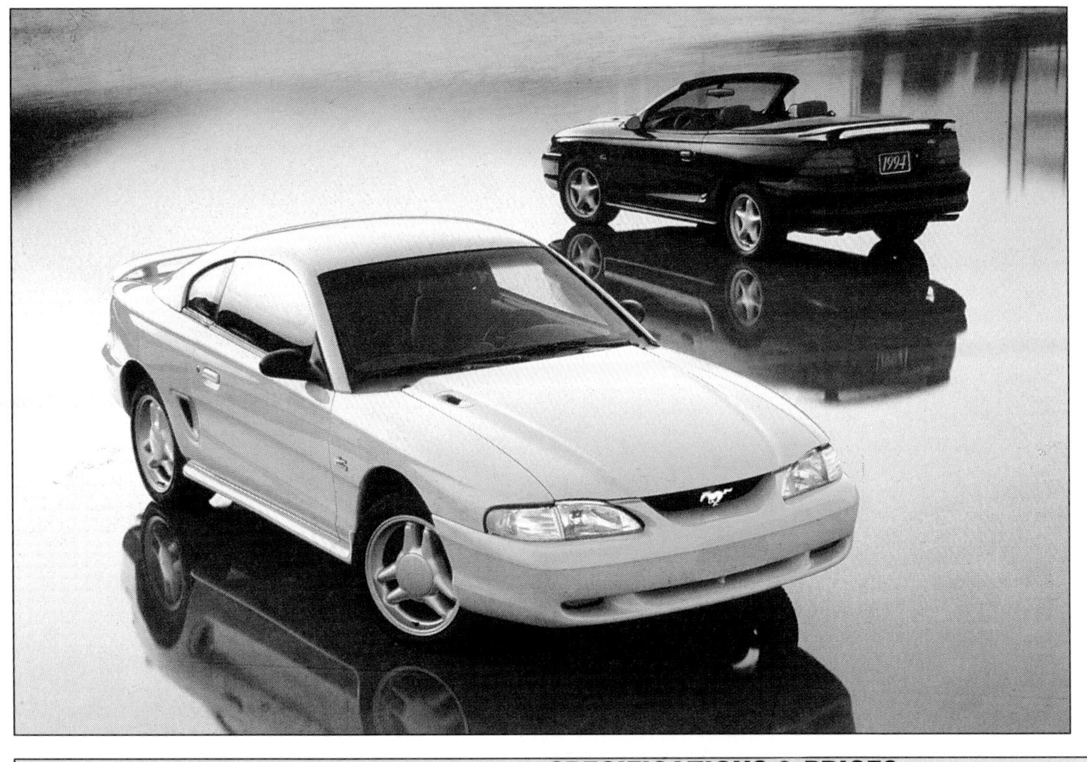

SPECIFICATIONS & PRICES

Model	Version/Trim	Body/ Seats	Interior volume cu.ft.	Trunk volume cu ft.	Drag coef. Cd	Wheel-base in	Lgth x Wdthx Hght. inx inx in	Curb weight lb	Towing capacity std. lb	Susp. type ft/rr	Brake type ft/rr	Steering system	Turning diameter ft	Steer. turns nber	Fuel. tank gal.	Standard tire size	Standard powertrain	PRICES US. $ 1994
FORD		**Total warranty and antipollution: 3 years / 36,000 miles;corrosion perforation: 5 years / unlimited.**																
Mustang		2 dr. cpe.4	83	11	0.34	101.2	181.4x71.8x52.9	3064	1 000	i/r	dc/dc	pwr.r&p.	38.4	2.38	15.4	205/55R15	V6/3.8/M5	**$13,365**
Mustang		2 dr. con.4	76	8.5	0.38	101.2	181.4x71.8x52.8	3245	1 000	i/r	dc/dc	pwr.r&p.	38.4	2.38	15.4	205/65R15	V6/3.8/M5	**$20,160**
Mustang	GT	2 dr. cpe.4	83	11	0.35	101.2	181.4x71.8x52.9	3197	1 000	i/r	dc/dc	pwr.r&p.	38.4	2.38	15.4	225/55ZR16	V8/5.0/M5	**$17,280**
Mustang	GT	2 dr. con.4	76	8.5	0.39	101.2	181.4x71.8x52.8	3289	1 000	i/r	dc/dc	pwr.r&p.	38.4	2.38	15.4	225/55ZR16	V8/5.0/M5	**$21,970**
Mustang	Cobra	2 dr. cpe.4	83	11	0.35	101.2	181.4x71.8x52.9	3223	1 000	i/r	dc/dc	pwr.r&p.	38.4	2.38	15.4	245/45ZR17	V8/5.0/M5	**NA**

See page 393 for complete 1995 Price List

Another machine to please you...

Ford has indeed a choice of models to satisfy everyone. Besides the classic Mustang, loyal to its legend, sports coupe fans can express their passion in a different style. Pure shape, front-wheel drive and a V6 engine are given a whole other perspective.

The Probe is derived from the Mazda 626/MX-6 of which it shares the platform and the assembly line. Mazda took charge of the drive train, the engineering, and the manufacturing, while Ford handled body styling and interior. It is offered in a base version with a 4 cylinder 2.0L or in a GT with a 2.5 L V6 with a manual or automatic transmission. The optional equipment list, as you can see at the right, is long even on a GT.

STRONG POINTS

• Safety: 90%

The general structure has been reinforced to allow effective handling and good inpact resistance. Two standard equipment air bags maximize safety.

• Satisfaction: 85%

Buyers are very satisfied with this sporty and popular worry-free coupe, to the point that 80% of the owners would again make the same purchase. The only point that remains to be watched is the wiring harness.

• Noise level: 80%

It is low in normal driving. Even with a 4 cylinder, it can produce an agreeable sporty tones when you step on it with a V6.

• Technical: 75%

The Probe coupe is of modern conception. Its steel unit body has spectacular lines whose aerodynamic finess is average. Weight has been trimmed by 176 lbs. A more compact McPherson front suspension helps lower the hood height. At the rear, the same type of suspension unit is controlled by 4 links. The rear wheels do introduce a steering action. Stabilizer bar sizes are the same for both versions. The brakes are disc and drum on a Probe and 4-wheel discs on the Probe GT. Either way the ABS is an extra cost option.

• Performance: 75%

While the V6 is outstanding, the 4 cylinder is not bad, because the power-to-weight ratio remains high. It offers power at all rpm. Takeoffs and accelerations are well-muscled, particularly above 4500 rpm.

• Handling: 75%

In its changeover, the Probe has become more fun and more safe to drive. In tight turns its handling remains neutral, while in a straight line or wide curves it is very stable. All this holds more for the GT than for the base model which is less powerful and not as well shoed.

• Fuel consumption: 75%

It is reasonable, even on a V6 , and hovers around 26 miles per gallon.

• Suspension: 70%

While sporty, its comfort is civilized and, in spite of its firmness, it effectively absorbs small road defects, but reacts more harshly to tar strips.

DATA

Category: front-wheel drive sport coupes.
Class : 3S

HISTORIC

Introduced in : 1988
Modified in : 1990: V6 3.0L engine;1993: body.
Made in: Flat Rock, Michigan, U.S.A.

RATINGS

Safety: 90 %
Satisfaction: 85 %
Depreciation: 54 %
Insurance: 7.0 % ($690)
Cost per mile: $ 0.47

NUMBER OF DEALERS

In U.S.A.: 5,200 Ford-Lincoln-Mercury

SALES ON THE AMERICAN MARKET

Model	1992	1993	Result
Probe	63,659	90,435	+29.7 %

MAIN COMPETITORS

EAGLE Talon, CHEVROLET Camaro, DODGE Stealth base, FORD Mustang, HONDA Prelude, MAZDA MX-6, NISSAN 240 SX, PONTIAC Firebird, TOYOTA Celica, VOLKSWAGEN Corrado.

EQUIPMENT

FORD Probe	base	GT
Automatic transmission:	O	O
Cruise control:	O	O
Power steering:	S	S
ABS brake:	O	O
Air conditioning:	O	O
Air bag:	S	S
Leather trim:	-	-
AM/FM/ Radio-cassette:	O	O
Power door locks:	O	O
Power windows:	O	O
Tilt steering column:	O	O
Dual power mirrors:	O	O
Intermittent wipers:	O	O
Light alloy wheels:	-	S
Sunroof:	O	O
Anti-theft system:	O	O

S : standard; O : optional; -: not available

COLORS AVAILABLE

Exterior: Black, White, Mandarin, Red, Blue, Green,Titanium.
Interior: Green, Opal Gray, Black, Saddle.

MAINTENANCE

First revision: 5,000 miles
Frequency: 6 months/ 5,000 miles
Diagnostic plug: Yes

WHAT'S NEW IN 1995 ?

-Redesigned rear headlamps and bumpers.
-New directional aluminum wheels.
-Door moldings touch-ups.
-New differential ratio for the 2.0L engine.

Model/ versions *: standard	ENGINE Type / timing valve / fuel system	Displacement cu/in	Power bhp @ rmn	Torque lb.ft @ rpm	Compres. ratio	TRANSMISSION Driving wheels / transmission	Final ratio	PERFORMANCE Acceler. 0-62 mph s	Stand. 1/4 mile s	Stand. 5/8 mile s	Braking 62-0 mph ft	Top speed mph	Lateral acceler. G	Noise level dBA	Fuel economy mpg city	hwy	Gasoline type / octane
base	L4* 2.0 DOHC-16-EFI	122	118 @ 5500	127 @ 4500	9.0 :1	front - M5*	4.11	9.9	17.0	31.2	144	112	0.80	65	30.0	40.0	R 87
						front - A4	3.77	11.0	17.8	32.1	148	109	0.80	65	29.0	38.5	R 87
GT	V6* 2.5 DOHC-24-EFI	153	164 @ 5600	160 @ 4000	9.2 :1	front - M5*	4.39	7.9	15.8	28.2	148	131	0.86	64	24.5	32.5	R 87
						front - A4	4.16	9.0	16.3	29.4	154	125	0.86	64	23.5	32.5	R 87

Probe

• **Conveniences: 70%**
In addition to a large glove box, you will find long pockets in the doors and a combination storage and armrest on the center console.
On the other hand, it is deplorable that the rear windows don't open and that the seat belt is not height-adjustable.

• **Quality/fit/finish: 70%**
The assembly tolerances and the care given to the fit and finish are superior to other products of the same make but the fabrics and plastics look cheap in some colors.

• **Trunk: 75%**
It is vast and can be increased still further by lowering the seat back. But in an effort to achieve rear body stiffness a high lift-over interferes with access.

• **Price/equipment: 60%**
The base model is inexpensive but its equipment lacks many amenities even in a GT whose price is higher.

• **Driver's compartment: 60%**
The best seating position is comfortable, but not easy to find. Visibility is excellent all around, except through the narrow backlight. Dashboard styling leaves much to be desired, but instrumentation is complete and easy to read and the main controls are typical of Japanese cars.
The stick shift is fast and precise while the pedal assembly allows heel and toeing. The steering wheel has a very disagreeable feel.

• **Insurance: 60%**
The premium is more reasonable for married people over 25. Younger buyers are heavily penalized.

• **Access: 60%**
It is more touchy to get into the rear seat where the roof is low and legspace is symbolic.

• **Seats: 50%**
Their appearance is flattering but they do not offer sufficient lateral support and they also lack padding in the side wings.

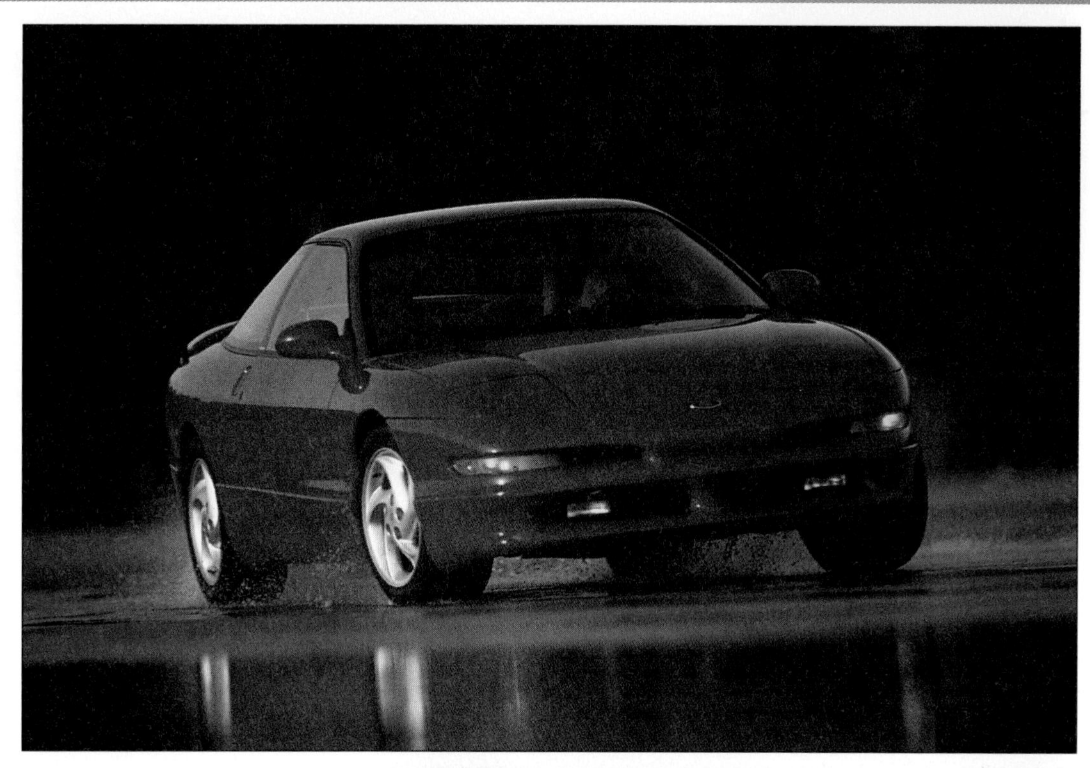

• **Depreciation: 50%**
Barely higher than average on a V6. The 4 cylinders loose more than 50% and take longer to resell.

WEAK POINTS

• **Interior space: 20%**
As in sporty 2+2 's there is more space in the front than in the back where leg-and headroom is lacking and only little children will feel at ease.

• **Braking: 40%**
It is deceiving because it lacks both bite and efficiency. The long stoppping distances are not worthy of a real sports car and in an urgent situation the stops are only linear with the optional ABS. The latter does things both bizarre and disquieting.

• **Steering: 45%**
Power steering assist is fast and well-balanced but vague in the center. On a wet road you note the slight torque steer which is not as pronounced as on the older models.

CONCLUSION

• **Overall rating: 64.0%**
Since its update, the Probe coupe has again found the favor of sports car buyers and its sales rose 35% in one year alone. Attractive in its lines and V6 engine it is a machine designed to please one's self ego. ☺

OWNER'S SUGGESTIONS

-A more resistant paint
-Rear windows that can open
-Fewer options
-More rear seat room
-More accessible trunk
-Better finish
-Better steering wheel design
-More typical instrument panel
-More technical front seats and a "Recaro" style

SPECIFICATIONS & PRICES

Model	Version/Trim	Body/ Seats	Interior volume cu.ft.	Trunk volume cu ft.	Drag coef. Cd	Wheel-base in	Lgth x Wdthx Hght. inx inx in	Curb weight lb	Towing capacity std. lb	Susp. type ft/rr	Brake type ft/rr	Steering system	Turning diameter ft	Steer. turns nber	Fuel. tank gal.	Standard tire size	Standard powertrain	PRICES US. $ 1994
FORD		**Total warranty and antipollution: 3 years / 36,000 miles;corrosion perforation: 5 years / unlimited.**																
Probe	base/SE	3dr.cpe.2+2	80.8	18	0.33	102.8	178.9x69.8x51.6	2689	NR	i/i	dc/dr	pwr.r&p.	35.8	2.9	15.4	195/65R14	L4/2.0/M5	**$13,685**
Probe	GT	3dr.cpe.2+2	80.8	18	0.34	102.8	178.9x69.8x51.8	2866	NR	i/i	dc/dc	pwr.r&p.	35.8	2.9	15.4	225/50VR16	V6/2.5/M5	**$16,015**

See page 393 for complete 1995 Price List

Ranger / B-Series

Another cluster on the Ford Award List...

The Ranger is by far the most popular compact pickup in the North American market because with its Mazda clone it totals more than 400,000 units sold. Because it is very conventional, you have to study it carefully to understand the buying motivation.

MAZDA B-Series

Manufactured by Ford, these identical pickups differ by their appearance, trim and equipment. They exist in 2- and 4-wheel drive with a short and long pickup bed and standard or extended cab. At Ford the trims are S, Sport, Custom, XLT and STX and at Mazda B2300, B3000, B4000 in base or SE. Three engines: a 4-cylinder 2.3L, a V6 3.0 and 4.0L. Four cylinders are standard on the 4X2 and the 4.0 L V6 on the 4X4. The 5-speed manual transmission is standard and the 4-speed automatic is optional. Rear wheel ABS braking is standard on all models.

STRONG POINTS

• Safety: **80%**
The presence of two air bags, reinforcing bars in the doors and improvement in structural integrity under impact give good occupant protection.

• Satisfaction: **80%**
It remains high despite oil leaks from the differential, and problems with the computer modules that control fuel and ignition.

• Cargo space: **70%**
With 1,650 pounds maximum load capacity, these mini-trucks cannot pretend to substitute themselves for the F-Series but their tow capacity is superior to that of most of their Japanese competition. Width between the wheel wells does not allow for 4X8 sheets unless you install supports in the openings provided for this purpose.

• Price/equipment: **70%**
These trucks are not the least expensive in their category but their price is justified by a stout build and honest reliability. Also their equipment is correct, even on base models.

• Quality/fit/finish: **65%**

DATA

Category: rear-wheel drive or AWD compact pickups.
Class : utility

HISTORIC

Introduced in : 1983
Modified in : 1986: Supercab & V6 2.9L engine; 1990: V6 4.0L.
Made in: Louisville, Kentucky, Twin Cities, Minnesota, Edison, New Jersey, U.S.A.

RATINGS

Safety: 75 %
Satisfaction: 80 %
Depreciation: 55 %
Insurance: 9 % ($655)
Cost per mile: $ 0.36

NUMBER OF DEALERS

In U.S.A.: 5,200 Ford-Lincoln-Mercury

SALES ON THE AMERICAN MARKET

Model	1992	1993	Result
Ranger	247,777	340,184	+27.2 %
B-Series	49,686	47,207	-5.0 %

MAIN COMPETITORS

DODGE Dakota, CHEVROLET-GMC S-10 Sonoma, ISUZU, MITSUBISHI Mighty Max, NISSAN Hardbody, TOYOTA pickup & T100.

EQUIPMENT

FORD Ranger MAZDA B-Series	XL base	XL Sport SE	XLT	STX	Splash
Automatic transmission:	O	O	O	O	O
Cruise control:	O	O	O	O	O
Power steering:	O	S	S	S	S
ABS brake:	S	S	S	S	S
Air conditioning:	O	O	O	O	O
Air bag:	S	S	S	S	S
Leather trim:	-	-	-	-	-
AM/FM/ Radio-cassette:	O	O	O	O	O
Power door locks:	-	O	O	O	O
Power windows:	-	O	O	O	O
Tilt steering column:	O	O	O	O	O
Dual power mirrors:	O	O	O	O	O
Intermittent wipers:	S	S	S	S	S
Light alloy wheels:	O	O	O	O	S
Sunroof:	O	O	O	O	O
Anti-theft system:	O	O	O	O	O

S : standard; O : optional; -: not available

COLORS AVAILABLE

Exterior: Blue , Red , Gray , Mocha, Yellow, Green , Black, White.
Interior: Green , Red, Graphite, Blue, White, Black, Silver, Saddle Brown.

MAINTENANCE

First revision: 5,000 miles
Frequency: 6 months/ 5,000 miles
Diagnostic plug: Yes

WHAT'S NEW IN 1995 ?

-Redesigned grill and instrument panel.
-Air bag added to the passenger side.
-New interior trim styling.
-Seat belts height adjustable.

	ENGINE					TRANSMISSION		PERFORMANCE									
Model/ versions *: standard	Type / timing valve / fuel system	Displacement cu/in	Power bhp @ rmn	Torque lb.ft @ rpm	Compres. ratio	Driving wheels / transmission	Final ratio	Acceler. 0-62 mph s	Stand. 1/4 mile s	Stand. 5/8 mile s	Braking 62-0 mph ft	Top speed mph	Lateral acceler. G	Noise level dBA	Fuel economy mpg city	hwy	Gasoline type / octane
1)	L4* 2.3 SOHC-8-SFI	140	112 @ 4800	135 @ 2400	9.2 :1	rear - M5*	-	12.6	19.0	33.5	138	97	0.69	70	28.0	32.0	R 87
						rear - A4	-	13.2	19.4	34.2	148	94	0.69	69	27.0	30.0	R 87
2)	V6* 3.0 OHV-12-SFI	182	145 @ 4800	165 @ 3000	9.3 :1	rear - M5	-	9.5	17.3	31.2	157	100	0.71	68	23.0	30.0	R 87
						rear - A4	-	10.8	18.0	32.0	167	97	0.71	68	22.0	29.0	R 87
3)	V6* 4.0 OHV-12-SFI	245	160 @ 4200	220 @ 3000	9.0 :1	rear- M5	-	NA									
						rear - A4	-	NA									

1) * Std cab. 4x2, 4x4 & base cab. 2)* 4x4 auto et 4x4. 3) option all models.

FORD Ranger

The ruggedness and durability of these utility machines place them at the head of their category in domestic manufacture. An effort is still needed to achieve a finish quality approaching passenger car level.

• **Technical: 60%**
These trucks are built on a ladder-type steel frame with seven cross members (six on the 4X4) on which the steel body is retained. The front suspension, same in both traction types, is independent, with twin I-beams, while at the rear the rigid rear axle is leaf-spring mounted. ABS is only used on the rear wheels. On the 4X4s, front wheel engagement is controlled by a simple switch. Without a central differential, this part-time 4-wheel drive can only be used on slippery surfaces.

• **Access: 60%**
It is easy to enter even the 4X4 whose ground clearance is higher.

• **Suspension: 60%**
Softer on the 4X2 and stiffer on the 4X4, it offers acceptable comfort under the proper load.

• **Driver's compartment: 60%**
Visibility is better than the driving position because the bench seat will not move back far enough and the steering column is moved too far toward the driver. The instrument panel is complete and both the instruments and controls are well laid out except for the window switches which are placed too low.

• **Performance: 55%**
The base engine is recommended only to those who use utility vehicles as a second car. Solely the V6 allows carrying or pulling heavy loads and their efficiency is not economical. Luxury versions are equipped with a 4.0 L engine offering sports car performance.

• **Seats: 50%**
The bench or jump seats of the extended cab are uncomfortable and only bucket seats offer decent comfort.

• **Steering: 50%**
Well-calibrated, it is slightly vague in the center which complicates straight line driving on freeways, especially under cross winds.

• **Depreciation: 50%**
The resale value stays at a reasonable level, given the good reputation for strength and reliability of these vehicles. However, it varies with condition and previous use.

• **Fuel consumption: 50%**
Relatively economical with a 4-cylinder and stick shift, it reaches considerable numbers with the V6 engines depending on load and terrain.

WEAK POINTS

• **Handling: 20%**
As on most utility vehicles, the rear is unstable and begins to swing out at the slightest road imperfection. To this is added a rebound action on 4X4s shoed with big rubber. The long wheelbase adds stability at the expense of maneuverability.

• **Conveniences: 20%**
One cannot say that storage abounds since it is limited to a glove box.

• **Interior space: 40%**
The standard cab accepts only two people. Two children will fit at the back of an extended cab, but space there is extremely limited and useful mostly for baggage.

• **Noise level: 40%**
Reasonable with V6 engines, it is less comfortable with 4-cylinders which work hard and lack discretion.

• **Braking: 40%**
Normal stops don't create problems, but sudden ones in the rain produce bad surprises because the front wheels lack ABS, lock up fast and the path becomes uncertain.

• **Insurance: 40%**
The rate for trucks is higher even in the best case for the simplest models and double those of a small family car.

CONCLUSION

• **Overall rating: 53.0%**
The success of the Ranger truck and its Mazda clone did not happen by chance. Value is the keyword that explains this result.

SPECIFICATIONS & PRICES

Model	Version/Trim	Body/ Seats	Wheel-base in	Lgth x Wdthx Hght. inx inx in	Curb weight lb	Towing capacity std. lb	Susp. type ft/rr	Brake type ft/rr	Steering system	Turning diameter ft	Steer. turns nber	Fuel. tank gal.	Standard tire size	Standard powertrain	PRICES US. $ 1994
FORD		**Total warranty and antipollution: 3 years / 36,000 miles;corrosion perforation: 5 years/ unlimited.**													
Ranger 4X2	regular	2 dr. p-u.2	107.9	184.2x69.4x63.9	2908	1649	i/r	dc/dr/ABS	bal.	-	-	16.4	195/70R14	L4/2.3/M5	**$9,389**
Ranger 4X2	long	2 dr. p-u.2	113.9	196.1x69.4x64.0	2950	1649	i/r	dc/dr/ABS	pwr.bal.	-	-	16.4	195/70R14	L4/2.3/M5	**$11,111**
Ranger 4X2	Supercab	2 dr. p-u.2+2	125.1	198.2x69.4x64.1	3210	1550	i/r	dc/dr/ABS	pwr.bal.	-	-	19.8	195/70R14	L4/2.3/M5	**$12,100**
Ranger 4X4	regular	2 dr. p-u.2	109.4	184.3x69.4x67.4	3252	1530	i/r	dc/dr/ABS	pwr.bal.	-	-	16.4	215/75R15	L4/2.3/M5	**$13,858**
Ranger 4X4	long	2 dr. p-u.2	114.2	196.2x69.4x67.6	3278	1530	i/r	dc/dr/ABS	pwr.bal.	-	-	16.4	215/75R15	L4/2.3/M5	**$15,426**
Ranger 4X4	Supercab	2 dr. p-u.2+2	125.4	198.2x69.4x67.5	3556	1429	i/r	dc/dr/ABS	pwr.bal.	-	-	19.8	215/75R15	V6/3.0/M5	**$16,262**
MAZDA		**General warranty: 3 years / 36,000 miles;perforation corrosion: 5 years / 36,000 miles; antipollution: 6 years / 100, 000 miles.**													
B2300 4X2	regular	2 dr. p-u.2	107.9	184.2x69.4x63.9	2908	1649	i/r	dc/dr/ABS	bal.	-	-	16.4	195/70R14	L4/2.3/M5	**$9,390**
B2300 4X2	long	2 dr. p-u.2	114.0	196.1x69.4x64.0	2950	1649	i/r	dc/dr/ABS	pwr.bal.	-	-	16.4	195/70R14	L4/2.3/M5	**$12,020**
B3000 4X2	Cab Plus SE	2 dr. p-u.2+2	125.1	198.2x69.4x64.1	3210	1550	i/r	dc/dr/ABS	pwr.bal.	-	-	19.8	195/70R14	V6/3.0/M5	**$13,170**
B3000 4X4	regular	2 dr. p-u..2	109.4	196.2x69.4x67.6	3278	1530	i/r	dc/dr/ABS	pwr.bal.	-	-	16.4	215/75R15	L4/2.3/M5	**$14,435**
B3000 4X4	Cab Plus SE	2 dr. p-u.2+2	125.4	198.2x69.4x67.5	3556	1429	i/r	dc/dr/ABS	pwr.bal.	-	-	19.8	215/75R15	V6/3.0/M5	**$15,495**

See page 393 for complete 1995 Price List

Taurus - Sable

Champions in all categories...

Since 1986 the Taurus Sable team has exceeded all popularity standards to set itself at the head of the sales hit parade in North America for 2 years. For once, Ford took the chance to elbow its traditional clientele and has brilliantly won its bet.

While waiting for the 1996 Taurus/Sable, the 1995 remain the same in sedans and 4-door stations wagons with V6 3.6 L and 3.8 L and a 4-speed automatic in trims with GL, SE and LX for the Taurus and GS, LS for the Sable. The sporty SHO, a Ford exclusive, is offered with a double overhead cam of 3.0 L in a stick shift, and 3.2 L in an automatic.

STRONG POINTS

• Safety: 80%

In spite of the two air bags the driver area could use more impact protection as the stiffness of the body is beginning to be dated. ABS is only offered as an option, except on the SHO.

• Satisfaction: 80%

With time, it has reached a very enviable level which explains in part the sales success. These models inspire confidence.

• Insurance: 80%

Proportionally, they are not expensive to insure, especially the SHO which is quite a bargain.

• Trunk: 80%

Well-proportioned, it can accept much baggage, and only its high sill complicates the access.

• Access: 80%

There are no problems at the front seats, but large-bodied people will find it less easy to reach the rear seats through the narrower doors.

• Technical: 75%

The styling has not aged, but the aerodynamics are no longer up-to-date. The steel unit body receives an independant 4-wheel suspension mixed brakes and optional ABS. Four wheel discs and ABS are standard on the SHO.

• Driver's compartment: 70%

The most convenient driving position is not easy to find since the seat cushion design is not ideal. Visibility is satisfactory except toward the rear of the wagon where the rear window is narrow. The dashboard is ergonomic but analog instrumentation is preferable to the digital one which cannot be read in sunlight. The different controls are well laid out except for those of the radio, which are too low, and those on the armrest, which are invisible at night.

• Suspension: 70%

Its amplitude allows it to easily absorb larger defects, but it lacks the cushioning that is the charm of American cars - its shock settings have the firmness of a European car.

• Seats: 65%

Those of the SHO are exemplary because you can adjust lateral and lumbar support with considerable precision while the other versions lack sculpturing and the padding is on the firm side.

• Handling: 60%

Less supple than the average American suspension, the Taurus-Sable do not roll as much on models equipped with 15 inch wheels or

DATA

Category: front-wheel drive mid-size sedans and wagons.
Class : 6

HISTORIC

Introduced in : 1986
Modified in : 1988: V6 3.8L. 1989: SHO; 1991: body.
Made in: Atlanta, Georgie & Chicago, Illinois, U.S.A.

RATINGS

Safety: 80 %
Satisfaction: 78 %
Depreciation: 66 %
Insurance: 5.0 % ($585, SHO: $785)
Cost per mile: $ 0.45

NUMBER OF DEALERS

In U.S.A.: 5,200 Ford-Lincoln-Mercury

SALES ON THE AMERICAN MARKET

Model	1992	1993	Result
Taurus	409,751	360,448	-12.1 %
Sable	115,075	120,977	+4.9 %

MAIN COMPETITORS

BUICK Regal, CHEVROLET Lumina, DODGE Intrepid, EAGLE Vision, CHRYSLER Concorde, HYUNDAI Sonata, OLDSMOBILE Cutlass Supreme, PONTIAC Grand Prix, TOYOTA Camry.

EQUIPMENT

FORD Taurus	GL	LX	SE	SHO
MERCURY Sable	**GS**	**LS**	-	
Automatic transmission:	S	S	S	O
Cruise control:	O	O	O	O
Power steering:	S	S	S	S
ABS brake:	O	O	O	S
Air conditioning:	S	S	S	S
Air bag:	S	S	S	S
Leather trim:	O	O	O	S
AM/FM/ Radio-cassette:	O	O	O	S
Power door locks:	-	S	S	S
Power windows:	O	S	S	S
Tilt steering column:	S	S	S	S
Dual power mirrors:	S	S	S	S
Intermittent wipers:	S	S	S	S
Light alloy wheels:	-	O	S	S
Sunroof:	-	O	O	O
Anti-theft system:	O	O	O	O

S : standard; O : optional; -: not available

COLORS AVAILABLE

Exterior: Mocha, Silver, Plum, Blue, Green, Black, Emerald.
Interior: Opaque Gray, Chrystal blue, Cranberry, Mocha, Black.

MAINTENANCE

First revision: 5,000 miles
Frequency: 6 months/ 5,000 miles
Diagnostic plug: Yes

WHAT'S NEW IN 1995 ?

-New SE version designed to compete with the imports.
-Low fuel warning standard on a GL.
-Improved 3.0 L engine and AX4N transmission.

	ENGINE					TRANSMISSION		PERFORMANCE										
Model/ versions *: standard	Type / timing valve / fuel system	Displacement cu/in	Power bhp @ rmn	Torque lb.ft @ rpm	Compres. ratio	Driving wheels / transmission	Final ratio	Acceler. 0-62 mph s	Stand. 1/4 mile s	Stand. 5/8 mile s	Braking 62-0 mph ft	Top speed mph	Lateral acceler. G	Noise level dBA	Fuel economy mpg city	Fuel economy mpg hwy	Gasoline type / octane
1)	V6* 3.0 OHV-12-SFI	182	140 @ 4800	165 @ 3250	9.3 :1	front - A4	3.37	11.8	18.2	36.5	138	109	0.76	67	22.5	38.0	R 87
2)	V6* 3.0 DOHC-24-SFI	182	220 @ 6200	200 @ 4800	9.8 :1	front - M5	3.74	7.5	15.5	28.9	148	137	0.82	67	22.0	35.0	S 91
3)	V6* 3.2 DOHC-24-SFI	195	220 @ 6000	215 @ 4800	9.8 :1	front - A4	3.77	8.4	16.0	29.2	157	131	0.82	67	21.0	35.0	S 91
4)	V6* 3.8 OHV-12-SFI	232	140 @ 3800	215 @ 2200	9.0 :1	front - A4	3.19	9.6	17.0	34.1	141	112	0.76	68	22.5	35.5	R 87

1) * base 2) * SHO manual 3) *SHO Automatic, opt. SHO manual 4) option except SHO.

the SHO sport suspension. The traction under power of the SHO is sometimes affected by a tight turn and strong accelerations provoke high torque steer.

• **Conveniences: 60%**
Insufficient on a GL they are more numerous on the LX and the SHO. The rear of the station wagon, which can be set up in picnic fashion, is an interesting gadget.

• **Performance: 60%**
The 3.8 L engine is the best adapted to the Taurus Sable, especially for those who travel with a full load or tow. The transmission is efficient but does not provide sufficient engine braking. The SHO constitutes an interesting high performance family sedan. Its automatic transmission version, introduced last year with a 3.2 L V6, considerably widens the buyer group for this model.

• **Quality/fit/finish: 60%**
Their level is very honest and the assembly gives an impression of sturdiness. Also fit and finish receive greater care. However, some rattles are still there.

• **Steering: 60%**
It is precise, with a good ratio, but the assist is too high which makes it sensitive. Also, the large turning diameter does not help maneuverability.

• **Noise level: 50%**
There is still progress to be made in terms of soundproofing, because for a car of this class, one still hears much rolling noise.

• **Interior space: 50%**
It is difficult to consider these cars as 6-seaters. Three people will not be comfortable on a bench seat in the front or in the rear. Four adults or 2 adults and 3 children would be comfortable.

• **Price/equipment: 50%**
Slighly more costly than their GM or Chrysler rivals, these models have the advantage of a station wagon and full equipment in LX and LS.

FORD Taurus

MERCURY Sable

WEAK POINTS

• **Braking: 40%**
The braking distances are long, pad life is average and stability is dubious without ABS on all models, except the SHO which handles in a very safe fashion during unforeseen stops.

• **Depreciation: 40%**
It remains higher than average because of the large number of vehicles on the market.

• **Fuel consumption: 45%**
It is average for its category but is higher than for the last, most spacious Chrysler LH.

CONCLUSION

• **Overall rating: 62.75%**
In spite of the lively competition in this category, the Taurus-Sable have every chance of making it through 1995 without any problem, while waiting for the unveiling of the following models which will replace them. ☺

OWNERS' SUGGESTIONS

- More powerful brakes and standard ABS
- More lively but economic engines (except SHO)
- Higher comfort level
- Improved tires
- Better finish

SPECIFICATIONS & PRICES

Model	Version/Trim	Body/ Seats	Interior volume cu.ft.	Trunk volume cu ft.	Drag coef. Cd	Wheel-base in	Lgth x Wdthx Hght. inx inx in	Curb weight lb	Towing capacity std. lb	Susp. type ft/rr	Brake type ft/rr	Steering system	Turning diameter ft	Steer. turns nber	Fuel. tank gal.	Standard tire size	Standard powertrain	PRICES US. $ 1994
FORD		**Total warranty and antipollution: 3 years / 36,000 miles;corrosion perforation: 5 years / unlimited.**																
Taurus	GL	4dr.sdn.5	103	18	0.32	106	192.0x71.1x54.1	3117	2 000	i/i	dc/dr	pwr.r&p.	38.6	2.7	16	205/65R15	V6/3.0/A4	**$16,140**
Taurus	GL	4dr.wgn.5	104	38	0.36	106	193.1x71.1x55.5	3285	2 000	i/i	dc/dr	pwr.r&p.	38.6	2.7	16	205/65R15	V6/3.0/A4	**$17,220**
Taurus	SE	4dr.sdn.5	103	18	0.32	106	192.0x71.1x54.1	3117	2 000	i/i	dc/dr	pwr.r&p.	38.6	2.7	16	205/65R15	V6/3.8/A4	**NA**
Taurus	LX	4dr.sdn.5	103	18	0.32	106	192.0x71.1x54.1	3192	2 000	i/i	dc/dr	pwr.r&p.	38.6	2.7	16	205/65R15	V6/3.0/A4	**$18,785**
Taurus	LX	4dr.wgn.5	103	38	0.36	106	193.1x71.1x55.5	3388	2 000	i/i	dc/dr	pwr.r&p.	38.6	2.7	16	205/65R15	V6/3.8/A4	**$20,400**
Taurus	SHO	4dr.sdn.5	103	18	0.32	106	192.0x71.1x54.1	3307	2 000	i/i	dc/dc	pwr.r&p.	40.0	2.5	18.4	215/60R16	V6/3.0/M5	**$24,715**
MERCURY		**Total warranty and antipollution: 3 years / 36,000 miles;corrosion perforation: 5 years / unlimited.**																
Sable	GS	4dr.sdn.5	100	18	0.33	106	192.2x71.1x54.1	3144	2 000	i/i	dc/dr	pwr.r&p.	38.6	2.7	16	205/65R15	V6/3.0/A4	**$17,740**
Sable	GS	4dr.wgn.5	104	38	0.36	106	193.2x71.1x55.5	3291	2 000	i/i	dc/dr	pwr.r&p.	38.6	2.7	16	205/65R15	V6/3.0/A4	**$18,900**
Sable	LTS	4dr.sdn.5	100	18	0.33	106	192.2x71.1x54.1	3126	2 000	i/i	dc/dr	pwr.r&p.	38.6	2.7	16	205/65R15	V6/3.8/A4	**NA**
Sable	LS	4dr.sdn.5	100	18	0.33	106	192.2x71.1x54.1	3186	2 000	i/i	dc/dr	pwr.r&p.	38.6	2.7	16	205/65R15	V6/3.0/A4	**$20,000**
Sable	LS	4dr.wgn.5	104	38	0.36	106	193.2x71.1x55.5	3331	2 000	i/i	dc/dr	pwr.r&p.	38.6	2.7	16	205/65R15	V6/3.0/A4	**$21,100**

See page 393 for complete 1995 Price List

Thunderbird - Cougar XR7

Specialized...

These coupes of purely American inspiration still reach sales of over 120,000 units per year. Their format remains difficult to export because it implies fuel hungry drive trains. The not-really sporty handling gains little driving pleasure.

MERCURY Cougar XR7

Starting from identical platforms and drive train, these two coupes offer different and opposite faces. The first is sold in versions LX and SC (Super Coupe) when the second is uniquely the XR7. The naturally-aspirated 3.8L V6 is in a base Cougar and T-Bird while the SC is fitted with a Roots blower. The 4.6 L V8 is offered as an option on the LX and XR7.

STRONG POINTS

• **Safety:** 90%
The very stiff structure is good at absorbing impacts and the two front seat air bags available as standard provide good occupant protection.

• **Satisfaction:** 80%
The T-Bird and Cougar are not problem cars and the defects which plagued the first buyers have been corrected. The exhaust system remains fragile, but 85% of the owners would again make the same purchase.

• **Insurance:** 75%
The premium on these cars is not as high as that of other spicier and more exotic sports machines.

• **Driver's compartment:** 70%
Seated low, the driver sees better from the T-Bird than from the Cougar which has a massive C-pillar. The appearance of the dashboard has been improved since last year by the inclusion of a more ergonomic center console. The instruments are legible and more practical.

• **Handling:** 70%
The 4-wheel independent suspension on the base model delivers safe handling but no pleasure due to average-quality tires. With its adjustable suspension and larger stabilizer bars, the SC corners flatter but

DATA

Category: mid-size rear-wheel drive coupes.
Class : 6

HISTORIC

Introduced in : 1955
Modified in : 1988: actual model;1991: V8 5.0L engine.
Made in: Lorain, Ohio, U.S.A.

RATINGS

Safety:	90 %
Satisfaction:	80 %
Depreciation:	56 %
Insurance:	5.5 % ($720)
Cost per mile:	$ 0.45

NUMBER OF DEALERS

In U.S.A.: 5,200 Ford-Lincoln-Mercury

SALES ON THE AMERICAN MARKET

Model	1992	1993	Result
T-Bird	84,186	122,415	+31.3 %
Cougar	54,557	73,362	+25.7 %

MAIN COMPETITORS

BUICK Regal, CHEVROLET Lumina & Camaro, OLDSMOBILE Cutlass Supreme, PONTIAC Grand Prix & Firebird.

EQUIPMENT

FORD Tunderbird MERCURY Cougar	LX XR7	SC
Automatic transmission:	S	O
Cruise control:	S	S
Power steering:	S	S
ABS brake:	O	S
Air conditioning:	S	S
Air bag:	S	S
Leather trim:	-	S
AM/FM/ Radio-cassette:	S	S
Power door locks:	S	S
Power windows:	S	S
Tilt steering column:	S	S
Dual power mirrors:	S	S
Intermittent wipers:	S	S
Light alloy wheels:	O	S
Sunroof:	O	O
Anti-theft system:	O	O

S : standard; O : optional; -: not available

COLORS AVAILABLE

Exterior: Green, Red, Plum, Mocha, Blue, Black, Gray, Silver, White, Pink Iris, Jewel Green.
Interior: Titanium, Crystal Blue, Red Burry, Mocha, Black.

MAINTENANCE

First revision:	5,000 miles
Frequency:	6 months/ 5,000 miles
Diagnostic plug:	Yes

WHAT'S NEW IN 1995 ?

-A wider selection of exterior colors.
-Steering ratio varies with speed but is only installed on the V8 and 4.6L V6.

Model/ versions *: standard	ENGINE Type / timing valve / fuel system	Displacement cu/in	Power bhp @ rmn	Torque lb.ft @ rpm	Compres. ratio	TRANSMISSION Driving wheels / transmission	Final ratio	PERFORMANCE Acceler. 0-62 mph s	Stand. 1/4 mile s	Stand. 5/8 mile s	Braking 62-0 mph ft	Top speed mph	Lateral acceler. G	Noise level dBA	Fuel economy mpg city	hwy	Gasoline type / octane
base	V6* 3.8 OHV-12-SFI	232	140 @ 3800	215 @ 2400	9.0 :1	rear -A4*	3.27	11.5	18.1	32.8	157	106	0.78	66	22.5	33.0	R 87
SC/XR7	V6* 3.8C OHV-12-SFI	232	230 @ 4400	330 @ 2500	8.5 :1	rear -M5*	2.73	7.6	15.7	29.2	167	137	0.82	66	21.0	32.0	S 91
						rear -A4	3.27	8.8	16.4	31.0	164	131	0.82	66	21.0	32.0	S 91
option	V8* 4.6 OHV-16-SFI	279	205 @ 4500	265 @ 3200	9.0 :1	rear -A4	3.08	9.0	16.8	30.3	180	115	0.78	66	21.0	32.0	R 87

its weight and bulk deprive it of agility and spontaneity.

• **Steering: 70%**
It is precise and fairly fast with good control over the power assist. Maneuverability is penalized by the size of the wheelbase and the turning diameter.

• **Suspension 70%**
It is more comfortable on base models than on the SC which totally feeds back road conditions.

• **Noise level: 70%**
Most mechanical sounds are snuffed out by the insulation but those caused by wind and rolling dominate.

• **Technical: 65%**
These large coupes have a steel unit body with independent suspension at all four wheels. The brakes are mixed disc and drum on base models but the four discs mounted standard on the SC are an option on the XR7 Cougar. On the SC, the shock damping is adjustable set in "automatic" or firm depending on the driving style. Body proportions are harmonious, with efficient lines and a favorable drag coefficient but total drag is influenced by a large frontal area.

• **Performance: 60%**
You need the supercharged V6 or the V8 power to liven up these heavy duty pre-historic Mastodons. Too bad that Ford did not deem it useful to replace the SC engine with a DOHC, 32-valve, 4.6L V8 which equips the Mark VIII and develops nearly 300 hp.

• **Trunk: 60%**
Its volume is sufficient but it lacks height, especially if the sound amplifier or cassette changer is going to remain there.

• **Quality/fit/finish: 60%**
There is no doubt about the rugged build of these models but some of the finish and appearance details, as well as the quality of some materials, leave to be desired.

• **Seats: 60%**
Well-padded, they provide a good grip but on the base models lateral and lumbar supports are not as effective as on the SC whose seat side wings are motorized.

• **Conveniences: 55%**
Storage consists of door pockets, a center console bin and a glove box of good volume.

• **Price/equipment: 50%**
The price of these coupes is not unreasonable if you are content with a base motor with a V8 engine, since the original equipment is interesting. The SC addresses itself to a buying group which is more interested in image than in the pure performance that a Taurus SHO can deliver for nearly the same price, together with, as a bonus, some obvious practical benefits.

• **Interior space: 50%**
It has nothing to do with the volume of these vehicles. The front seats are very comfortable but head and legroom is lacking at the back of the ThunderBird where the roof slopes more steeply than that of a Cougar.

WEAK POINTS

• **Braking: 30%**
Their efficiency and life are mediocre. The discs and drums can't handle the weight and the front wheels block rapidly in panic stops. Braking is more effective and stable with 4 discs and ABS offered as standard in the SC which makes it safer to use.

• **Fuel consumption: 30%**
Because of the weight and bulk of these vehicles, their efficiency has nothing economical. The naturally aspirated V6 is less thirsty, but the supercharged version drinks as much as the V8.

• **Access: 40%**
It's difficult to settle into the the back seats, which lack space, and the long heavy doors are hard to close.

• **Depreciation: 45%**
It is relatively strong because the loss of new car sales has dragged down that of the new cars which lose more than average.

CONCLUSION

• **Overall rating: 60.0%**
The more time goes by and the more those models are suffering from their oversize shape without mentioning the poor efficiency and the lack of pleasure when driving. ☺

FORD Thunderbird

MERCURY Cougar XR7

OWNERS' SUGGESTIONS

- Manual 5-speed option on base models.
- More favorable axle ratio for acceleration.
- Seat with better support.
- Lower fuel consumption.
- Better balanced braking.
- More realistic format, weight and price.
- Better performance.

SPECIFICATIONS & PRICES

Model	Version/Trim	Body/ Seats	Interior volume cu.ft.	Trunk volume cu ft.	Drag coef. Cd	Wheel-base in	Lgth x Wdthx Hght. inx inx in	Curb weight lb	Towing capacity std. lb	Susp. type ft/rr	Brake type ft/rr	Steering system	Turning diameter ft	Steer. turns nber	Fuel. tank gal.	Standard tire size	Standard powertrain	PRICES US. $ 1994
FORD			**Total warranty and antipollution: 3 years / 36,000 miles;corrosion perforation: 5 years / unlimited.**															
Thunderbird	LX	2dr.cpe.5	101	15	0.31	113	200.3x72.7x52.5	3750	2 000	i/i	dc/dr	pwr.r&p.	36.6	2.75	18	205/70R15	V6/3.8/A4	**$16,830**
Thunderbird	LX	2dr.cpe.5	101	15	0.31	113	200.3x72.7x52.5	3794	2 000	i/i	dc/dr	pwr.r&p.	36.6	2.75	18	205/70R15	V8/4.6/A4	**NA**
Thunderbird	SC	2dr.cpe.5	101	15	0.34	113	200.3x72.7x53.0	3761	2 000	i/i	dc/dc/ABS	pwr.r&p.	36.6	2.75	18	225/60ZR16	V6C/3.8/M5	**$22,240**
MERCURY			**Total warranty and antipollution: 3 years / 36,000 miles;corrosion perforation: 5 years / unlimited.**															
Cougar	XR7	2dr.cpe.5	102	15	0.36	113	199.8x72.7x52.5	3563	2 000	i/i	dc/dr	pwr.r&p.	36.6	2.75	18	205/70R15	V6/3.8/A4	**$16,260**
Cougar	XR7	2dr.cpe.5	102	15	0.36	113	199.8x72.7x52.5	3626	2 000	i/i	dc/dr	pwr.r&p.	36.6	2.75	18	205/70R15	V8/4.6/A4	**NA**

See page 393 for complete 1995 Price List

A separate niche...

Pressed to put on the market a product more competitive than the Aerostar or the Access in an attempt to damn up the success of the Chrysler mini-vans, Ford and Nissan have generated this hybrid which has created a market niche more car than utility vehicle...

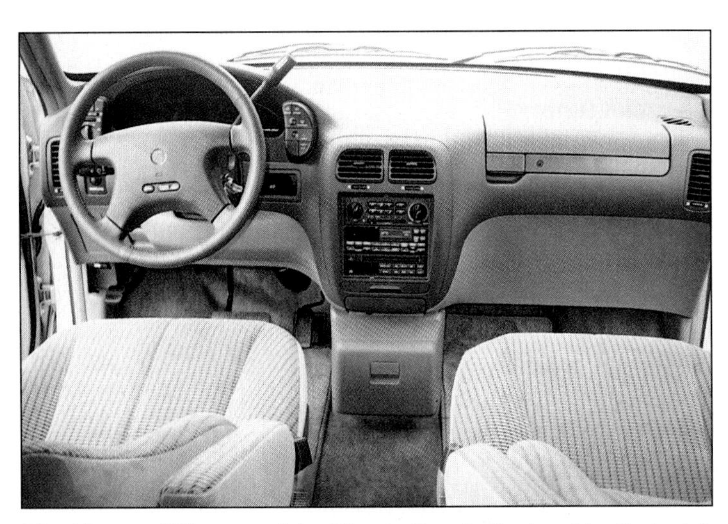

The Mercury Villager and the Nissan Quest offer two identical mini-vans with minor exceptions in detail and equipment. The short wheelbase Villager has been seconded by a long wheelbase Windstar to offer a product range almost equivalent to that of Chrysler. The Villager is offered in GS and LS trims, while the Quest comes in XE and GXE. Both are equiped with a 3.0 L V6 Nissan and a 4 speed automatic transmission with electronic computer control.

STRONG POINTS

• Safety: 90%
The very rigid structure, unfortunately, imposes a high weight penalty and cuts performance. Only one air bag is installed on the driver's side but motorized seat belts will remain until a second air bag has been installed up front. ABS is standard on all models except the Quest XE.

• Insurance: 85%
From this point of view the Villager-Quests are bargains because the premium is not higher than that of other vans which are less costly but have higher rates.

• Satisfaction: 80%
Some youthful, growing-up problems existed in 1993 centered mostly on the engine, fuel injection, transmission, body and some equipment. However, 80% of the owners polled said they were ready to make the same purchase again.

• Quality/fit/finish: 80%
Japanese participation is no stranger to the tight assembly specifications and materials quality which give the Villager-Quest a flattering and finished appearance as well as a strong impression of durability.

• Suspension: 80%
Thanks to a well-calculated wheel travel it efficiently absorbs road defects and oddly, it is the front which sounds loudest when crossing tar strips.

DATA

Category: rear-wheel drive passenger vans.
Class : utility

HISTORIC

Introduced in : 1993
Modified in : -
Made in: Avon Lake, Ohio, U.S.A.

RATINGS

Safety:	90 %
Satisfaction:	82 %
Depreciation:	30 % (2 years)
Insurance:	4.5 % ($585)
Cost per mile:	$ 0.42

NUMBER OF DEALERS

In U.S.A.: 5,200 Ford-Lincoln-Mercury, 1,100 Nissan

SALES ON THE AMERICAN MARKET

Model	1992	1993	Result
Villager	-	71,567	
Quest	-	44,602	

MAIN COMPETITORS

CHEVROLET Astro, CHRYSLER Town & Country, DODGE Caravan, DODGE Colt wagon, EAGLE Summit, FORD Aerostar, PLYMOUTH Voyager, MAZDA MPV, NISSAN Axxess, TOYOTA Previa.

EQUIPMENT

MERCURY Villager	GS	LS	Nautica		
NISSAN Quest			XE	GXE	
Automatic transmission:	S	S	S	S	S
Cruise control:	O	S	S	S	S
Power steering:	O	S	S	S	S
ABS brake:	S	S	S	S	S
Air conditioning:	S	S	S	S	S
Air bag:	S	S	S	S	S
Leather trim:	-	O	S	-	O
AM/FM/ Radio-cassette:	S	S	S	S	S
Power door locks:	-	S	S	S	S
Power windows:	-	S	S	S	S
Tilt steering column:	S	S	S	S	S
Dual power mirrors:	O	S	S	S	S
Intermittent wipers:	S	S	S	S	S
Light alloy wheels:	-	-	S	O	S
Sunroof:	-	O	O	-	O
Anti-theft system:	O	O	O	O	O

S : standard; O : optional; -: not available

COLORS AVAILABLE

Exterior: Silver, Black, Granite, White, Blue, Green, Brown, Red, Champagne, Bronze.
Interior: Gray, Beige, Blue, Garnet.

MAINTENANCE

First revision:	5,000 miles
Frequency:	6 months/ 5,000 miles
Diagnostic plug:	Yes

WHAT'S NEW IN 1995 ?

-New body colors and different combinations.

	ENGINE					TRANSMISSION		PERFORMANCE										
Model/ versions *: standard	Type / timing valve / fuel system	Displacement cu/in	Power bhp @ rmn	Torque lb.ft @ rpm	Compres. ratio	Driving wheels / transmission	Final ratio	Acceler. 0-62 mph s	Stand. 1/4 mile s	Stand. 5/8 mile s	Braking 62-0 mph ft	Top speed mph	Lateral acceler. G	Noise level dBA	Fuel economy mpg city	hwy	Gasoline type / octane
base	V6 3.0 SOHC-8-SEFI	181	151 @ 4800	174 @ 4400	9.0 :1	front -A4	3.861	12.0	18.6	34.7	148	112	0.75	68	21.0	30.0	R 87

damper units. The Villager-Quest enters curves readily. During the turn, they exhibit a certain amount of roll and an easily controlled understeer. Straight line stability is satisfactory. Both makes offer a more sporty suspension option.

• Noise level: 50%
Wind and rolling noises, as well as all the complaints from the hard working engine, maintain an average sound level.

• Depreciation: 50%
It is average and ranges about 30% for the second year.

• Performance: 40%
They are soft considering the high weight-to-power ratio and some passing maneuvers require considerable space. The last version of the Maxima engine which developes 190 hp instead of 151 improves this point.

• Price/equipment: 45%
These vans are expensive, but their standard equipment is relatively complete.

• Interior space: 45%
Thanks to a slightly longer wheelbase than their competitors, these vans theoretically offer more interior space. Unfortunately, the thickness of the seats reduces the useful volume and one feels cramped, particularly with captains seats.

• Conveniences: 80%
The hiding spaces are numerous, ample, and well-distributed through the interior. The backs of the rear seats form pallets when folded down and a net limits the displacement of small objects in the trunk area. For those who do not need 7 seats, an ingenious system allows them to be rearranged in 14 different ways. The last seat bench is installed on two rails which extend the full length of the cabin. This allows you to move it forward on demand and it can fold back on itself to take the least possible space.

• Access: 70%
It is easy to get in through the front or side doors and at the rear the hatchback is easily handled and has well-designed counter weighting. However, it is less easy to circulate between seats because the benches are very close to each other.

• Seats: 70%
Their padding is good but there is little wrap and they lack in both height and length. On the other hand, they are equipped with a headrest on all seats, sufficiently rare to be worth mentioning.

• Technical: 70%
The platform and the main drive train components are derived from the Maxima, but only the engine and transmision are imported from Japan. A McPherson transmission is used up front and a Hotchkiss at the rear which consists of a rigid axle and two leaf springs. Braking is mixed on all versions, except the Quest GXE which has 4 discs and a Bosch ABS. The ABS is standard at Ford Mercury but with rear brake drums.

•Driver's compartment: 65 %
The driver's seat does not offer sufficient lateral support but visibility is good and the rear-view mirrors are amply sized. The modern dashboard is fitted with complete and easy-to-read instrumentation, either digital or analog. On the other hand, ergonomics are not ideal because some switches on either side of the instrument panel are hard to reach and confusing, while the radio and AC controls are placed too low on the center console.

• Steering: 50%
It is vague at the center and its power assist is a bit strong and sensitive. The ratio is normal, but too large a turning diameter limits its maneuverability in spite of a short wheelbase.

• Handling: 50%
It is honest in spite of the simplicity of the rear suspension and of the softness of the spring-

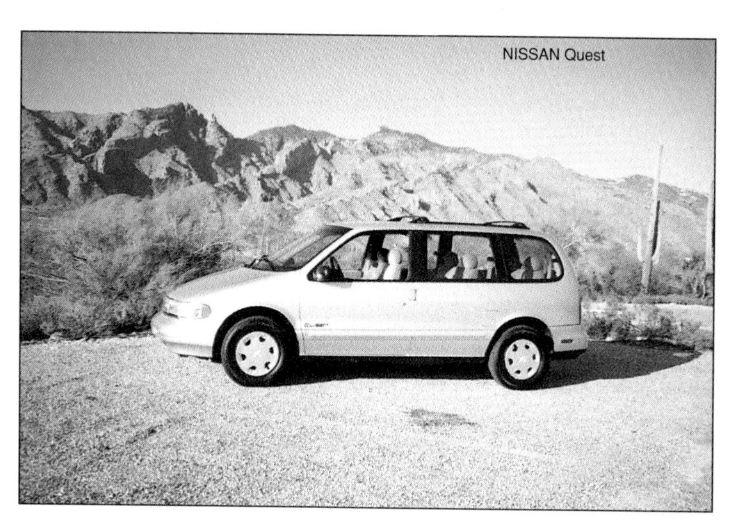

WEAK POINTS

• Fuel consumption: 30%
Reaching an average of 19 miles per gallon it is not particularly economical because the large body weight. and low engine power give it an unfavorable weight-to-power ratio of 27 lbs/hp.

• Trunk: 30%
When the seven seats are occupied, there is only enough room for two small suitcases behind the last bench.

• Braking: 40%
Its lack of efficiency and fade resistance in panic stops give long stopping distances which the ABS makes straight but the brake pads begin to smoke if one insists.

CONCLUSION

• Overall rating: 60%
Interesting, because of their original styling and design and their deluxe connotations, these vans seem to have discovered a new market niche where their prices match them to buyers who seek out spacious luxury rather than utility. :-|

SPECIFICATIONS & PRICES

Model	Version/Trim	Body/ Seats	Interior volume cu.ft.	Trunk volume cu ft.	Drag coef. Cd	Wheel-base in	Lgth x Wdthx Hght. inx inx in	Curb weight lb	Towing capacity std. lb	Susp. type ft/rr	Brake type ft/rr	Steering system	Turning diameter ft	Steer. turns nber	Fuel. tank gal.	Standard tire size	Standard powertrain	PRICES US. $ 1994
MERCURY			Total warranty & antipollution: 3 years / 36,000 miles;corrosion perforation: 5 years / unlimited.															
Villager	GS	4dr. van. 7	107.7	14	0.36	112.2	189.9x73.7x67.6	3990	3500	i/r	dc/dr/ABS	pwr.r/p.	40	3.0	20	205/75R15	V6/3.0/A4	**$18,375**
Villager	LS	4dr. van. 7	107.7	14	0.36	112.2	189.9x73.7x67.6	4074	3500	i/r	dc/dr/ABS	pwr.r/p.	40	3.0	20	205/75R15	V6/3.0/A4	**$23,155**
Villager	Nautica	4dr. van. 7	107.7	14	0.36	112.2	189.9x73.7x67.6	4074	3500	i/r	dc/dr/ABS	pwr.r/p.	40	3.0	20	205/75R15	V6/3.0/A4	**$24,635**
NISSAN			Total warranty & antipollution: 3 years / 36,000 miles;corrosion perforation: 5 years / unlimited.															
Quest	XE	4dr. van. 7	107.7	14	0.36	112.2	189.9x73.7x67.6	3871	3500	i/r	dc/dr	pwr.r/p.	40	3.0	20	205/75R15	V6/3.0/A4	**$18,529**
Quest	GXE	4dr. van.7	107.7	14	0.36	112.2	189.9x73.7x67.6	3964	3500	i/r	dc/dc/ABS	pwr.r/p.	40	3.0	20	215/70R15	V6/3.0/A4	**$23,039**

See page 393 for complete 1995 Price List

Ten years later...

Ten years after the arrival of Chrysler Minivans, Ford has finally created a product that is really competitive. However, in spite of its qualities, the Windstar already seems well behind the next generation that is about to make its debut.

Unveiled at the January 1994 Detroit show, the Windstar Mini-van will allow Ford to rank with honors in this category, which, with the Aerostar will not leave just good memories. Available in only one body, with a long wheelbase, it is offered in a GL version with a V6 3.0 L, or as an LX with a 3.8 L engine and a 4 speed automatic.

STRONG POINTS

• **Safety:** 90%
The rigid structure with numerous reinforcements is paid for by a high weight. The presence of air bags and standard ABS further enhances occupant protection.

• **Interior space:** 90%
The vast interior can accommodate 7 people on seats and benches in a conventional layout. Except for removing the benches, the interior is not modular, as in the Villager, which is too bad, because the system was original and clever. The back of the middle bench seat folds over to form a table on the LX version.

• **Insurance:** 90%
The Windstar is inexpensive to insure, increasing its appeal to large families, for whom budget is a major concern.

• **Access:** 90%
It's easy to get into, no matter which door is used, since the ground clearance is lower than average and the large doors have wide openings. However, there is too little space between the seats to allow moving around with ease.

• **Suspension:** 80%
Its comfort is more akin to that of a car rather than that of a utility, because it is smooth and well-dampened. It outclasses competitive models in terms of efficiency in that it never reacts violently to road imperfections.

• **Technical:** 75%
The makeup of the Windstar is far from revolutionary since its platform derives from that of the Taurus. Well-framed out, the body is rigid but heavy. A McPherson type front suspension was modified to allow a turning diameter one meter smaller than its competition. At the rear there are 2 helical springs and a drawn torsional axle that consists of 2 lengthwise leaves covered with a section that acts as an integral stabilizer bar. The ABS system has 3 channels and four sensors and the power steering is a version derived from the Lincoln Continental. While the lines offer no originality, they are harmonious and favorable from an aerodynamic view point with a 0.35 coefficient, comparable to that of many cars.

• **Quality/fit/finish:** 70%
General assembly is as careful as the finish, but some materials do not offer the best appearance. Carpets and fabrics are fluffy and look cheap.

DATA

Category: rear-wheel drive passenger vans.
Class : utility

HISTORIC

Introduced in : 1994
Modified in : 1995: GL V6 3.0L.
Made in: Oakville, Ontario, Canada.

RATINGS

Safety: 90 %
Satisfaction: NA
Depreciation: NA
Insurance: 4.0 % ($ 585)
Cost per mile: $ 0.42

NUMBER OF DEALERS

In U.S.A.: 5,200 Ford-Lincoln-Mercury

SALES ON THE AMERICAN MARKET

Model	1992	1993	Result
Windstar	New model		

MAIN COMPETITORS

CHEVROLET Astro & Lumina, DODGE-PLYMOUTH Caravan-Voyager, FORD Aerostar, PONTIAC Trans Sport, MAZDA MPV, MERCURY Villager, NISSAN Quest, TOYOTA Previa, VOLKSWAGEN Eurovan.

EQUIPMENT

FORD Windstar	GL	LX
Automatic transmission:	S	S
Cruise control:	O	S
Power steering:	S	S
ABS brake:	S	S
Air conditioning:	O	S
Air bag:	S	S
Leather trim:	-	-
AM/FM/ Radio-cassette:	O	O
Power door locks:	O	O
Power windows:	-	S
Tilt steering column:	O	O
Dual power mirrors:	O	O
Intermittent wipers:	S	S
Light alloy wheels:	S	S
Sunroof:	-	S
Anti-theft system:	O	O

S : standard; O : optional; - : not available

COLORS AVAILABLE

Exterior: Green, Plum, Indigo, Blue, White, Opal, Champagne, Bronze Crimson.
Interior: Royal blue, Opal gray, Mocha.

MAINTENANCE

First revision: 5,000 miles
Frequency: 6 months
Diagnostic plug: Yes

WHAT'S NEW IN 1995 ?

- **New model introduced during 1994.**
- **Base GL version with a 3.0 L V6 and simplified equipment.**

	ENGINE					TRANSMISSION		PERFORMANCE										
Model/ versions *: standard	Type / timing valve / fuel system	Displacement cu/in	Power bhp @ rmn	Torque lb.ft @ rpm	Compres. ratio	Driving wheels / transmission	Final ratio	Acceler. 0-62 mph s	Stand. 1/4 mile s	Stand. 5/8 mile s	Braking 62-0 mph ft	Top speed mph	Lateral acceler. G	Noise level dBA	Fuel economy mpg city	Fuel economy hwy	Gasoline type / octane
GL	V6*3.0 OHV-12-SFI	182	147 @ 5000	170 @ 3250	9.2 :1	front-A4	NA										
LX	V6*3.8 OHV-12-SFI	232	155 @ 4000	220 @ 3000	9.0 :1	front-A4	3.37	10.8	17.6	31.8	128	108	0.73	68	19.0	28.0	R 87

• Seats: 70%
The front ones are better designed and here the lateral and lumbar supports as well as the padding are excellent. At the rear, the backs and the cushions of the bench seats are short and there are no headrests.

• Storage: 70%
Enough space remains behind the last bench seat to accept a reasonable amount of baggage but by sliding it forward you can liberate more space.

• Driver's compartment: 60%
The steering column is too long which complicates the search for the best driving position and the dashboard is more aesthetic than ergonomic especially in what concerns the excessively low center console. Visibility is satisfactory but the rear windshield wiper is off center and does not have an intermittent action. Controls and gauges are typically Ford, easy-to-use, especially the well-located handbrake that takes little space.

• Handling: 60%
It is more stable in straight lines and in long curves. In tight cornerings a large roll interferes with handling. However, on bad pavement, or while crossing road joints, the rear axle does not jump around, as with most competitive models.

• Steering: 60%
Fast and well-modulated, it still lacks precision at the center but allows good maneuverability, even with the extended wheelbase.

• Braking: 50%
Its efficiency is remarkable because most stops can be carried out in distances under 131 ft with ABS which is well above average.

• Performance: 50%
They are more favorable with a 3.8 L than with the 3.0 L because of the high vehicle weight. Take-offs are crisper than passings because the upper transmission ratios are rather long.

• Price/equipment: 50%
While the base 3.0 L engine has almost exactly what it takes, the LX with a 3.8 L is better equipped and does not cost more than its Chrysler equivalent, which should help market penetration.

• Fuel consumption: 50%
With an average of 23 miles per gallon for the 3.8L, it is lower than that of its competition with the same displacement.

WEAK POINTS

• Conveniences: 30%
The glove box is not very large, and you find no door pockets. However the seat backs have bellows-style pockets and the LX uses a box between the seats.

• Noise level: 40%
As with most unit bodies, the sound level remains high and resonances travel through the interior, especially from the rear suspension.

CONCLUSION

• Overall rating: 62.25%
Too bad that the Windstar arrived so late because in a few months Chrysler will unveil its new model which, according to rumors, will age its competition by 10 years. By its philosophy, more than by its makeup, it addresses those who are in search of a large comfortable station wagon rather than a multiuse minivan. 😐

OWNER'S SUGGESTIONS

-Less anonymous body style
-More practical storage
-Bench seat headrests
-Less vague steering
-Intermittent rear-windshield wiper

SPECIFICATIONS & PRICES

Model	Version/Trim	Body/ Seats	Interior volume cu.ft.	Trunk volume cu ft.	Drag coef. Cd	Wheel-base in	Lgth x Wdthx Hght. inx inx in	Curb weight lb	Towing capacity std. lb	Susp. type ft/rr	Brake type ft/rr	Steering system	Turning diameter ft	Steer. turns nber	Fuel. tank gal.	Standard tire size	Standard powertrain	PRICES US. $ 1994
FORD		**Total warranty & antipollution: 3 years / 36,000 miles;corrosion perforation: 5 years / unlimited.**																
Windstar	GL	4dr. van. 7	-	16	0.35	118.3	201.1x75.4x68.0	3699	2000	i/r	dc/dr/ABS	pwr.r/p.	40.7	2.8	20	215/70R15	V6/3.0L/A4	**$19,590**
Windstar	LX	4dr. van. 7	-	16	0.35	118.3	201.1x75.4x68.0	3699	2000	i/r	dc/dr/ABS	pwr.r/p.	40.7	2.8	20	215/70R15	V6/3.8L/A4	**$23,760**

See page 393 for complete 1995 Price List

BUICK Century
OLDSMOBILE Ciera

Force of habit...

For the past dozen years these two models are part of the daily scene, so well integrated that you no longer see them. However, when going through them in detail, one notices that in spite of the good service they can still offer, they have gotten older.

The longevity of these 2 models testifies to their popularity, because in the last 12 years they have been sold with the same appearance. Even though their range has been extremely simplified, it continues to be offered in the form of sedans and 4-wheel wagons in Special or Custom trim for the Century and SL for the Ciera. The base engine has become a 2.2 L 4 cylinder, while the better-equipped versions benefit from the 3.1 L V6. The first is supplied with a 3-speed automatic and the second with an electronically controlled 4-speed.

STRONG POINTS

- **Satisfaction:** 80%
With the help of the years and some recall campaigns, their reliability has reached a satisfactory level but owners still complain of a lack of assembly care that generates small nuisance problems.
- **Interior space:** 70%
The inside volume of the Century/Ciera can accommodate, depending on the type of seats, up to 6 people while the wagons offer up to eight seats using the 3 benches. This constitutes, together with the price, the major attractions of these models.
- **Access:** 70%
It's easy to get into these cars whose doors are amply sized and open adequately.
- **Suspension:** 70%
The "Dynaride" Buick Century soaks up road defects better than that of the Ciera, which is too soft and reacts brutally to major road defects.
- **Trunk:** 70%
Huge and easily accessible, it's hard to use because the trunk floor is not flat which does not help with baggage layout. Storage space in the station wagon is immense, but you often have to load in height, which detracts from visibility.
- **Price/equipment:** 60%
While the base models are real work horses, the Custom and SL provide a more luxurious equipment and are more pleasant to use.
- **Insurance:** 60%
Considering their format, the rates for these models are not higher than that of smaller-sized vehicles and they are a buy for large families.
- **Fuel consumption:** 60%
In spite of their age, the 2 engines have excellent efficiency and their fuel consumption remains excellent compared to some of their more modern competitors.
- **Technical:** 60%
The steel unit body of these simplistic cars is outfitted with a McPherson suspension up front and a rigid axle at the back. The brakes are mixed but the ABS is standard. Very classical, their lines are not up-to-date, as confirmed by a drag coefficient of 0.38.
- **Noise level:** 60%
It is generally low at constant speed and only gets higher during fast

DATA

Category: mid-size front-wheel drive sedans and wagons.
Class : 6

HISTORIC

Introduced in : 1982
Modified in : -
Made in: Oklahoma City, U.S.A.

RATINGS

Safety: 80 %
Satisfaction: 82 %
Depreciation: 65 %
Insurance: 5.0 % ($ 585)
Cost per mile: $ 0. 45

NUMBER OF DEALERS

In U.S.A.: 3,000 Buick, 3,100 Oldsmobile.

SALES ON THE AMERICAN MARKET

Model	1992	1993	Result
Ciera/	117,292	143,699	+ 18.4 %
Century	114,274	116,034	+ 1.6 %

MAIN COMPETITORS

CHRYSLER Concorde-Intrepid, EAGLE Vision, FORD Taurus, GM W-Series: Regal-Lumina-Cutlass-Grand-Prix, HONDA Accord, HYUNDAI Sonata V6, MAZDA 626, MERCURY Sable, TOYOTA Camry V6.

EQUIPMENT

BUICK Century OLDSMOBILE Ciera	Special	Custom SL
Automatic transmission:	S	S
Cruise control:	S	S
Power steering:	S	S
ABS brake:	S	S
Air conditioning:	S	S
Air bag:	S	S
Leather trim:	-	-
AM/FM/ Radio-cassette:	O	S
Power door locks:	S	S
Power windows:	O	O
Tilt steering column:	S	S
Dual power mirrors:	O	S
Intermittent wipers:	S	S
Light alloy wheels:	O	O
Sunroof:	O	O
Anti-theft system:	O	O

S : standard; O : optional; - : not available

COLORS AVAILABLE

Exterior: Blue, White, Silver, Driftwood, Black, Gray, Cherry, Amethyst.
Interior: Blue, Mole, Red, Gray.

MAINTENANCE

First revision: 5,000 miles
Frequency: 6 months
Diagnostic plug: Yes

WHAT'S NEW IN 1995 ?

- 3.1 L V6 engine and 4-speed automatic standard on the Custom.
- Safety interlock between the brake and the shift lever.
- Dextron transmission lubricant good for 100,000 miles.
- New exterior/interior colors, new wheel covers.

ENGINE					TRANSMISSION		PERFORMANCE										
Model/ versions *: standard	Type / timing valve / fuel system	Displacement cu/in	Power bhp @ rmn	Torque lb.ft @ rpm	Compres. ratio	Driving wheels / transmission	Final ratio	Acceler. 0-62 mph s	Stand. 1/4 mile s	Stand. 5/8 mile s	Braking 62-0 mph ft	Top speed mph	Lateral acceler. G	Noise level dBA	Fuel economy mpg city	hwy	Gasoline type / octane
base	L4*2.2 SOHC-8-EFI	133	120 @ 5200	130 @ 4000	9.0 :1	front - A3	3.06	12.8	18.8	34.6	136	100	0.76	68	27.0	32.0	R 87
option	V6 3.1 OHV-12-MPFI	191	160 @ 5200	185 @ 4000	9.6 :1	front - A4*	2.97	11.5	17.6	33.3	144	112	0.78	66	25.0	31.0	R 87

acceleration or on bad pavement. Above the legal speed limit, wind noise sets in around the windshield and the side-view mirrors.

• **Safety: 60%**
There is an air bag on the driver's side, but the structure does not offer enough stiffness to resist impacts and the doors are not fitted with reinforcing beams. Fortunately, ABS stabilizes braking and the occupants receive 3-point seat belts.

• **Quality/fit/finish: 60%**
Even though they are not as strictly controlled as the latest models, assembly is good and the finish simple and ordinary. The new interior trim looks less low priced than in the preceding models but the plastics don't look rich at all.

• **Handling: 55%**
Buick's Dynaride suspension is better at controlling body roll and motions than the Oldsmobile that is really too soft. These cars primarily understeer when pushed into the turn and the front end feels heavy, the steering is vague, and the stock undersized rubber is overworked.

• **Steering: 50%**
Fast and precise, its assist is too high which makes it light and sensitive, and it requires a certain amount of attention to keep it going straight. Maneuverability is normal and matches the turning diameter.

• **Seats: 50%**
The lateral and lumbar supports on the new Century are better than those of the Ciera that have little shape. The padding, cushion sizes, and seat backs are adequate.

BUICK Century

BUICK Century

OLDSMOBILE Ciera SL

WEAK POINTS

• **Conveniences: 30%**
The storage space is reduced to its lowest common denominator which means a mini-glove box.

• **Depreciation: 35%**
These models resell fairly rapidly when they are well-equipped, but their resale value suffers from age and outmoded looks.

• **Performance: 40%**
Take-off and acceleration with a V6 engine allow a less frustrating use than the base 4 cylinder that is very anemic and has a weight/power ratio which is distinctly less favorable.

• **Braking: 40%**
It's assistance is too strong which complicates simple slow downs. Panic stops are more stable since the ABS has been installed as standard. While the stopping distances remain long, due to lack of bite at the break pads, their fade-resistance is normal.

CONCLUSION

• **Overall rating: 54.0%**
These models remain the choice of those who already know them and who don't need more for low cost travel, but want space and a certain comfort. These cars' main defects are their qualities... ☹

OWNER'S SUGGESTIONS

-More careful finish
-Better anti-corrosion protection
-Larger stabilizer bars on the Ciera
-Fewer oil leaks
-Avoid the 4 cylinder

SPECIFICATIONS & PRICES

Model	Version/Trim	Body/ Seats	Interior volume cu.ft.	Trunk volume cu ft.	Drag coef. Cd	Wheel-base in	Lgth x Wdthx Hght. inx inx in	Curb weight lb	Towing capacity std. lb	Susp. type ft/rr	Brake type ft/rr	Steering system	Turning diameter ft	Steer. turns nber	Fuel. tank gal.	Standard tire size	Standard powertrain	PRICES US. $ 1994
BUICK		**General warranty: 3 years / 36,000 miles; antipollution: 5 years / 50,000 miles; perforation corrosion: 6 years / 100,000 miles.**																
Century	Special	4dr. sdn.6	97	16	0.38	104.9	189.1x69.4x54.2	2985	2000	i/r	dc/dr	pwr.r/p.	38.4	3.04	16.5	185/75R14	L4/2.2/A3	**$15,800**
Century	Special	4dr.wgn.6/8	96	41	0.40	104.9	189.1x69.4x54.2	3128	2000	i/r	dc/dr	pwr.r/p.	38.4	3.04	16.5	185/75R14	L4/2.2/A3	**$16,650**
Century	Custom	4dr. sdn.6	97	16	0.38	104.9	189.1x69.4x54.2	2994	2000	i/r	dc/dr	pwr.r/p.	38.4	3.04	16.5	185/75R14	V6/3.1/A4	**$17,000**
OLDSMOBILE		**General warranty: 3 years / 36,000 miles; antipollution: 5 years / 50,000 miles; perforation corrosion: 6 years / 100,000 miles.**																
Ciera	SL	4dr. sdn.6	97	16	0.38	104.9	190.3x69.5x54.1	2930	2000	i/r	dc/dr	pwr.r/p.	38.0	3.05	16.5	185/75R14	L4/2.2/A3	**$16,070**
Ciera	Cruiser SL	4dr.wgn.6/8	96	41	0.37	104.9	190.3x69.5x54.1	3225	2000	i/r	dc/dr	pwr.r/p.	38.0	3.05	16.5	185/75R14	V6/3.1/A4	**$17,570**

See page 393 for complete 1995 Price List

Strongbox for the road...

In the past these big cars testified to the wealth of their owners and the health of the U.S. Times have changed and the symbol is not entirely the same. Today the Caprice is used as a fast police car for writing tickets. How to lower the taxes and fill the coffers...

Even after the passage of time, the lines of the Roadmaster and Caprice continue to disturb the public and Chevrolet has just finished touching up this model to lighten up the body appearance and improve visibility. Cadillac Fleetwood is available as a base 4-door sedan or a Brougham, derived from the same platform and drive train. The Caprice and Roadmaster are offered as 4-doors in Classic, LS for the Caprice sedan and a sports Impala SS. The Roadmaster can be a standard model or a Limited. As with Chevrolet, Buick wagons come in only one version.

STRONG POINTS

• Trunk: 100%
Its theoretical volume is enormous but it is not easy to make use of it with the tormented floor shape due to a spare tire enthroned in the center.

• Interior space: 90%
Not really proportional to the outside volume, it seats 6 people comfortably when there is a front bench seat.

• Safety: 80%
In spite of some reinforcements, the structure of these cars does not offer the best protection in case of collision. The presence of 2 air bags in the front seats, ABS and 3-point seat belt installed standard improves the score.

• Noise level: 80%
It is very low at cruise, particularly on the better insulated Fleetwood and Roadmaster. However, above 60 mph the road and tire noise gets higher.

• Satisfaction: 80%
Curiously the Buick rating is less flattering than that of the Caprice and Fleetwood which is very good.

• Insurance : 80%
They have the same rating and the premium varies with vehicle price

BUICK Roadmaster Estate

DATA

Category: full-size rear-wheel drive sedans and wagons.
Class : 6 & 7

HISTORIC

Introduced in : 1969
Modified in : 1977, 1990: Brougham, 1993: Fleetwood.
Made in: Arlington, Texas & Lakewood, GA, U.S.A.

RATINGS

Safety:	70 %	
Satisfaction:	80 % (75% Roadmaster)	
Depreciation:	65 %	60 % (Buick &Cadillac)
Insurance:	4.0 %($ 585-$ 654-$ 870)	
Cost per mile:	$ 0.54	$ 0.64 (Cadillac)

NUMBER OF DEALERS

In U.S.A.: 3,000 Buick, 4,466 Chevrolet-Geo,1,600 Cadillac.

SALES ON THE AMERICAN MARKET

Model	1992	1993	Result
Roadmaster	44,801	32,018	- 28.6 %
Fleetwood	-	27,372	-
Caprice	88,972	82,646	- 7.1 %

MAIN COMPETITORS

Caprice & Roadmaster: BUICK le Sabre, CHRYSLER New Yorker, FORD Crown Victoria / MERCURY Grand Marquis, OLDSMOBILE 88 PONTIAC Bonneville. **Fleetwood**: Lincoln Town Car. **Impala SS**: CHRYSLER LHS

EQUIPMENT

BUICK Roadmaster	base	Ltd		
CHEVROLET Caprice	Classic	LS		
CHEVROLET Impala			SS	
CADILLAC Fleetwood				base
Automatic transmission:	S	S	S	S
Cruise control:	-	S	S	S
Power steering:	S	S	S	S
ABS brake:	S	S	S	S
Air conditioning:	S	S	S	S
Air bag:	S	S	S	S
Leather trim:	-	-	S	S
AM/FM/ Radio-cassette:	O	S	S	S
Power door locks:	S	S	S	S
Power windows:	O	S	S	S
Tilt steering column:	S	S	S	S
Dual power mirrors:	S	S	S	O
Intermittent wipers:	S	S	S	S
Light alloy wheels:	-	O	O	S
Sunroof:	-	-	O	O
Anti-theft system:	S	S	S	S

S : standard; O : optional; - : not available

COLORS AVAILABLE

Exterior: Silver, Amethyst, Blue, Black, White, Yellow, Red, Gray, Cherry.
Interior: Gray, Blue, Beige, Red, White, Black, Antelope.

MAINTENANCE

First revision: 5,000 miles
Frequency: 6 months
Diagnostic plug: Yes

WHAT'S NEW IN 1995 ?

- New design of the quarter glass (Caprice/Impala).
- New outside rear-view mirrors, new colors for the body and trim.
- Dextron transmission oil good for 100,000 miles.
- New seats (Roadmaster).

Model/ versions *: standard	ENGINE Type / timing valve / fuel system	Displacement cu/in	Power bhp @ rmn	Torque lb.ft @ rpm	Compres. ratio	TRANSMISSION Driving wheels / transmission	Final ratio	PERFORMANCE Acceler. 0-62 mph s	Stand. 1/4 mile s	Stand. 5/8 mile s	Braking 62-0 mph ft	Top speed mph	Lateral acceler. G	Noise level dBA	Fuel economy mpg city	hwy	Gasoline type / octane
1)	V8* 4.3 OHV-16-SFI	265	200 @ 5200	235 @ 2400	9.9 :1	rear-A4*	2.93	11.0	17.8	32.0	141	112	0.78	65	19.0	30.0	R 87
2)	V8* 5.7 OHV-16-SFI	350	260 @ 4800	330 @ 3200	10.0 :1	rear-A4*	2.56	9.5	16.8	30.5	148	124	0.80	65	17.0	25.0	R 87
3)	V8* 5.7 OHV-16-SFI	350	260 @ 4800	330 @ 3200	10.5 :1	rear-A4*	3.08	7.5	15.6	28.5	131	136	0.82	66	18.0	26.0	R 87
4)	V8* 5.7 OHV-16-MPFI	350	260 @ 5000	335 @ 2400	10.5 :1	rear-A4*	2.56	10.4	17.6	32.3	157	112	0.76	64	18.0	26.0	R 87

1) base Caprice 2) base Roadmaster & Caprice Wagon, opt. Caprice. 3) Impala SS 4)Cadillac Fleetwood

CHEVROLET Impala SS

CADILLAC Fleetwood

and ranges between $585 and $850 American dollars.

• **Suspension: 75%**
Too soft, that of the Chevrolet afflicts the base Caprice with large body motions while that of the Fleetwood and Roadmaster is better coordinated and stable. That of the Impala SS (based on police option) represents, in our mind, the best compromise. This option is also available as an RPO (regular production option) for the Caprice. Unfortunately, the rigid rear axle reacts harshly to large road defects.

• **Access: 75%**
The doors are sufficiently long and the available space is ample but the door opening angle remains insufficient.

• **Performance: 70%**
The base 4.3 L V8 on the Caprice cannot equal the 5.7 L V8. The V8 confers surprising performance to these heavy cars. Take off and performance on the Impala SS are exotic enough to please pony car owners.

• **Quality/fit/finish: 60%**
The assembly procedures and finish of these models have not progressed sufficiently and some adjustments are approximate; materials quality is average.

• **Seats: 60%**
They provide better support on the Roadmaster, Impala SS and Fleetwood, thanks to their numerous adjustments, than on the base Caprice where they are insipid and uncomfortable.

• **Driver's compartment: 50%**
Short drivers find it difficult to get comfortable because of the steering column length. Visibility has improved for a 3/4 rear unlike the Roadmaster and Fleetwood. This year the outside rear-view mirrors are better-sized but the dashboards are short of instrumentation.

• **Technical: 50%**
The steel body is mounted on a perimeter chassis which dates from antiquity but provides superior sound and vibration proofing from the drive train. In spite of their size these models offer a satisfactory aerodynamic finesse (0.32). Front suspension is independent while the rear uses a rigid Salisbury axle controlled by a 4 bar linkage and helical springs. The Caprice base engine is a 4.3 L V8 with a 200 hp output while the 5.7 L 260 hp engine is standard on the Impala SS, Roadmaster and Fleetwood and is coupled to a 4-speed automatic. The ABS is standard but only the Impala SS has rear wheel discs.

• **Handling: 50%**
The "Dynaride" Roadmaster suspension delivers better results than that of the Caprice and Fleetwood. The higher roll and pitch on these models cause a lack of traction and aplomb in tight corners. The Impala SS does particularly well, thanks to its stouter and firmer springs and dampers as well as the larger diameter stabilizer bars.

WEAK POINTS

• **Steering: 40%**
Even though its assistance is lower than in the past, it remains sensitive and calls for a certain amount of attention, but its precision and directional stability are satisfactory. It does not inform much on the state of the road, and its feedback is insufficient. In spite of a moderate turning ratio and steering diameter their maneuverability in urban driving is delicate. Caprice taxi drivers do not seem to have this kind of a problem...

• **Braking: 40%**
The super power of its booster makes control a little touchy and the ABS kicks in too soon in panic stops so that you have to work hard to stabilize the large mass. Stopping distances are remarkable when you take the weight into account. The front end dives in exaggerated fashion and fade-resistance is mediocre.

• **Price/equipment: 40%**
Less expensive than the competition, if they were sold by weight, these cars are well-equipped, even those of the base Caprice whose equipment is now plusher.

• **Conveniences: 40%**
Storage is a little more generous than in the past but still not proportional to the bulk.

• **Fuel consumption: 40%**
Acceptable in normal use, it rises rapidly when the cars are loaded or towing heavy trailers.

• **Depreciation: 40%**
The base Caprice loses more of its value than the more prestigious and luxurious Roadmaster and Fleetwood.

CONCLUSION

• **Overall rating: 62.0%**
These models address themselves to a specific buyer group which does not want to accept the mini-van fad. They may offer space, comfort and comparable equipment for a saving in fuel and superior maneuverability, but big car fans are different.

SPECIFICATIONS & PRICES

Model	Version/Trim	Body/ Seats	Interior volume cu.ft.	Trunk volume cu ft.	Drag coef. Cd	Wheel-base in	Lgth x Wdthx Hght. inx inx in	Curb weight lb	Towing capacity std. lb	Susp. type ft/rr	Brake type ft/rr	Steering system	Turning diameter ft	Steer. turns nber	Fuel. tank gal.	Standard tire size	Standard powertrain	PRICES US. $ 1994
BUICK		**General warranty: 3 years / 36,000 miles; antipollution: 5 years / 50,000 miles; perforation corrosion: 6 years / 100,000 miles.**																
Roadmaster		4dr. sdn.6	114.7	21	0.33	115.9	215.8x78.1x55.9	4211	4978	i/r	dc/dr/ABS	pwr.r/p.	38.9	3.17	23	235/70R15	V8/5.7/A4	**$24,184**
Roadmaster Limited		4dr. sdn.6	114.7	21	0.33	115.9	215.8x78.1x55.9	4244	4978	i/r	dc/dr/ABS	pwr.r/p.	38.9	3.17	23	235/70R15	V8/5.7/A4	**$26,584**
Roadmaster Estate		4dr.wgn.6/8	115	54.7	0.35	115.9	217.5x79.9x60.3	5004	4978	i/r	dc/dr/ABS	pwr.r/p.	39.5	3.17	21	225/75R15	V8/5.7/A4	**$25,784**
CADILLAC		**General warranty: 4 years / 50,000 miles; antipollution: 5 years / 50,000 miles; perforation corrosion: 6 years / 100,000 miles.**																
Fleetwood base		4dr. sdn.6	124.8	21	0.36	121.5	225.1x78.0x57.1	4367	7000	i/r	dc/dr/ABS	pwr.bal.	42.6	3.2	23	235/70R15	V8/5.7/A4	**$33,990**
Fleetwood Brougham		4dr. sdn.6	124.8	21	0.36	121.5	225.1x78.0x57.1	4367	7000	i/r	dc/dr/ABS	pwr.bal.	42.6	3.2	23	235/70R15	V8/5.7/A4	**$35,670**
CHEVROLET		**General warranty: 3 years / 36,000 miles; antipollution: 5 years / 50,000 miles; perforation corrosion: 6 years / 100,000 miles.**																
Caprice	Classic	4dr. sdn.6	114.7	20.4	0.32	115.9	214.5x77.5x55.7	4061	4978	i/r	dc/dr/ABS	pwr.r/p.	39.9	3.17	23	215/75R15	V8/4.3/A4	**$19,153**
Caprice	Classic LS	4dr. sdn.6	114.7	20.4	0.32	115.9	214.5x77.5x55.7	4061	4978	i/r	dc/dr/ABS	pwr.r/p.	39.9	3.17	23	215/75R15	V8/4.3/A4	**$21,593**
Caprice	Wagon	4dr.wgn.6/8	115.4	54.7	0.35	115.9	217.3x79.6x60.9	4473	4978	i/r	dc/dr/ABS	pwr.r/p.	39.9	3.17	21	225/75R15	V8/5.7/A4	**$21,338**
Impala	SS	4dr. sdn.5	114.2	20.4	0.33	115.9	214.1x77.5x54.7	4037	4978	i/r	dc/dc/ABS	pwr.r/p.	43.9	3.06	23	255/50ZR17	V8/5.7.A4	**$21,098**

See page 393 for complete 1995 Price List

BUICK Park Avenue, CADILLAC DeVille, OLDSMOBILE 98

Threatened?

While Cadillacs still managed to maintain their status of typically American luxury cars, Buicks and Oldsmobiles seemed to loose ground to the unfurling of new models from Japan. Fortunately the high prices of the Japanese energetically brake their progress.

BUICK Park Avenue

These large 4-door sedans show once again that GM knows how to do the best. The H platform includes the Olds 98 Regency available as the Elite. The Buick Park Avenue comes in standard or Ultra, and the Cadillac De Ville in standard or Concours. This year it was Buick's right to a couple of aesthetic plus points while the others remained practically unchanged. A new, more powerful 3.8 L V6 is standard on the Park Ave and the 98.

STRONG POINTS

- **Safety:** 90%
While the Buick and Oldsmobile structures do not resist impacts as efficiently as the Cadillac, they benefit from a high rating, thanks to 2 air bags and a standard ABS.
- **Trunk:** 90%
Its huge capacity is not fully useful because of the inconvenient spare tire location and the complex floor shape.
- **Interior space:** 85%
These models offer ample space, particularly the De Ville where five or six people can travel at ease and interior dimensions are ample in all measurements.
- **Reliability:** 85%
With the time and the over abundance of attention that GM brought to some details, the rating of these cars reaches and passes that of some Japanese.
- **Insurance:** 80%
Even though the premium is ample, especially for the De Villes, their percentage is favorable compared to the price.
- **Suspension:** 80%
Its softness gently rocks the occupants during freeway travel but

DATA

Category: front-wheel drive luxury sedans.
Class : 7

HISTORIC

Introduced in : 1972:Park Avenue-98; 1970: De Ville.
Modified in : 1976:Park Avenue-98; 1984: De Ville.
Made in: Detroit-Hamtramck, MI (Cadillac) Lake Orion, MI (Oldsmobile), Wentzville, MO (Buick).

RATINGS

	Buick	Cadillac	Olds.
Safety:	90 %	90 %	90 %
Satisfaction:	82 %	86 %	87 %
Depreciation:	70 %	65 %	62.5 %
Insurance:	3.6 % ($ 720)	4.0 % ($ 925)	3.8 % ($ 720)
Cost per mile:	$ 0.52	$ 0.62	$ 0.52

NUMBER OF DEALERS

In U.S.A.: 3,000 Buick, 1,600 Cadillac, 3,100 Oldsmobile.

SALES ON THE AMERICAN MARKET

Model	1992	1993	Result
Park Avenue	59,836	59,533	- 0.5 %
De Ville	127,766	115,650	- 9.5 %
98	35,597	23,072	- 25.2 %

MAIN COMPETITORS

CHRYSLER New Yorker & LHS, LINCOLN Continental.

EQUIPMENT

	BUICK Park Avenue base	BUICK Park Avenue Ultra	CADILLAC De Ville base	CADILLAC De Ville Concours	OLDSMOBILE 98 Elite
Automatic transmission:	S	S	S	S	S
Cruise control:	S	S	S	S	S
Power steering:	S	S	S	S	S
ABS brake:	S	S	S	S	S
Air conditioning:	S	S	S	S	S
Air bag:	S	S	S	S	S
Leather trim:	O	O	O	O	O
AM/FM/ Radio-cassette:	S	S	S	S	S
Power door locks:	S	S	S	S	S
Power windows:	S	S	S	S	S
Tilt steering column:	S	S	S	S	S
Dual power mirrors:	S	S	S	S	S
Intermittent wipers:	S	S	S	S	S
Light alloy wheels:	S	S	S	S	S
Sunroof:	O	O	O	O	O
Anti-theft system:	S	S	S	S	S

S : standard;-O : optional; - : not available

COLORS AVAILABLE

Exterior: Gray, Mocha,Green, Blue, White, Black, Cherry, Beige, Red, Amethyst.
Interior: Graphite, Beige, Red, White, Black, Neutral, Mole, Light Gray.

MAINTENANCE

First revision: 3,000 miles
Frequency: 6,000 miles
Diagnostic plug: Yes

WHAT'S NEW IN 1995 ?

- **New 205 hp 3.8 L V6 engine.**
- **Dextron automatic transmission fluid good for 100,000 miles.**
- **Cosmetic touch ups, front and rear (Park Avenue).**
- **Standard traction control (Cadillac).**
- **Aluminum rims, more powerful engine and new suspension adjustment (Concours).**

Model/ versions *: standard	ENGINE Type / timing valve / fuel system	Displacement cu/in	Power bhp @ rmn	Torque lb.ft @ rpm	Compres. ratio	TRANSMISSION Driving wheels / transmission	Final ratio	PERFORMANCE Acceler. 0-62 mph s	Stand. 1/4 mile s	Stand. 5/8 mile s	Braking 62-0 mph ft	Top speed mph	Lateral acceler. G	Noise level dBA	Fuel economy mpg city	Fuel economy hwy	Gasoline type / octane
1)	V6* 3.8 OHV-12-TPFI	231	205 @ 5200	230 @ 4000	9.4 :1	front - A4*	2.97	10.5	17.0	31.2	151	112	0.75	66	19.0	27.0	R 87
2)	V6* C 3.8 OHV-12-TPFI	231	225 @ 5000	275 @ 3200	8.5 :1	front - A4*	2.97	8.0	15.8	29.3	144	124	0.77	65	17.0	27.0	M 89
3)	V8* 4.6 DOHC-32-EFI	278	275 @ 5600	300 @ 4000	10.3 :1	front - A4*	3.11	8.4	16.0	8.5	138	124	0.75	65	16.0	25.0	R 87
4)	V8* 4.9 OHV-16-TPFI	300	200 @ 4100	275 @ 3000	9.5 :1	front - A4*	2.73	9.5	16.3	30.0	157	112	0.76	64	16.0	25.0	R 87

1) * Park Avenue, 98.   2) option 98 Elite, Ultra   3) * De Ville Concours   4) * De Ville

OLDSMOBILE 98 Regency Elite

jostles them harshly on small poorly maintained roads. De Villes with electronically controlled shock absorbers or those of the Elite which are a little firmer result in less pronounced body motions.

• Access: 80%
Long doors and seats with little contour allow even tall people to get in without contortions.

• Technical: 80%
All based on nearly identical platforms, these front wheel drives differ by a longer wheelbase and V8 engines on the de Ville. The 98 and Park Avenue are equipped with the new 3.8 L V6, which gives 30 hp more than the previous engine. On the 98 Elite there is an optional Roots type supercharger which gains an extra 20 hp and substantially more torque. The standard De Ville uses the 4.9 L, engine while the Concours and the 4.6 L Northstar deliver 270 hp. They have a 4-wheel independent suspension. The wheel camber of the De Ville varies with road condition and a control maintains a constant ground clearance. The brakes are mixed on the Buick and Oldsmobile and Ford discs on the De Ville and ABS is standard. Their steel unit body has enough aerodynamic finesse to reach a drag coefficient of 0.34.

• Performance: 70%
Thanks to the new V6, the takeoff and acceleration of the Park Avenue and Olds 98 are sharper and spicier with the supercharger. Even though the 4.9 L of the De Ville offers a favorable power-to-weight ratio (18.8 lbs/hp), the Northstar of the Concours gives it wings with 14.7 lbs/hp.

• Quality/fit/finish: 70%
While constant improvement can be seen, it is still not to the level of Japanese or German standards for fit or appearance of the materials.

• Noise level: 70%
Low in cruise while the engines are quiet, only tire slap on the road and wind noise beyond some speed disturb the quiet of this cocoon. The Northstar can be heard during acceleration but without any disagreeable results.

• Driver's compartment: 60%
Driver's seat does not offer a comfortable driver's position because of the lack of support and the length of the steering column. Digital instrumentation is very simple but distracting, and some controls behind the steering wheel are difficult to reach. Also, the steering wheel rim doesn't offer a good grip.

• Seats: 60%
In spite of generous padding, their lateral support is nil and lumbar support is only available as an option.

• Steering: 60%
Variable as a function of speed, its power assist is still a little too strong and calls for some attention. Guidance is precise and steering ratio normal, but in town maneuverability suffers from too large a turning diameter and sizeable outside car dimensions.

• Handling: 50%
Excessive roll and pitch in these models, except for the super-

CADILLAC De Ville Concours

charged Elite and Concours whose shocks absorber spring assembly are adjustable. They are also fitted with larger stabilizer bars.

WEAK POINTS

• Price/equipment: 20%
Expensive, but well-equipped, these cars have seen their sales drop due to strong competition. Only the Cadillac seems to offer enough class and tech content to justify its selection.

• Conveniences: 30%
Storage spaces are not numerous and consist most often of a glove box, a hiding space built into the center console and poorly accessible door pockets.

• Braking: 40%
ABS stabilized, the mixed brakes work hard, considering the major weight of these cars and 4-wheel discs on the Cadillac are not enough to shorten the stopping distances. In both instances, the fade- resistance is only average.

• Fuel consumption: 40%
It is only economical with a base V6 of the Buick-Oldsmobile because the supercharged V6's and the V8 are more thirsty. Particularly, the Northstar penalizes high performance with a heavy hand.

• Depreciation: 40%
While the Buick looses more value than the Cadillac, the Oldsmobile brings a better resale price with no apparent explanation.

CONCLUSION

• Overall rating: 65.5%
GM is beginning to find some competition in the segment where it reigned for a long time. Newcomers from many sources are a temptation for a demanding buyer group. ☺

SPECIFICATIONS & PRICES

Model	Version/Trim	Body/ Seats	Interior volume cu.ft.	Trunk volume cu ft.	Drag coef. Cd	Wheel-base in	Lgth x Wdthx Hght. inx inx in	Curb weight lb	Towing capacity std. lb	Susp. type ft/rr	Brake type ft/rr	Steering system	Turning diameter ft	Steer. turns nber	Fuel. tank gal.	Standard tire size	Standard powertrain	PRICES US. $ 1994
BUICK		**General warranty: 3 years / 36,000 miles; antipollution: 5 years / 50,000 miles; perforation corrosion: 6 years / 100,000 miles.**																
Park Ave.	base	4dr. sdn.6	109	20.2	0.34	110.7	205.9x74.0x55.1	3591	2000	i/i	dc/dr/ABS	pwr.r/p.	40	2.97	18	205/70R15	V6/3.8/A4	**$27,164**
Park Ave.	Ultra	4dr. sdn.6	109	20.2	0.34	110.7	205.9x74.0x55.1	3642	2000	i/i	dc/dr/ABS	pwr.r/p.	40	2.97	18	2125/70R15	V6C/3.8/A4	**$31,864**
CADILLAC		**General warranty: 4 years / 50,000 miles; antipollution: 5 years / 50,000 miles; perforation corrosion: 6 years / 160 000km.**																
De Ville	Sedan	4dr. sdn.6	117	20	0.35	113.8	209.7x76.6x56.3	3757	1000	i/i	dc/dc/ABS	pwr.r/p.	40.7	2.97	20	215/70R15	V8/4.9/A4	**$32,990**
De Ville	Concours	4dr. sdn.6	117	20	0.35	113.8	209.7x76.6x56.3	3984	1000	i/i	dc/dc/ABS	pwr.r/p.	40.7	2.97	20	225/60HR16	V8/4.6/A4	**$37,990**
OLDSMOBILE		**General warranty: 3 years / 36,000 miles; antipollution: 5 years / 50,000 miles; perforation corrosion: 6 years / 100,000 miles.**																
98	Regency Elite	4dr. sdn.5/6	109	20.1	0.34	110.7	205.7x74.6x54.8	3514	2000	i/i	dc/dr/ABS	pwr.r/p.	40	2.97	18	205/70R15	V6/3.8/A4	**$28,270**

See page 393 for complete 1995 Price List

BUICK Riviera

Intermediate "Bourgeois" coupe...

There is no crowd in this category of intermediate coupes of which the majority originates in America. Sufficiently luxurious without costing too much, they are performance-minded without being really sporty. The Riviera occupies a special niche where it is finally unique.

After a year's absence from the market, the Riviera coupe has returned with entirely new forms, inside and out. However the drive train has not evolved much and the venerable V6 of the 3.8 L has just finished celebrating its 35th birthday. The naturally aspirated version is rated at 205 hp and is sold only in the U.S.

STRONG POINTS

• Safety: 90%
All vehicles recently updated have seen their structure benefit from a serious increase in rigidity, to satisfy the U.S. National Highway Traffic Administration. Also, standard front air bags plus an ABS bring its rating near maximum.

• Insurance: 90%
Considering its price, the Riviera costs less to insure than some more modest vehicles, thanks to its ratio of 3.6%.

• Satisfaction: 80%
If you rely on the rating of the previous model and take into account that the drive train is directly inherited from the previous one, you should not have any worries about reliability. However, it is better to be careful during the first production year.

• Performance: 80%
The supercharged version gives the best takeoffs and accelerations, thanks to a favorable weight-to-power ratio (16.7 lbs/hp). However, the character of this coupe does not lend itself to sporty use and it is in normal driving that you can best appreciate the power and torque range of the "silky" engine together with the well-spaced transmission ratios.

• Steering: 80%
Average in ratio, it benefits from a well-modulated power assist, which varies with speed, based on magnetic pickup signals. In spite of a reasonable turning diameter, maneuverability is encumbered by the imposing bulk of this model.

•Trunk: 80%
Utilization of its vast volume is penalized by its sill height and by the narrowness of the opening.

• Access: 80%
It is as easy to take a seat at the rear as at the front, thanks to long doors which open wide and to sufficient height under the roof.

• Interior space: 75%
Well-proportioned interior dimensions allow four people to be transported in full comfort, with a fifth in a pinch.

• Quality/fit/finish: 75%
Just as the Aurora, the Riviera benefits from the new GM policy which has noticeably increased the standards for assembly, fit and quality of materials. At the limit, you sometimes get the impression that the creators got inspired by their German rivals. You should note that the sound system is very average, in spite of the prestigious brand, and

DATA

Category: mid-size front-wheel drive coupes.
Class : 7

HISTORIC

Introduced in : 1962
Modified in : 1966, 1976, 1978, 1986: front-wheel drive & V6 engine, 1995.
Made in: Hamtramck-Detroit, Michigan, U.S.A.

RATINGS

Safety: 90 %
Satisfaction: 80 % (previous model)
Depreciation: 65 % (previous model)
Insurance: 3.6 % ($ 858)
Cost per mile: $ 0.45

NUMBER OF DEALERS

In U.S.A.: 3,000 Buick

SALES ON THE AMERICAN MARKET

Model	1992	1993	Result
Riviera	11,476	3,443	- 70.0 %

MAIN COMPETITORS

ACURA Legend coupe, FORD Thunderbird SC, MERCURY Cougar XR7.

EQUIPMENT

BUICK Riviera	Supercharged
Automatic transmission:	S
Cruise control:	S
Power steering:	S
ABS brake:	S
Air conditioning:	S
Air bag:	S
Leather trim:	O
AM/FM/ Radio-cassette:	S
Power door locks:	S
Power windows:	S
Tilt steering column:	S
Dual power mirrors:	S
Intermittent wipers:	S
Light alloy wheels:	S
Sunroof:	O
Anti-theft system:	S

S : standard; O : optional; - : not available

COLORS AVAILABLE

Exterior: Beige, Amethyst, Green, Black, White, Red, Cherry, Blue, Gray.
Interior: Graphite, Red, Beige, Jade, Blue.

MAINTENANCE

First revision: 3,000 miles
Frequency: 6,000 miles
Diagnostic plug: Yes

WHAT'S NEW IN 1995 ?

-Exterior and interior totally updated

ENGINE					TRANSMISSION			PERFORMANCE									
Model/ versions *: standard	Type / timing valve / fuel system	Displacement cu/in	Power bhp @ rmn	Torque lb.ft @ rpm	Compres. ratio	Driving wheels / transmission	Final ratio	Acceler. 0-62 mph s	Stand. 1/4 mile s	Stand. 5/8 mile s	Braking 62-0 mph ft	Top speed mph	Lateral acceler. G	Noise level dBA	Fuel economy mpg city	hwy	Gasoline type / octane
1993	V6* 3.8 OHV-12-SFI	231	170 @ 4800	220 @ 3200	8.5 :1	front - A4*	3.06	9.5	16.8	30.7	125	121	0.79	67	23.0	31.0	R 87
1995	V6* 3.8 OHV-12-SFI	231	205 @ 5200	230 @ 4000	9.4 :1	front - A4	3.05	NA									
option	V6C 3.8 OHV-12-SFI	231	225 @ 5000	275 @ 3200	8.5 :1	front - A4	3.05	8.0	16.4	29.5	144	124	0.79	66	19.0	29.0	M 89

the lack of depth is noticeable, in spite of the many available adjustments.

• Driver's compartment: 75%
The driver easily finds the most comfortable seating position, thanks to the numerous adjustments. The absence of the "B" pillar improves the lateral visibility, but the front and rear corners are far away, invisible from the driver's seat: hard to locate.

• Technical: 75%
Conservative, it has picked up numerous elements from the preceding model, beginning with the engine, whose latest version is more powerful. The suspension is independent and 4-wheel disc brakes plus ABS are standard. In spite of its very streamlined appearance, the body only reaches a very average Cd of 0.34.

• Seats: 75%
Their contouring is superior to that of the previous model which allows better support, and by a slightly harder padding for better lumbar support to delay fatigue during long trips.

• Conveniences: 75%
The numerous storage areas include: a glove box, a center console container, and door pockets of ample size.

• Suspension: 70%
It provides a better comfort at cruise speeds on freeways than on bad pavement, at which time it reacts in a fairly uncivilized way.

• Noise level: 70%
For the most part it remains very low, thanks to a quiet drive train, high body stiffness and good soundproofing.

• Handling: 65%
Suspension softness is better for straight lines than for tight turns that close in. Cornering hard, the inside wheel lightens, traction is lost, and the weight and bulk take away all agility.

• Fuel consumption: 60%
Reasonable in cruising, it climbs rapidly when the mission is to better know the engine limits.

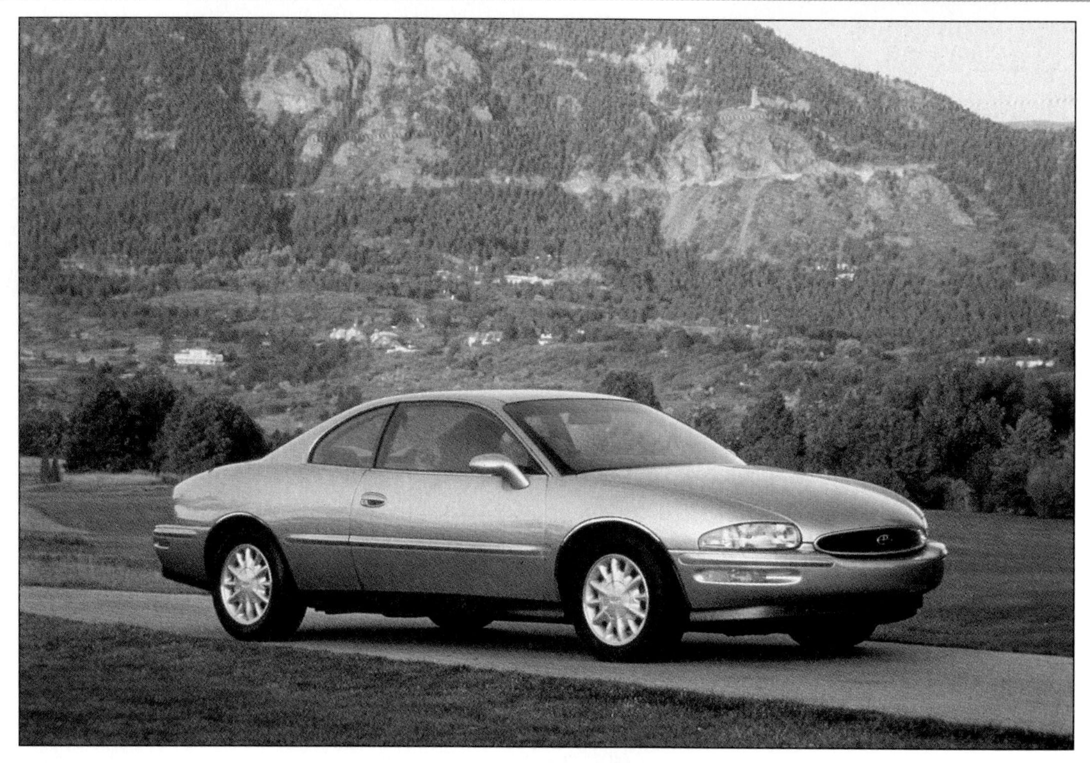

WEAK POINTS

• Price/equipment: 30%
In spite of the originality of its design and the richness of its equipment, the Riviera is not given away. This segment is not crowded because the Japanese are not active and the Americans are represented by just Ford and Buick. The equipment is rather complete, since only leather trim, CD players, and heated seats are optional.

• Depreciation: 35%
Based on the old model, it is as high as that of the intermediate sedans and also greater than that of its competitors. Next year we will see if the new model improves on this tendency.

• Braking: 40%
While it is relatively effective in a straight line, its fade resistance is nothing to write home about since the disc pads and brake efficiency go up in smoke when tested hard.

CONCLUSION

Overall rating: 62.5%
More loyal to its radical style and cosy comfort than performance or sporty handling, the Riviera will still be a grand "bourgeois " coupe. ☺

SPECIFICATIONS & PRICES

Model	Version/Trim	Body/ Seats	Interior volume cu.ft.	Trunk volume cu ft.	Drag coef. Cd	Wheel-base in	Lgth x Wdthx Hght. inx inx in	Curb weight lb	Towing capacity std. lb	Susp. type ft/rr	Brake type ft/rr	Steering system	Turning diameter ft	Steer. turns nber	Fuel. tank gal.	Standard tire size	Standard powertrain	PRICES US. $ 1994
BUICK		**General warranty: 3 years / 36,000 miles; antipollution: 5 years / 50,000 miles; perforation corrosion: 6 years / 100,000 miles.**																
Riviera	1994 base	2dr.cpe.5	99	14	0.36	108	198.2x73.1x52.9	3503	2000	i/i	dc/dcABS	pwr.r/p.	39.4	2.81	19	205/70R15	V6/3.8/A4	**$22,767**
Riviera	1995 base	2dr.cpe.5	99.5	17.4	0.34	113.8	207.2x75.0x55.2	3752	2000	i/i	dc/dcABS	pwr.r/p.	39.0	3.05	75.7	225/60R16	V6/3.8/A4	**$27,632**
Riviera	Supercharged	2dr.cpe.5	99.5	17.4	0.34	113.8	207.2x75.0x55.2	3787	2000	i/i	dc/dcABS	pwr.r/p.	39.0	3.05	75.7	225/60R16	V6C/3.8/A4	**$28,732**

See page 393 for complete 1995 Price List

Sumo dancers...

These popular coupes with spectacular lines perpetuate the tradition of the line they carry but do not know as much success as the preceding models. They remain alone in the market with this imposing combination of bulk, weight and engine.

The Camaro and Firebird coupes were freshened up in 1993; they remain practically identical, except for some cosmetic details. Base models are equipped with a 3.4 L V6 with a 5-speed manual transmission, while the Z28-Formula disposes of a 5.7 L V8 rated at 275 hp and a manual 6-speed. The same engine also powers the Firebird Trans Am, available exclusively with a 6-speed manual and a shorter axle ratio. For 1994 each coupe is also available in a convertible, whose drive train is identical.

STRONG POINTS

• Safety: 80%
Even though their steel unit body has been stiffened to gain handling and impact resistance, it is still not sufficient because body sounds can be heard while crossing road defects, especially with convertibles. Happily, two air bags and ABS are standard.

• Handling: 75%
The improved body stiffness, effective damping, and well-calibrated tires gain good handling in the corners, even tight ones. Traction, even on damp but smooth pavement, is satisfactory. On rough pavement the suspension bounces, acts jerky, and is hard to guide in a straight line.

• Performance: 75%
The majority of buyers will be content with a V6 engine whose pickups and accelerations are as ordinary as the power-to-weight ratio (20.7 lbs/hp). But it provides good sensations at a lower cost. The distinctly more muscular (12.8 lbs/hp) V8 will bring one to Heaven or Traffic Court depending on the cruise speed setting.

• Satisfaction: 75%
Here is wishing that the new generation offers better quality than the quite mediocre preceding one.

DATA

Category: rear-wheel drive sports coupes.
Class : S

HISTORIC

Introduced in : 1966 (Camaro) 1967 (Firebird)
Modified in : 1979, 1981, 1993.
Made in: Ste-Thérèse, Québec, Canada.

RATINGS

Safety: 90 %
Satisfaction: 75 %
Depreciation: 65 %
Insurance: 9.1 % ($ 927) V8: 8.3 % ($ 995)
Cost per mile: $ 0.47

NUMBER OF DEALERS

In U.S.A.: 4,466 Chevrolet-Geo, 2,953 Pontiac.

SALES ON THE AMERICAN MARKET

Model	1992	1993	Result
Camaro	56,909	68,773	+17.3 %
Firebird	22,501	26,893	+16.4 %

MAIN COMPETITORS

ACURA Integra, CHRYSLER Sebring, DODGE Avenger & Stealth, EAGLE Talon, FORD Mustang & Probe, HONDA Prelude, MAZDA MX-6, NISSAN 240SX, TOYOTA Celica & VOLKSWAGEN Corrado.

EQUIPMENT

CHEVROLET Camaro PONTIAC Firebird	base base	Z28 Formula	 Trans AM
Automatic transmission:	O	O	O
Cruise control:	O	O	S
Power steering:	S	S	S
ABS brake:	S	S	S
Air conditioning:	O	S	S
Air bag:	S	S	S
Leather trim:	O	O	O
AM/FM/ Radio-cassette:	S	S	S
Power door locks:	O	O	S
Power windows:	O	O	S
Tilt steering column:	S	S	S
Dual power mirrors:	S	S	S
Intermittent wipers:	S	S	S
Light alloy wheels:	O	S	S
Sunroof:	-	-	-
Anti-theft system:	O	O	O

S : standard; O : optional; - : not available

COLORS AVAILABLE

Exterior: Purple, White, Silver, Lime, Green, Red, Black, Blue.
Interior: Fabric: Beige, Graphite, Medium Gray; Leather: Red, White.

MAINTENANCE

First revision: 3,000 miles
Frequency: 6,000 miles
Diagnostic plug: Yes

WHAT'S NEW IN 1995 ?

- Traction control offered with the 2 transmissions (Pontiac).
- All-weather tires on the Trans Am (P245/50ZR16).
- Motorized aerial.
- New style wheel covers and alloy rims.
- New body covers.
- Optional CD player (Pontiac).

ENGINE						TRANSMISSION		PERFORMANCE										
Model/ versions *: standard	Type / timing valve / fuel system	Displacement cu/in	Power bhp @ rmn	Torque lb.ft @ rpm	Compres. ratio	Driving wheels / transmission	Final ratio	Acceler. 0-62 mph s	Stand. 1/4 mile s	Stand. 5/8 mile s	Braking 62-0 mph ft	Top speed mph	Lateral acceler. G	Noise level dBA	Fuel economy mpg city	hwy	Gasoline type / octane
1)	V6* 3.4 OHV-12-SFI	207	160 @ 4600	200 @ 3600	9.0 :1	rear - M5*	3.23	9.8	16.7	30.5	138	109	0.80	68	19.0	28.0	R 87
						rear - A4	3.23	10.5	17.2	31.2	144	106	0.80	68	17.0	24.0	R 87
2)	V8 5.7 OHV-16-SFI	350	275 @ 5000	325 @ 2000	10.5 :1	rear - M6*	3.42	6.7	15.0	26.5	148	131	0.85	70	17.0	24.0	R 87
						rear - A4	3.23	7.8	15.7	27.2	154	124	0.85	70	17.0	25.0	R 87
3)	V8 5.7 OHV-16-MPFI	350	275 @ 5000	325 @ 2400	10.5 :1	rear - M6*	3.42	6.7	14.9	25.6	138	131	0.85	70	17.0	25.0	R 87

1) base Camaro-Firebird 2) Camaro Z28 &Firebird Formula 3) Firebird Trans Am

• Driver's compartment: 70%
The driver can rapidly find an acceptable position, but the steering column is long and the manual transmission shifter is too far back while the parking brake is more practical.
Gauges are numerous and easy to read. Visibility is mediocre, because of low seating and the dashboard is high which makes it hard to evaluate the size of the vehicle. Also, the "C" pillar creates a major blind spot, especially on the convertible, where the back light is narrower. Finally, the side-view mirrors are very small.

• Steering: 70%
It is sufficiently fast, direct, precise, good enough to make "inspired" driving . As to the steering wheel, it is pleasant to use and its thickness provides an effective grip.

• Suspension: 70%
On the base models, it surprises by its good manners on smooth surfaces, except for the Z28-Formula-Trans Am, which is distinctly more unstable in spite of the quality of the standard equipment De Carbon shocks.

• Technical: 60%
Identical, these coupes have exactly the same dimensions as the previous version, but the body is longer, larger, and taller. Front suspension is independent, while at the rear, the Salisbury rigid axle is controlled by a four link suspension.

• Braking: 60 %
Standard brakes are mixed and 4-wheel discs come with V8 power; ABS is standard on both. Steering is with a power assisted rack-and-pinion and the 16 inch wheels are fitted with Goodyear Eagle GA tires.

• Quality/fit/finish: 60%
In spite of some improvements, it is still far from the quality of the Japanese competition. The appearance of the new dashboard is more technical, but it is built in plastic that looks econo-minded. As for the trim, it looks less bleak than before but is still not very rich. The inside trim of the convertible top has been fully simplified and it carries no insulation, just trim cloth.

• Seats: 60%
Up front they provide sufficient lateral and lumbar support and their padding is almost soft. However, the floor hump on the passenger side above the catalytic converter will be the source of cramps and heat on long trips. The rear seat room is so cramped that it will only hold little kids. Inside rear space is much reduced, the seat back is too vertical and access is acrobatic.

• Conveniences: 60%
The storage space consists of a glove box of acceptable volume, a hollow in the center console, and small pockets placed too far back in the doors. The convertible top is easy to take down or raise, thanks to electric power assist.

• Price/equipment: 50%
Even though proposed at popular prices, taking into account their performance and standard equipment, these coupes are too bulky. They have not encountered as much success as the more compact Ford Mustang.

• Access: 50%
You can reach the seats without any problems, thanks to the long doors which open wide, also the roof line is relatively high.

• Fuel consumption: 50%
It is not too exaggerated if you consider the bulk of these coupes and their engines, but compared to the appetite of an Eagle Talon it is not very economical.

PONTIAC Formula décapotable

WEAK POINTS

• Noise level: 30%
The sounds of the V6 engine are more discrete than those of the V8, which growl with a dull roar and transmits as many decibels as vibrations. However, at cruise speed it is more discrete and tolerable.

• Interior space: 35%
It is far from being proportional to the volume taken up, but space is better organized than in the past. Only 2 adults and 2 children can be accommodated.

• Depreciation: 35%
It is strong and compares to the overall American products whose bulk and fuel consumption are not fully adapted to our era.

• Braking: 40%
Effective and easier to modulate, it is sometimes brutal initially; the ABS stabilizes it well in panic stops. However, the stopping distances are long and the wheels see little untimely lockups.

• Trunk: 40%
It consists of a hole and a platform, located behind the rear bench seat. The seat back can be folded to increase the storage space, however it is not easy to reach, because of its tall sill. Also the driver will find it difficult to reach and operate the remote trunk opening, which is located in the glove box.

• Insurance: 40%
The young will find the tariff quite steep because the premium for the V6 is as strong as for the V8.

CONCLUSION

• Overall rating: 56.0%
These coupes have had more success in the U.S. than in Canada, where fuel is more expensive, and the market is directed to compact models such as the Ford Mustang, which is seen everywhere.

OWNER'S SUGGESTIONS

-A more compact format
-Superior finish
-A more rigid unit body (convertible)
-A more practical trunk as in other sports coupes
-Manual 6-speed virtually unusable (needs improvement)
-Rear seats should become more useful

SPECIFICATIONS & PRICES

Model	Version/Trim	Body/ Seats	Interior volume cu.ft.	Trunk volume cu ft.	Drag coef. Cd	Wheel-base in	Lgth x Wdthx Hght. inx inx in	Curb weight lb	Towing capacity std. lb	Susp. type ft/rr	Brake type ft/rr	Steering system	Turning diameter ft	Steer. turns nber	Fuel. tank gal.	Standard tire size	Standard powertrain	PRICES US. $ 1994
CHEVROLET		**General warranty: 3 years / 36,000 miles; antipollution: 5 years / 50,000 miles; perforation corrosion: 6 years / 100,000 miles.**																
Camaro	base	3dr.cpe. 2+2	82	13	0.32	101.1	193.2x74.1x51.3	3161	1000	i/r	dc/dr/ABS	pwr.r/p.	40.7	2.67	15.5	215/60R16	V6/3.4/M5	**$13,499**
Camaro	base	2dr.con. 2+2	80.5	7.5	0.36	101.1	193.2x74.1x52.0	3324	1000	i/r	dc/dr/ABS	pwr.r/p.	40.7	2.67	15.5	215/60R16	V6/3.4/M5	**$18,745**
Camaro	Z28	3dr.cpe. 2+2	82	13	0.32	101.1	193.2x74.1x51.3	3294	1000	i/r	dc/dc/ABS	pwr.r/p.	40	2.28	15.5	235/55R16	V8/5.7/M6	**$16,999**
Camaro	Z28	2dr.con. 2+2	80.5	7.5	0.36	101.1	193.2x74.1x52.0	3500	1000	i/r	dc/dc/ABS	pwr.r/p.	40	2.28	15.5	235/55R16	V8/5.7/M6	**$22,075**
PONTIAC		**General warranty: 3 years / 36,000 miles; antipollution: 5 years / 50,000 miles; perforation corrosion: 6 years / 100,000 miles.**																
Firebird	base	3dr.cpe. 2+2	82	13	0.32	101.1	193.2x74.1x52.0	3230	1000	i/r	dc/dr/ABS	pwr.r/p.	37.9	2.67	15.5	215/60R16	V6/3.4/M5	**$14,099**
Firebird	base	2dr.con. 2+2	80.5	7.5	0.36	101.1	193.2x74.1x52.7	3344	1000	i/r	dc/dr/ABS	pwr.r/p.	37.9	2.67	15.5	215/60R16	V6/3.4/M5	-
Firebird	Formula	3dr.cpe. 2+2	82	13	0.32	101.1	193.2x74.1x52.0	3371	1000	i/r	dc/dc/ABS	pwr.r/p.	37.7	2.28	15.5	235/55R16	V8/5.7/M6	**$18,249**
Firebird	Formula	2dr.con. 2+2	80.5	7.5	0.36	101.1	193.2x74.1x52.7	3490	1000	i/r	dc/dc/ABS	pwr.r/p.	37.7	2.28	15.5	235/55R16	V8/5.7/M6	-
Firebird	Trans Am	3dr.cpe. 2+2	82	13	0.32	101.1	197.1x74.1x51.6	3347	1000	i/r	dc/dc/ABS	pwr.r/p.	37.7	2.28	15.5	245/50ZR16	V8/5.7/M6	**$20,009**
Firebird	Trans Am	2dr.con. 2+2	80.5	215	0.36	101.1	197.1x74.1x52.7	3609	1000	i/r	dc/dc/ABS	pwr.r/p.	37.7	2.28	15.5	245/50ZR16	V8/5.7/M6	-

See page 393 for complete 1995 Price List

Oldsmobile Aurora

Palace revolution...

"No excuses" has been the key note punch line during the design of these luxury sedans ,capable of comparing themselves, without blushing, to the characteristics of their German or Japanese competitors. Taking off in a good direction, GM will find that quality pays in the long run...

The Aurora is the first car of the new era at General Motors. After the monstrous reorganization, the inevitable mentality change goes in place slowly but surely. First studied without a particular label, the Aurora was finally given to Oldsmobile. Of international caliber, this deluxe 4-door sedan intends to fight the Japanese and German cars equipped with 6 cylinder engines in a format that is slightly lower at a price that is equal or less.

STRONG POINTS

• Safety: **90%**
In this aspect, Aurora benefits from a very solid unit body that offers good resistance to frontal or lateral impact, two air bags protect the front seat occupants and the shoulder harness is adjustable in height. However, it is curious to note that with a car of this caliber, the seat belt locks are not better integrated with the seats. On the active side, the ABS and the traction control, offered standard, stabilize takeoffs and stops, no matter what the road conditions.

• Insurance: **90%**
The Aurora does not cost more to insure than the convertible GT Mustang, and its rate is lower than that of a simple 318 BMW.

• Technical: **80%**
Starting with a platform very close to the De Ville, Cadillac engineers created a unique vehicle in which the slogan was "no excuses." The engine is placed transversely between the drive wheels, the suspension is independent and 4-wheel disc brakes are used together with traction control and ABS. A tubular cage protects the interior and has been designed with a natural frequency of 25 KHz (cycles per second). Natural frequency is a measure of body stiffness and the way to control interior noises. This unit body has an average 0.34 drive coefficient in spite of slick lines. To reduce weight, the hood is aluminum and the fuel tank nylon. The new 4.0 L V8 which develops 250 hp has been designed especially for the Aurora and was the object of 14 patents.

• Steering: **80%**
Precise and fast, its assist is a little too strong which makes it too light and calls for extra attention. Maneuverability is good, taking into account a large steering diameter, an average ratio and extensive overhang.

• Quality/fit/finish: **80%**
Surprising for GM, it practically reaches an international standard. The leather trim, the appliqué wood and the carpets look pricey, the fits are strictly adhered to and the quality of the dashboard is beyond reproach.

• Storage: **80%**
Here, GM surprises by the number and volume of cleverly designed storage areas. For instance, the central arm rest is remarkably outfitted with integral cup holders. Unfortunately, they did not use any more imagination than they needed when designing the really impractical fuel filler cap.

DATA

Category: front-wheel drive luxury sedan.
Class : 7

HISTORIC

Introduced in : 1994
Modified in : -
Made in: Orion Township, MI, U.S.A.

RATINGS

Safety:	90 %
Satisfaction:	NA %
Depreciation:	NA %
Insurance:	4.0 % ($ 925)
Cost per mile:	$ 0.61

NUMBER OF DEALERS

In U.S.A.: 3,100 Oldsmobile.

SALES ON THE AMERICAN MARKET

Model	1992	1993	Result
Aurora	not on the market at that time.		

MAIN COMPETITORS

AUDI A8, BMW 540, INFINITI Q45, LEXUS LS400, MERCEDES BENZ 400E.

EQUIPMENT

ODLSMOBILE Aurora	base
Automatic transmission:	S
Cruise control:	S
Power steering:	S
ABS brake:	S
Air conditioning:	S
Air bag:	S
Leather trim:	S
AM/FM/ Radio-cassette:	S
Power door locks:	S
Power windows:	S
Tilt steering column:	S
Dual power mirrors:	S
Intermittent wipers:	S
Light alloy wheels:	S
Sunroof:	S
Anti-theft system:	S

S : standard; O : optional; - : not available

COLORS AVAILABLE

Exterior: Light Gray, White, Green,Teal, Blue, Black, Garnet, Cherry Red,Champagne, Mocha, Purple.
Interior: Graphite, Pearl, Blue, Teal, Mushroom.

MAINTENANCE

First revision:	3,000 miles
Frequency:	6,000 miles
Diagnostic plug:	Yes

WHAT'S NEW IN 1995 ?

- New model, sold by Oldsmobile but manufactured by Cadillac.
- Front wheel drive, 4.0 L V8 engine, automatic transmission, traction regulator and 4-wheel disc brakes with ABS.

	ENGINE				TRANSMISSION			PERFORMANCE									
Model/ versions *: standard	Type / timing valve / fuel system	Displacement cu/in	Power bhp @ rmn	Torque lb.ft @ rpm	Compres. ratio	Driving wheels / transmission	Final ratio	Acceler. 0-62 mph s	Stand. 1/4 mile s	Stand. 5/8 mile s	Braking 62-0 mph ft	Top speed mph	Lateral acceler. G	Noise level dBA	Fuel economy mpg city	hwy	Gasoline type / octane
Aurora	V8* 4.0 DOHC-32-SFI	244	250 @ 5600	260 @ 4400	10.3 :1	front - A4	3.48	9.5	16.7	30.6	157	134	0.80	65	19.0	27.0	M 89

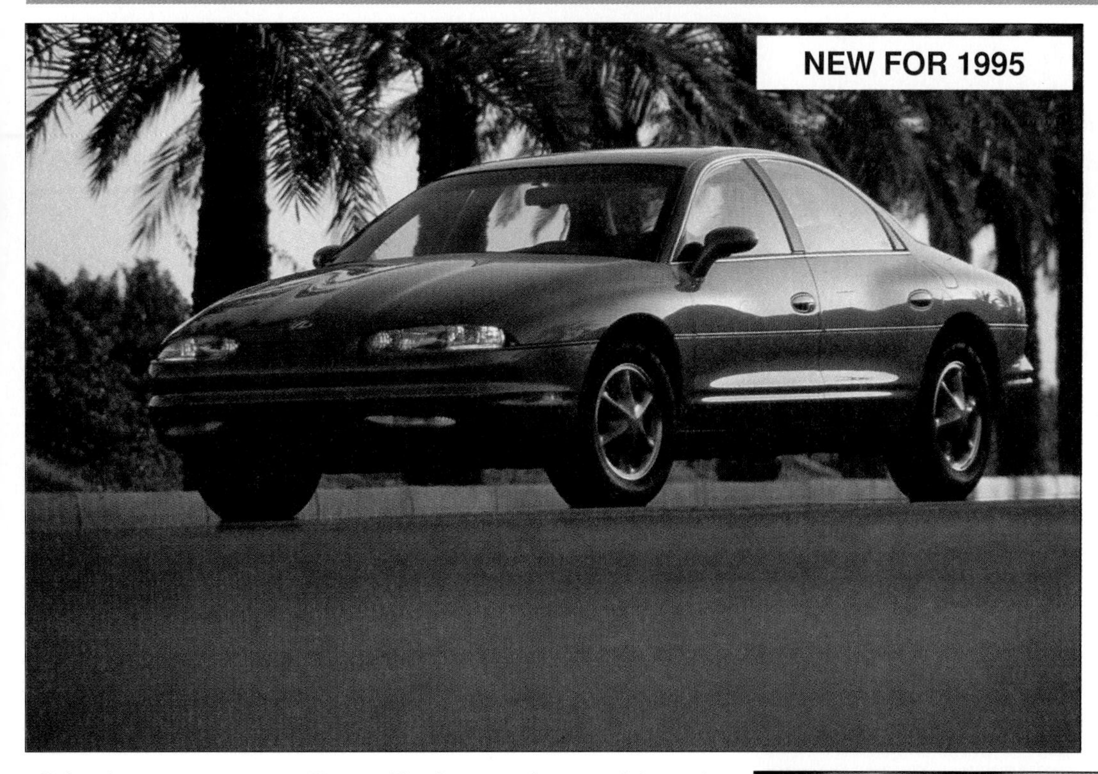

• **Driver's compartment: 75%**
If its style lacks a little in its simplicity, the ergonomics are effective. The main controls are well-located, except for the switches located at the bottom of the dash and instruments that are difficult to read even in the daytime. The center console is well-oriented towards the driver, and the displays visible in spite of their unusual form. One finds little electronics at this level, because the odometer is mechanical and of an old style. It is unexcusable that there is no shift indicator among the gauges which forces the driver to take his eyes off the road.

• **Interior space: 75%**
It allows five people to sit comfortably, the dimensions are ample for the rear legroom, but rear height is limited for tall people.

• **Access: 75%**
Thanks to ample length and door opening angle, it is easy to take place without contortions.

• **Trunk: 70%**
It is large and deep, but a narrow opening complicates its use, and bulky cases are hard to fit in.

• **Handling: 70%**
The Aurora enjoys good dynamic stability in straight lines or in wide curves and its moderate roll does not perturb a tight turn. It sets up easily and cleanly.

• **Noise level: 70%**
A quiet engine is coupled to excellent soundproofing that only allows rolling noises to come through when the pavement is poor. Airstreams along the body are only heard at very high and prohibited speeds.

• **Seats: 70%**
Their cushions and back are well-proportioned, but they lack the shape to offer good lateral support, while the lumbar support and the headrest are satisfactory, both at the front and the rear.

• **Suspension: 70%**
Most of the time it reacts with much smoothness and good manners. However, front wheel travel is limited and on larger obstacles the suspension bottoms.

• **Performance: 65%**
In spite of a favorable weight-to-power ratio (15.8 lbs/hp) standing starts are more exciting than accelerations, which should be

under 9 seconds from 0 to 62.5 mph to be in the mid-range of this category.

• **Fuel consumption: 50%**
It is reasonable for a car of this weight (3,966 pounds) and displacement (4.0 L) because we have recorded numbers lower than that of government agencies, particularly in the city.

WEAK POINTS

• **Price/equipment: 22%**
Even if it is distinctly less expensive than most of its competitors, it stays out of reach for the crowd. Its equipment is very complete, but we have detected the little details that make one think that GM has dipped into its large store of parts to equip the car and some out-of-style elements, such as the seat belts, cut into the modern feel of the vehicle.

• **Braking: 30%**
The efficiency is just very average, since stopping distances are long, in part because the ABS reacts in a disagreeable way by hardening and brutally raising the pedal toward the end of the stop. Fade-resistance is poor and the pads smoke, releasing characteristic and nauseating odor.

CONCLUSION

• **Overall rating: 69.0%**
Well-targeted, the Aurora changes the perception we have had of GM products because technical achievement, assembly, finish, and the value of the products used are extremely competitive and of world class. Let us simply wish that many others will follow, inspired by the new philosophy. ☺

SPECIFICATIONS & PRICES

Model	Version/Trim	Body/ Seats	Interior volume cu.ft.	Trunk volume cu ft.	Drag coef. Cd	Wheel-base in	Lgth x Wdthx Hght. inx inx in	Curb weight lb	Towing capacity std. lb	Susp. type ft/rr	Brake type ft/rr	Steering system	Turning diameter ft	Steer. turns nber	Fuel. tank gal.	Standard tire size	Standard powertrain	PRICES US. $ 1994
OLDSMOBILE		**General warranty: 3 years / 36,000 miles; antipollution: 5 years / 50,000 miles; perforation corrosion: 6 years / 100,000 miles.**																
Aurora	base	4dr. sdn. 5	102	16	0.32	113.8	205.4x74.4x55.4	3966	2000	i/i	dc/dc/ABS	pwr.r/p.	41	3.1	20	235/60R16	V8/4.0/A4	**$31,370**

See page 393 for complete 1995 Price List

BUICK Le Sabre , OLDSMOBILE 88 Royale
PONTIAC Bonneville

Classics...

These large sedans still draw a wide number of buyers who have backed away from the large American classic sedans. While they are not technical overachievers, they are just as well-equipped as their Japanese competitors and a freshening up would do them a world of good.

Positioned between the intermediate and luxury cars, these 4-door sedans are offered at Buick as Le Sabre Custom or Limited, at Oldsmobile as 88 Royale in the standard or an LS, or at Pontiac as Bonneville SE or SSE. While the Le Sabre and the 88 are equipped with the ancient 3.8 L 170 hp V6, the Bonneville SE inherits from its latest version a 205 hp V6. The SSE tops them all with 225 hp from a supercharged engine.

STRONG POINTS

• Safety: 90%
Even though the structure of these cars begins to be dated and does not benefit from the latest stiffening techniques of recent models, their impact protection index is excellent, thanks to two air bags and ABS.

• Satisfaction: 80%
It is highest on the Pontiacs which attain the Japanese percentages, while the Buicks and Oldsmobiles are manufactured at another plant that needs to catch up.

• Trunk: 80%
Too bad that the tortured shape and the ill-placed spare tire do not allow getting the most of it, but the lid opens high for ease of access.

• Technical: 75%
These three models share the same platform and some drive train components while the Le Sabre and 88 Royale differentiate from each other by body details and glass. In terms of style, the Bonneville is in a separate class, which explains why it is assembled in a separate plant. The steel unit body receives an independent suspension at all 4 wheels, but the braking remains mixed, even on a Bonneville SSE whose performance is fairly high.

• Suspension: 75%
On freeways at cruising speed, these cars are at their best, otherwise the overly soft suspension shakes up the occupants as soon as the pavement gets worse. There is no shortage of roll and pitch and it's not rare for the suspension to bottom because its wheel travel is limited. The sports version, offered as an option, is firmer and allows distinctly more precise control.

• Quality/fit/finish: 70%
In spite of the modifications brought into the assembly line and in the presentation of some of the detailing, these cars still have a few wrinkles of their era and cannot yet compare to more recent models.

• Interior space: 70%
The interior of these cars accomodates 5 occupants in comfort and a 6th can tag along for a short trip.

• Driver's compartment: 70%
Driver's position is effective and easy to find in spite of the excessive steering column length. Visibility and the instrumentation are satisfactory, even on standard models. On the Pontiac, the gauges at the right of the dashboard are virtually invisible, concealed by the driver's hands.

DATA

Category: full-size front-wheel drive sedans.
Class : 6

HISTORIC

Introduced in : 1969
Modified in : 1976: reduced size; 1986, 1987: front-wheel drive.
Made in: Lake Orion MI, U.S.A.

RATINGS

Safety: 80 %
Satisfaction: 85 % Bonneville : 75 %
Depreciation: 65 % Bonneville : 60 %
Insurance: Le Sabre : 4.5 % ($ 650) 88 : 5.5 % ($ 720) Bonneville SSE :4.0 % ($ 785)
Cost per mile: $ 0.52 Bonneville SSE: $ 0.55

NUMBER OF DEALERS

In U.S.A.: 3,000 Buick, 3,100 Oldsmobile, 2,953 Pontiac.

SALES ON THE AMERICAN MARKET

Model	1992	1993	Result
Le Sabre	138,409	149,299	+7.3 %
88	71,303	70,396	-1.3 %
Bonneville	97,944	88,591	- 9.6 %

MAIN COMPETITORS

CHEVROLET Caprice, CHRYSLER Concorde, Intrepid, New Yorker, EAGLE Vision, FORD Crown Victoria, MERCURY Grand Marquis.

EQUIPMENT

BUICK Le Sabre	**Custom**	**Limited**
OLDSMOBILE 88 Royale	**base**	**LS**
PONTIAC Bonneville	**SE**	**SSE**
Automatic transmission:	S	S
Cruise control:	O	S
Power steering:	S	S
ABS brake:	S	S
Air conditioning:	S	S
Air bag:	S	S
Leather trim:	-	O
AM/FM/ Radio-cassette:	O	S
Power door locks:	S	S
Power windows:	S	S
Tilt steering column:	S	S
Dual power mirrors:	S	S
Intermittent wipers:	S	S
Light alloy wheels:	S	S
Sunroof:	-	O
Anti-theft system:	S	S

S : standard; O : optional; - : not available

COLORS AVAILABLE

Exterior: Gray, Beige, Green, Blue, Red, Black, White, Cherry, Amethyst.
Interior: Red, Beige, Blue, Gray.

MAINTENANCE

First revision: 3,000 miles
Frequency: 6,000 miles
Diagnostic plug: Yes

WHAT'S NEW IN 1995 ?

- **Dexron automatic transmission lubricant, good for 100,000 miles.**
- **New 205 hp 3.8 L V6 on the Bonneville SE.**
- **New automatic air conditioning on the Bonneville SE.**

Model/ versions *: standard	ENGINE Type / timing valve / fuel system	Displacement cu/in	Power bhp @ rmn	Torque lb.ft @ rpm	Compres. ratio	TRANSMISSION Driving wheels / transmission	Final ratio	PERFORMANCE Acceler. 0-62 mph s	Stand. 1/4 mile s	Stand. 5/8 mile s	Braking 62-0 mph ft	Top speed mph	Lateral acceler. G	Noise level dBA	Fuel economy mpg city	hwy	Gasoline type / octane
1)	V6* 3.8 OHV-12-SFI	231	170 @ 4800	225 @ 3200	9.0 :1	front - A4*	2.84	10.5	17.5	31.2	144	112	0.78	67	19,0	28.0	R 87
2)	V6* 3.8 OHV-12-SFI	231	205 @ 5200	230 @ 4000	9.4 :1	front - A4	2.84	9.0	16.7	30.0	148	118	0.78	67	19.0	28.0	R 87
3)	V6C*3.8 OHV-12-MPFI	231	225 @ 5000	275 @ 3200	9.0 :1	front - A4*	3.06	8.0	16.2	29.6	157	124	0.78	67	17.0	27.0	R 87

1) * Le Sabre, 98 2) * Bonneville 3) * Bonneville SSEi

- **Handling: 70%**
Neutral most of the time, an average driver will not often be confronted by understeering except when going too fast on a tight curve, at which time weight and roll become an important factor.
- **Steering: 70%**
It does not give you much information about the road and remains light because of excessive power assist, even though its steering ratio is good. Maneuverability suffers from the long wheelbase and large steering diameter.
- **Access: 70%**
No problem getting into these cars with generous proportions, even for big bodies on the way back from dinner. Here the doors are ample-sized and open fully.
- **Seats: 70%**
Well-proportioned and padded, they lack shape for efficient lateral support except on the Bonneville SSE buckets.
- **Noise level: 70%**
Generally quiet, the old Le Sabre and 88 engines produce disagreeable vibrations in some ranges. The tires hammer hard over tar strips and wind shows its presence above 60 mph.
- **Insurance: 70%**
The Le Sabre and 88 premiums cost less than those of the Bonneville, considered more exotic and riskier by insurers.
- **Fuel consumption: 60%**
It is favorable, considering the weight and engine displacement of these cars.
- **Conveniences: 70%**
The glove box is not giant, but it is complemented by a small box and a tub in the center console.
- **Performances: 60%**
The weight-to-power ratio, which is very average with the old engine, is superior on the new V6 or its supercharged version. Its accelerations are more exotic, without penalizing the efficiency.

OLDSMOBILE 88

PONTIAC Bonneville SLE

BUICK Le Sabre

Standing starts and accelerations are clean but the driving remains antiseptic, even on the SSE which lacks character.
- **Braking: 50%**
In spite of their high weight, these cars retain drums on the rear brakes, even on the SSE whose performances are higher. The spongy pedal does not allow modulating decelerations with precision. While cold stops are relatively short, temperature causes their fade-resistance to go up in smoke.

WEAK POINTS

- **Depreciation: 30%**
It is major because this type of machine loses fans and support each year to the benefit of real intermediates, often more effective and more spacious as well as less bulky on the outside.
- **Price/equipment: 40%**
Midway between compact and luxury cars, the price of these models seems justified by their volume, their relatively complete equipment and well-finished interior.

CONCLUSION

- **Overall rating: 66.5%**
The latest New Yorker (Chrysler) is gaining on the aging GMs' which are in need of being rapidly renewed. The Bonneville is favored by its superior quality and its personalized aesthetics. ☺

OWNERS' SUGGESTIONS
-Disc brakes at the rear
-A more rigid body
-Less injector and electronic problems
-A more reliable catalytic converter
-Better maneuverability
-A more modern look

SPECIFICATIONS & PRICES

Model	Version/Trim	Body/ Seats	Interior volume cu.ft.	Trunk volume cu ft.	Drag coef. Cd	Wheel-base in	Lgth x Wdthx Hght. inx inx in	Curb weight lb	Towing capacity std. lb	Susp. type ft/rr	Brake type ft/rr	Steering system	Turning diameter ft	Steer. turns nber	Fuel. tank gal.	Standard tire size	Standard power train	PRICES US. $ 1994
BUICK		**General warranty: 3 years / 36,000 miles; antipollution: 5 years / 50,000 miles; perforation corrosion: 6 years / 100,000 miles.**																
Le Sabre	Custom	4 dr.sdn.6	109	17	0.32	110.8	200.0x74.1x55.7	3441	2000	i/i	dc/dr/ABS	pwr.r/p.	40.7	2.97	18	205/70R15	V6/3.8/A4	**$21,080**
Le Sabre	Limited	4 dr.sdn.6	109	17	0.32	110.8	200.0x74.1x55.7	3441	2000	i/i	dc/dr/ABS	pwr.r/p.	40.7	2.97	18	205/70R15	V6/3.8/A4	**$24,640**
OLDSMOBILE		**General warranty: 3 years / 36,000 miles; antipollution: 5 years / 50,000 miles; perforation corrosion: 6 years / 100,000 miles.**																
88	Royale	4 dr.sdn.6	108	17	NA	110.8	200.4x74.1x55.7	3399	2000	i/i	dc/dr/ABS	pwr.r/p.	40.7	2.97	18	205/70R15	V6/3.8/A4	**$21,120**
88	Royale LS	4 dr.sdn.6	108	17	NA	110.8	200.4x74.1x55.7	3428	2000	i/i	dc/dr/ABS	pwr.r/p.	40.7	2.97	18	205/70R15	V6/3.8/A4	**$23,220**
PONTIAC		**General warranty: 3 years / 36,000 miles; antipollution: 5 years / 50,000 miles; perforation corrosion: 6 years / 100,000 miles.**																
Bonneville	SE	4 dr.sdn.6	109	18	NA	110.8	199.5x74.4x55.7	3446	2000	i/i	dc/dr/ABS	pwr.r/p.	38.7	2.79	18	215/65R15	V6/3.8/A4	**$20,424**
Bonneville	SSE	4 dr.sdn.6	109	18	NA	110.8	201.1x74.4x55.7	3587	1000	i/i	dc/dc/ABS	pwr.r/p.	40.4	2.86	18	215/60R16	V6/3.8/A4	**$25,884**

See page 393 for complete 1995 Price List

CHEVROLET Cavalier
PONTIAC Sunfire

The Model T's of our era...

If they are not the most-sold cars in North America, the Cavalier-Sunbird-Sunfire are still very popular. They are more affordable than imports which are generally more modest. Their renewed appearance will enhance their chances.

For 1995, the Cavaliers and Sunfires have their body and interior totally renewed based on the platform and base mechanics of the old model. At Chevrolet, the range includes a base coupe or Z24, a base sedan or LS and a Z24 convertible. At Pontiac, the Sunfire is a sedan, a coupe or a convertible SE plus a GT coupe. The station wagon versions will become available in 1996. The base engine is a 2.2 L 4 cylinder coupled to a manual 5-speed or a 3-speed automatic, optional 4-cylinder 2.3 L DOHC Quadfour delivering 145 hp with manual 5-speeds or automatic 4 shifts will replace the old 3.1 L V6.

STRONG POINTS

• Safety: **90%**
The Cavalier/Sunfire have been updated to the customs of the day and will now be equipped with 2 front seat air bags and ABS. The new structure, which has been considerably reinforced already complies with the 1997 safety requirements.

• Price/equipment: **90%**
The price will constitute the first advantage of these popular cars. However, the succinct equipment tends to the essential, and you have to go search for the little extras in the options list.

• Steering: **80%**
With the base engine, the torque from the previous models seems to have disappeared, but as we have not been able to drive the model with a Quad4, this point remains in suspension. Precise and with a good ratio, it retains good maneuverability, thanks to a normal turning diameter.

• Satisfaction: **70%**
Let us wish that the new models are more reliable than those they replace, whose ratings were typical of American cars of the 80s'.

• Fuel consumption: **70%**
It is reasonable, in spite of the engine displacement, which confers on this car an appreciable economy aspect.

• Suspension: **70%**
The comfort it provides is remarkable; that of the model we tried did not have brutal road reactions on bad pavement and made good use of the large wheel travel.

• Quality/fit/finish: **60%**
The exterior fit seems to be more precise and the interior is more carefully done. Trim appearance is less econobox than on some imports.

• Driver's compartment: **60%**
It is made more pleasant by the successful dashboard design, which merges well with the center console. The controls are set up in a more international way and are more rational than before (such as the automatic transmission shifter, as original as it is practical, and the thick steering wheel rim with a good hand grip). One needs to get used to the location of the electric window switches, on the center console.

DATA

Category: compacts, front -wheel drive coupes and sedans.
Class : 3

HISTORIC

Introduced in: 1981
Modified in: 1982: convertible; 1985: V6 engine.1995: body
Made in: Lansing, MI, Lordstown, OH, U.S.A.

RATINGS

	Cavalier	Sunbird
Safety:	90 %	
Satisfaction:	74 %	70 %
Depreciation:	62 %	55.5 %
Insurance:	7.5 % ($520-$785)	
Cost per mile:	$0.32	

NUMBER OF DEALERS

In U.S.A.: 4,466 Chevrolet-Geo, 2,953 Pontiac.

SALES ON THE AMERICAN MARKET

Model	1992	1993	Result
Cavalier	212,374	273,617	+ 32.4 %
Sunbird	72,563	93,223	+ 32.2 %

MAIN COMPETITORS

CHEVROLET Corsica, DODGE-PLYMOUTH Neon, FORD-MERCURY Escort-Tracer, FORD-MERCURY Tempo-Topaz, HONDA Civic, HYUNDAI Elantra, MAZDA Protegé, SATURN, SUBARU Impreza, TOYOTA Corolla, VOLKSWAGEN Golf/Jetta.

EQUIPMENT

CHEVROLET Cavalier PONTIAC Sunfire	base SE	LS	Z24 GT
Automatic transmission:	O	S	O
Cruise control:	O	O	O
Power steering:	S	S	S
ABS brake:	S	S	S
Air conditioning:	O	S	S
Air bags (2):	S	S	S
Leather trim:	-	-	-
AM/FM/ Radio-cassette:	O	O	S
Power door locks:	O	O	O
Power windows:	O	O	O
Tilt steering column:	O	O	S
Dual power mirrors:	O	S	S
Intermittent wipers:	S	S	S
Light alloy wheels:	-	O	S
Sunroof:	O	O	O
Anti-theft system:	-	-	-

S : standard; O : optional; - : not available

COLORS AVAILABLE

Exterior: Blue Aqua, White, Red, Black, Green, Autumn Wood, Orchid.
Interior: Blue Aqua, Red, Neutral, Graphite, White.

MAINTENANCE

First revision: 3,000 miles
Frequency: 6,000 miles
Diagnostic plug: Yes

WHAT'S NEW IN 1995 ?

- **Exterior/interior entirely redesigned.**
- **Two air bags standard.**
- **One key for the entire vehicle.**

Model/ versions *: standard	ENGINE Type / timing valve / fuel system	Displacement cu/in	Power bhp @ rmn	Torque lb.ft @ rpm	Compres. ratio	TRANSMISSION Driving wheels / transmission	Final ratio	PERFORMANCE Acceler. 0-62 mph s	Stand. 1/4 mile s	Stand. 5/8 mile s	Braking 62-0 mph ft	Top speed mph	Lateral acceler. G	Noise level dBA	Fuel economy mpg city	hwy	Gasoline type / octane
base	L4* 2.2 SOHC-8-MFI	132	120 @ 5200	120 @ 4000	9.0 :1	front - M5*	3.58	11.0	17.7	35.2	138	103	0.73	68	25.0	36.0	R 87
						front - A3	3.18	12.2	18.5	36.0	144	100	0.73	68	23.0	33.0	R 87
option	L4* 2.3 DOHC-16-MFI	138	150 @ 6000	145 @ 4800	9.5 :1	front - M5*	3.91	NA									
						front - A4	3.94	NA									

NEW FOR 1995

CHEVROLET Cavalier

Gauges clustered under a large shade are easy to decipher.

• **Access**: 60%

It's rather easy to take a seat in the back of the coupe or the convertible as long as you have a small build, but it's easier in the sedans whose rear doors have a greater length and a sufficient opening angle.

• **Seats**: 60%

Their padding could be more generous, but their shape provides an effective lateral hold and lumbar support. Things are better up front than at the back, where the three padded ribs of the bench seat lack contouring.

• **Performance**: 60%

Disappointing because with a base engine whose weight-to-power ratio is a mediocre 22 lbs/hp, accelerations are more laborious than takeoffs. The engine remains rough and noisy and vibrates since it lacks the balance shafts such as the ones on the Quadfour.

• **Technical:** 60%

The steel unit body uses a McPherson front suspension and a semi-independent axle at the rear. Brakes are mixed but ABS and power assist are standard.

• **Trunk**: 50%

Large rather than deep, its volume can be increased by moving the rear bench seat, but access is hampered by the narrowness of the opening.

• **Brakes**: 50%

Their effectiveness is only average because stopping distances are rather long with the standard ABS. Panic stops are now straight, but there are intermittent lockups and the pedal gets firmer and more difficult to modulate.

• **Noise level**: 50%

The base engine, the only one that we were able to test, emits a fair amount of noises and vibrations, especially with the automatic.

• **Conveniences**: 50%

Storage spaces consist of a large glove box and a deep and lockable center console box, but door pockets are fairly small.

• **Handling**: 50%

It seems to have progressed in the sense that the latest Cavalier-Sunfire are more stable in a tight turn. Roll is less pronounced than on earlier models and the appearance of understeer is delayed. We were unable to lay our hands on the sports version by deadline to check its progress.

• **Interior space**: 50%

The sedan is roomier than the coupe. On the convertible, rear legroom is scarce, especially when the front seats have been moved back.

WEAK POINTS

• **Insurance**: 40%

Taking into account their price, these cars have a high insurance cost because the insurers know that they are sold in industrial quantities,which guarantees them a comfortable bundle.

• **Depreciation**: 40%

The Cavalier loses more than Sunbird-Sunfire, but it is mostly the large number on the market which increases their depreciation.

CONCLUSION

• **Overall rating**: 60.5%

You can't hold anything against these cars which, for a large number of owners, are a safe way to get around on a reasonable budget. 😐

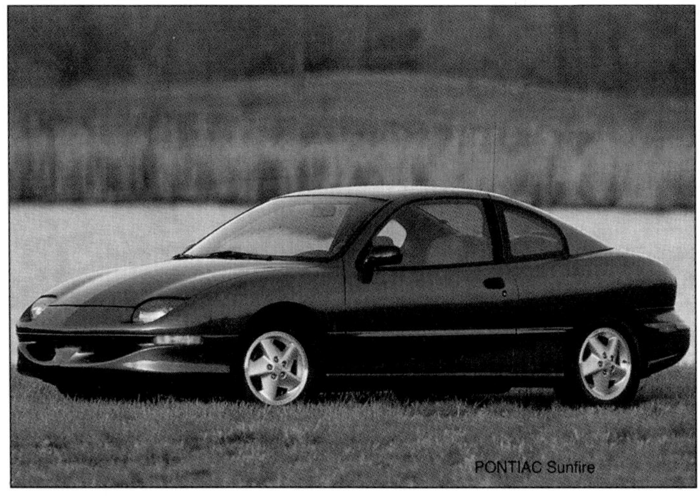

PONTIAC Sunfire

SPECIFICATIONS & PRICES

Model	Version/Trim	Body/ Seats	Interior volume cu.ft.	Trunk volume cu ft.	Drag coef. Cd	Wheel-base in	Lgth x Wdthx Hght. inx inx in	Curb weight lb	Towing capacity std. lb	Susp. type ft/rr	Brake type ft/rr	Steering system	Turning diameter ft	Steer. turns nber	Fuel. tank gal.	Standard tire size	Standard power train	PRICES US. $ 1994
CHEVROLET		**General warranty: 3 years / 36,000 miles; antipollution: 5 years / 50,000 miles; perforation corrosion: 6 years / 100,000 miles.**																
Cavalier		2 dr.cpe.4	84	12.4	0.39	104.1	180.3x67.4x53.2	2617	NR	i/si	dc/dr/ABS	pwr.r/p.	35.6	2.88	15.2	195/70R14	L4/2.2/M5	**$8,970**
Cavalier		4 dr.sdn.5	88	13.2	0.38	104.1	180.3x67.4x54.8	2676	1000	i/si	dc/dr/ABS	pwr.r/p.	35.6	2.88	15.2	195/70R14	L4/2.2/M5	**$9,120**
Cavalier	RS	4 dr.sdn.5	88	13.2	0.38	104.1	180.3x67.4x54.8	2736	1000	i/si	dc/dr/ABS	pwr.r/p.	35.6	2.88	15.2	195/65R15	L4/2.2/A3	**$11,440**
Cavalier	Z24	2 dr.cpe.5	84	12.4	0.39	104.1	180.3x67.4x53.2	2789	NR	i/si	dc/dr/ABS	pwr.r/p.	35.6	2.88	15.2	205/55R16	L4/2.2/M5	**$13,994**
Cavalier	Z24	2 dr. con.4	76	10.4	0.42	104.1	180.3x67.4x53.9	2837	NR	i/si	dc/dr/ABS	pwr.r/p.	35.6	2.88	15.2	195/65R15	L4/2.3/A3	**$19,995**
PONTIAC		**General warranty: 3 years / 36,000 miles; antipollution: 5 years / 50,000 miles; perforation corrosion: 6 years / 100,000 miles.**																
Sunfire	SE	con.. 2 p.4	76	10.4	0.42	104.1	181.9x68.4x53.9	2833	NR	i/si	dc/dr/ABS	pwr.r/p.	35.6	2.88	15.2	195/70R14	L4/2.2/A3	-
Sunfire	SE	2 dr.cpe.5	87.8	12.4	0.39	104.1	181.9x68.4x53.2	2679	NR	i/si	dc/dr/ABS	pwr.r/p.	35.6	2.88	15.2	195/70R14	L4/2.2/M5	-
Sunfire	SE	4 dr.sdn.5	92	13.1	0.38	104.1	181.7x67.9x54.8	2723	1000	i/si	dc/dr/ABS	pwr.r/p.	35.6	2.88	15.2	195/70R14	L4/2.2/M5	-
Sunfire	GT	3 dr. cpe.5	87.8	12.4	0.39	104.1	181.9x68.4x53.2	2828	NR	i/si	dc/dr/ABS	pwr.r/p.	35.6	2.88	15.2	195/70R14	L4/2.3/M5	-

See page 393 for complete 1995 Price List

A new status...

While the Eldorado remains very American, the Seville has given Cadillac added status. Happily, its image has been traded up from car supplier to retirees, limousines, and funeral cars to the status of creator of deluxe cars of international caliber.

CADILLAC Seville

The success of these two Cadillacs has allowed their builder to change image and to compete against high-pitched Europeans in matters of deluxe-top-performance cars. The Eldorado and the Seville are offered in standard and touring versions, equipped with the famous Northstar 4.6 L V8 with 32 valves, which puts out 270 hp standard or 295 in touring.

STRONG POINTS

• **Technical**: 90%

These models share the same platform and power train. Their steel unit body uses an independent suspension, disc brakes at four wheels, an ABS and a traction control. The self-adjusting suspension is available on the base SLS which also receives Michelin tires. The STS are supplied with Goodyear Eagles. The conservative lines which Pininfarina designed for the Allante provide an effective aerodynamic coefficient and low drag considering the car's bulk.

• **Safety**: 90%

In case of an impact, front-seat occupants seem better protected than the ones in the rear seats, which is why the safety does not reach a maximum rating in spite of the presence of two air bags and good structural rigidity.

• **Satisfaction**: 80%

It could improve still further to equal that of the Lexus which, according to J.D. Power's studies, is the best current reference for quality as perceived by the buyers.

• **Driver's compartment**: 80%

These cars address themselves to a different buyer group, more familiar with deluxe imports, who will not be let down by the conservative Audi-like look of the dashboard. The analog instrumentation is

DATA

Category: front-wheel drive luxury coupes and sedans.
Class : 7

HISTORIC

Introduced in: 1966: Eldorado; 1975: Seville.
Modified in: Eldorado: 1971, 1975, 1979, 1986, 1991.
Seville: 1979,1985, 1988, 1991.
Made in: Hamtramck, Detroit, MI, U.S.A.

RATINGS

Safety:	90 %
Satisfaction:	82 %
Depreciation:	65 %
Insurance:	3.7 % ($1,000-$1,065)
Cost per mile:	$0.94

NUMBER OF DEALERS

In U.S.A.: 1,600 Cadillac.

SALES ON THE AMERICAN MARKET

Model	1992	1993	Result
Eldorado	27,527	22,982	- 16.6 %
Seville	41,152	35,280	- 14.3 %

MAIN COMPETITORS

BMW 540, CADILLAC De Ville, INFINITI Q45, LEXUS LS400 & SC400, LINCOLN Continental & Mark VIII, MERCEDES-BENZ E420.

EQUIPMENT

CADILLAC Eldorado CADILLAC Seville	base SLS	Touring Cpe STS
Automatic transmission:	S	S
Cruise control:	S	S
Power steering:	S	S
ABS brake:	S	S
Air conditioning:	S	S
Air bags (2):	S	S
Leather trim:	S	S
AM/FM/ Radio-cassette:	S	S
Power door locks:	S	S
Power windows:	S	S
Tilt steering column:	S	S
Dual power mirrors:	S	S
Intermittent wipers:	S	S
Light alloy wheels:	S	S
Sunroof:	O	O
Anti-theft system:	S	S

S : standard; O : optional; - : not available

COLORS AVAILABLE

Exterior: Green, White, Black, Beige, Garnet, Red, Mocha, Amethyst, Blue,Argile.
Interior: Black, Taupe, Beige, Capuccino, Cherry, Blue, Argile.

MAINTENANCE

First revision:	3,000 miles
Frequency:	6,000 miles
Diagnostic plug:	Yes

WHAT'S NEW IN 1995 ?

- **A few different cosmetic and equipment details.**
- **Dextron III automatic transmission fluid guaranteed for 100,000 miles.**

Model/ versions *: standard	ENGINE Type / timing valve / fuel system	Displacement cu/in	Power bhp @ rmn	Torque lb.ft @ rpm	Compres. ratio	TRANSMISSION Driving wheels / transmission	Final ratio	PERFORMANCE Acceler. 0-62 mph s	Stand. 1/4 mile s	Stand. 5/8 mile s	Braking 62-0 mph ft	Top speed mph	Lateral acceler. G	Noise level dBA	Fuel economy mpg city	Fuel economy hwy	Gasoline type / octane
1)	V8* 4.6 DOHC-32-EFI	279	275 @ 5600	300 @ 4000	10.3 :1	front - A4*	3.11	8.0	16.4	29.0	154	124	0.80	66	16.0	25.0	S 91
2)	V8* 4.6 DOHC-32-EFI	279	300 @ 6000	295 @ 4400	10.3 :1	front - A4*	3.71	7.7	16.0	28.6	157	137	0.80	66	16.0	25.0	S 91

1) * Eldorado & Seville SLS 2) * Eldorado Touring & Seville STS

easy to read, with dials lit up from behind. Controls are those of an American car, with complexity in setting delays for the headlights and interior lights. Visibility would be better from 3/4 rear if the "C" pillar wasn't as thick and if the outside rear-view mirrors were larger.

• **Quality/fit/finish**: **80%**
The interior is conservative, directly inspired by Audi. However, appliqués of "Zebrano" wood are a little showy, and the quality of some trim items looks inexpensive or lacks the quiet elegance indispensable in this category. In this sense some redesign will be necessary.

• **Steering**: **80%**
Its assistance varies with speed and the ratio as well as its precision is that of a sports car. However the body size and the large steering diameter limit maneuverability.

• **Insurance**: **80%**
Its rate is fairly reasonable but the size of the sticker price results in a hefty premium.

• **Suspension**: **80%**
The SLS suspension is smooth on good pavement but becomes a noisy package when the pavement turns bad. That of the STS is firmer in all conditions and can become uncomfortable if the rhythm is not moderated.

• **Performance**: **80%**
The arrival of the Northstar rocks the tranquil image that Cadillac built up over the years. Takeoffs and pickups of the 4.6 L V8 are well-muscled. There is ample midrange torque as well as power. You keep getting an almost linear increase in power along with rpm thanks to a near flat torque curve.

• **Noise level**: **70%**
Soundproofing is impressive and the drive train reasonably discrete. The engine makes itself heard during strong accelerations, but this is not necessarily disagreeable. At times you perceive some airstream activity around the dash or the sliding roof.

• **Conveniences: 70%**
Air conditioning efficiency is fierce but storage spots are not numerous. There is a lack of pockets at the bottom of the doors, and a storage bin could be included in the rear center armrest as in the Aurora.

• **Access**: **70%**
It is no easier to reach the rear seats of a Seville than those of an Eldorado because the doors are relatively narrow and space is limited.

• **Handling: 70%**
The suspension of the SLS version is more unctuous than that of the firmer STS. In both instances, the body motions are well-controlled by an automatic device that can detect an increase in lateral acceleration and use this information to control body roll. Traction and side bite are superior.

• **Seats**: **70%**
They provide effective lateral hold and have better shape and padding than before. A sophisticated lumbar support allows good adjustment to fit a broad spectrum of people. Unfortunately, the leather trim is very slippery.

• **Interior space**: **70%**
Only four people will be able to appreciate the charm of the interior, which does not exactly relate to the exterior bulk of these vehicles. The sliding roof interferes with the available interior height.

• **Trunk**: **60%**
Without being immense, it will hold a sufficient amount of baggage, which, thanks to a low sill, will be handled without difficulty.

WEAK POINTS

• **Price/equipment: 20%**
The Northstar engine gives a new dimension to these cars which compare favorably to their Japanese or German competition. The STS version is better-equipped and includes rear-view mirrors with electrochemical color change.

• **Fuel consumption**: **30%**
The weight and performance are respectable and translate themselves into a husky fuel consumption that dips under 17 mpg.

• **Braking: 30%**
Panic stops provoke a major nose dip, and in spite of ABS you can note light wheel blockage while pedal effort becomes difficult to gauge. Otherwise the stops are straight and padding fade-resistance is above average.

• **Depreciation: 30%**
It remains a little higher than that of its competition because Cadillac's reputation has not reached that of a Lexus or a Mercedes.

CONCLUSION

• **Overall rating: 66.5%**
Cadillac has taken the high road to reach a younger buyer group and to interest purchasers who are looking elsewhere for high-performance luxury cars. ☺

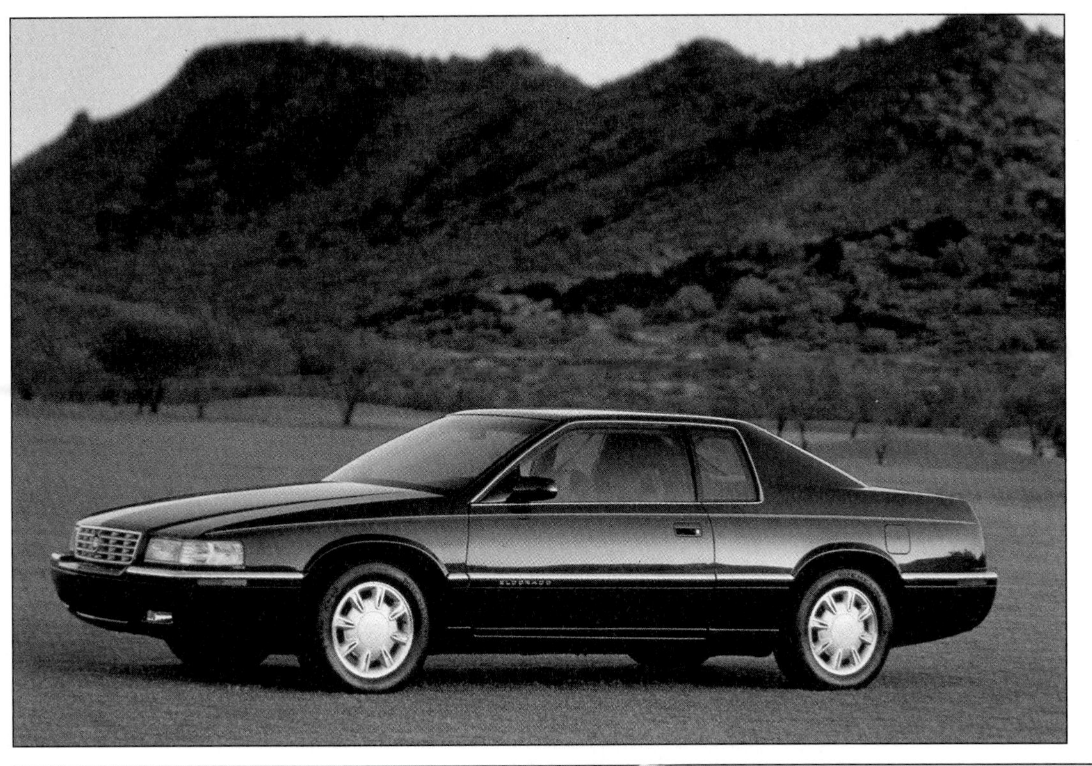

SPECIFICATIONS & PRICES

Model	Version/Trim	Body/ Seats	Interior volume cu.ft.	Trunk volume cu ft.	Drag coef. Cd	Wheel-base in	Lgth x Wdthx Hght. inx inx in	Curb weight lb	Towing capacity std. lb	Susp. type ft/rr	Brake type ft/rr	Steering system	Turning diameter ft	Steer. turns nber	Fuel. tank gal.	Standard tire size	Standard power train	PRICES US. $ 1994
CADILLAC		**General warranty: 4 years / 50,000 miles; antipollution: 5 years / 50,000 miles; perforation corrosion: 6 years / 100,000 miles.**																
Eldorado	base	2 dr.cpe.5	100	15	0.33	108.0	202.2x75.5x53.6	3774	1000	i/i	dc/dc/ABS	pwr.r/p.	40.4	2.65	20	225/60R16	V8/4.6/A4	**$37,690**
Eldorado	Touring	2 dr.cpe.5	100	15	0.33	108.0	202.2x75.5x53.6	3818	1000	i/i	dc/dc/ABS	pwr.r/p.	40.4	2.65	20	225/60ZR16	V8/4.6/A4	**$40,990**
Seville	SLS	4 dr.sdn.5	105	14.4	0.33	111.0	204.0x74.2x54.5	3891	1000	i/i	dc/dc/ABS	pwr.r/p.	41.7	2.65	20	225/60R16	V8/4.6/A4	**$41,430**
Seville	STS	4 dr.sdn.5	105	14.4	0.33	111.0	204.0x74.2x54.5	3891	1000	i/i	dc/dc/ABS	pwr.r/p.	41.7	2.65	20	225/60ZR16	V8/4.6/A4	**$45,330**

See page 393 for complete 1995 Price List

Between two...

General Motors is the only builder to offer intermediate vans. For increased comfort, only the extended body versions have been retained, and the new front end design gives these vehicles a more defined and sympathetic allure.

The Astro-Safari are the only intermediate vans offered on the market. They fit between the Sportvan/Rally and the Lumina/Trans Sport. The only available engine is a 4.3 L V6 offered in 165 and 190 hp versions with electronically controlled 4-speed automatic and 2- or 4-wheel drive. This year the old extended body is the only one available. The trims are SC, SL, and LT at Chevrolet and SL, SLE, and SLT at GMC.

STRONG POINTS

• Interior space: 90%
It is not really proportional to the bulk of these vehicles. While width is interesting, it lacks height, and there is little space to circulate between seats and even less legroom up front.

• Steering: 80%
Precise, well-modulated, with a good ratio, it doesn't have much feedback and doesn't inform about the state of the road. Maneuverability of the rear-wheel drive is superior to the 4WD, which has a larger turning diameter.

• Satisfaction: 75%
With time, and some needed design changes, it has gained good points, but there is still something to be done because a certain amount of owners would not again buy the same vehicle.

• Access: 70%
The doors do not open sufficiently to reach the front seats with ease, but the rear bumper is a good foot step and the rear doors open "ranch" style. These are original and practical plus-points.

• Seats: 70%
It is only with the CL/SLE trims that one can start pretending to be well-seated in these vehicles because those that equip the base vehicle are as poorly shaped as they are padded.

DATA

Category: mid-size rear-wheel drive or AWD vans.
Class: utilities

HISTORIC

Introduced in: 1985
Modified in: 1990: stretch version & AWD; 1995: short wheel base withdraw.
Made in: Baltimore, MD, U.S.A.

RATINGS

Safety: 70 %
Satisfaction: 75 %
Depreciation: 67 %
Insurance: 5.5 % ($585- $700)
Cost per mile: $0.42

NUMBER OF DEALERS

In U.S.A.: 4,446 Chevrolet-Geo.

SALES ON THE AMERICAN MARKET

Model	1992	1993	Result
Astro	115,815	125,567	+ 8.0 %
Safari	38,873	46,054	+ 25.6 %

MAIN COMPETITORS

CHEVROLET Lumina, CHRYSLER Town & Country, DODGE Caravan/Grand Caravan, FORD Aerostar, Villager & Windstar, PONTIAC Trans Sport, MAZDA MPV, NISSAN Quest, PLYMOUTH Voyager/Grand Voyager, TOYOTA Previa, VW Eurovan.

EQUIPMENT

CHEVROLET Astro GMC Safari	CS SL	CL SLE	LT SLT
Automatic transmission:	S	S	S
Cruise control:	O	O	S
Power steering:	S	S	S
ABS brake:	S	S	S
Air conditioning:	S	S	S
Air bag (left):	S	S	S
Leather trim:	-	-	O
AM/FM/ Radio-cassette:	O	O	S
Power door locks:	O	O	S
Power windows:	O	O	S
Tilt steering column:	O	O	S
Dual power mirrors:	O	S	S
Intermittent wipers:	S	S	S
Light alloy wheels:	O	S	S
Sunroof:	-	-	-
Anti-theft system:	-	-	-

S : standard; O : optional; - : not available

COLORS AVAILABLE

Exterior: White, Silver, Gray, Blue, Teal, Black, Brown, Beige, Red.
Interior: Gray, Blue, Garnet, Beige.

MAINTENANCE

First revision: 3,000 miles
Frequency: 6,000 miles
Diagnostic plug: Yes

WHAT'S NEW IN 1995 ?

- **Available only with the extended body.**
- **New front end design (grill, fender, hood).**
- **V6 4.3 L 190 hp standard.**
- **Enriched equipment.**
- **Air conditioning without CFC.**
- **Dextron III automatic transmission fluid guaranteed for 100,000 miles.**

ENGINE					TRANSMISSION			PERFORMANCE									
Model/ versions *: standard	Type / timing valve / fuel system	Displacement cu/in	Power bhp @ rmn	Torque lb.ft @ rpm	Compres. ratio	Driving wheels / transmission	Final ratio	Acceler. 0-62 mph s	Stand. 1/4 mile s	Stand. 5/8 mile s	Braking 62-0 mph ft	Top speed mph	Lateral acceler. G	Noise level dBA	Fuel economy mpg city	hwy	Gasoline type / octane
base	V6 4.3 OHV-12-CPI	262	190 @ 4400	260 @ 3400	9.1 :1	re./4 - A4*	3.73	11.0	17.9	32.3	157	106	0.67	68	15.0	19.0	R 87

• **Suspension: 70%**
It is when loaded and traveling down a freeway in good condition that these vehicles provide the best comfort. Traveling empty on side roads, the wheels do much jumping and vehicle path can be affected.

• **Driver's compartment: 60%**
Seated high, the driver benefits from an excellent visibility toward the front and the sides, and the outside rear-view mirrors are well-sized. Rear visibility is limited by the small windows which also lack a defroster. There is a windshield washer on side hinged doors while the ranch doors correct these problems. The instrumentation has the gadget look, lacks precision and remains hard to read when the lighting is weak.

• **Insurance: 60%**
The premium is not that much higher than for the compact models, taking into account their load capacity and superior traction.

• **Safety: 60%**
The structure of these utility vehicles resists collisions fairly well and the driver has an air bag. Unfortunately, other occupants are less well-protected.

• **Storage: 60%**
There is enough space behind the last bench seat to pack in a sufficient volume of baggage.

• **Technical: 50%**
The steel Astro van unit body is mounted on a ladder-type chassis derived from the S15 pickups. Even though the front end has been redesigned with less flowing lines, the aerodynamic coefficient didn't suffer too much. The front suspension is independent, even on a 4X4, while at the rear the rigid axle is supported by 2 leaf springs. Mixed braking is fitted with ABS on all 4 wheels.

• **Quality/fit/finish: 50%**
It leaves much to be desired because after a period of time the trim and the plastics rapidly lose some of their freshness, and the body is full of rattles.

• **Performance: 60%**
The torque of the V6 engine is well utilized by the transmission. Traction depends on tire quality. Takeoff is livelier (more brutal in some instances) than passing because the accelerator is more sensitive. In 2WD and winter driving, this can produce some surprises.

• **Noise level: 50%**
Empty, the body resonates and the tires pound when crossing tar strips.

WEAK POINTS

• **Fuel consumption: 30%**
It is high in all circumstances because of the weight and large engine displacement.

• **Braking: 35%**
The ABS on the 4 wheels makes the stops linear but the distances are fairly long and the pedal feel hardens rapidly, which makes it difficult to control the deceleration with precision.

• **Price/equipment: 40%**
More expensive than compact pickups, these vehicles are not better equipped and the investment can only be justified for carrying loads or towing a heavy trailer.

• **Depreciation: 40%**
It is higher than average because the budget is closer to that of a utility than of a station wagon.

• **Handling: 40%**
The 2WD version is distinctly understeering, especially empty, and traction is poor, even on dry pavement. The transfer of 35% of torque to the front wheels balances out the all-wheel drive to good advantage and makes winter use safer, especially in heavy snow areas.

• **Conveniences: 40%**
The interior does not have much storage space other than a tray on the engine cover. One would like to find door pockets or a ceiling console that is really practical.

CONCLUSION

• **Overall rating: 56.0%**
The Astro/Safari remains more of a utilitarian than family van aimed at professionals who are looking for a compromise between compact and conventional workhorse. 😐

OWNERS' SUGGESTIONS

-An economical Diesel engine
-Better handling when running empty
-Less accelerator sensitivity
-More self-assured winter driving
-More front space
-Better manufacturing and finish quality

SPECIFICATIONS & PRICES

Model	Version/Trim	Body Seats	Drag coef. Cd	Wheel-base in	Lgth x Wdthx Hght. inx inx in	Curb weight lb	Towing capacity std. lb	Susp. type ft/rr	Brake type ft/rr	Steering system	Turning diameter ft	Steer. turns nber	Fuel. tank gal.	Standard tire size	Standard power train	PRICES US. $ 1994
CHEVROLET / GMC		General warranty: 3 years / 36,000 miles; antipollution: 5 years / 50,000 miles; perforation corrosion: 6 years / 100,000 miles.														
Astro / Safari	2x4 long CS / SL	4dr.van.2/8	0.38	111.0	189.8x77.4x75.9	3997	5500	i/r	dc/dr/ABS	pwr.bal.	39.5	3.10	27	215/75R15	V6/4.3/A4	$16,525
Astro / Safari	2x4 long CL / SLE	4dr.van.5/8	0.38	111.0	189.8x77.4x75.9	4123	5500	i/r	dc/dr/ABS	pwr.bal.	39.5	3.10	27	215/75R15	V6/4.3/A4	$16,827
Astro / Safari	2x4 long LT / SLT	4dr.van.5/8	0.38	111.0	189.8x77.4x75.9	4175	5500	i/r	dc/dr/ABS	pwr.bal.	39.5	3.10	27	215/75R15	V6/4.3/A4	-
Astro / Safari	4x4 long CS / SL	4dr.van.2/8	0.38	111.0	189.8x77.4x75.9	4310	5000	i/r	dc/dr/ABS	pwr.bal.	39.5	2.67	27	215/75R15	V6/4.3/A4	$18,854
Astro / Safari	4x4 long CL / SLE	4dr.van.5/8	0.38	111.0	189.8x77.4x75.9	4431	5000	i/r	dc/dr/ABS	pwr.bal.	39.5	2.67	27	215/75R15	V6/4.3/A4	$19,156
Astro / Safari	4x4 long LT / SLT	4dr.van.5/8	0.38	111.0	189.8x77.4x75.9	4471	5000	i/r	dc/dr/ABS	pwr.bal.	39.5	2.67	27	215/75R15	V6/4.3/A4	-

See page 393 for complete 1995 Price List

BUICK Skylark, CHEVROLET Beretta/Corsica, PONTIAC Grand Am, OLDSMOBILE Achieva

Show off...

GM has always bet on the appearance of its vehicles and looked for buyers less sensitive to road performance or quality. This philosophy has backfired because today it's the results that count and those of the models above have compromises.

These cars share the same format and the major part of their drive train. Chevrolets are manufactured in one plant and the other three makes share the same assembly line in another plant. The Skylark is sold in the following trims: Custom, Limited (sedan) or Grand Sport. The Achieva is an S or SC (coupe) and SL (sedan), the Grand Am is SE or GT and the base Chevrolets are standard or Z26 for the Beretta. The standard engine is a 2.3 L with a manual 5-speed or an automatic 4-speed. The most exciting models have the multi-valve Quad4 rated at 150 hp (Grand Am & Achieva SC) or the optional 160 hp 3.1 L V8.

STRONG POINTS

• Technical: 70%
Their steel unit body uses an independent McPherson style front suspension and a semi-independent torsion type axle at the rear. Brakes are mixed; ABS and variable power steering are standard equipment. While the Grand Am lines are the most spectacular, those of the Skylark are more controversial while the Achieva and Corsica go by totally unnoticed. Without setting any records, their aerodynamic coefficients are acceptable.

• Seats: 70%
The lateral and lumbar supports are only effective on the sport model where their contouring and padding are sufficient.

• Steering: 70%
Fast and direct on a good road, it becomes imprecise on bad pavement with poor guidance from the front end. The reduced steering diameter and a lower number of steering wheel turns lock-to-lock have improved maneuverability.

• Suspension: 70%
Because of its softness, the ride of the standard models is superior to that of the sports coupes on the freeway. The sports coupe suspension

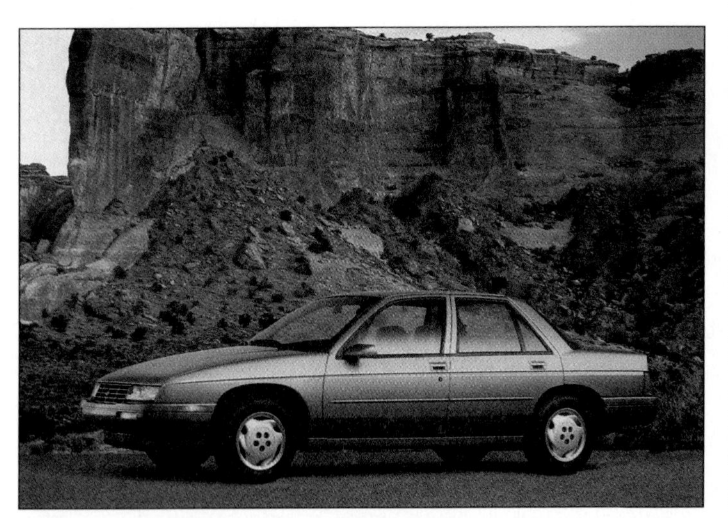

DATA

Category: front-wheel drive compact sedans and coupes.
Class : 4

HISTORIC

Introduced in: 1985 (2-door),1986 (4-door) L: 1988.
Modified in: 1990: 2.2 L engine & V6 3.1L. 1991: Achieva.
Made in: Wilmington DE & N. Lansing, MI, U.S.A.

RATINGS

	sedans	coupes
Safety:	90 %	
Satisfaction:	65 %	72%
Depreciation:	60 %	62 %
Insurance:	6.6 % ($520)	7.5 % ($660-$720)
Cost per mile:	$0.40	$0.46

NUMBER OF DEALERS

In U.S.A.: 3,000 Buick, 4,466 Chev-Geo, 3,100 Olds, 2,953 Pontiac.

SALES ON THE AMERICAN MARKET

Model	1992	1993	Result
Beret/Cors.	166,625	171,794	+ 3.1 %
Achieva	71,705	53,473	- 25.5 %
Grand Am	210,332	214,760	+ 2.1 %
Skylark	71,805	53,473	- 25.6 %

MAIN COMPETITORS

sedans: CHRYSLER Cirrus, DODGE Stratus, FORD Contour, HONDA Accord, HYUNDAI Sonata, MAZDA 626, MERCURY Mystique, NISSAN Altima, SUBARU Legacy, TOYOTA Camry, VOLKSWAGEN Passat.
coupes: CHRYSLER Sebring, DODGE Avenger, FORD Probe, HONDA Prelude, MAZDA MX-6, NISSAN 240SX,TOYOTA Celica, VOLKSWAGEN Corrado.

EQUIPMENT

BUICK Skylark	**Cust.**	**Ltd**	**Gran Sport**	
CHEVROLET Corsica/Beret.	**base**			**Z26**
OLDS. Achieva	**S**	**SC**	**SL**	
PONTIAC Grand Am	**-**		**SE**	**GT**
Automatic transmission:	O	S	S	S
Cruise control:	O	O	S	S
Power steering:	S	S	S	S
ABS brake:	S	S	S	S
Air conditioning:	O	S	S	S
Air bag (left):	S	S	S	S
Leather trim:	-	-	-	-
AM/FM/ Radio-cassette:	O	O	S	S
Power door locks:	S	S	S	S
Power windows:	O	O	O	S
Tilt steering column:	O	O	S	S
Dual power mirrors:	O	S	S	S
Intermittent wipers:	S	S	S	S
Light alloy wheels:	-	O	S	S
Sunroof:	-	-	O	O
Anti-theft system:	-	-	-	-

S : standard; O : optional; - : not available

COLORS AVAILABLE

Exterior: Silver, White, Blue, Black, Green, Red, Gray.
Interior: Graphite, Beige, Red, Blue.

MAINTENANCE

First revision: 3,000 miles
Frequency: 6,000 miles
Diagnostic plug: Yes

WHAT'S NEW IN 1995 ?

- **New Quad4 150 hp 2.3 L with balanced shafts.**
- **Dextron transmission fluid, good for 100,000 miles.**
- **New power steering pump.**
- **New tubular rear axle with springs centered on axis of the wheels.**

ENGINE					TRANSMISSION					PERFORMANCE							
Model/ versions *: standard	Type / timing valve / fuel system	Displacement cu/in	Power bhp @ rmn	Torque lb.ft @ rpm	Compres. ratio	Driving wheels / transmission	Final ratio	Acceler. 0-62 mph s	Stand. 1/4 mile s	Stand. 5/8 mile s	Braking 62-0 mph ft	Top speed mph	Lateral acceler. G	Noise level dBA	Fuel economy mpg city	hwy	Gasoline type / octane
1)	L4* 2.2 OHV-8-MFI	133	120 @ 5200	130 @ 4000	9.0 :1	front - M5*	3.83	10.0	16.8	31.0	138	100	0.77	68	25.0	34.0	R 87
						front-A3-A4	3.18	11.7	17.7	31.8	144	97	0.77	68	25.0	31.0	R 87
2)	L4* 2.3 DOHC-16-MFI	138	150 @ 6000	145 @ 4800	9.5 :1	front - M5*	3.94	9.5	16.6	30.6	154	112	0.75	67	23.0	35.0	R 87
						front - A3/A4	2.93	10.4							22.0	32.0	R 87
3)	V6* 3.1 OHV-12-MFI	191	155 @ 5200	185 @ 4000	9.6 :1	front-A3-A4	2.97	9.5	17.7	30.4	150	109	0.80	67	21.0	29.0	R 87

1) *Corsica/Beretta 2)* Grand Am & Achieva SC 3) * Beretta Z26 & Gran Sport, optional on other models.

OLDSMOBILE Achieva

is firm enough to stir up the occupants on the bad roads

• Safety: 70%
The structure of the 4-door models resists impacts better than the 2-door. The driver's protection has been improved by installing an air bag but that of the other occupants is dubious.

• Fuel consumption: 65%
That of the standard engine is normal while the QuadFour does not lack appetite when pushed to its limits.

• Satisfaction: 60%
Very average because most owners are disappointed by a lack of quality and little problems which penalize the resale value.

• Noise level: 60%
The 4-cylinder engines remain noisy and vibrate during accelerations while the V6 is quieter. Better insulation would allow a reduction of rolling noise.

• Price/equipment: 60%
Prices are tempting, but equipment is not always complete and the quality is often questionable.

• Trunk: 60%
Deeper and longer than before, its high sill does not allow easy baggage handling.

• Quality/fit/finish: 60%
The materials' quality is as mediocre as the care devoted to finish and interior trim. The interior styling of the Skylark is one of the most delirious ever encountered.

• Access: 60%
It is easier to reach the front seats than the rear ones because the doors are narrow and the front seat offers little space.

• Driver's compartment: 60%
The driver's position is old-fashioned with the pedals off-centered and the steering column too long. Visibility is better toward the front than 3/4 rear or toward the back because of the thickness of the back light pillars and the small rear-view mirrors.

• Performance: 60%
The weight-to-power ratio of the standard models does not allow them to post anything but mediocre performance. They are improved with the Quad4 and the V6, whose torque will seduce performance fans. The manual transmission is not well-staged and shifts are rough. Fortunately, the 4-speed automatic offers more glowing alternatives.

• Handling: 50%
While it has progressed, much depends on tire quality and the sports level of the suspension. The standard one is very soft and generates body motions proportional to the roll, while that of the GT and SC is firmer and more stable.

• Interior space: 50%
The volume of the interior only allows seating for 4 people and the rear seats offer little user comfort, more like catching a case of claustrophobia.

WEAK POINTS

• Braking: 40%
Stopping distances are long, friction materials lack bite. The ABS helps balance panic stops in spite of some small but ill-timed wheel lock-ups.

• Conveniences: 40%
Practical storage space is scarce. In richer versions, the center console storage compensates for the small size of the glove box.

• Depreciation: 40%
The Grand Am loses less value and sells faster than the Skylark, Achieva or Corsica.

• Insurance: 45%
These cars are relatively high to insure, especially the Grand Am, which is perceived as a sports car.

CONCLUSION

• Overall rating: 58.0%
Seduced by their allure, the owners of these models become quickly disenchanted with them in use. 😐

OWNERS' SUGGESTIONS

- More effective braking
- Improved quality
- Higher performance tires
- Improved handling
- More storage room
- More modern look (Corsica)

SPECIFICATIONS & PRICES

Model	Version/Trim	Body/ Seats	Interior volume cu.ft.	Trunk volume cu ft.	Drag coef. Cd	Wheel-base in	Lgth x Wdthx Hght. inx inx in	Curb weight lb	Towing capacity std. lb	Susp. type ft/rr	Brake type ft/rr	Steering system	Turning diameter ft	Steer. turns nber	Fuel. tank gal.	Standard tire size	Standard power train	PRICES US. $ 1994
BUICK		**General warranty: 3 years / 36,000 miles; antipollution: 5 years / 50,000 miles; perforation corrosion: 6 years / 100,000 miles.**																
Skylark	Custom	2 dr.cpe.5	89.5	13.3	0.32	103.4	189.2x68.7x53.5	2888	1000	i/si	dc/dr/ABS	pwr.r/p.	35.3	2.33	15.2	195/70R14	L4/2.3/A3	**$13,734**
Skylark	Custom	4 dr.sdn.5	89.5	13.3	0.32	103.4	189.2x68.7x53.5	2941	1000	i/si	dc/dr/ABS	pwr.r/p.	35.3	2.33	15.2	195/70R14	L4/2.3/A3	**$13,734**
Skylark	Limited	2 dr.cpe.5	89.5	13.3	0.32	103.4	189.2x68.7x53.5	2888	1000	i/si	dc/dr/ABS	pwr.r/p.	35.3	2.33	15.2	195/70R14	L4/2.3/A3	**NA**
Skylark	Limited	4 dr.sdn.5	89.5	13.3	0.32	103.4	189.2x68.7x53.5	2941	1000	i/si	dc/dr/ABS	pwr.r/p.	35.3	2.33	15.2	195/70R14	L4/2.3/A3	**$16,334**
Skylark	Gran Sport	2 dr.cpe.5	89.5	13.3	0.32	103.4	189.2x68.7x53.5	3005	1000	i/si	dc/dr/ABS	pwr.r/p.	35.3	2.33	15.2	205/55R16	V6/3.1/A4	**$18,434**
Skylark	Gran Sport	4 dr.sdn.5	89.5	13.3	0.32	103.4	189.2x68.7x53.5	3058	1000	i/si	dc/dr/ABS	pwr.r/p.	35.3	2.33	15.2	205/55R16	V6/3.1/A4	**$18,434**
CHEVROLET		**General warranty: 3 years / 36,000 miles; antipollution: 5 years / 50,000 miles; perforation corrosion: 6 years / 100,000 miles.**																
Corsica	base	4 dr.sdn.5	91.6	379	0.36	103.4	183.4x68.5x54.2	2745	1000	i/i	dc/dr/ABS	pwr.r/p.	35.3	2.33	15.2	195/70R14	L4/2.2/M5	**$13,315**
Beretta	base	2 dr.cpe.5	90	382	0.36	103.4	187.3x68.2x53.0	2755	1000	i/i	dc/dr/ABS	pwr.r/p.	35.3	2.33	15.2	195/70R14	L4/2.2/M5	**$12,585**
Beretta	Z26	2 dr.cpe.5	90	382	0.33	103.4	187.3x68.2x53.0	2989	NR	i/i	dc/dr/ABS	pwr.r/p.	38.6	2.33	15.2	205/60R15	V6/3.1/A4	**$15,310**
OLDSMOBILE		**General warranty: 3 years / 36,000 miles; antipollution: 5 years / 50,000 miles; perforation corrosion: 6 years / 100,000 miles.**																
Achieva	S	2 dr.cpe.5	91	14	0.33	103.4	189.2x67.5x52.2	2716	1000	i/si	dc/dr/ABS	pwr.r/p.	35.3	2.88	15.2	205/70R14	L4/2.3/M5	**$14,210**
Achieva	S	4 dr.sdn.5	89.8	14	0.33	103.4	189.2x67.5x52.2	2780	1000	i/si	dc/dr/ABS	pwr.r/p.	35.3	2.88	15.2	205/55R16	L4/2.3/M5	**$14,310**
Achieva	SC	2 dr.cpe.5	91	14	0.33	103.4	189.2x67.5x52.2	2844	1000	i/si	dc/dr/ABS	pwr.r/p.	35.3	2.88	15.2	205/55R16	L4/2.3/M5	**$17,710**
Achieva	SL	4 dr.sdn.5	89.8	14	0.33	103.4	189.2x67.5x52.2	2877	1000	i/si	dc/dr/ABS	pwr.r/p.	35.3	2.88	15.2	205/55R16	L4/2.3/M5	**$17,710**
PONTIAC		**General warranty: 3 years / 36,000 miles; antipollution: 5 years / 50,000 miles; perforation corrosion: 6 years / 100,000 miles.**																
Grand Am	SE	2 dr.cpe.5	90.7	13.3	0.34	103.4	186.9x67.5x52.2	2824	1000	i/si	dc/dr/ABS	pwr.r/p.	35.3	2.5	15.2	195/70R14	L4/2.3/M5	**$12,514**
Grand Am	SE	4 dr.sdn.5	90.7	13.3	0.34	103.4	186.9x67.5x52.2	2881	1000	i/si	dc/dr/ABS	pwr.r/p.	35.3	2.5	15.2	195/70R14	L4/2.3/M5	**$12,614**
Grand Am	GT	2 dr.cpe.5	90.7	13.3	0.34	103.4	186.9x67.5x52.2	2888	NR	i/si	dc/dr/ABS	pwr.r/p.	35.3	2.5	15.2	205/55R16	L4/2.3/M5	**$15,014**
Grand Am	GT	4 dr.sdn.5	90.7	13.3	0.34	103.4	186.9x67.5x52.2	2940	NR	i/si	dc/dr/ABS	pwr.r/p.	35.3	2.5	15.2	205/55R16	L4/2.3/M5	**$15,114**

See page 393 for complete 1995 Price List

Mission accomplished...

With time, Saturns confirm what their builder claimed when they were first launched nearly three years ago. As Americans, they are original, reliable and sufficiently economical to seed doubt, among dyed in the wool Japanese car fanatics.

The Saturns continue their mission of being GM's broadside attack against Japanese compacts. The range consists of three models: 2-door coupe SC1 and SC2, 4-door sedan sold in three versions, SL, SL1 and SL2: and a 4-door station wagon SW1 and SW2. The engine is an 85 hp 1.9 L, and on the 2 versions, the same 1.9 L equipped with double overhead cams delivers 125 hp. A 5-speed manual transmission is standard and a 4-speed automatic is offered as an option.

STRONG POINTS

• **Satisfaction:** 90%
Excellent results. The first owners were surprised by the reliability of the Saturn, which does not give many problems.

• **Safety:** 90%
It improves with the delivery of a second front air bag, while their structure is good at resisting impacts.

• **Price/equipment:** 80%
The price of the standard models is more attractive than that of the more luxurious ones. The latter flirt dangerously with those of the competition which offers a long-standing reliability and a larger dealership network. Much vital equipment is optional, which is quite discouraging.

• **Technical:** 80%
Even though controversial, their lines are original and their aerodynamic is efficient with a drag coefficient that varies between 0.32 and 0.33. The Saturns' build is inspired by that inaugurated on the Fiero and later picked up on AP vans. The platform is based on a steel chassis-cage of which some sections are galvanized. This structure serves as a support for steel body panels at the hood, roof and trunk and thermo-formed plastic at the fenders and doors. On the sedan, the roof and trunk are also plastic. The suspension is independent at all four wheels and the braking is mixed: 4-wheel discs and ABS are optional.

• **Depreciation:** 70%
It is greater than average, no doubt because of the originality of these models. However, the dealer network is sparse.

• **Fuel consumption:** 70%
It holds around 28 mpg in normal driving but can drop to 24, when driving the SC2 with a heavy foot.

• **Suspension:** 70%
Its firmness reminds you of vintage Volkswagens. The unyielding shock control and lack of wheel travel often cause bottoming on poor pavement.

• **Quality/fit/finish:** 70%
The general appearance is flattering, both inside and out, and assembly, just like the finishing, is carefully done. The plastic body panels are well-adjusted and the finish is very brilliant.

• **Driver's compartment:** 70%
The driver can find a convenient position, thanks to combined seat and steering column adjustments, but the latter is still too long to be perfect. Visibility is good in spite of the windshield post thickness, and the

DATA

Category: compact front-wheel drive coupes and sedans.
Class : 3

HISTORIC

Introduced in: 1990: U.S.A; 1991: Canada.
Modified in: 1993: wagon and base coupe.
Made in: Spring Hill, Tennessee, U.S.A.

RATINGS

Safety: 90 %
Satisfaction: 90 %
Depreciation: 30 % (SC), 44 % (SL) 2 years
Insurance: SL/SC:12.0 % ($535) SL2/SC2: 7.7 % ($585)
Cost per mile: $0.32

NUMBER OF DEALERS

In U.S.A.: 300

SALES ON THE AMERICAN MARKET

Model	1992	1993	Result
Saturn	196,126	229,356	+ 14.5 %

MAIN COMPETITORS

Saturn SL: ACURA Integra, CHEVROLET Cavalier, DODGE-PLYMOUTH Neon, EAGLE Summit, FORD-MERCURY Escort-Tracer, HONDA Civic, HYUNDAI Elantra, MAZDA Protegé, PONTIAC Sunfire, SUBARU Impreza, TOYOTA Corolla, VOLKSWAGEN Jetta.
Saturn SC: HONDA del Sol, HYUNDAI Scoupe, MAZDA MX-3, TOYOTA Paseo.

EQUIPMENT

SATURN	SL/SC1SL1/SW1	SL2/SW2	SC2
Automatic transmission:	O	O	O
Cruise control:	O	O	O
Power steering:	O	S	S
ABS brake:	S	S	S
Air conditioning:	O	S	S
Air bags (2):	S	S	S
Leather trim:	-	O	O
AM/FM/ Radio-cassette:	O	O	O
Power door locks:	O	S	S
Power windows:	O	S	S
Tilt steering column:	O	O	O
Dual power mirrors:	S	S	S
Intermittent wipers:	O	S	S
Light alloy wheels:	O	O	S
Sunroof:	-	O	O
Anti-theft system:	-	-	-

S : standard; O : optional; - : not available

COLORS AVAILABLE

Exterior: White, Navy Blue, Plum, Green, Black, Red, Gold.
Interior: Tan, Gray, Black.

MAINTENANCE

First revision: 3,000 miles
Frequency: 6,000 miles
Diagnostic plug: Yes

WHAT'S NEW IN 1995 ?

- **Two air bags and ABS standard.**
- **New front end for the SC1 and SC2 coupes.**
- **Redesigned dashboard.**
- **Adjustable steering column.**

Model/ versions *: standard	ENGINE Type / timing valve / fuel system	Displacement cu/in	Power bhp @ rmn	Torque lb.ft @ rpm	Compres. ratio	TRANSMISSION Driving wheels / transmission	Final ratio	PERFORMANCE Acceler. 0-62 mph s	Stand. 1/4 mile s	Stand. 5/8 mile s	Braking 62-0 mph ft	Top speed mph	Lateral acceler. G	Noise level dBA	Fuel economy mpg city	hwy	Gasoline type / octane
1)	L4* 1.9 SOHC-8-MFI	116	100 @ 5000	115 @ 2400	9.3 :1	front - M5*	4.06	11.6	18.0	31.9	144	100	0.76	68	28.0	37.0	R 87
						front - A4	4.06	12.5	18.6	32.6	151	94	0.76	68	26.0	35.0	R 87
2)	L4* 1.9 DOHC-16-MFI	116	124 @ 5600	122 @ 4800	9.5 :1	front - M5*	4.06	8.0	16.0	29.0	148	119	0.80	68	25.0	34.0	R 87
						front - A4	4.06	8.8	16.7	29.8	141	112	0.80	68	23.0	32.0	R 87

1) SL,SL1,SC1 2) SC2 & SL2

dashboard has been redesigned and is more ergonomical than before.

• Handling: 70%
It is clean with a moderate roll, and body motions are well-controlled by simple but efficient damping. The understeering characteristics do not show up unless the driver goes in too deep into a corner, but this is easily controlled.

• Steering: 70%
The power-assisted one is more pleasant thanks to its better precision, modulation and faster reaction, while the manual steering is heavy, with many turns lock-to-lock and a pain for urban parking.

• Conveniences: 70%
The storage spaces include a large-sized glove box, door pockets, a recess in the center console and a flat tray under the dash on the 2 versions . The shoulder harness belts do not have a height adjustment, and there is no indicator for the automatic transmission shift adjuster on the dash.

• Access: 60%
Big folks will find it hard to reach the rear seats. The doors are narrow and the backs of the front seats provide little room.

• Seats: 60%
With little wrap around on standard models, they're better shaped on the other versions and provide good lateral support in spite of their Germanic padding.

• Performance: 60%
Those of the standard engine are anemic, but torque is sufficient to insure decent passing, while that of the multivalve is more playful, thanks to a well-balanced weight-to-power ratio. Standing starts and passing are crisper. However, both engines share the peculiarity of becoming rough, noisy and vibrating at high rpm.

• Trunk: 50%
Though you can enlarge it by lowering the rear bench seat, it is not very voluminous, and this even on a station wagon. However, access is facilitated by a sill which comes down to bumper level.

SATURN SW1

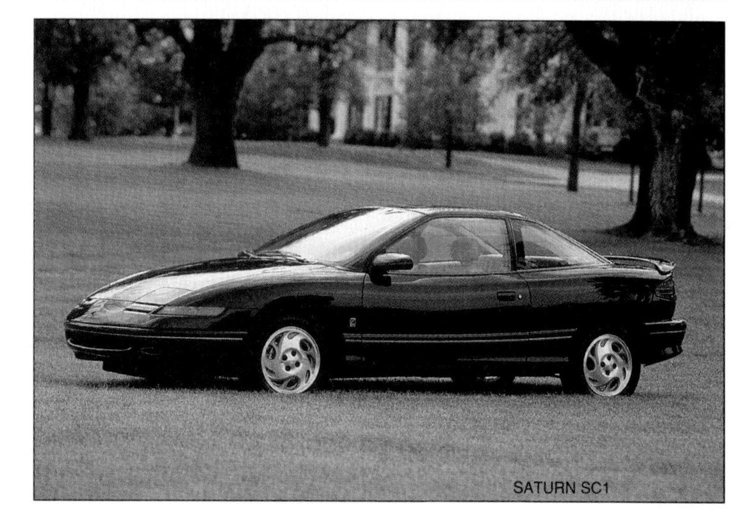

SATURN SC1

WEAK POINTS

• Insurance: 30%
The Saturns are more expensive to insure than others due to their structure. It is costlier to repair in case of a serious scrape.

• Interior space: 40%
The interior accepts four grown-ups in the sedan and station wagon, but the limited seat height earmarks it for children's use.

• Noise level: 40%
Soundproofing is one of the points in need of improvement on these models whose mechanical and rolling noises are high.

• Braking: 40%
Even though progressive and easy to modulate in normal circumstances, it shows up brutally during panic stops because the front wheels rapidly lock up and the trajectory becomes uncertain. ABS and 4-wheel discs are indispensable in stopping the SL2 and SC2 with their higher performance, but light wheel lockups and pedal pulsing are more disagreeable than dangerous.

CONCLUSION

• Overall rating: 64.0%
The arrival of Chrysler's Neon, which has more numerous sales points, will give the Saturn a run for its money. Saturn models remain one of the valid choices if you can control the optional equipment. ☺

OWNERS' SUGGESTIONS

-Less options
-New appearance
-ABS standard
-More storage volume in the station wagons
-Engines with more performance, less vibrations and noise
-Better soundproofing

SPECIFICATIONS & PRICES

Model	Version/Trim	Body/ Seats	Interior volume cu.ft.	Trunk volume cu ft.	Drag coef. Cd	Wheel-base in	Lgth x Wdthx Hght. inx inx in	Curb weight lb	Towing capacity std. lb	Susp. type ft/rr	Brake type ft/rr	Steering system	Turning diameter ft	Steer. turns nber	Fuel. tank gal.	Standard tire size	Standard power train	PRICES US. $ 1994
SATURN		**General warranty: 3 years / 36,000 miles; antipollution: 5 years / 50,000 miles; perforation corrosion: 6 years / 100,000 miles.**																
SC1		3 dr cpe. 4	76.4	11	0.32	99.2	173.2x67.5x50.6	2281	1000	i/i	dc/dr/ABS	pwr.r/p.	36	3.0	12.8	175/70R14	L4/1.9/M5	**$11,695**
SC2		3 dr cpe.4	76.4	11	0.31	99.2	170.9x67.5x50.6	2359	1000	i/i	dc/dr/ABS	pwr.r/p.	36	2.67	12.8	195/60R15	L4/1.9/M5	**$12,895**
SL		4 dr.sdn. 5	88.7	12	0.33	102.4	176.3X67.6X52.5	2324	1000	i/i	dc/dr/ABS	r/p.	37	4.0	12.8	175/70R14	L4/1.9/M5	**$9,995**
SL1		4 dr.sdn. 5	88.7	12	0.33	102.4	176.3X67.6X52.5	2326	1000	i/i	dc/dr/ABS	pwr.r/p.	37	3.0	12.8	175/70R14	L4/1.9/M5	**$10,795**
SL2		4 dr.sdn. 5	88.7	12	0.33	102.4	176.3X67.6X52.5	2405	1000	i/i	dc/dr/ABS	pwr.r/p.	37	2.67	12.8	195/60R15	L4/1.9/M5	**$11,795**
SW1		4 dr wgn. 5	90	24	0.32	102.4	176.3X67.6X52.5	2372	1000	i/i	dc/dr/ABS	pwr.r/p.	37	3.0	12.8	175/70R14	L4/1.9/M5	**$11,695**
SW2		4 dr wgn.5	90	24	0.32	102.4	176.3X67.6X52.5	2447	1000	i/i	dc/dr/ABS	pwr.r/p.	37	2.67	12.8	195/60R15	L4/1.9/M5	**$12,595**

See page 393 for complete 1995 Price List

Especially stylish ...

Very spectacular, with audacious lines, the GM minivans please especially by their style because from a practical point of view, others do better in offering long-body versions that allow seven-person seating and enough baggage space.

The minivans Chevrolet Lumina, Pontiac Trans Sport and Oldsmobile Silhouette share their drive train with the W family. The differences between them are only details in appearance and equipment. The Chevrolet Lumina is offered in standard or LS, the Pontiac Trans Sport in SE, while the Silhouette exists in a unique version sold only in the United States. The standard engine is a 3.1 L V6 with 3-speed automatic, while the 3.8 L V6 and the 4-speed are offered as options.

STRONG POINTS

• Safety: 80%
The structure of these vehicles is good at resisting impacts, an air bag now protects the driver and the doors have beamed reinforcements. The location of the taillights above the collision zone, well within view of other drivers, remains one of the innovations of these vans.

• Satisfaction: 80%
It is improved and is on a level playing field with the Caravan-Voyager which set the standards in this matter.

• Technical: 80%
These compact minivans distinguish themselves by a structure consisting of a galvanized steel platform topped by a steel cage to support thermo-formed plastic side panels and doors. The front suspension is independent, while at the rear there is a rigid axle connected by control arms and by a transversal Panhard rod. The brakes are mixed, but the ABS is standard. Their aerodynamic finesse is efficient and last year's front-end design changes reduce the front overhang.

• Insurance: 80%
In spite of their special construction, more finicky to repair, it compares with other compact vans.

• Quality/fit/finish: 70%
It is improved, but the immense dashboard lacks stiffness and the

OLDSMOBILE Silhouette

DATA

Category: front-wheel drive compact vans.
Class : utilities

HISTORIC

Introduced in: 1990
Modified in: 1992: V6 3.8L.engine 1994: touch-ups
Made in: North Tarrytown, NY, U.S.A.

RATINGS

	Lumina	Trans Sport
Safety:	80 %	
Satisfaction:	82 %	
Depreciation:	58 %	53 %
Insurance:	5.0 % ($585)	
Cost per mile:	$0.42	

NUMBER OF DEALERS

In U.S.A.: 4,466 Chevrolet-Geo, 3,100 Oldsmobile, 2,953 Pontiac.

SALES ON THE AMERICAN MARKET

Model	1992	1993	Result
Lumina	43,781	41,444	- 5.4 %
Trans Sport	29,881	28,324	- 5.3 %
Silhouette	14,831	11,860	- 20.0 %

MAIN COMPETITORS

CHEVROLET Astro, CHRYSLER Town & Country, DODGE Caravan, FORD Aerostar & Windstar, PLYMOUTH Voyager, MAZDA MPV, MERCURY Villager, NISSAN Quest, TOYOTA Previa, VW Eurovan.

EQUIPMENT

CHEVROLET Lumina	base	LS	
OLDSMOBILE Silhouette			base
PONTIAC Trans Sport		SE	
Automatic transmission:	S	S	S
Cruise control:	O	O	O
Power steering:	S	S	S
ABS brake:	S	S	S
Air conditioning:	O	O	S
Air bag (left):	S	S	S
Leather trim:	-	O	O
AM/FM/ Radio-cassette:	O	O	O
Power door locks:	O	O	S
Power windows:	O	O	S
Tilt steering column:	O	O	S
Dual power mirrors:	O	S	S
Intermittent wipers:	S	S	S
Light alloy wheels:	S	S	S
Sunroof:	O	O	O
Anti-theft system:	O	O	O

S : standard; O : optional; - : not available

COLORS AVAILABLE

Exterior: Silver, White, Blue, Black, Hed.
Interior: Light Gray, Blue, Ruby.

MAINTENANCE

First revision: 3,000 miles
Frequency: 6,000 miles
Diagnostic plug: Yes

WHAT'S NEW IN 1995 ?

- Safety interlock between the brake pedal and the shift selector.
- Dextron automatic transmission fluid good for 100,000 miles.
- New body colors.
- Traction control available on Trans Sport and SE Lumina.

	ENGINE					TRANSMISSION		PERFORMANCE									
Model/ versions *: standard	Type / timing valve / fuel system	Displacement cu/in	Power bhp @ rmn	Torque lb.ft @ rpm	Compres. ratio	Driving wheels / transmission	Final ratio	Acceler. 0-62 mph s	Stand. 1/4 mile s	Stand. 5/8 mile s	Braking 62-0 mph ft	Top speed mph	Lateral acceler. G	Noise level dBA	Fuel economy mpg city	hwy	Gasoline type / octane
base	V6* 3.1 OHV-12-EFI	191	120 @ 4200	175 @ 2200	8.5 :1	front - A3*	3.06	12.9	18.6	36.8	144	94	0.70	68	19.0	23.0	R 87
option	V6* 3.8 OHV-12-SFI	231	170 @ 4800	225 @ 3200	9.0 :1	front - A4*	3.06	11.5	18.4	35.6	154	109	0.72	68	17.0	22.0	R 87

PONTIAC Trans Sport

doors emit annoying noises.

• Suspension: 70%
The damping offers a good compromise between comfort and handling. The rear suspension gives good tracking and jumps around less than that of the Caravan/Voyager.

• Conveniences: 70%
The glove box is minuscule, but there are several other storage areas and more cup holders (16) than there could ever be occupants. When folded down, the rear seat backs form a loading platform.

• Interior space: 70%
It is better at the front seats than at the back where space is at a premium. The inside height decreases toward the back. The reduced distance between the seats, which are low, is inadequate for a comfortable position. You sit elbow to elbow, tight, narrow, and compressed.

• Driver's compartment: 60%
The position of the driver takes getting used to and makes some maneuvers touchy. A lack of distance feel is caused by the windshield being moved too far forward. In spite of the huge glass area, visibility is not perfect. Also, it is limited in height by the tilt of the windshield and the swept windshield wiper area. In addition, during bad weather, the triangular lateral windows quickly get dirty, and can be cleaned only from the outside. The thickness of the center post creates a major no-vision angle, but the outside rear-view mirrors have been upsized. The seat backs obstruct the back light and the rear windshield wiper is not centered. The dashboard is just as fussy and the instrumentation is out of the bottom of a black hole. Some controls, such as the gear shift lever which is too short and the switches which are laid out around the gauge assembly, are impractical to use.

• Seats: 60%
Small but well-shaped, they include no headrest but are easy to adjust and manipulate. Their weight is reasonable and the mountings are fairly simple. The modularity of the seats is limited and the standard bench seat on a five-seater is more practical since children can comfortably stretch out on it.

• Price/equipment: 60%
After some prior sales of models with little equipment, the sticker price is comparable to that of the direct competition.

• Depreciation: 60%
It is less than normal as the demand stays high.

• Performance: 60%
Lymphatic with a standard 3.1 L, with the high weight of these machines, it is a good boy sent out to do a man's job. The optional 3.8 L allows more acceptable starts and passings.

• Steering: 50%
It has a good ratio and power assist but lacks return and is imprecise at the center. In addition, maneuverability is penalized by the large steering diameter.

• Handling: 50%
It is one of the healthiest among the domestic vans, and the roll and body motions are well-controlled.

CHEVROLET Lumina

• Noise level: 50%
Drive train resonance maintains a higher interior noise level than that of vehicles with more compartments. The 3.8 L is much more discreet than the 3.1 L.

• Fuel consumption: 50%
The 3.8 L offers the best efficiency package and consumes less than the 3.1 L.

WEAK POINTS

• Access: 40%
The doors are heavy and hard to manipulate, especially in the winter when the sliding door is exasperating. The electric power option of the slide door is very interesting, but safety is not absolute because it is easy to get your fingers caught. You get better access to the rear when the median center seat is lowered.

• Storage: 40%
It is non-existent when the seats are in place and you have to sacrifice the last two to get some more space.

• Braking: 40%
The spongy pedal does help modulate the decelerations, and during panic stops the distances are long because the friction surfaces lack bit. However, thanks to the ABS, the trajectories are linear and stable.

CONCLUSION

• Overall rating: 61.0%
These vehicles enjoy a deserved success as much for their design and original lines as for the competitive prices at which base models with little equipment are sold. 😐

OWNERS' SUGGESTIONS

-Better brake fade-resistance
-Less dangerous doors
-Electric sliding doors with better safety features
-More comfortable pivot-type front seats
–Rear seat headrests

SPECIFICATIONS & PRICES

Model	Version/Trim	Body Seats	Drag coef. Cd	Wheel-base in	Lgth x Wdthx Hght. inx inx in	Curb weight lb	Towing capacity std. lb	Susp. type ft/rr	Brake type ft/rr	Steering system	Turning diameter ft	Steer. turns nber	Fuel. tank gal.	Standard tire size	Standard power train	PRICES US. $ 1994
CHEVROLET		**General warranty: 3 years / 36,000 miles; antipollution: 5 years / 50,000 miles; perforation corrosion: 6 years / 100,000 miles.**														
Lumina	base/Cargo	4 dr.van.2/7	0.34	109.8	194.2X73.9X65.7	3516	2000	i/r	dc/dr/ABS	pwr.r/p.	43.1	3.05	20	205/70R15	V6/3.1/A3	**$17,015**
Lumina	LS	4 dr.van.5/7	0.34	109.8	194.2X73.9X65.7	3593	3000	i/r	dc/dr/ABS	pwr.r/p.	43.1	3.05	20	205/70R15	V6/3.8/A4	**$20,477**
PONTIAC		**General warranty: 3 years / 36,000 miles; antipollution: 5 years / 50,000 miles; perforation corrosion: 6 years / 100,000 miles.**														
Trans Sport	SE	4 dr.van.5/7	0.34	109.8	192.2x74.6X65.7	3523	2000	i/r	dc/dr/ABS	pwr.r/p.	43.1	3.05	20	205/70R15	V6/3.1/A3	**$17,469**
Trans Sport	SE	4 dr.van.5/7	0.34	109.8	192.2x74.6X65.7	3659	3000	i/r	dc/dr/ABS	pwr.r/p.	43.1	3.05	20	205/70R15	V6/3.8/A4	**$20,671**
OLDSMOBILE		**General warranty: 3 years / 36,000 miles; antipollution: 5 years / 50,000 miles; perforation corrosion: 6 years / 100,000 miles.**														
Silhouette	base	4 dr.van.5/7	0.34	109.8	194.7x73.9X65.7	3633	3000	i/r	dc/dr/ABS	pwr.r/p.	43.1	3.05	20	205/70R15	V6/3.8/A4	**$20,635**

See page 393 for complete 1995 Price List

GM W1-Series

BUICK Regal / CHEVROLET Monte-Carlo OLDSMOBILE Cutlass Supreme PONTIAC Grand Prix coupes

Usurped identity...

The names of the coupes remind of a glorious period of the American automobile, the times when the engine had heart and no anti-pollution system. Today the sad-sack heirs are a shabby shadow of their former selves. Only the names remain.

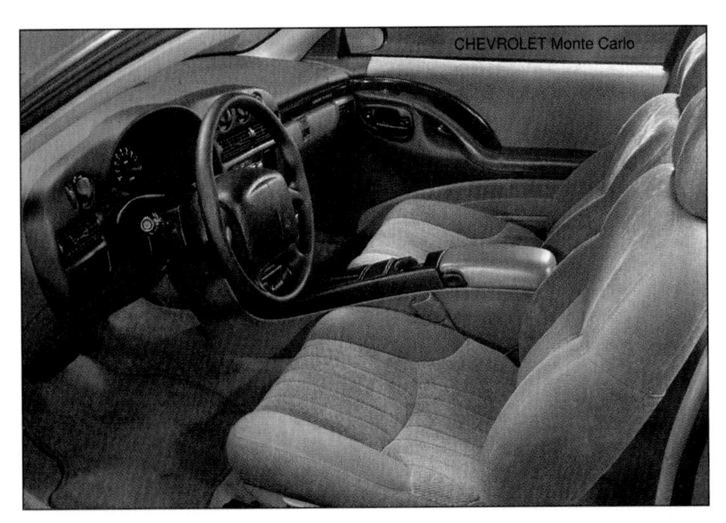

CHEVROLET Monte Carlo

These coupes with prestigious names perpetuate North American automotive tradition and the era when winning on Sunday sold cars on Monday. If this remains true in some parts of the Unites States, elsewhere the market laws are playing hard ball. The Buick Regal is sold as custom or Grand sport, the Pontiac Grand Prix as SE or GTP and the Olds Cutlass Supreme S or SL convertible. The news of the year is the return of Monte Carlo at Chevrolet. It is based on the old Lumina with trims in LS and Z34.

STRONG POINTS

• **Safety:** **90%**
It is considerable, thanks to the presence of two standard air bags, of a structure that resists impacts well and of a roll bar in the Cutlass Supreme convertible.

• **Steering:** **70%**
It is more direct, except on the Regal, the power assist is more powerful than before with some honest precision. In spite of their size, these coupes are maneuverable and enjoy a reasonable turning diameter.

• **Suspension:** **70%**
The standard suspension is smooth and comfortable on good pavement but stiffer and harsher on Sports versions, which react noticeably on bad asphalt.

• **Satisfaction:** **70%**
In this department the Regal stays ahead of its teammates, but its lead is down to 8%.

• **Technical:** **70%**
The coupes share the platform and drive train of sedans with the same name. They have steel unit bodies, independent suspensions, and disc brakes on all 4 wheels. Their more or less personal lines do not shine in terms of aerodynamic efficiency because the average Cd is

DATA

Category: mid-size front-wheel drive coupes.
Class : 5

HISTORIC

Introduced in: 1988 & 1989: Lumina; 1995: Monte Carlo
Modified in: -
Made in: Kansas City, Kansas, U.S.A. & Oshawa,Ont., Canada.

RATINGS

Safety: 90 %
Satisfaction: Lumina/G-Prix: 75 % Regal/Cutlass: 80 %
Depreciation: 60 % Regal: 52 %
Insurance: 5.5 % ($585-$650-$785)
Cost per mile: $0.46

NUMBER OF DEALERS

In U.S.A.: 3,000 Buick, 4,466 Chev-Geo, 3,100 Olds, 2,953 Pontiac.

SALES ON THE AMERICAN MARKET

Model	1992	1993	Result
Regal	91,972	87,647	- 4.7 %
Lumina	218,114	219,683	+0.7 %
Cutlass Sup	80,195	89,217	+10.2 %
Grand Prix	103,517	113,134	+8.5 %

MAIN COMPETITORS

FORD Thunderbird, MERCURY Cougar XR7, TOYOTA Camry.

EQUIPMENT

BUICK Regal		**Cust.**	**Gran Sport**
CHEVROLET Monte Carlo	**LS**	**Z34**	
OLDS. Cut. Supreme		**SL**	**Conv.**
PONTIAC Grand Prix		**SE**	**GTP**
Automatic transmission:	S	S	S
Cruise control:	S	S	S
Power steering:	S	S	S
ABS brake:	S	S	S (optional forPontiac)
Air conditioning:	S	S	S
Air bags (2):	S	S	S
Leather trim:	-	O	O
AM/FM/ Radio-cassette:	O	S	S
Power door locks:	S	S	S
Power windows:	S	S	S
Tilt steering column:	S	S	S
Dual power mirrors:	S	S	S
Intermittent wipers:	S	S	S
Light alloy wheels:	O	S	S
Sunroof:	-	O	O
Anti-theft system:	S	S	S

S : standard; O : optional; - : not available

COLORS AVAILABLE

Exterior: Silver, Black, White, Blue, Driftwood, Gray, Green, Red,
Interior: Blue, Gray, Beige, Red.

MAINTENANCE

First revision: 3,000 miles
Frequency: 6,000 miles
Diagnostic plug: Yes

WHAT'S NEW IN 1995 ?

- Lumina coupe replaced by the new Monte Carlo.
- New body colors and interior details.
- New dashboard that includes 2 air bags (Regal/Cutlass).
- Power steering assist increases as a function of speed on the 3.4 L V6.

ENGINE					TRANSMISSION			PERFORMANCE										
Model/ versions *: standard	Type / timing valve / fuel system	Displacement cu/in	Power bhp @ rmn	Torque lb.ft @ rpm	Compres. ratio	Driving wheels / transmission	Final ratio	Acceler. 0-62 mph s	Stand. 1/4 mile s	Stand. 5/8 mile s	Braking 62-0 mph ft	Top speed mph	Lateral acceler. G	Noise level dBA	Fuel economy mpg city	Fuel economy hwy	Gasoline type / octane
1)	V6* 3.1 OHC-12-MFI	191	160 @ 5200	185 @ 4000	9.6 :1	front - A4*	3.33	9.8	17.2	31.0	138	109	0.78	67	19.0	29.0	R 87
2)	V6 3.4 DOHC-24-SFI	207	210 @ 4000	215 @ 4000	9.2 :1	front - M5*	3.43	8.8	16.4	29.3	141	124	0.80	66	17.0	26.0	R 87
3)	V6 3.8 OHC-12-SFI	231	170 @ 4800	225 @ 3200	9.0 :1	front - A4*	3.06	9.5	17.0	31.4	144	115	0.79	66	19.0	29.0	R 87

1) base 2) * Grand Prix GTP & Monte Carlo Z34 3) Regal Gran Sport

BUICK Regal / CHEVROLET Monte-Carlo OLDSMOBILE Cutlass Supreme PONTIAC Grand Prix coupes

GM W1-Series

OLDSMOBILE Cutlass Supreme convertiblle

0.37. The base engine is a 3.1 L V6 for which you can substitute a double overhead cam 3.4 L or a not very sporty 3.8 L at Buick. A 4-speed automatic is standard on all models.

• **Insurance: 70%**
There is a major difference between the premium of the standard model and that of its more luxurious version (50%).

• **Conveniences: 70%**
Storage spots are in sufficient numbers both on the simple and on the luxurious models.

• **Trunk: 70%**
Its volume is ample. Even though it lacks height, the flat and regular contours allow full use.

• **Quality/fit/finish: 70%**
Finishing improves constantly. The materials are less tacky, but a few details still need correcting.

• **Performance: 65%**
The 24 valve 3.4 L V6 provides the best performance, while the 3.1 L gives an honest output with no excitement. Buick's 3.8 L is not more powerful, but it offers more torque within a useful rpm range.

• **Interior space: 65%**
Four adults will be at ease in these coupes even though headroom is more limited at the rear because of the roof shape. On the convertible, width is lacking because the top storage takes up space.

CHEVROLET Monte Carlo

NEW FOR 1995

• **Seats: 65%**
Well-padded, their contouring does not wrap around the body as well as those of the Grand Prix GPT.

• **Noise level: 65%**
The drive train is silent (except during strong accelerations), and it is the rolling and wind noises that take over due to lack of soundproofing.

• **Driver's compartment: 60%**
Their dashboard is not very inspiring and the digital instrumentation not easy to read, depending on the ambient light level. The commands which are specific to the car are poorly accessible, such as the ones for the AC, and take getting use to. Visibility is good except with the top up on the Cutlass Supreme convertible and the driving position is comfortable but low.

• **Handling: 60%**
Suspension softness on these models results in major roll and pitch. Those who want a more sports-like handling will opt for a heavy-duty suspension fitted for the Z34 and GTP. It allows flatter cornering and easier entry into the turns as well as more straight-line stability.

• **Access: 50%**
Easy at the front, it is more complicated to slide in at the back since the front seats intrude on rear entry space.

• **Price/equipment: 50%**
Better equipped than in the past, the simpler models are a better choice than big guns such as the GTP.

• **Fuel consumption: 50%**
It is reasonable, taking into account the engine displacement and the weight.

WEAK POINTS

• **Depreciation: 40%**
That of the Buick Regal is smaller by 8% than that of other models.

• **Braking: 40%**
The stopping distances are long, in spite of discs at all 4 wheels, but the ABS allows straight stops.

CONCLUSION

• **Overall rating: 63.0%**
Roomy and practical, the only part of being sporty is in their name because they don't stir up any passion. 😐

PONTIAC Grand Prix GTP

SPECIFICATIONS & PRICES

Model	Version/Trim	Body/ Seats	Interior volume cu.ft.	Trunk volume cu ft.	Drag coef. Cd	Wheel-base in	Lgth x Wdthx Hght. inx inx in	Curb weight lb	Towing capacity std. lb	Susp. type ft/rr	Brake type ft/rr	Steering system	Turning diameter ft	Steer. turns nber	Fuel. tank gal.	Standard tire size	Standard power train	PRICES US. $ 1994
BUICK		**General warranty: 3 years / 36,000 miles; antipollution: 5 years / 50,000 miles; perforation corrosion: 6 years / 100,000 miles.**																
Regal	Custom	3dr.cpe. 5/6	95.3	15.6	0.36	107.5	193.9x72.5x53.3	3261	1000	i/i	dc/dc/ABS	pwr.r/p.	36.7	3.05	17	205/70R15	V6/3.1/A4	**$18,324**
Regal	Gran Sport	3dr.cpe. 5	95	15.6	0.34	107.5	193.9x72.5x53.3	3329	1000	i/i	dc/dc/ABS	pwr.r/p.	36.7	3.05	17	225/60R16	V6/3.8/A4	**$20,324**
CHEVROLET		**General warranty: 3 years / 36,000 miles; antipollution: 5 years / 50,000 miles; perforation corrosion: 6 years / 100,000 miles.**																
Monte Carlo	LS	3dr.cpe. 5/6	96	15.7	0.34	107.5	200.7x72.5x53.8	3307	1000	i/i	dc/dc/ABS	pwr.r/p.	36.7	2.60	17	205/70R15	V6/3.1/A4	**$16,875**
Monte Carlo	Z34	3dr.cpe. 5/6	96	15.7	0.33	107.5	200.7x72.5x53.8	3437	1000	i/i	dc/dc/ABS	pwr.r/p.	36.7	2.26	17	225/60R16	V6/3.4/A4	**$19,310**
OLDSMOBILE		**General warranty: 3 years / 36,000 miles; antipollution: 5 years / 50,000 miles; perforation corrosion: 6 years / 100,000 miles.**																
Cutlass SupremeSL		3dr.cpe. 5/6	96	15.5	0.35	107.5	193.9x71.0x53.3	3285	1000	i/i	dc/dc/ABS	pwr.r/p.	36.7	2.60	17	205/70R15	V6/3.4/A4	**$17,670**
Cutlass Supr Convt.		3dr.cpe. 5	101.5	12.1	NA	107.5	193.9x71.0x54.3	3629	1000	i/i	dc/dc/ABS	pwr.r/p.	36.7	2.30	17	225/60R16	V6/3.4/A4	**$25,470**
PONTIAC		**General warranty: 3 years / 36,000 miles; antipollution: 5 years / 50,000 miles; perforation corrosion: 6 years / 100,000 miles.**																
Grand Prix SE		3dr.cpe. 5/6	95.2	14.9	0.35	107.5	194.8x71.9x52.8	3243	1000	i/i	dc/dc/ABS	pwr.r/p.	36.7	2.60	17	215/60R16	L4/3.1/A4	**$16,670**
Grand Prix GTP		3dr.cpe. 5/6	95.2	14.9	0.33	107.5	194.8x71.9x52.8	3243	1000	i/i	dc/dc/ABS	pwr.r/p.	36.7	2.60	17	225/60R16	V6/3.4/A4	**$19,465**

See page 393 for complete 1995 Price List

Sclerosis...

For the past six years, Taurus/Sable raked the market, and for the past two years, the LH seduced in wily ways, while the GM W slep with clenched fists, allowing itself to lose precious parts of the market. It is time that the wind of renewal blows on the sleepers.

BUICK Regal LTD

These models lose ground faced with an aggressive competition, for they have not been redesigned for six years. The standard power trains of all these 4-door sedans is a 3.1 L. V6 with an automatic 4-speed. The finishes are Custom, Limited and Gran Sport at Buick, Standard and LS at Chevrolet, SL at Oldsmobile and, finally, SE and GT at Pontiac.

STRONG POINTS

- **Safety:** 90%
Thanks to additional reinforcements, their structure resists better in impact tests and the presence of two standard air bags gives them an excellent coefficient in spite of their age.
- **Insurance:** 80%
Reasonable on the base models, the premium of the sports versions is equivalent to that of the coupes.
- **Satisfaction:** 75%
As coupes, the Buick sedans seem to satisfy more owners than do the other three makes.
- **Steering:** 70%
Direct with a positive assistance, it is precise and allows good maneuverability, thanks to a normal steering diameter.
- **Suspension:** 70%
Too soft when it is not sports oriented, it weaves on bad pavements. Otherwise, it is firm and reacts more harshly to road defects.
- **Interior space:** 70%
The interior of these cars comfortably accepts four to five people and disposes of enough space at the rear for legs and head.
- **Access:** 70%
It does not pose any problems at the front, but at the rear the narrower doors interfere with the seating of large people.

DATA

Category: mid-size front-wheel drive sedans.
Class : 5

HISTORIC

Introduced in: 1988: Regal, Cutlass Sup., & 1989-1995 (Lumina).
Modified in: 1994: ABS and air bag standard.
Made in: Kansas City, Kansas, U.S.A. & Oshawa,Ont., Canada.

RATINGS

Safety: 90 %
Satisfaction: Lumina/G-Prix: 75 % Regal/Cutlass: 80 %
Depreciation: 60 % Regal: 52 %
Insurance: 5.5 % ($585-$650-$785)
Cost per mile: $0.46

NUMBER OF DEALERS

In U.S.A.: 3,000 Buick, 4,466 Chev-Geo, 3,100 Olds, 2,953 Pontiac.

SALES ON THE AMERICAN MARKET

Model	1992	1993	Result
Regal	91,972	87,647	- 4.7 %
Lumina	218,114	219,683	+0.7 %
Cutlass Sup	80,195	89,217	+10.2 %
Grand Prix	103,517	113,134	+8.5 %

MAIN COMPETITORS

BUICK Century, CHRYSLER Concorde & Intrepid, EAGLE Vision, FORD Taurus, HONDA Accord, MAZDA 626, MERCURY Sable, OLDSMOBILE Ciera, SUBARU Legacy, TOYOTA Camry.

EQUIPMENT

BUICK Regal	Custom	Ltd	Gran Sport
CHEVROLET Lumina	base	LS	
OLDS. Cut. Supreme	S	SL	
PONTIAC Grand Prix	SE	GT	
Automatic transmission:	S	S	S
Cruise control:	S	S	S
Power steering:	S	S	S
ABS brake:	S	S	S
Air conditioning:	S	S	S
Air bags (2):	S	S	S
Leather trim:	-	O	O
AM/FM/ Radio-cassette:	O	S	S
Power door locks:	S	S	S
Power windows:	S	S	S
Tilt steering column:	S	S	S
Dual power mirrors:	S	S	S
Intermittent wipers:	S	S	S
Light alloy wheels:	O	S	S
Sunroof:	-	O	O
Anti-theft system:	S	S	S

S : standard; O : optional; - : not available

COLORS AVAILABLE

Exterior: Silver, Black, White, Blue, Driftwood, Gray, Green, Red.
Interior: Blue, Gray, Beige, Red.

MAINTENANCE

First revision: 3,000 miles
Frequency: 6,000 miles
Diagnostic plug: Yes

WHAT'S NEW IN 1995 ?

- Lumina redesigned with the same power train.
- New body colors and interior detailing.
- New dash to include two air bags (Regal - Cutlass).
- Power assist has a function of speed on a V6 3.4L.

Model/ versions *: standard	ENGINE Type / timing valve / fuel system	Displacement cu/in	Power bhp @ rmn	Torque lb.ft @ rpm	Compres. ratio	TRANSMISSION Driving wheels / transmission	Final ratio	PERFORMANCE Acceler. 0-62 mph s	Stand. 1/4 mile s	Stand. 5/8 mile s	Braking 62-0 mph ft	Top speed mph	Lateral acceler. G	Noise level dBA	Fuel economy mpg city	hwy	Gasoline type / octane
1)	V6* 3.1 OHC-12-MFI	191	160 @ 5200	185 @ 4000	9.6 :1	front - A4*	3.33	9.8	17.2	31.0	138	109	0.78	67	19.0	29.0	R 87
2)	V6 3.4 DOHC-24-SFI	207	210 @ 4000	215 @ 4000	9.2 :1	front - M5*	3.43	8.8	16.4	29.3	141	124	0.80	66	17.0	26.0	R 87
3)	V6 3.8 OHC-12-SFI	231	170 @ 4800	225 @ 3200	9.0 :1	front - A4*	3.06	9.5	17.0	31.4	144	115	0.79	66	19.0	29.0	R 87

1) base 2) option 3) Regal Gran Sport

BUICK Regal / CHEVROLET Lumina OLDSMOBILE Cutlass Supreme PONTIAC Grand Prix sedans

GM W 2-Series

• Technical: 70%
These cars share the same platform and Power train components with the preceding coupes; only their body and interiors differ. They offer a steel unit body with an independent suspension and disc brakes on all four wheels. While not displeasing, their lines are dated because they have been seen too long.

• Seats: 60%
The two standard versions offer neither lateral hold nor lumbar support. It is difficult to seat a third person in the center of the bench seat because the shoulder harness buckles are in the way. At the rear, the seat is low and the cushion is too short.

• Noise level: 60%
Low at cruise speed, it is amplified during accelerations and the tires strongly pound tar strips as the speed increases.

• Trunk: 60%
It is vast and its regular forms make it easy to use in spite of a high sill on the Lumina.

• Quality/fit/finish: 70%
Body squeaks are not absent after a few thousand miles and the artificial wood of the Lumina lacks class. However, other plastic items and trim have a more flattering appearance than in the past. Assembly could be more careful and some elements are not always well-aligned.

• Driver's compartment: 60%
While a little low, the driver is well-seated and visibility is good from all angles. However, the dashboard is unnecessarily complicated, the digital instrumentation not always legible, and some controls still out of reach.

• Performance: 60%
Ordinary with a 3.1 L, they are sharper with a 3.4 L which provides a better weight-to-power ratio. As to the 3.8 L, its torque allows pulling higher loads.

OLDSMOBILE Cutlass Supreme

CHEVROLET Lumina

PONTIAC Grand Prix

• Handling: 55%
It is typical of standard range American cars with a suspension that is too soft. You have to rely on options offering better quality tires, shocks and springs to control roll and gain a more precise and inspired handling.

• Fuel consumption: 50%
It is economical with a 3.1 L but the temptations of the 3.4 raise the tab.

• Price/equipment: 50%
With nearly complete standard equipment, the price of these models is better justified and constitutes a great argument in their favor.

WEAK POINTS

• Braking: 40%
Standard on all models, the ABS stabilizes panic stops but stopping distances are still too long.

• Conveniences: 40%
Not numerous, storage is only present on the more luxurious versions.

• Depreciation: 40%
It is weaker than for coupes, where the demand is more sustained.

CONCLUSION

• Overall rating: 60%
Sales numbers for these sedans are all in decline and in need of renewed beauty care. 😐

OWNER'S SUGGESTIONS

- Improved quality and finish.
- More powerful heating.
- Firmer suspension.
- Better quality tires.
- Winter-proofed windshield wiper.

SPECIFICATIONS & PRICES

Model	Version/Trim	Body/ Seats	Interior volume cu.ft.	Trunk volume cu ft.	Drag coef. Cd	Wheel-base in	Lgth x Wdthx Hght. inx inx in	Curb weight lb	Towing capacity std. lb	Susp. type ft/rr	Brake type ft/rr	Steering system	Turning diameter ft	Steer. turns nber	Fuel. tank gal.	Standard tire size	Standard Power train	PRICES US. $ 1994
BUICK		**General warranty: 3 years / 36,000 miles; antipollution: 5 years / 50,000 miles; perforation corrosion: 6 years / 100,000 miles.**																
Regal	Custom	4dr.sdn.5/6	100	15.7	0.35	107.5	193.7x72.5x54.5	3336	1000	i/i	dc/dc/ABS	pwr.r&p.	36.7	3.05	17.1	205/70R15	V6/3.1/A4	$18,624
Regal	Limited	4dr.sdn.5/6	100	15.7	0.35	107.5	193.7x72.5x54.5	3463	1000	i/i	dc/dc/ABS	pwr.r&p.	36.7	3.05	17.1	205/70R15	V6/3.1/A4	$20,124
Regal	Gran Sport	4dr.sdn.5	100	15.7	0.34	107.5	193.7x72.5x54.5	3406	1000	i/i	dc/dc/ABS	pwr.r&p.	36.7	3.05	17.1	225/60R16	V6/3.8/A4	$20,624
CHEVROLET		**General warranty: 3 years / 36,000 miles; antipollution: 5 years / 50,000 miles; perforation corrosion: 6 years / 100,000 miles.**																
Lumina	base	4dr.sdn.5/6	101	15.7	0.33	107.5	200.9x71.9x55.2	3329	1000	i/i	dc/dc/ABS	pwr.r&p.	36.7	2.60	17.1	205/70R15	V6/3.1/A4	$15,305
Lumina	LS	4dr.sdn.5/6	101	15.7	0.33	107.5	200.9x71.9x55.2	3373	1000	i/i	dc/dc/ABS	pwr.r&p.	36.7	2.26	17.1	225/60R16	V6/3.4/A4	$16,515
OLDSMOBILE		**General warranty: 3 years / 36,000 miles; antipollution: 5 years / 50,000 miles; perforation corrosion: 6 years / 100,000 miles.**																
Cutlass Supreme	S	4dr.sdn.5/6	100	15.5	0.36	107.5	193.7x71.0x54.8	3369	1000	i/i	dc/dc/ABS	pwr.r&p.	37.4	2.60	16.6	205/70R15	V6/3.1/A4	$17,770
Cutlass Supreme	SL	4dr.sdn.5	100	15.5	0.36	107.5	193.7x71.0x54.8	3417	1000	i/i	dc/dc/ABS	pwr.r&p.	37.4	2.26	16.6	225/60R16	V6/3.4/A4	$19,823
PONTIAC		**General warranty: 3 years / 36,000 miles; antipollution: 5 years / 50,000 miles; perforation corrosion: 6 years / 100,000 miles.**																
Grand Prix	SE	4dr.sdn.5/6	99.5	15.5	0.34	107.5	194.9x71.9x54.8	3318	1000	i/i	dc/dc/ABS	pwr.r&p.	36.7	2.60	16.5	205/70R15	V6/3.1/A4	$16,254
Grand Prix	GT	4dr.sdn.5/6	99.5	15.5	0.33	107.5	194.9x71.9x54.8	3318	1000	i/i	dc/dc/ABS	pwr.r&p.	39.0	2.26	16.5	225/60R16	V6/3.4/A4	$19,291

See page 393 for complete 1995 Price List

Megalomania...

Worshipped or put down, the Corvette has nourished the same debate year after year. Less popular because of its price and less refined than its Japanese rivals, it has seen its sales increase in a significant way, which proves that the myth is not dead, because it is still running fast.

The Corvette has been an integral part of the history of American values; with time, it became an institution that everyone contests but no one wants to question, for fear of provoking something irreparable. This model represented what General Motors knew best in terms of rocket travel. It is equipped with a 5.7 L. V8 and a stanrard 4-speed automatic or a manual 6-speed at no extra cost, the only one available on the ZR-1.

STRONG POINTS

• Performance: 100%

The power and the generous torque of the big V8, plus the German 6-speed stick shift produce good straight line numbers. The little key on the ZR-1 dashboard opens the door to the land of marvels and speeding tickets, since its engine offers a superior power curve that does not seem to have an end. However, its clutch is not very progressive and its gear shift, which gives a fight, take away a little pleasure from those in search of the absolute.

• Handling: 90%

Unidirectional tires and a rigid suspension, allow the Corvette to corner flat and stay neutral for a long time. This works best on a dry road with perfect asphalt, or traction drops off and the trajectories become uncertain.

• Safety: 80%

There is much to be done for it to become optimal in spite of the impact absorbent fiber glass body. The safety hoop of the coupe crushes easily and is absent on the convertible. Fortunately, two air bags protect the occupants. Really needed, is an integral hard top roof, stiffening up a lighter platform.

• Technical: 80%

The Corvettes still have a platform chassis, topped by a steel rib structure to support the polyester resin and fiberglas body. The 4-wheel independent suspension includes a fiberglas transverse rear spring with a strong roll-resistance in relation to its vertical ride rate. Vented disc brakes are controlled by a Bosch ABS which also serves as a traction control. It lightly brakes (10%) the wheel which lacks traction. The electronic content of this model constantly increases. The ignition and fuel controls are very sophisticated and there is also a device for controlling tire pressure.

• Seats: 70%

They maintain and support perfectly and accept with equal pleasurability people of all sizes.

• Insurance: 70%

The Corvettes are not a bargain to insure, especially on the standard model. The amount of the premium directly reflects the anxiety of the insurers.

• Satisfaction: 70%

Little problems persist in spite of the years of experience, but that does not discourage the fans and fanatics, so sales keep increasing.

DATA

Category: rear-wheel drive coupes, convertibles and GT.
Class : S & GT

HISTORIC

Introduced in: 1953, 1963, 1967, 1983.
Modified in: 1986: convertible; 1990: ZR-1.
Made in: Bowling Green, Kentucky, U.S.A.

RATINGS

		LT1	ZR-1	
Safety:		90 %	90 %	
Satisfaction:		78 %	76 %	
Depreciation:		49 %	62 %	
Insurance:	($1,340)	5.4 %	3.4 %	($1,500)
Cost per mile:		$0.83	$1.00	

NUMBER OF DEALERS

In U.S.A.: 4,466 Chevrolet

SALES ON THE AMERICAN MARKET

Model	1992	1993	Result
Corvette	19,819	20,487	+3.3

MAIN COMPETITORS

Corvette LT1: ACURA Legend cpe, DODGE Stealth R/T Turbo, MAZDA RX-7, NISSAN 300 ZX, SUBURU SVX, TOYOTA Supra.
Corvette: ZR-1: ACURA NS-X, BMW 850i, JAGUAR XJS, MERCEDES BENZ SL350 & 500, PORSCHE 911& 928.

EQUIPMENT

CHEVROLET Corvette	cpe	Convertible	ZR1
Automatic transmission:	O/no charge	O/no charge	S
Cruise control:	S	S	S
Power steering:	S	S	S
ABS brake:	S	S	S
Air conditioning:	S	S	S
Air bag (2):	S	S	S
Leather trim:	O	O	S
AM/FM/ Radio-cassette:	S	S	S
Power door locks:	S	S	S
Power windows:	S	S	S
Tilt steering column:	S	S	S
Dual power mirrors:	S	S	S
Intermittent wipers:	S	S	S
Light alloy wheels:	S	S	S
Sunroof:	-	-	-
Anti-theft system:	S	S	S

S : standard; O : optional; - : not available

COLORS AVAILABLE

Exterior: White, Yellow, Black, Green, Pink, Red, Blue, Purple.
Interior: Gray, Black, Beige, Red, White, Yellow.

MAINTENANCE

First revision: 3,000 miles
Frequency: 6 months/6,000 miles
Diagnostic plug: Yes

WHAT'S NEW IN 1995 ?

- More powerful de Carbon electronic shock control.
- Grill redesign.
- "Flat-proof tires" (optional).
- Last year of ZR-1 production.

	ENGINE					TRANSMISSION		PERFORMANCE										
Model/ versions *: standard	Type / timing valve / fuel system	Displacement cu/in	Power bhp @ rmn	Torque lb.ft @ rpm	Compres. ratio	Driving wheels / transmission	Final ratio	Acceler. 0-62 mph s	Stand. 1/4 mile s	Stand. 5/8 mile s	Braking 62-0 mph ft	Top speed mph	Lateral acceler. G	Noise level dBA	Fuel economy mpg city	hwy	Gasoline type / octane
base LT-1	V8* 5.7 OHV-16-SFI	350	300 @ 5000	340 @ 4000	10.5 :1	rear-A4*	2.59	6.2	14.6	29.8	148	149	0.89	70	17.0	27.0	M 89
						rear-M6	2.51	5.8	13.5	29.4	144	155	0.89	70	17.0	28.0	M 89
ZR1 LT5	V8* 5.7 DOHC-32-MPFI	350	405 @ 5800	385 @ 5200	11.0 :1	rear-M6*	3.45	5.2	12.8	28.8	128	171	0.87	70	17.0	24.0	S 91

• Steering: 70%
At normal speed it is direct and precise, but maneuverability is reduced by a large steering diameter and body bulk. On a high speed track, the front end shows its age and becomes less stable and the steering lightens up due to lack of chassis stiffness.

• Braking: 60%
Effective, fade-free and easy to modulate, the ABS system is affected by repeated rebounds of the wheels which makes the system unpredictable and increases stopping distances.

• Quality/fit/finish: 60%
The assembly, finish and quality of some materials date from another era. The lack of quality shows in the roof of the convertible which is not tight and in the electric circuit that is not worthy of such a car.

• Driver's compartment: 60%
The latest Corvette interior update has been beneficial to the seats, which are more effective and less complicated to adjust, so even when seated low, the driver is comfortable. However, visibility remains limited because of the massive dashboard, the height of the body belt line, the enormous blind spot caused by the B pillar on the coupe and that caused by the convertible cover. The main controls are well laid out, the steering wheel and the gear shift are well in hand. The dashboard remains a major problem, anything but ergonomic. Instrumentation is illegible in the sunshine, especially with the convertible top down.

• Depreciation: 50%
While the value of the ZR-1 drops dramatically, that of the ordinary Corvette holds up well.

WEAK POINTS

• Price/equipment: 00%
Though the equipment is relatively complete, the quality price ratio is unfavorable when the Corvette is compared to its sophisticated Japanese rivals.

• Interior space: 10%
In spite of its enormous proportions, the Corvette interior remains tight and one feels taken between the doors and the transmission tunnel. The dimensions are limited in all directions.

• Trunk: 30%
The Corvette is still not a practical car because it is just as difficult to put baggage into the coupe or the convertible. On the former, the sill is high and the opening is distant and on the latter, access to storage is through the seat back.

• Noise level: 30%
Exhilarating at low speeds, the engine maintains a road level of noise and vibrations which soon becomes tiring.

• Fuel consumption: 30%
It rarely goes over 19 mpg and as a result, is never economical.

• Access: 40%
One should be in shape and well-trained to rapidly get in without damage, or simply be young...

• Suspension: 40%
On the freeway everything goes well, but if the asphalt is deteriorated, the occupants receive a massage that has nothing medical about it.

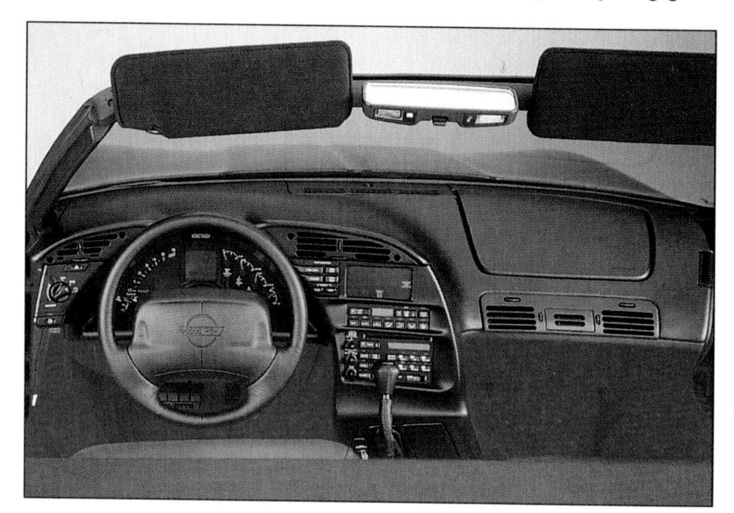

CONCLUSION

• Overall rating: 56.5%
Since they started talking about it, no one has hoped for a radical revamping of the American myth on four wheels. 😐

OWNER'S SUGGESTIONS

-Really effective anti-theft device
-Better quality sound system
-More rigid chassis
-A finish that is less "kit car"

SPECIFICATIONS & PRICES

Model	Version/Trim	Body/ Seats	Interior volume cu.ft.	Trunk volume cu ft.	Drag coef. Cd	Wheel-base in	Lgth x Wdthx Hght. inx inx in	Curb weight lb	Towing capacity std. lb	Susp. type ft/rr	Brake type ft/rr	Steering system	Turning diameter ft	Steer. turns nber	Fuel. tank gal.	Standard tire size	Standard Power train	PRICES US. $ 1994
CHEVROLET		**General warranty: 3 years / 36,000 miles; antipollution: 5 years / 50,000 miles; perforation corrosion: 6 years / 100,000 miles.**																
Corvette	LT1	2dr.cpe.2	48.7	12.5	0.33	96.2	178.5x70.7x46.3	3203	NR	i/i	dc/dc/ABS	pwr.r&p.	40	2.32	20	255/45ZR17a	V8/5.7/A4	**$36,285**
Corvette	LT1	2dr.con.2	48.7	6.6	0.35	96.2	178.5x70.7x47.3	3360	NR	i/i	dc/dc/ABS	pwr.r&p.	40	2.32	20	285/40ZR17r	V8/5.7/A4	**$43,060**
Corvette	ZR1	2dr.cpe.2	48.7	12.5	0.33	96.2	178.5x70.7x46.3	3514	NR	i/i	dc/dc/ABS	pwr.r&p.	40	2.32	20	275/40ZR17 315/35ZR17 (re.)	V8/5.7/M6	**$67,813**

See page 393 for complete 1995 Price List

CHEVROLET pick-up S-10 & Blazer
GMC pick-up Sonoma & Jimmy

With equal weapons...

Chevrolet and GMC pursue their modernization of the utility and sports utility compact ranges by equipping the Blazer-Jimmy with a chassis inaugurated last year on the pickups, and by redesigning the body in a more contemporary style. The competition better hang on.

After updating its pickups last year, GM applied the same treatment to its sport utility Blazer and Jimmy for 1995. The body has been fully redesigned and the chassis received substantial improvements aimed at increasing the stiffness it needed. These vehicles share the same Power train and are cloned for GMC as Sonoma and Jimmy. Proof without a doubt of its maturity, the Blazer name has been given to a new vehicle. The larger one is christened Tahoe. The pickups are available in 2- and 4-wheel drive with a longer or short box, normal or extended cab and standard or LS trims at Chevrolet or SL, SLS and SLE at GMC. The Blazer-Jimmy can be 2- or 4-door, 2- or 4-wheel drive with finishes LS or LT at Chevrolet, SL, SLS, SLE or SLT at GMC. For 1995, the standard pickup engine is a 2.2 L. 4-cylinder and that of the Blazer-Jimmy a 4.3 L. V6.

STRONG POINTS

• **Safety:** 70%
An air bag and improved structural rigidity substantially improve occupant safety.

• **Satisfaction:** 70%
The sports-utility give more satisfaction than pickups which trailed by 10%. Owner's complain mostly of differentials, transmissions and suspensions.

• **Conveniences:** 70%
The glove box in the latest dashboard is practical, since it can serve as a small desk and holds pocket change. However, the shoulder belts are not adjustable.

• **Quality/fit/finish:** 70%
The new vehicles are better assembled and the finish is superior, while some of the materials do not look particularly rich.

DATA

Category: rear-wheel drive or 4 WD all-purpose and vans.
Class : utility

HISTORIC

Introduced in: 1982: S-10 pick-up; 1983: Blazer S-10
Modified in: 1983, 1988, 1989, 1991.
Made in: Moraine, OH, Linden, NJ, Schreveport, LO, U.S.A.

RATINGS

		S-10-Sonoma	Blazer-Jimmy
Safety:		80%	80 %
Satisfaction:		65 %	75 %
Depreciation:		60 %	53 %
Insurance:	($525)	8.5 %	5.5 % ($585)
Cost per mile:		$0.38	$0.51 (4x4)

NUMBER OF DEALERS

In U.S.A.: 4,466 Chevrolet

SALES ON THE AMERICAN MARKET

Model	1992	1993	Result
Blazer S-10	147,742	167,421	+11.8 %
S-10	191,982	183,700	- 4.4 %

MAIN COMPETITORS

Blazer-Jimmy: FORD Explorer, ISUZU Rodeo & Trooper, JEEP Cherokee & Grand Cherokee, NISSAN Pathfinder, TOYOTA 4Runner.
S-10-Sonoma: DODGE Dakota, FORD Ranger, MAZDA B, NISSAN Hardbody& TOYOTA.

EQUIPMENT

CHEVROLET S-10 / **GMC Sonoma S** / **CHEVROLET Blazer** / **GMC Jimmy**	base SL	LS SLE	base SL	2dr. SLE	4dr. SLT	4dr. LT
Automatic transmission:	O	O	O	O	O	O
Cruise control:	-	O	S	O	O	S
Power steering:	S	S	S	S	S	S
ABS brake:	S	S	S	S	S	S
Air conditioning:	S	S	S	S	S	S
Air bag (left):	S	S	S	S	S	S
Leather trim:	-	-	-	-	O	S
AM/FM/ Radio-cassette:	O	O	S	O	O	S
Power door locks:	O	O	S	O	O	S
Power windows:	O	O	S	O	O	S
Tilt steering column:	O	O	S	O	O	S
Dual power mirrors:	-	O	S	O	O	S
Intermittent wipers:	S	S	S	S	S	S
Light alloy wheels:	-	O	S	O	O	S
Sunroof:	-	-	-	-	-	-
Anti-theft system:	O	O	O	O	O	O

S : standard; O : optional; - : not available

COLORS AVAILABLE

Exterior: White, Gray, Black, Blue, Green, Red, Kaki, Metallic Gray, Purple.
Interior: Blazer:Gray, Beige, Blue, Graphite.
S-10: Light Gray, Charcoal, Blue, Beige.

MAINTENANCE

First revision: 3,000 miles
Frequency: 6 months/6,000 miles
Diagnostic plug: Yes

WHAT'S NEW IN 1995 ?

- **New models redesigned (Jimmy and Blazer).**
- **Air bag on the driver's side.**
- **Dextron automatic transmission fluid guaranteed 100,000 miles.**

Model/ versions *: standard	ENGINE Type / timing valve / fuel system	Displacement cu/in	Power bhp @ rmn	Torque lb.ft @ rpm	Compres. ratio	TRANSMISSION Driving wheels / transmission	Final ratio	PERFORMANCE Acceler. 0-62 mph s	Stand. 1/4 mile s	Stand. 5/8 mile s	Braking 62-0 mph ft	Top speed mph	Lateral acceler. G	Noise level dBA	Fuel economy mpg city	hwy	Gasoline type / octane
1)	L4* 2.2 OHV-8-MPFI	133	118 @ 5200	130 @ 2800	9.0 :1	rear-M5*	2.73	14.0	19.7	39.0	171	90	NA	68	26.0	37.0	R 87
						rear-A4	4.11	15.2	20.8	40.3	177	85	NA	68	22.0	33.0	R 87
2)	V6 4.3 OHV-12-EFI	262	155 @ 4000	235 @ 2400	9.1 :1	re./4 -M5*	2.73	9.5	16.9	31.2	168	109	NA	67	21.0	33.0	R 87
						re./4 -A4	3.08	10.7	17.5	32.3	180	106	NA	67	20.0	30.0	R 87
3)	V6 4.3 OHV-12-MFI	262	195/191@ 4500	260 @ 3400	9.1 :1	four-M5*	3.42	10.4	17.1	31.8	174	106	0.68	67	20.0	30.0	R 87
						four-A4	3.42	11.2	17.8	33.0	183	100	0.68	67	19.0	28.0	R 87

1) base S-10-Sonoma 2) base Blazer-Jimmy, option S-10-Sonoma 3) std Blazer-Jimmy 4x4 & opt-S-10-Sonoma

• Driver's compartment: 70%
The new dashboard is better laid out, with more logical equipment and a better design look. The gauges are legible, the main controls accessible, except for that of the parking brake which is still foot-operated. Visibility is good, the gear shift lever is well laid out but the steering column is too long for a comfortable driving position.

• Technical: 70%
These utilities are based on a ladder-type chassis to which is mounted a body with a number of hot dipped galvanized panels. It uses a part-time 4-wheel drive with front wheel hubs than can be engaged with a lever or with an optional electronic control on the dashboard. The front suspension is independent and at the rear, there is a rigid axle with leaf springs. The brakes are mixed, with ABS on the rear wheels of 4-cylinder pickups and with a Blazer-Jimmy and V6 pickups.

• Steering: 70%
Fairly precise, its assistance is well-modulated on the Blazer-Jimmy and on the pickups, and their ratio is shorter.

• Performance: 60%
Laborious with a 4-cylinder, whose power and torque are geared to small jobs, they are more interesting with a 4.3 L. V6, especially in its 195 hp version. If anything, there is a little more power than the traction can stand in a sharp corner. With this engine, the Blazer-Jimmy can now be on equal footing with the big 6 in-line of the Grand and little Cherokees.

• Handling: 60%
It is noticeably improved by the substantial increase in rigidity, allowing it to confront difficult terrain without fear. However, road handling also benefits in terms of precision and cornering ability, less vague than on preceding models.

• Seats: 60%
The bench-type seats do not maintain enough and lumbar support is poor on the bucket seat.

• Price/equipment: 60%
In spite of their technical update, they remain affordable compared to their direct competitors but their equipment must still be complemented by numerous options.

• Insurance: 60%
The Blazer-Jimmy cost distinctly less to insure than the pickups and logically it should be the other way around.

• Suspension: 60%
Comfortable on the parkways, it is less jumpy on small bumpy roads, thanks to better spring and shock calibrations, especially on the rear axle which is now more stable, even running with an empty pickup.

• Noise level: 50%
Comfortable at constant interstate speed, it is only disturbed by engine noise during strong accelerations and by wind and rolling noise which is ever present.

• Interior Space: 50%
Even though the interiors are larger, you can still seat not more than 2 people up front in the pickups, while the extended cab is better for baggage than for passengers. Space is more generous at the back of a 4-door Jimmy than on the 2-door, which are less in demand.

• Storage: 50%
Its volume is more practical in the extended cabs pickups and in the Blazer.

• Access: 50%
To seat yourself at the back of a 2-door Blazer in the extended cab is not as easy as up front, because space is limited.

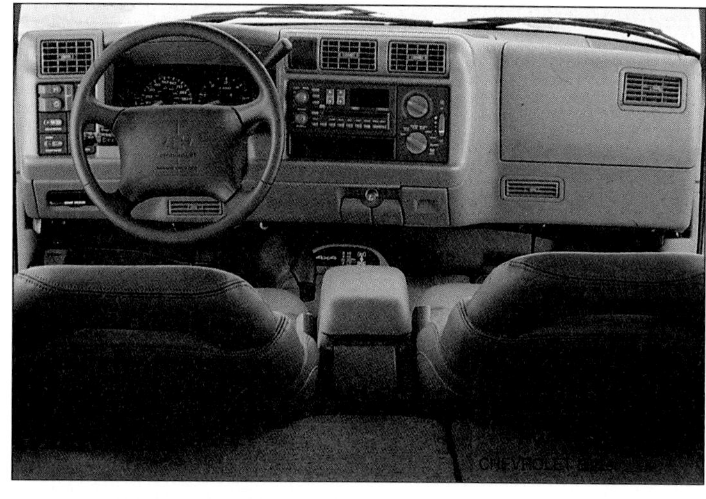

CHEVROLET S-10

NEW FOR 1995

CHEVROLET Blazer

WEAK POINTS

• Fuel consumption: 40%
It is never really economical, but it increases in strong proportions under load or rough terrain.

• Depreciation: 40%
It is stronger for utilities than for passenger cars, whose usage is different.

• Braking: 40%
ABS has brought a much better stability during panic stops, but stopping distances remain long and the power assist is too high and hard to modulate with the excess sensitivity of the pedal.

CONCLUSION

• Overall rating: 58.5%
It is not in vain that Chevrolet has improved its pickups and sports utilities, because many owners had begun to look elsewhere for the safety and handling they were unable to offer. 😐

SPECIFICATIONS & PRICES

Model	Version/Trim	Body/ Seats	Wheel-base in	Lgth x Wdthx Hght. inx inx in	Curb weight lb	Towing capacity std. lb	Susp. type ft/rr	Brake type ft/rr	Steering system	Turning diameter ft	Steer. turns nber	Fuel. tank gal.	Standard tire size	Standard Power train	PRICES US. $ 1994
CHEVROLET-GMC	**General warranty: 3 years / 36,000 miles; antipollution: 5 years / 50,000 miles; perforation corrosion: 6 years / 100,000 miles.**														
S10-Sonoma 4x2	std regular	2dr.p-u.2/3	108.3	189.0x67.9x62.1	2822	1653	i/r	dc/dr/ABSre.	pwr.bal.	36.9	2.75	20	205/75R15	L4/2.2/M5	**$9,806**
S10-Sonoma 4x2	std extd.	2dr.p-u.2/3	117.9	204.9x67.9x62.1	2875	1715	i/r	dc/dr/ABSre.	pwr.bal.	36.9	2.75	20	205/75R15	L4/2.2/M5	-
S10-Sonoma 4x2	cab.extd.	2dr.p-u.4/5	122.9	203.3x67.9x61.9	3082	1459	i/r	dc/dr/ABSre.	pwr.bal.	41.6	2.75	20	205/70R15	V6/4.3/A4	**$12,113**
S10-Sonoma 4x4	std regular	2dr.p-u.2/3	108.3	189.0x67.9x63.8	3355	1653	i/r	dc/dr/ABS	pwr.bal.	36.9	2.75	20	205/75R15	L4/2.2/M5	**$15,638**
S10-Sonoma 4x4	std extd.	2dr.p-u.2/3	117.9	204.9x67.9x63.8	3437	1600	i/r	dc/dr/ABS	pwr.bal.	36.9	2.75	20	205/75R15	L4/2.2/M5	-
S10-Sonoma 4x4	cab.extd.	2dr.p-u.4/5	122.9	203.3x67.9x64.6	3604	1415	i/r	dc/dr/ABS	pwr.bal.	41.6	2.75	20	205/75R15	V6/4.3/A4	**$16,613**
Blazer-Jimmy 4x2	base	3dr.wgn.4	100.5	174.7x67.8x66.0	3507	1988	i/r	dc/dr/ABS	pwr.bal.	34.8	3.38	19	205/75R15	V6/4.3/M5	**$15,022**
Blazer-Jimmy 4x2	base	5dr.wgn.6	107.0	181.2x67.8x65.9	3600	1850	i/r	dc/dr/ABS	pwr.bal.	36.6	3.38	20	205/75R15	V6/4.3/A4	-
Blazer-Jimmy 4X4	base	3dr.wgn.4	100.5	170.3x67.8x66.0	3867	1420	i/r	dc/dr/ABS	pwr.bal.	35.2	2.97	19	205/75R15	V6/4.3/M5	**$16,905**
Blazer-Jimmy 4X4	base	5dr.wgn.6	107.0	193.3x67.8x65.9	4072	1219	i/r	dc/dr/ABS	pwr.bal.	39.5	2.97	20	205/75R15	V6/4.3/A4	**$18,287**

See page 393 for complete 1995 Price List

Towing time...

GM is looking at Ford prance to the front of the sales honor list, while waiting to improve its big utilities so as to unseat its rival. The arrival of the Dodge Ram has not changed much in the market place of the two bigger ones, because its sales have not yet become significant.

GM remains second in sales of its utilities in North America. The Chevrolet GMC family includes full-size pickups of which there are 80 different versions shared between the 4X2 (C1500-2500-3500) and 4X4 (K1500-2500-3500) on the normal or extended wheelbase and with a standard or extended 2-door multi-seaters 2- or 4-door to which are added the Tahoe and the Suburban. This year, a 4-door version of the Tahoe makes it into a little Suburban. Four engines are offered in the pickups, a 4.3 L V6 and 3 V-8s of 5.0 L, 5.7 L and 7.4 L running on gasoline. There are also two Diesels of 6.5 L, one naturally aspirated and the other turbo-charged. The box on the pickups takes three shapes: short with step sides and short with straight sides and single or dual drive wheels. The Blazer Yukon are wagons with 2 doors and 4-wheel drive while the Suburban is a 4-door wagon with 2- or 4-wheel drive. The main finish level of these vehicles are Cheyenne and Silverado at Chevrolet or SL and SLE at GMC.

STRONG POINTS

Technical:
The Chevrolet GMC utilities begin with the same mechanical base, with minor differences in finish and appearance details. Based on a galvanized steel ladder chassis, it has five cross members, an independent front suspension, plus a rigid rear axle with leaf springs. The macho aesthetics of their body are not very aerodynamic because the Cd varies between 0.45 for Blazer Suburban and 0.50 for the pickups. The standard ABS acts on all four wheels for all vehicles in the range.

Safety:
If the structural resistance remains average, the driver is now protected by an air bag and the other occupants are relatively well-shielded.

Interior space:
The simple pickup cab can accept three people per bench, double that with the extended cab. The Tahoe or the 4-door cab is the triplet, taking in nine people on three bench seats.

Storage:
Baggage space is very limited in the simple pickup cab, but there is more storage in the extended cab, the Blazer and the Suburban, which has an immense storage area even when nine people are seated.

Quality/fit/finish:
The interior is very flattering, but some materials have the utility look and some adjustments are vague.

Driver's compartment:
The driver lacks lateral support on the bench seats, which have little shape. Individual bucket seats on some models do not even have a reclining back rest. Visibility is good all around, even on the Suburban whose roof supports are thick. Rear-view mirrors are ample. The new dashboard appearance improves both aesthetics and ergonomics as

DATA

Category: rear-wheel drive and 4x4 all-purposes and vans.
Class : utility

HISTORIC

Introduced in: 1936:C/K pick-ups & Suburban; 1970: Blazer.
Modified in: 1937, 1948, 1955, 1988, 1992.
Made in: Fort Wayne, IN, Janesville, WI, Pontiac & Flint, MI,U.S.A.

RATINGS

	C/K	Blazer	Suburban
Safety:	80 %	70 %	70 %
Satisfaction:	82 %	80 %	83 %
Depreciation:	55 %	65 %	55 %
Insurance:	6.4 %	5.0 %	4.3 % ($585)
Cost per mile:	$0.55	$0.55	$0.60

NUMBER OF DEALERS

In U.S.A.: 4,466 Chevrolet

SALES ON THE AMERICAN MARKET

Model	1992	1993	Result
C/K	455,250	544,373	+16.4%
Suburban	56,839	82,615	+32.3 %

MAIN COMPETITORS

C/K: DODGE Ram, FORD F-Series.
Blazer: DODGE Ramcharger, FORD Bronco.
Suburban: none

EQUIPMENT

CHEVROLET C/K, Tahoe, Subur. GMC Sierra, Yukon, Suburban	Cheyenne SL	Silverado SLE
Automatic transmission:	S	S
Cruise control:	O	S
Power steering:	S	S
ABS brake:	S	S
Air conditioning:	O	O
Air bag (left):	S	S
Leather trim:	-	-
AM/FM/ Radio-cassette:	O	O
Power door locks:	O	O
Power windows:	O	O
Tilt steering column:	O	O
Dual power mirrors:	O	O
Intermittent wipers:	S	S
Light alloy wheels:	O	O
Sunroof:	-	-
Anti-theft system:	-	-

S : standard; O : optional; - : not available

COLORS AVAILABLE

Exterior: Blue, Green,Black, Red, Silver, Beige, White, Brown.
Interior: Light Gray, Dark Blue, Red, Beige.

MAINTENANCE

First revision: 3,000 miles
Frequency: 6 months/6,000 miles
Diagnostic plug: No

WHAT'S NEW IN 1995 ?

- Driver's side air bag (C/K).
- ABS brakes on all four wheels as standard equipment.
- New dashboard appearance (C/K).
- Remote control keys for door opening and closing offered as an option.
- The Blazer becomes the Tahoe.

	ENGINE					TRANSMISSION		PERFORMANCE										
Model/ versions *: standard	Type / timing valve / fuel system	Displacement cu/in	Power bhp @ rmn	Torque lb.ft @ rpm	Compres. ratio	Driving wheels / transmission	Final ratio	Acceler. 0-62 mph s	Stand. 1/4 mile s	Stand. 5/8 mile s	Braking 62-0 mph ft	Top speed mph	Lateral acceler. G	Noise level dBA	Fuel economy mpg city	hwy	Gasoline type / octane
base	V6* 4.3 OHV-12-EFI	262	160 @ 4000	235 @ 2400	9.1 :1	rear - M5*	3.08	13.7	19.3	36.8	187	94	0.68	68	19.0	29.0	R 87
						rear - A4	3.08	14.5	21.0	36.2	174	87	0.68	68	20.0	28.0	R 87
option	V8 5.0 OHV-16-EFI	305	175 @ 4200	265 @ 2800	9.1 :1	rear./4 -A4	3.42	13.3	18.8	36.6	180	100	0.68	67	18.0	25.0	R 87
	V8 5.7 OHV-16-EFI	350	200 @ 4000	310 @ 2400	9.1 :1	rear./4 -A4*	3.42	12.6	18.7	35.4	200	106	0.68	67	16.0	23.0	R 87
	V8D 6.5 OHV-16-MI	380	155 @ 3600	275 @ 1700	21.5 :1	rear./4 -A4*	3.73	15.2	20.5	37.3	197	94	0.67	68	19.0	26.0	D
	V8TD 6.5 OHV-16-MI	380	190 @ 3400	385 @ 1700	21.5 :1	rear./4 -A4*	3.73	14.4	19.7	36.5	203	100	0.67	68	20.0	27.0	D
	V8 7.4 OHV-16-EFI	454	230 @ 3600	385 @ 1600	7.9 :1	rear./4 -A4*	4.10	12.5	18.5	35.0	197	106	0.68	67	13.0	19.0	R 87

it is better organized.

Performance:
In spite of their size, these vehicles are relatively easy to drive. Empty, their power and torque are not lacking. The base V6 and a 5.0 L. V8 are sufficient for light work and leisure but for heavy duty you need the gasoline 5.7 or 7.4, very thirsty. The naturally aspirated and turbo Diesels are more economical.

Handling:
On a good road, the responses are clean and the roll is limited. On bad pavement the rear axle jumps and wanders producing substantial departures from the intended path. The 2-door Tahoe has a tendency to wander, because of its short wheelbase, while the Suburban and 4-door Tahoe are more stable and heavier.

Steering:
Still too much power assist, too many turns lock-to-lock, with poor return, the maneuverability of these cars is not ideal considering their bulk and large steering diameter.

Braking:
Its excess assistance makes pedal control delicate and the stopping distances are very long. Also, the ABS does not eliminate some wheel locks which provoke disquieting shudders.

WEAK POINTS

Access:
It is satisfactory, except toward the rear seats of the extended cab of the pickup or the Tahoe 2-door as well as toward the third bench of the Tahoe 4-door and Suburban.

Seats:
Their ergonomics are not perfect, even on the deluxe versions.

CHEVROLET-GMC K1500

NEW FOR 1995

CHEVROLET Tahoe

Suspension:
Almost soft on parkways, it is a little more vague on bad pavement or off road, especially with heavy duty 4X4 parts and heavy tires.

Noise level:
Very low with V8 engines, it is a little higher with the Diesel without it being really disturbing and when the speed increases, whistles from the wind and tires do the same.

Conveniences:
The glove box is standard and the other storage depends on the option group.

Price/equipment:
When the equipment is acceptable, the price is less tolerable.

Insurance:
The pickups are distinctly costlier to insure than the Tahoe and the Suburban.

Fuel consumption:
It is strong in all instances except for the diesels.

Satisfaction:
Owners affirm that the durability of this vehicle is not yet equivalent to that of Ford or Dodge.

Depreciation:
Often rudely driven, the pickups quickly lose value and the Suburban sells better than the less versatile Tahoe.

CONCLUSION

The strength, quality and reliability of these utilities should improve to equal the competition and to convince some buyers to return to GM because if aesthetics are important they are the least of things on the utility. 😐

SPECIFICATIONS & PRICES

Model	Version/Trim	Body/ Seats		Wheel-base in.	Lgth x Wdthx Hght. inx inx in	Curb weight lb	Towing capacity std. lb	Susp. type ft/rr	Brake type ft/rr	Steering system	Turning diameter ft	Steer. turns nber	Fuel. tank gal.	Standard tire size	Standard Power train	PRICES US. $ 1994
CHEVROLET-GMC 1500-Series		**General warranty: 3 years / 36,000 miles; antipollution: 5 years / 50,000 miles; perforation corrosion: 6 years / 100,000 miles.**														
C-Sierra	Cheyenne-SL reg. short	2dr.p-u.3	4x2	117.5	194.5x77.1x70.4	3827	1735	i/r	dc/dr/ABS.	pwr.bal.	41.3	3.04	25	225/75R15	V6/4.3/M5	**$14,267**
C-Sierra	Cheyenne-SL reg. extd	2dr.p-u.3	4x2	131.5	213.1x76.8x70.4	4012	2565	i/r	dc/dr/ABS.	pwr.bal.	44.4	3.04	34	225/75R15	V8/5.0/M5	**$16,094**
C-Sierra	Silverado-SLE cab.short	2dr.p-u.3	4x2	141.5	218.0x77.1x70.6	4321	2588	i/r	dc/dr/ABS.	pwr.bal.	52.9	3.04	25	225/75R15	V8/5.0/M5	-
C-Sierra	Silverado-SLE cab. extd	2dr.p-u.3/5	4x2	155.5	237.0x76.8x70.6	4127	3567	i/r	dc/dr/ABS.	pwr.bal.	54.5	3.04	34	225/75R15	V8/5.0/M5	-
K-Sierra	Cheyenne-SL reg. short	2dr.p-u.3	4x4	117.5	194.5x77.1x73.8	4138	2565	i/r	dc/dr/ABS.	pwr.bal.	40.3	2.88	25	225/75R16	V6/4.3/M5	**$16,709**
K-Sierra	Cheyenne-SL reg. extd	2dr.p-u.3	4x4	131.5	213.1x76.8x73.8	4288	3177	i/r	dc/dr/ABS.	pwr.bal.	52.9	2.88	34	225/75R16	V8/5.0/A4	**$18,415**
K-Sierra	Silverado-SLE cab. short	2dr.p-u.3/5	4x4	141.5	218.0x77.1x73.8	4435	5120	i/r	dc/dr/ABS.	pwr.bal.	52.1	2.88	25	225/75R16	V8/5.9/A4	-
K-Sierra	Silverado-SLE cab. extd	2dr.p-u.3/5	4x4	155.5	237.0x76.8x73.8	4568	4795	i/r	dc/dr/ABS.	pwr.bal.	52.9	2.88	34	225/75R16	V8/5.0/A4	-
Tahoe-Yukon	Cheyenne-SL	2dr.wgn.5	4x4	111.5	188.5x77.1x72.4	4747	1559	i/r	dc/dr/ABS.	pwr.bal.	41.5	2.88	30	245/75R16	V8/5.7/A4	-
Tahoe-Yukon	Silverado-SLE	4dr.wgn.5	4x4	111.5	188.5x77.1x72.4	4747	1559	i/r	dc/dr/ABS.	pwr.bal.	41.5	2.88	30	245/75R16	V8/5.7/A4	-
Suburban	Cheyenne-SL	4dr.wgn.6/9	2x4	131.5	220.0x76.7x70.2	4689	2121	i/r	dc/dr/ABS.	pwr.bal.	45.8	3.04	42	245/75R15	V8/5.7/A4	-
Suburban	Silverado-SLE	4dr.wgn.6/9	4x4	131.5	220.0x76.7x70.2	5176	2357	i/r	dc/dr/ABS.	pwr.bal.	47.9	2.88	42	245/75R16	V8/5.7/A4	-

See page 393 for complete 1995 Price List

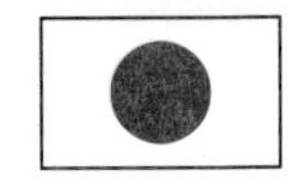

Slow evolution...

There was an era where Honda was the initiator and did not hesitate to innovate to go forward. Today, this firm has aged considerably, at least mentally, because it is only with much difficulty that it completes its range of products and motorization. Is there anybody home?

The Honda Accord remains the family car par excellence of our era in North America, and the success it encounters comes from its homogeneous design in terms of Power train and aesthetics. It has prudently evolved, beginning with the comments from its conservative clients, which probably explains the timidity of its motions. The 1994 range includes: a 2-door coupe in LX and EX-R, a sedan in LX, EX and EX-R and a 4-door wagon in EX only. In 1995, the revolution has finally arrived with a 2.7 L. V6 which is offered as an option on the EX and EX-R versions.

STRONG POINTS

• Safety: **90%**
The body has been reinforced to offer a structural resistance that is 21% higher in beaming and 49% higher in torsion than the previous model. Reinforcements have been installed in the doors and rocker panels to better resist lateral impact. Padding has been added to the doors to protect the hips and shoulders of the occupants. Two air bags and knee protectors are installed in the front seats of all models.

• Satisfaction: **90%**
Its degree has not improved as much as on some of its competitors, which easily reach 95%.

• Quality/fit/finish: **80%**
If assembly and finish are strict, the sheet metal seems light and some of the fabrics and plastics are not very flattering and resemble those of the Civic.

• Driver's compartment: **80%**
Thanks to seat and steering column adjustments, the driver is comfortably installed and visibility is not disturbed by pillars, which are thicker. Also, the outside rear-view mirrors are amply sized. Redesigned, the dashboard is more ergonomic but with no imagination or personality.

• Technical: **80%**
The unit body of the latest Accord is steel, with 80% of the panels galvanized to improve its corrosion resistance. The 4-wheel independent suspension is as on the preceding version and braking remains mixed on the LX-EX and 4-discs and third generation ABS on the EX-R. While the lines are refined up front, the rear is more bulbous and the aerodynamics remain conservative, with a 0.34 drag coefficient. The 4-cylinder 2.2 L. engines develops 130 hp on the LX and EX or 145 hp in the EX-R; the optional V6 develops 170 hp. The transmissions are manual 5-speed standard and 4-speed automatic as an option.

• Fuel consumption: **80%**
At a reasonable speed, it remains moderate on the 4-cylinder but above allowed speeds it climbs rapidly.

• Conveniences: **80%**
In sufficient numbers, the storage spaces include a large glove box, door pockets and hollows in the center console.

DATA

Category: front-wheel drive compact coupes, sedans and wagons.
Class : 4

HISTORIC

Introduced in: 1981
Modified in: 1985, 1989. 1991: wagon; 1994: body; 1995: V6.
Made in: Marysville, OH,U.S.A.

RATINGS

Safety: 90 %
Satisfaction: 90 %
Depreciation: 56 %
Insurance: 5.7 % ($585)
Cost per mile: $0.40

NUMBER OF DEALERS

In U.S.A.: 1,001

SALES ON THE AMERICAN MARKET

Model	1992	1993	Result
Accord	393,477	330,030	- 16.2 %

MAIN COMPETITORS

BUICK Skylark, DODGE-PLYMOUTH Spirit-Acclaim, Cirrus-Stratus,FORD Contour-Mystique MAZDA 626, MITSUBISHI Galant, NISSAN Altima & Maxima, OLDSMOBILE Achieva, PONTIAC Grand Am, SUBARU Legacy, TOYOTA Corolla & Camry, VOLKSWAGEN Passat & Jetta.

EQUIPMENT

HONDA Accord	LX	EX	EX-R
Automatic transmission:	O	O	O
Cruise control:	O	S	S
Power steering:	O	S	S
ABS brake:	S	S	S
Air conditioning:	S	S	S
Air bag (2):	S	S	S
Leather trim:	-	-	O
AM/FM/ Radio-cassette:	S	S	S
Power door locks:	-	S	S
Power windows:	-	S	S
Tilt steering column:	S	S	S
Dual power mirrors:	S	S	S
Intermittent wipers:	S	S	-
Light alloy wheels:	-	O	S
Sunroof:	O	O	O
Anti-theft system:	-	O	O

S : standard; O : optional; - : not available

COLORS AVAILABLE

Exterior: Black, White, Burgundy Red, Cashmere, Gray, Green, Malachite.
Interior: Gray, Taupe, Ivory, Jade.

MAINTENANCE

First revision: 4,000 miles
Frequency: 4,000 miles
Diagnostic plug: Yes

WHAT'S NEW IN 1995 ?

- Two new models EX and EX-R with a V6.
- Some cosmetic redesign.

		ENGINE				TRANSMISSION		PERFORMANCE									
Model/ versions *: standard	Type / timing valve / fuel system	Displacement cu/in	Power bhp @ rmn	Torque lb.ft @ rpm	Compres. ratio	Driving wheels / transmission	Final ratio	Acceler. 0-62 mph s	Stand. 1/4 mile s	Stand. 5/8 mile s	Braking 62-0 mph ft	Top speed mph	Lateral acceler. G	Noise level dBA	Fuel economy mpg city	hwy	Gasoline type / octane
EX-LX	L4* 2.2 SOHC-16-MFI	131	130 @ 5300	139 @ 4200	8.8 :1	front - M5*	4.062	9.0	16.8	29.2	128	131	0.78	65	30.0	41.0	R 87
						front - A4	4.133	10.2	17.6	30.8	134	125	0.78	65	27.0	39.0	R 87
EX-R	L4* 2.2 SOHC-16-MFI	131	145 @ 5600	147 @ 4500	8.8 :1	front - M5	4.062	8.5	15.8	28.0	121	138	0.80	64	29.0	40.0	R 87
option EX-EX-R	V6* 2.7 SOHC-24-MFI	163	170 @ 5600	165 @ 4500	-	front - A4	NA										

• **Access** 70%
It is easy to take seat in these cars whose doors are well-designed, even on the coupe, and where thc back of the front seats provides enough access to the rear bench seats.

• **Seats:** 70%
They provide good lateral and lumbar support, but the firmness of their padding and the narrowness of the seat (trying to get an impression of space) make trips more tiring.

• **Suspension:** 70%
It is rather soft on the EX and LX version and becomes firmer on the EX-R, closer to European settings.

• **Noise level:** 70%
It is lower thanks to greater body stiffness and more advanced soundproofing. However, the rolling noises are noticeable and those due to airstreams begin to appear beyond 60 mph.

• **Steering:** 70%
Precise and with a good ratio, its excessively strong assistance penalizes road holding in a cross wind and a tighter turning radius would favor maneuverability.

• **Performance:** 60%
It spite of its extra 5 hp, the standard engine has problem handling car weight, while the VTEC of the EX-R accelerates more crisply. The V6, which we were unable to test, will eliminate these short comings.

• **Handling:** 60%
The greater rigidity of the body and the effective suspension calibration as well as ample sized tires produce neutral road handling in most instances it will understeer when pushed to the limit. The larger tires and stabilizer bar of the EX-R allow it to maintain its neutral feel longer.

• **Braking:** 60%
While progressive and easy to modulate in normal use, the stopping distances on the LX-EX arelonger. Standard 4-wheel discs and ABS on the EX-R are needed to obtain shorter and better-balanced stops.

• **Interior space:** 60%
The reduction of some dimensions, such as wheelbase length, has been compensated by the increase in width and height to give a more useful interior volume. Four people will be at ease, but a fifth only in a panic.

• **Insurance:** 60%
The Accord is not more costly to insure than most of the competition in its category.

• **Trunk:** 50%
More spacious than before, the storage capacities of the three bodies are ample and easy to load, thanks to their simple and rational shapes and lowering the bench seat back, allows it to handle bulky objects.

• **Price/equipment:** 50%
Honda still practices a policy of high price, which doesn't always help it in terms of sales, which have seriously dropped. This is hard to explain for vehicles manufactured in North America.

WEAK POINTS

• **Depreciation:** 40%
It increases because of the great number of used Hondas that are encountered on the market.

CONCLUSION

• **Overall rating:** 68.5%
The latest Accord disappoints by its lack of dash. Let us hope that the long awaited V6 accomplishes a miracle. ☺

SPECIFICATIONS & PRICES

Model	Version/Trim	Body/ Seats	Interior volume cu.ft.	Trunk volume cu ft.	Drag coef. Cd	Wheel-base in	Lgth x Wdthx Hght. inx inx in	Curb weight lb	Towing capacity std. lb	Susp. type ft/rr	Brake type ft/rr	Steering system	Turning diameter ft	Steer. turns nber	Fuel. tank gal.	Standard tire size	Standard Power train	PRICES US. $ 1994
HONDA		**General warranty: 3 years / 36,000 miles; Power train: 5 years / 60,000 miles.**																
Accord	LX	2dr.cpe.5	90	13	0.32	106.9	184.0x70.1x54.7	2789	2000	i/i	dc/dr	pwr.r&p.	36.1	3.11	17	185/70R14	L4/2.2/M5	**$17,030**
Accord	LX	4dr.sdn.5	94.8	13	0.33	106.9	184.0x70.1x55.1	2811	2000	i/i	dc/dr	pwr.r&p.	36.1	3.11	17	185/70R14	L4/2.2/M5	**$17,230**
Accord	EX	4dr.sdn.5	94.8	13	0.33	106.9	184.0x70.1x55.1	2877	2000	i/i	dc/dr	pwr.r&p.	36.1	3.11	17	185/70R14	L4/2.2/M5	**$19,750**
Accord	EX	4dr.wgn.5	96.6	25.7	0.32	106.9	184.0x70.1x55.1	-	2000	i/i	dc/dr	pwr.r&p.	36.1	3.11	17	195/60R15	L4/2.2/M5	**$20,750**
Accord	EX-R	2dr.cpe.5	90	13	0.32	106.9	184.0x70.1x54.7	2965	2000	i/i	dc/dc/ABS	pwr.r&p.	36.1	3.11	17	195/60R15	L4/2.2/M5	**$20,600**
Accord	EX-R	4dr.sdn.5	94.8	13	0.33	106.9	184.0x70.1x55.1	3009	2000	i/i	dc/dc/ABS	pwr.r&p.	36.1	3.11	17	195/60R15	L4/2.2/M5	**$21,550**
Accord	EX-V6	4dr.sdn.5	-	-	-	-	-	-	-	i/i	dc/dc/ABS	pwr.r&p.					V6/2.7/A4	-
Accord	EX-R V6	4dr.sdn.5																

See page 393 for complete 1995 Price List

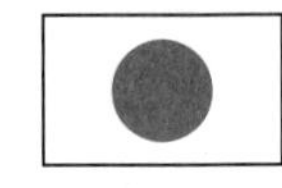

Disgrace...

The latest Honda generations have not known the success of their predecessors. This is the case with the latest Accord, but also of the Civic, which was the cherished baby of the public. High prices and politics in the company have led buyers to other purchases.

HONDA Civic Hatchback

In 1995 the Honda Civic range includes: 3 bodies, a 3-door sedan in DX finish, a 2-door coupe in DX trim, Si, and a 4-door sedan LX or EX. The DX, CX, and LX receive a 102 hp 1.5 L engine with manual or stick shift while the Sports Si and the EX sedan have a 125 hp 1.6 L with the 2 usual transmissions.

STRONG POINTS

• Satisfaction: 90%
The reliability of the Civic has made its reputation and explains in part the high price which their builder keeps. Owners continue to complain of corrosion problems.

• Safety: 80%
As on most subcompacts, the Civics are vulnerable to collisions, considering the size of their structure and the thinness of sheet metal. They have been substantially improved to satisfy American impact tests. This obligation has imposed that the hatchback opens in 2 parts, so the lower portion of the body can better support the bumper in case of rear impact. Fortunately, two standard air bags protect the front seats (std on the LX).

• Fuel consumption: 80%
It continues to be economical, since the average of the range is around 35 miles per gallon.

• Technical: 80%
The steel unit body includes 4 wheel independent suspension and mixed brakes without ABS and the steering assist is only available on the CX hatchback. Their simple, streamlined shape has an air drag that varies between 0.31 and 0.32.

• Driver's compartment: 80%
The most convenient driving position is easy to find and the seat offers effective lateral and lumbar support in spite of the thin padding.

DATA

Category: front-wheel drive sub-compact sedans.
Class : 3S

HISTORIC

Introduced in: 1972
Modified in: 1980, 1984, 1988, 1992.
Made in: Alliston, Ontario, Canada.

RATINGS

Safety: 70 %
Satisfaction: 90 %
Depreciation: 59 %
Insurance: 9.4 % ($585)
Cost per mile: $0.30

NUMBER OF DEALERS

In U.S.A.: 1,001

SALES ON THE AMERICAN MARKET

Model	1992	1993	Result
Civic	218,958	168,880	- 32.1 %

MAIN COMPETITORS

DODGE-PLYMOUTH Colt, EAGLE Summit, FORD Escort, HYUNDAI Accent (4dr.) MAZDA 323-Protegé, NISSAN Sentra, TOYOTA Tercel, VOLKSWAGEN Golf.

EQUIPMENT

HONDA Civic Hbk	DX			
HONDA Civic sedan			LX	EX
HONDA Civic Coupe	DX	SI		
Automatic transmission:	O	-	O	O
Cruise control:	O	S	O	S
Power steering:	-	-	-	O
ABS brake:	S	S	O	S
Air conditioning:	S	S	S	S
Air bag (2):	S	O	S	-
Leather trim:	-	-	-	-
AM/FM/ Radio-cassette:	O	S	O	S
Power door locks:	-	-	-	S
Power windows:	-	-	-	S
Tilt steering column:	S	S	S	S
Dual power mirrors:	S	S	O	S
Intermittent wipers:	S	S	S	S
Light alloy wheels:	-	S	-	-
Sunroof:	-	S	-	-
Anti-theft system:	-	-	-	-

S : standard; O : optional; - : not available

COLORS AVAILABLE

Exterior: Black, Red, White, Green, Blue, Gray.
Interior: Black, Blue, Gray, Beige.

MAINTENANCE

First revision: 4,000 miles
Frequency: 4,000 miles
Diagnostic plug: Yes

WHAT'S NEW IN 1995 ?

- No major change.

Model/ versions *: standard	ENGINE Type / timing valve / fuel system	Displacement cu/in	Power bhp @ rmn	Torque lb.ft @ rpm	Compres. ratio	TRANSMISSION Driving wheels / transmission	Final ratio	PERFORMANCE Acceler. 0-62 mph s	Stand. 1/4 mile s	Stand. 5/8 mile s	Braking 62-0 mph ft	Top speed mph	Lateral acceler. G	Noise level dBA	Fuel economy mpg city	hwy	Gasoline type / octane
DX-LX	L4* 1.5 SOHC-16-EFI	91	102 @ 5900	98 @ 5000	9.2 :1	front - M5*	3.88	9.4	16.4	30.0	148	112	0.80	68	40.0	52.0	R 87
						front - A4	4.33	10.5	17.3	31.5	161	106	0.80	68	33.0	47.0	R 87
Si &Ex	L4* 1.6 SOHC-16-EFI	97	125 @ 6600	106 @ 5200	9.2 :1	front - M5*	4.25	8.8	16.2	28.8	151	118	0.81	68	34.0	46.0	R 87

Instrumentation is well laid out, and easy to read. The main controls are conventional and the steering wheel offers a good grip, thanks to a thick rim.

• Price/equipment: 80%
In spite of their limited equipment, the Honda's are always a little more expensive than the others, a little surprising since they are manufactured in Canada.

• Suspension: 70%
It does not count among the softest ones but absorbs road irregularities because its damping is more consistent and the wheel travel more generous.

• Quality/fit/finish: 70%
Even though the assembly and finish remain strict, some fabrics and plastics do not have the best appearance. The very light sheet metal does not inspire confidence.

• Performance: 70%
All Hondas suffer from the same problem. Their engine revs at the slightest accelerator motion but their torque is "hollow" and the noise which accompanies the acceleration gives the impression of going faster than the stop watch. The latest models have less performance than the old ones which were lighter. The manual shift selection is fast and precise but the automatic down shifts slowly to save a little fuel.

• Handling: 70%
The Civic holds the road well, thanks to its suspension, with fork-shaped arms and tires wihich are amply sized and well-adapted. They are great in a tight turn and remain neutral long enough so the average driver does not see the understeer that characterizes them. While you no longer note high torque steer during strong accelerations, you can frequently find traction loss of the driving wheels, even on dry pavement.

• Seats: 70%
Well-shaped, they provide good lumbar and lateral support but their firm padding becomes painful on long trips.

• Steering: 70%
Precise and well-modulated, it continues to have too high a ratio. Assisted or manual, maneuverability is excellent and the steering diameter is short.

• Conveniences: 60%
The storage consists of a good size glove box and a storage compartment on the center console, because the very practical door pockets disappear on the cheap versions.

• Access: 50%
Acceptable on the coupe and on the sedan, whose interior height is 1.1 inches larger than the hatchback, where it is more difficult to reach the bench seat because of the small space freed by the front seats.

• Noise level: 50%
Lacking in insulation material, it is tolerable at cruise speed but you soon get tired of it during accelerations.

WEAK POINTS

• Trunk: 40%
The rear bench seat now can be lowered to increase storage capacity. The 2-part hatchback helps stiffness but not baggage handling. If you lean on the back of the body, while handling baggage, dry cleaning bill increases.

HONDA Civic sedan

• Braking: 40%
Progressive and easy to modulate during normal deceleration, it becomes irregular in a panic stop, because the front wheels lock up immediately, the pedal becomes hard, and the trajectories both long and unstable. The ABS is not offered, even as an option on any of these models.

• Interior space: 40%
The increase in the main dimensions has improved the Civics' interior space but advantages are more evident on a coupe and the sedan whose interior height is greater than on the hatchback.

• Insurance: 40%
Taking into account their format and prices, the Civics are fairly expensive to insure because they address themselves to a youthful buyer group.

• Depreciation: 40%
Just like the Accord, the used Civics begin to lose more of their value than before because there are many of them on the market.

CONCLUSION

• Overall rating: 63.4%
With time, the Civic has lost more of what made its charm, which is an affordable price and performances which made their driving fun. It is this evolution in the wrong direction which explains the drop in sales and buyers wandering away.

OWNER'S SUGGESTIONS

- More powerful heater
- More complete equipment
- Stiffer body
- Effective soundproofing
- Better general quality
- More comprehensive warranty
- Better corrosion resistance

SPECIFICATIONS & PRICES

Model	Version/Trim	Body/ Seats	Interior volume cu.ft.	Trunk volume cu ft.	Drag coef. Cd	Wheel-base in	Lgth x Wdthx Hght. inx inx in	Curb weight lb	Towing capacity std. lb	Susp. type ft/rr	Brake type ft/rr	Steering system	Turning diameter ft	Steer. turns nber	Fuel. tank gal.	Standard tire size	Standard Power train	PRICES US. $ 1994
HONDA		**General warranty: 3 years / 36,000 miles; Power train: 5 years / 60,000 miles.**																
Civic	DX	3dr.sdn.4/5	77	13.3	0.31	101.2	160.2x66.9x53.0	2147	NR	i/i	dc/dr	r&p.	32.2	3.88	11.9	175/70R13	L4/1.5/M5	**$10,800**
Civic	DX	2dr.cpe.4/5	81	12	0.31	103.1	172.8x66.9x53.3	2227	NR	i/i	dc/dr	r&p.	32.8	3.58	11.9	175/70R13	L4/1.5/M5	**$11,220**
Civic	Si	2dr.cpe.4/5	81	12	0.31	103.1	172.8x66.9x53.3	2396	NR	i/i	dc/dr	r&p.	32.8	3.58	11.9	185/60R14	L4/1.6/M5	**$13,170**
Civic	LX	4dr.sdn.4/5	85	13	0.32	103.1	172.8x66.9x54.1	2286	1000	i/i	dc/dr	r&p.	32.8	3.58	11.9	175/70R13	L4/1.5/M5	**$12,950**
Civic	EX	4dr.sdn.4/5	85	13	0.32	103.1	172.8x66.9x54.1	2350	1000	i/i	dc/dr	r&p.	32.8	3.58	11.9	175/60R14	L4/1.5/M5	**$15,740**

See page 393 for complete 1995 Price List

Another missed chance...

The destiny of big companies is cyclic after a brilliant period of conquest, Honda is in the era of bad discussions and dubious compromises. A new direction and better planning are essential to insure the future. If that is late in coming...

The coupe-convertible del Sol is an original idea which consists of offering two vehicles in one. If the style of the front part offers similarity to other models of the range, the rear breaks radically with the hatchback of the previous model by distinguishing itself with a Targa-style roll bar on which a light aluminum alloy roof panel (weight 22 lbs) comes to rest. It can be stored in the trunk, which is separate from the interior, on an articulated cradle to which it locks. This allows trunk access, even when the roof is stored there. Finally the backlight disappears electrically to give the impression of a convertible and to favor air flow toward the rear. The del Sol is offered in versions base, S, Si and VTEC. The first is offered with a 1.5 L engine already in the Civic, the second, equipped with a 1.6 L, has a variable intake and develops 125 hp. It is found on the Civic Si and the VTEC which produces 160 hp.

STRONG POINTS

• Safety: **90%**
The body shell and the doors include reinforcement to insure stiffness, always a problem on this type of car. Finally, two air bags complete the issue.

• Satisfaction: **90%**
The first buyers have not yet complained of major problems and one must wait another year to find out if this percentage can hold.

• Quality/fit/finish: **80%**
The assembly, finish and quality of the materials used is very strict and the presentation is impeccable on all scores.

• Technical: **80%**
The steel body shell has been the object of advanced aerodynamic study to eliminate most airstreams when the roof is removed. The

DATA

Category: front-wheel drive sport coupes-convertibles.
Class : S3

HISTORIC

Introduced in: 1992
Modified in: -
Made in: Saitama, Japan.

RATINGS

Safety: 90 %
Satisfaction: 90 %
Depreciation: 40 % (2 years)
Insurance: 6.4 % ($725)
Cost per mile: $0.33

NUMBER OF DEALERS

In U.S.A.: 1,001

SALES ON THE AMERICAN MARKET

Model	1992	1993	Result
del Sol	NA	86,699	

MAIN COMPETITORS

EAGLE Talon Esi, HYUNDAI Scoupe, MAZDA Miata & MX3, SATURN SC, TOYOTA Paseo VW Corrado & Cabrio.

EQUIPMENT

HONDA Civic del Sol	Si	VTEC
Automatic transmission:	O	O
Cruise control:	O	O
Power steering:	O	O
ABS brake:	S	S
Air conditioning:	S	S
Air bag (left):	S	S
Leather trim:	-	-
AM/FM/ Radio-cassette:	S	S
Power door locks:	-	-
Power windows:	S	S
Tilt steering column:	S	S
Dual power mirrors:	S	S
Intermittent wipers:	S	S
Light alloy wheels:	S	S
Sunroof:	S	S
Anti-theft system:		

S : standard; O : optional; - : not available

COLORS AVAILABLE

Exterior: Black, Red, White, Blue, Green.
Interior: Black.

MAINTENANCE

First revision: 4,000 miles
Frequency: 4,000 miles
Diagnostic plug: Yes

WHAT'S NEW IN 1995 ?

- No major change.

ENGINE						TRANSMISSION		PERFORMANCE									
Model/ versions *: standard	Type / timing valve / fuel system	Displacement cu/in	Power bhp @ rmn	Torque lb.ft @ rpm	Compres. ratio	Driving wheels / transmission	Final ratio	Acceler. 0-62 mph s	Stand. 1/4 mile s	Stand. 5/8 mile s	Braking 62-0 mph ft	Top speed mph	Lateral acceler. G	Noise level dBA	Fuel economy mpg city	hwy	Gasoline type / octane
Si	L4* 1.6 SOHC-16-MFI	97	125 @ 6600	106 @ 5200	9.2 :1	front - M5*	4.250	9.5	16.4	29.5	144	118	0.80	68	34.0	46.0	R 87
						front - A4	4.333	10.7	16.8	31.4	138	112	0.80	68	31.0	43.0	R 87
VTEC	L4* 1.6 DOHC-16-MFI	97	160 @ 7600	112 @ 7000	10.2 :1	front - M5*	4.266	8.5	15.6	29.7	125	131	0.84	69	31.0	40.0	M 89

suspension is independent at all 4 wheels, using double A-frames. Brakes are mixed on the S and 4 disc on the Si which receives a stabilizer bar at the front as well as an ABS system on the VTEC.

• **Performance: 75%**
They are deceiving, because the weight of the del Sol is a little higher than that of other Civics. The accelerations have nothing thunderous about them, and this, even with the 150 hp engine. Contrary to the Miata, which seems to go very fast when there is nothing there, the del Sol goes very fast without it appearing or feeling so. That is its problem, because it does not produce any real driving fun.

• **Driver's compartment: 70%**
The driver is well-seated, no matter what his size. The bucket seat provides effective lateral hold and support, but visibility is limited by the giant side blind spot created by the B pillar. The dashboard is very aesthetic and its elements well-rounded. A wide number of easy-to-read gauges are included into a kidney shaped cluster. Switches disposed around the rim are easy to reach. It would have been better to invert the controls of the radio and AC. The trunk opening from the inside of the driving compartment was omitted to protect the trunk's contents while the top is off.

• **Fuel consumption: 70%**
It is very modest, because in general it is close to 30 mpg.

• **Handling: 70%**
The front wheel drive cuts back on its sports character and the softness is accented by the suspension being connected to an insufficient body panel thickness which results in a vague and unstable directional feel, which has more to do with design than steering. This phenomena accelerates the sharp understeer and deprives the del Sol of its agility and fun character which characterized CRX driving.

• **Seats: 70%**
Thin and rather firm, they provide effective lateral and lumbar support thanks to their great contour.

• **Suspension: 70%**
Its softness is better appreciated on bad pavement, where the reactions are more civilized than the average of its competition.

• **Price/equipment: 70%**
You have to accept the evidence that the del Sol coupe does not sell as well as the Miata, in spite of the originality of its concept and a character that is both exotic and multifaceted. As to its equipment it is as expected, thin, lean and mean on the most affordable models.

• **Conveniences: 70%**
The storage areas include a small glove box, a cassette container on the center tunnel and two hollows with a key lock, between the seats. It is easy for even one person working alone to remove the roof and install it without damage into its trunk cradle. It is not complicated, but still calls for some practice.

• **Access: 60%**
One can easily climb on board in spite of low height as long as the body is sufficiently limber.

• **Depreciation: 60%**
As a convertible, the del Sol has lost less in its first year than comparable convertibles.

• **Steering: 60%**
Too overratioed to handle panic corrections, it is imprecise when manual on the S and a little better assisted on the Si, the VTEC and automatics. It generally gives a bizarre sensation of filtering the reactions from the front end.

• **Insurance: 65%**
The del Sol does not cost anymore to insure than most of its competitors.

• **Brakes: 50%**
Progressive and easy to modulate in normal use, it shows a near dangerous instability in panic stops because without ABS, the front wheels block rapidly and make the vehicle path sinusoidal. Only the VTEC can pretend to have braking at the level of its potential.

WEAK POINTS

• **Interior space: 10%**
Two people will be very much at ease thanks to sufficient cabin length and width.

• **Trunk: 40%**
It is relatively large for this type of vehicle because you can store two average-sized suitcases and two small ones, even when the roof is in the trunk.

• **Noise level: 40%**
It equals that of a coupe at normal speeds and under strong acceleration it is not too hard on the eardrums any more than when the roof is open.

CONCLUSION

• **Overall rating: 64.5%**
Honda did not have the right to spoil a good idea like that. Let's hope that Mazda takes it up soon while adding the necessary correction to transform it like the Miata and the MX-3 into a success. ☺

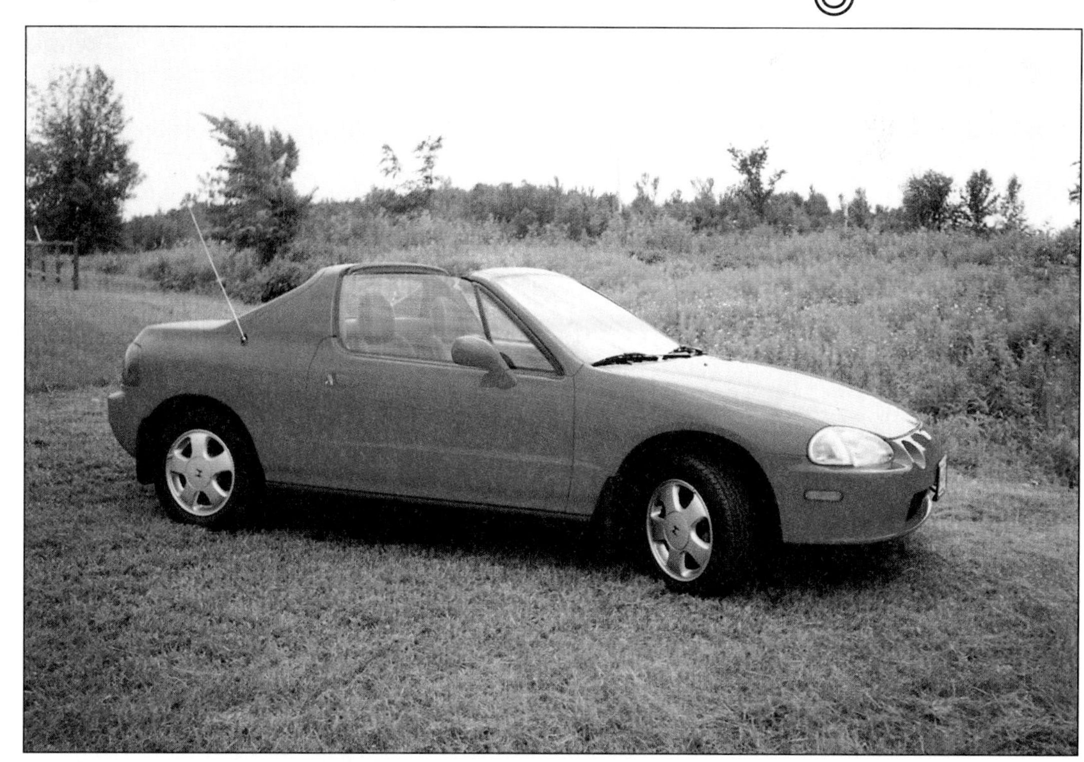

SPECIFICATIONS & PRICES

Model	Version/Trim	Body/ Seats	Interior volume cu.ft.	Trunk volume cu ft.	Drag coef. Cd	Wheel-base in	Lgth x Wdthx Hght. inx inx in	Curb weight lb	Towing capacity std. lb	Susp. type ft/rr	Brake type ft/rr	Steering system	Turning diameter ft	Steer. turns nber	Fuel. tank gal.	Standard tire size	Standard Power train	PRICES US. $ 1994
HONDA		General warranty: 3 years / 36,000 miles; Power train: 5 years / 100 000 km; corrosion perforation: 5 years / kilométrage illimité.																
Civic del Sol Si		2dr.cpe. 2	NA	8.3-10.6		93.3	157.7x66.7x49.4	2396	NR	i/i	dc/dc	pwr.r&p.	30.2	3.6	11.9	185/60R14	L4/1.6/M5	$16,100
Civic del Sol VTEC		2dr.cpe. 2	NA	8.3-10.6		93.3	157.7x66.7x49.4	2427	NR	i/i	dc/dc/ABS	pwr.r&p.	32.8	3.6	11.9	195/60R14	L4/1.6/M5	$17,500

See page 393 for complete 1995 Price List

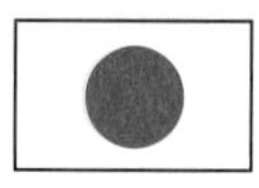

"Feelings..."

The Prelude coupe surprised the entire world by its unusual but original lines. It was however the lack of sports feel which characterized the first models. Actually, only the SR-V is capable of reflecting the sporty name, even then with many reservations.

The Prelude is another Honda product that has not been able to preserve the popularity acquired by the preceding model. Same motive, same punishment and it will soon be necessary to ask which models are still selling best at the Japanese builder. The Prelude is only offered in two versions in 1995: SR with a 2.3 L. double overhead cam developing 160 hp and VTEC whose variable intake timing Honda has specialized in, which delivers 190 hp.

STRONG POINTS

• **Satisfaction: 90%**
The reliability is excellent and the aging is slower than on other Japanese cars.

• **Safety: 90%**
In spite of body stiffening, its impact resistance is not ideal but two air bags mounted standard on all versions and 3-point seat belts allow a good protection level.

• **Technical: 80%**
The steel unit body has an independent suspension, 4-wheel disc brakes and an ABS installed standard on the SR and SR-V. In spite of its taut lines, the aerodynamic finesse is conservative and the Cd is 0.33 (0.40 is old-fashioned and 0.30 is excellent). The interesting 4-cylinder 2.3 L. engine has an electronic PGM-FI injection system derived from those used by competive engines of the marque. It includes two balance shafts to cancel most vibrations inherent in this type of engine.

• **Quality/fit/finish: 80%**
The interior presentation is very faded as often happens with Japanese builders, and some materials lack class on a car of this price. However, the build and finish are without reproach and carried out with

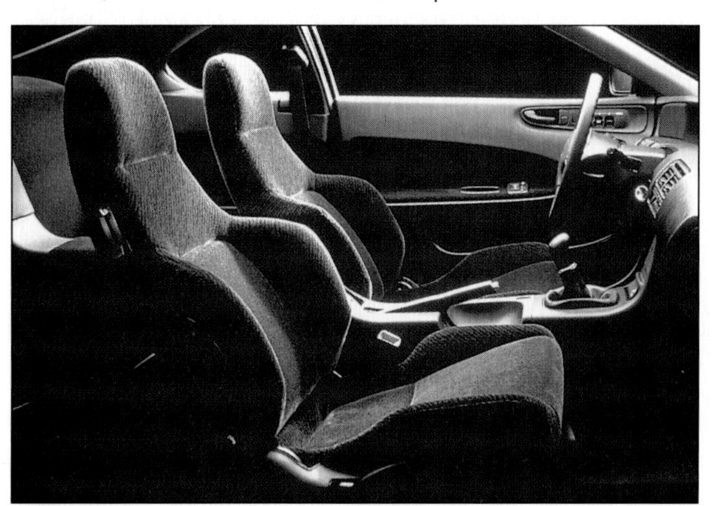

DATA

Category: front-wheel drive sport coupes.
Class : S

HISTORIC

Introduced in: 1979
Modified in: 1983, 1987 & 1991.
Made in: Sayama, Japan.

RATINGS

Safety: 90 %
Satisfaction: 92 %
Depreciation: 57 %
Insurance: 6.3 % ($785)
Cost per mile: $0.43

NUMBER OF DEALERS

In U.S.A.: 1,001

SALES ON THE AMERICAN MARKET

Model	1992	1993	Result
Prélude	36,040	22,540	- 37.5 %

MAIN COMPETITORS

ACURA Integra GS-R, CHEVROLET Camaro V6, EAGLE Talon, FORD Probe & Mustang, MAZDA MX-6 Mystere, NISSAN 240SX, PONTIAC Firebird V6, TOYOTA Celica, VOLKSWAGEN Corrado.

EQUIPMENT

HONDA Prelude	SR	SR-V
Automatic transmission:	O	-
Cruise control:	S	S
Power steering:	S	S
ABS brake:	S	S
Air conditioning:	S	S
Air bag (2):	S	S
Leather trim:	-	S
AM/FM/ Radio-cassette:	S	S
Power door locks:	S	S
Power windows:	S	S
Tilt steering column:	S	S
Dual power mirrors:	S	S
Intermittent wipers:	S	S
Light alloy wheels:	S	S
Sunroof:	S	S
Anti-theft system:	S	S

S : standard; O : optional; - : not available

COLORS AVAILABLE

Exterior: Black, Red, Silver, Blue, Green.
Interior: Black, Gray.

MAINTENANCE

First revision: 4,000 miles
Frequency: 4,000 miles
Diagnostic plug: Yes

WHAT'S NEW IN 1995 ?

-Standard "S" model eliminated from the lineup.

Model/ versions *: standard	ENGINE Type / timing valve / fuel system	Displacement cu/in	Power bhp @ rmn	Torque lb.ft @ rpm	Compres. ratio	TRANSMISSION Driving wheels / transmission	Final ratio	PERFORMANCE Acceler. 0-62 mph s	Stand. 1/4 mile s	Stand. 5/8 mile s	Braking 62-0 mph ft	Top speed mph	Lateral acceler. G	Noise level dBA	Fuel economy mpg city	hwy	Gasoline type / octane
SR	L4* 2.3 DOHC-16-MPFI	138	160 @ 5800	156 @ 4500	9.8 :1	front - M5*	4.266	8.2	15.8	28.1	135	125	0.80	66	27.0	34.0	R 87
						front - A4	4.428	9.4	16.7	29.4	138	118	0.80	66	26.0	34.0	R 87
SR-V	L4* 2.2 DOHC-16-MPFI	132	190 @ 6800	158 @ 5500	10.0 :1	front - M5*	4.266	7.7	15.2	27.6	131	131	0.80	66	26.0	33.0	M 89

the greatest of care.

• Seats: 80%
Real bucket seats which provide lateral and lumbar support but whose padding lacks softness.

• Performance: 75%
They are good and the numbers obtained correspond to that of a sports coupe but pleasurability is not part of the package because it is obtained at high revs and one notices constant lack of power below 4000 rpm. A V6 engine such as the one that is fitted to the Accord this year would allow the Prelude to better hold up against its old enemy the VW Corrado.

• Suspension: 70%
That of the SR offers the best compromise, because the one of the SR-V is too stiff, The body movements are very limited, even on a bad road, where the reaction is never disagreeable.

• Handling: 70%
The Prelude reactions are never harsh and it remains neutral most of the time, with the understeering tendency postponed by good control of body motions. The precision is a strong point because you can position the front toe the nearest millimeter.

• Steering: 70%
It gives the impression to function in steps, jitters at low speeds, light and vague as the speed increases. However, its ratio and steering diameter give it good maneuverability.

• Fuel consumption: 70%
It is reasonable with both engines, they offer excellent mileage which generally averages around 25 mpg.

• Driver's compartment: 60%
The driver will complain of not being able to adjust the height and the angle of the seat cushion. However, the contours provide excellent lateral and lumbar support. Visibility is penalized by the strong windshield angle, the narrow back light and the thickness of the C pillar. The dashboard remains as the most controversial element of the latest Prelude where the instrumentation is half analog and half digital, illegible in the sun and disposed far from the driver eyes and some switches are laid out at random.

• Brakes: 50%
In spite of its progressive modulation and balanced action, stopping distances are too long for a car of this weight and the ABS is late in reacting, which produces initial wheel lock, more annoying than dangerous.

• Noise level: 50%
Remaining low at cruise speeds and during strong accelerations, because soundproofing is effective and the Power train discreet.

• Conveniences: 50%
Storage space is insufficient with a minuscule glove box, missing door pockets and the center console which is of little use.

• Price/equipment: 50%
Honda has a reputation of being costlier than its competition and the problem is to know if it delivers enough for the difference.

• Insurance: 50%
The Prelude costs a little less to insure than other coupes but depending on the age and the record, the cost can be hefty.

WEAK POINTS

• Interior space: 20%
The Prelude coupe will only accept two adults comfortably because at the rear it is difficult to move in even for children, Depending on the size of the driver, the seat back of the front seat can touch the rear seat bench and you can't even sit sideways because the center console is in the way. Even the sunroof does its bit because it lowers the height and interferes with tall people.

• Trunk: 25%
Even though very limited, its volume is sufficient to accept baggage for two people traveling over a weekend. In addition, two trap doors behind the bench seat back allow you to store skis or other long objects.

• Access; 40%
In spite of the door length, the low height and contours of the bucket seats complicate getting in.

• Depreciation: 40%
That of the latest model is higher than that of the old one because of the lack of interest and the premium.

CONCLUSION

• Overall rating: 60.5%
Only the V6 engine could save the Prelude from indifference, providing it has the promised number of horsepower and that the rest of the vehicle, meaning handling and steering, will be readapted to provide strong driving pleasure without having to reach 100 mph on the track.

SPECIFICATIONS & PRICES

Model	Version/Trim	Body/ Seats	Interior volume cu.ft.	Trunk volume cu ft.	Drag coef. Cd	Wheel-base in	Lgth x Wdthx Hght. inx inx in	Curb weight lb	Towing capacity std. lb	Susp. type ft/rr	Brake type ft/rr	Steering system	Turning diameter ft	Steer. turns nber	Fuel. tank gal.	Standard tire size	Standard Power train	PRICES US. $ 1994
HONDA Prelude		**General warranty: 3 years / 36,000 miles; Power train: 5 years 100 000 km.**																
Prelude	SR	2dr.cpe.2+2	72.3	8.6	0.33	100.4	174.8x69.5x50.8	2866	NR	i/i	dc/dc/ABS	pwr.r&p.	35.8	2.9	15.9	205/55R15	L4/2.3/M5	**$21,400**
Prelude	SR/V	2dr.cpe.2+2	72.3	8.6	0.33	100.4	174.8x69.5x50.8	2932	NR	i/i	dc/dc/ABS	pwr.r&p.	35.8	2.9	15.9	205/55R15	L4/2.3/M5	**$24,500**

See page 393 for complete 1995 Price List

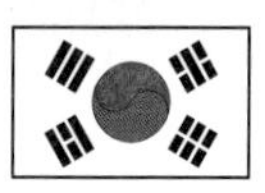

Double or nothing...

After the progress attained by the Sonata and the Elantra, Hyundai should consolidate its position in the group of minimum cars, by giving the Excel a replacement capable of confronting a well-targeted competition. If users' satisfaction goes up, the bet will have been won.

After encountering many mishaps, the Excel enjoyed a certain amount of success last year and saw its sales go up a little. It is in this favorable context that the Accent comes to take its place in the market where small popular cars are less and less numerous. and where a certain need has occurred. The change of name, the new body and numbers improvements represent that many chances for the newcomer. The Accent is proposed in a 2-door coupe and a 4-door sedan in versions L and GL equipped with a 4-cylinder 1.5 L with fuel injection and a manual 5-speed or an automatic with 4-speed offered as an option.

STRONG POINTS

• Price/equipment: 90%
The Accent comes with ample equipment and the options are not very costly. Their price starts below $10,000, a bargain compared to their Japanese competitors.

• Fuel consumption: 90%
It is economical, because it rarely blows 35 mpg with an automatic.

• Safety: 80%
The structure of the new Accent has been reinforced to match the 1997 American laws; unfortunately, air bags and ABS are not standard, as though life was an option.

• Driver's compartment: 70%
The driver rapidly finds the best position even though the steering column is not adjustable. Visibility is satisfactory but the wide C-pillar blocks the 3/4 view on the coupe; rear-view mirrors are well-sized. The dashboard is simple and well-organized. It contains enough instrumentation with small dials, and the commands are easy to reach and manipulate, even those below the gauge block.

• Steering: 70%
The standard manual steering still has a very high number ratio and is less firm than on the Excel. Assisted, fast, more precise and well-modulated, it makes driving more pleasant. However in both instances, maneuverability suffers from too large a steering diameter.

• Technical: 70%
The Accent does not derive, as the Excel, from the platform of the old Mitsubishi Mirage, it is 100% Hyundai and its sympathetic allure offers a good aerodynamic efficiency since its drag coefficient reaches a low 0.31. The steel unit body has a 4-wheel independent suspension, braking is mixed and ABS is optional.

• Handling: 60%
It is improved on the GL, whose tires are more consistent than those of the L. If the understeering tendency is marked, it occurs later and is easily controlled.

DATA

Category: front-wheel drive sub-compact sedans.
Class : 3S

HISTORIC

Introduced in: 1995
Modified in: -
Made in: Ulsan, South Korea.

RATINGS

Safety: 85 %
Satisfaction: 60 %(Excel)
Depreciation: 62% (Excel)
Insurance: 8.2% ($445)
Cost per mile: $0.30

NUMBER OF DEALERS

In U.S.A.: 500

SALES ON THE AMERICAN MARKET

Model	1992	1993	Result
Excel	42,324	43,498	+2.7 %

MAIN COMPETITORS

FORD Aspire, GEO Metro, HONDA Civic, MITSUBISHI Mirage, NISSAN Sentra, SATURN SL1, SUBARU Justy, SUZUKI Swift, TOYOTA Tercel.

EQUIPMENT

HYUNDAI Accent	L	GL
Automatic transmission:	O	O
Cruise control:	-	-
Power steering:	O	S
ABS brake:	-	O
Air conditioning:	-	-
Air bag (2):	O	O
Leather trim:	-	-
AM/FM/ Radio-cassette:	O	S
Power door locks:	-	-
Power windows:	-	-
Tilt steering column:	O	S
Dual power mirrors:	-	-
Intermittent wipers:	S	S
Light alloy wheels:	-	-
Sunroof:	-	O
Anti-theft system:	-	-

S : standard; O : optional; - : not available

COLORS AVAILABLE

Exterior: Green, Silver, Red, Blue.
Interior: Graphite, Mocha.

MAINTENANCE

First revision: 3,000 miles
Frequency: 6,000 miles
Diagnostic plug: No

WHAT'S NEW IN 1995 ?

-All new model which replaces the Excel.
-Alpha engine designed by Hyundai.
-Air bags and ABS optional.

	ENGINE					TRANSMISSION		PERFORMANCE										
Model/ versions *: standard	Type / timing valve / fuel system	Displacement cu/in	Power bhp @ rmn	Torque lb.ft @ rpm	Compres. ratio	Driving wheels / transmission	Final ratio	Acceler. 0-62 mph s	Stand. 1/4 mile s	Stand. 5/8 mile s	Braking 62-0 mph ft	Top speed mph	Lateral acceler. G	Noise level dBA	Fuel economy mpg city	Fuel economy mpg hwy	Gasoline type / octane
base	L4* 1.5 SOHC-12-MPFI	91	92 @ 5500	97 @ 2700	10:1	front - M5*	3.65	11.5	17.7	32.8	138	109	0.76	70	35.0	47.0	R 87
						front - A4	3.65	13.5	18.8	34.8	147	103	0.76	70	33.0	47.0	R 87

The reactions are predictable and surprise by their liveliness on the manual version; fun to drive on a twisty road.

• Suspension: 60%
Its comfort is very acceptable, for a small car, because larger wheel travel allows it to better soak up pavement defects. Also, the damping is more effective.

• Access: 60%
It is easier to get into the front seats than into the back ones where the sedan doors do not leave enough headroom. Tall people will find rear seats uncomfortable.

• Seats: 60%
Their padding is firm and none too thick. This increases fatigue on long trips. Thanks to their good shape, they do provide good lumbar and lateral support.

• Quality/fit/finish: 60%
That there is progress is clear because the plastics do not look as inexpensive, same for the fabrics, whose colors bring a little fantasy. Adjustments are precise and the finish more careful than on the Excel.

• Conveniences: 60%
The storage consists of a normal glove box, very small door pockets and a hollow in the console. The front shoulder harness has a height adjustment and headrests are fitted to the GL bench seat.

• Satisfaction: 60%
We hope that the owners of the new models will be more satisfied than those of the Excel, who only liked it 60%.

• Interior space: 50%
Normal for a car of this size, the space is well-allocated and there is enough headroom at the rear. Rear legroom depends on the front seat position.

• Trunk: 50%
It is more useful on the coupe, where it can be increased by folding the bench seat forward. Both are accessible thanks to a large opening and a sill that reaches down to the bumper.

NEW FOR 1995

WEAK POINTS

• Noise level: 30%
It remains high in all conditions because lack of insulation allows free entry of power train and drive noises, as well as strong frontal wind noise.

• Insurance: 40%
Lack of confidence in the marque translates into a substantially higher premium than those of other models in this category

• Braking: 40%
Its efficiency and fade-resistance are very average, due to lack of friction material quality. On the other hand, it is easy to modulate with precision and smooth progression. During a panic stop, you have a long distance in which to reflect on the smoothness. Wheel lock-up appears early without ABS, while ABS, by its nature, extends the stops. However, with it, the trajectory is more straight-lined and stable.

• Performance: 40%
The Alpha engine borrowed from the Elantra is more powerful so that you may, with a stick shift, pass crisply; weight-to-power ratio is average. Shifts are fast and precise, easily handled with a smooth clutch that makes driving convenient. However, it is not the same with the automatic, which constantly downshifts at the least grade change, hurting comfort and fuel consumption.

• Depreciation: 40%
The Accent loses more of its value because of its dubious reliability, especially after warranty expiration.

CONCLUSION

• Overall rating: 59.0%
After the renewal of the Aspire, Metro and Swift the arrival of the Accent again launches and economy cars. The latest Hyundai marks a net progress in relation to the Excel because it has better handling is greater fun to drive and has a better finish. Too bad that from a safety point of view you have to put your hand back in your pocket. 😐

SPECIFICATIONS & PRICES

Model	Version/Trim	Body/ Seats	Interior volume cu.ft.	Trunk volume cu ft.	Drag coef. Cd	Wheel-base in	Lgth x Wdthx Hght. inx inx in	Curb weight lb	Towing capacity std. lb	Susp. type ft/rr	Brake type ft/rr	Steering system	Turning diameter ft	Steer. turns nber	Fuel. tank gal.	Standard tire size	Standard Power train	PRICES US. $ 1994
HYUNDAI		**General warranty: 3 years / 36,000 miles; Power train: 5 years / 60,000 miles; corrosion perforation: 5 years / unlimited.**																
Accent	L	3dr.sdn.4	88	16.2	0.31	94.5	161.5x63.8x54.9	2101	1000	i/i	dc/dr	r/p.	31.8	3.87	11.9	155/80R13	L4/1.5/M5	**NA**
Accent	L	4dr.sdn.4	84.2	10.7	0.31	94.5	162.1x63.8x54.9	2105	1000	i/i	dc/dr	r/p.	31.8	3.87	11.9	155/80R13	L4/1.5/M5	**NA**
Accent	GL	3dr.sdn.4	88	16.2	0.31	94.5	161.5x63.8x54.9	2115	1000	i/i	dc/dr	pwr.r/p.	31.8	2.93	11.9	175/70R13	L4/1.5/M5	**NA**
Accent	GL	4dr.sdn.4	84.2	10.7	0.31	94.5	162.1x63.8x54.9	2119	1000	i/i	dc/dr	pwr.r/p.	31.8	2.93	11.9	175/70R13	L4/1.5/M5	**NA**

See page 393 for complete 1995 Price List

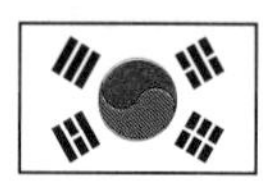

A good card...

To finish conjuring up the Pony syndrome, Hyundai has rebuilt its range with patience and ferocity in spite of the chronic technical lack of success which affected its sales.

Today the builder holds better cards and the Elantra is the master card. The Elantra represents one third of the Korean builders' sales and its popularity is growing. This 3-volume, 4-door sedan is offered in GL and GLS finishes. The GL comes with a 1.6 Mitsubishi and a manual 5-speed; an optional 1.8 L Alpha is fitted standard on the GLS with a manual transmission. An automatic 4-speed, which can run in "sport" or "normal", is optional in both engines.

STRONG POINTS

• Price/equipment: 80%
For what it costs, the Elantra offers more than the average of its competitors, especially the GLS whose equipment is complete and which constitutes a better buy.

• Satisfaction: 80%
The number of satisfied clients increases continually and by the merest of coincidences so do the sales. Many owners complain of the thin paint.

• Safety: 80%
The structural rigidity is acceptable. However, only the GS benefits from the driver-side air bag, optional is a more efficient braking with ABS.

• Fuel consumption: 80%
Even though normal, it is higher than that of the Corolla, Sentra and Civic.

• Access: 70%
Getting into Elantra poses no more problems up front than at the back, since the doors are amply sized and open wide.

• Seats: 70%
Their shape gives ample hold while cornering and good lumbar support, but the padding is thin and firm.

• Suspension: 70%
On a car of this category, it surprises by its softness. However, the lack of wheel travel causes it to react brutally on a bad road.

• Driver's compartment: 70%
The dashboard is ergonomic, with well laid out gauges and the main controls within hand's reach. With the exception of switches located on each side of the instrument cluster, these are difficult to reach. Even though low, the driving position is more comfortable on the GLS whose steering column and lumbar support are adjustable. Visibility is satisfactory, in spite of center pillar thickness, which creates a major dead spot.

• Technical: 70%
The steel unit body has an independent 4-wheel suspension of the McPherson type up front and traction bars at the rear. Only the GLS has gas shocks as an option. Braking is mixed, but the GLS can be fitted with 4-wheel disc brakes and ABS. Steering with power assist is standard on the GLS. The Alpha engine is a modern, double overhead

DATA

Category: front-wheel drive compact sedans.
Class: 3

HISTORIC

Introduced in: 1991 (1.6L)
Modified in: 1992: 1.8L engine; 1994: security elements.
Made in: Ulsan, South Korea.

RATINGS

Safety: 80 %
Satisfaction: 78 %
Depreciation: 45 %
Insurance: 7.5 % ($545)
Cost per mile: $0.32

NUMBER OF DEALERS

In U.S.A.: 500

SALES ON THE AMERICAN MARKET

Model	1992	1993	Result
Elantra	32,042	35,874	+ 10.7 %

MAIN COMPETITORS

CHEVROLET Cavalier , DODGE-PLYMOUTH Colt, EAGLE Summit, FORD-Escort, HONDA Civic 4dr, MAZDA Protegé, NISSAN Sentra, PONTIAC Sunfire, SATURN SL1/SL2, SUBARU Impreza, TOYOTA Corolla, VW Jetta.

EQUIPMENT

HYUNDAI Elantra	GL	GLS
1.8L Alpha engine	O	S
Automatic transmission:	-	O
Cruise control:	O	O
Power steering:	O	S
ABS brake:	O	O
Air conditioning:	-	S
Air bag:	-	O
Leather trim:	-	-
AM/FM/ Radio-cassette:	S	S
Power door locks:	-	O
Power windows:	-	S
Tilt steering column:	-	S
Dual power mirrors:	-	S
Intermittent wipers:	S	S
Light alloy wheels:	-	O
Sunroof:	-	O
Anti-theft system:	-	-

S : standard; O : optional; - : not available

COLORS AVAILABLE

Exterior: Silver, Gray, Green, Red, Blue, White, Black, Rose Petal, Satin Pearl, Navy Blue.
Interior: Dark Gray, Cabernet Red.

MAINTENANCE

First revision: 4,000 miles
Frequency: 3 months/4,000 miles
Diagnostic plug: No

WHAT'S NEW IN 1995 ?

- **Pollution control improvement.**
- **Removal of all asbestos based gaskets and insulation.**
- **New aquamarine body color.**

ENGINE					TRANSMISSION			PERFORMANCE									
Model/ versions *: standard	Type / timing valve / fuel system	Displacement cu/in	Power bhp @ rmn	Torque lb.ft @ rpm	Compres. ratio	Driving wheels / transmission	Final ratio	Acceler. 0-62 mph s	Stand. 1/4 mile s	Stand. 5/8 mile s	Braking 62-0 mph ft	Top speed mph	Lateral acceler. G	Noise level dBA	Fuel economy mpg city	hwy	Gasoline type / octane
GL	L4*1.6 DOHC-12-MPFI	97.3	113 @ 6000	102 @ 5000	9.2 :1	front - M5*	NA	12.5	18.5	35.5	148	103	0.78	68	26.0	38.0	R 87
GLS	L4* 1.8 DOHC-12-MPFI	112	124 @ 6000	116 @ 4500	9.2 :1	front - M5*	NA	9.5	16.4	30.8	141	109	0.80	67	26.0	37.0	R 87
GL/GLS	L4 1.8 DOHC-12-MPFI	112	124 @ 6000	116 @ 4500	9.2 :1	front - A4	NA	10.8	17.0	32.0	138	106	0.80	67	27.0	37.0	R 87

cam with 12 valves and has a balance shaft to neutralize most vibrations inherent in this type of engine.

• Quality/fit/finish: 60%
While the presentation is austere, the adjustments are more closely held and the detailing has improved. However, the appearance of some materials still looks low dollar.

• Steering: 60%
It lacks precision and has a dead-center feel. However, its assistance is well-modulated and ratio is average. Maneuverability could be improved; steering diameter is high and torque steer can be felt during strong acceleration.

• Noise level: 50%
It remains above average, due to lack of insulating material. Tire, wind and engine noises can be heard to your heart's content.

• Interior: 50 %
The cabin easily accommodates 4 adults and there is ample room, especially in height.

• Trunk: 50%
It may lack a little height but its volume is sufficient and it can be increased in size by moving the rear seat bench. However, the opening remains narrow.

• Performance: 50%
Timid with the base engine, they are more interesting with the Alpha, whose superior torque and power make driving more fun. Even with an automatic transmission, standing starts and acceleration are crisper and allow fast passing.

• Handling: 50%
The independent suspensions allow satisfactory road holding in most circumstances, as long as the tires are of good quality. If one pushes a little in the corners, the limits of the simplistic suspension is quickly discovered: the Elantra quickly understeers. Be careful on a wet road, because it is easy to lose control. As the turns get tighter, forward traction becomes a problem in spite of the tires working hard and trajectories become vague as soon as pavement deteriorates since traction becomes intermittent.

• Conveniences: 50%
Storage is more practical in the GLS than in the GL which must content itself with a reasonable-sized glove box.

WEAK POINTS

• Insurance: 30%
As on most cars whose value is uncertain, the Hyundai premium is a little higher than that of other compacts.

• Depreciation: 40%
The Elantra loses its value more rapidly than the average of other compacts and the GL more than the GLS.

• Braking: 40%
Its efficiency is as irregular as its stability because it lacks the 4 discs and the optional ABS of the GLS. Otherwise the front wheels block rapidly during panic stops and distances can vary from single to double, depending on heat and fade.

CONCLUSION

• Overall rating: 60.0%
Among the compacts, the Elantra GLS constitutes a valid choice by its price-equipment ratio and it is more convenient use. 😐

OWNERS' SUGGESTIONS

-Air bags and ABS as standard
-Eliminate the 1.6 L engine
-Better quality paint and accessories
-Fewer body noises
-Less options

SPECIFICATIONS & PRICES

Model	Version/Trim	Body/ Seats	Interior volume cu.ft.	Trunk volume cu ft.	Drag coef. Cd	Wheel-base in	Lgth x Wdthx Hght. inx inx in	Curb weight lb	Towing capacity std. lb	Susp. type ft/rr	Brake type ft/rr	Steering system	Turning diameter ft	Steer. turns nber	Fuel. tank gal.	Standard tire size	Standard Power train	PRICES US. $ 1994
HYUNDAI		**General warranty: 3 years / 36,000 miles; Power train: 5 years / 60,000 miles;corrosion perforation: 5 years; antipollution: 5 years / 36,000 miles.**																
Elantra	base	4dr.sdn.4/5	90	12	0.34	98.4	172.8x66.4x54.5	2474	1000	i/i	dc/dr	r/p.	36.1	3.0	13.7	175/65R14	L4/1.6/M5	**$9,749**
Elantra	GLS	4dr.sdn.4/5	90	12	0.34	98.4	172.8x66.4x54.5	2564	1000	i/i	dc/dr	pwr.r/p.	36.1	3.0	13.7	185/60HR14	L4/1.8/M5	**$10,959**
Elantra	GLS	4dr.sdn.4/5	90	12	0.34	98.4	172.8x66.4x54.5	2615	1000	i/i	dc/dr/ABS	pwr.r/p.	36.1	3.0	13.7	185/65HR14	L4/1.8/A4	**$11,684**

See page 393 for complete 1995 Price List

Reprieve...Life support?

The Scoupe is in its last year under its present form because its replacement has already shown the tip of its nose. One would have thought that with its sympathetic allure and attractive price it would have known better success. Fact is the Hyundai image has not yet washed off the shadows of the past.

The Scoupe coupe, derived from the old Excel, addresses itself to a young buyer group. They do not have any of the prejudices of their elders and are ready to take off in their first sports rocket: new, inexpensive and with a good guarantee. The Scoupe is offered in three versions: base, LS and turbo, equipped with a 4-cylinder 1.5 L Alpha, in naturally aspirated version or turbocharged. It comes with a standard manual 5-speed or an optional automatic, except on the turbo.

STRONG POINTS

• Fuel consumption: 85%
Normal for an engine of this displacement, it constitutes one of the main attractions of this vehicle.

• Price/equipment: 80%
Affordable to buy and inexpensive to maintain (at least within the limits of the warranty), these models offer enough for the price, but while the equipment of the LS and Turbo is relatively complete, a rear windshield washer is missing from both models.

• Satisfaction: 75%
It is high, just like the reliability, which should encourage the Korean builder.

• Driver's compartment: 70%
The driver sits well, especially in the versions that benefit from an adjustable steering column. Visibility towards the rear is excellent, even when the spoiler is installed on the trunk lid. The dashboard is simple and functional, with just what is needed in terms of instruments and easy-to-use controls, except for the overdrive switch; poorly positioned on the automatic shift selectors, and whose operation is capricious; and those of the door glass which are unfindable at night.

• Suspension: 75%
Its softness allows it to better absorb road defects on the standard and LS models than on the stiffer Turbo whose wheel displacement is

DATA

Category: front-wheel drive sub-compact sports coupes.
Class: S3

HISTORIC

Introduced in: 1991
Modified in: 1992: turbo engine.
Made in: Ulsan, South Korea.

RATINGS

Safety: 70 %
Satisfaction: 75 %
Depreciation: 57 %
Insurance: base: 7.8 % ($520) turbo: 7.9 % ($655)
Cost per mile: $0.42

NUMBER OF DEALERS

In U.S.A.: 500

SALES ON THE AMERICAN MARKET

Model	1992	1993	Result
Scoupe	16,987	14,004	- 17.6 %

MAIN COMPETITORS

CHEVROLET Cavalier, FORD Escort GT, GEO Metro GSi, HONDA Civic SI & Del Sol, MAZDA MX3, PONTIAC Sunfire, SATURN SC1, SUZUKI Swift GT, TOYOTA Paseo.

EQUIPMENT

HYUNDAI Scoupe	base	LS	Turbo
Automatic transmission:	O	O	O
Cruise control:	-	-	-
Power steering:	-	S	S
ABS brake:	-	S	S
Air conditioning:	-	-	O
Air bag:	-	O	O
Leather trim:	S	S	S
AM/FM/ Radio-cassette:	-	O	S
Power door locks:	-	O	O
Power windows:	-	S	S
Tilt steering column:	-	O	S
Dual power mirrors:	S	S	S
Intermittent wipers:	-	S	S
Light alloy wheels:	-	-	S
Sunroof:	-	-	O
Anti-theft system:	-	-	-

S : standard; O : optional; - : not available

COLORS AVAILABLE

Exterior: Silver, Gray, Blue, Green, White, Black, Red, Fuschia,Yellow.
Interior: Dark Gray, Navy Blue.

MAINTENANCE

First revision: 4,000 miles
Frequency: 3 months/4,000 miles
Diagnostic plug: No

WHAT'S NEW IN 1995 ?

-Improvements in the Alpha engine, in term of cam drive and ignition, to make it quieter and more reliable.
-Improvement of the transmission whose new electronic control adapts shifts to the driving style. A new mechanism improves the drive-neutral-reverse shift.

Model/ versions *: standard	ENGINE Type / timing valve / fuel system	Displacement cu/in	Power bhp @ rmn	Torque lb.ft @ rpm	Compres. ratio	TRANSMISSION Driving wheels / transmission	Final ratio	PERFORMANCE Acceler. 0-62 mph s	Stand. 1/4 mile s	Stand. 5/8 mile s	Braking 62-0 mph ft	Top speed mph	Lateral acceler. G	Noise level dBA	Fuel economy mpg city	hwy	Gasoline type / octane
base-LS	L4* 1.5 SOHC-12 MPFI	89	92 @ 5500	97 @ 4000	10.0 :1	front - M5*	4.32	11.5	18.0	33.7	148	103	0.78	66	33.0	49.0	R 87
						front - A4	4.37	12.3	19.3	35.0	138	100	0.78	66	30.0	48.0	R 87
Turbo	L4* 1.5T SOHC-12 MPFI	89	115 @ 5500	123 @ 4500	7.5 :1	front - M5*	4.32	8.5	16.0	28.4	144	125	0.84	67	31.0	43.0	R 87

limited.

• Safety: 60%
The Scoupe structure should be stiffened to meet present standards and air bags need to be standard equipment for it to gain in ratings.

• Access: 60%
It is not as acrobatic toward the rear seats as some competing coupes but it will not be easy for big folks.

• Seats: 60%
They give good lateral support, but the lumbar support is inadequate since their padding is thin and their mounts are fragile.

• Noise level: 60%
Reasonable with a standard engine, it is a little higher on a Turbo, because the engines growl on acceleration. The windows whistle as the speed increases and the hatchback sounds off on some pavements.

• Technical: 60%
This Scoupe is derived from the Excel, from which it takes the platform and the main drivetrain elements. For the last two years, it disposes of the Alpha engine in standard, naturally aspirated version or of the Turbo which equips the LS Scoupe. Agreeable, the lines of the Scoupe do not set any new aerodynamic records since the drag coefficient varies between 0.35 and 0.36. The steel unit body has an independent McPherson suspension and a torsional axle in the back. Braking is mixed, even on the LS Turbo Scoupe, and ABS is standard only on the LS and Turbo.

• Quality/fit/finish: 60%
The interior presentation is not the most attractive, but the finish is honest, even though some materials and accessories look econo.

• Steering: 60%
It is more precise, better ratioed and modulated with power assist than on the manual. However, it is also sensitive to the high torque of the Turbo engine and tends to lighten up with speed.

• Performance: 50%
They are livelier, thanks to short and close ratios of the manual transmission, but the standard engine lacks a little horsepower to achieve the same in an automatic. Standing starts and passings with a turbo engine are definitely more muscled, and at the limit one could say that the power arrives in one big package which the chassis has trouble handling.

• Handling: 50%
Satisfactory on good pavement, it leaves much to be desired on a bad road, where the narrow tires tend to lose contact with the road and there are losses of traction in corners when the car leans and the inside wheels lighten up. Suspension softness provokes body motions, amplifies the roll and brings on understeer. Suspension is a little stiffer on the Turbo, but you need the driving ability to use the available power since on bad pavement the driving wheels become schizophrenic (go bonkers and patter on the pavement).

• Conveniences: 50%
Storage is more numerous on the LS and Turbo than on standard models.

WEAK POINTS

• Interior space: 30%
Good at the front seats, it is less so at the rear, where there is little legroom when the front seats are backed up. Also, the roof shape limits the height.

• Trunk: 30%
Its volume is not very big when the rear seats are occupied, but it can be increased by tipping the back of the bench. However, loading baggage is complicated by a high sill and a narrow opening. Would you consider pouring in liquid baggage?

• Insurance: 30%
The insurers are already suspicious of the Hyundai and if, to boot, it is sporty, the premium climbs in an unreal manner.

• Depreciation: 40%
This coupe loses a little less value than the Excel, but more than the average in its category.

• Braking: 40%
It remains a major problem because, in spite of progressive modulation, it is terribly short on efficiency, balance and fade-resistance in a panic situation. It is difficult to accept that a model which reaches a 125 mph speed does not have 4-wheel discs as standard.

CONCLUSION

• Overall rating: 56.0 %
In spite of some restrictions, the Scoupe is a good first used car, if not for any other reason than the guarantee level, which could spare a young budget but may torpedo it later.

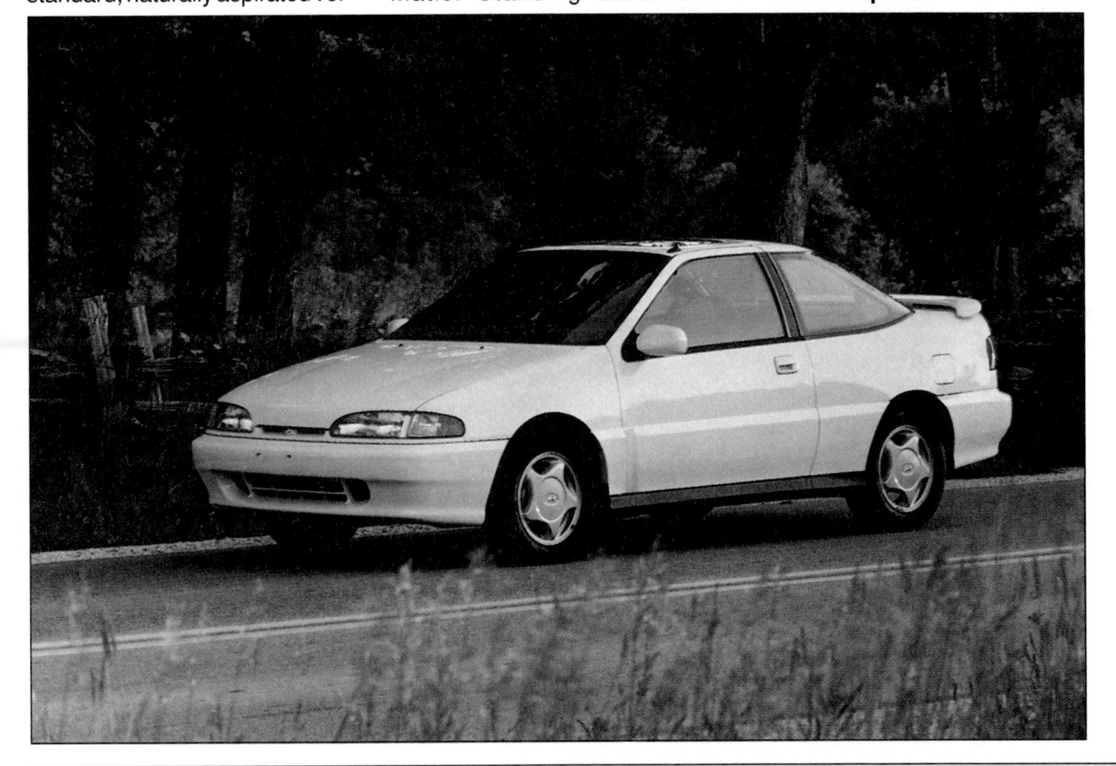

OWNERS' SUGGESTIONS

- Better quality paint
- Higher performance tires
- Brakes/heating efficiency
- Better soundproofing
- Less body noise
- Reduced window whistles
- Better padded seats

SPECIFICATIONS & PRICES

Model	Version/Trim	Body/ Seats	Interior volume cu.ft.	Trunk volume cu ft.	Drag coef. Cd	Wheel-base in	Lgth x Wdthx Hght. inx inx in	Curb weight lb	Towing capacity std. lb	Susp. type ft/rr	Brake type ft/rr	Steering system	Turning diameter ft	Steer. turns nber	Fuel. tank gal.	Standard tire size	Standard Power train	PRICES US. $ 1994
HYUNDAI		**General warranty: 3 years / 36,000 miles; Power train: 5 years / 60,000 miles; corrosion perforation: 5 years / unlimited.**																
Scoupe	base	2dr.cpe.4	80	9.0	0.35	93.8	165.9x64.0x50.0	2119	1000	i/i	dc/dr	r/p.	32.1	2.75	11.9	175/70R13	L4/1.5/M5	**$9,799**
Scoupe	LS	2dr.cpe.4	80	9.0	0.35	93.8	165.9x64.0x50.0	2165	1000	i/i	dc/dr	pwr.r/p.	32.1	2.75	11.9	185/60R14	L4/1.5/M5	**$11,099**
Scoupe	LS Turbo	2dr.cpe.4	80	9.0	0.36	93.8	165.9x64.0x50.0	2438	1000	i/i	dc/dr	pwr.r/p.	32.1	2.75	11.9	185/60R14	L4T/1.5/M5	**$11,899**

See page 393 for complete 1995 Price List

HYUNDAI

Sonata

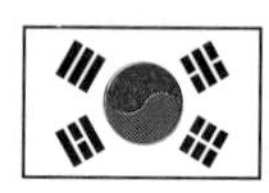

The Korean Camry...

Renewed in early 1994 with a quasi-identical content and form, the Sonata relates to models that are distinctly more costly. However, its success remains modest because the buyers are suspicious of too good a bargain. On the technical side, its renewal only adds to its value.

The Sonata is the second-best seller after the Elantra. It's up to Hyundai to offer a very spacious car with good equipment (GLS) at the price of a model in a lower category. Redefined at the end of 1994, it is a 3-volume, 4-door sedan, offered in GL and GLS. The standard GL engine is a 2.0 L 4 cylinder coupled to a standard transmission, and the option is a 3.0 L V6 with an automatic, standard on a GLS.

STRONG POINTS

• Safety: 80%
The unit body is sufficiently rigid to respect collision standards and the doors are reinforced against lateral impact. However, only the occupants of the GLS are protected by 2 front-seat air bags. Four-wheel disc brakes and ABS can be ordered on the GLS.

• Access: 80%
The length of the doors and their door angle allow front or rear entry without any difficulty.

• Satisfaction: 75%
Very satisfied owners are numerous, and intelligent application of the warranty allow leveling out the little problems.

• Price/equipment: 70%
The Sonata GL is one of the least expensive of its category while still properly equipped, and the only options on the GL are the AC, the ABS, the sliding roof and a CD player.

• Interior space: 70%
After the price, it is the main appeal of the Sonata because its interior dimensions are those of an intermediate, and 5 people can ride without difficulty, enjoying ample head and legroom in the rear seats.

• Technical: 70%
The steel unit body has simple shapes (designed by Giugiaro and an inspiration to some of its direct competitors) but effective, with a drag coefficient that stays at 0.32. The independent 4-wheel suspensions have been redesigned to improve stability. Steering with power assist is standard and braking is mixed on all versions. The 2.0 L DOHC with 4 cylinders and 16 valves, as well as the V6, have gained power and torque.

• Quality/fit/flnlsh: 70%
In constant improvement, they still have not reached the level of the Japanese because the fabrics, plastics and some accessories still look "econo-box."

• Insurance: 70%
The Sonata is just as penalized as the other Hyundais on this point, but fortunately its price keeps the premium within easy limits.

• Interior space: 70%
Thanks to seat and steering column adjustments, it's easy to find the most comfortable position and the driver receives good lateral and lumbar support. Also, the padding is more consistent. Visibility is

DATA

Category: front-wheel drive mid-size sedans.
Class: 5

HISTORIC

Introduced in: 1989
Modified in: 1990: V6 3.0L; 1992: 2.0L; 1994: renewed.
Made in: Ulsan, South Korea

RATINGS

Safety: 90 %
Satisfaction: 75 %
Depreciation: 62 % (V6:65%)
Insurance: 6.0 % ($520)
Cost per mile: $0.45

NUMBER OF DEALERS

In U.S.A.: 500

SALES ON THE AMERICAN MARKET

Model	1992	1993	Result
Sonata	17,196	15,420	- 10.4 %

MAIN COMPETITORS

BUICK Skylark, CHEVROLET Corsica, CHRYSLER Cirrus, DODGE Stratus, FORD Contour-Taurus, HONDA Accord, MAZDA 626, MERCURY Mystique-Sable, NISSAN Altima, OLDSMOBILE Achieva, PONTIAC, Grand Am, SUBARU Legacy, TOYOTA Camry, VOLKSWAGEN Passat.

EQUIPMENT

HYUNDAI Sonata	GL	GLS
Automatic transmission:	O	S
Cruise control:	O	S
Power steering:	S	S
ABS brake:	O	O
Air conditioning:	O	O
Air bag:	O	S
Leather trim:	-	-
AM/FM/ Radio-cassette:	S	S
Power door locks:	O	S
Power windows:	O	S
Tilt steering column:	S	S
Dual power mirrors:	O	S
Intermittent wipers:	S	S
Light alloy wheels:	O	O
Sunroof:	O	O
Anti-theft system:	O	O

S : standard; O : optional; - : not available

COLORS AVAILABLE

Exterior: Sandal Wood, Titanium, Navy Blue, Gray, Green, Red, White, Black.
Interior: Dark Gray, Beige.

MAINTENANCE

First revision: 4,000 miles
Frequency: 3 months/4,000 miles
Diagnostic plug: No

WHAT'S NEW IN 1995 ?

- Better general soundproofing.
- Stiffness improved by the addition of numerous reinforcements.
- Rear bench back folds in a 60/40 ratio.

Model/ versions *: standard	ENGINE Type / timing valve / fuel system	Displacement cu/in	Power bhp @ rmn	Torque lb.ft @ rpm	Compres. ratio	TRANSMISSION Driving wheels / transmission	Final ratio	PERFORMANCE Acceler. 0-62 mph s	Stand. 1/4 mile s	Stand. 5/8 mile s	Braking 62-0 mph ft	Top speed mph	Lateral acceler. G	Noise level dBA	Fuel economy mpg city	hwy	Gasoline type / octane
base	L4* 2.0 DOHC-16-MFI	122	137 @ 5800	129 @ 4000	9.0 :1	front - M5*	4.322	11.0	17.4	31.2	170	118	0.78	68	26.0	38.0	R 87
						front - A4	4.007	12.8	19.0	33.5	141	115	0.78	68	26.0	37.0	R 87
GLS V6	V6* 3.0 SOHC-12-MFI	181	142 @ 5000	168 @ 2500	8.9 :1	front - A4*	3.958	10.0	17.5	30.7	151	124	0.78	67	21.0	28.0	R 87

excellent from all angles, and the instruments are both numerous and easy to read. Some of the dashboard switches are out of comfortable reach.

• Suspension: 70%
It is sufficiently smooth to provide a decent comfort for as long as the pavement is good. Otherwise, it reacts harshly as soon as the pavement deteriorates.

• Fuel consumption: 60%
It is realistic of a car of this format, but the 4-cylinder is distinctly more economical than the V6.

• Conveniences: 60%
Storage spaces are practical, good-sized, and well-distributed throughout the interior.

• Performance: 60%
They are quite honorable and the weight of these models remains reasonable. The 2.0 L engine shows ample proof of power at high revs and of good lower rpm torque. However, the automatic transmission goes better on a V6, whose power and torque are more comfortable.

• Handling: 60%
This is the point that improved the most and one can say that, in spite of the roll, the body motions are better controlled. The damping is more consistent and wheel travel is more generous and better absorbed road unevenness than in the preceding model. There is too much that remains to be done to improve drive wheel traction in a tight turn.

• Seats: 60%
They are amply sized and well-formed and the padding has also improved.

• Steering: 60%
Soft and fast, its assistance has been improved to give it more precision and feedback. Its steering ratio is average, and it gives in good maneuverability thanks to its favorable turning diameter.

• Trunk: 50%
Its volume is not proportional to the size of the Sonata, but it can be increased by lowering the 60/40 seat back.

• Noise level: 50%
It remains acceptable at cruise speed and is well fed by wind, roll and power train noises and sometimes by body creaks.

WEAK POINTS

• Depreciation: 35%
Distinctly stronger than that of Japanese cars, it is, however, comparable to some good cars of American manufacture.

• Braking: 40%
Easy to modulate during normal deceleration, it becomes unpredictable in panic stops because the front wheels lock up rapidly, stopping distances are long and fade resistance is just average which incites one to strongly recommend 4-wheel discs and ABS be offered as an option.

CONCLUSION

• Overall rating: 63.0%
The Sonata continues to offer an excellent ratio of price/format/equipment, but you have to expect some problems at resale, because Hyundai's reputation is not yet strongly established. :-|

OWNERS' SUGGESTIONS

-Manual transmission and fuel pump more reliable
-Better fade resistance
-Better defogging in the winter
-Improved radio quality
-More effective windshield washer
-More legible oil gauge
-Better seat and belt adjustment
-Less noisy windshield wipers
-Dealer hari kari kit

SPECIFICATIONS & PRICES

Model	Version/Trim	Body/ Seats	Interior volume cu.ft.	Trunk volume cu ft.	Drag coef. Cd	Wheel-base in	Lgth x Wdthx Hght. inx inx in	Curb weight lb	Towing capacity std. lb	Susp. type ft/rr	Brake type ft/rr	Steering system	Turning diameter ft	Steer. turns nber	Fuel. tank gal.	Standard tire size	Standard Power train	PRICES US. $ 1994
HYUNDAI		**General warranty: 3 years / 36,000 miles; Power train: 5 years / 60,000 miles;perforation corrosion: 5 years / unlimited.**																
Sonata	GL	4dr.sdn.5	101	13.2	0.32	106.3	185.0x69.7x55.3	2769	1000	i/i	dc/dr	pwr.r/p.	34.6	3.1	17.2	195/70R14	L4/2.0/M5	**$12,799**
Sonata	GL V6	4dr.sdn.5	101	13.2	0.32	106.3	185.0x69.7x55.3	2921	1000	i/i	dc/dc	pwr.r/p.	34.6	3.1	17.2	195/70R14	V6/3.0/A4	**$14,369**
Sonata	GLS	4dr.sdn.5	101	13.2	0.32	106.3	185.0x69.7x55.3	2820	1000	i/i	dc/dr	pwr.r/p.	34.6	3.1	17.2	195/70R14	L4/2.0/A4	**$14,199**
Sonata	GLS V6	4dr.sdn.5	101	13.2	0.32	106.3	185.0x69.7x55.3	2972	1000	i/i	dc/dc	pwr.r/p.	34.6	3.1	17.2	195/70R14	V6/3.0/A4	**$15,769**

See page 393 for complete 1995 Price List

Progenitor...

The arrival on the market of models with strong European connotations such as the Ford Contour/ Mystique or the Cirrus/Stratus makes one realize how far in advance of its time the G20 was with its concept of a luxurious sports compact in advance of its time. From this point on, the students may pass the master.

For the last three years, the G20 Infiniti did not find the success that it merits on our side of the Atlantic . Europeans, however difficult in the matter of imports, have reserved for it an unusual reception. One must say that the Primera is not sold as a full-dress deluxe car but as a popular model in several versions, including a station wagon which is unknown here. For 1995 the G20 remains in the form of a 4-door sedan in standard trim or "t" the latter disposes of a firmer suspension, a spoiler, wheels and tires which give it a more sportslike handling.

STRONG POINTS

• Safety: 90%
The G20 steel unit body offers good collision resistance, while the restraint devices include, among other things, two air bags which protect the driver and the passengers.

• Satisfaction: 90%
Its reliability never created major problems and the respect of the warranty as well as dealership reception has impressed most owners.

• Technical: 80%
The steel unit body is characterized by lines which are ordinary overall, but pleasant, balanced and effective with excellent streamlining. The drag coefficient is just 0.30. The only engine available is a 2.0 L, 4-cylinder with a double overhead cam. It develops 140 hp and has eight counterweights on the crank to reduce vibrations and improved bearing life.The manual 5-speed is standard and so is the limited slip differential with a viscous coupling, while the automatic transmission is optional. The suspension is independent, and the disc brakes on all four wheels are assisted by ABS.

• Steering: 80%
Its precision, modulation, speed and good grip at the steering wheel rim are appreciated. It's too bad that maneuverability is only average

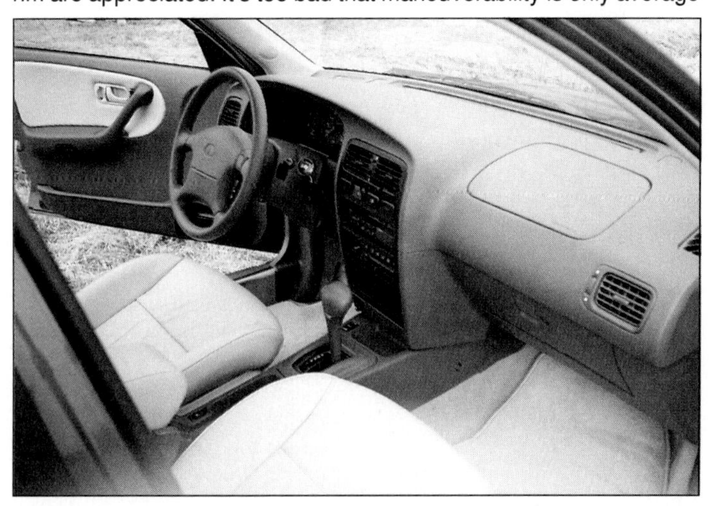

DATA

Category: front-wheel drive compact luxury sedans.
Class: 7

HISTORIC

Introduced in: 1991
Modified in: 1994: G20t version
Made in: Tochigi, Japan.

RATINGS

Safety: 90 %
Satisfaction: 90 %
Depreciation: 58 %
Insurance: 4.6 % ($720)
Cost per mile: $0.62

NUMBER OF DEALERS

In U.S.A.: 150

SALES ON THE AMERICAN MARKET

Model	1992	1993	Result
G20	14,592	16,533	+ 11.8 %

MAIN COMPETITORS

ACURA Integra, BMW 318i, HONDA Accord, MAZDA 626 Cronos, NISSAN Altima, VW Jetta GLX.

EQUIPMENT

INFINITI	G20	G20t
Automatic transmission:	O	O
Cruise control:	S	S
Power steering:	S	S
ABS brake:	S	S
Air conditioning:	S	S
Air bag:	S	S
Leather trim:	O	S
AM/FM/ Radio-cassette:	S	S
Power door locks:	S	S
Power windows:	S	S
Tilt steering column:	S	S
Dual power mirrors:	S	S
Intermittent wipers:	S	S
Light alloy wheels:	S	S
Sunroof:	S	S
Anti-theft system:	S	S

S : standard; O : optional; - : not available

COLORS AVAILABLE

Exterior: Beige, White, Black, Blue, Crimson,Emerald.
Interior: Fabric: Gray, Beige. Leather: Beige, Black.

MAINTENANCE

First revision: 7,500 miles
Frequency: 6 months/7,500 miles
Diagnostic plug: Yes

WHAT'S NEW IN 1995 ?

- New model G20t.

Model/ versions *: standard	Type / timing valve / fuel system	Displacement cu/in	Power bhp @ rmn	Torque lb.ft @ rpm	Compres. ratio	Driving wheels / transmission	Final ratio	Acceler. 0-62 mph s	Stand. 1/4 mile s	Stand. 5/8 mile s	Braking 62-0 mph ft	Top speed mph	Lateral acceler. G	Noise level dBA	Fuel economy mpg city	Fuel economy mpg hwy	Gasoline type / octane
G20	L4* 2.0 DOHC-16-SFI	120	140 @ 6400	132 @ 4800	9.5 :1	front - M5*	4.058	9.0	16.8	29.4	138	128	0.81	68	28.0	41.0	R 87
						front - A4	4.072	9.8	17.4	30.2	144	125	0.81	68	26.0	37.0	R 87
G20t	L4* 2.0 DOHC-16-SFI	120	140 @ 6400	132 @ 4800	9.5 :1	front - M5	4.058	9.0	16.8	29.4	141	128	0.81	68	28.0	41.0	R 87
						front - A4	4.072	9.7	17.3	30.2	151	125	0.81	68	26.0	37.0	R 87

because the turning diameter is a bit large.

• **Quality/fit/finish: 80%**
The interior of the G20 is presented in a luxurious fashion, and its leather trim in contrasting multicolors as well as the wood appliqués are in good taste. The quality of the materials and the assembly and finishing care are evident.

• **Driver's compartment: 80%**
The driver is perfectly well-seated thanks to combined adjustments of the seats and steering column. Visibility is excellent from all angles, even from the rear where the spoiler does not obstruct the visibility. Controls and gauges are easy to use.

• **Seats: 80%**
They are comfortable in spite of some firmness, and the contour of the seat back efficiently maintains the back with a good lumbar support. Unfortunately, the leather trim is slippery.

• **Suspension: 80%**
Its consistence favors handling without penalizing comfort because wheel travel allows them to soak up road defects without excess harshness.

• **Fuel consumption: 80%**
It is not ruinous and holds around 28 mpg which allows a 375-mile cruise range.

• **Insurance: 80%**
Taking into account its luxury and sports possibilities, the G20 premium is reasonable.

• **Conveniences: 70%**
The glove box is not very large, but it is complimented by door pockets and the storage in the back of the front seats.

• **Access: 70%**
Getting into a G20 poses absolutely no problem because the doors are large and have an ample opening angle.

• **Handling: 70%**
The manual G20 is great fun to drive on switch backs and twisty roads because it is more agile and precise to guide than a number of supposed sports coupes. Road holding in a turn remains neutral most of the time. On a slippery road or in a very tight turn, traction remains optimal, thanks to well-calibrated pneumatic equipment and a limited slip differential. The firmer suspension of the "t" allows roll to be reduced still further.

• **Performance: 60%**
An increase in weight led to an unfavorable weight/power ratio that has become very average (21 lbs/hp). If the standing starts and the passings are livelier with a stick shift than with the automatic, it is because of the gear ratio spacing. The driving is just as pleasant in both because the engine is powerful at high rpm and benefits from low rpm torque. A 2.5 L V6 would not be too much to face the music.

• **Trunk: 60%**
As roomy as it is accessible, its shape is regular and the cutout at the back simplifies baggage handling.

• **Interior space: 50%**
In spite of its compact format, the interior can accept 4 adults who enjoy adequate head and legroom even at the rear seats.

• **Noise level: 50%**
Low at constant cruising speed, thanks to good soundproofing, it rises when the engine is pushed. Wind and rolling noise increase with speed.

WEAK POINTS

• **Braking: 40%**
Progressive during average decelerations, it remains very stable during panic stops, but the distances are relatively long and the friction materials lack bite.

• **Price/equipment: 40%**
The strength of the yen brings the price of this model in the $20,000 range, where choice is not lacking. Its equipment is complete, and it only lacks heated seats.

• **Depreciation: 40%**
The anonymity of this model plays it a bad hand because its resale value is weaker than that of its competitors, and it's not always easy to find an enlightened fan as a second buyer.

CONCLUSION

• **Overall rating: 68.5%**
Even though it is easy on the eyes and pleasant to drive, the G20 is getting a few wrinkles around the eyes. Its line, trim and performances date to the preceding generation. The clock is ticking... fast. ☺

OWNERS' SUGGESTIONS

-Heated front seats
-More dealers
-Turbo engine
-Better soundproofing

SPECIFICATIONS & PRICES

Model	Version/Trim	Body/ Seats	Interior volume cu.ft.	Trunk volume cu ft.	Drag coef. Cd	Wheel-base in	Lgth x Wdthx Hght. inx inx in	Curb weight lb	Towing capacity std. lb	Susp. type ft/rr	Brake type ft/rr	Steering system	Turning diameter ft	Steer. turns nber	Fuel. tank gal.	Standard tire size	Standard Power train	PRICES US. $ 1994
INFINITI		General warranty: 4 years / 60,000 miles; Power train & antipollution: 6 years / 100,000 miles; corrosion perforation: 7 years / unlimited .																
G20	base	4dr.sdn. 5	89	14.1	0.30	100.4	174.8x66.7x54.7	2892	1000	i/i	dc/dc/ABS	pwr.r/p.	10.8	2.6	60	195/60HR14	L4/2.0/M5	$21,750
G20	base	4dr.sdn. 5	89	14.1	0.30	100.4	174.8x66.7x54.7	2941	1000	i/i	dc/dc/ABS	pwr.r/p.	10.8	2.6	60	195/60HR14	L4/2.0/A4	$22,750
G20	"t"	4dr.sdn. 5	89	14.1	0.30	100.4	174.8x66.7x54.7	2954	1000	i/i	dc/dc/ABS	pwr.r/p.	10.8	2.6	60	195/60HR14	L4/2.0/M5	NA
G20	"t"	4dr.sdn. 5	89	14.1	0.30	100.4	174.8x66.7x54.7	3003	1000	i/i	dc/dc/ABS	pwr.r/p.	10.8	2.6	60	195/60HR14	L4/2.0/A4	NA

See page 393 for complete 1995 Price List

Styling excess...

The J30 attracts attention, mostly by its exaggerated forms which are an apology to roundness. A cult of a style long past. It effects the same panoply of luxury as a number of its colleagues and the choice can only relate to the question of taste.

The deluxe sedans represent approximately half the world of automobiles in numbers of models, but are far from accounting for half the sales. The number is explained by the considerable profits they generate for the car makers and dealers and the image they project of more discreet models. The J30 Infinity tries to attract a well-off clientele which makes two way trips without much baggage. It is offered in the form of a 4-door sedan in a standard J30 or a J30t for "touring". Identified by the spoiler on the trunk lid, the BBS rims, a firmer suspension and rear-wheel steer based on the HICAS principle, inherited, like the rest of the drivetrain, from the 300 ZX.

STRONG POINTS

• Safety: 90%
Structural rigidity is excellent, and it has been reinforced at numerous points. There are also 2 standard air bags, 3-point seat belts and the ones up front include a pretensioning device and padding designed to minimize collision injuries. All of this confers on the J30, a high safety rating.

• Quality/fit/finish: 90%
Fine leather trim, appliqués of rosewood and zebra, all carefully varnished, the quality of the plastics and carpets endow the J30 with class worthy of the most luxurious and reminds by small detail qualities of the Jaguar.

• Satisfaction: 90%
J.D. Power Research proves that Japanese deluxe cars are distinctly less capricious than their European or American competition.

• Driver's compartment: 80%
The rational and ergonomic dashboard is very well organized. Everything is in place and effective, with the exception of the switches controlling the trunk opening and fuel tank flap, located under the armrest of the left door. The best driving position is easy to find by combining seat and steering column adjustments. Visibility would be total if the "C" pillar was thinner and the outer rear-view mirrors were a bit larger.

• Technical: 80%
Identical, in terms of power train, the two J30 versions are equipped with a 210 HP engine derived from the 300ZX with a variable intake, direct ignition and a SOFIS fuel injection which gives optimum throttle response. The 7-position selector for the electronically controlled automatic transmission can function in a manual or fully automatic mode and the limited slip differential uses a viscous coupling. The steel unit body carries sub-frames for the front and rear suspension as a way of filtering out noises and vibrations. Up front the suspension is of a McPherson type, while at the rear, it uses mechanical arms like those of the Q45. Its very rounded lines attract attention, but the aerodynamic values are very average.

DATA

Category: rear-wheel drive luxury sedans.
Class: 7

HISTORIC

Introduced in: 1992
Modified in: -
Made in: Tochigi, Japan.

RATINGS

Safety: 90 %
Satisfaction: 91 %
Depreciation: 40 % (2 years)
Insurance: 3.8 % ($925)
Cost per mile: $0.62

NUMBER OF DEALERS

In U.S.A.: 150

SALES ON THE AMERICAN MARKET

Model	1992	1993	Result
J30	13,820	21,537	+ 35.9 %

MAIN COMPETITORS

ACURA Legend, ALFA ROMEO 164, AUDI 100 & A6, BMW 5-Series, LEXUS ES300, LINCOLN Continental, MERCEDES-BENZ C Class, SAAB 9000, VOLVO 960.

EQUIPMENT

INFINITI	J30	J30t
Automatic transmission:	S	S
Cruise control:	S	S
Power steering:	S	S
ABS brake:	S	S
Air conditioning:	S	S
Air bag:	S	S
Leather trim:	S	S
AM/FM/ Radio-cassette:	S	S
Power door locks:	S	S
Power windows:	S	S
Tilt steering column:	S	S
Dual power mirrors:	S	S
Intermittent wipers:	S	S
Light alloy wheels:	-	S
Sunroof:	-	-
Anti-theft system:	S	S

S : standard; O : optional; - : not available

COLORS AVAILABLE

Exterior: Crimson, Blue, Black, Ivory, Beige, Gray, Silver Crystal.
Interior: Gray, Beige, Black, White.

MAINTENANCE

First revision: 7,500 miles
Frequency: 6 months/7,500 miles
Diagnostic plug: Yes

WHAT'S NEW IN 1995 ?

- No major change.

ENGINE					TRANSMISSION				PERFORMANCE									
Model/ versions *: standard	Type / timing valve / fuel system	Displacement cu/in	Power bhp @ rmn	Torque lb.ft @ rpm	Compres. ratio	Driving wheels / transmission	Final ratio	Acceler. 0-62 mph s	Stand. 1/4 mile s	Stand. 5/8 mile s	Braking 62-0 mph ft	Top speed mph	Lateral acceler. G	Noise level dBA	Fuel economy mpg city	hwy	Gasoline type / octane
J30	V6* 3.0 OHC-24-MPSFI	181	210 @ 6400	193 @ 4800	10.5 :1	rear - A4*	3.917	8.7	16.3	29.5	138	131	0.78	65	22.0	30.0	M 89
J30t	V6* 3.0 OHC-24-MPSFI	181	210 @ 6400	193 @ 4800	10.5 :1	rear - A4*	3.917	8.7	16.3	29.5	138	131	0.80	65	22.0	30.0	M 89

- **Insurance:** **80%**
The modest insurance premium, in relation to the cost of the J30, is explained by the fact that it addresses itself to an older and more conservative clientele that does not ruin insurers.
- **Seats:** **80%**
Beautiful and efficient, they provide good side hold in turns and excellent back support, however their padding could be a little softer.
- **Suspension:** **80%**
Soft and smooth on parkways, it doesn't react badly to rough pavement in spite of the limited suspension travel, and the difference in ride between the base J30 and the J30T is imperceptible.
- **Noise level:** **80%**
The soundproofing efficiency and the quiet power train keep it at a low level.
- **Conveniences:** **70%**
The small size of the glove box is compensated by a container in the center console, door pockets and bellows into the cushions built into the back of the front seat.
- **Performance:** **70%**
Depending on the transmission shift mode, they can be sports-minded or peaceful because the weight-to-power ratio is favorable (18.3 lbs/hp). Being able to shift manually adds to the driving pleasure, but numerical results position the J30 in the average of 3.0 L V6's.
- **Handling:** **70%**
The "t" version is more agile on a twisty coarse, thanks to stiffer stabilizer bars which limit the roll of the standard J30 which also understeers earlier. The HICAS control adds rear wheel steer, improves stability in turns and fast lane changes.
- **Steering:** **60%**
Precise and with a good ratio, it is very sensitive and its assistance is too strong. While the turning diameter is reasonable, maneuverability leaves much to be desired.

INFINITI J30

INFINITI J30t

- **Depreciation:** **50%**
The J30, as the other Infinitis, has an above-average value loss because of the lack of identity of the make and exaggerate appearance which does not please everyone.
- **Fuel consumption:** **50%**
It is in the average of 3.0 L V6's and 21 mpg is considered reasonable.
- **Access:** **50%**
It's more delicate to get in at the back where the doors are narrow and the roof has too much curvature to allow tall people to enter without too much damage. Up front there is no problem.

WEAK POINTS

- **Price/equipment:** **10%**
The strength of the yen, the very complete equipment, and the great luxury of the J30s make them too expensive for what they have to offer. The practical side is doubtful and they lack practical gadgets, such as an on-board computer for the driver's seat and rear-view mirrors, to be at the level of their competitors.
- **Interior space:** **40%**
It is deceiving because the rear seats are particularly limited and two people will feel cramped, particularly when the front seats are moved too far back.
- **Trunk:** **40%**
The ovoid shape of the tail deck favors neither its volume nor its capacity, and in spite of its regular shape, it doesn't hold much. High sill complicates access.
- **Braking:** **40%**
In spite of their advanced technology, the brakes are not among the more effective, and stopping distances are never under 130 feet. The onset of the ABS is irregular and allows incipient wheel lockup, which is more alarming than really dangerous.

CONCLUSION

- **Overall rating:** **65.0%**
Excessive by their lines, not truly sporty in spite of their mechanical advances, nor really practical in light of their reduced volumes, they are part of the market curiosities from which it is better to hold yourself aloof. ☺

SPECIFICATIONS & PRICES

Model	Version/Trim	Body/ Seats	Interior volume cu.ft.	Trunk volume cu ft.	Drag coef. Cd	Wheel-base in	Lgth x Wdthx Hght. inx inx in	Curb weight lb	Towing capacity std. lb	Susp. type ft/rr	Brake type ft/rr	Steering system	Turning diameter ft	Steer. turns nber	Fuel. tank gal.	Standard tire size	Standard Power train	PRICES US. $ 1994
-																		
INFINITI		**General warranty: 4 years / 60,000 miles; Power train & antipollution: 6 years / 60,000 miles; corrosion perforation: 7 years / unlimited .**																
J30	base	4dr.sdn. 4	86.5	10.1	0.35	108.7	191.3x69.7x54.7	3527	2000	i/i	dc/dc/ABS	pwr.r/p.	36.1	2.93	19	215/60R15	V6/3.0/A4	**$36,950**
J30	t	4dr.sdn. 4	86.5	10.1	0.34	108.7	191.3x69.7x54.7	3578	2000	i/i	dc/dc/ABS	pwr.r/p.	36.1	2.93	19	215/60R15	V6/3.0/A4	**$39,250**

See page 393 for complete 1995 Price List

Zero at the tiller, we are sinking...

The flagship of the Infiniti range navigates in waters made difficult by ferocious competition from better-established models. At the stage where this make is, the simplest would be to enroll the witness protection program, that is to disappear.

Last year, Infinity redesigned the front end of the Q45, giving it a real Jaguar style grill, which finally gave it the identity it was lacking. In the first commercials this wasn't too bad because you couldn't see the car, but in the dealers showroom potential customers were turned off by the absence of a grill and by the Texas-belt-buckle-style emblem. This 3-volume 4-door version was unique for Canada and to this was added the model T for the American market.(Ford's rights to that name expired a while back). Its drivetrain consists of a 4.5 L V8 and a 4-speed automatic.

STRONG POINTS

• Safety: 90%
Thanks to reinforcements in the door and roof and to padding that protects the occupants' knees at the front seats, 3-point seat belts, and 2 air bags, the hefty Infiniti easily attains the maximum score.

• Technical: 90%
The steel unit body has 4-wheel independent suspension, McPherson up front and an axle guided by multiple arms at the rear. The new grill has completely changed the Q45 appearance which resembles more closely what people expect in this segment. The massive body offers a remarkably low 0.30 drag coefficient. The center of interest in this model resides in the 4.5 L V8 with an aluminum block and heads developing 278 hp. It is the most powerful V8 of the Japanese industry, and only the latest Northstar from GM or the 4.6 L FourCam from Ford can ram it and sink it. The limited slip differential with a viscous coupling and traction control coupled to ABS is offered as an option. It reduces engine power when slippage is detected, independently of the driver's accelerator control.

• Satisfaction: 90%
It is practically total because owners only mention minor incidents which the quality of service at Infiniti soon resolves.

• Steering: 80%
It is as precise and fast as its assistance is well-modulated and also allows you to feel the road. Its steering diameter of over 36 ft slighty penalizes the average maneuverability and the body is a little bulky.

• Driver's compartment: 80%
Combined steering column and seat adjustments help the driver to quickly become comfortable. Both configurations can fit into memory. Visibility is excellent from all angles, but the outside rear-view mirrors are too small. Some controls are not the most practical, such as seat adjustments on the left door or the rear-view control hidden by the steering wheel. Analog instruments are numerous, easy-to-interpret and laid out in conventional fashion.

• Quality/fit/finish: 80%
Assembly and finish are strict, but the interior lacks the warmth one finds in British cars, and the leather and wood trims are too faded.

DATA

Category: rear-wheel drive luxury sedans.
Class: 7

HISTORIC

Introduced in: 1989
Modified in: 1994: front grill.
Made in: Tochigi, Japan.

RATINGS

Safety: 90 %
Satisfaction: 90 %
Depreciation: 63.2%
Insurance: 2.8 % ($1,200)
Cost per mile: $0.94

NUMBER OF DEALERS

In U.S.A.: 150

SALES ON THE AMERICAN MARKET

Model	1992	1993	Result
Q45	12,216	12,294	+ 0.7 %

MAIN COMPETITORS

AUDI V8, BMW 7-Series, CADILLAC De Ville & Seville, LEXUS LS400, LINCOLN Continental, Town Car, MERCEDES E-300, SAAB 9000, VOLVO 960.

EQUIPMENT

INFINITI	Q45 & Q45t
Automatic transmission:	S
Cruise control:	S
Power steering:	S
ABS brake:	S
Air conditioning:	S
Air bag:	S
Leather trim:	S
AM/FM/ Radio-cassette:	S
Power door locks:	S
Power windows:	S
Tilt steering column:	S
Dual power mirrors:	S
Intermittent wipers:	S
Light alloy wheels:	S
Sunroof:	S
Anti-theft system:	S

S : standard; O : optional; - : not available

COLORS AVAILABLE

Exterior: Silver, White, Beige, Gray, Black, Crimson, Blue, Wildberry.
Interior: Beige, Blue, Gray, White, Mole.

MAINTENANCE

First revision: 7,500 miles
Frequency: 6 months/7,500 miles
Diagnostic plug: Yes

WHAT'S NEW IN 1995 ?

-No major change.

	ENGINE				TRANSMISSION			PERFORMANCE									
Model/ versions *: standard	Type / timing valve / fuel system	Displacement cu/in	Power bhp @ rmn	Torque lb.ft @ rpm	Compres. ratio	Driving wheels / transmission	Final ratio	Acceler. 0-62 mph s	Stand. 1/4 mile s	Stand. 5/8 mile s	Braking 62-0 mph ft	Top speed mph	Lateral acceler. G	Noise level dBA	Fuel economy mpg city	hwy	Gasoline type / octane
Q45	V8* 4.5 DOHC-32-MPSFI	274	278 @ 6000	292 @ 4000	10.2 :1	rear - A4*	3.538	7.5	15.0	25.5	128	149	0.79	65	20.0	29.0	M 89

• **Performance: 80%**
High on torque and power, the engine provides good takeoff and strong passing power considering the large weight to be moved. The Q45 suppresses most of its direct competitors on this point, including the Lexus, because its V8 is one of the most powerful on the market. While it "pulls" long, the automatic transmission is well-adapted and the electronically controlled traction in turns is without reproach.

• **Noise level: 80%**
The extremely quiet engine allows tire noises to be heard more effectively as they patter on bad pavement.

• **Insurance: 80%**
The rate is one of the lowest in the market and the premium is not exaggerated compared to the price of this car.

• **Convenience: 80%**
Storage consists of a spacious glove box, door pockets, bellows in back of the front seats and a storage box in the center console.

• **Access: 80%**
It poses no problem because the doors open wide and the steering column rises automatically as soon as the engine stops.

• **Seats: 70%**
Their lateral support is more effective than their ability to hold up your back. The padding is hard and the leather trim is very artificial to the touch, which is not pleasant.

• **Suspension: 70%**
It efficiently absorbs road defects, is never harsh, but it lacks the cushiony feel that makes the charm of big cars.

• **Handling: 70%**
With minimum roll in a turn, and a traction hard to fault for a sedan of this size, the Q45 shows a self-assurance and balance that is found more often on sports cars, but it lacks the agility and charm of a BMW.

• **Interior space: 65%**
It is proportional to its bulk and seats five. If the width and height are generous, there is a lack of rear legroom when the front seats are at the end of their run.

• **Trunk: 60%**
As tall as it is deep, it offers a huge amount of room, and its access is eased by a truck lid that comes down to the bumpers.

• **Braking: 60%**
In spite of its imposing mass, the Q45 stops in a reasonable distance. Its brakes demonstrate balance and fade resistance and the pedal effort and the precision with which it can be modulated are satisfying.

WEAK POINTS

• **Price/equipment: 00%**
Even though competitive, it doesn't attract as many fans as Jaguar, Mercedes, or BMW, as people from Infiniti predicted. Average at this price level, the equipment is very complete but lacks a few electronic gadgets, an area in which the Japanese are normally very strong.

• **Fuel consumption: 35%**
Reasonable if you take into account the weight displacement and compare it to some anemic V6s. It is due to the finesse of the body lines, which allows it a cruise range of 375 miles on the highway at constant and legal speed.

• **Depreciation: 40%**
It is strong because of a lack of identity of the marque and the problems which resulted from a failed publicity campaign during launch, as well as the fact that the economic climate is not very favorable to this type of car.

CONCLUSION

• **Overall rating: 69.0%**
In spite of its new appearance, the Q45 did not break any more hearts than before its plastic surgery. It is the heroine of a story which started badly and which will end up about that way. ☺

SPECIFICATIONS & PRICES

Model	Version/Trim	Body/ Seats	Interior volume cu.ft.	Trunk volume cu ft.	Drag coef. Cd	Wheel-base in	Lgth x Wdthx Hght. inx inx in	Curb weight lb	Towing capacity std. lb	Susp. type ft/rr	Brake type ft/rr	Steering system	Turning diameter ft	Steer. turns nber	Fuel. tank gal.	Standard tire size	Standard Power train	PRICES US. $ 1994
INFINITI		**General warranty: 4 years / 60,000 miles; Power train: 6 years / 60,000 miles; corrosion perforation: 7 years / unlimited .**																
Q45	base	4dr.sdn. 5	96	14.5	0.30	113.2	199.8x71.9x56.3	4039	2000	i/i	dc/dc/ABS	pwr.r/p.	37.4	2.6	22.5	215/65VR15	V8/4.5/A4	**$49,950**
Q45t	base	4dr.sdn. 5	96	14.5	0.30	113.2	199.8x71.9x56.3	4083	2000	i/i	dc/dc/ABS	pwr.r/p.	37.4	2.6	22.5	215/65VR15	V8/4.5/A4	**$54,950**

See page 393 for complete 1995 Price List

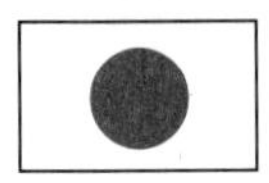

Losing speed...

Things are not going at their best for Isuzu's products. The Amigo is a gadget and the Rodeo, the only valid product, has lost most of its power of seduction. With a gain in price, it is no longer a bargain.

ISUZU Pick-up

Only the Rodeo is sold in Canada in 1995 and Amigo, as well as Isuzu pickups, are sold solely in the U.S. These vehicles share the same chassis with the Trooper as well as some mechanical components. It is a conventional 4-wheel drive, easy to engage. The Rodeo has a 4-door station wagon with two trim levels: S and LS, while the Amigo is a 2-door convertible in finishes S and XS and the pickups have a box in the cab: short or long. These different vehicles are equipped with 4- or 6-cylinder engines and manuals or optional automatics.

STRONG POINTS

• Satisfaction: 80%
Their reliability has never given any grave concerns, but the owners complain of an insufficient number of dealers and the rarity of some replacements parts.

• Seats: 70%
Their contours provide effective lateral and lumbar supports, but their padding is on the firm side with bucket seats, while the bench seats are too flat to offer good comfort.

• Suspension: 70%
Comfortable on a good road thanks to its softness, it is less agreeable when crossing large road defects because it rears and bucks, especially when empty.

• Steering: 70%
Setup more for offroad than freeways, it has too many turns lock-to-lock and is not precise. However, a short turning diameter gains good maneuverability.

• Driver's compartment: 70%
The excellent driver's position comes from a good relationship between the steering wheel, seat and pedals. A tall driving position gains

DATA

Category: 2-or 4-wheel drive multipurpose vehicles.
Class: Utility

HISTORIC

Introduced in: 1990
Modified in: -
Made in: Lafayette, Indiana, U.S.A.

RATINGS

Safety: 50 %
Satisfaction: 78 %
Depreciation: 60 %
Insurance: 9.5 % ($585)
Cost per mile: $0.36

NUMBER OF DEALERS

In U.S.A.: NA

SALES ON THE AMERICAN MARKET

Model	1992	1993	Result
Pick-up	31,387	30,731	- 2.1 %
Rodeo	45,257	48,903	+ 7.5 %

MAIN COMPETITORS

Rodeo: CHEVROLET Blazer, FORD Explorer, GMC Jimmy, JEEP Cherokee & Grand Cherokee, NISSAN Pathfinder, TOYOTA 4Runner.
Pick-up: DODGE Dakota, FORD Ranger, CHEVROLET S-10, GMC Sonoma, MAZDA B, MITSUBISHI Mighty Max, NISSAN Hardbody, TOYOTA.

EQUIPMENT

ISUZU Rodeo	S	LS
Automatic transmission:	O	O
Cruise control:	S	S
Power steering:	O	O
ABS brake:	O	O
Air conditioning:	O	S
Air bag:	S	S
Leather trim:	-	-
AM/FM/ Radio-cassette:	S	S
Power door locks:	-	S
Power windows:	-	S
Tilt steering column:	S	S
Dual power mirrors:	-	S
Intermittent wipers:	-	S
Light alloy wheels:	O	S
Sunroof:	-	
Anti-theft system:	-	

S : standard; O : optional; - : not available

COLORS AVAILABLE

Exterior: Red, White, Metallic Gray, Blue, Black, Tin, Pink (Amigo).
Interior: Gray, Beige.

MAINTENANCE

First revision: 3,000 miles
Frequency: 6,000 miles
Diagnostic plug: No

WHAT'S NEW IN 1995 ?

- New emission control device (Rodeo).
- Mirror with a cover added to the sun visors (Rodeo).
- Black tailgate handle for both Rodeo models.

Model/ versions *: standard	ENGINE Type / timing valve / fuel system	Displacement cu/in	Power bhp @ rmn	Torque lb.ft @ rpm	Compres. ratio	TRANSMISSION Driving wheels / transmission	Final ratio	PERFORMANCE Acceler. 0-62 mph s	Stand. 1/4 mile s	Stand. 5/8 mile s	Braking 62-0 mph ft	Top speed mph	Lateral acceler. G	Noise level dBA	Fuel economy mpg city	hwy	Gasoline type / octane
1)	L4* 2.6 SOHC-8-MPFI	156	120 @ 4600	150 @ 2600	8.3 :1	rear./4-M5*	4.30	13.8	19.3	38.8	154	97	NA	69	21.0	28.0	R 87
						rear. - A4	4.10	15.0	19.8	40.3	167	94	NA	69	20.0	27.0	R 87
2)	V6* 3.2 OHC-12-MPFI	193	175 @ 5200	188 @ 4000	9.3 :1	rear./4 - M5	4.30	12.5	18.0	35.4	135	94	0.74	66	18.0	24.0	R 87
						rear./4 - A4	4.10	13.7	19.4	38.5	144	87	0.72	67	19.0	25.0	R 87

1) base Pick-up 3) Rodeo & Pick-up

good visibility and the rear-view mirrors are large. The main dials are easy to interpret, but the rocker switches under the gauge panel are not practical. You have to turn your eyes away from the road to identify them while it would be so simple to put them under the steering wheel. You could search the instrument panel in vain for an indicator that would tell you the gear in which the automatic is running.

• Conveniences: 60%
Storage is well-distributed, except on the standard pickup, and the interiors of the Rodeo and Amigo are well-presented.

• Insurance: 60%
It is a little costlier than average to insure these utility vehicles whose premium remains affordable.

• Access: 60%
It is easier to climb into a 2-wheel drive than in the taller 4X4, but the doors open wide even at the rear of the Rodeo/Amigo which has 2- part hatchback.

• Technical: 60%
A chassis with five or seven cross members serves as the base for these three vehicles with different wheelbases. The Rodeo chassis is identical to that of the extended cabs and the Trooper. The front of their body and front doors are identical and built of steel. The suspensions have double arms and torsion bars plus an antiroll bar up front. While at the rear, the rigid axle is controlled by leaf springs. Braking now has discs on all 4 wheels and a mechanical ABS on the rear wheels.

• Quality/fit /finish: 60%
It is honest, but the plastics on the dashboard and the seat covering are not fancy.

• Interior space: 50%
There is more room in a Rodeo than an Amigo with a shorter wheelbase or in the extended cab, whose auxiliary seats can only be used in an emergency.

• Trunk: 50%
The spare tire hinged on the tailgate frees up ample baggage space. Folding down the rear seat bench of the Rodeo/Amigo you can double the available volume.

• Noise level: 50%
Silent on the road when equipped with a V6, these vehicles are noisier during acceleration, especially on a 4-cylinder which is very generous with decibels.

• Price/equipment: 50%
Less competitive because there is little in the standard versions. People who like comfort will have to pay for the options.

WEAK POINTS

• Fuel consumption: 30%
It is high, no matter which engine is considered, because it can reach between 18 and 23 mpg on the road and more off road, where the engine works harder.

• Performance: 30%
Heavy and shoe-box shaped, these cars are under-engined and have unfavorable weight-to-power ratios. Their takeoffs and accelerations are acceptable but the engines always give the impression of working hard.

• Handling: 40%
From this point of view, the vehicles do not do too badly, in spite of their weight, the height of the center of gravity, and the combination of leaf springs and rigid rear axle. The suspensions are fairly soft and the roll they generate is never dangerous, considering the speed reached. In spite of clean reactions, it is better to lift your foot before tight turns or changing lanes because these vehicles detest slaloms.

• Braking: 40%
Effective when cold, they weaken as soon as the linings heat up and sporadically block the front wheels. Stopping distances are long and the ABS uniquely on the rear wheels has the only virtue of keeping the trajectory straight.

• Safety: 40%
In spite of door reinforcements to add lateral resistance, the structure offers only an average resistance to impact testing, while the driver protection is insured by the presence of an air bag. That of the other occupants leaves much to be desired.

• Depreciation: 40%
The uncertainty connected with this make doesn't give it a high exchange value.

CONCLUSION

• Overall rating: 55.0%
The Rodeo remains the most valid product of the family because it is equipped with the more powerful engine. The Amigo is an amusing, but marginal, vehicle and the pickups have to deal with too strong a competition. ☹

ISUZU Rodeo

SPECIFICATIONS & PRICES

Model	Version/Trim		Body Seats	Wheel-base in	Lgth x Wdthx Hght. inx inx in	Curb weight lb	Towing capacity std. lb	Susp. type ft/rr	Brake type ft/rr	Steering system	Turning diameter ft	Steer. turns nber	Fuel. tank gal.	Standard tire size	Standard Power train	PRICES US. $ 1994
ISUZU		**Warranty: 3 years / 36,000 miles; Power train 5 years / 60,000 miles ; perforation 6 years / 100,000 miles.**														
Rodeo	S	2x4	5dr.wgn. 5	91.7	164.2x70.1x65.2	3973	4500	i/r	dc/dc/ABS	re.pwr.bal.	37.7	3.6	22	225/75R15	V6/3.2/M5	**$15,089**
Rodeo	LS	4x4	5dr.wgn. 5	91.7	165.7x70.1x65.7	3999	4500	i/r	dc/dc/ABS	re.pwr.bal.	37.7	3.2	22	245/70R16	V6/3.2/M5	**$25,019**
ISUZU		**Warranty: 3 years / 36,000 miles; Power train 5 years / 60,000 miles ; perforation 6 years / 100,000 miles.**														
Pick-up S	4x2	short deck	2dr.p-u.2	105.6	177.3x66.6x62.4	2700	1000	i/r	dc/dr/ABS	re. bal.	35.4	5.0	14	195/75R14	L4/2.3/M5	**$9,399**
Pick-up XS	4x2	long deck	2dr.p-u.2	119.2	193.8x66.6x62.4	3000	1000	i/r	dc/dr/ABS	re. pwr.bal.	39.4	3.4	19.8	195/75R14	L4/2.6/M5	**$10,809**
Pick-up XS	4x2	Super cab.	2dr.p-u.2	119.2	193.8x66.6x66.3	3029	2000	i/r	dc/dr/ABS	re. pwr.bal.	39.4	3.4	19.8	225/75R15	L4/2.3/M5	**$12,709**
Pick-up XS	4x2	Super cab.	2dr.p-u.2	119.2	193.8x66.6x66.2	3329	2000	i/r	dc/dc/ABS	re.pwr.bal.	39.4	3.6	19.8	225/75R15	L4/2.6/M5	-

See page 393 for complete 1995 Price List

Too much or too little?

Since it has swapped its tenure as a woods runner for a "cruiser" in finer quarters, the Trooper hasn't stopped losing ground. Its new vocation doesn't seem to pay off or it may not be luxurious enough to attract the sector the Land Rover has raked over for years.

Over the years the Trooper, which started as a rustic vehicle, has become a luxury sports utility whose rates have climbed in exaggerated fashion. No change in 1995 and it retains the form of a 5-door wagon with 4-wheel drive and a standard automatic or an optional 5-speed. The only engine available is a 3.2 L V6 which develops 175 hp in a standard S version and 190 hp in the more luxurious LS.

STRONG POINTS

• Interior space: 80%
It is excellent, since height and length are generous, and its sufficient width will allow 5 people to get in with ease.

• Satisfaction: 80%
Not a big problem for this vehicule whose owners appreciate the reliability but deplore the rarity and cost of replacement parts.

• Insurance: 80%
Comparing it to other vehicles of the same type, the premium is affordable because the rates are lower.

• Driver's compartment: 80%
The steering wheel is placed near the dashboard and allows a comfortable driving position whose height gains visibility and this, in spite of sizable door posts. The instruments are clear and well laid out but some controls are particular to the Trooper, such as the large knobs that control the headlights and windshield wipers, and take getting used to.

• Storage: 80%
The space behind the rear bench seat is large, because the spare tire has been moved outdoors. Storage becomes immense when the bench seat is folded, and the flat floor plus the 2-part tailgate, which opens full width, make it easy to load.

• Quality/fit/finish: 70%
The construction is strong and the finishing is done with care. However, the plastics of the dash in the velvet trim could have an appearance that is a little less synthetic.

• Access: 70%
It is easy to get in, thanks to numerous hand holds provided for the purpose. At the rear, storage access is convenient, thanks to asymmetric doors and bumpers which act as a footstep.

• Seats: 70%
Well-shaped, they provide ample lateral and lumbar supports, but their padding is firm and the velvet that covers them has a rough texture.

• Conveniences: 70%
For once there is storage everywhere. The glove box and door pockets are huge and the center console is laid out in a practical way. The cab has a number of practical and uncommon details which make one wonder why other builders are not as sharp.

• Suspension: 70%
Its smoothness makes the charm of this vehicle, which handles like a limousine on the freeway or on bad roads. Its generous wheel travel allows it to absorb large bumps.

DATA

Category: 2-or 4-wheel drive all-terrain multipurpose vehicles.
Class: utility

HISTORIC

Introduced in: 1981 (Big Horn, Japan) imported since 1987.
Modified in: 1992: totally renewed.
Made in: Fujisawa, Japan.

RATINGS

Safety: 50 %
Satisfaction: 83 %
Depreciation: 63 %
Insurance: 4.2 %($655)
Cost per mile: $0.48

NUMBER OF DEALERS

In U.S.A.: NA

SALES ON THE AMERICAN MARKET

Model	1992	1993	Result
Trooper	12,956	21,786	+ 40.6 %

MAIN COMPETITORS

CHEVROLET Blazer, FORD Explorer, GMC Jimmy, JEEP Cherokee & Grand Cherokee, LAND ROVER Discovery, NISSAN Pathfinder, TOYOTA 4Runner.

EQUIPMENT

ISUZU Trooper	XS	LS
Automatic transmission:	O	S
Cruise control:	S	S
Power steering:	S	S
ABS brake:	-	-
Air conditioning:	S	S
Air bag:	S	S
Leather trim:	-	-
AM/FM/ Radio-cassette:	S	S
Power door locks:	S	S
Power windows:	S	S
Tilt steering column:	-	-
Dual power mirrors:	S	S
Intermittent wipers:	S	S
Light alloy wheels:	O	S
Sunroof:	-	
Anti-theft system:	-	

S : standard; O : optional; - : not available

COLORS AVAILABLE

Exterior: White, Green, Gray, Red, Black, Blue, Mica, Silver.
Interior: Brown, Gray, Beige.

MAINTENANCE

First revision: 3,000 miles
Frequency: 5,000 miles
Diagnostic plug: Yes

WHAT'S NEW IN 1995 ?

- No major change.

	ENGINE					TRANSMISSION		PERFORMANCE										
Model/ versions *: standard	Type / timing valve / fuel system	Displacement cu/in	Power bhp @ rmn	Torque lb.ft @ rpm	Compres. ratio	Driving wheels / transmission	Final ratio	Acceler. 0-62 mph s	Stand. 1/4 mile s	Stand. 5/8 mile s	Braking 62-0 mph ft	Top speed mph	Lateral acceler. G	Noise level dBA	Fuel economy mpg city	hwy	Gasoline type / octane
S	V6* 3.2 SOHC-12-MFI	193	175 @ 5200	188 @ 4000	9.3 :1	rear./4 - M5	4.555	13.9	19.5	38.5	144	87	0.73	66	18.0	23.0	R 87
						rear./4 - A4*	4.555	14.8	21.0	39.7	138	94	0.71	67	17.0	24.0	R 87
LS	V6* 3.2 DOHC-24-MFI	193	190 @ 5600	195 @ 3800	9.8 :1	rear./4 - A4*	4.555	13.0	18.3	35.5	144	100	0.73	64	18.8	20.0	R 87

• **Steering:** 60%
Its assistance is well-proportioned and is acceptable, but the number of turn, lock-to-lock, takes away from its spontaneity and maneuverability.
• **Depreciation:** 60%
It has increased appreciably with the arrival of numerous competitive models.
• **Noise level:** 60%
The soundproofing keeps out all drivetrain and wind noise, which reminds one of the lack of body streamlining. As to the engine, it only signals its presence during strong accelerations.
• **Technical:** 60%
The unit body is steel, mounted on a seven cross member chassis. The independent front suspension uses torsion bars, while the rigid rear axle is fitted with helical springs. A four bar linkage prevents torque lift at the rear during accelerations to improve comfort. Underbody shields are installed standard, to protect the radiator, gas tank, oil pan, transmission and transfer case. Four-wheel disc brakes with ABS on the rear wheels are standard and 4-wheel ABS is an option on the LS. The single V6 has a SOHC, while the more powerful DOHC with 24 valves is unusual for this type of vehicle.
• **Safety:** 60%
The trooper body has only average impact resistance and only a driver's-side air bag is standard. Fortunately, there are door reinforcements, to resist lateral impacts.

WEAK POINTS

• **Performance:** 20%
It is very mediocre, especially with an automatic which needs 23 seconds to accelerate from zero to 62 mph and the maximum speed hardly exceeds 90 mph. A three-day clock is optional.
• **Fuel consumption:** 30%
Under-powered, it burns more fuel than average in its category.
• **Handling:** 40%
The Trooper has good straight-line stability, and on a sweeping curve is less affected by the pavement than by crosswinds. It is safer to slow down when you come to tight turns or need a lane change because the tall center of gravity and suspension softness result in impressive body movements. Traction in the straight and on tight curves depends quite a bit on the quality of the tires.
• **Braking:** 40%
Stopping distances are very long and hot, with ABS acting on all 4 wheels and are shorter when the ABS only acts on the rear wheel and doesn't keep the front wheels from locking up. In both instances, fade resistance is satisfactory but the front dives in an impressive manner during panic stops.
• **Price/equipment:** 40%
Less favorable than in the past, the price reflects the change of philosophy which made the Trooper more luxurious than useful, even on the base model.

CONCLUSION

• **Overall rating:** 61.0%
Too luxurious to go roam in the woods, it is not enough to make the Range Rover uneasy. The Trooper has neglected those who made it a success which risks losing them. 😐

SPECIFICATIONS & PRICES

Model	Version/Trim	Body/ Seats	Max Payload lb	Wheel base in	Lgth x Wdthx Hght. inx inx in	Curb weight lb	Towing capacity std. lb	Susp. type ft/rr	Brake type ft/rr	Steering system	Turning diameter ft	Steer. turns nber	Fuel. tank gal.	Standard tire size	Standard power train	PRICES US. $ 1994
ISUZU		**Warranty: 3 years / 36,000 miles; Power Train 5 years / 60,000 miles ; perforation 7 years / 100,000 miles.**														
Trooper	LS 4x4	5dr.wgn.5	1206	108.6	183.5x68.7x72.8	4070	5000	i/r	dc/dc/ABS	re.pwr.bal.	38.0	3.6	22.5	245/70R16	V6/3.2/A4	**$28,200**

See page 393 for complete 1995 Price List

Return to sources...

In the search for its roots, Jaguar does not hesitate to step back in what concerns its car style, to better go forward with the search for ever higher performance, which has always been the base of the Coventry creations; for the great pleasure of everyone.

Desirous to preserve the traditional character of its cars, Jaguar has redrawn the front and rear sections for 1995. Up front, we find round head lights and the curves of the older XJ6/XJ12, while at the rear, the cutout at the trunk opening and its taillights remind of the Mercedes S-Series. The XJ6/XJ12 are 3-volume, 4-door sedans sold in several trim levels: XJ6, Vanden Plas, XJR and XJ12. The engine of the first 3 is the in-line 4.0 L 6 cylinder, naturally aspirated or supercharged (XJR) with a ZF 4 speed automatic of German origin; the latest is the new 6.0 L V12, also available on the XJS.

STRONG POINTS

- **Safety:** 90%

The Jaguar structure resists impact relatively well and the front seat occupants are protected by air bags, which constitutes a serious rise to standards.

- **Insurance:** 90%

In spite of its low index, the fabulous price of these cars leads to a high premium.

- **Quality/fit/finish:** 85%

Massive sheet metal, thick chrome, high-quality leather, and lacquered marquetry have made the renown of these British cars.

- **Technical:** 80%

The steel unit body has taken back its traditional appearance, so now its aerodynamic drag coefficient is only 0.37. The independent suspension includes antidive and antilift geometry; also, the spring, damper and stabilizer bar rates are up. Four wheel disc brakes have ABS and the differential includes a limited slip. The 6-cylinder in-line engine is a classic of the make. This year it offers a supercharged version with an output of 322 hp (fully equal to a mildly improved small block Chevy) and also more than the V12. There is a new and original

JAGUAR XJ12

DATA

Category: rear-wheel drive luxury sedans.
Class: 7

HISTORIC

Introduced in: 1986
Modified in: 1990: 4.0L; 1993: V12. 1995: aesthetics & XJR
Made in: Browns Lane, Coventry, England.

RATINGS

Safety: 90 %
Satisfaction: 73 %
Depreciation: XJ-6: 71.0 % XJ12: 68 %
Insurance: XJ-6: 3.0 % ($1,270) XJ12: 2.8 % ($1,300)
Cost per mile: $0.96

NUMBER OF DEALERS

In U.S.A.: 129

SALES ON THE AMERICAN MARKET

Model	1992	1993	Result
XJ6	6,299	8,990	+ 30.0 %

MAIN COMPETITORS

XJ6: ACURA Legend, AUDI V8, BMW 735i, INFINITI Q45, LEXUS LS400, MERCEDES 300 & S.
XJ12: BMW 750iL, MERCEDES BENZ 600SEL.

EQUIPMENT

JAGUAR XJ6	base	Vanden Plas
JAGUAR XJ12	base	
Automatic transmission:	S	S
Cruise control:	S	S
Power steering:	S	S
ABS brake:	S	S
Air conditioning:	S	S
Air bag:	S	S
Leather trim:	S	S
AM/FM/ Radio-cassette:	S	S
Power door locks:	S	S
Power windows:	S	S
Tilt steering column:	S	S
Dual power mirrors:	S	S
Intermittent wipers:	S	S
Light alloy wheels:	S	S
Sunroof:	S	S
Anti-theft system:	S	S

S : standard; O : optional; - : not available

COLORS AVAILABLE

Exterior: Metallic Gray, White, Black, Navy Blue, Bordeaux, Topaz, Bronze Pink, Jade, Red.
Interior: Black, Tan, Blue, Charcoal, Oatmeal, Parchment, Coffee.

MAINTENANCE

First revision: 10,000 miles
Frequency: 6 months
Diagnostic plug: Yes

WHAT'S NEW IN 1995?

- Cosmetic retouches to the front and rear.
- New version with a supercharged engine (XJR).
- Power steering varies as a function of speed.
- Steering wheel with leather and wood (except XJ6).

Model/ versions *: standard	ENGINE Type / timing valve / fuel system	Displacement cu/in	Power bhp @ rmn	Torque lb.ft @ rpm	Compres. ratio	TRANSMISSION Driving wheels / transmission	Final ratio	PERFORMANCE Acceler. 0-62 mph s	Stand. 1/4 mile s	Stand. 5/8 mile s	Braking 62-0 mph ft	Top speed mph	Lateral acceler. G	Noise level dBA	Fuel economy mpg city	hwy	Gasoline type / octane
XJ6 & VP	L6* 4.0 DOHC-24-EFI	242	245 @ 4700	289 @ 4000	10.0 :1	rear - A4*	3.58	8.2	15.7	28.5	164	134	0.75	66	21.0	30.0	S 91
XJR	L6*C 4.0 DOHC-24-EFI	242	322 @ 5000	378 @ 3050	8.5 :1	rear - A4*	3.58	7.2	14.5	27.8	148	149	0.80	67	17.0	28.0	S 91
V12	V12* 6.0 SOHC 24-EFI	366	313 @ 5350	353 @ 3750	11.0 :1	rear - A4*	3.58	8.0	15.4	28.2	151	143	0.80	65	14.0	21.0	S 91

electronically controlled transmission, which can be shifted automatically on the right side or manually on the left side of the "U"-shaped shifter gate. You can also select a sport or normal mode for the electronic control. The new V12 has an aluminum block and head and a new cam drive as well as a forged steel crank. Outside dimensions are reduced to fit under the hood which was originally projected for a 4-cylinder.

• Performances: 80%
It is curious to note how close the 6 and 12 cylinder naturally aspirated engines are in car performance, identical to within a few tenths of a second. Standing starts and accelerations are respectable, taking into account the high total weight. The XJ keeps a fast pace without difficulty and doesn't disdain becoming sports car-like as long as one is willing to play tunes on the gear shift lever. The supercharged version is faster because it needs just a little over 7 seconds to go from 0 to 60 mph.

• Suspension: 80%
The happy compromise between the springs, dampers and the tires endows these cars with a smooth and comfortable ride, effectively cushioning out the road defects.

• Steering: 80%
Precise, with a fast ratio, its variable power assist produces a good directional stability and requires less driver attention. Regrettably, the large steering diameter interferes with maneuverability.

• Access: 80%
Thanks to well-sized doors, it is not difficult to slide in.

• Satisfaction: 75%
It climbs, year after year, to reach a rate which could be qualified as satisfactory.

• Noise level: 70%
Ride and mechanical noises are better silenced than airstream noises around the windshield, which increase with speed .

• Seats: 70%
They would be more comfortable if the seats were longer and their back rests taller. Up front their side rolls are less prominent and the curvature more effective.

• Driver's compartment: 60%
It is more spectacular than ergonomic because except for effective lateral and lumbar support, the driving position is imperfect. Visibility is satisfactory even though the rear-view mirrors are small, but it is difficult to evaluate the ends of the fenders during parking. Front seat controls have been relocated to the left side of the seats, digital and analog gauges are well-matched and easy to read, but it takes an intensive course before you can use the computer.

• Handling: 60%
These sedans prefer freeways to small twisty roads in a state of decomposition. There, the suspension loses its unperturbable phlegm and you need to play the steering wheel to stay the course. In a turn, they are neutral, up to the moment where they oversteer brutally, especially if the pavement is wet.

• Interior space: 60%
The interior will only accept 4 occupants, who will enjoy sufficient space in length; but the tall will complain of a lack in height, even up front.

• Storage: 60%
Large door pockets compensate for a small glove box, and the center console is poorly laid out.

• Trunk: 50%
Easily accessible, thanks to a cutout in its lid, the volume is not immense, but it receives a sufficient amount of baggage in spite of its tormented forms, and the spare tire encumbers what is left.

WEAK POINTS

• Price/equipment: 00%
Jaguar pricing is even more overvalued than that of its higher-tech competitors, but the equipment is complete and luxurious, offering fancier trim, privacy curtains, walnut burl trays and sheepskin carpets.

JAGUAR XJ6

JAGUAR XJR

• Fuel consumption: 30%
Considering the weight and displacement of these models, it is not surprising that they don't go over 19 mpg.

• Depreciation: 30%
The Jaguars are among the cars that lose the most of their value, but this tendency should change with time.

• Braking: 40%
For cars of this weight, the stopping distances are long. Yet, the stops are also stable and easy to modulate, even though the ABS sometimes reacts bizarrely allowing the wheels to lock up for a fraction of a second.

CONCLUSION

• Overall rating: 64.0%
Coming back to more traditional aesthetics and power train components, capable of giving the greater excitement, the British builder takes the path which allows to sell better. Ford representatives have placed their bets on this to put Jaguar back on track. 😐

SPECIFICATIONS & PRICES

Model	Version/Trim	Body/ Seats	Interior volume cu.ft.	Trunk volume cu ft.	Drag coef. Cd	Wheel-base in	Lgth x Wdthx Hght. inx inx in	Curb weight lb	Towing capacity std. lb	Susp. type ft/rr	Brake type ft/rr	Steering system	Turning diameter ft	Steer. turns nber	Fuel. tank gal.	Standard tire size	Standard power train	PRICES US. $ 1994
JAGUAR		**Warranty: 4 years / 50,000 miles; corrosion: 6 years / unlimited.**																
XJ 6	Sovereign	4dr.sdn.4	93	11.1	0.37	113.0	197.8x70.8x52.7	4081	NR	i/i	dc/dc//ABS	pwr.r/p.	40.8	2.8	23.2	225/60ZR16	L6/4.0/A4	**$51,750**
XJ 6	Vanden Plas	4dr.sdn.4	93	12	0.37	113.0	197.8x70.8x52.7	4105	NR	i/i	dc/dc//ABS	pwr.r/p.	40.8	2.8	23.2	225/60ZR16	L6/4.0/A4	**$59,400**
XJ 6	XJR	4dr.sdn.4	93	11.1	0.37	113.0	197.8x70.8x52.7	4215	NR	i/i	dc/dc//ABS	pwr.r/p.	40.8	2.8	23.2	255/45ZR17	L6C/4.0/A4	**NA**
XJ12		4dr.sdn.4	93	12	0.37	113.0	197.8x70.8x52.7	4420	NR	i/i	dc/dc//ABS	pwr.r/p.	40.8	2.8	23.2	225/60ZR16	V12/6.0/A4	**$71,750**

See page 393 for complete 1995 Price List

Relic...

The coming year will bring the discovery of the XJS replacement, which has largely earned its retirement. According to the first published documents, the new Jaguar coupes and convertibles have the proportions and streamlined form of the ancient XK-E, in vogue today.

In 1995 the XJS is offered as a coupe or convertible 2+2, 4.0 or 6.0. The first are equipped with a 6-cylinder 4.0 L, which is also fitted to the XJ6, a 4-speed automatic or an optional 5-speed. These vehicles can also receive the new 6.0 L V12, inaugurated in the XJ12 with an automatic only. The electronic shift which controls the latter, allows, depending on mood and circumstances, to select manually or automatically, and in either "normal" or "sport" mode.

STRONG POINTS

- **Safety:** **90%**
The good structural impact resistance and the presence of two air bags have considerably improved occupant safety.
- **Insurance:** **90%**
Their index is low, but cost brings the premium to a high level that still seems justified.
- **Technical:** **80%**
The basic design of the coupes and convertibles is classic: front engine, rear-wheel drive with 4-wheel independent suspension. At the rear, two coil over shock suspension units; the drivetrain is mounted on a sub-frame with mounts that isolate the body from vibrations and noise. There are four disc brakes and the ones at the back are no longer inboard on both sides of the differential but are within the wheels for ease of maintenance. To Pininfarina is due the general shape of the XJS, which has been updated several times by the Jaguar designers. The steel unit body is specially reinforced on the convertible to increase platform rigidity. In spite of the body curves, the aerodynamics really aged and the drag coefficient approaches 0.40.
- **Quality/fit/finish:** **80%**
Ford has set to improving some materials and components as well as

DATA

Category: rear-wheel drive GT coupes and convertibles.
Class: GT

HISTORIC

Introduced in: 1975
Modified in: 1988: convertibles.
Made in: Browns Lane, Coventry, England.

RATINGS

Safety: 90 %
Satisfaction: 65 %
Depreciation: 59 % (con) 65 % (cpe)
Insurance: 3.0 % ($1,200)
Cost per mile: $0.80

NUMBER OF DEALERS

In U.S.A.: 129

SALES ON THE AMERICAN MARKET

Model	1992	1993	Result
XJS	2,382	3,744	+ 36.4 %

MAIN COMPETITORS

ACURA NSX, BMW 8-Series, LEXUS SC400, MERCEDES SL, PORSCHE 928.

EQUIPMENT

JAGUAR XJS	4.0	6.0
Automatic transmission:	S	S
Cruise control:	S	S
Power steering:	S	S
ABS brake:	S	S
Air conditioning:	S	S
Air bag:	S	S
Leather trim:	S	S
AM/FM/ Radio-cassette:	S	S
Power door locks:	S	S
Power windows:	S	S
Tilt steering column:	S	S
Dual power mirrors:	S	S
Intermittent wipers:	S	S
Light alloy wheels:	S	S
Sunroof:	S	S
Anti-theft system:	S	S

S : standard; O : optional; - : not available

COLORS AVAILABLE

Exterior: Metallic Gray, White, Black, Navy Blue, Bordeaux, Topaz, Bronze Pink, Jade, Red.
Interior: Black, Tan, Blue, Charcoal, Oatmeal, Parchment, Coffee.

MAINTENANCE

First revision: 10,000 miles
Frequency: 6 months
Diagnostic plug: Yes

WHAT'S NEW IN 1995?

- New wheel rims.
- New sound system and a radio with a removable front.
- New more powerful 4.0 L engine.
- Redesigned interior.
- New variable assistance steering.

	ENGINE					TRANSMISSION		PERFORMANCE										
Model/ versions *: standard	Type / timing valve / fuel system	Displacement cu/in	Power bhp @ rmn	Torque lb.ft @ rpm	Compres. ratio	Driving wheels / transmission	Final ratio	Acceler. 0-62 mph s	Stand. 1/4 mile s	Stand. 5/8 mile s	Braking 62-0 mph ft	Top speed mph	Lateral acceler. G	Noise level dBA	Fuel economy mpg city	hwy	Gasoline type / octane
4.0	L6*4.0 DOHC-24-EFI	242	237 @ 4700	282 @ 4000	10.0 :1	rear - M5	3.54	8.2	15.8	28.7	144	143	0.80	67	17.0	28.0	S 91
						rear - A4	3.54	8.8	16.5	29.5	151	137	0.80	67	21.0	30.0	S 91
6.0	V12 6.0 SOHC-24- EFI	366	301 @ 5400	351 @ 2800	11.0 :1	rear - A4*	3.54	7.8	15.5	28.2	151	150	0.80	66	14.0	22.0	S 91

the care given to finishing details.

• **Performance: 80%**
In spite of a weight that approaches two tons, takeoffs and accelerations of the naturally aspirated 4.0 L are already very respectable, but the supercharged version of the XJR does even better thanks to its 322 hp. In contrast, the V12 does not do as well as the latter, based on a stop watch, because its power is lower. It has on the other hand, a silky feel, very agreeable to those who prefer comfort to adrenaline.

• **Handling: 70%**
More at ease on the freeway where straight-line stability, as on wide curves, is good, these big coupes are not as effective on a twisty path where their weight and size take away all agility. It is better not to tempt fate in the rain, when you can lose the rear without warning and it is hard to recover. With its stiffer sports suspension, the 4.0 L coupe is sharper than the V12, which is dedicated to comfort.

• **Driver's compartment: 70%**
Recent design changes have considerably improved driver comfort, including dashboard layout and the location and readability of the many gauges. The shape of the 3-speed gear shift has remained just as irrational. The steering wheel does not offer a good grip, is angled and also lacks a telescopic adjustment. Visibility from 3/4 rear is penalized by excess B pillar thickness, the narrowness of the quarter-panel glass, and the rear-view mirrors are still too small. Finally, the seats are not worthy of such refined production because, in spite of their appearance, they provide neither lateral nor lumbar support. Seat controls are now more rational to use and are positioned next to the seat.

• **Satisfaction: 70%**
In spite of the constant redesign they endure, there is increasing difficulty for these cars to stand comparison with their more recent European and Japanese rivals.

• **Steering: 70%**
As on the XJ6/12, the power steering has an assist that varies with speed. It provides more positive assistance and a better directional stability. Even though fairly fast, the large steering diameter hurts maneuverability.

• **Suspension: 70%**
That of V12 engines is smoother on freeways than the 4.0L sports coupe which is better at limiting roll. Their common characteristic is the ability to vigorously shake up the occupants on a bad road.

• **Noise level: 60%**
In spite of the soundproofing which creates a velvety feel, mechanical noises remain perceptible, as much with a 4.0 L engine as with a 6.0 L, to which are added wind and rolling noises as soon as speed increases.

• **Seats: 60%**
Marvelous to look at, but their shape does not provide sufficient lateral hold, their rolls are aggressive and the padding firm.

• **Conveniences: 60%**
Storage spaces include a small glove box, door pockets and a receptacle which forms part of the center console.

• **Access: 50%**
The length and door opening angle are small. It is not easy to slide into the front compartment, not to speak of rear seats, which will only accept baggage.

WEAK POINTS

• **Price/equipment: 00%**
In spite of their prestigious status and very complete equipment, Jaguar prices are exaggerated because their technical achievement has long been surpassed.

• **Fuel consumption: 20%**
Don't expect much more than 17 mpg when rapidly moving the two tons of this vehicle.

• **Interior space: 30%**
Considering its outside bulk, the XJS is not very roomy because it is a defacto a 2-seater with very limited room in all directions, particularly in the back.

• **Trunk: 30%**
In spite of the space taken up by the spare tire, the volume reserved for baggage is sufficient in a car of this size.

• **Depreciation: 40%**
Resale value of the Jaguar is an elusive number, and you better consider this investment as sentimental.

• **Braking: 40%**
The ABS stabilizes perfectly for panic stops, but the distances are much too long and the fade-resistance mostly mediocre.

CONCLUSION

• **Overall rating: 58%**
The Jaguar will reveal during the '95 model year the successor to the XJS. It is great timing because of the two models built by Jaguar, it is the most controversial in terms of philosophy, overall design and lines. The performance, passé by today's standards, doesn't match the price requested. 😐

SPECIFICATIONS & PRICES

Model	Version/Trim	Body/ Seats	Interior volume cu.ft.	Trunk volume cu ft.	Drag coef. Cd	Wheel-base in	Lgth x Wdthx Hght. inx inx in	Curb weight lb	Towing capacity std. lb	Susp. type ft/rr	Brake type ft/rr	Steering system	Turning diameter ft	Steer. turns nber	Fuel. tank gal.	Standard tire size	Standard power train	PRICES US. $ 1994
JAGUAR		**Warranty: 4 years / 50,000 miles; corrosion: 6 years / unlimited.**																
XJS	4.0	2dr.cpe.2+2	77	9.4	0.38	102.0	191.2x69.4x48.7	3805	NR	i/i	dc/dc/ABS	pwr.r/p.	42.7	2.76	24	225/60ZR16	L6/4.0/M5	**$51,950**
XJSC	4.0	2dr.con.2	NA	9.4	0.39	102.0	191.2x69.4x48.7	4021	NR	i/i	dc/dc/ABS	pwr.r/p.	42.7	2.76	20.7	225/60ZR16	L6/4.0/M5	**$59,950**
XJS	6.0	2dr.cpe.2+2	77	9.4	0.38	102.0	191.2x69.4x48.7	4052	NR	i/i	dc/dc/ABS	pwr.r/p.	42.7	2.76	24	225/60ZR16	V12/6.0/A4	**$69,950**
XJSC	6.0	2dr.con.2	NA	9.4	0.39	102.0	191.2x69.4x48.7	4306	NR	i/i	dc/dcABS	pwr.r/p.	42.7	2.76	20.7	225/60ZR16	V12/6.0/A4	**$79,950**

See page 393 for complete 1995 Price List

Passed by...

Theoretically, the Cherokee remains an efficient all-use vehicle at the level of its 4-wheel drive and its 6-cylinder engine, but sales began to suffer from the competition of the Grand Cherokee and from the fact that its lack of quality makes its use frustrating and costly.

It is offered in 1995 with a 2-or 4-wheel drive in the form of 3-and 5-door wagons in 3 finish levels: SE, Sport, and Country.

STRONG POINTS

• Storage space: 80%
The presence of a full-sized spare tire limits its inside volume, but there remains enough space to shelter a volume equivalent to that of a large sedan. In addition, the bench can disappear to liberate more space or to transport bulky objects.

• Technical: 75%
The Cherokee is based on a steel ladder frame to which is bolted the body. The suspensions consist of a rigid axle with helical springs at the front and leaf springs at the rear. The brakes are mixed and the 4-wheel ABS is only offered as an option. The characteristic lines are far from streamlined, so the drag coefficient varies between 0.51 and 0.52. The standard engine is a 2.5 L 4-cylinder, with a manual 5-speed standard and the 3-speed automatic optional. The in-line 4.0 L 6-cylinder engines are fitted to the Sport and Country with the same choice of transmissions. Full-time or part-time 4WD and 2WD are optional on all versions.

• Satisfaction: 70%
Reliability is not always without reproach, because parts, just like maintenance, are very expensive, and some owners have been discouraged by the frequency of small and large annoyances.

• Safety: 60%
While the structure, which dates from the 80s' has only an average impact resistance, reinforcements have been installed in the doors and an air bag is fitted to the steering wheel hub.

• Seats: 60%
The short cushion and the low seat-back height make them uncomfortable on long trips, especially in the back, where the bench seat back is short and lacking headrests.

• Suspension: 60%
In spite of its rusticity, it is comfortable on the freeways, but it begins to jump at the slightest pavement unevenness.

• Steering: 60%
Its power assist and ratio are too high, which make it sensitive and imprecise, but the maneuverability is good, thanks to a compact format and a short steering diameter.

• Quality/fit/finish: 50%
In spite of some improvements, the assembly quality, the materials, and the care brought to finishing are from another era (not the good old times).

• Driver's compartment: 50%
The Cherokee is not the most comfortable car to drive because the steering column is too long and the seat doesn't hold in turns and fails to support the back. Visibility is good, but the outside rear-view mirrors

DATA

Category: 2-or 4-WD all-terrain multipurpose vehicles.
Class: utility

HISTORIC

Introduced in: 1962, 1984.
Modified in: 1985: 2WD; 1987: L6 4.0L engine;1989: ABS re.
Made in: Toledo, Ohio, U.S.A.

RATINGS

Safety: 60 %
Satisfaction: 70 %
Depreciation: 55 %
Insurance: 7.1 % ($655)
Cost per mile: $0.48

NUMBER OF DEALERS

In U.S.A.: 1,029 Eagle-Jeep.

SALES ON THE AMERICAN MARKET

Model	1992	1993	Result
Cherokee	128,960	125,443	- 2.8 %

MAIN COMPETITORS

CHEVROLET Blazer S-10, FORD Explorer, ISUZU Rodeo & Trooper, GMC Jimmy, NISSAN Pathfinder, SUZUKI Sidekick 4 dr.,TOYOTA 4Runner.

EQUIPMENT

JEEP Cherokee	SE	Sport	Country
	O	O	O
Automatic transmission:	O	O	O
Cruise control:	O	O	O
Power steering:	O	S	S
ABS brake:	O	S	S
Air conditioning:	S	S	S
Air bag:	-	-	O
Leather trim:	O	O	S
AM/FM/ Radio-cassette:	S	S	S
Power door locks:	-	O	O
Power windows:	O	O	S
Tilt steering column: -	O	O	S
Dual power mirrors:	O	S	S
Intermittent wipers:	O	O	O
Light alloy wheels:	-		
Sunroof:	-		
Anti-theft system:			

S : standard; O : optional; - : not available

COLORS AVAILABLE

Exterior: Red, Blue, Beige, White, Green, Black.
Interior: Dark Sand, Charcoal.

MAINTENANCE

First revision: 7,500 miles
Frequency: 6 months
Diagnostic plug: Yes

WHAT'S NEW IN 1995 ?

- **Airbags on the driver's side.**
- **The Country is offered only as a 4-door.**
- **Automatic 3-speed transmission available as an option with a 2.5 L engine.**
- **New body colors.**
- **Front seats with reclining back, standard.**

	ENGINE					TRANSMISSION		PERFORMANCE										
Model/ versions *: standard	Type / timing valve / fuel system	Displacement cu/in	Power bhp @ rmn	Torque lb.ft @ rpm	Compres. ratio	Driving wheels / transmission	Final ratio	Acceler. 0-62 mph s	Stand. 1/4 mile s	Stand. 5/8 mile s	Braking 62-0 mph ft	Top speed mph	Lateral acceler. G	Noise level dBA	Fuel economy mpg city	hwy	Gasoline type / octane
base	L4* 2.5 OHC-8-MPFI	151	130 @ 5250	149 @ 3250	9.2 :1	rear./4 - M5*	3.48	13.5	18.8	35.2	170	94	0.68	69	24.0	30.0	R 87
						rear./4 - A3	3.73	14.5	19.5	36.8	174	91	0.68	68	23.0	28.0	R 87
option	L6* 4.0 OHC-12-MPFI	243	190 @ 4750	225 @ 4000	8.8 :1	rear./4 - M5*	2.42	9.0	16.6	30.5	167	106	0.70	69	20.0	28.0	R 87
						rear./4 - A4	3.55	10.5	17.0	31.3	164	103	0.70	68	19.0	28.0	R 87

are too small and placed too far back to be effective. The controls are typically American, such as a foot-operated parking brake, very impractical in some off road maneuvers. Not very numerous on the SE, the gauges are laid out well but easier to read in the day than at night.

• Performance: 50%
The standard 4-cylinder is recommended only for 2-wheel drive and stick shift, just for family use. A transfer case necessitates a 6-cylinder whose weight-to-power ratio allows fast passing and ample freeway cruise speeds.

• Handling: 50%
With quality tires,It comes closer to that of a car than of a utility machine. One should not trust the sometimes excessive suspension softness which accents the roll and the acceleration of the 6-cylinder on a wet road.

• Interior space: 50%

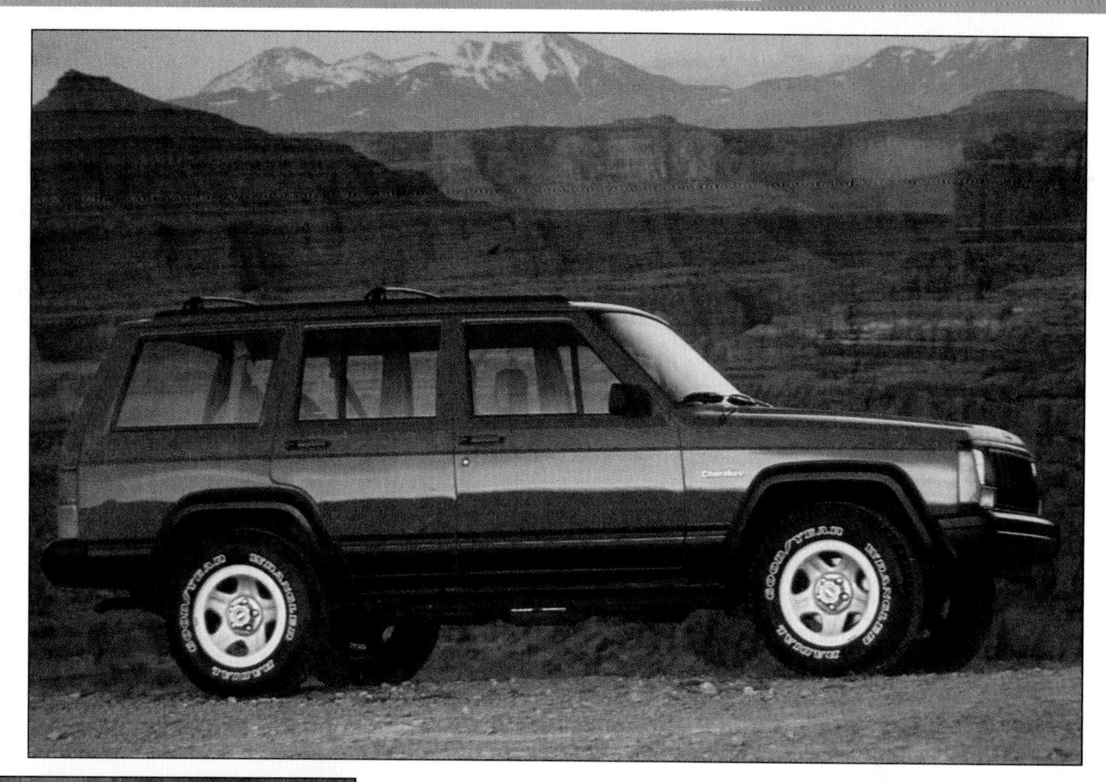

The compact dimensions of the Cherokee are paid for by a reduced usable space, comparable to that of a compact. The cab lacks length and width, but inside height is adequate.

• Access: 50%
Narrow doors, high ground clearance, and small inside dimensions complicate access to cab or baggage, and the tall floor doesn't help.

• Noise level: 50%
Soundproofing is more effective on the luxury versions, where it is better at snuffing out rolling and engine compartment noises. The 6-cylinder is more discreet than the 4, which noisily manifests its presence during takeoffs and accelerations.

• Price/equipment: 50%
Whatever the versions selected, it is expensive because its equipment, which is never complete, leads to the costly options list, still too long for a vehicle at the end of its career.

WEAK POINTS

• Fuel consumption: 30%
It was never economical, no matter what the engine or the type of transmission, and becomes gargantuan in difficult terrain.

• Braking: 30%
Panic stops lack efficiency because the distances are long and the excessive assist prevents precise modulation. The ABS, offered as an option, is not a luxury, because without it the vehicle path in a panic stop is part of fantasy land.

• Conveniences: 40%
The practical side was never the strong point of the Cherokee, whose storage areas have never been sufficient.

• Insurance: 40%
As on all vehicles with specific risks, the premium is higher than average.

• Depreciation: 40%
It has accelerated since the appearance of the Grand Cherokee and because of its chronic lack of quality.

CONCLUSION

• Overall rating: 52.5%
While obsolescent and costly, the Cherokees are compact and have multiple uses, but the more refined Grand Cherokee steals most of its clients. ☹

OWNERS' SUGGESTIONS

- -A better general quality
- -More efficientsound proofing
- - A more refines suspension
- -Better finishing
- -Improved reliability
- -Better fuel consumption

SPECIFICATIONS & PRICES

Model	Version/Trim	Body/ Seats	Drag coef. Cd	Wheel-base in	Lgth x Wdthx Hght. inx inx in	Curb weight lb	Towing capacity std. lb	Susp. type ft/rr	Brake type ft/rr	Steering system	Turning diameter ft	Steer. turns nber	Fuel. tank gal.	Standard tire size	Standard power train	PRICES US. $ 1994
JEEP	General warranty: 3 years / 36,000 miles; surface corrosion 1 year / 12,000 miles; perforation 7 years / 100,000 miles; road assistance 3 years / 36,000 miles.															
Cherokee 4X2	SE	3dr.wgn.5	0.51	101.4	167.0x67.7x63.8	2890	4550	i/r	dc/dr/ABS	re.pwr.bal.	35.8	3.4	20.2	215/75R15	L4/2.5/M5	$12,827
Cherokee 4X2	Sport	3dr.wgn.5	0.51	101.4	167.0x67.7x63.8	2932	4850	r/r	dc/dr/ABS	re.pwr.bal.	35.8	3.4	20.2	225/75R15	L6/4.0/M5	$14,484
Cherokee 4X2	Country	3dr.wgn.5	0.51	101.4	167.0x67.7x63.8	2998	4850	i/r	dc/dr/ABS	re.pwr.bal.	35.8	3.4	20.2	225/70R15	L6/4.0/M5	$16,771
Cherokee 4X4	SE	3dr.wgn.5	0.52	101.4	167.0x67.7x63.8	3058	4550	i/r	dc/dr/ABS	re.pwr.bal.	35.8	3.4	20.2	215/75R15	L4/2.5/M5	$13,387
Cherokee 4X4	Sport	5dr.wgn.5	0.52	101.4	167.0x67.7x63.8	3064	4850	r/r	dc/dr/ABS	re.pwr.bal.	35.8	3.4	20.2	225/75R15	L6/4.0/M5	$17,479
Cherokee 4X4	Country	5dr.wgn.5	0.52	101.4	167.0x67.7x63.8	3102	4850	i/r	dc/dr/ABS	re.pwr.bal.	35.8	3.4	20.2	225/70R15	L6/4.0/M5	$19,266

See page 393 for complete 1995 Price List

Charmer...

The Grand Cherokee knows increased popularity among those who are already fans of the make and those which cross over from the competition, the latter will be better organized this year with the freshening up of the Blazer and Explorer. This will be the mother of all sales contests...

The Grand Cherokee knows wide success in spite of the poor publicity coming from the tests of J.D. Power, which claim a dubious reliability and a higher than average number of problems. It is offered as a 5-door wagon with 2-or 4-wheel drive in the SE, Laredo Limited and new for this year, the Orvis for off road use only. The engines are 4.0 L 6-cylinders in-line or the 5.2 L V8 with a 4-speed automatic.

STRONG POINTS

• Technical: 80%
The Grand Cherokee unit body is extremely rigid, has a reduced height but retains the large ground clearance. The 2 axles are rigid with helical springs and disc brakes at all 4 wheels. Stiff connections between the axles and the body, thanks to multiple arms, is the handling and comfort secret of this vehicle. You can opt for a full-time 4-wheel drive in "Quadra Track" or the "Select Track". The Grand Cherokee 4x4 is equipped standard with a part-time "Command Track". More modern and better shaped, the body lines have an average drag coefficient of 0.34, normal for this type of vehicle.

• Safety: 80%
The Grand Cherokee is the first off-road vehicle which comes through standard with an ABS on all 4 wheels and an air bag on the driver's side, while the occupants are maintained by 3-point seat belts. The structure includes solid bumpers and resists impacts well, but NHTSA test indicates that the and passengers are not as well protected as the driver. Removable headrests will now become available for the rear seat bench.

• Satisfaction: 80%
In spite of the J.D. Power reports, the number of very satisfied users is rising, which could indicate that the problems mentioned are not that important.

• Suspension: 80%
The quality of its damping and the wheel travel allow very civilized comfort, because in spite of the presence of vulgar rigid axles, the wheels are well able to absorb large bumps.

• Storage: 80%
The seat bench back can be folded down in 2/3 and 1/3 proportions to increase total holding volume, which is penalized by the spare tire and wheel. Jeep installs it inside to facilitate its access off road and to keep it clean.

• Quality/fit/finish: 70%
The assembly, appearance of the materials and finishing care are more evident on these models than on other jeeps.

• Access: 70%
In spite of its large ground clearance, the length of the doors and the cab size ease climbing in, but handles for hanging on to are not numerous.

• Seats: 70%
Their lack of curvature does not provide enough side hold, and their

DATA

Category: 2-or 4-WD all-terrain multipurpose vehicles.
Class: utility

HISTORIC

Introduced in: 1992
Modified in: 1993: Gd Wagoneer & V8 5.2L engine.
Made in: Jefferson Avenue, Detroit, U.S.A..

RATINGS

Safety: 80 %
Satisfaction: 80 %
Depreciation: 35 % (2 years)
Insurance: 6.5 % ($655)
Cost per mile: $0.48

NUMBER OF DEALERS

In U.S.A.: 1,029 Eagle-Jeep.

SALES ON THE AMERICAN MARKET

Model	1992	1993	Result
G. Cherokee	86,859	217,232	+150 %

MAIN COMPETITORS

CHEVROLET Blazer S-10, FORD Explorer, ISUZU Rodeo & Trooper, JEEP Cherokee, GMC Jimmy, LAND ROVER Range & Discovery, MITSUBISHI Montero, NISSAN Pathfinder, TOYOTA 4Runner & Land Cruiser.

EQUIPMENT

Grand Cherokee	SE	Laredo	Ltd	Orvis
Automatic transmission:	S	S	S	S
Cruise control:	S	S	S	S
Power steering:	S	S	S	S
ABS brake:	S	S	S	S
Air conditioning:	S	S	S	S
Air bag:	S	S	S	S
Leather trim:	O	O	S	S
AM/FM/ Radio-cassette:	S	S	S	S
Power door locks:	S	S	S	S
Power windows:	S	S	S	S
Tilt steering column:	S	S	S	S
Dual power mirrors:	S	S	S	S
Intermittent wipers:	S	S	S	S
Light alloy wheels:	O	S	S	S
Sunroof:	-			
Anti-theft system:	-			

S : standard; O : optional; - : not available

COLORS AVAILABLE

Exterior: Black, White, Green, Red, Brown, Blue.
Interior: Fabric: Medium Quartz, Champagne, Driftwood; Leather: Beige, Black, White.

MAINTENANCE

First revision: 7,500 miles
Frequency: 6 months
Diagnostic plug: Yes

WHAT'S NEW IN 1995 ?

- **New "Orvis" version designed for off-road use.**
- **Limited model offered in 2-wheel drive.**
- **Disc brakes and ABS offered on all 4 wheels, standard.**
- **Electric door locks, standard.**
- **Remote lock system, standard.**
- **Automatic transmission, standard.**

	ENGINE					TRANSMISSION		PERFORMANCE										
Model/ versions *: standard	Type / timing valve / fuel system	Displacement cu/in	Power bhp @ rmn	Torque lb.ft @ rpm	Compres. ratio	Driving wheels / transmission	Final ratio	Acceler. 0-62 mph s	Stand. 1/4 mile s	Stand. 5/8 mile s	Braking 62-0 mph ft	Top speed mph	Lateral acceler. G	Noise level dBA	Fuel economy mpg city	hwy	Gasoline type / octane
1)	L6* 4.0 OHV-12SFI	242	190 @4750	225 @ 4000	8.8 :1	re./4 - M5*	3.55	10.7	17.0	30.1	131	118	0.75	67	17.0	24.0	R 87
2)	V8* 5.2 OHV-16-SFI	318	220 @ 4800	285 @ 3600	9.1 :1	re./4 - A4*	3.73	9.0	16.8	29.0	134	118	0.74	67	16.0	23.0	R 87

1) base 2) option

cushion, as well as the height of the seat backs, has been gnawed away to accent the impression of space, especially at the back.

• **Driver's compartment: 70%**
The dashboard is complicated, for no reason, and not very ergonomic because the center console is pulled away instead of projected. The steering column is too long and forces the driver to sit too far from the dashboard. There is no shift selector position indicator among the instrumentation. Visibility is excellent, thanks to the size of the glass surface and to ample side-view mirrors.

• **Interior: 70%**
In relation to the little Cherokee, the cab is less spacious but has gained length and height. Curvature at the sides has preserved hip and shoulder room.

• **Performance: 70%**
The 4.0 L 6-cylinder allows very respectable takeoffs and accelerations but it loads the little Cherokee. The automatic transmission downshifts often and frequently to keep up when passing. The V8 has lots of torque, quietly delivered, but it has great thirst.

• **Noise level: 60%**
The rigidity of the unit body and its soundproofing give it a quiet worthy of a deluxe car. However, the engine sounds off during strong accelerations and the suspension takes due notice of road joints.

• **Depreciation: 60%**
The high demand and a good reputation keep the bidding high and you will have to wait for another year to make a good deal.

• **Insurance: 60%**
The premium for the SE 4x2 is more affordable than that of a 4x4 Limited.

• **Handling: 60%**
The Grand Cherokee in 4-wheel drive is one of the most secure in its category. Because of its equipment and handling in a curve, it tends to make you forget its utilitarian nature since its reactions are those of an automobile. The quality of the tires influence the handling, but the ideal compromise between road and off road doesn't exist, and you will have to choose the playground.

• **Steering: 60%**
Calculated to ease handling in difficult terrain, its excessively strong assistance makes it vague in the center and calls for numerous corrections on the road, particularly in a crosswind.

• **Braking: 50%**
The stopping distances and the straight line path, as well as the fade-resistance, are remarkable, considering the vehicle's weight.

• **Conveniences: 50%**
The practical side is not perfect because storage places are more numerous than useful.

WEAK POINTS

• **Fuel consumption: 20%**
The V8 offers exceptional pulling power (nearly 3 tons, as an option), but its fuel consumption stays about 16 miles per gallon. While the 6-cylinder "contents" itself with 17 mpg.

• **Price/equipment: 40%**
The average price of the Grand Cherokee is at $23,000, expensive for being fashionable, but at least at this price, the equipment has been enriched for this year.

CONCLUSION

• **Overall rating: 64%**
The Grand Cherokee knows a considerable success because of its style, many uses and safe handling even though 90% of its owners do not make full use of it.

SPECIFICATIONS & PRICES

Model	Version/Trim	Body/ Seats	Drag coef. Cd	Wheel-base in	Lgth x Wdthx Hght. inx inx in	Curb weight lb	Towing capacity std. lb	Susp. type ft/rr	Brake type ft/rr	Steering system	Turning diameter ft	Steer. turns nber	Fuel. tank gal.	Standard tire size	Standard power train	PRICES US. $ 1994
JEEP	**General warranty: 3 years / 36,000 miles; surface corrosion 1 year / 12,000 miles; perforation 7 years / 100,000 miles; road assistance 3 years / 36,000 miles.**															
Gd Cherokee 4x2	SE	4dr.wgn. 5	0.44	105.9	179.0x70.9x64.7	3569	3000	r/r	dc/dc/ABS	pwr.bal.	36.7	3.3	23	215/75R15	L6/4.0/A4	**$21,156**
Gd Cherokee 4x2	Laredo	4dr.wgn. 5	0.44	105.9	179.0x70.9x64.7	3593	5000	r/r	dc/dc/ABS	pwr.bal.	36.7	3.3	23	215/75R15	L6/4.0/A4	-
Gd Cherokee 4x2	Limited	4dr.wgn. 5	0.44	105.9	179.0x70.9x64.7	3651	5000	r/r	dc/dc/ABS	pwr.bal.	36.7	3.3	23	225/70R15	L6/4.0/A4	-
Gd Cherokee 4x4	SE	4dr.wgn. 5	0.44	105.9	179.0x70.9x64.7	3675	3000	r/r	dc/dc/ABS	pwr.bal.	36.7	3.2	23	215/75R15	L6/4.0/A4	**$22,096**
Gd Cherokee 4x4	Laredo	4dr.wgn. 5	0.44	105.9	179.0x70.9x64.7	3704	5000	r/r	dc/dc/ABS	pwr.bal.	36.7	3.2	23	215/75R15	L6/4.0/A4	-
Gd Cherokee 4x4	Limited	4dr.wgn. 5	0.44	105.9	179.0x70.9x64.7	3946	5000	r/r	dc/dc/ABS	pwr.bal.	36.7	3.2	23	225/70R15	V8/5.2/A4	**$29,618**
Gd Cherokee 4x4	Orvis	4dr.wgn. 5	0.33	105.9	179.0x70.9x64.7	3957	5000	r/r	dc/dc/ABS	pwr.bal.	36.7	3.2	23	245/70R15	V8/5.2/A4	-

See page 393 for complete 1995 Price List

Kidnapped...

The WranglerJeep flourishes best in the shopping streets of Center City, where it remains a flashy sight, rather than in the middle of nature for which it was originally intended. It is not surprising that it has never been rationalized, even if its replacement is slow making the scene.

The YJ doesn't have much to do with its ancestors which were agile, compact and go-anywhere machines. It is sold as a convertible with 2 steel half doors or as a hardtop with full doors and removable windows. Its trim levels are S, SE and Sahara. The engine of the S and SE is a 2.5 L 4 cylinder, while the Sahara receives an in-line 6-cylinder 4.0 L. The manual 5-speed is standard and the old 3-speed automatic is optional on all models.

STRONG POINTS

• Satisfaction: 80%
It has improved with the years to reach an enviable level, but the long wait for the arrival of a more realistic new model has the fans turning more and more towards the competition.

• Price/equipment: 80%
The YJ is too expensive for what it has to offer because its equipment is rudimentary and options are limited.

• Steering: 80%
Rapid and precise, its assistance is too high, which makes it light and forces a constant watch on the road at all times, especially in a crosswind. It is short, has a tight turning radius and good maneuverability, but excessive width limits its progress off road.

• Braking: 60%
The ABS, optional on the SE and the Sahara, is not another gadget, but a safety item which should be standard on the S since sudden stops lack stability, especially on a wet road. Brake materials lack bite and fade-resistance, and stopping distances are much too long.

• Insurance: 60%
It is expensive, because this type of vehicle represents a higher risk, especially in the hands of people who are not too reasonable.

DATA

Category: 2-or 4-WD all-terrain mnultipurpose vehicles.
Class: utility

HISTORIC

Introduced in: 1952
Modified in: 1975: CJ7; 1985: Wrangler.
Made in: Toledo, Ohio, U.S.A.

RATINGS

Safety: 40 %
Satisfaction: 80 %
Depreciation: 51 %
Insurance: 6.8 % ($655)
Cost per mile: $0.48

NUMBER OF DEALERS

In U.S.A.: 1,029 Eagle-Jeep

SALES ON THE AMERICAN MARKET

Model	1992	1993	Result
Wrangler	49,724	65,648	+ 24.3 %

MAIN COMPETITORS

SUZUKI Sidekick, GEO Tracker, LAND ROVER Defender 90.

EQUIPMENT

JEEP Wrangler	S	SE/Sport	Sahara
Automatic transmission:	O	O	O
Cruise control:	-	-	-
Power steering:	O	S	S
ABS brake:	-	O	O
Air conditioning:	O	O	S
Air bag:	-	-	-
Leather trim:	-	-	-
AM/FM/ Radio-cassette:	O	O	S
Power door locks:	-	-	-
Power windows:	-	-	-
Tilt steering column:	-	-	-
Dual power mirrors:	O	O	O
Intermittent wipers:	O	O	S
Light alloy wheels:	O	O	O
Sunroof:	-		
Anti-theft system:	-		

S : standard; O : optional; - : not available

COLORS AVAILABLE

Exterior: Black, Red, White, Blue,Blue Green, Beige, Green.
Interior: Charcoal, Spice.

MAINTENANCE

First revision: 7,500 miles
Frequency: 6 months
Diagnostic plug: No

WHAT'S NEW IN 1995 ?

- Renegade is eliminated.
- Rio Grande trim is optional on the S version.
- New body colors.

ENGINE						TRANSMISSION		PERFORMANCE										
Model/ versions *: standard	Type / timing valve / fuel system	Displacement cu/in	Power bhp @ rmn	Torque lb.ft @ rpm	Compres. ratio	Driving wheels / transmission	Final ratio	Acceler. 0-62 mph s	Stand. 1/4 mile s	Stand. 5/8 mile s	Braking 62-0 mph ft	Top speed mph	Lateral acceler. G	Noise level dBA	Fuel economy mpg city	hwy	Gasoline type / octane
1)	L4* 2.5 OHV-8-MPSFI	151	123 @ 5250	139 @ 3250	9.2 :1	rear/4 - M5*	4.11	12.8	19.2	35.5	164	90	0.72	72	21.0	25.0	R 87
						rear/4 - A3	3.73	13.6	20.5	36.8	154	87	0.72	72	20.0	24.0	R 87
2)	L6* 4.0 OHV-12-MPSFI	243	180 @ 4750	220 @ 4000	8.8 :1	rear/4 - M5*	3.07	10.0	17.5	32.4	157	100	0.72	70	19.0	25.0	R 87
						rear/4 - A3	3.07	11.2	18.8	34.0	151	97	0.72	70	18.0	22.0	R 87

1) * S,SE. 2) * Sahara opt. SE/Sport

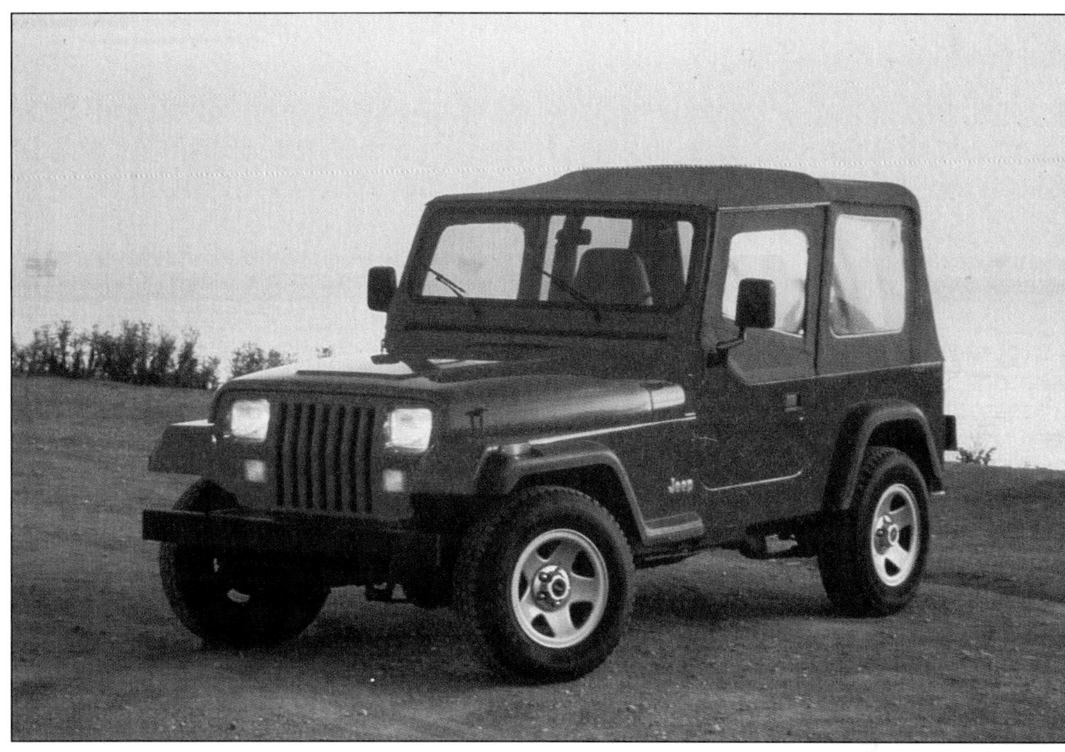

• **Quality/fit/finish: 50%**
They are those of a utility machine, and sheet metal fit is vague and the material ordinary. However, the appearance of upscale versions is less radical than that of the S. Handling the soft top remains complicated.

• **Driver's compartment: 50%**
Recentering the seats and the reclining backs have considerably improved the driving position. Visibility is more favorable with a hardtop than with a convertible, whose dead spots are annoying and numerous. The outside rear-view mirrors are installed in bizarre ways. The main controls are accessible, except those for the radio and foot parking brake. Yet there is room to put a hand brake right on the center tunnel, where it would be very useful for off-road maneuvers. The instrumentation is complete, but the large gate bezels are easier to read than little ones lined up at the right.

• **Performances: 50%**
The 6-cylinder with a manual transmission puts wings on a jeep, which few obstacles can stop in difficult terrain, while on the road it sometimes has too much power. As to the 4-cylinder, it is distinctly less brilliant, even with a manual transmission.

• **Seats: 50%**
Their lateral hold is not ideal, but their padding insures a more comfortable back support than in the past.

• **Depreciation: 50%**
It is low, considering the popularity of this model with the younger set and fans of the great outdoors.

WEAK POINTS

• **Storage: 10%**
Practically nonexistent when the bench seat is occupied. Then it is only accessible from the back of the vehicle, and even then with complex manipulations with either the hard or the softtop.

• **Conveniences: 20%**
The practical side is unknown because the glove box won't conceal more than a pair of mittens, and the optional lockable container mounted between the seats is essential.

• **Interior space: 30%**
In spite of its respectable overall size, space is counted by the cubic inches inside the Wrangler, especially at the back where the width and length are limited, which makes for temporary seating.

• **Noise level: 30%**
Lower with a hardtop than with a soft one, the radio is always superfluous and long trips painful.

• **Handling: 30%**
Even though the stability was improved in the course of the last redesign, it remains controlled by the short wheelbase, the high ground clearance and the suspension harshness. The rebound of the tires on the Saraha incites you to slow down in tight turns and to avoid rapid lane changes.

• **Technical: 40%**
The Wrangler consists of a ladder-type chassis with 4 cross members to which rigid front and rear axles are connected by steel leaf springs. The body also mounts on chassis. Most mechanical elements are those of the Cherokee but specially adapted. The body design is more concerned with its historical side than with aerodynamic finesse, about equal to that of a container. The transmission and transfer box are set up for part-time 4-wheel drive and can also be switched to low or high range. This is far from the sophistication of the Cherokee.

• **Access: 40%**
Easier to get into the front than the back where space is limited and a footrest would be useful.

• **Suspension: 40%**
The short wheelbase, the leaf springs, and damping firmness combine to toss around the occupants. Trips over 100 miles are hard on the bones.

• **Fuel consumption: 40%**
It is never economical, and it would be interesting to offer a diesel for those who really work with a jeep.

• **Safety: 40%**
It is not famous, since the unit body is not perfectly impact-resistant and the front-seat occupants are not protected by air bags. The tubular roll cage reinforcing the structure in case of a roll over is standard.

CONCLUSION

• **Overall rating: 44.5%**
The real Jeep offers a lifestyle, a nonconforming, rational, practical way of getting around. Just like all other fashionable items, it costs more than it is worth. ☹

OWNERS' SUGGESTIONS

-Lighter more compact models
-More cargo space
-A top that is easier to manipulate
-Better general quality
-A more rational dashboard.

SPECIFICATIONS & PRICES

Model	Version/Trim	Body/ Seats	Drag Coef. Cd	Max Payload lb	Wheel-base in	Lgth x Wdthx Hght. inx inx in	Curb weight lb	Towing capacity std. lb	Susp. type ft/rr	Brake type ft/rr	Steering system	Turning diameter ft	Steer. turns nber	Fuel. tank gal.	Standard tire size	Standard power train	PRICES US. $ 1994
-																	
JEEP		**General warranty: 3 years / 36,000 miles; surface corrosion 1 year / 12,000 miles; perforation 7 years / 100,000 miles; road assistance 3 years / 36,000 miles**															
Wrangler	S	2dr.con.2	0.65	1382	93.4	151.9x66.0x71.9	2943	1000	r/r	dc/dr	bal.	32.8	4.2	15	205/75R15	L4/2.5/M5	**$11,390**
Wrangler	SE	2dr.con.4	0.65	330	93.4	151.9x66.0x71.9	3082	1000	r/r	dc/dr	pwr.bal.	32.8	4.2	15	215/75R15	L4/2.5/M5	**$14,454**
Wrangler	Sahara	2dr.con.4	0.65	330	93.4	151.9x66.0x71.9	3082	2000	r/r	dc/dr	pwr.bal.	32.8	3.6	20	225/75R15	L6/4.0/M5	-

See page 393 for complete 1995 Price List

Range Rover-Discovery

Democracy...

Land Rover is better perceived in North America through the Range Rover than by its other highly specialized vehicles. The arrival of the Discovery, which will attack Blazer, Explorer and Grand Cherokee, will allow its image to become more democratic and will sell more cars...

In spite of the push from American and Japanese products which hold the high road, the Range Rover pursues its career in the niche at the top of the range. Range Rover is imported in the form of a 5-door wagon as a standard County or as a County LWB with a long wheelbase, whose equipment is more luxurious and more complete. It is rejoined this year by the Discovery based on the same power train but whose finish and prices are more populist. The standard Range and the Discovery share the 3.9 L V8, while the Range LWB receives the 4.2 L with a 4-speed automatic. The report which follows concerns only the Range Rover.

STRONG POINTS

• Storage **90%**
It holds an impressive amount of baggage even when the bench is used. The opening at the 2-part tailgate does not ease the access.

• Safety: **80%**
Two air bags protect the front-seat occupants and the unit body offers good impact resistance, plus a roll cage.

• Technical: **80%**
Its design is robust, based on a steel chassis with five cross members which accepts two rigid axles and helical springs on the standard and suspension air bags on the LWB. The load leveler maintains the rear ride height regardless of load, and the adjustable pneumatic damping on the LWB version is electronically controlled, thanks to eight sensors located around the body. The full-time 4-wheel drive uses a differential with a viscous coupling to allocate the power between the front and rear drivetrain and to ensure the locking between the axles without any need for driver intervention. For reasons of compatibility with the automatic, the center differential is driven by a chain rather than a shaft. Also, to reduce weight and corrosion risk, the outside panels of the roof, fenders and doors are aluminum. The body, whose lines have been modernized, is not an aerodynamic miracle, and its style does not stray too far from that of its predecessors, which accounts for a good part of its success.

• Quality/fit/finish: **80%**
Interior design is flattering, with typically English leather trim, and the wood appliqué is craftsmanlike. However, while the overall design seems sturdy, some accessories are fragile, such as the dome lights and ancient switches. It is not rare to see the upper part of the tailgate open abruptly after a rear suspension rebound.

• Access: **80%**
In spite of the high sill and the limited door opening angle, it is not hard to get in, thanks to other sizes which are ample.

• Seats: **70%**
They are better designed up front than at the back and the rear bench seat lacks depth to liberate a little more legroom.

DATA

Category: 2-or 4-WD all-terrain multipurpose vehicles.
Class: utility

HISTORIC

Introduced in: 1970: Range 2-door; 1993 Discovery (imported in 1995).
Modified in: 1981: Range 4-door; 1986: injection; 1993: suspension.
Made in: Solihull, England.

RATINGS

Safety: 80 %
Satisfaction: 75 %
Depreciation: 52 %
Insurance: 3.8 % ($1,340)
Cost per mile: $1.00

NUMBER OF DEALERS

In U.S.A.: 84

SALES ON THE AMERICAN MARKET

Model	1992	1993	Result
Land Rover	-	4,907	

MAIN COMPETITORS

CHEVROLET-GMC Suburban, FORD Explorer, JEEP Grand Cherokee, TOYOTA Land Cruiser.

EQUIPMENT

	LAND ROVER Discovery base	LAND ROVER Range Rover base	LAND ROVER Range Rover LWB
Automatic transmission:	O	S	S
Cruise control:	S	S	S
Power steering:	S	S	S
ABS brake:	S	S	S
Air conditioning:	S	S	S
Air bag:	S	S	S
Leather trim:	O	S	S
AM/FM/ Radio-cassette:	S	S	S
Power door locks:	S	S	S
Power windows:	S	S	S
Tilt steering column:	S	S	S
Dual power mirrors:	S	S	S
Intermittent wipers:	S	S	S
Light alloy wheels:	S	S	S
Sunroof:	O	O	S
Anti-theft system:	S	-	-

S : standard; O : optional; - : not available

COLORS AVAILABLE

Exterior: White, Green, Black, Bronze, Blue.
Interior: Tan.

MAINTENANCE

First revision: 3,000 miles
Frequency: 6 months/6,000 miles
Diagnostic plug: No

WHAT'S NEW IN 1995 ?

- **Optional air suspension (Range Rover).**
- **New Discovery model.**

Model/ versions *: standard	ENGINE Type / timing valve / fuel system	Displacement cu/in	Power bhp @ rmn	Torque lb.ft @ rpm	Compres. ratio	TRANSMISSION Driving wheels / transmission	Final ratio	PERFORMANCE Acceler. 0-62 mph s	Stand. 1/4 mile s	Stand. 5/8 mile s	Braking 62-0 mph ft	Top speed mph	Lateral acceler. G	Noise level dBA	Fuel economy mpg city	Fuel economy hwy	Gasoline type / octane
base	V8* 3.9 OHV-16-MPFI	240	182 @ 4750	232 @ 3100	9.35 :1	all-A4*	3.54	10.8	18.2	35.7	180	109	0.69	66	15.0	20.0	M 89
LWB	V8* 4.2 OHV-16-MPFI	261	200 @ 5500	251 @ 3250	8.95 :1	all-A4*	3.54	9.9	17.4	33.2	164	112	0.71	66	15.0	20.0	M 89
Discovery	V8* 3.9 OHV-16-MPFI	240	182 @ 4750	232 @ 3100	9.35 :1	all-M5*	3.54	NA									
						all-A4	-	NA									

Range Rover-Discovery

LAND ROVER Discovery

LAND ROVER Range Rover

LAND ROVER Range Rover

• Conveniences: 70%
The storage is well-distributed at the front of the cab where the glove box, a dashboard tray and a lock box in the center console are amply sized. At the rear, a compartmented divider closes off the storage area and allows carrying numerous items.

• Suspension: 70%
Helical springs are better at soaking up bumps and ensure improved comfort, compared to leaf springs used by its competitors, rare at this level of price and quality.

• Driver's compartment: 70%
The driver sits high and enjoys good visibility through the large glazed area. The dashboard is neat and well-organized. Some controls and accessories are neither conventional nor within hand's reach. At last the steering wheel is a little larger and the pedal assembly off-centered.

• Satisfaction: 60%
The reliability still does not inspire total confidence and the dealer network is still embryonic.

• Insurance: 60%
The premium on the Range Rover is close to that of a Jaguar, which says it all.

• Steering: 60%
Its very high assistance is compounded by a lack of center feel. A high number of turns, lock-to-lock, which is marvelous off road, penalizes maneuverability, and its turning diameter is a bit too large.

• Interior space: 50%
The interior is vast and luminous and can accept five people thanks to its well-proportioned sizing. However, the wheelbase of the long LWB earns larger legroom at the rear seats than on the standard County.

• Handling: 50%
Four wheel drive allows excellent cornering traction and its proverbial versatility lets it switch from turnpikes, where it can run head to head against many cars, to little backwoods roads where it makes many so-called specialized vehicles look ridiculous. Off road, its aptitude for crossing obstacles is surprising, thanks to its ground clearance and large entry and exit angles as well as good skid plate protection. On the road, its ability to stay the course is less edifying because it is sensitive to pavement and crosswind. The air suspension introduced last year is more effective at modifying the ground clearance than to improving road holding.

• Performance: 50%
The V8 engine is not too much for getting this 2-ton machine to move at car speeds on the road, but thanks to enough gear reduction, takeoffs and passings are sufficient. The engine is smooth and the transmission ratios well-staged, which allows it to keep the pace.

• Noise level: 50%
In spite of some tire, transmission and wind noises, the sound level remains positive, thanks to an efficient soundproofing.

WEAK POINTS

• Price/equipment: 00%
One must want to be distinguished at any price to face the sticker for this deluxe truck which makes you pay full tab for its style, equipment and versatility.

• Fuel consumption: 10%
It is never economical and can easily drop to 14 miles per gallon.

• Braking: 30%
It surprises by the precise modulation and secure balance with the perfected ABS system. Its efficiency is mediocre and the stopping distances are long in a panic stop because the weight is too high for the small brakes.

• Depreciation: 40%
It fluctuates a lot because potential buyers do not rush in and it is not rare to lose more than 60% at the end of three years.

CONCLUSION

• Overall rating: 57.5%
The Discovery will change the elitist's perception of the Land Rover and will allow it to sell more cars than can be produced with the Range. The relationship with BMW will certainly not detract from this maneuver.

SPECIFICATIONS & PRICES

Model	Version/Trim	Body/ Seats	Box Length ft	Drag coef. Cd	Wheel-base in	Lgth x Wdthx Hght. inx inx in	Curb weight lb	Towing capacity std. lb	Susp. type ft/rr	Brake type ft/rr	Steering system	Turning diameter ft	Steer. turns nber	Fuel. tank gal.	Standard tire size	Standard power train	PRICES US. $ 1994
LAND ROVER			General warranty: 3 years / 70 000 km; antipollution: 5 years / 50,000 miles; perforation corrosion: 6 years / unlimited.														
Range Rover		County	5dr.wgn.5	-	100.0	175.1x71.4x70.8	4409	2976	r/r	dc/dc/ABS	pwr.s&w.	39.4	3.375	23.4	205R16	V8/3.9/A4	**$46,900**
Range Rover	LWB	County	5dr.wgn.5	-	108.0	183.0x71.4x70.8	4575	2976	r/r	dc/dc/ABS	pwr.s&w.	39.4	3.375	23.4	205R16	V8/4.2/A4	**$50,200**
Discovery		base	5dr.wgn.5/7	-	100.0	178.7x70.6x77.4	4378	2976	r/r	dc/dc/ABS	pwr.s&w.	39.4	3.3	23.4	235/70R16	V8/3.9/A4	-

See page 393 for complete 1995 Price List

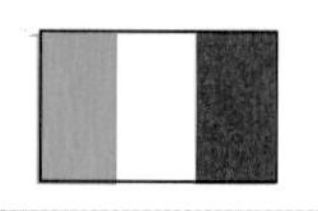

Happy anniversary...

After 25 years of tribulations, incertitude and anxieties, the small Bologna builder celebrates its anniversary with great pomp by producing the fastest, most powerful car in a history which has given us machines that are exceptional and unforgettable

To celebrate its 25th anniversary, Lamborghini releases a special SE-Series of the Diablo, which is one of the fastest production cars in the world: 212 mph. As its ancestors, the Miura and the Countach, they were designed by Marcello Gandini. It is a 2-door coupe with two seats and 2- or 4-wheel drive.

STRONG POINTS

• Performance: 100%
With nearly 500 hp for its 3,525 pounds, the V12 acts as a catapult. The standing starts as well as passings are fantastic as long as traction is perfect, which explains the presence of a 4-wheel drive. The gear shift of the manual transmission is slow and hard, while the clutch pedal is very civilized. Torque is abundant and you can accelerate from 40 mph in fifth without groans from the engine, which saves you a few shifts.

• Handling: 100%
It is especially at the level of handling that the 4-wheel drive justifies its presence, because it does not improve handling in a turn where the car remains basically understeering. At the limit, the rear end takes off as dryly as on the standard version. You have to also note the slowness with which the viscous coupling reacts in a slalom. On the dry, the car holds by itself, whether the turns are wide or tight, thanks to its huge tires; while in a straight line, stability deteriorates over 160 mph because of the negative air loading and this calls for a rear spoiler offered as an option.

• Technical: 90%
Its spectacular aesthetics include doors that swing out, a centrally located 12-cylinder engine, and a transmission which intrudes into the cab within the center tunnel. This allows a weight distribution of 49/51 front to rear. Unlike the Bugatti EB110, the Lamborghini does not draw from high-tech aircraft technology. The chassis is simply tubular, the body is composed of hand-fitted aluminum panels and only the front and rear bumpers, hood and trunk and center tunnel are molded with carbon fibers. The independent suspensions have their antidive and antilift geometry, and the electronically adjusted Koni shock absorbers stiffen as a function of speed. Below 80 mph, the suspension is soft but it progressively firms up at 120, 150 mph and beyond. The brakes use competition style vented discs and do not include ABS.
In the 4-wheel drive version, the center tunnel accommodates the drive shaft, which runs the front drivetrain. It only transmits from 0 to 25% of the engine power so that the Diablo maintains its rear wheel feel and pleasurability, even when traction conditions are precarious. The engine, at the fine point of technology, develops 492 hp standard and to 525 hp on the SE in rear wheel drive. The manual transmission is a 5-speed ZF with a competition clutch.

• Quality/fit/finish: 90%
The assembly and finish of the Diablo are more strict than those of the Countach thanks to manufacturing techniques brought in by Chrysler.

DATA

Category: rear-wheel drive or AWD GT coupes.
Class: exotic

HISTORIC

Introduced in: 1990
Modified in: 1994: VT; 1995:
Made in: Santa Agata Bolognese, Bologna, Italy.

RATINGS

Safety: NA
Satisfaction: NA
Depreciation: NA
Insurance: 3.0 % ($7,150)
Cost per mile: $3.25

NUMBER OF DEALERS

In U.S.A.: 14

SALES ON THE AMERICAN MARKET

Model	1992	1993	Result
Diablo	156	156	=

MAIN COMPETITORS

BUGATTI EB110, FERRARI F512M , VECTOR W8.

EQUIPMENT

LAMBORGHINI	Diablo VT
Automatic transmission:	-
Cruise control:	-
Power steering:	S
ABS brake:	-
Air conditioning:	S
Air bag:	-
Leather trim:	S
AM/FM/ Radio-cassette:	S
Power door locks:	S
Power windows:	S
Tilt steering column:	S
Dual power mirrors:	s
Intermittent wipers:	S
Light alloy wheels:	S
Sunroof:	-
Anti-theft system:	-

S : standard; O : optional; - : not available

COLORS AVAILABLE

Exterior: Red, Yellow, White, Black, Metallic Gray, Blue.
Interior: Tan, White, Black.

MAINTENANCE

First revision: 2,000 miles
Frequency: 6 months/6,000 miles
Diagnostic plug: Yes

WHAT'S NEW IN 1995 ?

- A convertible will be introduced in the course of the year.
- Special limited edition SE will underscore the 25th birthday of the marque.

ENGINE					TRANSMISSION			PERFORMANCE										
Model/ versions *: standard	Type / timing valve / fuel system	Displacement cu/in	Power bhp @ rmn	Torque lb.ft @ rpm	Compres. ratio	Driving wheels / transmission	Final ratio	Acceler. 0-62 mph s	Stand. 1/4 mile s	Stand. 5/8 mile s	Braking 62-0 mph ft	Top speed mph	Lateral acceler. G	Noise level dBA	Fuel economy mpg city	hwy	Gasoline type / octane
Diablo VT	V12 5.7 DOHC-48-MPFI	348	492 @ 7000	428 @ 5200	10.0:1	rear-M5*	3.83	5.8	13.8	20.7	164	186	0.92	77	10.0	17.0	S 91
Diablo SE	V12 5.7 DOHC-48-MPFI	348	525 @ 7100	417 @ 5900	10.0:1	rear-M5	3.83	4.0 (builder's datas)				206	0.95	79	8.0	16.5	S 91

• **Access:** 70%
In spite of its low height, it is not difficult to get in because the doors liberate sufficient space and the interior is large.
• **Steering:** 70%
That of the VT is power-assisted and has a little higher number ratio than the conventional Diablo, which is direct and precise like a scalpel. In both instances, maneuverability is limited by the very low ground clearance and the large steering diameter.
• **Seats:** 70%
Their shape provides good lateral hold, but lumbar support is not adjustable and the soft padding soon makes it uncomfortable. Why not a real Recaro?
• **Suspension:** 70%
Surprise, it is almost comfortable, and substantially less hard than we imagined, and on the VT the shock absorber adjustments add to driving pleasure.
• **Safety:** 70%
The Diablo unit body resists impact tests, but at the moment when this was written it is unknown if air bags will finally be installed in the '95 version.
• **Driver's compartment:** 60%
Since our last test, the dashboard became more ergonomic and lower, and some controls have been repositioned. Front view is less obstructed and toward 3/4 or rear visibility is very limited. In spite of an adjustable steering column which moves with the gauge panel, a comfortable driving position is hard to find. The seat back does not recline, and controls on the console are far from the hand. It is hard to understand why on this type of car, the pilot is not coddled a little better.
• **Depreciation:** 60%
It will fluctuate until the recession ends.

WEAK POINTS

• **Price/equipment:** 00%
It is very exagerated, even if the technical level is sophisticated. In addition, the general equipment is reduced to a strict minimum.
• **Fuel consumption:** 00%
It is at the level of the performance, and the fuel tank gives a range of 250 miles (in smooth driving).
• **Trunk:** 20%
Located between the front wheels it has replaced the spare tire which was replaced by a flat-tire canister.
• **Interior space:** 30%
The inside is fairly open, thanks to the dashboard shape, and the size is impressive, but big folks will encounter a low ceiling.
• **Noise level:** 30%
The song of the engine becomes terrifying above 125 mph and has no problem masking those from the tires and wind.
• **Conveniences:** 30%
They are rare because storage space is symbolic and you will find neither cup holders nor courtesy mirrors...
• **Insurance:** 50%
Around $7,000 is the Diablo premium for a year.
• **Braking:** 50%
Of a competition style, it is effective but difficult to modulate, and the pedal is hard. The 1.6 tons of Diablo stops in less than 165 ft from 60 mph, but an ABS would bring more security on rainy days.
• **Satisfaction:** 50%
No one who is normally constituted will confess to having had problems with his preferred investment, and the dealer has amnesia, so...

LAMBORGHINI Diablo SE

CONCLUSION

• **Overall rating:** 55%
Suffering and pleasure are intermittently intermixed on the Diablo, which is a masochist's dream come true.

LAMBORGHINI Diablo SE

SPECIFICATIONS & PRICES

Model	Version/Trim	Body/ Seats	Interior volume cu.ft.	Trunk volume cu ft.	Drag coef. Cd	Wheel-base in	Lgth x Wdthx Hght. inx inx in	Curb weight lb	Towing capacity std. lb	Susp. type ft/rr	Brake type ft/rr	Steering system	Turning diameter ft	Steer. turns nber	Fuel. tank gal.	Standard tire size	Standard power train	PRICES US. $ 1994
LAMBORGHINI		**Warranty: 2 years / 7,500 miles.**																
Diablo	VT	2dr.cpe. 2	NA	5.0	0.30	104.3	175.6x80.3x43.5	3737	NR	i/i	dc/dc	pwr.r/p.	42.7	3.0	26.4	235/40ZR17 335/35ZR17	V12/5.7/M5	**$239,00(**
Diablo	SE	2dr.cpe. 2	NA	5.0	0.30	104.3	177.4x80.3x43.5	3197	NR	i/i	dc/dc	pwr.r/p.	42.7	3.0	26.4	235/40ZR17 335/35ZR17	V12/5.7/M5	**NA**

See page 393 for complete 1995 Price List

The standard...

The phenomenon of leasing cars to enterprises or to professionals has given rise without precedence to middle class deluxe cars. An avalanche of models inundates this lucrative market. The Lexus' success sets the standards.

The ES300 is the most-sold car in the Lexus range, in which it constitutes the standard model. It is a 3-volume 4-door sedan sold in a unique finish. It is narrowly derived from the Camry from which it picks up the platform and power train as well as some elements of body and glazing. This year it received a few cosmetic retouches designed to reinforce its personality and to differentiate it a little more strongly from its roots.

STRONG POINTS

•Safety: **90%**
The structural rigidity is remarkable and it offers good impact resistance. The front seat occupants are protected by two air bags.

• Technical: **90%**
The ES300 takes its main dimensions from the Camry, except length and width which differ. Its advanced appearance gives it an honest aerodynamic finesse with a very good Cd of 0.32. The steel unit body has an independent suspension and disc brakes on all four wheels. The power train and the rear are mounted on cradles isolated from the body by rubber mounts. A large portion of the body sheet metal is galvanized and coated with a particularly effective insulating material. A 3.0 L. V6 introduced last year develops 188 hp and is coupled to a 4-speed automatic transmission with electronic controls.

• Satisfaction: **90%**
The number of very satisfied buyers leaves little place for those who are just average, middle or not at all.

• Quality/fit/finish: **90%**
The general presentation is anonymous but very refined, the assembly and finish are careful, and the materials are of excellent quality.

DATA

Category: front-wheel drive luxury sedans.
Class: 7

HISTORIC

Introduced in: 1992
Modified in: 1994: new engine.
Made in: Tahara, Japan.

RATINGS

Safety: 90 %
Satisfaction: 90 %
Depreciation: 48 %
Insurance: 3.8 % ($860)
Cost per mile: $0.62

NUMBER OF DEALERS

In U.S.A.: 167

SALES ON THE AMERICAN MARKET

Model	1992	1993	Result
ES300	39,652	35,655	-10.1 %

MAIN COMPETITORS

ACURA Vigor, AUDI A4 & A6, BMW 3-Series, INFINITI G-20, MAZDA Millenia, MERCEDES BENZ C280, NISSAN Maxima, SAAB 900 & 9000, VOLVO 850.

EQUIPMENT

LEXUS ES300 Automatic transmission:	base
Cruise control:	S
Power steering:	S
ABS brake:	S
Air conditioning:	S
Air bag:	S
Leather trim:	S
AM/FM/ Radio-cassette:	O
Power door locks:	S
Power windows:	S
Tilt steering column:	S
Dual power mirrors:	S
Intermittent wipers:	S
Light alloy wheels:	S
Sunroof:	S
Anti-theft system:	O

S : standard; O : optional; - : not available

COLORS AVAILABLE

Exterior: White, Black, Amethyst, Red, Metallic Beige, Quartz Pink, Emerald.
Interior: Black, Ivory, Gray, Taupe, Oak.

MAINTENANCE

First revision: 3,000 miles
Frequency: 3,000 miles
Diagnostic plug: Yes

WHAT'S NEW IN 1995 ?

-Grill, headlights and fog lights redesigned.

Model/ versions *: standard	ENGINE Type / timing valve / fuel system	Displacement cu/in	Power bhp @ rmn	Torque lb.ft @ rpm	Compres. ratio	TRANSMISSION Driving wheels / transmission	Final ratio	PERFORMANCE Acceler. 0-62 mph s	Stand. 1/4 mile s	Stand. 5/8 mile s	Braking 62-0 mph ft	Top speed mph	Lateral acceler. G	Noise level dBA	Fuel economy mpg city	hwy	Gasoline type / octane
ES 300	V6*3.0 DOHC-24-MPFI	181	188 @ 5200	203 @ 4400	10.5:1	front - A4	3.72	9.2	16.2	29.3	125	131	0.80	64	22.0	31.0	R 87

• Driver's compartment: 80%
The most comfortable driving position is easy to find, thanks to numerous seat and steering column adjustments. The dashboard, typically Lexus, provides more forward space. Its layout is ergonomic, with well laid out controls and digital gauges arranged conventionally. Visibility is excellent all around, the rear-view mirrors are well-dimensioned and the roof pillars fairly small.

• Access: 80%
It is easy to seat yourself in the front or rear of the ES300, and most main dimensions are ample.

•Seats: 80%
They resemble those of the other models of the make. The shape provides good lateral and lumbar support but the padding is relatively firm.

• Suspension: 80%
Soft enough to favor comfort but not so much as to detract from handling, it represents one of the best compromises on the market. The two cradles that isolate the drive train from the body attenuate vibrations and shake, and give you the impression of being in a car of much larger size.

• Insurance: 80%
Because the ES300 addresses itself mostly to reasonable people, its index is low and the premium affordable.

• Noise level: 80%
The Lexus has become the champion of soundproofing and is one of the most effective in the world. The ES300 has not escaped this principle. Drive train and rolling noise are well-attenuated and only small wind streams murmur around the windshield.

• Conveniences: 70%
Storage spaces are sufficient in number and volume and a trap in the rear armrest allows carrying long objects such as skis.

• Fuel consumption: 70%
More economical than the Camry or some competitive models such as the Mercedes C220 or the Nissan Maxima, it stays well within acceptable levels.

• Performance: 70%
Thanks to a favorable power-to-weight ratio (17.9 lbs/hp) the latest engine has improved both standing starts and passings which are crisper, but some competitors do distinctly better with equivalent design.

• Handling: 70%
It offers a very secure feeling in a straight line or in wide turns and the ES300 follows easily within tight curves. Its drivability is not more inspired than that of its cousin, but traction in curves is good and an understeering temperament does not show up till high speeds. Otherwise it remains neutral most of the time.

• Steering: 70%
Fast, with a somewhat positive assist, like that of the Camry, it is a little vague in the center and lightens up at high speeds.

• Trunk: 60%
Vast and simple in shape, it can easily accept much baggage but cannot be transformed, as on the Camry, by tipping the rear seat bench. The same does not go for the trunk which has a much higher sill than the Camry, complicating baggage handling.

• Interior space: 60%
Four people will be at ease on a long trip and the fifth can join when necessary. Head and legroom are sufficient, even though the seats are thicker than on the Camry.

• Brakes: 60%
Progressive and balanced, it remains effective because stopping distances stay under 130 ft in spite of the ABS which makes the path straight. Fade-resistance is satisfactory.

• Depreciation: 50%
A reputation for reliability and relatively modest maintenance cost makes the ES300 a good second hand purchase. That is the reason for the resale value staying above average and often better than those of more prestigious German models.

WEAK POINTS

• Price/equipment: 30%
Toyota products cost more because of the manufacturing quality and the value of the yen pushes them into the dangerous competitive zone. Even if the ES300 equipment is complete, it does not include many options and is overvalued by about 9%.

CONCLUSION

• Overall rating: 72.5%
The little Lexus has created a market segment in which it serves as a reference as much for its Japanese competitors as for Europeans, which find it very disturbing. Technically, very homogeneous, it provides a discreet luxury, an effective comfort and a trouble-free operation which explain its success. ☺

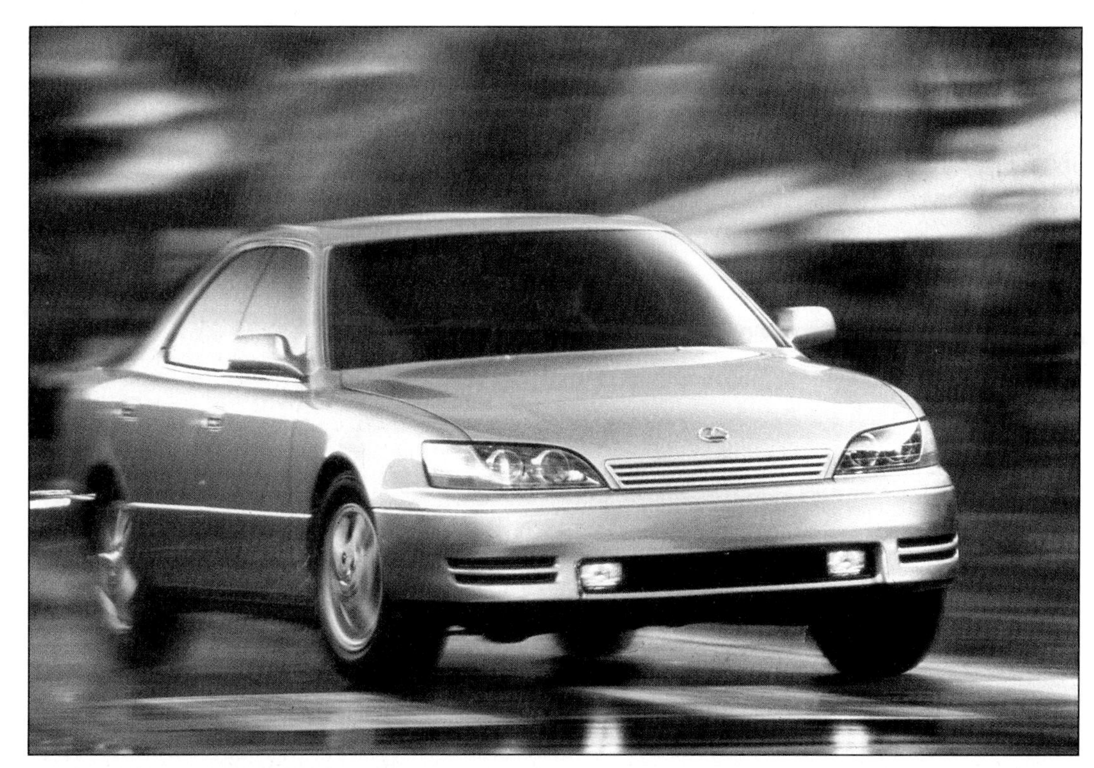

SPECIFICATIONS & PRICES

Model	Version/Trim	Body/ Seats	Interior volume cu.ft.	Trunk volume cu ft.	Drag coef. Cd	Wheel-base in	Lgth x Wdthx Hght. inx inx in	Curb weight lb	Towing capacity std. lb	Susp. type ft/rr	Brake type ft/rr	Steering system	Turning diameter ft	Steer. turns nber	Fuel. tank gal.	Standard tire size	Standard power train	PRICES US. $ 1994
LEXUS		**General warranty: 4 years / 50,000 miles; Power train: 6 years / 70,000 miles; corrosion perforation: 6 years / unlimited & road assistance.**																
ES300	base	4dr.sdn.5	95	14.3	0.32	103.1	187.8x70.0x53.9	3373	2000	i/i	dc/dcABS	pwr.r/p.	36.7	2.7	18.5	205/65R15	V6/3.0/A4	**$31,200**

See page 393 for complete 1995 Price List

Between two...

The GS300 was interposed two years ago between the two most significant models of the Lexus range. In spite of its original allure, it doesn't bring much because it has less performance than an ES300 and finally, not much more room. So what?

Since 1993, the GS300 fits between the ES300 and the LS400. The design of this 4-door 3-volume sedan is owed to Italian stylist Giugiaro. It's the first time that a car manufactured by Toyota was not designed in its own styling bureau, but was instead assigned on the basis of a competition The GS300 is available in only one version powered by an in-line 3.0 L 6 cylinder with a 4-speed automatic.

STRONG POINTS

• Safety: **90%**
Its unit body is very rigid and the front seat occupants are protected by 2 air bags and pre-tensionable belts. This device, triggered by the air bag sensors, tensions both the shoulder harness and the seat belt.

• Satisfaction: **90%**
Hidden quality was Toyota's philosophy and it pays off at the level of the buyer who always has the feeling of getting more for his money.

• Driver's compartment: **90%**
The best driving position is easy to discover, thanks to electrically controlled steering and seat adjustments, both simple and effective. Visibility is good in spite of the "C" pillar thickness. The dashboard is a model of ergonomics because the commands and controls are well laid out, with the exception of the well hidden cruise control switch.

• Access: **80%**
It is as easy to get a seat at the front as at the rear and the large doors open wide.

• Seats: **80%**
They gain a very effective lateral and lumbar support but their contouring could be more pronounced. Padding is firm - just right - allowing long trips without fatigue.

• Suspension: **80%**
Handling and comfort are good, with ample wheel travel to softly absorb most of the road defects. However, the front end has a tendency to rebound repeatedly on low amplitude ripples, giving the feeling of a corrugated road ride.

• Noise level: **80%**
The Lexus counts among the best insulated cars as of this moment, thanks to some very elaborate work.

• Quality/fit/finish: **80%**
The body does not attract attention because its presentation is discrete. The luxury is more evident for the interior, which has leather trim and varnished wood appliqués. Assembly is strict, finish is a mania and the quality of material above any suspicion.

• Steering: **80%**
Even though very precise, its power assist is very strong and makes it sensitive, while the steering ratio and turning diameter gain good maneuverability.

• Technical: **80%**
Unlike the ES300 which is a front wheel drive, the GS300 has a classic

DATA

Category: rear-wheeldrive luxury sedans.
Class : 7

HISTORIC

Introduced in: 1993
Modified in:
Made in: Tahara, Japan

RATINGS

Safety: 90 %
Satisfaction: 92 %
Depreciation: 44 % (2 years)
Insurance: 3.1 % ($1,000)
Cost per mile: $0.94

NUMBER OF DEALERS

In U.S.A.: 167

SALES ON THE AMERICAN MARKET

Model	1992	1993	Result
GS300	NA	19,164	

MAIN COMPETITORS

ACURA Legend, AUDI A6, BMW 5 Series, INFINITI J30, MERCEDES BENZ E Class, SAAB 9000, VOLVO 960.

EQUIPMENT

LEXUS GS300	base
Automatic transmission:	S
Cruise control:	S
Power steering:	S
ABS brake:	S
Air conditioning:	S
Air bag (2):	S
Leather trim:	S
AM/FM/ Radio-cassette:	S
Power door locks:	S
Power windows:	S
Tilt steering column:	S
Dual power mirrors:	S
Intermittent wipers:	S
Light alloy wheels:	S
Sunroof:	O
Anti-theft system:	O

S : standard; O : optional; - : not available

COLORS AVAILABLE

Exterior: White, Silver, Blue, Red, Black, Beige, Jade, Indigo.
Interior: Leather: Ivory, Gray, Spruce.

MAINTENANCE

First revision: 4,000 miles
Frequency: 4,000 miles
Diagnostic plug: Yes

WHAT'S NEW IN 1995 ?

- A few minor cosmetic changes.

Model/ versions *: standard	ENGINE Type / timing valve / fuel system	Displacement cu/in	Power bhp @ rmn	Torque lb.ft @ rpm	Compres. ratio	TRANSMISSION Driving wheels / transmission	Final ratio	PERFORMANCE Acceler. 0-62 mph s	Stand. 1/4 mile s	Stand. 5/8 mile s	Braking 62-0 mph ft	Top speed mph	Lateral acceler. G	Noise level dBA	Fuel economy mpg city	Fuel economy hwy	Gasoline type / octane
GS300	L6* 3.0 DOHC-24-MPFI	183	220 @ 5800	210 @ 4800	10.0 :1	rear - A4	4.083	9.7	17.7	31.7	125	136	0.80	65	17.5	25.5	M 89

architecture inherited from the Cressida. The engine is at the front and the drive wheels at the rear. The 3.0 L in-line 6 cylinder surprises by a displacement equal to that of the ES300. It has however, two overhead cams and develops 220 hp, for a weight of 3,660 pounds (16,5 lbs/hp). The automatic transmission is electronically controlled and identical to that which equips the 400 but the gear ratios are adapted to get the most out of the in-line 6 power. The steel unit body received a very successful styling look, as well as the better low drag coefficient of 0.31. The suspension is independent, the front discs are vented and the rear ones, which do not have to dissipate as much heat, are solid. The ABS system insures stability during stops and also acts as a traction control allowing maximum traction at all times. To better filter out vibrations coming from the engine, steering, differential, and suspension, the drive trains are mounted on cradles which isolate them from the body.

• Insurance: 80%
High price brings a premium to match, but it is not exaggerated.

• Conveniences: 80%
The front shoulder harness is adjustable in height and the Nakamishi sound system is of excellent quality. Air conditioning temperature control is precise to within 1/2 of a degree,. with precise distribution. However, the size and number of storage areas are calculated too tight.

• Interior space: 70%
The interior will only accept 4 grownups in comfort because the rear center tunnel, the bench seat shape and the shoulder harness arrangement prevent carrying a 5th passenger.

• Handling: 70%
Mainly oriented toward the comfort, the suspension presents a moderate roll but body motions are well-controlled and the GS300 posts a reassuring neutral steer that only manifests itself in extreme conditions.

• Performance: 60%
In spite of its torque and willingness, the engine doesn't develops enough power to give this sedan the performance it deserves. The 0 to 60 time painfully edges under 10 seconds and the accelerations are soft. The turbo version, marketed elsewhere, confers on this car a temperament that better justifies the price and differentiates itself from the ES300, which is not the case.

• Braking: 60%
It is excellent because the panic stopping distances are all below 40 meters, even after repeated stops, which test the brake fade. The small ABS wheel blockage is more annoying than dangerous.

• Trunk: 50%
The regular shape makes its use efficient, but the contents are limited and only one golf bag can be brought since storage will not extend toward the interior.

• Fuel consumption: 50%
It is average, considering car weight and cylinder displacement.

WEAK POINTS

• Price/equipment: 00%
Even though its equipment is very complete, the price requested from the GS300 is not proportional to performance when one compares the GS to more prestigious European competitors.

• Depreciation: 40%
It is larger than predicted, which surprises, considering the durability of the product.

CONCLUSION

• Overall rating: 69.5%
Remarkable for more than one point of view, the GS300 lacks character too much to justify its cost. ☺

SPECIFICATIONS & PRICES

Model	Version/Trim	Body/ Seats	Interior volume cu.ft.	Trunk volume cu ft.	Drag coef. Cd	Wheel-base in	Lgth x Wdthx Hght. inx inx in	Curb weight lb	Towing capacity std. lb	Susp. type ft/rr	Brake type ft/rr	Steering system	Turning diameter ft	Steer. turns nber	Fuel. tank gal.	Standard tire size	Standard power train	PRICES US. $ 1994
LEXUS		General warranty: 4 years / 50,000 miles; Power train: 6 years / 70,000 miles; corrosion perforation: 6 years / unlimited & road assistance.																
GS300		4dr.sdn. 4/5	104	13	0.31	109.4	194.9x70.7x55.1	3660	2000	i/i	dc/dcABS	pwr.r/p.	36.1	3.23	21.1	215/60VR16	L6/3.0/A4	$41,100

See page 393 for complete 1995 Price List

Continuity...

To affirm that the LS400 has been renewed, would be a little bold. Let's simply say that it has been refined in a number of ways. But that on the whole, its definition remains identical with what it was. It is always more difficult to do better when you have reached the peak.

Only the experts will be able to detect the details which differentiate the LS and SC from 94. They have been noticeably improved but the most significant changes are invisible. The LS 400 is a 3-volume 4-door sedan and the SC is a 3-volume 2-door coupe which share the same platform and power train within a few details. The SC300 takes the appearance of the 400 but borrows the drive train of the GS 300. The 3 models are marketed fully equipped.

STRONG POINTS

• Satisfaction: 95%
The Lexus comes in largely at the head of all owners' polls which say they are extremely satisfied with their acquisition.

• Quality/fit/finish: 90%
Quiet and in good taste, the presentation, assembly and finish are strict and the quality of the materials studied to prevent aging.

• Safety: 80%
It is optimal in terms of structural impact resistance which has been improved by a safety cage around the interior. There are two standard air bags, ABS and a traction control.

• Technical: 90%
In relation to the model it replaces, the LS has received a number of modifications since 90% of its components are new or have been improved and weight has been trimmed by 200 pounds. The wheel-base is longer, the engine suspension, and interior has been subtlely refined. The most visible modification has been at the level of the grill and hood whose design is different. The style of the LS remains a happy blend of Audi, BMW and Mercedes styles. It has the lowest aerodynamic drag in the market: 0.28 (lower number means more slipperiness). The appearance of the SC coupes does not look like anything known. Their steel unit body is double walled, and any exposed parts are made of galvanized steel. The suspension is independent and the ventilated disc brakes are used on all four wheels together with ABS and traction control. The engine is a V8 4.0 L, very modern, quiet, with ample power and torque. It connects to a 4-speed automatic with electronic management and can function in 2 modes, sport or economy. The final drive of the SC400 is shorter than that of the sedan to give it performance related to its vocation. The SC300 is equipped with an in-line 3.0 L 6 cylinder from the GS and connects to a manual 5-speed.

• Insurance: 90%
The very low rate is compensated by the very high price to give a premium to match, but justified.

• Driver's compartment: 90%
Visibility is good from all angles and the driver is comfortably seated even with no separate lumbar adjustment. The dashboard is a model of its kind with a simple and functional aesthetic presentation.

• Seats: 80%

DATA

Category: rear-wheel drive luxury sedans and coupes.
Class : 7

HISTORIC

Introduced in: 1990: LS400; 1991: SC400.
Modified in: 1993: SC300; 1995: body.
Made in: Tahara, Japan.

RATINGS

		LS	SC	
Safety:		90 %	90 %	
Satisfaction:		94 %	94 %	
Depreciation:		61 %	56 %	
Insurance:	($1,135)	2.8 %	2.9 %	($1,065)
Cost per mile:		$0.94	$0.94	

NUMBER OF DEALERS

In U.S.A.: 167

SALES ON THE AMERICAN MARKET

Model	1992	1993	Result
LS	32,561	23,783	- 27.0 %
SC	20,677	16,075	- 22.3 %

MAIN COMPETITORS

LS: AUDI A8, BMW 7 Series, CADILLAC Seville, INFINITI Q45, JAGUAR XJ6, LINCOLN Continental, MERCEDES 420SE, OLDSMOBILE Aurora.
SC: BMW 850, CADILLAC Eldorado Touring, JAGUAR XJS, LINCOLN Mark VIII, MERCEDES-BENZ SL, PORSCHE 928.

EQUIPMENT

LEXUS	LS400	SC400	SC300
Automatic transmission:	S	S	-
Cruise control:	S	S	S
Power steering:	S	S	S
ABS brake:	S	S	S
Air conditioning:	S	S	S
Air bag (2):	S	S	S
Leather trim:	S	S	S
AM/FM/ Radio-cassette:	S	S	S
Power door locks:	S	S	S
Power windows:	S	S	S
Tilt steering column:	S	S	S
Dual power mirrors:	S	S	S
Intermittent wipers:	S	S	S
Light alloy wheels:	S	S	S
Sunroof:	S	S	S
Anti-theft system:	S	S	S

S : standard; O : optional; - : not available

COLORS AVAILABLE

Exterior: White, Silver, Black, Champagne, Beige, Indigo, Taupe.
Interior: Leather: Ivory, Gray, Taupe, Black, Blue, Spruce.

MAINTENANCE

First revision: 4,000 miles
Frequency: 4,000 miles
Diagnostic plug: Yes

WHAT'S NEW IN 1995 ?

- Body face lift.
- Higher performance V8.
- Increased interior volume and trunk (LS).

Model/ versions *: standard	ENGINE Type / timing valve / fuel system	Displacement cu/in	Power bhp @ rmn	Torque lb.ft @ rpm	Compres. ratio	TRANSMISSION Driving wheels / transmission	Final ratio	PERFORMANCE Acceler. 0-62 mph s	Stand. 1/4 mile s	Stand. 5/8 mile s	Braking 62-0 mph ft	Top speed mph	Lateral acceler. G	Noise level dBA	Fuel economy mpg city	hwy	Gasoline type / octane
SC300	L6* 3.0 DOHC-24-SMPFI	183	225 @ 6000	210 @ 4800	10.0 :1	rear - M5	4.083	9.5	17.0	30.8	135	125	0.81	66	18.2	26.5	S 91
SC400	V8* 4.0 DOHC-32-MPEFI	242	250 @ 5600	260 @ 4400	10.0 :1	rear - A4	3.916	7.0	15.0	27.6	125	143	0.85	64	18.0	25.7	S 91
LS 400	V8* 4.0 DOHC-32-EFI	242	260 @ 5300	270 @ 4500	10.4 :1	rear - A4*	3.615	8.5	16.5	29.0	121	136	0.78	65	19.0	28.0	S 91

While their contours are not very pronounced, they hold and provide efficient lumbar support, and their padding is just right.

• Performance: 80%
With the V8 engine, the favorable weight-to-power ratio allows crisp takeoffs and accelerations while the 6 cylinder has more torque than occidental horse power. The well-padded interior helps you forget speed limits and it is good to keep an eye on the gauges from time to time to know what is going on.

• Handling SC: 80%
The SC corners flatter thanks to its noticeably firmer springs-damper. An adjustable suspension can be set in "comfort", "sport" or "automatic" allowing the handling to be adapted to different handling styles.

• Steering: 80%
Smooth and precise, its assistance varies as a function of speed, but remains too high and too light, requiring constant attention. That of the SC reacts faster, because its ratio is shorter.

• Access: 80%
It's easier to get into the back of an LS than an SC in spite of the right front seat automatically moving out of the way which compensates for the narrow door opening.

• Suspension: 80%
That of the LS is honeyed, thanks to its softness and the extent of its wheel travel. The optional pneumatic air lift is only needed under heavy load.

• Conveniences: 80%
Storage is well-distributed and the interior includes a number of refinements such as an electrically adjustable steering column, telescopic and angular.

• Braking: 60%
It is powerful and balanced during panic stops and the distances are remarkable for vehicles of this weight.

• Depreciation: 60%
It compares with some German ones, which augurs well.

LEXUS LS400

NEW FOR 1995

LEXUS LS400

LEXUS SC400

• Noise level: 60%
The remarkable soundproofing snuffs out all noise whether they come from the power train or roll, even at high speeds.

• Trunk LS: 60%
Even though lesser than that of its competitors, the LS trunk has sufficient capacity and its forms are regular.

• Handling LS: 60%
Crisp and well-balanced, the softness of the suspension is better controlled than before, same as with body movement.

• Interior space LS: 50%
A fifth adult can be temporarily seated at the back of the LS for a short trip, but it is really a four-seater.

WEAK POINTS

• Price/equipment: 00%
While the price of these two cars does not put them within reach of everyone, it is never the less competitive to those of its competitors if you compare their characteristics, manufacturing quality, and the richness of its equipment.

• Trunk: SC: 20%
That of the SC is equal or superior to that of its competitors.

• Fuel consumption: 40%
For those who respect traffic rules, the fuel consumption is equivalent to that of some V6s of lesser displacement and stays around 20 mpg.

• Interior space SC: 40%
As on most vehicles of this type, space is more limited in length than in height at the rear seats of the coupes, which have more seat height and allow two adults to travel in comfort.

CONCLUSION

• Overall rating: LS: 74.0%
SC: 70.0%

The Lexus LS and SC have no problem remaining in the lead because they begin to have an excellent reputation and their reliability makes green with envy the most aristocratic of their competitors. ☺

SPECIFICATIONS & PRICES

Model	Version/Trim	Body/ Seats	Interior volume cu.ft.	Trunk volume cu ft.	Drag coef. Cd	Wheel-base in	Lgth x Wdthx Hght. inx inx in	Curb weight lb	Towing capacity std. lb	Susp. type ft/rr	Brake type ft/rr	Steering system	Turning diameter ft	Steer. turns nber	Fuel. tank gal.	Standard tire size	Standard power train	PRICES US. $ 1994
LEXUS		**General warranty: 4 years / 50,000 miles; Power train: 6 years / 70,000 miles; corrosion perforation: 6 years / unlimited & road assistance.**																
LS400		4dr.sdn.5	97	14	0.28	112.2	196.7x72.0x55.9	3693	2000	i/i	dc/dcABS	pwr.r&p.	34.8	3.46	22.4	225/60VR16	V8/4.0/A4	**$51,200**
SC300		2 dr.cpe.2+2	85	9.3	0.30	105.9	191.1x70.5x52.6	3660	2000	i/i	dc/dc/ABS	pwr.r&p.	36.1	3.10	20.6	225/55VR16	L6/3.0/M5	**$38,000**
SC400		2 dr.cpe.2+2	85	9.3	0.30	105.9	191.1x70.5x52.6	3624	2000	i/i	dc/dc/ABS	pwr.r&p.	36.1	3.10	20.6	225/55VR16	V8/4.0/A4	**$45,100**

See page 393 for complete 1995 Price List

End of the series...

The new Lincoln Continental will arrive too late to be in these pages. Ford will profit from this model change to reenter it on the market invaded by foreign competition, which changes by culture and society have profoundly modified.

Entry model of the Lincoln range, one can never say that the Continental set a mark on its era because its sales have always been confidential. This 4-door 3-volume sedan is offered in two trim levels, Executive and Signature, which differ mostly by their equipment level because their drive train remains identical.

STRONG POINTS

• Safety: 90%
The impact resistance of the structure is satisfactory and the protection of the front occupants is insured by 2 air bags.

• Satisfaction: 80%
Without pretending to stand up to the competition of the German or Japanese cars, its reliability results in fewer owners' problems than in one era where it encountered many small nuisance problems. Its pneumatic suspension is complex and expensive to repair.

• Technical: 80%
The Continental was the first and only Lincoln to adopt front wheel drive and it shares high tech with the Mark VIII. Its steel unit body is provided with an independent suspension on all 4 wheels, including computer-controlled pneumatic dampers. The four wheel disc brakes include a standard ABS and the power assist to the rack-and-pinion varies as a function of speed. Its lines by some details are reminiscent of those of the Mercedes, Jaguar and BMW., but hold only an average aerodynamic coefficient with a Cd of 0.34. As is the case of the Q45 Infinity, its grill has been redesigned in a Jaguar style which seems reasonable since it is now family business.

• Noise level: 80%
Soundproofing snuffs out most noises coming from the power train. But those of the wind and tires persist depending on pavement and speed.

• Trunk: 80%
Its volume is well proportioned and it will hold a large amount of baggage. Its forms are regular and usable and the low sill facilitates access.

• Insurance: 80%
The premium is reasonable since the car is American and therefore easy to repair in the event of an accident and most of its owners are seasoned people with no surprises.

• Access: 80%
It is easy to get into this limousine whose doors are well-proportioned and whose height is comfortable.

• Seats: 70%
They are not comfortable on a long trip because their padding is too soft and their form is too evasive so they do not provide enough lateral hold or lumbar support. All things considered, velvet trim is preferable to leather, which is slippery, summer or winter.

DATA

Category: front-wheel drive luxury sedans.
Class : 7

HISTORIC

Introduced in: 1988
Modified in: 1989: multiple-point fuel injection.
Made in: Wixom, MI, U.S.A..

RATINGS

Safety: 90 %
Satisfaction: 77 %
Depreciation: 60 %
Insurance: 3.7 % ($985)
Cost per mile: $0.94

NUMBER OF DEALERS

In U.S.A.: 5,200 Ford-Lincoln-Mercury

SALES ON THE AMERICAN MARKET

Model	1992	1993	Result
Continental	38,458	31,746	- 17.5 %

MAIN COMPETITORS

ACURA Legend, AUDI A6, BMW 535i, CADILLAC De Ville & Seville, INFINITI J30, JAGUAR XJ6, LEXUS GS300, LINCOLN Town Car, MERCEDES E 320, VOLVO 900 Series.

EQUIPMENT

LINCOLN	Continental
Automatic transmission:	S
Cruise control:	S
Power steering:	S
ABS brake:	S
Air conditioning:	S
Air bag (2):	S
Leather trim:	S
AM/FM/ Radio-cassette:	S
Power door locks:	S
Power windows:	S
Tilt steering column:	S
Dual power mirrors:	S
Intermittent wipers:	S
Light alloy wheels:	S
Sunroof:	O
Anti-theft system:	O

S : standard; O : optional; - : not available

COLORS AVAILABLE

Exterior: Metallic Blue, Ivory Pearl, Black, Red, Beige, Green.
Interior: Leather :Tan.

MAINTENANCE

First revision: 5,000 miles
Frequency: 6 months/ 6,000 miles
Diagnostic plug: Yes

WHAT'S NEW IN 1995 ?

- New model presented at the beginning of 1995 (See the chapter on Innovations - page 39).

Model/ versions *: standard	ENGINE Type / timing valve / fuel system	Displacement cu/in	Power bhp @ rmn	Torque lb.ft @ rpm	Compres. ratio	TRANSMISSION Driving wheels / transmission	Final ratio	PERFORMANCE Acceler. 0-62 mph s	Stand. 1/4 mile s	Stand. 5/8 mile s	Braking 62-0 mph ft	Top speed mph	Lateral acceler. G	Noise level dBA	Fuel economy mpg city	Fuel economy hwy	Gasoline type / octane
1994	V6* 3.8 OHV-12-MPSFI	232	160 @ 4400	225 @ 3000	9.0 :1	front - A4*	3.37	10.0	16.7	30.8	154	115	0.75	66	21.0	33.0	R 87
1995	V8* 4.6 DOHC-32-SFI	281	260 @ 5750	265 @ 4750	9.85 :1	front - A4*	3.56	-	-	-	-	-	-	-	18.0	28.0	S 91

• **Suspension: 70%**
In spite of its sophistication it is not as effective as that of the Mark VIII and gains more comfort on the freeway where its automatic control does marvels, than on small, bumpy country roads where the body shows posts uncoordinated motions and shakes up the occupants.
• **Quality/fit/finish: 70%**
The interior presentation doesn't exactly correspond to the body style whose construction is robust because the gauges and imitation wood appliqués are tacky even though the materials are of good quality and the finish is careful. Some details like seat rails that appear on the rear seats deserve a little more attention.
• **Driver's compartment: 70%**
The dashboard is not unanimously approved, because confusing controls and digital instruments don't fit the classic image this car projects.

• **Steering: 60%**
Even though fast, precise and rather well-modulated, its maneuverability suffers from too large a turning diameter.
• **Braking: 60%**
In spite of 4-disc braking and an ABS system, its efficiency is only moderate, because stopping distances are long and the pedal is hard to modulate, the power assist being very strong.
• **Fuel consumption: 60%**
High marks in this department considering the weight and bulk: the Continental is relatively economical compared to some of its competition.
• **Interior space: 60%**
Five adults will be at ease in the car whose dimensions are ample, particularly in height.
• **Performance: 50%**
The engine lacks passing punch and the V8, such as the one on the new Mark VIII, would be welcome. The transmission is well-staged but it is slow to down shift and in some instances must be eased into it and engine braking is not very convincing.

WEAK POINTS

• **Price/equipment: 20%**
Compared to some imports costing approximately the same, the Continental is less well-equipped, less refined and less powerful.
• **Handling: 40%**
The great suspension softness does not control effectively. While all goes well on the freeway and the driving precision, rhythm and comfort suffer on small bumpy, twisty roads. There, the tires work hard and it's better to slow down to avoid sea sickness.
• **Conveniences: 40%**
The practical side is not one of the Continental's strong points because storage spaces are down to a minimum.
• **Depreciation: 40%**
It is stronger than average because these models age quickly and their budget for maintenance and repairs scares off the prospects.

CONCLUSION

• **Overall rating: 64.0%**
At the eve of the model being replaced, you can conclude that Ford did not give it a great chance for success by always supplying the wrong compromises. While the outside could tempt buyers of European models, the interior has always been designed to tempt the classic Lincoln buyers at the opposite. Finally, the power train has never been up to its requirements and has discouraged more than one person. One risks to loose everything by wanting to gain everything.

SPECIFICATIONS & PRICES

Model	Version/Trim	Body/ Seats	Interior volume cu.ft.	Trunk volume cu ft.	Drag coef. Cd	Wheel-base in	Lgth x Wdthx Hght. inx inx in	Curb weight lb	Towing capacity std. lb	Susp. type ft/rr	Brake type ft/rr	Steering system	Turning diameter ft	Steer. turns nber	Fuel. tank gal.	Standard tire size	Standard power train	PRICES US. $ 1994
LINCOLN		**General warranty: 3 years / 50,000 miles; corrosion perforation: 6 years / 100,000 miles.**																
Continental	Exe(1994)	4dr.sdn.6	104	19	0.34	109.0	205.1x72.3x55.4	3576	1000	i/i	dc/dcABS	pwr.r/p.	38.4	2.84	18.4	205/70R15	V6/3.8/A4	**$34,750**
Continental	Sig (1994)	4dr.sdn.6	104	19	0.34	109.0	205.1x72.3x55.4	3622	1000	i/i	dc/dcABS	pwr.r/p.	38.4	2.84	18.4	205/70R15	V6/3.8/A4	**$36,050**
Continental	1995	4dr.sdn.6	102.7	18.1	0.32	109.0	206.3x73.3x55.9	3969	1000	i/i	dc/dcABS	pwr.r/p.	41.1	2.86	18.0	225/60R16	V8/4.6/A4	**NA**

See page 393 for complete 1995 Price List

Unique...

The Mark VIII coupe has surprised as much by its resolutely futuristic line as by its performance and handling, unusual for a car built in North America. The sophistication of its technique and the brilliance of its engine don't always fit with its somewhat international format.

The Mark VIII coupe differs much from the model it replaces, as much by its innovative line as by its modernized power train. It exists only in a 3-volume 2-door with 4 seats in the unique version that one day should be followed by a convertible.

STRONG POINTS

• Safety: 90%

The Mark VIII coupe comes standard with two air bags, 3-point seat belts and a structure that resists impacts well which allows it to reach the maximum rating.

• Performance: 80%

The new 280 hp FourCam 4.6 L has instant throttle response and lives up to its rating. This American V8 is without a doubt the most brilliant of its generation, and gains for the Mark VIII a weight-to-power ratio of 1.48 lbs/hp. It is coupled to an automatic transmission whose ratios are well-staged, the shifts smooth and precise and the down-shifts allow a good power reserve and adequate engine braking.

• Steering: 80%

In spite of a fairly powerful assist, it is fast and precise and provides unusual maneuverability for a vehicle of this size whose steering diameter is very reasonable.

• Technical: 80%

The steel unit body, with the exception of the front hood, which is in molded plastic, receives a four wheel independent suspension. Up front there are long and short control arms and the rear is mounted on a sub-frame which includes cast aluminum hubs. Air shocks with computer controls damps the body motions in four axis. The control level is based on road condition and driving style. Four disc braking is handled by a Teves ABS tied into a traction control which has a dashboard on/off switch. The rack-and-pinion steering power assist varies as a function of speed. Body lines combine to produce a 0.33 drag coefficient, near the high end of efficiency. The grill, built by Davidson Textron is of chromed urethane applied in successive coats, which does not delaminate even when folded in two, will not rust and resists impacts at 20 degrees F as well as ultra violet light.

• Satisfaction: 80%

The commentary of the first owners are not numerous enough and it is too early to comment on the precision and reliability of the Mark VIII.

• Driver's compartment: 80%

It is easy to rapidly find the best driving position and the lateral and lumbar supports of the seats are effective. Visibility is better toward the front than the rear, and the "C" pillar is large while at the sides the rear-view mirrors are ridiculously small.

• Insurance: 80%

Its rate is no higher than that of an ordinary car, but the price isn't what

DATA

Category: rear-wheel drive luxury coupes.
Class : 7

HISTORIC

Introduced in: 1993
Modified in: -
Made in: Wixom, MI, U.S.A.

RATINGS

Safety: 90 %
Satisfaction: 78 % (Mark VII)
Depreciation: 40 % (2 years)
Insurance: 3.6 % ($1,100)
Cost per mile: $0.94

NUMBER OF DEALERS

In U.S.A.: 5,200 Ford-Lincoln-Mercury

SALES ON THE AMERICAN MARKET

Model	1992	1993	Result
Mark VIII	8,115	31,852	+ 392 %

MAIN COMPETITORS

ACURA Legend coupe, BUICK Riviera, CADILLAC Eldorado, JAGUAR XJS, LEXUS SC400.

EQUIPMENT

LINCOLN	Mark VIII
Automatic transmission:	S
Cruise control:	S
Power steering:	S
ABS brake:	S
Air conditioning:	S
Air bag (2):	S
Leather trim:	S
AM/FM/ Radio-cassette:	S
Power door locks:	S
Power windows:	S
Tilt steering column:	S
Dual power mirrors:	S
Intermittent wipers:	S
Light alloy wheels:	S
Sunroof:	O
Anti-theft system:	S

S : standard; O : optional; - : not available

COLORS AVAILABLE

Exterior: Mocha, Blue, Garnet, Burgundy, Cranberry, Eggplant, Green, Black, White, Opal Gray, Opal.
Interior: Mocha, Ébony, Seaweed, Opal Gray.

MAINTENANCE

First revision: 5,000 miles
Frequency: 6 months/6,000 miles
Diagnostic plug: Yes

WHAT'S NEW IN 1995 ?

- **Chromed exhaust tips.**
- **Redesigned dash.**
- **The mid-1995 edition has a more sporty appearance.**

Model/ versions *: standard	ENGINE Type / timing valve / fuel system	Displacement cu/in	Power bhp @ rmn	Torque lb.ft @ rpm	Compres. ratio	TRANSMISSION Driving wheels / transmission	Final ratio	PERFORMANCE Acceler. 0-62 mph s	Stand. 1/4 mile s	Stand. 5/8 mile s	Braking 62-0 mph ft	Top speed mph	Lateral acceler. G	Noise level dBA	Fuel economy mpg city	hwy	Gasoline type / octane
base	V8* 4.6 DOHC-32-SFI	281	280 @ 5500	285 @ 4500	9.85 :1	rear - A4*	3.07	7.8	16.0	28.5	138	125	0.80	65	18.0	28.0	S 91

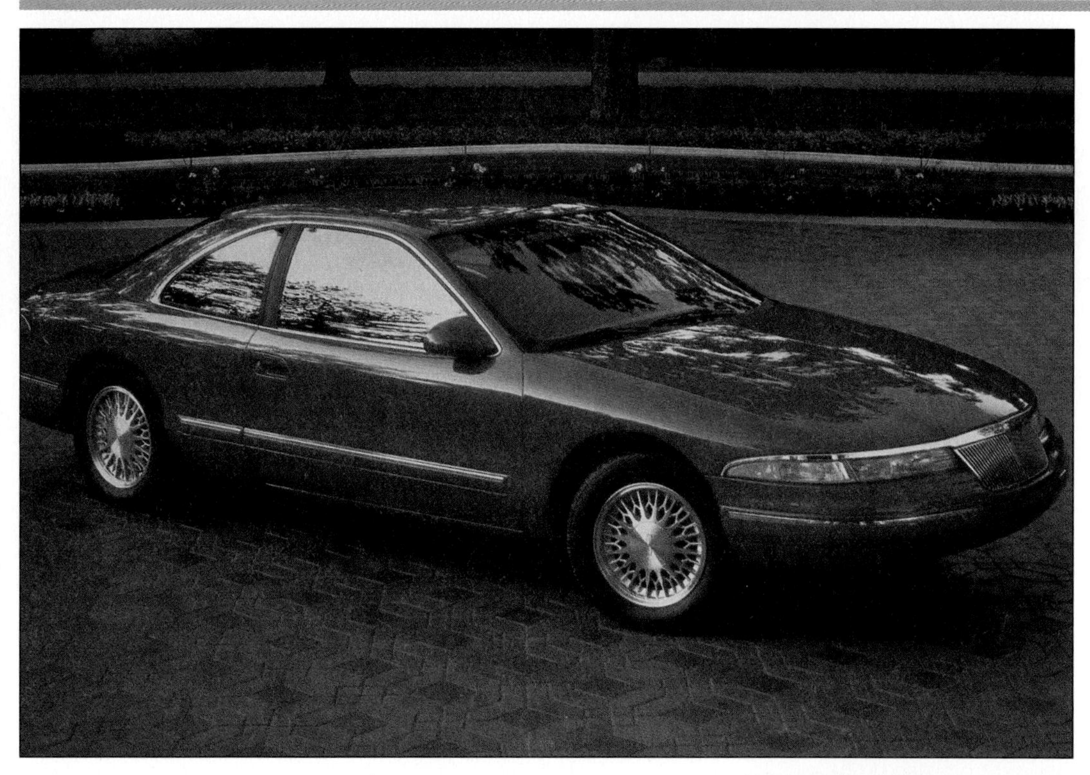

• Conveniences: 50%
They leave much to be desired because the glove box is as small as the door pockets and the center armrest is monopolized by a cellular phone.

• Braking: 50%
Even though it has to fight a large inertia, it is at the level of its performance, powerful and balanced, and its fade-resistance is surprising.

WEAK POINTS

• Price/equipment: 00%
In spite of its very complete equipment and the fact that it compares with the best in performance and comfort, the Mark VIII is costly for its average quality.

• Fuel consumption: 10%
It is reasonable for its performances but its fuel tank capacity limits the autonomy.

makes the premium shoot up like an arrow, the 280 hp V8 is..

• Seats: 80%
The shaped bench seat which forms the two buckets offers a lateral hold slightly better than that of the front seats, already very acceptable on an American car and their adequate padding allows it to make long fatigue-free trips.

• Suspension: 80%
Damping is effective, and smart, and the suspension amplitude is sufficient to absorb major bumps without penalizing occupants comfort.

• Noise level: 70%
Discrete in normal use, the engine allows you to hear its pleasant tones during strong acceleration when rolling noise is neutralized by efficient soundproofing and body finesse minimizes wind noise.

• Quality/fit /finish: 70%
The dashboard looks very ordinary and a few touches of wood, even false, would help warm up its somewhat severe appearance. Some elements look too cheap, such as the door handles, ashtray lids and cup carrier whose plastics are not flattering. The finish is not as careful as one would expect on a car of this price which is disappointing in relation to the whole.

• Handling: 70%
In spite of the available power, traction in turns is surprising even when traction control is not switched on. The imposing format and weight of the coupe do not harm its travel down a twisty path, thanks to excellent control of body motions by the pneumatic suspension. Roll and pitch are minimized and you will find real pleasure in driving fast without ever feeling that the situation can get out of hand at any moment or realizing that a car of such an imposing format is driven.

• Access: 70%
The large and heavy doors allow easy seating on a bench seat and the autoglide moves the seats forward as soon as the back is inclined.

• Interior space: 60%
Two normal sized adults can comfortably occupy the rear seats where they have enough space

in length and height but lack foot room under the front seats. There is no folding center armrest. Front seats are generous even if the size of the center tunnel and door arm rests limits their width.

• Trunk: 60%
Large and deep, it is not practical since it lacks height, has a narrow opening and its sill complicates baggage handling.

• Depreciation: 60%
This type of vehicle loses a little more than the average because prospects are not numerous.

CONCLUSION

• Overall rating: 65%
A little unusual in its styling, the Mark VIII coupe surprises as much by its brilliant performance as by the quality which needs instant improvement. ☺

SPECIFICATIONS & PRICES

Model	Version/Trim	Body/ Seats	Interior volume cu.ft.	Trunk volume cu ft.	Drag coef. Cd	Wheel-base in	Lgth x Wdthx Hght. inx inx in	Curb weight lb	Towing capacity std. lb	Susp. type ft/rr	Brake type ft/rr	Steering system	Turning diameter ft	Steer. turns nber	Fuel. tank gal.	Standard tire size	Standard power train	PRICES US. $ 1994
LINCOLN		**General warranty: 3 years / 50,000 miles; corrosion perforation: 6 years / 100,000 miles..**																
Mark VIII		2 dr.cpe. 5	97	14	0.33	113.0	206.9x74.6x53.6	3765	2000	i/i	dc/dc/ABS	pwr.r&p.	11.34	2.60	68.1	225/60R16	V8/4.6/A4	**$38,050**

See page 393 for complete 1995 Price List

My aunt is on a diet...

The Town Car, which has always seemed part of the family, constantly improves, to the point that car critics will find less to joke about. More refined, less prone to over eating, it only lacks a little handling for its roundness to make it presentable.

The Lincoln Town Car has nothing of the urban car. Quite the contrary, it is the most voluminous car of the Ford range. It is a 3-volume, 4-door limousine proposed in Executive, Signature, and Cartier trims. Only one engine is available, the 4.6 L V8 which is coupled to a 4-speed automatic with electronic controls.

STRONG POINTS

• Interior space: 95%
Its imposing dimensions allow it to claim the most spacious interior space on the market because it can easily accommodate up to 6 people.

• Safety: 90%
The occupants are well-protected by 2 air bags standard on all versions and the structure offers good collision resistance.

• Trunk: 90%
In spite of its enormous volume, it is difficult to utilize because of its tortured forms and the ridiculous minispare tire disposed in the middle.

• Access: 90%
Taking a seat in the Town Car doesn't cause any problems because the length and opening of the doors and roof clearance are ample.

• Noise level: 90%
It's one of the lowest on the market, thanks to the quietness of the drive train and the numerous rubber connections which filter out noises and vibrations and the enormous quantity of soundproofing materials.

• Satisfaction: 80%
It continues to progress, even though the owners still speak of electric circuit problems.

• Suspension : 80%
Cushiony on good pavement, it panics at the slightest road defect and mistreats its occupants without respect.

• Insurance: 80%
The premium of the Lincoln's seems reasonable compared to that of some of its German rivals.

• Technical: 70%
The Town Car has been created beginning with a platform which has a perimeter chassis of the Crown-Victoria and the Grand Marquis where the layout is very classic with the drive train up front and rear drive wheel with a rigid axle. Even though simplistic, this recipe is seasoned by gas shock absorbers and pneumatic springs whose leveling control is automatic. The 4.6 L V8 is a modern power plant which benefits from a single cam per cylinder bank. The 4-wheel disc brakes, with a standard ABS and the steering power assist, varies as a function of speed. More fluid than in the past, the lines favorably influence the aerodynamic coefficient which shows a surprising 0.36.

• Quality/fit/finish: 60%
In relation to old models, they have improved, but the assembly, fit, finish and material are behind those of German or Japanese counter-

DATA

Category: rear-wheel drive luxury limousines.
Class : 7

HISTORIC

Introduced in: 1980
Modified in: 1990: actual body.
Made in: Wixon, MI, U.S.A.

RATINGS

Safety: 90 %
Satisfaction: 83 %
Depreciation: 62 %
Insurance: 3.7 % ($985)
Cost per mile: $0.94

NUMBER OF DEALERS

In U.S.A.: 5,200 Ford-Lincoln-Mercury

SALES ON THE AMERICAN MARKET

Model	1992	1993	Result
Town Car	115,075	110,046	- 4.4 %

MAIN COMPETITORS

BMW 740i, CADILLAC Brougham, De Ville, Seville, INFINITI Q45, JAGUAR XJ6, LEXUS LS400, MERCEDES BENZ 420 SE.

EQUIPMENT

LINCOLN Town Car	Executive	Signature	Cartier
Automatic transmission:	S	S	S
Cruise control:	S	S	S
Power steering:	S	S	S
ABS brake:	S	S	S
Air conditioning:	S	S	S
Air bag (2):	S	S	S
Leather trim:	O	O	S
AM/FM/ Radio-cassette:	S	S	S
Power door locks:	S	S	S
Power windows:	S	S	S
Tilt steering column:	S	S	S
Dual power mirrors:	S	S	S
Intermittent wipers:	S	S	S
Light alloy wheels:	S	S	S
Sunroof:	O	O	O
Anti-theft system:	S	S	S

S : standard; O : optional; - : not available

COLORS AVAILABLE

Exterior: Burgundy, Silver, Cranberry, Blue, Black, Wildberry, Willow Green, Ivory.
Interior: Blue, Cranberry, Ebony, Titanium, Graphite, Ivory, Green.

MAINTENANCE

First revision: 5,000 miles
Frequency: 6 months
Diagnostic plug: Yes

WHAT'S NEW IN 1995 ?

- Cosmetic front end changes.
- Choice of degree of power assist.
- New dashboard.
- Tele-control of temperature and radio on the steering wheel (Cartier and Signature).
- New crystal headlights.

Model/ versions *: standard	ENGINE Type / timing valve / fuel system	Displacement cu/in	Power bhp @ rmn	Torque lb.ft @ rpm	Compres. ratio	TRANSMISSION Driving wheels / transmission	Final ratio	PERFORMANCE Acceler. 0-62 mph s	Stand. 1/4 mile s	Stand. 5/8 mile s	Braking 62-0 mph ft	Top speed mph	Lateral acceler. G	Noise level dBA	Fuel economy mpg city	hwy	Gasoline type / octane
base	V8* 4.6-SOHC-16-SMPFI	281	210 @ 4250	270 @ 3250	9.0 :1	rear - A4*	3.08	11.0	18.0	33.5	151	109	0.70	64	17.8	27.0	R 87

parts. The body posts a very successful international style. The interior disappoints by its American "Old Timers" style. The false wood and plastic blend badly with digital instrumentation placed on a car of this class. A good point is to have concealed the seat gliders at the rear, which have the tendency to scratch up the "Guccis".

• **Steering: 60%**
You need the soul of a master chauffeur, a good eye and patience to drive this big car in town because the size of its turning diameter and its ratio deprive it of any maneuverability. On the road, it's imprecise in the center which deprives it of any sensation coming from the pavement.

• **Seats: 60%**
They are not comfortable on a long trip because they are too soft, and you don't know how to seat yourself in them in comfort.

• **Driver's compartment: 60%**
The lack of lateral hold and lumbar support from the seat, which is not tightly curved enough, doesn't put the driver at ease and visibility is too sharply reduced from 3/4 rear. In spite of the reorganized dashboard, it is still not any more ergonomic because some controls are still out of the drivers reach,,

• **Performance: 50%**
In spite of the V8 power, it still has lots to do to move nearly 2 tons of this heavy weight. Takeoffs and accelerations are good but you have to force the transmission into a downshift to keep pace on the grade and the engine break is of no great help.

WEAK POINTS

• **Price/equipment: 10%**
Less expensive than a big Lexus or Infiniti, the Town Car offers superior roominess but its status is at another level, typically North-American. Its equipment is complete on all the versions, which differ only by their interior trim.

• **Handling: 20%**
The Town Car prefers the parkway to the little twisty roads, where it gives up on the first corner. Its suspension is too soft and the pneumatic controls are not effective enough to control body movements as on the Mark VIII.

• **Fuel consumption: 30%**
It is less catastrophic than before and holds around 19 mpg.

• **Braking: 40%**
The four discs succeed in stopping this imposing mass within acceptable distances and the ABS stabilize the vehicle's path during panic stops, but the front dives hard and unloads the driving wheels.

• **Depreciation: 40%**
It remains high and in countries in which fuel is expensive, other models are more realistic.

• **Conveniences: 40%**
Storage area is limited to a glove box and a storage cabinet in the front center armrest. The front shoulder harness lacks a height adjustment.

CONCLUSION

• **Overall rating: 61.0%**
The Town Car has evolved much on specific points such as its aerodynamic value, braking and engine output. It now needs to progress in handling precision and body motion control to be able to recruit a new breed of customers. 😐

OWNER'S SUGGESTIONS

-Less «boat-like» handling
-Better fuel economy
-A more competitive price.
-Less tacky dashboard appearance and interior trim which lacks class.
-Decor less "kitsch"
-Window tint standard

SPECIFICATIONS & PRICES

Model	Version/Trim	Body/ Seats	Interior volume cu.ft.	Trunk volume cu ft.	Drag coef. Cd	Wheel-base in	Lgth x Wdthx Hght. inx inx in	Curb weight lb	Towing capacity std. lb	Susp. type ft/rr	Brake type ft/rr	Steering system	Turning diameter ft	Steer. turns nber	Fuel. tank gal.	Standard tire size	Standard power train	PRICES US. $ 1994
LINCOLN		**General warranty: 3 years / 50,000 miles; corrosion perforation: 6 years / 100,000 miles..**																
Town Car	Executive	4 dr.sdn.6	116	22.5	0.36	117.4	218.9x76.7x56.9	4030	2000	i/r	dc/dc/ABS	pwr.bal.	40	3.4	20	215/70R15	V8/4.6/A4	**$34,750**
Town Car	Signature	4.dr.sdn.6	116	22.5	0.36	117.4	218.9x76.7x56.9	NA	2000	i/r	dc/dc/ABS	pwr.bal.	40	3.4	20	215/70R15	V8/4.6/A4	**$36,050**
Town Car	Cartier	4 dr.sdn.6	116	22.5	0.36	117.4	218.9x76.7x56.9	NA	2000	i/r	dc/dc/ABS	pwr.bal	40	3.4	20	225/60R16	V8/4.6/A4	**$38,100**

See page 393 for complete 1995 Price List

A new start?

Under GM which was its previous owner, Lotus did not find the hoped for resurgence, perhaps because its potential meant little to the American giant. Bought by Bugatti, which seems to manufacture more money than cars, Lotus will know a second chance.

In 1989 the Elan picked up the famous name of its progenitor, Lotus, the company of founder Collin Chapman. At the beginning of the project its was going to be a 2x2 coupe but practical considerations dictated a 2-seater convertible. Its main mechanical elements were borrowed from the old Isuzu Impulse/Storm including the engine and transmission assembly. It has a 1.6 L Turbo charged, 4-cylinder, with a manual 5-speed.

STRONG POINTS

• Performance: 90%
With a weight-to-power ratio of only 14.6 lbs/hp it is not surprising to see the Elan accelerate from 0 to 60 mph. This sums up Chapman's philosophy: he preferred to lighten up his car rather than pull ultimate horse power from an engine that became fragile. Passing is alive and allows fun driving even if with front wheel drive, pleasure is not as complete as with rear wheels that do the pushing. In spite of its commoner origins, the engine has nerve and forms , with the rest of the components, an assembly that surprises by its homogeneous ways.

• Handling: 90%
It constitutes Lotus' specialty and has been the object of special attention. The stability has been imperturbable on takeoff. In a straight line or in a curve, under braking or under the effect of violent acceleration, torque steer is minimal and guidance virtually perfect. The rear axle follows without problems, thanks to good quality tires.

• Technical: 80%
Those who have ever seen this model go by, and there are not many, do not know that its dimensions are slightly smaller than those of a Miata. To obtain maximum rigidity, the Lotus engineers have turned toward a steel "Chassis-beam" to which is bolted a floor plan of composite material which constitutes the bottom of the car. Numerous steel reinforcements are glued to it and the side beams are also riveted. The different fiberglass reinforced polyester resin body panels are injected under a vacuum to prevent gas bubbles. The line, which is agreeable to the eye, necessitated numerous aerodynamic studies to obtain the best drag coefficient and to reduce the turbulence always unwelcomed on a convertible. This explains why the windshield is very inclined, an extension of the front end of it. Suspension is independent and there are 4 wheel disc brakes without ABS and a power steering.

• Driver's compartment: 80%
More like a production vehicle than a kit car which is a bit the case of the Esprit. The dashboard is well-designed, with a massive gauge assembly that also includes most of the controls. It reminds you more of Japanese than of English cars. The bezels are well-grouped and easy to read, except for the tach which is not circular and visibility is only average, due to the high belt line and major blind spots created by the convertible top. The driving position is low but excellent and one

DATA

Category: front-wheel drive sport convertibles.
Class : S3

HISTORIC

Introduced in: 1989
Modified in: -
Made in: Norwich, Norfolk, England.

RATINGS

Safety: NA %
Satisfaction: NA %
Depreciation: NA %
Insurance: 4.2 % ($1,100)
Cost per mile: $0.80

NUMBER OF DEALERS

In U.S.A.: NA

SALES ON THE AMERICAN MARKET

Model	1992	1993	Result
Lotus	125	125	

MAIN COMPETITORS

ALFA ROMEO Spider, MAZDA Miata

EQUIPMENT

LOTUS Elan	S2
Automatic transmission:	-
Cruise control:	-
Power steering:	S
ABS brake:	-
Air conditioning:	O
Air bag (left):	-
Leather trim:	O
AM/FM/ Radio-cassette:	S
Power door locks:	S
Power windows:	S
Tilt steering column:	S
Dual power mirrors:	S
Intermittent wipers:	S
Light alloy wheels:	S
Sunroof:	S
Anti-theft system:	S

S : standard; O : optional; - : not available

COLORS AVAILABLE

Exterior: White, Red, Blue, Silver, Black, Yellow.
Interior: Gray

MAINTENANCE

First revision: 3,000 miles
Frequency: 6 months/5,000 miles
Diagnostic plug: No

WHAT'S NEW IN 1995 ?

- Lotus will be imported in North America by Bugatti which is building a network of dealers in the United States.

	ENGINE					TRANSMISSION		PERFORMANCE										
Model/ versions *: standard	Type / timing valve / fuel system	Displacement cu/in	Power bhp @ rmn	Torque lb.ft @ rpm	Compres. ratio	Driving wheels / transmission	Final ratio	Acceler. 0-62 mph s	Stand. 1/4 mile s	Stand. 5/8 mile s	Braking 62-0 mph ft	Top speed mph	Lateral acceler. G	Noise level dBA	Fuel economy mpg city	hwy	Gasoline type / octane
Elan S2	L4* 1.6T-DOHC-16-MPFI	97	155 @ 6000	146 @ 4200	8.2: 1	front - M5*	3.83	7.2	15.4	28.6	131	130	0.88	70	22.3	26.9	S 91

wishes that the seat back would support laterally as well as the cushion.

• Steering: 80%
It is responsible for a large part of the driving pleasure, very incisive, precise and also very fast, with just 2.6 turns lock-to-lock and positively modulated.

• Safety: 70%
In spite of the platform rigidity, the structure of the convertible is always more vulnerable to impacts when there is no rollover bar. An air bag will be installed in the steering wheel.

• Quality/fit/finish: 70%
It is surprising, because the small makes are rarely able to reach the big ones on this point. Assembly and finish are careful and the leather trim is more flattering. The plastic of the dashboard is too shiny.

• Seats: 70%
While insufficiently shaped, they provide average hold and their padding is sufficient to be comfortable.

• Insurance: 70%
The premium on this little toy is at the level of the price and reflects the uncertainty of the insurers as to what it would take to put it back together after an accident.

• Fuel consumption: 70%
Reasonable in relation to its performance and strong for its displacement it will depend on the the right foot of the driver.

• Conveniences: 60%
They consist of a small glove box, door pockets and recesses in the center console.

• Access: 60%
It is not too difficult to get in, even when the convertible top is in place because the doors open sufficiently, but it is better not to be too big.

• Braking: 50%
Even though the bite and fade-resistance are good, one would expect shorter stopping distances, especially without the ABS, since the pedal is sensitive and does not allow precise modulation.

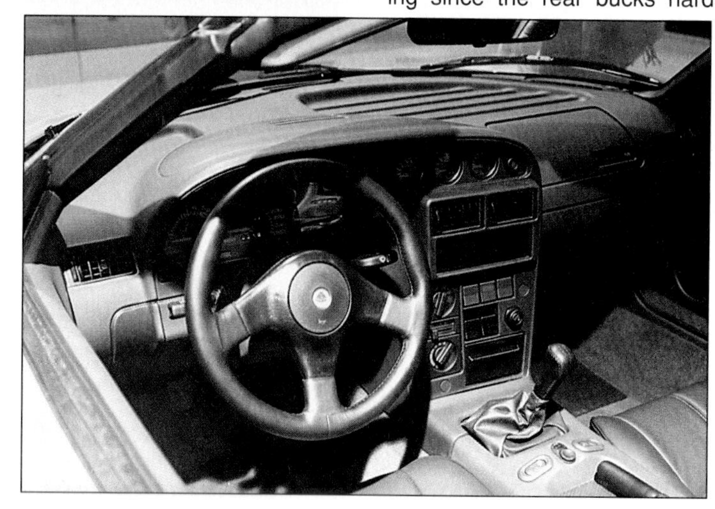

WEAK POINTS

• Price/equipment: 00%
You need a lot of passion or the cult of the English make to accept spending that much money for this type of toy, whose resale value will be most uncertain and depends on the success of the commercial network Bugatti is putting in place in North America.

• Interior space: 10%
Even though the occupants have enough room in length and height, the tall belt line and low seating produce a disagreeable claustrophobic feel which disappears when you open the roof.

• Trunk: 20%
It is disposed vertically since the rear part has no overhang and will only accept soft bags of small size.

• Noise level: 30%
It is not only the exhaust system which purrs happily to keep up the noise level but also rolling noises and those of the wind and from under the hood. Rolling noises are not modest, especially on pavement in poor shape. However, above 80 mph any conversation, audio noises or sirens become nearly inaudible.

• Suspension: 40%
The Lotus have never been comfort monsters, however that of the Elan is not as hard as one could imagine. Along the parkway everything goes well but tar strips and expansion joints are disturbing since the rear bucks hard while crossing them, as though it was surprised.

CONCLUSION

• Overall rating: 60.5%
Too bad that this little car is as expensive in any part of its budget as it is original and fun to drive, well in keeping with the tradition of the famous make whose colors it carries with a flourish. 😐

SPECIFICATIONS & PRICES

Model	Version/Trim	Body/ Seats	Interior volume cu.ft.	Trunk volume cu ft.	Drag coef. Cd	Wheel-base in	Lgth x Wdthx Hght. inx inx in	Curb weight lb	Towing capacity std. lb	Susp. type ft/rr	Brake type ft/rr	Steering system	Turning diameter ft	Steer. turns nber	Fuel. tank gal.	Standard tire size	Standard power train	PRICES US. $ 1994
LOTUS		**General warranty :1 year/ unlimited: perforation 8 years / unlimited.**																
Elan	S2	2 dr.con.2	53	7	0.34	88.6	149.7x68.3x48.4	2249	NR	i/i	dc/dc	pwr.r&p.	34.8	2.6	12.2	205/45VR16	L4/1.6/M5	**NA**

See page 393 for complete 1995 Price List

Chapman's spirit...

The Esprit symbolizes an entire era of the past where Lotus rhymed with competitions, victories, and success. Colin Chapman, for nearly 20 years, was the great magician of the automobile. When he disappeared, he took all his secrets with him, with the spirit that animated the British office.

The Lotus Esprit doesn't date from yesterday, because this year it celebrates its 15th birthday. It was designed by Giugiaro and was then redrawn by the GM's stylists, when they owned it. Lotus is an exotic 2-seater coupe, offered in a unique version which has the peculiarity of being equipped with a 4-cylinder engine, something rare in a segment where V8s and V12s roam. The only transmission is a manual 5-speed.

STRONG POINTS

• Performance: 100%
This vehicle was designed to bite into the asphalt and it loves doing that. As proof, it only needs 5 seconds to go from 0 to 62, like a Diablo. Its weight-to-power ratio is 11.2 lbs/hp, which explains this result. Urban driving is not its forte because there is no power under 2500 rpm after which the turbo kicks in. The 4-cylinder does not leave from lower rpm, where it lacks torque. The rpm climb of the tach is very exciting, accompanied as it is by a competition car-style growl. Now you know why James Bond swapped his Aston Martin for a Lotus which would in turn transform itself into a submarine. That's why the streamlined styling offers good aerodynamic efficiency: despite its large tires and numerous appendages its drag coefficient stays at 0.32.

• Handling: 90%
It continues to be excellent, in spite of the age of its conception. One must say that at that time Formula 1 engineers were at the drawing board. Not surprising that it stays glued to the road in all circumstances and demonstrates great neutrality. Its centrally located engine is there for a purpose, so its ample tires and near absence of roll allow it to go through hairpin turns with surprising aplomb.

• Insurance: 90%
Even though the premium represents a round sum, it is only a small percentage of its value.

• Technical: 80%
This mid-engined coupe is made of glass reinforced polyester resin injected under pressure. Kevlar fibers reinforce the roof, whose removable panels have an aircraft style honeycomb design. The beam-type chassis in galvanized steel is integrated into the structure which supports elaborate independent suspensions, front and rear. Disc brakes, vented at the front, are all ABS controlled and the steering assist is a function of speed.

• Safety: 80%
The resin shell has a reputation for offering a stiffness that remains stable with time and whose capacity to absorb impact is superior to that of steel. For its Atlantic crossing, the Esprit will be equipped with an air bag on the driver's side.

• Quality/fit/finish: 80%
In spite of refinements in production process, the Esprit is not as sharp as the Elan which precedes it. The material quality is good but some

DATA

Category: rear-wheel drive GT coupes.
Class : GT

HISTORIC

Introduced in: 1975
Modified in: 1980: Turbo; 1987: body.
Made in: Norwich, Norfolk, England.

RATINGS

Safety: 70 %
Satisfaction: NA %
Depreciation: NA %
Insurance: 3.0 % ($2,250)
Cost per mile: $1.43

NUMBER OF DEALERS

In U.S.A.: NA

SALES ON THE AMERICAN MARKET

Model	1992	1993	Result
Lotus	125	125	

MAIN COMPETITORS

ACURA NSX, BMW 850, CHEVROLET Corvette ZR-1, DODGE Stealth R/T Turbo, MERCEDES BENZ SL, PORSCHE 911.

EQUIPMENT

LOTUS Esprit	S4
Automatic transmission:	-
Cruise control:	-
Power steering:	S
ABS brake:	S
Air conditioning:	S
Air bag (left):	S
Leather trim:	S
AM/FM/ Radio-cassette:	S
Power door locks:	S
Power windows:	S
Tilt steering column:	-
Dual power mirrors:	S
Intermittent wipers:	S
Light alloy wheels:	S
Sunroof:	S
Anti-theft system:	S

S : standard; O : optional; - : not available

COLORS AVAILABLE

Exterior: White, Red, Blue, Silver, Black, Yellow.
Interior: Gray, Tan, Black.

MAINTENANCE

First revision: 3,000 miles
Frequency: 6 months/5,000 miles
Diagnostic plug: No

WHAT'S NEW IN 1995 ?

- Lotus will be imported in North America by Bugatti which is building a network of dealers in the United States.

Model/ versions *: standard	ENGINE Type / timing valve / fuel system	Displacement cu/in	Power bhp @ rmn	Torque lb.ft @ rpm	Compres. ratio	TRANSMISSION Driving wheels / transmission	Final ratio	PERFORMANCE Acceler. 0-62 mph s	Stand. 1/4 mile s	Stand. 5/8 mile s	Braking 62-0 mph ft	Top speed mph	Lateral acceler. G	Noise level dBA	Fuel economy mpg city	Fuel economy mpg hwy	Gasoline type / octane
Esprit	L4T* 2.2-DOHC-16-MPFI	133	264 @ 6500	261 @ 3900	8.0	rear - M5*	3.89	5.0	13.6	26.5	125	155	0.90	75	15.0	25.0	S 91

details deserve improvement. Its reliability has not always been exemplary and has known a number of electrical problems.

• **Steering: 80%**
Fven though it is not especialy direct, it is of great precision and its assistance is fairly firm, which is good safety insurance.

• **Driver's compartment: 70%**
It begins to be a little Rococo with its instrument cluster placed on a tablet and the ergonomics are imperfect because some controls, including radio and air conditioning, are far from the driver. Small-sized folks will be disadvantaged driving this car with fixed seats, which is to say the least, uncomfortable. Visibility is precarious from 3/4 rear and the pedals are so close to each other that you need narrow Italian shoes not to hang up on them. However, the steering and the shift lever fall well in hand and it is a pleasure to shift up and down with simple wrist motion.

• **Access: 70%**
It is not easy to take a seat in a Lotus in spite of the door length because its low height obliges you to slither into a sort of narrow tunnel at the end of which you find the pedals.

• **Seats: 70%**
They are not uncomfortable, it's simply that their adjustments do not allow them to adapt themselves to some morphologies, otherwise, they hold well, thanks to large side pads but offer no lumbar adjustment.

• **Braking: 60%**
The ABS is responsible for relatively long stopping distances, because this car used to stop in a shorter distance when not so equipped. One thing that does not change is the ephemeral fade-resistance as soon as the brakes are abused by sports-like driving.

• **Suspension: 50%**
It is closer to that of a competition device than that of a limousine and it is better to select one's itinerary to limit back pain but that forms an integral part of the sporting spirit.

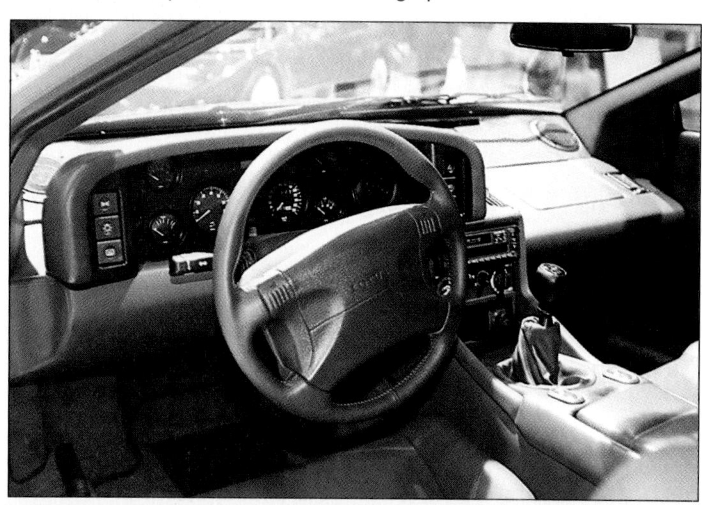

• **Fuel consumption: 50%**
Reasonable when the speed is, it climbs proportionally to the tach needle while the fuel gage descends vertiginously.

WEAK POINTS

• **Price/equipment: 00%**
In spite of the uncommon performance capability of the Esprit, it doesn't seem very reasonable to pay as much for this sympathetic little coupe as for an SL Mercedes which is a monument of technology.

• **Noise level: 10%**
At high speeds, this Lotus will render one deaf and dumb because the noise, which is not always melodious, exceeds the limits of comfort.

• **Interior space: 10%**
Wedged between a door and a central tunnel, which takes much space, one feels trapped and big sturdy bodies will know anxiety.

• **Trunk: 10%**
Purely symbolic, because one can only store a few things in it.

• **Conveniences: 20%**
Storage is limited to a small glove box and a pocket emptier located between the seat backs.

CONCLUSION

• **Overall rating: 56.5%**
Exotic without really being expensive, conserving no other value than sentimental, performing but unusable in the realistic context, the Esprit retains its allure, which has not gotten older, but that is the only thing.

SPECIFICATIONS & PRICES

Model	Version/Trim	Body/ Seats	Interior volume cu.ft.	Trunk volume cu ft.	Drag coef. Cd	Wheel-base in	Lgth x Wdthx Hght. inx inx in	Curb weight lb	Towing capacity std. lb	Susp. type ft/rr	Brake type ft/rr	Steering system	Turning diameter ft	Steer. turns nber	Fuel. tank gal.	Standard tire size	Standard power train	PRICES US. $ 1994
LOTUS		**General warranty :1 year/ unlimited: perforation 8 years / unlimited.**																
Esprit	S4	2 dr.cpe.2	49.4	5.3	0.32	95.3	172.0x73.5x45.3	2976	NR	i/i	dc/dc/ABS	pwr.r&p.	36.1	3.0	19.3	215/40ZR17 245/45ZR17 (re)	L4/2.2T/M5	**NA**

See page 393 for complete 1995 Price List

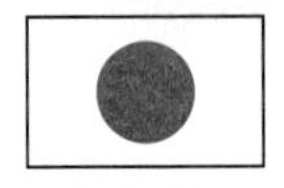

Expensive improvements...

The 323 and the Protegé are the most popular models, not without a cause, because their prices up until now were reasonable. This risks changing with the new models which have been redesigned and improved, but their prices have made a considerable jump. Disquieting...

Very unusual situation for the 323 and Protegé which have been completely redesigned for 1995. But the old versions, identified as S, will continue to be sold at the same time as the new ones for self evident price reasons. Since some of these old models are well-known, we have decided to reserve comments in the evaluation of the new born, the performances and characteristics of both generations are grouped in the tables. The latest 323 has been metamorphosed into a 3-door coupe with a sports look offered in GS with a 1.5 L and in LS with a 1.8 L. While the Protege remains a 4-door sedan whose format has increased, and is sold in LS or LX with a 1.8 L engine. The manual transmission is standard and the automatic 4-speed is optional.

STRONG POINTS

• Satisfaction: 85%
As a whole, the owners are very satisfied, but they advise of rapid clutch facing and brake pad wear as well as the flimflam quality of some accessories.

• Fuel consumption: 80%
If the new 1.8 L engine remains as it was, the new 1.5 L is very frugal and contents itself with 37 mpg in town and 50 mpg on the road which gives it an average of 40 mpg.

• Safety: 80%
The structure of those 2 models was further stiffened and reinforced. Also 2 air bags are offered as an option on the LS and GS and are standard on the ES/LS.

• Driver's compartment : 80%
Combined adjustments of the seat and steering column allow quickly finding the ideal driving position. The controls are well laid out (radio a little low) and the instruments are simple and easy to read. Their visibility is excellent in the Protegé, but less so toward the back of the coupe, whose forms are tortured, especially when fitted with a spoiler. One could complain that the two dashboards are different and the center console is recessed instead of pulled toward the driver as logic would have it.

• Price/equipment: 80%
The prices of the new models have nothing to do with the old ones, because they jump significantly up, which seems like a lot for popular cars whose equipment is only normal. This explains Mazda's decision to keep the old models with more reasonable lead prices.

• Suspension: 75%
It is the result of good compromises because its reactions are never harsh and it's only disagreeable on corrugated roads, where the wheels do their own natural frequency dance.

• Technical: 70%
While the 323 coupe is based on the old platform, the Protegé has seen its wheelbase, length and width increase. The steel unit bodies include 60% galvanized panels. The independent 4-wheel suspensions with stabilizer bars are those of preceding models with a few retouches. They are of the McPherson type at the front, with struts and double trapezoidal arms at the rear. Brakes are mixed and the ABS is optional on the LS/GS and standard on the ES/LS. If the 1.8 L engine remains identical, the 1.5 L is new and its fuel economy superior to that of the preceding 1.5 L.

DATA

Category: front-wheel drive compact sedans.
Class : 3S & 3

HISTORIC

Introduced in: 1977 :GLC
Modified in: 1989-1995 323-Protegé
Made in: Hiroshima, Japan.

RATINGS

		323	Protegé	
Safety:		80 %	80 %	
Satisfaction:		83%	83%	
Depreciation:		60 %	55 %	
Insurance:	($520)	7.5 %	6.5 %	($585)
Cost per mile:		$0.32	$0.32	

NUMBER OF DEALERS

In U.S.A.: 871

SALES ON THE AMERICAN MARKET

Model	1992	1993	Result
323	14,485	9,512	- 34.6 %
Protegé	53,474	81,642	+34.6 %

MAIN COMPETITORS

323: DODGE Colt, HONDA Civic Hbk, FORD Aspire-Escort, HYUNDAI Accent, NISSAN Sentra,TOYOTA Tercel, VOLKSWAGEN Golf.
Protegé: CHEVROLET Cavalier, DODGE-PLYMOUTH Colt-Neon, EAGLE Summit, FORD-MERCURY Escort-Tracer, HONDA Civic 4dr, HYUNDAI Elantra, PONTIAC Sunfire, SATURN SL1, TOYOTA Corolla, VW Jetta.

EQUIPMENT

MAZDA 323-Protegé	S	LX/GS	ES/LS
Automatic transmission:	O	O	O
Cruise control:	-	-	S
Power steering:	O	O	O
ABS brake:	-	O	S
Air conditioning:	-	S	S
Air bag (2):	O	S	S
Leather trim:	-	O	S
AM/FM/ Radio-cassette:	O	S	S
Power door locks:	-	-	S
Power windows:	-	-	S
Tilt steering column:	-	S	S
Dual power mirrors:	-	S	S
Intermittent wipers:	S	S	S
Light alloy wheels:	-	S	-
Sunroof:	-	-	-
Anti-theft system:	-	-	-

S : standard; O : optional; - : not available

COLORS AVAILABLE

Exterior: White, Red, Blue, Green, Emerald, Black, Beige, Ruby.
Interior: Gray, Taupe.

MAINTENANCE

First revision: 5,000 miles
Frequency: 5,000 miles
Diagnostic plug: No

WHAT'S NEW IN 1995 ?

- **Model fully redesigned.**

Model/ versions *: standard	ENGINE Type / timing valve / fuel system	Displacement cu/in	Power bhp @ rmn	Torque lb.ft @ rpm	Compres. ratio	TRANSMISSION Driving wheels / transmission	Final ratio	PERFORMANCE Acceler. 0-62 mph s	Stand. 1/4 mile s	Stand. 5/8 mile s	Braking 62-0 mph ft	Top speed mph	Lateral acceler. G	Noise level dBA	Fuel economy mpg city	Fuel economy mpg hwy	Gasoline type / octane
1)	L4* 1.6 SOHC-8-EFI	97	82 @ 5000	92 @ 2500	9.3 :1	front-M5*	4.10	11.0	17.5	35.0	128	106	0.75	69	28.0	37.9	R 87
						front-A4	3.74	12.5	18.4	34.4	138	100	0.75	69	24.6	35.0	R 87
2)	L4* 1.5 DOHC-16-EFI	91	92 @ 5000	96 @ 2500	9.4 :1	front-M5*	3.85	11.9	17.6	33.2	147	102	0.76	69	31.0	42.0	R 87
						front-A4	4.06	13.4	18.3	34.8	147	100	0.76	69	26.3	38.0	R 87
3)	L4* 1.8 DOHC-16-EFI	112	103 @ 5500	111 @ 4000	8.9 :1	front-M5*	3.62	8.5	15.4	29.0	135	115	0.78	69	27.2	36.8	R 87
						front-A4	3.48	9.4	16.0	29.7	141	112	0.78	69	24.0	32.7	R 87
4)	L4* 1.8 DOHC-16-EFI	112	122 @ 5500	117 @ 4000	9.0 :1	front-M5*	4.10	9.7	16.5	29.9	141	112	0.78	68	26.3	35.0	R 87
						front-A4	3.83	11.8	18.0	32.8	144	109	0.78	67	22.8	35.0	R 87

1) 323 base (Canada) 2) 323 GS (Canada) 3) Protegé S 4) Protegé LX/ES, 323 LS (Canada)

MAZDA 323

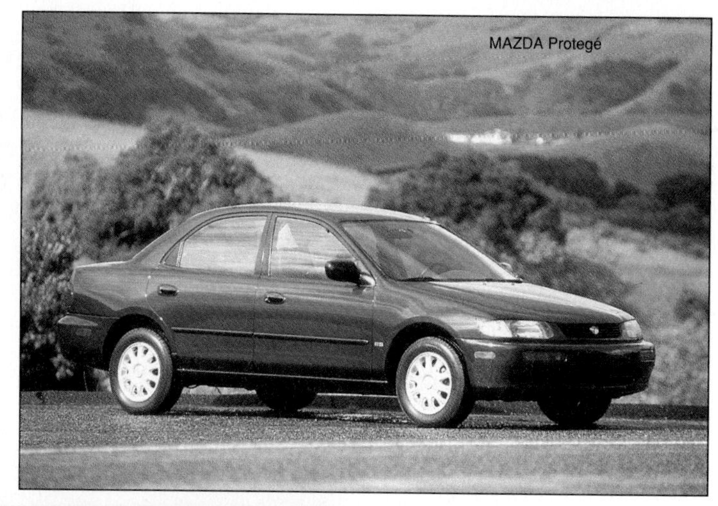
MAZDA Protegé

• **Steering: 70%**
It is now power-assisted on all versions and its degree of assistance varies as a function of speed. In practice, while it is precise, its assistance remains a little strong which makes it light, while the speed and maneuverability are only average, in spite of a normal steering diameter.

• **Quality/fit/finish: 70%**
The interior presentation is conventional and without fantasy, but the assembly quality, finish, and materials selected show care.

• **Access: 70%**
It does not cause any problem, particularly on the two models because the doors are amply sized, they open wide and the roof crown presents enough headroom.

• **Interior space: 65%**
It is as satisfactory in the 2 bodies, whose different dimensions are generous for this category. But especially in the Protegé whose main dimensions have been increased. The domed roof, a new tendency, gives much headroom without penalizing the aerodynamics.

• **Conveniences: 60%**
The most optimized storage (ES/LS) spaces include a little glove box, door pockets and two hollows in the dashboard however, a bad point, the shoulder harness and front seat belts are not adjustable in height.

• **Trunk: 60%**
It keeps the same volume as last year. However, it can be expanded by tipping the back of the seat bench and the opening on the Protegé is large.

• **Seats: 60%**
In spite of the firmness of their padding, their curvature provides an excellent lateral support, at hip level and across the back; also, the lumbar support is effective.

• **Insurance: 60%**
These small cars have a relatively high index and their premium attains an amount which is not negligible especially on the Protegé.

• **Handling: 50%**
The rigidity of the body, the efficient suspension geometry and the well-calibrated damping give these little cars a secure road handling, predictable and easy to control.

• **Performance: 50%**
These cars give the impression of going much faster than they really do; even though clean, standing starts and passings are ordinary, because the power increase just compensates for the weight increase.

WEAK POINTS

• **Braking: 30%**
Here, progress is less evident, because the facings lack bite, especially with ABS, which stabilizes the stopping path but lengthens it beyond the average.

• **Noise level: 40%**
In progress related to the previous models, it attracts attention by the impression of a lack of soundproofing and noisier engines than on the others. Wind noises are more discrete than rolling noises, which resonate in tune with the pavement.

• **Depreciation: 45%**
It is stronger on the 323 than on a Protegé, but is average for other models in this category.

CONCLUSION

• **Overall rating: 64%**
In spite of the value of the Yen, Mazda is charging full traffic for the improvements it has brought to these models stars whose improvements expanded in all directions.

SPECIFICATIONS & PRICES

Model	Version/Trim	Body/Seats	Interior volume cu.ft.	Trunk volume cu ft.	Drag coef. Cd	Wheel-base in	Lgth x Wdthx Hght. inx inx in	Curb weight lb	Towing capacity std. lb	Susp. type ft/rr	Brake type ft/rr	Steering system	Turning diameter ft	Steer. turns nber	Fuel. tank gal.	Standard tire size	Standard powertrain	PRICES US. $ 1994
MAZDA		**Sold in Canada only**																
323	S	3 dr.sdn.5	93	16.5	0.36	96.5	163.6x65.7x54.3	2194	1000	i/i	dc/dr	r&p.	31.5	4.3	13.2	155/80R13	L4/1.6/M5	-
323	GS	3 dr.sdn.5	91	15.7	0.33	98.6	163.6x67.3x55.3	2348	1000	i/i	dc/dr	pwr.r&p.	32.1	3.0	13.2	175/70R13	L4/1.5/M5	-
323	LS	3 dr.sdn.5	91	15.7	0.33	98.6	163.6x67.3x55.3	2480	1000	i/i	dc/dc	pwr.r&p.	32.1	3.0	13.2	185/65R14	L4/1.8/M5	-
		General warranty: 3 years / 50,000 miles; Power Train: 5 years / 60,000 miles; corrosion perforation: 5 years / unlimited.																
Protegé	S	4 dr.sdn.5	92	13	0.35	98.4	171.5x65.9x54.1	2326	1000	i/i	dc/dr	pwr.r&p.	32.1	3.0	14.5	175/70R13	L4/1.8/M5	**$8,995**
Protegé	LX	4 dr.sdn.5	95.5	13	0.32	102.6	174.8x67.3x55.9	2475	1000	i/i	dc/dc	pwr.r&p.	33.4	3.0	13.2	185/65R14	L4/1.8/M5	**$11,495**
Protegé	ES	4 dr.sdn.5	95.5	13	0.32	102.6	174.8x67.3x55.9	2524	1000	i/i	dc/dc	pwr.r&p.	33.4	3.0	13.2	185/65R14	L4/1.8/M5	**$13,195**

See page 393 for complete 1995 Price List

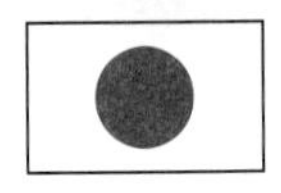

In the heart of the battle...

These compacts have been among the first to offer a V6 which has given them a serious advantage over their main rivals who stayed with 4 cylinders. This engine, which brought undeniable driving pleasure, is often missing these days from the "reasonable" cars.

Since their update, the 626 and MX 6 know a well-earned success, because they do not lack interest. These two models were created especially for the North-American market and are manufactured in Michigan with Ford's Probe, with which they share most powertrain elements. The 626s are 4-door, 3-volume sedans offered in versions DX, LX and SE while the MX 6 is a 3-volume coupe offered in finishes RS and LS.

STRONG POINTS

• Safety: **90%**
Two air bags and 3-point seat belts are standard equipment. These two cars attain the maximum score, thanks to the excellent rigidity of their structure.

• Seats: **80%**
They are as well-curved at the front as at the rear, and they maintain and support well with padding that is not too firm.

• Technical: **80%**
Their very rigid unit body includes two sub-frames which retain the suspension and mechanical elements. This characteristic offers many advantages both for handling, which is more precise, and for comfort, by suppressing a good part of the noise and vibration. In addition to the impression of robustness that stems from this type of construction, it is good for giving the occupants a feeling of very positive safety. Beginning with the same platform, each of these models has been adapted to offer a different personality. Their steel unit body has an independent McPherson suspension at all 4 wheels, with 2 trapezoidal arms at the rear and stabilizer bars front and rear. The standard braking is mixed, and four discs are offered with ABS at the high end of the range. The standard engine is a 4-cylinder 2.0 L while the 2.5 L V6 is mounted standard on the SE and LS. These power plants are at the head of today's technology. Their valve train includes 2 overhead cams and 4 valves per cylinder.V6 is one of the lightest in its category, thanks to its aluminum block and lightened accessories.

• Satisfaction: **80%**
Let us hope that these models will be less afflicted by reliability and materials quality problems.

• Performance: **70%**
They are virtually identical for both cars, whatever the powertrain selected. The V6 is one of the first of its generation to dispose of both horse power and torque at practically all rpm. Its climb in rpm rates is exciting and driving it gains a really sporty character.

• Quality/fit/finish: **70%**
The interior presentation is, as often happens on Japanese cars, rather dull because the plastics and trim are faded. However, the assembly is strict and the finish carefully done.

• Driver's compartment: **70%**
The dashboards are simple and rational and the well-grouped and

DATA

Category: front-wheel drive compact sedans and coupes.
Class : 4

HISTORIC

Introduced in: 1983 (front-wheel drive)
Modified in: 1986: Turbo;1987: reshaped;1988: MX-6; 1993:reshaped.
Made in: Flat Rock, MI, U.S.A.

RATINGS

Safety: 90 %
Satisfaction: 82 %
Depreciation: 62 % (626), 44% (2 years) MX-6
Insurance: 626 ($655) 5.4 % 6.1% ($725) MX-6
Cost per mile: $0.40

NUMBER OF DEALERS

In U.S.A.: 871

SALES ON THE AMERICAN MARKET

Model	1992	1993	Result
	57,065	71,377	+ 20.0 %
	26,785	29,068	+ 7.9 %

MAIN COMPETITORS

626 : BUICK Skylark, CHRYSLER Cirrus-Stratus HONDA Accord, MITSUBISH Galant, NISSAN Altima, OLDSMOBILE Achieva, PONTIAC Grand Am SUBARU Legacy, TOYOTA Corolla & Camry, VW Passat.
MX-6 : ACURA Integra, EAGLE Talon, FORD Probe, HONDA Prelude TOYOTA Celica, VW Corrado.

EQUIPMENT

MAZDA 626	DX	LX	ES
MAZDA MX-6		RS	LS
Automatic transmission:	O	O	O
Cruise control:	-	S	S
Power steering:	S	S	S
ABS brake:	O	O	S
Air conditioning:	-	S	S
Air bag (2):	S	S	S
Leather trim:	-	-	S
AM/FM/ Radio-cassette:	S	S	S
Power door locks:	-	S	S
Power windows:	-	S	S
Tilt steering column:	S	S	S
Dual power mirrors:	S	S	S
Intermittent wipers:	S	S	S
Light alloy wheels:	-	O	S
Sunroof:	-	O	S
Anti-theft system:	-	-	-

S : standard; O : optional; - : not available

COLORS AVAILABLE

Exterior: Artic Blue, Gold, Platinum Mica, Ruby, Black,Green, White, Red.
Interior: MX-6: Black, Beige. 626: Gray, Beige.

MAINTENANCE

First revision: 5,000 miles
Frequency: 5,000 miles
Diagnostic plug: Yes

WHAT'S NEW IN 1995 ?

- No major change.

Model/ versions *: standard	ENGINE Type / timing valve / fuel system	Displacement cu/in	Power bhp @ rmn	Torque lb.ft @ rpm	Compres. ratio	TRANSMISSION Driving wheels / transmission	Final ratio	PERFORMANCE Acceler. 0-62 mph s	Stand. 1/4 mile s	Stand. 5/8 mile s	Braking 62-0 mph ft	Top speed mph	Lateral acceler. G	Noise level dBA	Fuel economy mpg city	hwy	Gasoline type / octane
DX-LX	L4*2.0 DOHC-16-MFI	121	118 @ 5500	127 @ 4500	9.0 :1	front - M5*	4.11	10.2	16.8	30.2	138	112	0.78	65	30.0	43.0	R 87
						front - A4	3.77	11.3	17.6	30.8	144	109	0.78	65	26.0	39.0	R 87
ES	V6*2.5 DOHC-24-MPFI	152	164 @ 5600	160 @ 4800	9.2 :1	front - M5*	4.11	8.0	16.0	29.5	125	124	0.80	64	24.0	33.0	M 89
						front - A4	4.16	9.6	16.5	30.3	138	195	0.80	64	22.0	34.0	M 89

legible gauges as well as the main controls are within hand's reach. The switches at the bottom of the dash are the exception, invisible at night. Note the presence of the automatic transmission shift indicator, placed in the middle of the gauges. Visibility is good, in spite of the "C" pillar width on the 626, the lack of backlight height and the presence of a spoiler on the MX-6. The driver is more coddled in the coupe where he disposes of a real bucket seat but the locked container which is part of the center armrest is in the way of the right elbow.

• Steering: 70%
Precise and well-modulated, it suffers from a light torque steer during energetic accelerations and seems more direct and inspired on the MX-6.

• Access: 70%
It is not difficult to reach the rear seats of these two cars, even though the rear doors of the sedan are a little narrower.

• Suspension: 70%
Its damping is only average and the limited wheel travel makes it bottom out frequently on large bumps.

• Noise level: 70%
The 4-cylinder is a little noisier than the V6 whose discretion amazes on models of this class where wind or body noises are inaudible, even on bad pavement.

• Conveniences: 70%
The storage includes a medium-sized glove box, door pockets and hollows in the center console.

• Insurance: 65%
Its rate is in the average of this category.

• Handling: 65%
Even if it favors comfort, the softer suspension of the 626 retains an imperturbable stability in wide curves, tight turns or straight line, where it stays neutral for a long time before showing a tendency towards understeer. The weight balance and the rating of the suspension and tires make the MX-6 more sporty. It is less agile than Ford's Probe on a twisty path but more stable in a straight line or wide curb.

Mazda MX-6

Mazda 626

• Fuel consumption: 60%
The 4 cylinder is less thirsty than the V6 whose power is called upon more generously.

• Trunk: 55%
That of the 626 lacks a little height because of the lid lines but is wide and deep. Its capacity can be increased by lowering the seat back. That of the MX-6 is not as practical even though it can be increased toward the interior, because it has less height and a taller sill as well as a trunk opening that limits its use.

• Price/equipment: 50%
Except for the 626 DX, these cars are properly equipped and the prices are aimed at their Japanese rivals.

• Braking: 50%
They only match the performance level when equipped with 4 discs and ABS, with allows more effective stops. Otherwise, the front wheels tend to lock up early and the distances becomes longer. The owners complain of rapid wear and of break squeals.

WEAK POINTS

• Depreciation: 40%
It is strong, because of the mixed reputation of previous models and of Mazda's service.

• Interior space: 40%
The 628 interior accepts four adults in comfort and its main sizes have been well-calculated including ample legroom for the rear seats. In the coupe, the height is limited in the front by the sunroof and at the back by the roof shape.

CONCLUSION

• Overall rating: 66%
These cars don't lack in charm, in their own way, but their attraction is mainly in the pleasurability of their V6 engine. ☺

SPECIFICATIONS & PRICES

Model	Version/Trim	Body/ Seats	Interior volume cu.ft.	Trunk volume cu ft.	Drag coef. Cd	Wheel-base in	Lgth x Wdthx Hght. inx inx in	Curb weight lb	Towing capacity std. lb	Susp. type ft/rr	Brake type ft/rr	Steering system	Turning diameter ft	Steer. turns nber	Fuel. tank gal.	Standard tire size	Standard powertrain	PRICES US. $ 1994
MAZDA		**General warranty: 3 years / 50,000 miles; Power Train: 5 years / 60,000 miles; corrosion perforation: 5 years / unlimited.**																
626 DX	4dr.sdn.5	97	13.8	0.32	102.8		184.4x68.9x55.1	2743	1000	i/i	dc/dr	pwr.r/p.	34.8	2.9	15.8	195/65R14	L4/2.0/M5	**$14,255**
626 LX	4dr.sdn.5	97	13.8	0.32	102.8		184.4x68.9x55.1	2743	1000	i/i	dc/dr	pwr.r/p.	34.8	2.9	15.8	195/65R14	L4/2.0/M5	**$16,540**
626 ES	4dr.sdn.5	97	13.8	0.32	102.8		184.4x68.9x55.1	2906	1000	i/i	dc/dc/ABS	pwr.r/p.	34.8	2.9	15.8	205/55VR15	V6/2.5/M5	**$21,545**
MX-6 RS	2dr,cpe.4	79.6	12.4	0.31	102.8		181.7x68.9x51.6	2623	1000	i/i	dc/dr	pwr.r&p.	35.8	2.9	15.8	195/65R14	L4/2.0/M5	**$17,495**
MX-6 LS	2dr.cpe.4	79.6	12.4	0.31	102.8		181.7x68.9x51.6	2800	1000	i/i	dc/dc	pwr.r&p.	35.8	2.9	15.8	205/55VR15	V6/2.5/M5	**$21,495**

See page 393 for complete 1995 Price List

Maxima's version from Mazda...

By introducing the Millenia, Mazda got 2 shots from one stone. In one way it replaces its less-than-virtuous 929 by a more homogenous and original model, and on the other hand it saves part of its investment capital in a deluxe marque which aborted.

The Millenia, which originally was going to be part of the luxury range Mazda wanted to use as a response to the Acura Infiniti and Lexus will instead replace the 929 who never received the projected success. Financial problems decided otherwise, because a short while before its inauguration, the Amati was killed in the egg. Of the two models introduced to be sold under that label, only the Millenia survived and the other, a great luxury vehicle,with a V12, was tabled. Sold throughout the world, the Millenia is known as Eunos in Japan, Xedos in Europe and Amati in South America. It particularly addresses itself to male owners(80%), married (80%), earning an average of $75,000 American a year and 75% are self-employed living in the suburbs of a large metropolitan area. It is a front wheel drive deluxe sedan with 3 volumes, four doors, whose standard model is powered by the same 2.5 L V6 as the 626. The S has a 2.3 L which functions with a Miller cycle which allows it, for equal displacement, to develop more horse power with less fuel consumption and lower emissions. The base model is sold with cloth or leather trim.

STRONG POINTS

• Safety: **90%**
This sedan benefits from the latest development in matters of stiffness and structure and 2 air bags (standard) protect the front seats. The seats are as aesthetic as they are comfortable, a pleasant surprise on a Japanese car. They are well-shaped and padded to offer superb lateral hold even though they are not equipped with a lumbar support.

• Technical: **85%**
The Millenia is a steel unit body whose fluid lines have remarkable aerodynamics with a drag coefficient of 0.29 in the lowest end of the range. The four wheel independent suspension uses multiple arms at both ends and a variable power assist determined by the engine speed. The 4-wheel disc brakes have a standard ABS.It is the S engine which draws attention. It operates on a Miller cycle to obtain a very high expansion ratio but a low compression ratio thanks to a Lysholm compressor. The compression time is shortened ,which creates a fifth cycle allowing a better cylinder fill and a higher compression ratio to gain 1.5 times more torque and power for a given displacement and a 10 to 15% lower fuel consumption.

• Quality/fit/finish: **80%**
As with most Japanese products, the Millenia is assembled with much care, its finish is impeccable, down to the least details and the quality of the materials is without reproach. However, the presentation of the dashboard is questionable in that its style does not fit with the character of this model. The black matte plastic material looks gloomy.

• Insurance: **80%**
In spite of its reasonable rate, the premium is high in proportion to the price of these models.

DATA

Category: front-wheel drive luxury sedans.
Class : 7

HISTORIC

Introduced in: 1994
Modified in: -
Made in: Hofu, Japan.

RATINGS

Safety: 90 %
Satisfaction: NA %
Depreciation: NA %
Insurance: 3.6 % ($785) S: ($825)
Cost per mile: $0.60

NUMBER OF DEALERS

In U.S.A.: 871

SALES ON THE AMERICAN MARKET

Model	1992	1993	Result
Millenia	Not on the market at that time		

MAIN COMPETITORS

ACURA Vigor, AUDI A4, BUICK Riviera, CADILLAC De Ville, LEXUS ES300, LINCOLN Continental, MERCEDES C-Class, MITSUBISHI Diamante, NISSAN Maxima, OLDSMOBILE Aurora, SAAB 900, TOYOTA Avalon, VOLVO 850 & 940.

EQUIPMENT

Mazda Millenia	base fabric	base leather	S
Automatic transmission:	S	S	S
Cruise control:	S	S	S
Power steering:	S	S	S
ABS brake:	S	S	S
Air conditioning:	S	S	S
Air bag (2):	S	S	S
Leather trim:	-	S	S
AM/FM/ Radio-cassette:	S	S	S
Power door locks:	S	S	S
Power windows:	S	S	S
Tilt steering column:	S	S	S
Dual power mirrors:	S	S	S
Intermittent wipers:	S	S	S
Light alloy wheels:	S	S	S
Sunroof:	-	S	S
Anti-theft system:	S	S	S

S : standard; O : optional; - : not available

COLORS AVAILABLE

Exterior: Red, Blue, Metallic Gray, Champagne Gray, Green, White.
Interior: fabric: Gray. leather: Gray, Beige.

MAINTENANCE

First revision: 5,000 miles
Frequency: 5,000 miles
Diagnostic plug: Yes

WHAT'S NEW IN 1995 ?

- All new model launched in 1994 as a 95.

ENGINE					TRANSMISSION					PERFORMANCE								
Model/ versions *: standard	Type / timing valve / fuel system	Displacement cu/in	Power bhp @ rmn	Torque lb.ft @ rpm	Compres. ratio	Driving wheels / transmission	Final ratio	Acceler. 0-62 mph s	Stand. 1/4 mile s	Stand. 5/8 mile s	Braking 62-0 mph ft	Top speed mph	Lateral acceler. G	Noise level dBA	Fuel economy mpg city	Fuel economy hwy	Gasoline type / octane
Millenia	V6*2.5 DOHC-24-MPFI	152	170 @ 5800	160 @ 4800	9.2: 1	front - A4	4.176	9.3	16.7	29.8	148	125	0.80	65	23.0	33.0	S 91
Millenia S	V6*2.3 DOHC-24-MPFI	138	210 @ 5300	210 @ 3500	8.0: 1	front - A4	3.805	8.2	16.1	28.7	135	136	0.80	67	23.0	33.0	S 91

NEW FOR 1995

- **Steering: 70%**
Its power assist is well-modulated and the precision as well as the speed are great, but at some speeds it transmits road defects and maneuverability would be better if it turned a little tighter.
- **Handling: 70%**
The performance/handling ratio of the Millenia is great and reactions are precise. With them it is easy to enter the turn and follow a line, however some "thank you mam" pavement will make it lose its assurance.
- **Fuel consumption: 70%**
It is reasonable with a V6 and the 626 but climbs rapidly with a Miller engine whose temperament incites you to play more with the accelerator.
- **Driver's compartment: 70%**
It is relatively well-organized and its appearance relates to 929. The controls are distinctly more rational, but those on the center console would be easier to reach if the latter would be oriented towards the driver.
- **Performance: 70%**
Placid with the 2.5 L, it is more trouble handling the weight and more exciting with the Miller, which has nerve and whose take-offs and accelerations are tonic.
- **Access: 70%**
It is easier at the front than at the rear, and the roof curve forces you to lower the head to enter or leave. Let us point out that the mechanical access to the Miller cycle engine is touchy because the engine compartment overflows with technology.
- **Suspension: 70%**
Perfect on good pavement, but the front end expresses its disapproval of any road in poor condition.
- **Sound level: 70%**
It is higher with a Miller engine than with a base V6 which remains as discrete as on a 626.
- **Interior space: 60%**
It is acceptable as is, because while head and leg rooms is adequate, in front and back, there is nothing extra and a 5th person sitting in the back won't be a welcome sight.
- **Trunk: 60%**
Larger and more usable than the 929, it lacks height, especially when a CD player is installed.
- **Conveniences: 50%**
Storage areas do not abound, because the glove box and door pockets are small, the center box is adequate and the rear armrest is not used. At least the front shoulder harnesses have a height adjustment.

WEAK POINTS

- **Price/equipment: 25%**
The price of the S seems normal considering the advanced engine technology, but appears exaggerated on the base version with the cloth trim and is not competitive with the Maxima, its main rival.
- **Braking: 45%**
Its moderate efficiency does not square up with the performance of which the S is capable. Stopping distances are long with the ABS, which perfectly stabilizes the vehicle's trajectory but fade-resistance is much less adequate.

CONCLUSION

- **Overall rating: 68.0%**
The Millenia advantageously replaces the 929 from both the dynamic and ergonomic angles, but it is also more practical and economical. However Mazda is missing the boat by not offering a distinctly less expensive base version which would curtail the free run of the excellent Maxima that competes against it. ☺

SPECIFICATIONS & PRICES

Model	Version/Trim	Body/ Seats	Interior volume cu.ft.	Trunk volume cu ft.	Drag coef. Cd	Wheel-base in	Lgth x Wdthx Hght. inx inx in	Curb weight lb	Towing capacity std. lb	Susp. type ft/rr	Brake type ft/rr	Steering system	Turning diameter ft	Steer. turns nber	Fuel. tank gal.	Standard tire size	Standard powertrain	PRICES US. $ 1995
MAZDA		**General warranty: 3 years / 50,000 miles; Power Train: 5 years / 60,000 miles; corrosion perforation: 5 years / unlimited.**																
Millenia	fabric	4dr.sdn. 5	94	13.3	0.29	108.3	189.8x69.7x54.9	3216	NR	i/i	dc/dcABS	pwr.r/p.	37.4	2.9	18	205/65R15	V6/2.5/A4	**$25,995**
Millenia	leather	4dr.sdn. 5	94	13.3	0.29	108.3	189.8x69.7x54.9	3232	NR	i/i	dc/dcABS	pwr.r/p.	37.4	2.9	18	205/65R15	V6/2.5/A4	**$28,300**
Millenia	S	4dr.sdn. 5	94	13.3	0.29	108.3	189.8x69.7x54.9	3391	NR	i/i	dc/dcABS	pwr.r/p.	37.4	2.9	18	215/55R16	V6/2.3/A4	**$31,400**

See page 393 for complete 1995 Price List

End of romance...

The MPV has known a spectacular popularity during its introduction to the market. Mini-van fans saw it streamlined, sporty and less utilitarian than its colleagues. After several years and many frustrations you have to stop dreaming.

The MPV addresses itself to those who want a spacious vehicle but not the appearance of a delivery wagon. It is a short wheel based 2-volume with four doors, offering five or seven seats. The standard rear 2-wheel drive comes with a 4-cylinder 2.6 L with a 4-speed automatic or an optional 3.0 L V6 with electronic transmission and computer management which is standard on a 4-wheel drive. The trim levels are standard or LX.

STRONG POINTS

- **Safety:** 80%

It is adequate, thanks to its structural impact resistance and comes standard with a driver's side air bag and 3-point belts for all other occupants.

- **Quality/fit/finish:** 80%

The general appearance of the MPV is closer to that of an automobile than to a utility with its hinged side door, assembly, and careful finish.

- **Satisfaction:** 80%

Some owners complain of the rapid brake friction material wear, lack of spare parts, and the inability of some dealers to rapidly resolve problems.

- **Suspension:** 80%

In spite of a firm response, it is good at absorbing road defects, thanks to ample wheel travel.

- **Noise level:** 70%

It is comfortable on freeways at cruising speeds since rolling noises are well-filtered, but the engines signal their presence during acceleration.

- **Conveniences:** 70%

The conveniences include a minuscule glove box, practical door pockets and a drawer under the right front seat.

DATA

Category: rear-wheel drive or 4WD compact vans.
Class : utility

HISTORIC

Introduced in: 1988
Modified in: 1989: 4 WD version.
Made in: Hiroshima, Japan.

RATINGS

Safety: 90 %
Satisfaction: 77 %
Depreciation: 60 %
Insurance: 5.2 % ($655)
Cost per mile: $0.42

NUMBER OF DEALERS

In U.S.A.: 871

SALES ON THE AMERICAN MARKET

Model	1992	1993	Result
MPV	32,503	30,592	- 5.9 %

MAIN COMPETITORS

CHEVROLET Astro & Lumina, DODGE-PLYMOUTH Caravan-Voyager, FORD Aerostar & Windstar, MERCURY Villager, NISSAN Quest, PONTIAC Trans Sport, TOYOTA Previa, VW Eurovan.

EQUIPMENT

MAZDA MPV	2x4 base	2x4 LX	4x4 base	4x4 LX
Automatic transmission:	S	S	S	S
Cruise control:	-	S	-	S
Power steering:	S	S	S	S
ABS brake:	S	S	S	S
Air conditioning:	S	S	S	S
Air bag (left):	O	O	O	O
Leather trim:	-	O	-	0
AM/FM/ Radio-cassette:	S	S	S	S
Power door locks:	S	S	S	S
Power windows:	-	S	-	S
Tilt steering column:	S	S	S	S
Dual power mirrors:	S	S	S	S
Intermittent wipers:	S	S	S	S
Light alloy wheels:	-	S	S	S
Sunroof:	-	O	-	O
Anti-theft system:	-	-	-	-

S : standard; O : optional; - : not available

COLORS AVAILABLE

Exterior: White, Beige, Blue, Ruby, Green, Silver, Bordeaux.
Interior: Gray, Taupe.

MAINTENANCE

First revision: 5,000 miles
Frequency: 5,000 miles
Diagnostic plug: Yes

WHAT'S NEW IN 1995 ?

- No major change.

Model/ versions *: standard	ENGINE Type / timing valve / fuel system	Displacement cu/in	Power bhp @ rmn	Torque lb.ft @ rpm	Compres. ratio	TRANSMISSION Driving wheels / transmission	Final ratio	PERFORMANCE Acceler. 0-62 mph s	Stand. 1/4 mile s	Stand. 5/8 mile s	Braking 62-0 mph ft	Top speed mph	Lateral acceler. G	Noise level dBA	Fuel economy mpg city	hwy	Gasoline type / octane
base	L4* 2.6 SOHC-12-MPFI	159	121 @ 4600	140 @ 3800	8.4 :1	rear-A4*	3.91	12.7	19.3	36.6	180	100	0.71	66	21.0	30.0	R 87
V6 2x4	V6* 3.0 SOHC-18-MPFI	180	155 @ 5000	169 @ 4000	8.5 :1	rear-A4*	3.91	11.0	18.0	32.8	180	106	0.72	66	18.0	27.0	R 87
V6 4x4	V6* 3.0 SOHC-18-MPFI	180	155 @ 5000	169 @ 4000	8.5 :1	rear/4-A4*	4.10	11.8	18.4	35.5	177	106	0.72	66	17.0	24.0	R 87

• **Technical** 70%
The steel unit body is stiffened by an auxiliary chassis welded to the platform and the suspensions are of the McPherson type up front and there is a rigid axle at the rear, guided by length wise torsion bars plus a Panhard rod with stabilizer bars front and rear. The brakes are mixed with a standard ABS acting on the rear wheels. The MPV distinguishes itself by its rear wheel drive and a hinged rather than a sliding door. The lines are pleasant and efficient and the aerodynamic coefficient is favorable for this type of vehicle.

• **Access:** 70%
It offers the advantage of a solid mounting, exempt from noise vibrations, and a swinging sidedoor that is not the most practical because it cannot open wide in case of tight parking.

• **Insurance:** 70%
Its rate is normal and the premium is in the average of the category.

• **Interior space:** 60%
The front seats are among the more spacious, while room is tight between the bench seats when the MPV is set up for 7 people. The best arrangement is to remove one bench seat to have 5 readily accessible seats.

• **Driver's compartment:** 60%
The driver does not benefit from an ideal position because the pedals are too close to the front seat and the steering wheel is too far away. A telescopic steering post would help because the height adjustment is not sufficient. Visibility is excellent, but the dashboard is not very ergonomic since the gauge group is too high and the switches, which are located there, are unusual and difficult to reach.

• **Seats:** 60%
They are very firm, the armrests are too high and non-adjustable and they limit the width of the entry toward the front. At the rear, the middle bench seat is very heavy and hard to manipulate and the one at the end cannot be folded and is attached to the floor, limiting its versatility.

• **Steering:** 60%
It is well-modulated, but its large reduction makes it vague in the center on the 2WD which hurts its linearity and its maneuverability. The turning diameter is large.

• **Performances:** 50%
The large weight penalizes accelerations and take-offs in spite of a torque converter with a lock up clutch. The transmission operates in three modes, power, economy and cruise. The V6 is preferable to the anemic and not very economic 4-cylinder.

• **Handling:** 50%
On this plane, the MPV outclasses most of its competition, thanks to its more elaborate suspension which serves both comfort and handling. In spite of the height of the center of gravity, there is little roll, and traction in tight curves is hard to fault. However, in winter driving the stability of the 4WD will be more reassuring than that of the 2WD.

• **Price/equipment:** 50%
The standard 4X4 is less expensive than the Previa but it is not competitive with Chrysler, because its equipment is more complete and lacks only air conditioning and power equipment.

WEAK POINTS

• **Braking:** 20%
It is the weak point of this vehicle because its lack of bite and fade causes long stopping distances, fortunately straightlined, thanks to ABS acting on the rear wheels.

• **Fuel consumption:** 30%
The rigidity of the body is paid for by weight and high fuel consumption under all conditions.

• **Storage:** 40%
Its volume is virtually nil when the bench is occupied and many owners would like to see an extended version. When the seats are removed, you can see that the floor is not flat and the roof height is limited.

• **Depreciation:** 40%
The MPV begins to lose more value than its rivals in this category.

CONCLUSION

• **Overall rating:** 59.5%
The romance between the public and the MPV had a good beginning but is now over. Because one had to realize that it was not practical, not as reliable and certainly not as economical as imagined.

OWNER'S SUGGESTIONS

-More effective brakes
-Rear seats removable
-Extended body for additional baggage space
-Better performance
-More economical operation
-More comfortable seats
-More powerful engine

SPECIFICATIONS & PRICES

Model	Version/Trim		Body/ Seats	Drag coef. Cd	Wheel-base in	Lgth x Wdthx Hght. inx inx in	Curb weight lb	Towing capacity std. lb	Susp. type ft/rr	Brake type ft/rr	Steering system	Turning diameter ft	Steer. turns nber	Fuel. tank gal.	Standard tire size	Standard powertrain	PRICES US. $ 1994
MAZDA		General warranty: 3 years / 50,000 miles; Power Train: 5 years / 60,000 miles; corrosion perforation: 5 years / unlimited.															
MPV	2WD	base	4 dr.van.5	0.36	110.4	175.8x71.9x68.0	1632	2299	i/r	dc/dc/ABS	pwr.r&p.	39.0	3.9	15.8	195/75R15	L4/2.6/A4	**$18,195**
MPV	2WD	base	4 dr.van.7	0.36	110.4	175.8x71.9x68.0	1669	2600	i/r	dc/dc/ABS	pwr.r&p.	39.0	3.9	19.5	195/75R15	V6/3.0/A4	**$19,690**
MPV	2WD	LX	4 dr.van.7/8	0.36	110.4	175.8x71.9x68.0	1699	2600	i/r	dc/dc/ABS	pwr.r&p.	39.0	3.9	19.5	215/65R15	V6/3.0/A4	**$22,190**
MPV	4WD	base	4 dr.van.7	0.36	110.4	175.8x71.9x70.8	1819	2600	i/r	dc/dc/ABS	pwr.r&p.	39.7	3.6	19.8	215/65R15	V6/3.0/A4	**$23,395**
MPV	4WD	LX	4 dr.van.7	0.36	110.4	175.8x71.9x70.8	1833	2600	i/r	dc/dc/ABS	pwr.r&p.	39.7	3.6	19.8	215/65R15	V6/3.0/A4	**$26,945**

See page 393 for complete 1995 Price List

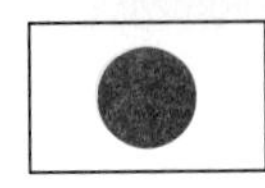

Chance smiles upon audacity...

The good initiative on the part of Mazda was to pick up the concept of a sports mini-coupe prematurely abandoned by Honda. It took a good dose of audacity to allow yourself to start up the smallest V6 in the industry. Why not a mini -rotary?

The MX-3 is successful because it corresponds exactly to what the public expects from a small popular sports coupe. It picks up precisely the formula Honda invented for its CRX, which was abandoned for obscure reasons. Based on the platform of the first 323 Protegé, its 2-volume 3-door body is offered in RS with a 4-cylinder 1.6 L, DOHC and 16 valves, which develops 105 hp. The GS comes with a 1.8 L V6 and DOHC with 4 valves that produces 130 hp.

STRONG POINTS

• Technical : 80%

The lines of the MX-3 don't fail to attract attention, especially in some colors, because its two extremities are too provocative. The back is squat, unlike the front, which is very slopping, with its inclined windshield and almond shaped-head lights. Its aerodynamic capabilities are ample, with a coefficient of 0.32 for the RS and 0.31 for the GS. The steel unit body has the suspension elements of the 323 Protegé, which is to say it is independent at all 4 wheels with McPherson type suspension units. However, the tracks have been widened and the geometry has been revised to improve comfort and handling. With its new cylinder head, the 4-cylinder is less wimpy than in the past but it is the 1.8 L V6 which will keep it as a star because it is the smallest V6 in the world.

• Satisfaction: 75%

The owners of the MX-3 are generally satisfied with their machine, but complain of the annoying disc brake pad noise and its rapid wear.

• Driver's compartment: 75%

Comfortably installed in a well-formed seat, the driver enjoys good visibility toward the front and sides, in spite of the lack of belt line height. Toward 3/4 rear, visibility is distinctly mediocre, because of the center pillar thickness and the narrowness of the backlight which is very limited and blots out in the rain.Dashboard is functional, with legible gauges and the controls remind us of those of the 323.

• Fuel consumption: 75%

The V6 is thirstier than the 4-cylinder, especially with an automatic, but it brings a pleasurability which allows output to be considered normal.

• Performance: 75%

In spite of the technical advance of the engine, the dynamic effect is not as extraordinary because the performances are insipid and the numbers compare to those of the Miata. In both instances, accelerations impress the ear more than the stop watch. The V6 is ahead of the 4-cylinder and its operation is exempt from vibrations.

• Handling: 75%

The body stiffness, the suspension geometry, and wheels moved to the 4 corners provide good road handling to the MX-3. There is less roll on the GS, whose suspension is firmer but remains neutral longer before the readily-controlled understeer put in an appearance. One can note once again, that the front wheel drive does not provide as

DATA

Category: front-wheel drive sport coupes.
Class : S3

HISTORIC

Introduced in: 1992
Modified in: 1994: DOHC1.6 L engine .
Made in: Hiroshima, Japan

RATINGS

Safety: 80 % (90 % with airbag)
Satisfaction: 75 %
Depreciation: (RS) 45 % 53 % (GS)
Insurance: RS ($585) 6.2 % 6.8 % GS ($655)
Cost per mile: $0.45

NUMBER OF DEALERS

In U.S.A.: 871

SALES ON THE AMERICAN MARKET

Model	1992	1993	Result
MX-3	27,674	23,441	- 15.3 %

MAIN COMPETITORS

ACURA Integra, EAGLE Talon, HONDA del Sol, HYUNDAI Scoupe, MAZDA Miata, SATURN SC, TOYOTA Paseo.

EQUIPMENT

MAZDA MX-3	RS	GS
L4 1.6L engine	S	-
V6 1.8L engine	-	S
Automatic transmission:	O	O
Cruise control:	-	S
Power steering:	O	O
ABS brake:	O	O
Air conditioning:	S	S
Air bag (2):	-	O
Leather trim:	-	-
AM/FM/ Radio-cassette:	S	S
Power door locks:	-	O
Power windows:	-	O
Tilt steering column:	S	S
Dual power mirrors:	S	S
Intermittent wipers:	S	S
Light alloy wheels:	O	S
Sunroof:	S	S
Anti-theft system:	-	-

S : standard; O : optional; - : not available

COLORS AVAILABLE

Exterior: White, Blue, Red, Silver, Black, Green.
Interior: Black.

MAINTENANCE

First revision: 5,000 miles
Frequency: 5,000 miles
Diagnostic plug: Yes

WHAT'S NEW IN 1995 ?

- No major change.

	ENGINE					TRANSMISSION		PERFORMANCE										
Model/ versions *: standard	Type / timing valve / fuel system	Displacement cu/in	Power bhp @ rmn	Torque lb.ft @ rpm	Compres. ratio	Driving wheels / transmission	Final ratio	Acceler. 0-62 mph s	Stand. 1/4 mile s	Stand. 5/8 mile s	Braking 62-0 mph ft	Top speed mph	Lateral acceler. G	Noise level dBA	Fuel economy mpg city	Fuel economy mpg hwy	Gasoline type / octane
RS	L4* 1.6 DOHC-16-MPFI	97	105 @ 6200	100 @ 3600	9.0 :1	front - M5*	4.11	9.8	16.8	29.7	131	112	0.82	68	34.0	47.0	R 87
						front - A4	3.83	11.0	17.2	32.0	134	106	0.82	68	30.0	43.0	R 87
GS*	V6* 1.8 DOHC-24-MPFI	112	130 @ 6500	115 @ 4500	9.2 :1	front - M5*	4.39	8.8	15.7	28.8	128	125	0.85	68	26.0	37.0	R 87
						front - A4	4.06	10.0	16.6	30.2	131	112	0.85	68	23.0	35.0	R 87

*(V6 engine only available in Canada)

much driving pleasure as the rear wheel drive of the Miata.

• Safety: 75%
Even though the unit body offers excellent rigidity, the optional furnishings such as the two airbags can vary the coefficient. It is odd that on a machine of this type, this essential safety element is only offered as an option.

• Steering: 70%
That of the GS is livelier and more precise than the one of the RS which is too overratioed and vague in the center.

• Conveniences: 70%
The storage includes a useful glove box, 2 door pockets, a lock box and a recess in the center console.

• Price/equipment: 70%
Since its engine is more powerful, the price of the RS has become more attractive than that of the GS and this in spite of the difference in equipment. However the option list remains long.

• Quality/fit/finish: 70%
The interior presentation is gray and contrasts with the sympathetic allure of the body. Assembly quality and finish are careful, the fabrics and plastics have a look and a "feel" touch that are econo. A simple little touch of color on the seats and doors would pleasantly distinguish the GS from the RS.

• Seats: 70%
They lack softness and the rear bench seat is less well-shaped than the front seat, which offers better support. It is however less insipid compared to other competing coupes.

• Insurance: 60%
Index is that of sports cars and reflects the performance and the clientele targeted by this coupe.

• Access: 60%
Tall people can't fit on the rear bench seat without pain because it is suited for young children.

• Suspension: 60%
More handling than comfort, because its strictness doesn't allow ignoring any road defect.

• Trunk: 60%
It offers more space than that of the MX-6 because it can be increased toward the interior by opening up the bench seat. However, its high sill, needed for stiffness, complicates its use.

• Depreciation: 50%
The resale value of these coupes holds up better than that of the sedans. Proportionately, that of the more affordable RS is higher than that of the GS, more exotic from a budget view point.

• Braking: 50%
The brakes have more bite and more efficiency on the GS equipped with 4 discs than on the RS whose stopping distances are long and their relative stability in panic stops does not necessarily require the presence of an ABS.

• Noise level: 50%
V6 is more discrete than the 4 cylinder, which loudly manifests itself during strong acceleration, while rolling noises reveal the lack of insulating material.

WEAK POINTS

• Interior space: 30%
While restrained, it is surprising for this type of vehicle, because it is equivalent to that of coupes in the higher category. Certainly one disposes of more space in front height than at the rear, where the ceiling is low because of the roof slope.

CONCLUSION

• Overall rating: 65.0%
Even though it is not a real performance rocket, the MX-3 is much fun to drive, practical to use on a reasonable budget and its allure is very sympathetic. ☺

SPECIFICATIONS & PRICES

Model	Version/Trim	Body/ Seats	Interior volume cu.ft.	Trunk volume cu ft.	Drag coef. Cd	Wheel-base in	Lgth x Wdthx Hght. inx inx in	Curb weight lb	Towing capacity std. lb	Susp. type ft/rr	Brake type ft/rr	Steering system	Turning diameter ft	Steer. turns nber	Fuel. tank gal.	Standard tire size	Standard powertrain	PRICES US. $ 1994
MAZDA		**General warranty: 3 years / 50,000 miles; Power Train: 5 years / 60,000 miles; corrosion perforation: 5 years / unlimited.**																
MX-3	RS	3dr.cpe.2+2	80	15.4	0.32	96.7	165.7x66.7x51.6	2414	NR	I/I	dc/dr	pwr.r&p.	32.8	3.1	13.2	P185/65R14	L4/1.6/M5	**$13,595**
	RS	3dr.cpe.2+2	80	15.4	0.32	96.7	165.7x66.7x51.6	2480	NR	I/I	dc/dr	pwr.r&p.	32.8	3.1	13.2	P185/65R14	L4/1.6/A4	**$14,345**
MX-3	GS	3dr.cpe.2+2	80	15.4	0.31	96.7	165.7x66.7x51.6	2581	NR	I/I	dc/dc	pwr.r&p.	32.8	2.7	13.2	P205/55R15	V6/1.8/M5	**$16,095**
	GS	3dr.cpe.2+2	80	15.4	0.31	96.7	165.7x66.7x51.6	2632	NR	I/I	dc/dc	pwr.r&p.	32.8	2.7	13.2	P205/55R15	V6/1.8/A4	**$16,845**

(V6 engine only available in Canada)

See page 393 for complete 1995 Price List

MAZDA MX-5 Miata

Road runner...

The Miata has opened the way of affordable pleasure cars. BMW and Mercedes have used it as an inspiration to create models that will be launched in 95 and 96. The winning formula consists of offering a simple, reliable, and authentic machine. In this case the contract is completed.

Another audacity from Mazda which decided to revive the roadsters of the 60's. It was a good idea because the Mazda has known ample success with a wide acceptance. This simple little car is the only one which allows tasting the joy of driving a sports car for a reasonable budget, without excess power of speed. It is a little two seater with a 1.8 L 4-cylinder, a manual 5-speed or a 4-speed automatic. Two option groups are offered, one favoring performance and the other luxurious equipment.

STRONG POINTS

• **Satisfaction:** 90%
Few problems other than fragile differentials. Owners love it but get tired at the end of two years of its lack of practical attributes.

• **Steering:** 80%
The manual is standard and doesn't correspond to the Miata temperament which prefers the livelier, more precise and more direct power steering.

• **Handling:** 80%
The chassis stiffness and good suspension geometry allow the Miata to corner flat, posting, when pushed sufficiently, a progressive oversteer that is fun to control.

• **Fuel consumption:** 80%
Its high weight (considering its size), explains the appetite for fuel.

• **Safety:** 70%
Considering the vulnerability of this type of structure, Miata gets an excellent rating, thanks to the standard air bag on the driver's side, but lacks a second one and especially the roll cage hoop to protect the occupants.

• **Technical:** 70%
Mazda has succeeded in building a production convertible with a

DATA

Category: rear-wheel drive sport convertibles.
Class : 3S

HISTORIC

Introduced in: 1989
Modified in: 1994: 128 hp engine
Made in: Hiroshima, Japan.

RATINGS

Safety: 80 %
Satisfaction: 92 %
Depreciation: 55 %
Insurance: 6.8% ($620)
Cost per mile: $0.43

NUMBER OF DEALERS

In U.S.A.: 871

SALES ON THE AMERICAN MARKET

Model	1992	1993	Result
Miata	24,964	21,588	- 13.6 %

MAIN COMPETITORS

HONDA Civic del Sol.

EQUIPMENT

MAZDA Miata	base
Automatic transmission:	O
Cruise control:	O
Power steering:	O
ABS brake:	O
Air conditioning:	O
Air bag (left):	S
Leather trim:	O
AM/FM/ Radio-cassette:	S
Power door locks:	-
Power windows:	O
Tilt steering column:	O
Dual power mirrors:	S
Intermittent wipers:	S
Light alloy wheels:	O
Sunroof:	O
Anti-theft system:	-

S : standard; O : optional; - : not available

COLORS AVAILABLE

Exterior: White, Red, Blue, Black, Green.
Interior: Balck, Tan.

MAINTENANCE

First revision: 5,000 miles
Frequency: 5,000 miles
Diagnostic plug: Yes

WHAT'S NEW IN 1995 ?

- No major change.

		ENGINE				TRANSMISSION					PERFORMANCE						
Model/ versions *: standard	Type / timing valve / fuel system	Displacement cu/in	Power bhp @ rmn	Torque lb.ft @ rpm	Compres. ratio	Driving wheels / transmission	Final ratio	Acceler. 0-62 mph s	Stand. 1/4 mile s	Stand. 5/8 mile s	Braking 62-0 mph ft	Top speed mph	Lateral acceler. G	Noise level dBA	Fuel economy mpg city	hwy	Gasoline type / octane
Miata	L4* 1.8 DOHC-16-MPFI	112	128 @ 6500	110 @ 6000	9.0 :1	rear - M5*	4.10	8.5	15.7	28.8	138	118	0.85	73	26.0	34.0	R 87
						rear - A4	4.10	9.7	16.5	30.0	144	112	0.85	73	26.0	34.0	R 87

sufficient rigidity to obtain sports handling. The steel unit body has been reinforced by two aluminum rails which connect the sub-frames for the powertrain at the front with another sub-frame for the rear axle. The suspension with twin "A" frames is independent on all four wheels, with high pressure shocks and stabilizer bars at the front and rear. There are disk brakes on all 4 corners, but the ABS is optional, just like the powersteering. A 1.8 L DOHC with 16 valves offers 9 hp more than the old 1.6 L which improves the weight-to-power ratio and brings it to 18.3 lbs/hp. Unfortunately, in spite of its bulbous forms, the aerodynamics are not ideal, but better with a hard top than a convertible.

• Quality/fit/finish: 70%
Solidly built, the Miata receives strict finishing, but the interior is very black. The quality of the trim is very evident with the deluxe option whose price is to match.

• Performances: 70%
The new engine has more pep and the accelerations and power are better used by a fast and precise manual with good shifting. The advantage of the Miata is to allow having fun without danger within the allowable speed limits.

• Seats: 70%
They hold well and provide good kidneys support but the padding is fairly harsh.

• Driver's compartment: 60%
The Miata is a real sports roadster. The bucket seat offers good hold but a tall driver will be unhappy with the non-adjustable steering column. Otherwise, the commands are simple and well laid out and the instruments, in sufficient numbers, are easy to consult. Perfect when the roof is removed, the visibility is more favorable with a hard top than with a convertible top afflicted by major blind spots. The form and location of the non-adjustable (standard) outside rear-view mirrors do not facilitate the view toward the back.

• Access: 60%
It is fairly easy for those who measure less than 5ft5inches...

• Insurance: 60%
The high price and rate result in a large premium. Taking into account the Miata characteristics, you might want to insure it only for the good season...

• Suspension: 55%
Acceptable on the freeway, it jumps around on little bumpy roads and after two hours on this treatment, a pause will always be welcome.

• Price/equipment: 50%
A little more equipment would justify the relatively high price of the Miata, especially if you compare it to the more versatile Civic del Sol.

WEAK POINTS

• Trunk: 00%
It won't hold much, mostly soft bags, because it has little protection and is encumbered by the battery and the mini-spare tire. Those two accessories could be relocated elsewhere to gain space.

• Noise level: 10%
The Miata exhaust doesn't purr as agreeably as those of the English of another era and noise from wind, tires and power train doesn't contain anything melodious.

• Interior space: 20%
Two people of average height will find in the Miata a minimal space because its size is figured too closely and tall or stout folks will suffer from claustrophobia.

• Conveniences: 40%
Storage consists of a small glove box and a lock box within the center console, to which are added, this year, door pockets and holders in the seat backs. You can use a helping hand when installing the hardtop, while the convertible top is easily manipulated with one hand. Its locks are simple and weather proofing excellent.

• Braking: 40%
Stable and fade-free, its stopping distances are a little long and its power assist a little too high, which makes it hard to modulate it with precision. Finally, the ABS offered as an option, only has the advantages of retaining the use of steering during panic stops.

• Depreciation: 45%
Higher than average, it remains within the normal order of things.

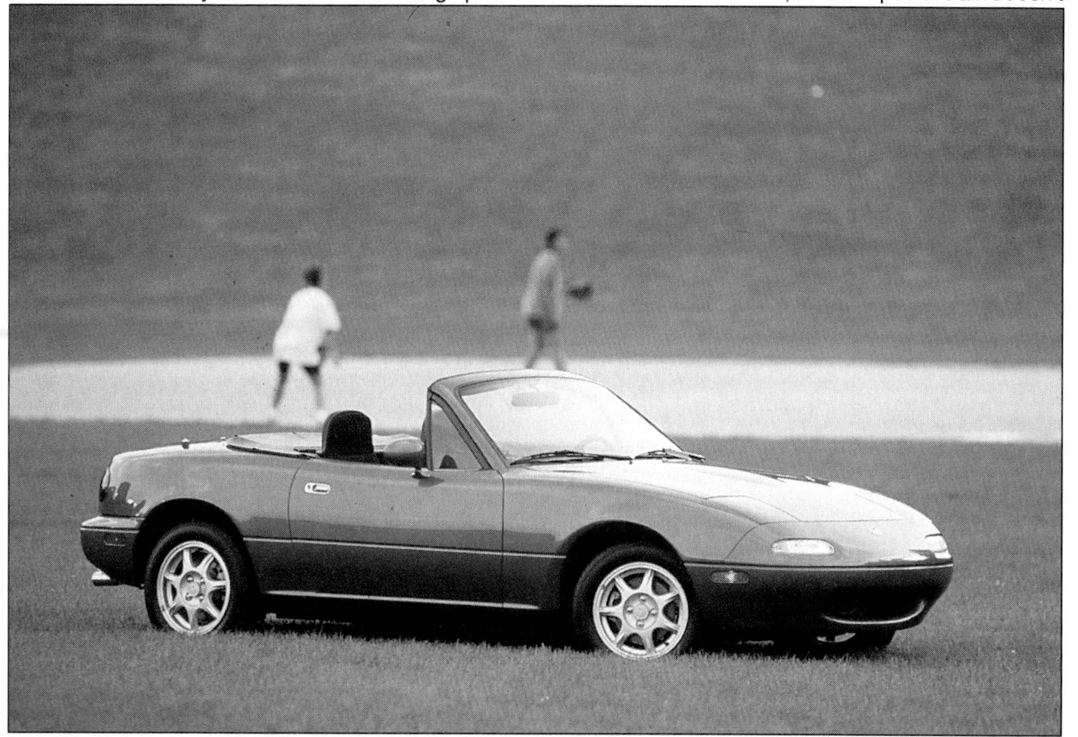

CONCLUSION

• Overall rating: 56.5%
The Miata is a fun road runner. One sits inside and savors. the taste of the road, not too fast, but with sensations. It would however be good if it evolved a little faster, simply so the fans would not think it is fossilized.

OWNER'S SUGGESTIONS

- -The 1.8 L V6 of the MX-3
- -Better side defrosting.
- -Better original tires
- -Less bulky headlights
- -A stronger battery
- -Less fragile differential
- -More economical operation
- -Stronger fabrics/plastics
- -Less austere interior
- -Storage and trunk more practical

SPECIFICATIONS & PRICES

Model	Version/Trim	Body/ Seats	Interior volume cu.ft.	Trunk volume cu ft.	Drag coef. Cd	Wheel-base in	Lgth x Wdthx Hght. inx inx in	Curb weight lb	Towing capacity std. lb	Susp. type ft/rr	Brake type ft/rr	Steering system	Turning diameter ft	Steer. turns nber	Fuel. tank gal.	Standard tire size	Standard powertrain	PRICES US. $ 1994
MAZDA		General warranty: 3 years / 50,000 miles; Power Train: 5 years / 60,000 miles; corrosion perforation: 5 years / unlimited.																
Miata	soft top	2 dr.con.2	NA	3.2	0.38	89.2	155.4x66.0x48.2	2295	NR	i/i	dc/dc	r&p.	30.2	3.33	12.7	185/60R14	L4/1.8/M5	**$16,650**
Miata	soft top	2 dr.con.2	NA	3.2	0.38	89.2	155.4x66.0x48.2	2302	NR	i/i	dc/dc	r&p.	30.2	3.33	12.7	185/60R14ec	L4/1.8/A4	**$17,500**

See page 393 for complete 1995 Price List

The Corvette syndrome...

One day a highly-placed Mazda director imagined that his firm, after its 24 hours of Le Mans success, should have in its stable an exotic vehicle that would reflect its glory. That is how the RX-7 became the deluxe, inaccessible go-cart whose sales have declined terribly.

During its last recast, the RX-7 inaugurated an inspiration line very different from that of the preceding models. At the same time it abandoned its popular character to rejoin the Corvette in the club of unaffordable exotics. It continues to be sold equipped with rotary engines of 1.3 L with 2 turbochargers and one heat exchanger. It delivers 255 hp at the rear wheels through a manual 5-speed transmission and a Torsen limited slip differential.

STRONG POINTS

• Performances: 100%
The acceleration of the RX-7 are notable, but we have not been as impressed as with the models tested the first year. The accelerations are muscled and the transmission draws the best available, which is somewhere between limits of the tach and State trooper tolerance.

• Technical: 90%
The line of the RX-7 doesn't owe anything to anyone and includes subliminal messages recalling Jaguar, Lotus, A.C. Bristol, while others more modern suggest the Stealth, the NSX and ... the Miata. Aerodymanic efficiency is excellent, since its low 0.29 drag coefficient is achieved in spite of big tires and large openings in the grill. The body, both rigid and light, partakes of both the unit body and the tubular chassis. The structure is in steel, with the exception of the engine hood, which is aluminum to insure exceptional torsional rigidity. Mazda has picked up the solution used on the Miata to connect the engine to the differential by 2 lengthwise aluminum reinforcements. The engine, disposed in the front center, provides a good mass balance. The independant 4-wheel suspension is based on forged aluminum arms, coil springs, and stabilizer bars. Induced geometry changes, front and rear, improve stability.

• Safety: 90%
The excellent structural rigidity plus 2 air bags protect the occupants

DATA

Category: rear-wheel drive sport coupes.
Class : GT

HISTORIC

Introduced in: 1978
Modified in: 1985: body; 1986:Turbo;1987 convertible; 1993: body.
Made in: Hiroshima, Japan.

RATINGS

Safety: 90 %
Satisfaction: 80 %
Depreciation: 34% (2 years)
Insurance: 3.9 % ($1,065)
Cost per mile: $0.80

NUMBER OF DEALERS

In U.S.A.: 871

SALES ON THE AMERICAN MARKET

Model	1992	1993	Result
RX-7	6,006	5,062	- 15.7 %

MAIN COMPETITORS

CHEVROLET Corvette, DODGE Stealth, NISSAN 300 ZX.

EQUIPMENT

MAZDA RX-7	base
Automatic transmission:	-
Cruise control:	S
Power steering:	S
ABS brake:	S
Air conditioning:	S
Air bag (2):	S
Leather trim:	S
AM/FM/ Radio-cassette:	S
Power door locks:	S
Power windows:	S
Tilt steering column:	S
Dual power mirrors:	S
Intermittent wipers:	S
Light alloy wheels:	S
Sunroof:	-
Anti-theft system:	-

S : standard; O : optional; - : not available

COLORS AVAILABLE

Exterior: Silver, Red, Blue, White, Black.
Interior: Leather or fabric: Black.

MAINTENANCE

First revision: 5,000 miles
Frequency: 5,000 miles
Diagnostic plug: Yes

WHAT'S NEW IN 1995 ?

- No major change.

	ENGINE					TRANSMISSION		PERFORMANCE										
Model/ versions *: standard	Type / timing valve / fuel system	Displacement cu/in	Power bhp @ rmn	Torque lb.ft @ rpm	Compres. ratio	Driving wheels / transmission	Final ratio	Acceler. 0-62 mph s	Stand. 1/4 mile s	Stand. 5/8 mile s	Braking 62-0 mph ft	Top speed mph	Lateral acceler. G	Noise level dBA	Fuel economy mpg city	hwy	Gasoline type / octane
base	R2T* 1.3-MPFI	80	255 @ 6500	217 @ 5000	9.0 :1	rear - M5*	3.29	5.5	14.3	25.7	115	156	0.97	69	20.0	32.0	S 91

in case of collision.

• **Quality/fit/finish: 80%**
The outside appearance is flattering, while the inside is discouraging and the finish very ordinary for a car of this price. The presentation is a mix of movie land with incurving shapes and retroview with instrumentation inspired in Ye Olde English.

• **Driver's compartment: 80%**
It is very ergonomic and the steering wheel falls right into the hands. The gear shift can be moved rapidly from the wrist. Analog and digital gauges are numerous and well-grouped: the large tach enthroned on a dash is graduated to 9 000 rpm with an 8 000 rpm red line, next to a speedometer graduated to 200 mph, something not seen every day! Some commands are not as rational ,such as switches on the console and those for the air conditioning (not automated at this price), which trip the light fantastic for no reason. Satisfactory toward the front, visibility is hindered toward 3/4 rear by the thickness of the "C" pillar and toward the rear by the narrow backlight which is very sloped and becomes blind at the least drop of rain, because its wiper (optionally standard model) is not very effective. In addition, it is limited by the presence of a rear spoiler.

• **Satisfaction: 80%**
The reliability is a more certain guarantee of buyers' contentment than some visits to the dealer.

• **Handling: 80%**
Road holding is exemplary if one has a smooth flowing driving style. Otherwise, reactions become brutal and in inexperienced hands cause skids, particularly on wet or poorly maintained roads. However, in addition to the Torsen differential, standard on these cars, a traction control based on ABS allows balancing out harsh starts, which generate weaving and some more or less elegant traces on wet pavement.

• **Steering: 80%**
The front wheels and the steering communicate, rather harshly, the state of the road. The power assist is firm and the steering ratio unusual for a car of this type. It is without a doubt to avoid fast lane changes that it is not more direct.

• **Access: 70%**
It is easy to slide in since the shape of the roof and the seats generate sufficient space.

• **Insurance: 70%**
Sports car rates and the high price will call for logging extra hours.

• **Seats: 70%**
The seats are particularly disappointing because they are ordinary in their form, which leaves something to be desired ergonomically. They are also ordinary in their workmanship. A real Recaro would be welcome

• **Braking: 60%**
It is efficient and balanced, especially on a wet road and the pedal effort is easily controlled with precision.

WEAK POINTS

• **Trunk: 10%**
There is more space on the base model, equipped with a more conventional sound system than on the premium because the Bose loud speakers occupy a good 1/4 of its content and even a medium-sized suitcase would have problems fitting there. One could however, utilize the available space behind the seats to store any excess baggage.

• **Interior space: 20%**
Two people can be squeezed into the RX-7 interior, which lacks height and width because of the center console size.

• **Noise level: 30%**
It is rich in mechanical and rolling noises and the remarkable Bose sound system offered as an option is not enough to cover the vehicle sound effects.

• **Price/equipment: 40%**
The RX-7 has ceased to be a popular sports coupe because its high tech and its complete equipment have brought its price to the outer confines of reason. Except for amortizing the development costs of the turbo rotary engine, one can't find justification for that high a price.

• **Fuel consumption: 40%**
With comparable performance, it is higher, in spite of displacement, than that of a 300ZX.

• **Depreciation: 40%**
It is stronger because the economic climate does not predispose to this type of expenses.

• **Suspension: 50%**
It does not spare anything to the occupants, which clearly know the intentions of the car as soon as the asphalt goes bad and imposes lifting the foot if one doesn't want to arrive exhausted.

• **Conveniences: 50%**
The glove box and the door pockets don't hold much, but you can make use of two high volume rear lockers that can be closed with a key.

CONCLUSION

• **Overall rating: 61.5%**
RX-7 is no longer within every one's reach because following Chevrolet's bad example with the Corvette, Mazda has cut itself off from many potential clients. 😐

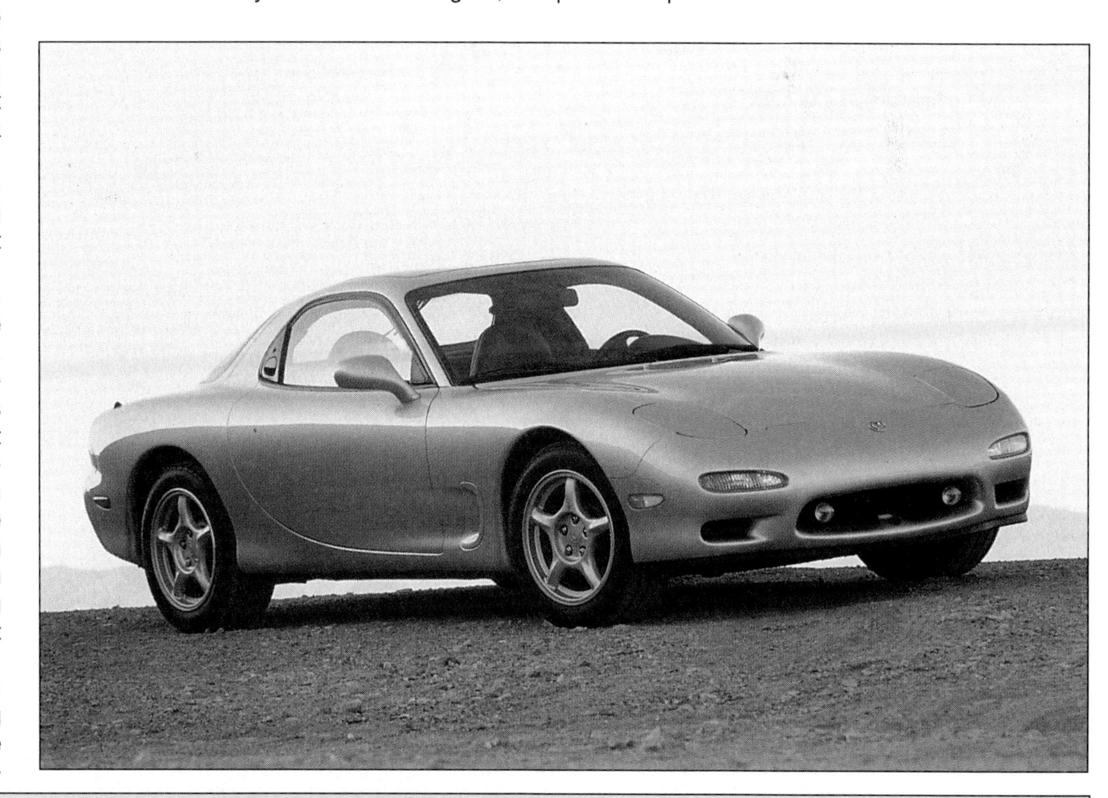

SPECIFICATIONS & PRICES

Model	Version/Trim	Body/ Seats	Interior volume cu.ft.	Trunk volume cu ft.	Drag coef. Cd	Wheel-base in	Lgth x Wdthx Hght. inx inx in	Curb weight lb	Towing capacity std. lb	Susp. type ft/rr	Brake type ft/rr	Steering system	Turning diameter ft	Steer. turns nber	Fuel. tank gal.	Standard tire size	Standard powertrain	PRICES US. $ 1994
MAZDA		**General warranty: 3 years / 50,000 miles; Power Train: 5 years / 60,000 miles; corrosion perforation: 5 years / unlimited.**																
RX-7	base	3 dr.cpe.2	-	3.5	0.29	95.4	168.5x68.9x48.4	2835	NR	i/i	dc/dc/ABS	pwr.r&p.	35.4	2.9	20	225/50R16	R2T/1.3/M5	**$36,000**
RX-7	Premium	3 dr.cpe.2	-	3.5	0.29	95.4	168.5x68.9x48.4	2886	NR	i/i	dc/dc/ABS	pwr.r&p.	35.4	2.9	20	225/50R16	R2T/1.3/M5	**$38,000**

See page 393 for complete 1995 Price List

Attention customers... special bargain

Many people still have not heard of a $35,000 canadian dollars Mercedes, which is, the price of the C220 Special Edition, sold only in Canada. Its equipment has been slightly simplified, its qualities remain intact and form an excellent buy.

The latest little Mercedes is a year old. It is a 4-door sedan with 3 volumes, offered in two versions, equipped with gasoline engines. The C220 is a 2.2 L 4-cylinder and the C280 is a 2.8 L 6-cylinder. These engines have 2 overhead cams and 4 valves per cylinder with only 4-speed automatics. The Special Edition exists only in Canada.

STRONG POINTS

• Safety: 100%
The Mercedes structure is among the most impact resistant, thanks to uncommon rigidity in all directions. Two air bags protect the front seat occupants and the automatic retraction seats belts gain for it the maximum rating.

• Quality/fit/finish: 100%
Mercedes builds its cars as no other builder in the world. The assembly is robust, the materials of great quality and the meticulous finish explains, as much as the value of the Deutsch Mark, the size of the price. For once, fabric trim breaks up the monotony of the interior.

• Satisfaction: 90%
The dealer network is not very expanded, but the guarantee is as solid as the cars. Maintenance and replacement parts are distinctely more costly than average.

• Driver's compartment: 80%
Comfortable driving position is easy to find, now that the steering column is adjustable; steering wheel diameter remains large. The seat provides effective hold and visibility is good from all angles with amply-sized rear-view mirrors. Gauges fall naturally under the eyes but some controls remain unique to Mercedes and call for getting used to.

• Steering: 80%
It is soft, precise, well-modulated, and its reduction, higher than average, doesn't really hurt maneuverability.

• Technical: 80%
The steel unit body has been considerably reworked to improve interior space, a main handicap of the preceding model. The line has been refined by giving more windshield inclination but it posts all the attributes of the make, including the famous grill that is enthroned up front. Unfortunately, the lines are not as effective as before because the drag coefficient has climbed to 0.35 (more drag). The 4-wheel independant suspension has struts and "A" frame lower control arms and an antidive as well as a negative off set. At the rear, a multiple arm linkage has been perfected to include antilift and antidive and hydropneumatic shocks are used. The disc brakes on all 4 wheels and the ABS are standard.

• Seats: 80%
Their form provides ideal support and their padding, less hard than before, allows long fatigue-free trips.

• Suspension: 70%
Not hard or soft, It absorbs the road defects well, thanks to an excellent

DATA

Category: rear-wheel drive luxury compact sedans.
Class : 7

HISTORIC

Introduced in: 1984 (190 2.3L)
Modified in: 1989: 2.6L; 1990: 2.0L diesel; 1993: C-Class.
Made in: Sindelfingen & Brement, Germany.

RATINGS

Safety: 100 %
Satisfaction: 90 %
Depreciation: 20 % (1year)
Insurance: C-220 ($1,065) 5.2 % 3.7 ($1,065) C-280
Cost per mile: $0.94

NUMBER OF DEALERS

In U.S.A.: 380

SALES ON THE AMERICAN MARKET

Model	1992	1993	Result
C-Class	-	14,773	

MAIN COMPETITORS

ACURA Vigor, AUDI 90, BMW 3-Series, INFINITI G20, LEXUS LS300, MAZDA 929, MITSUBISHI Diamante, SAAB 900, VOLVO 850.

EQUIPMENT

MERCEDES-BENZ	C220	C280
Automatic transmission:	O	S
Cruise control:	S	S
Power steering:	S	S
ABS brake:	S	S
Air conditioning:	O	O
Air bag (2):	S	S
Leather trim:	S	S
AM/FM/ Radio-cassette:	S	S
Power door locks:	S	S
Power windows:	S	S
Tilt steering column:	S	S
Dual power mirrors:	O	S
Intermittent wipers:	S	S
Light alloy wheels:	-	
Sunroof:	-	
Anti-theft system:	-	

S : standard; O : optional; - : not available

COLORS AVAILABLE

Exterior: Black, Red, Ivory, Blue, White, Gray, Green, Silver, Taupe.
Interior: Black, Blue, Burgundy, Beige, Gray, Palomino, Green.

MAINTENANCE

First revision: 3,000 miles
Frequency: 7,500 miles
Diagnostic plug: Yes

WHAT'S NEW IN 1995 ?

- No major change.

		ENGINE				TRANSMISSION					PERFORMANCE							
Model/ versions *: standard	Type / timing valve / fuel system	Displacement cu/in	Power bhp @ rmn	Torque lb.ft @ rpm	Compres. ratio	Driving wheels / transmission	Final ratio	Acceler. 0-62 mph s	Stand. 1/4 mile s	Stand. 5/8 mile s	Braking 62-0 mph ft	Top speed mph	Lateral acceler. G	Noise level dBA	Fuel economy mpg city	Fuel economy mpg hwy	Gasoline type / octane
C220	L4* 2.2 DOHC-16-EFI	134	147 @ 5500	155 @ 4000	10.0 :1	rear - A4	3.07	10.5	17.2	31.8	125	130	0.78	66	27.0	37.0	R 87
C280	L6* 2.8 DOHC-24-EFI	171	158 @ 5800	162 @ 4600	10.0 :1	rear - A4	2.87	9.0	16.4	29.3	131	143	0.80	64	24.0	33.0	M 89

damping quality.

• **Handling:** 70%
The excellent rear suspension allows following a curve without any problem, regardless of the radius without upsetting its balance. Its ability to hold a straight line is beyond reproach because it is insensitive to cross wind. The neutral steering gives way to a progressive oversteer if you approach a tight turn at excessive speeds, or on a slippery road where the limited slip controls traction.

• **Conveniences:** 70%
The storage areas include a large glove box, door pockets, and a locker in the center console. Those who suffer from allergies will appreciate the pollen and dust filters in the AC.

• **Noise level:** 70%
Extreme body rigidity and effective soundproofing, don't allow any rolling or powertrain noises.

• **Performance:** 65%
If one considers that this car is in the class of a Jetta, but weighs 441 pounds more, one will not be surprised that the 147 hp 2.2 L is necessary to allow it to accelerate from 0 to 62mph in 10.5 seconds while the 6-cylinder only asks for 8.5 seconds with an automatic transmission, the only one available.

• **Access:** 65%
In spite of the extension of the interior, it is always touchy to get in the back, because rear legroom is limited.

• **Insurance:** 60%
With the image of the car, the premium is not given away and collision repairs can be costly.

• **Fuel consumption:** 60%
It remains reasonable, considering the weight and respectable performance of these cars.

• **Braking:** 60%
It is very powerful, and panic stops are short, with seemingly unlimited stability and fade-resistance.

WEAK POINTS

• **Interior space:** 30%
Even though the inside volume has been slightly improved, it will always be difficult for a tall person to get in comfortably in the back, where length lacks. It is mostly at shoulder level and in height that sizes were improved.

• **Trunk:** 40%
With slightly more volume than before, it differs by the 60/40 rear bench rest which can fold down to liberate extra space. The locks are intelligently placed on the trunk side for added safety. Its sill comes down to bumper height for improved access.

• **Price/equipment:** 40%
Mercedes has decided to sell a good number of these cars in their most popular finish. For this it had to get undressed, just a little. However, if one takes into consideration the equipment and the strength of the company, it is a feat that should pay off.

• **Depreciation:** 50%
The Mercedes depreciates today as much as other much less noble models and sells for 50% of their nominal value after 3 years or 36,000 miles of use.

CONCLUSION

• **Overall rating:** 68.0%
While the ecomonic climate doesn't allow the Stuttgart builder to carpet our roads with its little models, it shouldn't get discouraged since the C-Class has many good arguments to convince a select few people to roll with Mercedes. ☺

SPECIFICATIONS & PRICES

Model	Version/Trim	Body/ Seats	Interior volume cu.ft.	Trunk volume cu ft.	Drag coef. Cd	Wheel-base in	Lgth x Wdthx Hght. inx inx in	Curb weight lb	Towing capacity std. lb	Susp. type ft/rr	Brake type ft/rr	Steering system	Turning diameter ft	Steer. turns nber	Fuel. tank gal.	Standard tire size	Standard powertrain	PRICES US. $ 1994
MERCEDES-BENZ		**Total warranty: 4 years / 50,000 miles.**																
C 220	S.E.	4dr.sdn.	84.7	15.2	0.32	105.9	177.4x67.7x56.1	3172	1000	i/i	dc/dc/ABS	pwr.bal.	35.1	3.5	16.4	195/65R15	L4/2.2/A4	**$22,900**
C 220		4dr.sdn.	84.7	15.2	0.32	105.9	177.4x67.7x56.1	3172	1000	i/i	dc/dc/ABS	pwr.bal.	35.1	3.5	16.4	195/65R15	L4/2.2/A4	**$29,900**
C 280		4dr.sdn.	84.7	15.2	0.32	105.9	177.4x67.7x56.1	3291	1000	i/i	dc/dc/ABS	pwr.bal.	35.1	3.5	16.4	195/65R15	L6/2.8/A4	**$34,900**

See page 393 for complete 1995 Price List

The last cartridges...

The Class E is on its last parade, because its successor will soon make its entry. It remains the most sold series and is at the median point of the German builders range. The next generation should bring many surprises, notably concerning the style.

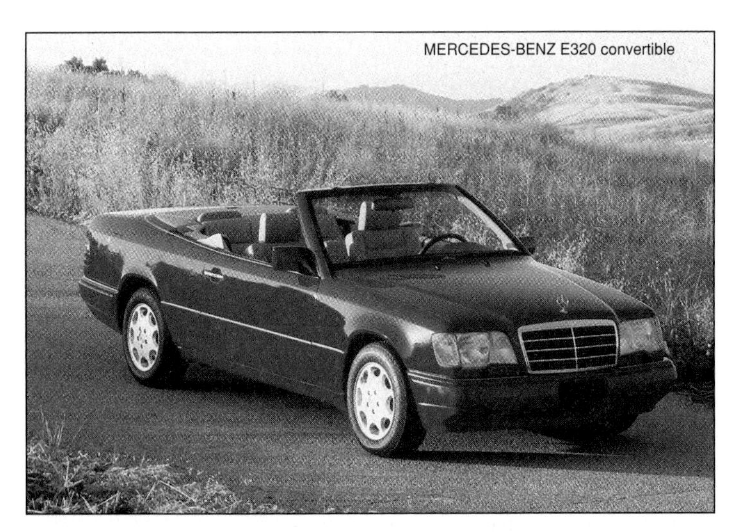

MERCEDES-BENZ E320 convertible

The Class E is in its last year under the form currently known with the new model to be introduced at the Geneva show in March 1995.
It offers four body styles: coupe, convertible, sedan, and station wagon with 6-cylinder Diesels of 3.0 L or gasoline with 3.2 L or V8 with 4-speed automatics.

STRONG POINTS

• Safety: 100%
Mercedes was one of the first car builders sensitized to the safety aspect of its vehicles and to elaborate avant garde techniques, well before this aspect was regimented. The Mercedes Benz bodies have been studied for a long time to resist all types of impacts and insure the best occupant protection by providing 2 air bags and retractible seat belts as standard. There is strong speculation that Mercedes will introduce air bags integrated into the doors to protect the occupants against lateral shocks and the next generation of E-Class will be so equipped.

• Technical: 90%
Compliments on the technical achievements of the German builder are not needed, the body of the E Class is steel, with an independant front suspension and 4 wheel discs with ABS standard. The rear suspension has five particularly effective arms that have been adapted to other models of the make. The latest cosmetic changes improved the aerodynamics whose drag varies between 0.31 and 0.34.

• Satisfaction: 90%
It is high because the guarantee is consistant but maintenance and repairs are particularly expensive.

• Quality/fit/finish: 90%
The assembly technique allied with the quality of materials used and the finishing care is in a class apart.

DATA

Category: rear-wheel drive luxury coupes,sedans, conv. and wagons.
Class : 7

HISTORIC

Introduced in: 1984: 300E; 1986: 300 TD & 260E;1987: 300 CE.
Modified in: 1987: 4Matic transmission; 1990: 24 valve engine.
Made in: Sindelfingen, Germany.

RATINGS

Safety: 100 %
Satisfaction: 95 %
Depreciation: 55 %
Insurance: 4.0 %
Cost per mile: $0.94

NUMBER OF DEALERS

In U.S.A.: 380

SALES ON THE AMERICAN MARKET

Model	1992	1993	Result
300	23,416	26,073	+ 10.2 %

MAIN COMPETITORS

ACURA Legend, ALFA ROMEO 164, AUDI A6, BMW 5-Series, INFINITI J30, LEXUS GS300, SAAB 9000, VOLVO 900-Series.

EQUIPMENT

MERCEDES-BENZ	E300	E320	E400
Automatic transmission:	S	S	S
Cruise control:	S	S	S
Power steering:	S	S	S
ABS brake:	S	S	S
Air conditioning:	S	S	S
Air bag (2):	S	S	S
Leather trim:	O	O	S
AM/FM/ Radio-cassette:	S	S	S
Power door locks:	S	S	S
Power windows:	S	S	S
Tilt steering column:	S	S	S
Dual power mirrors:	S	S	S
Intermittent wipers:	S	S	S
Light alloy wheels:	S	S	S
Sunroof:	-	O	S
Anti-theft system:	S	S	S

S : standard; O : optional; - : not available

COLORS AVAILABLE

Exterior: Black,Red,Ivory,Blue,White,Gray,Silver,Taupe,Garnet,Emerald.
Interior: Black, Blue, Beige, Gray, Palomino, Parchment, Saddle.

MAINTENANCE

First revision: 3,000 miles
Frequency: 7,500 miles
Diagnostic plug: Yes (except E300D)

WHAT'S NEW IN 1995 ?

-Disappearance of the E500 version.

	ENGINE					TRANSMISSION		PERFORMANCE										
Model/ versions *: standard	Type / timing valve / fuel system	Displacement cu/in	Power bhp @ rmn	Torque lb.ft @ rpm	Compres. ratio	Driving wheels / transmission	Final ratio	Acceler. 0-62 mph s	Stand. 1/4 mile s	Stand. 5/8 mile s	Braking 62-0 mph ft	Top speed mph	Lateral acceler. G	Noise level dBA	Fuel economy mpg city	Fuel economy hwy	Gasoline type / octane
E300D	L6* 3.0 SOHC-10-MFI	183	147 @ 4600	273 @ 2400	22.0:1	rear - A4*	2.65	12.5	19.0	33.0	131	118	0.77	67	26.0	37.0	D
E320	L6* 3.2 DOHC-24-MFI	195	217 @ 5500	310 @ 3750	10.0 :1	rear - A4*	3.06	8.5	16.2	28.6	128	137	0.80	66	23.0	33.0	S 91
E420	V8*4.2 DOHC-32-MFI	256	275 @ 5700	400 @ 3900	11.0 :1	rear - A4*	2.65	7.5	15.4	27.8	134	150	0.83	66	21.0	32.0	S 91

• **Driver's compartment: 80%**
The best driving position is easy to find, thanks to simple and effective seat and column adjustments. However, the steering wheel diameter remains too large. While the gauges are well laid out, some controls and switches are peculiar to Mercedes and call for getting used to. Numerous builders have adapted the large round buttons to set the air conditioning, which the Stuggart firm was the first to use.

• **Performance: 80%**
Placid with a Diesel engine, they are very spectacular with a 4.2 L V8 which transforms this car into an Autobahn eater, disguised as an ordinary sedan. As to the E320, their takeoffs and accelerations rival some of the best.

• **Steering: 80%**
Fast, precise and well-modulated, it provides virtual perfection and gains for these cars of respectable weight and size, surprising agility and maneuverability.

• **Insurance: 80%**
No miracles from this point of view: in spite of the low index, the exaggerated prices produce depressing premiums.

• **Access: 80%**
The size of the exits or of the available space is calculated, even on coupes and convertibles to allow reaching the rear seats with a minimum of disgraceful contortions.

• **Suspension: 70%**
Relatively firm, the shock damping is very consistent and allows the suspension to absorb road defects while efficiently controlling body motions.

• **Noise level: 70%**
It has improved thanks to the increased unit body rigidity, which resonates less. It also benefited from extensive soundproofing development. All this still allows the characteristic tire hamering on road joints to filter in, and wind

MERCEDES-BENZ E320

noises around the rear-view mirrors also increase when speed goes up.

• **Conveniences: 70%**
Storage includes a large glove box, door pockets, a housing on the console and a little storage box built in the rear center armrest.

• **Seats: 70%**
The lateral hold and the lumbar support are improved, even with leather trim, which is often more slippery than perforated vinyl, christened M-B Tex, which imitates it and is very durable.

• **Handing: 70%**
It is one of the most secure because it remains neutral at the speeds practiced in North America. Be it in a straight line or in a wide curve, these models find their own best line and you hardly notice a loss of traction in tight turns.

• **Braking: 60%**
It impresses by the initial bite, but stopping distances are long because of the weight and the ABS. The action is progressive, but a strong assist makes modulation somewhat delicate.

• **Interior space: 60%**
Four people can be seated in normal use and a fifth in an emergency. Dimensions are well-calculated in all directions, they leave enough head and legroom except in the convertible where rear legroom is more limited.

• **Trunk: 60%**
Of regular shape, it is at the same time spacious and easily accessible, thanks to the generous cutout at the sill.

WEAK POINTS

• **Price/equipment: 00%**
In spite of prices better adjusted to the Japanese competition, these cars remain targeted for privileged buyers. The prices are justified by the manufacturing quality and the strength of the German money facing the dollar. They are also justified by very complete and sophisticated equipment which includes as standard (in addition to the items noted in the table): headlight washers and wipers, heated seats, electrically controlled headrests, an antitheft device and electrically adjusted steering column.

• **Fuel consumption: 50%**
Only the diesel is really economical and thirst of the gasoline engine increases with the speed.

• **Depreciation: 50%**
The most effective way is to sell these cars early with a low mileage or to wear them down to a frazzle, which risks taking a while.

CONCLUSION

• **Overall rating: 70.0%**
Even on the eve of its replacement, the E-Class is, in spite of its age, a range of classic and exceptional models which justify the investment. ☺

SPECIFICATIONS & PRICES

Model	Version/Trim	Body/ Seats	Interior volume cu.ft.	Trunk volume cu ft.	Drag coef. Cd	Wheel-base in	Lgth x Wdthx Hght. inx inx in	Curb weight lb	Towing capacity std. lb	Susp. type ft/rr	Brake type ft/rr	Steering system	Turning diameter ft	Steer. turns nber	Fuel. tank gal.	Standard tire size	Standard powertrain	PRICES US. $ 1994
MERCEDES BENZ		**Total warranty: 4 years / 50,000 miles; corrosion perforation; 5 years / unlimited.**																
E300	diesel	4dr.sdn.5	93	14.6	0.31	110.2	187.2x68.5x56.3	3483	2000	i/i	dc/dc/ABS	pwr.bal.	37.1	3.1	23.8	195/65R15	L6/3.0/A4	**$40,000**
E320	gas	4dr.sdn.5	93	14.6	0.31	110.2	187.2x68.5x56.3	3527	2000	i/i	dc/dc/ABS	pwr.bal.	37.1	3.1	18.5	195/65R15	L6/3.2/A4	**$42,500**
E320	gas	2dr.cpe.5	82.8	14.5	0.31	106.9	183.9x68.5x54.9	3549	2000	i/i	dc/dc/ABS	pwr.bal.	36.1	3.3	18.5	195/65R15	L6/3.2/A4	**$61,600**
E320	gas	2dr.con.5	82.8	8.3	0.33	106.9	183.9x68.5x54.8	3990	NR	i/i	dc/dc/ABS	pwr.bal.	36.1	3.3	18.5	195/65R15	L6/3.2/A4	**$77,300**
E320	gas	4dr.wgn.5	94	42.3	0.34	110.2	188.2x68.5x59.9	3748	2000	i/i	dc/dc/ABS	pwr.bal.	37.1	3.1	19.0	195/65R15	L6/3.2/A4	**$46,200**
E420	gas	4dr.sdn.5	93	14.6	0.31	110.2	187.2x68.5x56.3	3748	2000	i/i	dc/dc/ABS	pwr.bal.	37.1	3.1	18.5	195/65R15	V8/4.2/A4	**$51,000**

See page 393 for complete 1995 Price List

Superlatives...

Never has the industry gone so far in matters of technology applied to volume production. The Mercedes S represents what man knows best in individual transport. Of course, the price matches, but without exaggeration.

The top of the Mercedes range did not get a unanimous decision when it was recast in 1992. Most critics considered it superlative and pretentious in an ecology era prone to end waste. For 1995, the retouches brought a tendency to be less showy and more discreet. Most bodies of the different versions are identical except the S350 with a shorter wheelbase. With its V12 engine, it can pretend to supplant the Rolls Royce/Bentley because it collects that many refinements. Lower in the range, you find gasoline 5.0L and 4.2L V8's or in-line 6-cylinder 3.2L and a 3.5L Turbo Diesel. All transmissions are 4-speed automatics.

STRONG POINTS

• Technical: 100%

The advent of the S-Class became the occasion for an avalanche without precedent of technical innovations and deluxe equipment. The steel unit body offers lines that cleverly conceal its enormous proportions. The appearance is refined and the aerodynamic coefficient remains at a remarkable 0.31. These lines are rapidly getting older and the cutout at the trunk opening is not very elegant. Mercedes was often the originator in aesthetic matters and should have pushed the streamlining of the sedan just as it did on the coupe that followed.

• Safety: 100%

The rigidity of the Mercedes largely surpasses the government standards and adding two air bags at the front seats and retractable seat belts easily gives them the maximum rating.

• Quality/fit/finish: 100%

The construction of these models involved refined technology and finishing as well as quality. All of the materials are above average. With its fine leathers and its rare wood appliqués, the interior shows much

DATA

Category: rear-wheel drive luxury coupes and sedans.
Class : 7

HISTORIC

Introduced in: 1992
Modified in: 1995: cosmetic touch-ups.
Made in: Sindelfingen, Germany.

RATINGS

	L6	V8	V12
Safety:	100 %	100 %	100 %
Satisfaction:	90 %	93 %	93 %
Depreciation:	43 %	45 %	40%
Insurance:	($1,135) 3.2 %	2.8 %($1,800)	5.2 % ($5,400)
Cost per mile:	$1.10	$1.25	$1.45

NUMBER OF DEALERS

In U.S.A.: 380

SALES ON THE AMERICAN MARKET

Model	1992	1993	Result
S-Class	18,788	15,633	- 16.8 %

MAIN COMPETITORS

AUDI A8, BMW 7-Series, INFINITI Q45, JAGUAR XJ6/12/XJR, LEXUS LS 400.

EQUIPMENT

MERCEDES-BENZ	S350	S320	S420	S500	S600
Automatic transmission:	S	S	S	S	S
Cruise control:	S	S	S	S	S
Power steering:	S	S	S	S	S
ABS brake:	S	S	S	S	S
Air conditioning:	S	S	S	S	S
Air bag (2):	S	S	S	S	S
Leather trim:	S	S	S	S	S
AM/FM/ Radio-cassette:	S	S	S	S	S
Power door locks:	S	S	S	S	S
Power windows:	S	S	S	S	S
Tilt steering column:	S	S	S	S	S
Dual power mirrors:	S	S	S	S	S
Intermittent wipers:	S	S	S	S	S
Light alloy wheels:	S	S	S	S	S
Sunroof:	NC	NC	NC	NC	NC
Anti-theft system:	S	S	S	S	S

S : standard; O : optional; - : not available

COLORS AVAILABLE

Exterior: Black,Red,Ivory,Blue,White,Gray,Silver,Taupe,Garnet,Emerald.
Interior: Black, Blue, Beige, Gray, Palomino, Parchment, Saddle.

MAINTENANCE

First revision: 3,000 miles
Frequency: 7,500 miles
Diagnostic plug: Yes

WHAT'S NEW IN 1995 ?

- Aesthetic touch-ups of the grill, bumper, lights, and rear lights.
- Cupholders in front and in the back.
- Remote control for the trunk opening.
- S-320 version shares the same body as the other S family.
- Integrated cellular phone in option.

Model/ versions *: standard	ENGINE Type / timing valve / fuel system	Displacement cu/in	Power bhp @ rmn	Torque lb.ft @ rpm	Compres. ratio	TRANSMISSION Driving wheels / transmission	Final ratio	PERFORMANCE Acceler. 0-62 mph s	Stand. 1/4 mile s	Stand. 5/8 mile s	Braking 62-0 mph ft	Top speed mph	Lateral acceler. G	Noise level dBA	Fuel economy mpg city	hwy	Gasoline type / octane
S350	L6* 3.5 TD SOHC-12-IM	210	148 @ 4000	229 @ 2200	22.0 :1	rear - A4*	2.82	13.5	19.0	36.8	141	112	0.72	67	26.0	37.0	D
S320	L6* 3.2 DOHC-24-MPFI	195	228 @ 5800	232 @ 3750	10.0 :1	rear - A5*	3.46	9.7	16.6	30.5	144	119	0.72	67	21.0	32.0	M 89
S420	V8* 4.2 DOHC-32-SMPFI	256	275 @ 5700	295 @ 3900	11.0 :1	rear - A4*	2.82	8.8	16.2	29.6	141	137	0.78	67	19.0	27.0	S 91
S500	V8* 5.0 DOHC-32-SMPFI	303	315 @ 5600	347 @ 3900	10.0 :1	rear - A4*	2.65	7.8	16.0	28.7	144	144	0.79	66	18.0	26.0	S 91
S600	V12* 6.0 DOHC-48-SMPFI	365	389 @ 5200	420 @ 3800	10.0 :1	rear - A4*	2.64	7.0	14.4	25.4	144	162	0.76	64	15.0	22.0	S 91

class and is less austere than in the past.

• Access: 90%
Entering does not present any inconvenience with either the long or the short body.

• Seats: 90%
Well-shaped and adjustable in every sense, even in the rear, the padding is not too firm and they maintain efficiently, thanks to their side rolls, and different pneumatic cushions take care of the lumbar zones. The cushion is rather short to accent the impression of space.

• Suspension: 90%
It offers one of the best compromises, neither too hard or too soft and its reactions are never harsh.

• Noise level: 90%
At 60 mph, the sound level meter drops to 62 decibels. the lowest level in world production held jointly with Rolls Royce. Thanks to double glazing, outside noises are repulsed and the body damps out those that could come from the power or drive trains.

• Conveniences: 90%
It would take several pages to describe them all, as many as there are, extending from double glazing the windows to charcoal filters for dust and pollens, plus headrests and adjustable shades.

• Satisfaction: 90%
It is at the expiration of the warranty, that one realizes the cost of manpower and parts.

• Driver's compartment: 80%
The best driving position is rapidly found, thanks to adjustments at all cardinal points of the seats, the headrest and of the steering column. Up to three different sets of adjustments can be held in memory. Commands are identical to those of other models of the make and the gauges are disposed symmetrically in relation to the speedometer, easy to read at all times, except for the fuel gauge which is masked by the driver's hand. Visibility is good from all angles, but the raised rear section leads you to use the small guides, extending from the rear fenders when backing up. The inside rear-view mirror is electrically adjusted from the same controls that handle the outside ones.

• Performance: 80%
Driving these cars calls for a particular awareness because they move silently, in complete isolation, removing all sense of speed or danger. There is no common measure between Turbo Diesel and V12 performance, or those of the 6-cylinder and the V8. The S600 accelerates like a Ferrari 456 in spite of its 2.3 tons, while the V8's offer the best output.

• Steering: 80%
Its assistance is still too strong, which detracts from guidance on a wet road and removes all pavement feel. It does remain precise and rapid, and confers to this technological tank an amazing maneuverability.

• Interior space: 80%
Five people will be very comfortable, except on the S600 where the rear seats are separated by a console. All other dimensions are monumental in every way.

• Insurance: 70%
Even though their rates are among the lowest, the premium is proportional to the cost of these cars and is listed at $5,400 for the S600.

• Handling: 70%
Extremely safe on dry roads where control can be effected with considerable precision. However, in the rain, the weight and some loss of traction can often induce easily controlled slides.

• Trunk: 70%
Even if its vast, volume is not proportional to the bulk of these cars. A low sill height accross its opening eases access.

• Brakes: 60%
To stop this car from 60 mph in less than 150 ft without any loss in efficiency or fade merits respect.

• Depreciation: 60%
It has improved, but remains tied to the astronomical cost, the level of the sums involved, maintenance and the scarcity of clients.

WEAK POINTS

• Price/equipment: 00%
Whatever the price of these cars, it is largely justified by their technical content, which endows them with very safe operation and superb enjoyment. This is refined equipment at its best.

• Fuel consumption: 10%
Except for the Turbo Diesel, which has reasonable consumption, the others develop a thirst to match the performance.

CONCLUSION

• Overall rating: 75.0%
The S-Class represents the "nec plus ultra" in automatic technology of our era. All of the science largely merits a small financial effort. ☺

SPECIFICATIONS & PRICES

Model	Version/Trim	Body/ Seats	Interior volume cu.ft.	Trunk volume cu ft.	Drag coef. Cd	Wheel-base in	Lgth x Wdthx Hght. inx inx in	Curb weight lb	Towing capacity std. lb	Susp. type ft/rr	Brake type ft/rr	Steering system	Turning diameter ft	Steer. turns nber	Fuel. tank gal.	Standard tire size	Standard powertrain	PRICES US. $ 1994
MERCEDES BENZ		**Total warranty: 4 years / 50,000 miles; corrosion perforation; 5 years / unlimited.**																
S350		4dr.sdn.5	108	15.7	0.32	119.7	201.3x74.3x58.4	4608	2000	i/i	dc/dc/ABS	pwr.bal.	40	3.1	26.4	225/60R16	L6TD/3.5/A4	**$70,600**
S320		4dr.sdn.5	108	15.7	0.32	123.6	205.2x74.3x58.4	4603	2000	i/i	dc/dc/ABS	pwr.bal.	41	3.1	26.4	225/60R16	L6/3.2/A5	**$70,600**
S420		4dr.sdn.5	112	15.7	0.32	123.6	205.2x74.3x58.4	4694	2000	i/i	dc/dc/ABS	pwr.bal.	41	3.1	26.4	235/60R16	V8/4.2/A4	**$79,500**
S500		4dr.sdn.5	112	15.7	0.32	123.6	205.2x74.3x58.4	4753	2000	i/i	dc/dc/ABS	pwr.bal.	41	3.1	26.4	235/60R16	V8/5.0/A4	**$95,300**
S600		4dr.sdn.5	112	15.7	0.32	123.6	205.2x74.3x58.4	5022	2000	i/i	dc/dc/ABS	pwr.bal.	41	3.1	26.4	235/60ZR16	V12/6.0/A4	**$130,300**

See page 393 for complete 1995 Price List

Peak of refinement...

The coupes and convertible Mercedes in the S Class are even more refined than the sedans, which says it all if you consider the compactness of their format. The innovations they bring are so numerous and complex that they fully justify the price at which they are sold.

The SL coupes and convertibles are the sports cars of the S-Class range from which they pick up mechanical and technical innovations. The SL is a soft top convertible which is delivered with a removable hard top. It is equipped with a 3.2L 6-cylinder, a 5.0L V8 or a V12 borrowed from the S600. While the 6-cylinder is seconded by a 5-speed automatic, unique in the world, the second and third engines are offered with a more conventional 4-speed automatic. The SC derived from the S600 sedan is not available with a 6-cylinder engine.

STRONG POINTS

• Safety: 100%
The structural rigidity of these vehicles is exemplary, especially on the convertible whose architecture is generally more vulnerable. Two air bags, retractable seat belts and pads to protect the knees are part of the occupant safety system. In case of a collision, the convertible's retractable roll cage rises into place in a fraction of a second, if the base position of the vehicle is disturbed.

• Technical: 100%
The Mercedes represents the summit of technology. The complex sequence of operations which raises or lowers the convertible top of the SL is the best example. Fifteen jacks, eleven rotary vanes and seventeen switches are essential so the driver does not have to go through this laborious exercise. The rollbar cage hidden behind the driver goes into place a fraction of a second after sensors detect a pronounced change in body inclination. The sensors for the brake antilock assist the differential, allowing the automatic lockup to go from 35 to 100%. This device, associated with the ASR, antislip slow down the wheel that lost traction by applying the corresponding disc brake. The damping of the suspension is also controlled by an electronic computer, which takes into account the speed, the vertical acceleration of the wheel and the body, the steering angle and the weight. Damping rates automatically vary from soft to firm. The aerodynamics are effective and the Cd reaches 0.31 with the hard top or 0.34 with the soft top. The coupe is as sophisticated as some of its equipment comes from the S600 sedan.

• Driver's compartment: 100%
A very well-seated driver enjoys the best visibility on the coupe. The convertible is more limited to the 3/4 rear by the soft top. While the dashboard is richly enhanced, some controls need getting used to.

• Quality/fit/finish: 90%
The assembly precision, the care and the quality of the materials used justify the price of these exotic models whose interior is most flattering but whose exterior is subdued.

• Satisfaction: 90%
It is difficult not to be satisfied after spending this much money for a car which clearly broadcasts the owner's social status.

DATA

Category: rear-wheel drive sport luxury coupes and convertibles.
Class : GT

HISTORIC

Introduced in: 1971:L6 3.8L.; 1989: L6 3.0L & V8 5.0L.
Modified in: 1985: V8 5.0L; 1987: V8 5.6L.;1993: V12 6.0L.
Made in: Sindelfingen, Germany.

RATINGS

	SL	SC
Safety:	100 %	100 %
Satisfaction:	92 %	93 %
Depreciation:	45 %	50 %
Insurance:	5.0 %	5.0 % ($4,670-$5,800)
Cost per mile:	$1.25	$1.35

NUMBER OF DEALERS

In U.S.A.: 380

SALES ON THE AMERICAN MARKET

Model	1992	1993	Result
SL	4,861	4,787	- 1.6 %

MAIN COMPETITORS

ACURA NS-X, BMW 850i, LEXUS SC400, FERRARI F-355, JAGUAR XJS, PORSCHE 911& 928.

EQUIPMENT

MERCEDES BENZ	SL320	SL500	SL600	SC500	SC600
Automatic transmission:	S	S	S	S	S
Cruise control:	S	S	S	S	S
Power steering:	S	S	S	S	S
ABS brake:	S	S	S	S	S
Air conditioning:	S	S	S	S	S
Air bag (2):	S	S	S	S	S
Leather trim:	S	S	S	S	S
AM/FM/ Radio-cassette:	S	S	S	S	S
Power door locks:	S	S	S	S	S
Power windows:	S	S	S	S	S
Tilt steering column:	S	S	S	S	S
Dual power mirrors:	S	S	S	S	S
Intermittent wipers:	S	S	S	S	S
Light alloy wheels:	S	S	S	S	S
Sunroof:	-	-	-	O	O
Anti-theft system:	S	S	S	S	S

S : standard; O : optional; - : not available

COLORS AVAILABLE

Exterior: Black,Red,Ivory,Blue,White,Gray,Silver,Taupe,Garnet,Emerald.
Interior: Black, Blue, Beige, Gray, Palomino, Parchment, Saddle.

MAINTENANCE

First revision: 3,000 miles
Frequency: 7,500 miles
Diagnostic plug: Yes

WHAT'S NEW IN 1995 ?

- No major change.

Model/ versions *: standard	ENGINE Type / timing valve / fuel system	Displacement cu/in	Power bhp @ rmn	Torque lb.ft @ rpm	Compres. ratio	TRANSMISSION Driving wheels / transmission	Final ratio	PERFORMANCE Acceler. 0-62 mph s	Stand. 1/4 mile s	Stand. 5/8 mile s	Braking 62-0 mph ft	Top speed mph	Lateral acceler. G	Noise level dBA	Fuel economy mpg city	hwy	Gasoline type / octane
SL320	L6* 3.2 SOHC-24-SMPFI	195	228 @ 5600	232 @ 3750	10.0 :1	rear - A5*	3.69	8.8	16.4	31.0	128	144	0.80	66-70+	21.0	32.0	S 91
SL500	V8* 5.0 DOHC-32 SMPFI	303	315 @ 5600	347 @ 3900	10.0 :1	rear - A4*	2.65	7.0	15.5	26.6	138	150	0.82	66-70+	19.0	27.0	S 91
SL600	V12* 6.0 SOHC-48-IEM	365	389 @ 5200	420 @ 3800	10.0 :1	rear - A4*	2.65	6.5	14.8	25.7	148	156	0.81	66-70+	17.0	23.0	S 91
S500	V8* 5.0 DOHC-32-SMPFI	303	315 @ 5600	347 @ 3900	10.0 :1	rear - A4*	2.65	7.5	16.1	28.0	141	156	0.74	64	18.0	26.0	S 91
S600	V12* 6.0 SOHC-48-SMPFI	365	389 @ 5200	420 @ 3800	10.0 :1	rear - A4*	2.64	6.8	15.6	27.0	148	156	0.77	63	15.0	23.0	S 91

+ hard-top-soft-top

• **Performance: 90%**
They differ much, depending on the motor because their weight is high. Thus, in spite of the power available, the SL 6-cylinder distinctly lacks the torque needed for crisp passing power. The driving is pleasant only with the V8 or even more so with the V12. The 500 SL or SC are more homogeneous, since the V8 torque allows powerful acceleration. The V12 drives directly into Nirvana because it seems to have no limit and it allows posting exotic times considering the imposing bulk of these vehicles.

• **Seats: 90%**
They provide effective lateral and lumbar support, but the leather trim is very slippery and their padding very firm.

• **Suspension: 80%**
Thanks to an electrical control, shock damping instantaneously receives the best setting between comfort and handling to the point that you have trouble feeling that a sports vehicle is being driven.

• **Access: 80%**
The doors and sizes are well-proportioned, allowing for easy seating.

• **Handling: 80%**
These two cars offer an exceptional performance in the climate of laid back safety. Certainly the weight and the electronics tend to make the reactions a little antiseptic but the performance is so self-assured that no one will complain. In a curve, oversteering follows a long neutral feel and is easy to control, because traction is extraordinary.

• **Steering: 80%**
Not the fastest, but its precision and ratio allow sharp reactions and good maneuverability.

• **Insurance: 75%**
The modest index, multiplied by the price results in a horribly costly premium.

• **Conveniences: 70%**
On the SL, the AC is effective even when the top is down and in case of rain, maintaining a speed of 80 mph is enough to keep the interior dry. Storage spaces are not numerous but of sufficient size.

MERCEDES-BENZ SC

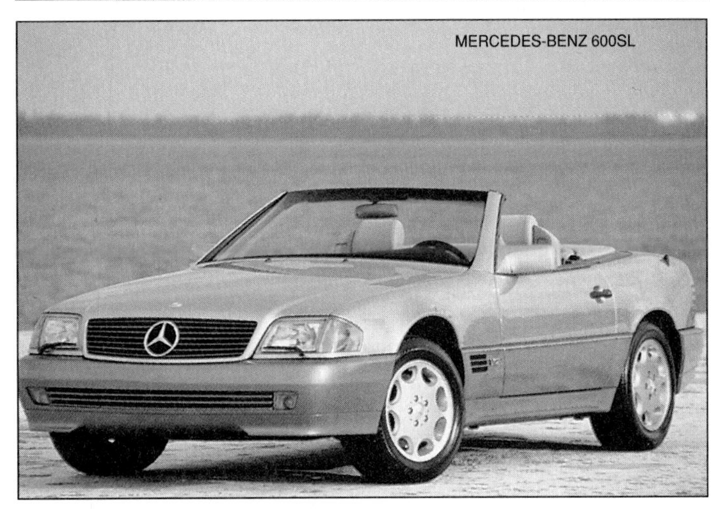

MERCEDES-BENZ 600SL

• **Depreciation: 60%**
In the current state of the economy, these vehicles are not the best investment, except for stature reasons.

• **Noise level: 60%**
These cars are as well-insulated as the sedans, even the convertible, which is quite a feat.

WEAK POINTS

• **Price/equipment: 00%**
These technological marvels are costly, but their equipment includes a thousand new refinements which sometimes surpass the imagination, such as the antiwind net of the convertible.

• **Fuel consumption: 20%**
It is high under all conditions, which surprises no one.

• **Interior space: 30%**
The SL offers only two seats while the coupe has four real seats and the interior sizes are well-dimensioned, which seems realistic for cars that are as bulky and weigh near two tons.

• **Trunk: 35%**
In both instances, its volume is proportional to the number of occupants.

CONCLUSION

• **Overall rating: 69.5%**
Not really sports cars, these two uncommon vehicles have the advantage of extreme sophistication and deliver exceptional power and safety. ☺

SPECIFICATIONS & PRICES

Model	Version/Trim	Body/ Seats	Interior volume cu.ft.	Trunk volume cu ft.	Drag coef. Cd	Wheel-base in	Lgth x Wdthx Hght. inx inx in	Curb weight lb	Towing capacity std. lb	Susp. type ft/rr	Brake type ft/rr	Steering system	Turning diameter ft	Steer. turns nber	Fuel. tank gal.	Standard tire size	Standard powertrain	PRICES US. $ 1994
MERCEDES BENZ		**Total warranty: 4 years / 50,000 miles; corrosion perforation; 5 years / unlimited.**																
SL320		2dr.con.2	NA	8.1	0.32	99.0	176.0x71.3x51.3	4098	NR	i/i	dc/dc/ABS	pwr.bal.	35.4	3.0	21.1	225/55ZR16	L6/3.2/A5	**$85,200**
SL500		2dr.con.2	NA	8.1	0.32	99.0	176.0x71.3x51.3	4167	NR	i/i	dc/dc/ABS	pwr.bal.	35.4	3.0	21.1	225/55ZR16	V8/5.0/A4	**$99,500**
SL600		2dr.con.2	NA	8.1	0.32	99.0	178.0x71.3x50.6	4453	NR	i/i	dc/dc/ABS	pwr.bal.	35.4	3.0	21.1	225/55ZR16	V12/6.0/A4	**$120,100**
S500		2dr.cpe.5	NA	14.3	0.30	115.9	199.4x75.3x57.1	4689	NR	i/i	dc/dc/ABS	pwr.bal.	38.4	3.1	26.4	235/60HR16	V8/5.0/A4	**$99,800**
S600		2dr.cpe.5	NA	14.3	0.30	115.9	199.4x75.3x57.1	4951	NR	i/i	dc/dc/ABS	pwr.bal.	38.4	3.1	26.4	235/60ZR16	V12/6.0/A4	**$133,300**

See page 393 for complete 1995 Price List

Diamante

Too intelligent?

Mitsubishi is taken for the Japanese Mercedes because these cars are girdled with electronic equipment and technical refinements not to be found elsewhere in these quantities. In spite of all the benefits of high technology, does a car need all this fighter pilot hardware?

The Diamante forms the high end of the range at Mitsubishi. It is sold only in the United States in the network of its builder. In 1995, it continues to be offered in the form of a 4-door sedan LS or the station wagon because the ES is henceforth reserved for fleet service. The standard engine is a 3.0L V6 with SOHC on the station wagon and DOHC on the LS with a 4-speed automatic.

STRONG POINTS

• Safety: **90 %**
The structure is sufficiently rigid to satisfy current standards and the front seats are protected by two standard air bags.

• Technical: **90%**
The steel unit body has good aerodynamic efficiency in spite of lines complicated by typically Japanese styling, plus a coefficient which is only 0.30. The suspension is independent, with 4-wheel disc brakes. The electronics take up much of the space on this model particularly the LS where it controls ignition, carburation, ABS, 4-wheel steering, traction control, electronic damper control and the variable intake system. At warranty expiration, it automatically deducts your Diamante service payments.

• Satisfaction: **85%**
The owners love this car for the richness of its technology which impresses even more by its reliability.

• Quality/fit/finish: **80%**
The leather seat trim is superb, but the wood appliqué on the dashboard is phony, too bad. Otherwise, the assembly and finish are very careful and the grain of the plastic that trims the dash is as agreeable to the eye as to the touch.

• Suspension: **80%**
The driver can choose the damping mode or allow the computer to decide on it depending on the circumstances. In "sport" position it is relatively firm while in auto it will adjust damping as a function of speed or brake application but will otherwise remain softer. The computer will raise the car by 3 cm on a bad road or lower it by 1 cm on the freeway.

• Driver's compartment: **80%**
It reminds one of the Audi by its organization and by gauges grouped under a long sunshield. The gauges and commands are treated in a high tech style, customary to Mitsubishi. If in the beginning it lightly complicates comprehension, later, they are easier to use and certainly more impressive. The driver has soon settled in comfort and disposes of excellent visibility.

• Conveniences: **80%**
In addition to a good-sized glove box, storage includes door pockets, an enclosure in the center console and more pockets in the back.

• Access: **80%**
No problem taking place in seats of these models, which are more imposing than they appear and whose doors open wide.

DATA

Category: front-wheel drive luxury sedans.
Class : 7

HISTORIC

Introduced in: 1990
Modified in: -
Made in: Nagoya, Japan.

RATINGS

Safety:	90 %
Satisfaction:	83 %
Depreciation:	55 %
Insurance:	7.0 %
Cost per mile:	$0.64

NUMBER OF DEALERS

In U.S.A.: 500

SALES ON THE AMERICAN MARKET

Model	1992	1993	Result
Diamante	22,112	25,267	+ 12.5 %

MAIN COMPETITORS

ACURA Vigor, AUDI A6, BMW 3-Series, HONDA Accord V6, INFINITI J30, LEXUS ES300 & GS300, MAZDA 626 V6 & Millenia, NISSAN Maxima, SAAB 900, TOYOTA Camry V6, VOLVO 850.

EQUIPMENT

MITSUBISHI Diamante	LS	Wagon
Automatic transmission:	S	S
Cruise control:	S	S
Power steering:	S	S
ABS brake:	O	S
Air conditioning:	S	S
Air bag (2):	S	S
Leather trim:	O	S
AM/FM/ Radio-cassette:	S	S
Power door locks:	S	S
Power windows:	S	S
Tilt steering column:	S	S
Dual power mirrors:	S	S
Intermittent wipers:	S	S
Light alloy wheels:	S	S
Sunroof:	O	O
Anti-theft system:	S	S

S : standard; O : optional; - : not available

COLORS AVAILABLE

Exterior: White, Silver, Blue, Green, Amethyst, Gray, Beige.
Interior: Beige, Blue, Gray, Black.

MAINTENANCE

First revision:	3,000 miles
Frequency:	6 months/6,000 miles
Diagnostic plug:	Yes

WHAT'S NEW IN 1995 ?

- In the United States the ES is out of the catalog and available only in fleet use.

Model/ versions *: standard	ENGINE Type / timing valve / fuel system	Displacement cu/in	Power bhp @ rmn	Torque lb.ft @ rpm	Compres. ratio	TRANSMISSION Driving wheels / transmission	Final ratio	PERFORMANCE Acceler. 0-62 mph s	Stand. 1/4 mile s	Stand. 5/8 mile s	Braking 62-0 mph ft	Top speed mph	Lateral acceler. G	Noise level dBA	Fuel economy mpg city	hwy	Gasoline type / octane
ES	V6*3.0 SOHC-12-MPFI	181	175 @ 5500	185 @ 3000	10.0 :1	front - A4*	3.958	10.6	17.8	30.5	144	118	0.72	65	21.0	29.0	R 87
LS	V6*3.0 DOHC-24-MPFI	181	202 @ 6000	201 @ 3000	10.0 :1	front - A4*	3.958	9.2	17.0	29.8	151	131	0.74	65	21.0	28.0	M 89

• **Seats: 70%**
Their appearance is much more impressive when girdled in leather but their lateral support could be extended and their lumber support should be more efficient. At the back, the bench seat back disposes of really operational headrests, rare enough to be well worth mentioning.

• **Trunk: 60%**
Relatively small for a car of this size, it closely resembles that of the old Mazda 929 because it lacks length and cannot be increased in size.

• **Performance: 60%**
The performance is only average, taking into accountthe cylinder displacement, because the weight is high. Takeoffs and pickups are soft particularly with a standard engine, but higher with a 200 hp of the LS. Even then, there is nothing there to smoke the tires.

• **Insurance: 60%**
The rate for the Diamante is a little higher than that of its competitors due to its technical complexity.

• **Noise level: 60%**
The Diamante is well-soundproofed, with rolling and power train noises thoroughly muted and the aerodynamic finesse minimizes air flow sound around the windshield.

• **Fuel consumption: 60%**
The Diamante is not one of the thirstiest, considering its weight.

• **Steering: 60%**
In its desire to always help the driver, the steering is always so well electronically controlled that one begins to resent the disagreeable feeling that someone else is deciding for you and under some cornering conditions it is positively annoying.

WEAK POINTS

• **Price/equipment: 30%**
The Diamante is not the only one to offer luxury and comfort at a reasonable price in this category, but it alone supplies, in addition to complete equipment, an incredible amount of electronic servo-controls of which driving pleasure and safety can easily dispense with.

• **Braking: 40%**
Its efficiency is not awesome even though it is easy to modulate and is very stable when assisted by the ABS. Disc pads with more bite are more necessary on this model than some of the gadgets. There effectiveness fades away after three panic stops.

• **Interior space: 40%**
The Diamante interior will only accept four adults; a fifth can only travel there in an emergency and its volume is lesser than that of its competitors. Taller persons will complain of lacking head space at the front seats, especially with an openable sun roof; rear legroom is also at a premium.

• **Handling: 40%**
Its more interesting on the LS which rolls less than the family sedan and whose engine is more muscled. This allows it to get through tight turns with a more secure feeling. The traction control functions in a frustrating way, leaving the driver the roll of a steering wheel holder accessory rather than being in charge of accelerator control. Fortunately this function can be annulled with a switch.

• **Depreciation: 40%**
It is stronger than that of its rivals because of the cost of repairs and maintenance beyond its warranty.

CONCLUSION

• **Overall rating: 64.5%**
We are fortunately not yet in the era where cars can drive themselves. Please Mitsubishi, let us have the pleasure of driving a little longer. ☺

SPECIFICATIONS & PRICES

Model	Version/Trim	Body/ Seats	Interior volume cu.ft.	Trunk volume cu ft.	Drag coef. Cd	Wheel-base in	Lgth x Wdthx Hght. inx inx in	Curb weight lb	Towing capacity std. lb	Susp. type ft/rr	Brake type ft/rr	Steering system	Turning diameter ft	Steer. turns nber	Fuel. tank gal.	Standard tire size	Standard powertrain	PRICES US. $ 1994
MITSUBISHI		**General warranty: 3 years / 36,000 miles: Power Train 5 years / 60,000 miles; Bodyperforation 7 years / 100,000 miles.**																
Diamante	LS	4dr.sdn. 4/5	94	13.6	0.30	107.2	190.2x69.9x52.6	3604	1000	i/i	dc/dcABS	pwr.r/p.	36.7	3.15	19	205/65R15	V6/3.0/A4	**$32,825**
Diamante	Wagon	5dr.wgn.4/5	102	37.3	0.40	107.2	192.4x69.9x57.9	3638	1000	i/r	dc/dc	pwr.r&p.	36.7	3.15	18.8	205/65R15	V6/3.0/A4	**$25,850**

See page 393 for complete 1995 Price List

MITSUBISHI CHRYSLER

EAGLE Talon, MITSUBISHI Eclipse

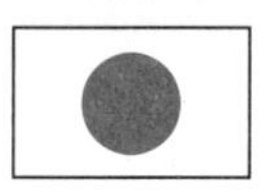

Gone for another tour...

The Chrysler group made the right bets by associating itself with Mitsubishi to concoct sports groups that respond perfectly to the expectations of the public. If the Stealth 3000GT address itself to a more affluent group, the Talon/Eclipse are more popular and enjoy a wide current success.

The Eagle Talon and the Mitsubishi Eclipse have been entirely re-thought for 1995. They are proposed as a 3-door coupe with front wheel drive, equipped with a 2.0 L DOHC naturally aspirated engine, shared with a Neon on the standard version, ESI for the Talon and RS/GS at Mitsubishi. A 2.0 L turbocharged Mitsubishi engine which was already fitted to the preceding model, can now be found in the Eclipse GS Turbo with 2-wheel drive and in the Talon TSi Turbo and Eclipse GSX with 4-wheel drive. The standard transmission is a 5-speed manual with the option of a 4-speed automatic. The Laser, which was sold by Plymouth, has been dropped.

STRONG POINTS

• Technical: 90%

These coupes are based on the Mitsubishi Galant platform and are manufactured in the Unites States at the Diamond Star Mitsubishi plant. Once more, the body style was created by Chrysler and Mitsubishi, which have modernized its lines while retaining its aggressive character. They have also improved its aerodynamics, which reached a remarkable 0.29 coefficient on all of their models.

• Safety: 90%

It improves considerably since its structure was extensively reinforced. These vehicles will henceforth carry 2 air bags as standard equipment and the 2 front shoulder harnesses have height adjustments.

• Satisfaction: 85%

Mitsubishi's strict controls and its reputation guarantee the quality and reliability of these models which have never caused large problems.

• Handling: 80%

The front wheel drives share the same efficiency with the 4-wheel drives and have an identical lateral acceleration coefficients, thanks to

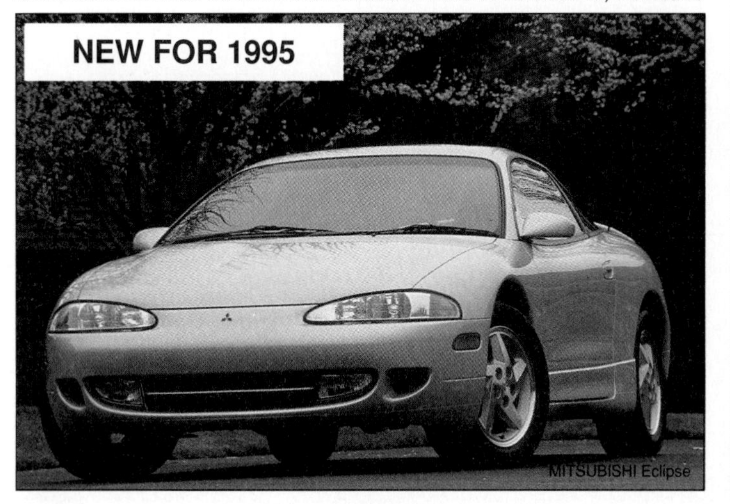

MITSUBISHI Eclipse

DATA

Category: front-wheel drive or AWD sport coupes.
Class : 3S

HISTORIC

Introduced in: 1990
Modified in: 1993: cosmetic touch-ups.1995: new model
Made in: Normal, Illinois, U.S.A..

RATINGS

		2WD	4WD
Safety:		90 %	90 %
Satisfaction:		95 %	95 %
Depreciation:		50 %	54 %
Insurance:	($725)	7.5 %	5.6 % ($885)
Cost per mile:		$0.44	$0.51

NUMBER OF DEALERS

In U.S.A.: 1,029 Eagle, 500 Mitsubishi

SALES ON THE AMERICAN MARKET

Model	1992	1993	Result
Talon	29,911	27,360	- 8.6 %
Eclipse	53,712	57,083	+6.0 %

MAIN COMPETITORS

ACURA Integra, FORD Mustang & Probe, HONDA Prelude & del Sol, MAZDA MX-3 & MX6, TOYOTA Celica & MR2, VOLKSWAGEN Corrado.

EQUIPMENT

EAGLE Talon MITSUBISHI Eclipse	DL RS	ESi GS	TSi GS-T	TSI AWD GSX
Automatic transmission:	O	O	O	O
Cruise control:	O	O	S	S
Power steering:	S	S	S	S
ABS brake:	O	O	O	O
Air conditioning:	O	O	S	S
Air bag (2):	S	S	S	S
Leather trim:	-	-	O	S
AM/FM/ Radio-cassette:	S	S	S	S
Power door locks:	-	O	O	O
Power windows:	-	O	S	S
Tilt steering column:	S	S	S	S
Dual power mirrors:	S	S	S	S
Intermittent wipers:	S	S	S	S
Light alloy wheels:	-	O	S	S
Sunroof:	O	O	O	O
Anti-theft system:	-	O	O	O

S : standard; O : optional; - : not available

COLORS AVAILABLE

Exterior: Black, Blue, Green, Red, Gray, White.
Interior: Brown, Gray.

MAINTENANCE

First revision: 3,000 miles
Frequency: 6 months/6,000 miles
Diagnostic plug: Yes

WHAT'S NEW IN 1995 ?

- This model is entirely new for 1995.

Model/ versions *: standard	ENGINE Type / timing valve / fuel system	Displacement cu/in	Power bhp @ rmn	Torque lb.ft @ rpm	Compres. ratio	TRANSMISSION Driving wheels / transmission	Final ratio	PERFORMANCE Acceler. 0-62 mph s	Stand. 1/4 mile s	Stand. 5/8 mile s	Braking 62-0 mph ft	Top speed mph	Lateral acceler. G	Noise level dBA	Fuel economy mpg city	Fuel economy hwy	Gasoline type / octane
1)	L4* 2.0 DOHC-16-MPSFI	122	140 @ 6000	131 @ 4800	9.6 :1	front - M5*	3.94	9.5	16.8	29.3	131	109	0.85	68	28.0	42.0	R 87
						front - A4	3.91	10.8	17.2	31.4	138	106	0.85	68	28.0	40.0	R 87
2)	L4* 2.0T DOHC-16-MPSFI	122	210 @ 6000	214 @ 3000	8.5 :1	front/4-M5*	4.93	7.0	15.6	27.8	125	125	0.85	69	24.0	32.0	S 91
	L4* 2.0T DOHC-16-MPSFI	122	205 @ 6000	220 @ 3000	8.5 :1	front/4-A4	4.42	8.2	16.3	28.4	131	118	0.85	69	22.0	28.0	S 91

1) Talon ESi/Eclipse/RS/GS 2) Talon TSi/Eclipse GS Turbo & GSX.

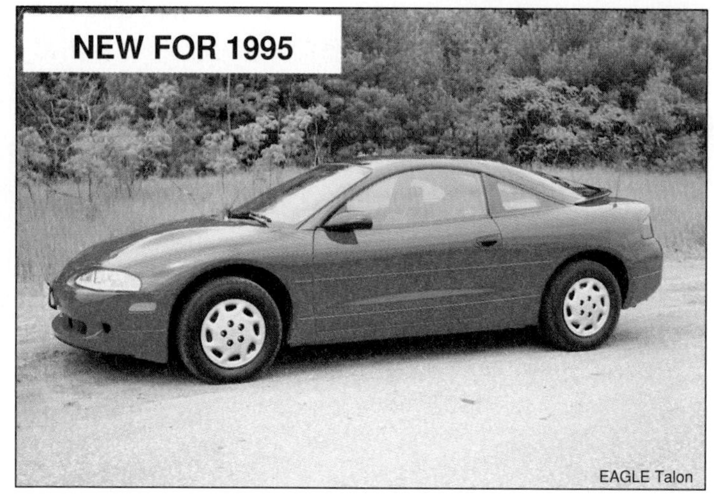

EAGLE Talon

their independent suspension. The difference is mostly in straight line traction ability as well as in the cornering traction when the inside wheels are unloaded on slippery pavement.

• Quality/fit/finish: 80%
The assembly is very homogenous and the finish, just like the quality of the materials, is more flattering than before.

• Driver's compartment: 80%
The interior has been entirely redesigned with the dash losing some of its ultra ordinary style to find a more normal and ergonomic configuration. The gauges are better laid out and easier to read. Also, the console protrudes slightly and is aimed toward the driver, putting controls within hands' reach. Those of the radio are low and behind the gear shift. Visibility is improved, both forward and at the side, while at the back the quarter panel window remains narrow.

• Seats: 80%
The effective bucket seats provide good side hold and lumbar support, while the bench is more symbolic.

• Performance: 70%
The 2.0 L engine on standard versions is full of good will and allows leisurely rides. Only the 2.0 L Turbo heartens the driving and instantly gets up in rpm as soon as the Turbo kicks in. It provides ample low and medium range torques with a relatively short response time. The 1st two stick shift gears are too short for acceleration. Traction in corners is never in short supply, no matter what the pavement quality.

• Steering: 70%
The manual version of the standard Mitsubishi is heavy and with too much ratio to pretend to be fast or precise and totally lacks any sort of sports feel. On the other hand it is fast and precise, with a power assist, but has a tendency to go light at high speeds. In both instances the somewhat large turning diameter penalizes maneuverability.

• Suspension: 70%
The sports coupes are no longer what they were, because comfort is now part of the standard equipment. In spite of some firmness it offers comfort equivalent to that of a compact sedan and becomes a little noisier when the pavement is seriously deficient.

• Fuel consumption: 65%
Essentially, it depends on how you drive, a "sport" oriented driver will get 21mpg and a more conservative driver will go to 25.

• Access: 65%
It is easier to take a seat up front, thanks to long doors that open wide; only a duck would be satisfied with the legroom in the back

• Depreciation: 55%
It has begun to increase due the larger competition but stays at an interesting level.

• Braking: 50%
Sufficient in normal conditions, the power assist, fade-resistance, and lack of bite show up in panic stops. Also, the rear wheels lock up rapidly. The optional ABS improves stability but has a tendency to lengthen the stopping distances.

• Noise level: 50%
While the Turbo sometimes makes itself heard with high pitched sounds, the naturally aspirated engines are quieter and road noises dominate while wind noise is very low.

• Conveniences: 50%
Storage includes a small glove box, door pockets, and a cabinet in the center console.

• Price/equipment: 50%
The old models enjoyed considerable success because their ratios of price/performance/quality was favorable. The 2-wheel drive versions have an army of competitors while the 4-wheel drives are unique in this price range. Only the TSi/RS/GSX are well-equipped, while the others have received a few small gratifications to justify the price increase.

• Insurance: 50%
As with all sports cars, the billing is upscale especially for a buyer under 25 years of age.

WEAK POINTS

• Trunk: 30%
The 2-wheel drive trunk is more spacious than the 4-wheel drive's which must carry rear axle and transmission components . One can however increase its capacity by laying down the rear of the bench seat but in both instances the tall sill is not practical.

• Interior space: 40%
Front seats are more generous than the rear ones, where head and legroom are very limited, good only for baggage.

CONCLUSION

• Overall rating: 65.0%
In the middle of a sea of sports coupes, these models stand out by the originality of their 4-wheel drive and good calibration of their format. The interior presents well and has attractive lines. ☺

SPECIFICATIONS & PRICES

Model	Version/Trim	Body/ Seats	Interior volume cu.ft.	Trunk volume cu ft.	Drag coef. Cd	Wheel-base in	Lgth x Wdthx Hght. inx inx in	Curb weight lb	Towing capacity std. lb	Susp. type ft/rr	Brake type ft/rr	Steering system	Turning diameter ft	Steer. turns nber	Fuel. tank gal.	Standard tire size	Standard powertrain	PRICES US. $ 1995
EAGLE		**General warranty: 3 years / 36,000 miles: Power Train 5 years / 60,000 miles; surface corrosion; 3 years; Body perforation 7 years / 100,000 miles.**																
Talon	ESi	3dr.cpe2+2	79.1	16.6	0.29	98.8	172.2x68.3x51.0	2835	NR	i/i	dc/dc	pwr.r&p.	38.0	2.4	15.8	195/70HR14	L4/2.0/M5	**$14,362**
Talon	TSi Turbo	3dr.cpe2+2	79.1	13.5	0.29	98.8	172.2x68.7x51.6	3240	NR	i/i	dc/dc	pwr.r&p.	38.0	2.4	15.8	215/55VR16	L4T/2.0/M5	**$17,978**
MITSUBISHI		**General warranty: 3 years / 36,000 miles: Power Train 5 years / 60,000 miles; Body perforation 7 years / 100,000 miles.**																
Eclipse	RS	3dr.cpe2+2	79.1	16.6	0.29	98.8	172.2x68.3x51.0	2723	NR	i/i	dc/dr	r&p.	38.0	2.41	15.8	195/70R14	L4/2.0/M5	**$12,099**
Eclipse	GS	3dr.cpe2+2	79.1	16.6	0.29	98.8	172.2x68.7x51.0	2822	NR	i/i	dc/dc	pwr.r&p.	38.0	2.41	15.8	205/55HR16	L4/2.0/M5	**$14,219**
Eclipse	GS Turbo	3dr.cpe2+2	79.1	16.6	0.29	98.8	172.2x68.7x51.0	2877	NR	i/i	dc/dc	pwr.r&p.	38.0	2.41	15.8	205/50HR16	L4/2.0/M5	**$18,689**
Eclipse	GSX	3dr.cpe2+2	79.1	13.5	0.29	98.8	172.2x68.3x51.6	3120	NR	i/i	dc/dc	pwr.r&p.	38.0	2.41	15.8	215/55VR16	L4T/2.0/M5	**$21,459**

See page 393 for complete 1995 Price List

Better armed...

In the United States, the Galant plays the roll of the second fiddle to the Honda Accord, The Mazda 626 and the Toyota Camry. It's enough to test them to become convinced that it has as much to offer on a technical plan and perhaps more in terms of finish and quality.

Right in the middle of the Mitsubishi range in the USA, the Galant finds itself between the Diamante and the Mirage. It had been totally renewed last year, in the form of a 3-volume 4-door sedan, which will be offered in 1995 in versions S, ES, LS and LS V6 replacing the preceding GS. The first 3 are equipped with a 2.4 L SOHC, while the V6 receives the new 2.5 L which is also used with the Chrysler Cirrus and Dodge Stratus. The manual 5-speed is standard, except on the LS V6 which receives a 4-speed automatic, offered as an option on the others.

STRONG POINTS

• Safety: 90%
Even though the weight has not increased much, the body has been seriously stiffened and 2 air bags protect the occupants in the front seats, while the others have 3-point belts.

• Satisfaction: 85%
The power trains are reliable, but American owners complain about their relationships with some dealers, concerning the application of the warranty.

• Quality/fit/finish: 80%
The assembly and the material used are of good quality and the fit, from some points of view, is more careful than that of the competition. Interior appearance is less sad than on the preceding model but doesn't project luxury. Only the LS allows the leather trim as an option.

• Technical: 80%
The steel unit body has seen its main dimensions increase in all directions and the aerodynamic finesse has improved with the coefficient dropped below 0.30. The independent 4-wheel suspension includes antidive at the front and antilift at the rear for improved pitch stability, while the stabilizer bars act laterally . The brakes are discs and drums on the S, ES, LS,and 4 discs are used on the LS V6, with ABS as an option on all models.

• Seats: 80%
They are well-designed, front and rear, (rare enough to be mentioned) and provide very effective lateral and lumbar support.

• Driver's compartment: 80%
The layout of the new dashboard is modern, but it does not shine by its originality and reminds one of the Colt. The best driving position is easy to find, thanks to different adjustments. The steering and the shift offer a good grip, visibility is good, the controls are without surprises and instrumentation is legible and well laid out.

• Noise level: 75%
It remains good thanks to the rigid body, the efficiency of the sound-proofing and the good engine balance which generates few vibrations.

• Performance: 70%
The 141 hp of the standard engine are not too much to launch the mass

DATA

Category: front-wheel drive & AWD compact sedans.
Class : 4

HISTORIC

Introduced in: 1983: rear-wheel drive.
Modified in: 1987: front-wheel drive and AWD;1993:reshaped.
Made in: Normal, IL, U.S.A.

RATINGS

Safety: 90 %
Satisfaction: 87 %
Depreciation: 55 %
Insurance: 6.5 % ($585)
Cost per mile: $0.42

NUMBER OF DEALERS

In U.S.A.: 500

SALES ON THE AMERICAN MARKET

Model	1992	1993	Result
Galant	23,758	18,351	- 32.8 %

MAIN COMPETITORS

BUICK Skylark, CHEVROLET Corsica, CHRYSLER Le Baron-Cirrus, DODGE-PLYMOUTH Spirit-Acclaim-Stratus, HONDA Accord, MAZDA 626, NISSAN Altima, OLDSMOBILE Achieva, PONTIAC Grand Am, SUBARU Legacy, TOYOTA Camry, VOLKSWAGEN Passat.

EQUIPMENT

MITSUBISHI Galant	S	ES	LS	LS V6
Automatic transmission:	O	O	O	S
Cruise control:	O	O	S	S
Power steering:	S	S	S	S
ABS brake:	O	O	O	O
Air conditioning:	O	S	S	S
Air bag (2):	S	S	S	S
Leather trim:	-	-	O	O
AM/FM/ Radio-cassette:	O	S	S	S
Power door locks:	O	S	S	S
Power windows:	O	S	S	S
Tilt steering column:	S	S	S	S
Dual power mirrors:	S	S	S	S
Intermittent wipers:	S	S	S	S
Light alloy wheels:	-	-	S	S
Sunroof:	-	-	S	S
Anti-theft system:	-			

S : standard; O : optional; - : not available

COLORS AVAILABLE

Exterior: Beige, Balck, Blue, Gray, Purple, Red, Silver, White.
Interior: Dark Blue, Dark Gray, Dark Brown.

MAINTENANCE

First revision: 4,000 miles
Frequency: 6 months/ 6,000 miles
Diagnostic plug: Yes

WHAT'S NEW IN 1995 ?

- The GS replaced by the LS V6.
- Front hood without the bump.
- Door trim and dashboard grey.

	ENGINE					TRANSMISSION		PERFORMANCE										
Model/ versions *: standard	Type / timing valve / fuel system	Displacement cu/in	Power bhp @ rmn	Torque lb.ft @ rpm	Compres. ratio	Driving wheels / transmission	Final ratio	Acceler. 0-62 mph s	Stand. 1/4 mile s	Stand. 5/8 mile s	Braking 62-0 mph ft	Top speed mph	Lateral acceler. G	Noise level dBA	Fuel economy mpg city	Fuel economy hwy	Gasoline type / octane
S/ES/LS	L4* 2.4 SOHC-16-MPFI	143	141 @ 5500	148 @ 3000	9.5 :1	front - M5*	4.32	8.8	16.6	30.2	141	115	0.79	66	28.0	37.0	R 87
						front - A4	4.35	9.7	17.4	31.0	148	112	0.79	65	26.0	35.0	R 87
LS V6	V6* 2.5 SOHC-24-MPFI	152	155 @ 5500	161 @ 4400	10.0 :1	front - A4	3.91	9.0	16.5	30.8	138	124	0.79	65	21.0	28.0	R 87

of this vehicle. Takeoffs are normal and accelerations a little too soft, defects which also affected the old model, because multivalve engines are generally short of low rpm torque. The V6 of the LS is not much more powerful but offers a higher torque.

• Steering: 70%
The power-assisted steering with electronic control is the only high tech component of the Galant. It is similar by some of its bizarre action, that of the Diamante and adds little for the complexity it involves.

• Access: 70%
While the rear doors are narrower than the front ones, it is not difficult to take a seat because they open wide.

• Suspension: 70%
Its smoothness provides freeway comfort and the ample wheel travel absorbs the road defects. Damping is consistant.

• Interior space: 65%
It is practically equal to those of the Honda Accord and the Mazda 626, with more generous dimensions, especially in height, even with a sliding roof. Four tall adults will be well at ease and a 5th can sit for a short trip.

• Fuel consumption: 65%
That of the 4-cylinder compares to the Nissan Altima while the ones of the V6 are equivalent to the 626.

• Handling: 65%
It's one of the most advanced points, because these cars are more neutral in a curve and understeering doesn't occur until latter. In the base version, the roll is considerable and becomes better controlled on the LS with its more elaborate suspension.

• Insurance: 65%
It's not one of the more expensive in this category. The premiums rose together with the popularity of these models.

• Conveniences: 60%
Thev consist of a sizable glove box, door pockets on the ES/LS and some hiding places in the dashboard.

• Price/equipment: 50%
They compare favorably with those of the Honda Accord and Mazda 626.

• Trunk: 50%
It's volume has increased in relation to the old model and the cutout in the opening makes it more accessible.

WEAK POINTS

• Braking: 40%
It poses no problem in normal use but is mediocre in models without rear wheel disc brakes or with ABS, since the front wheels block rapidly in panic stops, lengthening the distance, but making it straighter than before. The modulation on this pedal is of a rare precision, while the fade-resistance of the brake material is only average.

• Depreciation: 40%
Less popular than some of its competitors, the Galant loses a little more than average. The fact that Mitsubishi doesn't harness it with electronic gadgets which they flourish elsewhere in the world, will help them not lose value as before.

CONCLUSION

• Overall rating: 67.5%
The Galant is an interesting compact which competes better than its main rivals since it is equipped with a V6 which is becoming a standard in its group, where even Honda has acquired one. ☺

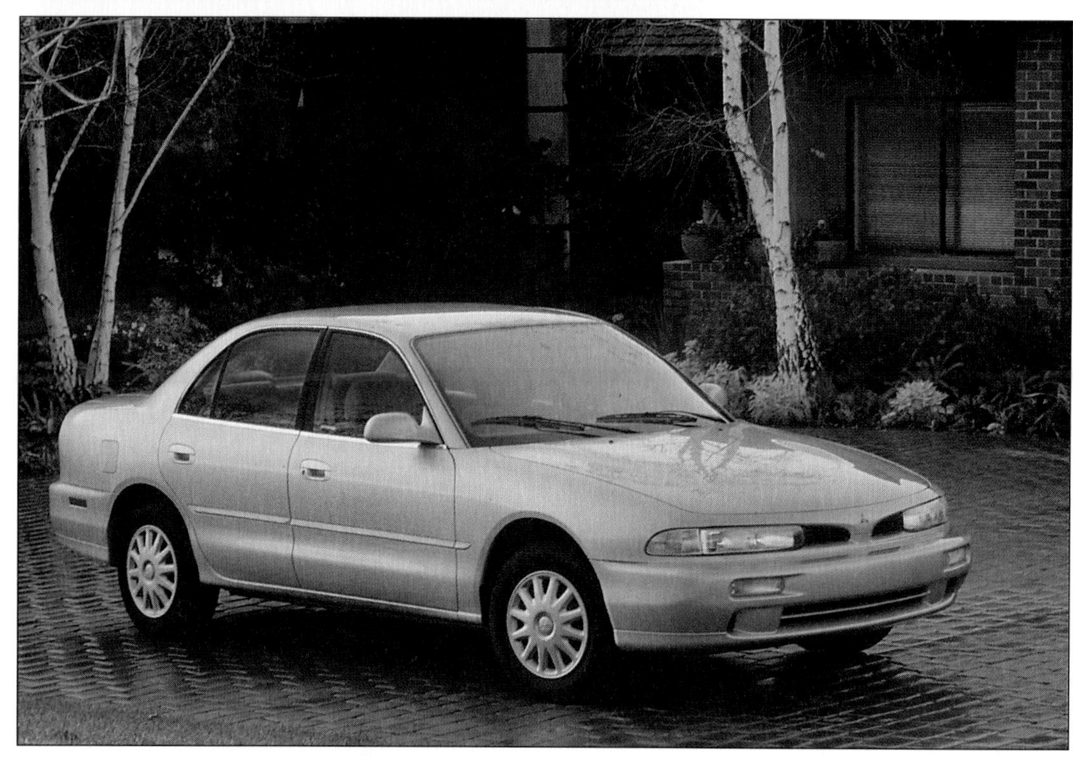

SPECIFICATIONS & PRICES

Model	Version/Trim	Body/ Seats	Interior volume cu.ft.	Trunk volume cu ft.	Drag coef. Cd	Wheel-base in	Lgth x Wdthx Hght. inx inx in	Curb weight lb	Towing capacity std. lb	Susp. type ft/rr	Brake type ft/rr	Steering system	Turning diameter ft	Steer. turns nber	Fuel. tank gal.	Standard tire size	Standard powertrain	PRICES US. $ 1994
MITSUBISHI		**General warranty: 3 years / 36,000 miles: Power Train 5 years / 60,000 miles; Bodyperforation 7 years / 100,000 miles.**																
Galant	S	4dr.sdn.5	97.3	12.5	0.29	108.5	187.0x68.1x53.1	2756	1000	i/i	dc/dr	pwr.r&p.	34.8	3.0	17	185/70R14	L4/2.4/M5	**$13,600**
Galant	ES	4dr.sdn.5	97.3	12.5	0.29	108.5	187.0x68.1x53.1	2866	1000	i/i	dc/dr	pwr.r&p.	34.8	3.0	17	185/70R14	L4/2.4/M5	**$17,880**
Galant	LS	4dr.sdn.5	94.4	12.5	0.29	108.5	187.0x68.1x53.1	2976	1000	i/i	dc/dr	pwr.r&p.	34.8	3.0	17	195/60R15	L4/2.4/M5	**$19,420**
Galant	LS V6	4dr.sdn.5	94.4	12.5	0.29	108.5	187.0x68.1x53.1	3042	1000	i/i	dc/dc/	pwr.r&p.	34.8	3.0	17	205/60R15	V6/2.5/A4	**$21,277**

See page 393 for complete 1995 Price List

Colt-Mirage-Summit

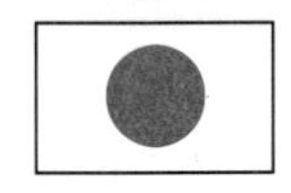

Politics...

Not to disturb the Neon, which Chrysler wants to carpet everywhere wall to wall, it was decided to retire the 4-door in the base model. It's too bad, because it was a more economical model, better made and more reliable than the Neon can be in its beginnings.

The Colts are the base models in the Dodge and Plymouth range, and are offered in standard or ES, while in the United States, Mitsubishi distributes them under the name of Mirage in E, ES, and LS versions. In 1995, the 4-door goes off market and only the engine remains. This is a 1.5 L coupled to a manual 5-speed with an automatic available as an option.

STRONG POINTS

• Fuel consumption: 100%
It is economical, because the weight and aerodynamics are favorable and the sequential fuel injection of these engines is relatively sophisticated for cars in this category.

• Safety: 90%
The Colt and Mirage reached maximum, thank to their impact proof structure and 2 air bags.

• Price/equipment: 90%
While the base coupe is offered at an attractive price, it comes with just about nothing and the sticker price grows fast if you dip into the long list of options.

• Satisfation: 80%
A high percentage of the owners of these popular models are satisfied. There have been just minor problems and the guarranty is not negligible. Their quality has improved, particulary in the corrosion area, where it doesn't appear as early as in some competing models.

• Driver's compartment: 70%
The driver's environment is simple and functional, and the assembly and the gauge panel and center console fit well in the dashboard, which visually is not intrusive. The main driving elements are well-located, and visibility is satisfactory from all angles.

• Suspension: 70%
It surprises agreeably by its smoothness, unusual in a car of this class. It readily absorbs small road defects and doesn't react too harshly to large wallows.

• Quality/fit/finish: 70%
The presentation of these vehicles is very simple, the fit and finish very homogenous, but the appearance of its plastics and fabrics looks bargain basement.

• Access: 70%
You get in without problems through doors that open wide and the seats move easily to allow access to the rear.

• Technical: 70%
The latest little Mitsubishi's are more elegant and refined; they have inherited the mechanical elements of preceding models. The shapes of the steel unit body are more rounded and aerodyanmically efficient, which brings the drag coefficient down to 0.29. The independent 4-wheel suspension includes a McPherson up front and multiple arms at the back. These front wheel drives are powered by 1.5 L 4-cylinder

DATA

Category: front-wheel drive sub-compact sedans and coupes.
Class : 3S

HISTORIC

Introduced in: 1979 (rear-wheel drive)
Modified in: 1985: front-wheel drive;1989 & 1993: new design.
Made in: Mizushima, Japan.

RATINGS

Safety: 90 %
Satisfaction: 85 %
Depreciation: 44 % (2 years)
Insurance: 8.4 % ($520)
Cost per mile: $0.30

NUMBER OF DEALERS

In U.S.A.: 1,887 Dodge, 1,029 Eagle, 500 Mitsubishi.

SALES ON THE AMERICAN MARKET

Model	1992	1993	Result
Colt	23,111	28,817	+ 19.9 %
Mirage	22,917	29,162	+ 21.5 %
Summit	6,885	19,436	+ 282 %

MAIN COMPETITORS

GEO Metro, HONDA Civic, HYUNDAI Accent, MAZDA 323, NISSAN Sentra, SUBARU Justy, SUZUKI Swift, TOYOTA Tercel & Corolla.

EQUIPMENT

DODGE-PLYMOUTH Colt MITSUBISHI Mirage	base S 2 dr.	ES ES 2 dr.	 LS 2 dr.
Automatic transmission:	O	O	O
Cruise control:	-	-	-
Power steering:	S	-	-
ABS brake:	O	O	O
Air conditioning:	-	O	O
Air bag (2):	S	S	S
Leather trim:	-	-	-
AM/FM/ Radio-cassette:	O	O	O
Power door locks:	-	-	-
Power windows:	-	-	-
Tilt steering column:	-	O	-
Dual power mirrors:	-	O	-
Intermittent wipers:	-	O	-
Light alloy wheels:	-	-	-
Sunroof:	-	-	-
Anti-theft system:	-	-	-

S : standard; O : optional; - : not available

COLORS AVAILABLE

Exterior: Blue, Green, Red, White.
Interior: Gray, Blue.

MAINTENANCE

First revision: 3,000 miles
Frequency: 6 months/6,000 miles
Diagnostic plug: Yes

WHAT'S NEW IN 1995 ?

- **Four-door models dropped.**
- **One available engine (1.5 L).**
- **Two air bags standard on all models.**
- **Adjustable steering column offered as an option on the ES.**
- **Optional outside rear-mirrors electrical on the ES.**

	ENGINE					TRANSMISSION		PERFORMANCE										
Model/ versions *: standard	Type / timing valve / fuel system	Displacement cu/in	Power bhp @ rmn	Torque lb.ft @ rpm	Compres. ratio	Driving wheels / transmission	Final ratio	Acceler. 0-62 mph s	Stand. 1/4 mile s	Stand. 5/8 mile s	Braking 62-0 mph ft	Top speed mph	Lateral acceler. G	Noise level dBA	Fuel economy mpg city	hwy	Gasoline type / octane
base	L4* 1.5 SOHC-12 IESPM	90	92 @ 6000	93 @ 3000	9.2 :1	front - M5*	2.91	12.3	18.5	34.7	144	103	0.75	68	40.0	51.0	R 87
						front - A4	3.600	13.5	19.2	35.6	141	100	0.75	68	35.0	43.0	R 87

Colt-Mirage-Summit

which develop 92 hp. The brakes are mixed with ABS optional on other models.

• Steering: 60%
You will do better to opt for the power assist whose fast and precise action gives these little cars a fun agility while a standard manual has too low a ratio and gets vague.

• Handling: 50%
These little sedans are agile in a turn in spite of the roll caused by suspension softness. While they understeer easily, they remain readily controllable, don't produce any underhanded reactions and surprise by their driving pleasure on little twisty roads which allows sports car playfulness.

• Interior space: 50%
In relation to the old models the latest have more head and legroom, mostly at the front, and overall volume has improved on the average by 3.5 cu.ft. in both bodies.

• Seats: 50%
Better dimensioned, they remain disapointing because their curvature is insufficient to provide adequate lateral support and their lumbar support is supplied by a padding as thin as it is hard.

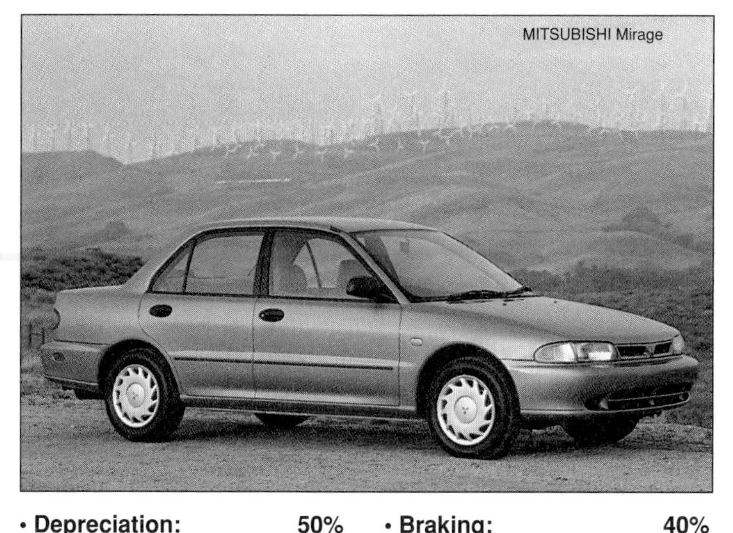

MITSUBISHI Mirage

• Depreciation: 50%
The vast Chrysler/Dodge/Plymouth dealership network is the performance bond that speaks of ease of maintance and service. This year their resale value is 55% of purchase cost 2 years ago.

WEAK POINTS

• Performance: 30%
Their diet has done well for these little cars, which are fun to drive and the engine has little inertia and loves to rev. It delivers a better performance with a manual than with an automatic whose ability to launch or accelerate is laborious.

DODGE Colt

MITSUBISHI Mirage

• Braking: 40%
While decelerations are progressive and easy to modulate, panic stops neccessitate an ABS since front wheel lockup extends stopping distances to 154 ft, which is very long, taking into account the weight of these cars.

• Trunk: 40%
The trunk, separated from the interior, is less practical than that of the old hatchbach because it lacks height and its opening is narrow. However, you can increase its volume by lowering the back of the bench seat.

• Noise level: 40%
Reasonable at cruise speed, the engines are noisy during accelerations and with less soundproofing, the standard models ring more on bad pavement.

• Conveniences: 40%
One can't say that the storage spots are very numerous but there is a large glove box and you can find pockets in the doors. Do consider a roof rack...

• Insurance: 45%
All things considered, these little cars cost more to insure than big ones and the average Colt index ranges around 9%.

CONCLUSION

• Overall rating: 60.5%
Colts sell in large numbers because they present one of the best combinations of "car-tool", pleasant to drive and are very economical. Pretty, amusing, practical, inexpensive and reliable, it has everything to please those on a small budget.

OWNERS' SUGGESTIONS

-Better soundproofing
-Better seat padding
-Standard equipment more complete
-Tire sizes increased to 165/80/13 on the base model
-An interior less insipid

SPECIFICATIONS & PRICES

Model	Version/Trim	Body/ Seats	Interior volume cu.ft.	Trunk volume cu ft.	Drag coef. Cd	Wheel-base in	Lgth x Wdthx Hght. inx inx in	Curb weight lb	Towing capacity std. lb	Susp. type ft/rr	Brake type ft/rr	Steering system	Turning diameter ft	Steer. turns nber	Fuel. tank gal.	Standard tire size	Standard powertrain	PRICES US. $ 1994
DODGE-PLYMOUTH-EAGLE	**General warranty: 3 years / 36,000 miles: Power Train 5 years / 60,000 miles; surface corrosion; 3 years; Body perforation 7 years / 100,000 miles**																	
Colt-Summit	base	2dr.sdn.5	87	10.5	0.29	96.1	171.1x66.1x51.6	2086	1000	i/i	dc/dr	r&p.	32.8	3.6	13.2	145/80R13	L4/1.5/M5	**$9,319**
Colt-Summit	ES	2dr.sdn.5	87	10.5	0.29	96.1	171.1x66.1x51.6	2108	1000	i/i	dc/dr	pwr.r&p.	32.8	3.6	13.2	155/80R13	L4/1.5/M5	**$12,298**
MITSUBISHI	**General warranty: 3 years / 36,000 miles: Power Train 5 years / 60,000 miles; Bodyperforation 7 years / 100,000 miles.**																	
Mirage	S	2dr.sdn.5	87	10.5	0.29	96.1	171.1x66.1x51.4	2083	1000	i/i	dc/dr	r&p.	32.8	3.6	13.2	145/80R13	L4/1.5/M5	**$9,179**
Mirage	ES	2dr.sdn.5	87	10.5	0.29	96.1	171.1x66.1x51.4	2105	1000	i/i	dc/dr	r&p.	32.8	3.6	13.2	155/80R13	L4/1.5/M5	**$10,579**
Mirage	LS	2dr.sdn.5	87	10.5	0.29	96.1	171.1x66.1x51.4	2127	1000	i/i	dc/dr	pwr.r&p.	33.4	3.2	13.2	175/70R13	L4/1.5/M5	**$12,129**

See page 393 for complete 1995 Price List

MITSUBISHI CHRYSLER

DODGE-EAGLE-MITSUBISHI-PLYMOUTH
Colt Vista - Expo - LRV - Summit Wagon

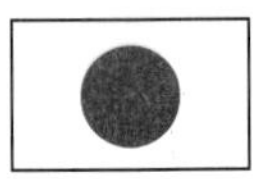

Condemned to dissapear...

The formula consisting of offering a vehicle which attempts to combine small engines with a station wagon that offers a larger volume, does not seem to be successful. The progression of mini-vans and the return of some station wagons sound taps for them.

The new Honda Odyssey, which will keep company with the Expo-LRV-VIsta-Summit is the only one of their kind to defend the idea of a vehicle situated between station wagons and mini-vans. Dodge and Plymouth distribute the Colt Vista, Eagle, and Summit Wagon. They are manufactured by Mitsubishi in Japan, which distributes them under its own network in the US under the names RLV and Expo, not available in Canada. You find them at Dodge, Eagle, Plymouth in standard front wheel drive or in SE with 4-wheel drive.

STRONG POINTS

• Safety: 90%
The impact resistance of the body is only average, but the belt line of the body, the doors and locks were highly reinforced to better resist impacts. This year, two front air bags are standard to allow an improvement in occupant protection.

• Satisfaction: 85%
Owners appreciate the reliabiltiy of these vehicles, which are easy to maintain but whose parts are expensive when not covered by the warranty.

• Driver's compartment: 80%
Thanks to the adjustments of the seat and steering column, the driver rapidly finds the most comfortable position and sits tall and straight, with excellent visibility. The dashboard is placed low, opening up cabin space and everything is well-organized from switches to gauges.

• Storage space: 80%
Its volume varies with the position of the benches, which can be folded or rapidly removed to provide a greater volume or transport bulkier objects.

• Technical: 80%

MITSUBISHI Expo

DATA

Category: front -wheel drive or AWD compact vans.
Class : 3

HISTORIC

Introduced in: 1983 (Chariot) 1985: imported in U.S.A. & Canada.
Modified in: 1992
Made in: Mizuchima, Japan.

RATINGS

Safety: 92 %
Satisfaction: 85 %
Depreciation: 54 %
Insurance: ($525) (2WD) 8.4 % ($585) (4WD)
Cost per mile: $0.42

NUMBER OF DEALERS

In U.S.A.: 1,887 Dodge, 1,822 Plymouth, 1,029 Eagle, 500 Mitsubishi.

SALES ON THE AMERICAN MARKET

Model	1992	1993	Result
Expo/LRV	16,558	-	

MAIN COMPETITORS

CHEVROLET Lumina (base), DODGE-PLYMOUTH Caravan-Voyager (base), MAZDA MPV (2.6L), NISSAN Axxess, TOYOTA Previa.

EQUIPMENT

DODGE-PLYMOUTH Colt EAGLE Summit Wagon MITSUBISHI	base DL	SE LX	AWD AWD	 LRV	 Expo	 AWD
Automatic transmission:	O	O	O	O	O	O
Cruise control:	O	O	O	O	O	S
Power steering:	S	S	S	S	S	S
ABS brake:	O	O	O	O	O	O
Air conditioning:	O	O	O	S	S	S
Air bag (2):	S	S	S	S	S	S
Leather trim:	-	-	-	-	-	-
AM/FM/ Radio-cassette:	O	O	O	O	O	O
Power door locks:	O	S	O	O	O	S
Power windows:	-	O	O	O	S	
Tilt steering column:	S	S	S	S	S	S
Dual power mirrors:	O	S	S	O	O	S
Intermittent wipers:	S	S	S	S	S	S
Light alloy wheels:	-	-	-	O	O	S
Sunroof:	-	-	-			
Anti-theft system:	-	-	-			

S : standard; O : optional; - : not available

COLORS AVAILABLE

Exterior: Blue, Red, White, Lime, Turquoise.
Interior: Gray, Blue.

MAINTENANCE

First revision: 3,000 miles
Frequency: 6 months/6,000 miles
Diagnostic plug: No

WHAT'S NEW IN 1995 ?

- Passenger side air bag.
- Redesigned grill and hood.
- Fabric-covered center console.
- Retractable cup holders.

	ENGINE					TRANSMISSION		PERFORMANCE									
Model/ versions *: standard	Type / timing valve / fuel system	Displacement cu/in	Power bhp @ rmn	Torque lb.ft @ rpm	Compres. ratio	Driving wheels / transmission	Final ratio	Acceler. 0-62 mph s	Stand. 1/4 mile s	Stand. 5/8 mile s	Braking 62-0 mph ft	Top speed mph	Lateral acceler. G	Noise level dBA	Fuel economy mpg city	hwy	Gasoline type / octane
1)	L4* 1.8 SOHC-16-MPSFI	112	119 @ 6000	116 @ 4500	9.5 :1	front -M5*	3.471	11.0	17.6	34.2	144	100	0.71	68	29.0	39.0	R 87
			113 @ 6000	116 @ 4500	9.5 :1	front -A4	2.980	12.7	18.8	36.0	148	94	0.71	68	28.0	38.0	R 87
2)	L4* 2.4 SOHC-16-MPSFI	144	136 @ 5500	145 @ 4250	9.5 :1	front -M5*	3.472	10.7	17.2	32.6	138	106	0.72	68	27.0	35.0	R 87
						front -A4	2.672	12.5	18.4	35.7	144	103	0.72	68	24.0	34.0	R 87
3)	L4* 2.4 SOHC-16-MPSFI	144	136 @ 5500	145 @ 4250	9.5 :1	all - M5*	3.285	12.0	18.0	34.8	141	103	0.70	68	25.0	32.0	R 87
						all - A4	3.029	13.2	19.5	37.5	144	100	0.70	68	22.0	30.0	R 87

1)* DL & SE 2)* DL opt.DE,SE 3)* AWD

The 2 steel unit bodies have a 4-wheel independent suspension with a McPherson front and 2 half axles with coil springs at the rear. The suspension elements in both types of traction are identical and only the spindle carriers are different, to allow the passage of the drive shaft. The front wheel drive version is provided with a standard 1.8 L SOHC (Single overhead Cam) head with 16 valves, which delivers 119 hp. As an option, it can receive the 2.4 L 16 valve which is standard on 4-wheel drive options. The rear wheels are powered through a central differential which directs the power through another differential with a viscous coupling, installed between the rear wheels.

• Access: 80%
It is excellent for all seats in the 5-door version, except for the last bench, reserved for children, which is more difficult to reach. It is less practical to seat 3 persons through the sliding doors of the short body.

• Suspension: 80%
Its smoothness is marvelous on a good road and is good at absorbing irregularities in the pavement but soon generates sea sickness on a twisty road because of its excessive softness which shakes up the body.

• Conveniences: 80%
Storage space consists of a large size glove box, door pockets, a small tray over the dashboard and a box within the center console.

• Price/equipment: 70%
Less expensive than some mini-vans, these hybrids cost more than station wagons of equivalent volume and as their equipment is reduced to a minimum, the options will bump up the bill.

• Seat: 70%
They are well-shaped and offer better support up front than at the rear, where the bench is flat.

• Quality/fit/finish: 70%
In spite of an austere interior appearance, the build is strict and the finish is carefully done but some of the plastics are not rich in appearance.

• Steering: 70%
Smooth and fast, it provides an amazing maneuverability in urban traffic, but becomes vague when speed increases.

• Fuel consumption: 60%
It is higher than average for vehicles with the same displacement and can reach 19 miles per gallon on a difficult road with 4-wheel drive, which is equal to that of a 3 L V6 and the range is limited to 200 miles at the most.

• Noise level: 50%
These engines sound off only during strong acceleration but are more discrete in cruise range, at which time rolling and wind noises dominate.

• Interior space: 50%
You can install 5 persons in the short body whose height gives an impression of space, but whose length is limited, while the Expo, sold in the US by Mitsubishi, accepts up to 7 people, thanks to 2 bench seats, between which space is limited.

• Insurance: 30%
The premium on these wagons is sometimes higher than that of equivalent value mini-vans.

WEAK POINTS

• Performance: 30%
They are not very explosive with the 1.8 L base version and one must move up to the 2.4 L to dispose of more energetic launch and acceleration. The same engine is less brilliant in 4-wheel drive, whose weight reaches 3,300 pounds. The transmission ratios are well-staged and gear shifts cause no problem.

• Handling: 35%
Stable on freeways, these vehicles are not wind-sensitive, except in gusts. On twisty roads, the suspension causes roll and early understeers in either 2 or 4-wheel drive. It's better to be a little careful on freeway exits than to find yourself in problems. Traction in a turn, where the inside wheels unload, is better in 4-wheel drive, coming out of tight turns than in front wheel drive where it encounters some traction loss.

• Braking: 40%
The optional 4 wheel disc brakes and the ABS are not a luxury because average stops are progressive and easy to modulate. In a panic stop, the distances are longer, since the front wheels block rapidly and the trajectories trip the light fantastic.

• Depreciation: 40%
It equals that of cars in this category, that is 20% a year.

CONCLUSION

• Overall rating: 65.0%
These hybrids are not ubiquitous or more economical and really not less expensive than real mini-vans. So? 😐

MITSUBISHI LRV

SPECIFICATIONS & PRICES

Model	Version/Trim	Body/ Seats	Interior volume cu.ft.	Trunk volume cu ft.	Drag coef. Cd	Wheel-base in	Lgth x Wdthx Hght. inx inx in	Curb weight lb	Towing capacity std. lb	Susp. type ft/rr	Brake type ft/rr	Steering system	Turning diameter ft	Steer. turns nber	Fuel. tank gal.	Standard tire size	Standard powertrain	PRICES US. $ 1994
DODGE-EAGLE Colt Vista-Summit Wagon General warranty: 3 years / 36,000 miles: Power Train 5 years / 60,000 miles; surface corrosion; 3 years; Body perforation 7 years / 100,000 miles																		
2WD	base	4dr.wgn.5	96.5	34.6	0.35	99.2	168.5x66.7x62.1	2734	1500	i/i	dc/dr	pwr.r&p.	33.4	3.1	14.5	185/75R14	L4/1.8/M5	**$13,114**
2WD	SE/FWD	4dr.wgn.5	96.5	34.6	0.35	99.2	168.5x66.7x62.1	2734	1500	i/i	dc/dr	pwr.r&p.	33.4	3.1	14.5	185/75R14	L4/1.8/M5	**$14,340**
4WD	AWD	4dr.wgn.5	96.5	34.6	0.35	99.2	168.5x66.7x62.6	3064	2000	i/i	dc/dr	pwr.r&p.	33.4	3.1	14.5	205/70R14	L4/2.4/M5	**$15,018**
MITSUBISHI General warranty: 3 years / 36,000 miles: Power Train 5 years / 60,000 miles; Bodyperforation 7 years / 100,000 miles.																		
Expo	LRV	4dr.wgn.5	96	35	0.35	99.2	168.5x66.7x64.4	2734	1500	i/i	dc/dr	pwr.r&p.	33.4	3.15	14.5	185/75R14	L4/2.4/M5	**$13,149**
Expo		5dr.wgn.7	98	44	0.35	107.0	177.4x66.7x62.6	2998	2000	i/i	dc/dr	pwr.r&p.	36.7	3.15	15.9	205/70R14	L4/2.4/M5	**$15,839**
Expo	LRV AWD	4dr.wgn.5	96	35	0.35	99.2	168.5x66.7x64.4	3053	2500	i/i	dc/dr	pwr.r&p.	33.4	3.15	14.5	185/75R14	L4/2.4/M5	**$16,949**
Expo	AWD	5dr.wgn.7	98	44	0.35	107.0	177.4x66.7x62.6	3263	2500	i/i	dc/dr	pwr.r&p.	36.7	3.15	15.9	205/70R14	L4/2.4/M5	**$17,299**

See page 393 for complete 1995 Price List

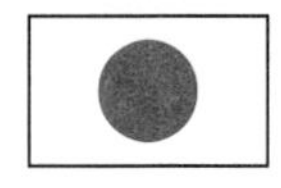

Street rat...

The Montero is faithful to the Mitsubishi philosophy, which always includes a sophisticated technical content. However it applies more to on-road safety and driving pleasure and to off road performance for weekend sport.

Well-known throughout the world under the Pajero name, the Montero is only sold in the United States as a 5-door station wagon in standard, RS, LS, and SR. It is outfitted with a 3.0 L V6 which develops 151 hp with a manual 5-speed on the standard RS. The LS and SR receive as standard a 4-speed automatic. The latter offered as an option, on the first two versions.

STRONG POINTS

• Safety: 90%

The Montero unit body stands up rather well during impact testing and the availability of air bags insures occupant protection in the front seats; we also note that rear seat headrests are standard equipment.

• Satisfaction: 80%

Seventy percent of Mitsubishi products' owners confirm that they would again make the same purchase. However, they find that spare parts are costly and hard to get.

• Technical: 80%

With a steel unit body, mounted on a 5-cross member chassis, the Montero offers an independent front suspension, a rigid rear axle and torsion bars. Dampers with 3 adjustment modes (soft-medium-hard) are offered as an option in the 2 top lines. The Active Trac WD 4-wheel drive can be shifted at will. It is available with a Multimode ABS. This unit allows shifting from 2 to 4-wheel drive while rolling, as long as the speed is less than 60 mph. You can also use the high range of the transfer case at any time, on any surface. Four wheel disc brakes are used, and the Multimode ABS acts whether in 2WD or 4WD. A recirculating ball bearing power steering is standard.

DATA

Category: rear-wheel drive or four-wheel drive all-purpose.
Class : utility

HISTORIC

Introduced in: 1970 Pajero,1983: U.S.A.: 1990: Raider, Canada.
Modified in: 1991: new body
Made in: Okasaki, Japan.

RATINGS

Safety: 90 %
Satisfaction: 85 %
Depreciation: 55 %
Insurance: 8.0 %
Cost per mile: 0.51 $

NUMBER OF DEALERS

In U.S.A.: 500

SALES ON THE AMERICAN MARKET

Model	1992	1993	Result
Montero	6,704	9,280	+ 27.8 %

MAIN COMPETITORS

CHEVROLET Blazer, FORD Explorer, GMC Jimmy, ISUZU Rodeo, JEEP Grand Cherokee, NISSAN Pathfinder, TOYOTA 4Runner.

EQUIPMENT

MITSUBISHI Montero	base	RS	LS	SR
Automatic transmission:	O	O	S	S
Cruise control:	-	O	S	S
Power steering:	O	O	O	O
ABS brake:	S	S	S	S
Air conditioning:	S	S	S	S
Air bag:	-	O	S	S
Leather trim:	-	-	O	O
AM/FM/ Radio-cassette:	O	S	S	S
Power door locks:	-	O	S	S
Power windows:	-	O	S	S
Tilt steering column:	S	S	S	S
Dual power mirrors:	-	S	S	S
Intermittent wipers:	-	S	S	S
Light alloy wheels:	S	S	S	S
Sunroof:	-	O	O	O
Anti-theft system:	-			

S : standard; O : optional; - : not available

COLORS AVAILABLE

Exterior: Beige, Blue, Green, Red, White
Interior: Gray, Blue.

MAINTENANCE

First revision: 3,000 miles
Frequency: 6 months/6,000 miles
Diagnostic plug: Yes

WHAT'S NEW IN 1995 ?

- 15-inch light alloy wheel and sunroof as standard equipment.
- New body colors.

	ENGINE					TRANSMISSION		PERFORMANCE									
Model/ versions *: standard	Type / timing valve / fuel system	Displacement cu/in	Power bhp @ rmn	Torque lb.ft @ rpm	Compres. ratio	Driving wheels / transmission	Final ratio	Acceler. 0-62 mph s	Stand. 1/4 mile s	Stand. 5/8 mile s	Braking 62-0 mph ft	Top speed mph	Lateral acceler. G	Noise level dBA	Fuel economy mpg city	hwy	Gasoline type / octane
LS	V6* 3.0 SOHC-12-MPFI	181	177 @ 5000	188 @ 4000	9.0 :1	rear/4 M5	4.636	14.0	18.7	35.4	141	103	0.70	67	20.0	19.0	R 87
						rear/4 A4	4.636	14.7	19.2	36.8	148	100	0.70	67	17.0	19.0	R 87
SR	V6* 3.5 SOHC-12-MPFI	213	214 @ 5000	228 @ 3000	9.5 :1	rear/4 A4	4.636	NA									

• Quality/fit/finish: 80%
The trim, exterior as well as interior, is more that of a deluxe car than a utility. The build and finish are meticulous, but the large use of plastic lacks class.

• Driver's compartment: 80%
The driver sits high and his position is made comfortable by a very short adjustable steering column, which makes the wheel very near the dashboard. This allows more latitude in driver's position for off road. Visibility is excellent except toward the rear where the headrests, the spare tire and the jump seats, when raised, close off the back light. The dashboard carries many gauges and you can receive, as an option, an altimeter, and inclinometer, and a thermometer, which measures inside and outside temperatures. Your pilot's certificate is in the mail.

• Suspension: 70%
The adjustable damping amply justifies its presence, but the suspension rate, set up for freeways, harshly shakes the occupants on little or poorly maintained roads and paths.

• Conveniences: 70%
Storage spaces are numerous and practical and hand grips are well-located. Many clever details testify to Mitsubishi's development.

• Access: 70%
Getting into the Montero really doesn't pose any problems, since the doors swing well out of the way and those at the back allow access to the storage area along it's full width. Footrests would be welcome to ease entry and exit.

• Seats: 60%
They provide superior support at the front than at the back, where the bench is almost flat. Their padding is generous, but firm, and the leather trim is particularly slippery.

• Interior Space: 60%
The Montero interior is longer and taller than is it large. It allows up to 6 people, thanks to a folding seat, installed in the storage area. Swing it out of the way, when not needed.

• Storage: 60%
Its volume is distinctly larger when the folding bench seats are not used.

• Steering: 60%
Its high ratio and many turns lock-to-lock make it vague on the road but it is agreeable to use off road. There, its power assist is well-modulated and its turning diameter, reasonable for this format, allows good maneuverability.

• Depreciation: 50%
The Montero resells well, more because of its comfort and equipment than for its off road aptitude.

• Insurance: 50%
Normal for this type of vehicle, the premium can reach high sums for better-equipped models.

WEAK POINTS

• Performance: 20%
The weight-to-power ratios are unfavorable (26.5 lbs/hp) so the Montero is no road rocket, particularly with an automatic. Take offs and accelerations are more akin to those of a diesel with ample low end torque.

• Handling: 30%
On the road the Montero handles like a large deluxe station wagon. Its overly soft suspension makes it roll in tight turns which is hardly dangerous, considering the modest performance; optional adjustable damping controls body motions. Off road, its extended length and modest ground clearance limit its ability to clear obstacles. Agility is limited by the lack of suspension travel, but mostly by the absence of a standard limited slip differential.

• Braking: 40%
The ABS is effective at stabilizing panic stops, but it is not very helpful off road where it complicates some maneuvers, especially in reverse, on slippery terrain. On the road, its stopping distances are long and the fade-resistance average.

• Noise level: 40%
The drive train is discrete at cruise and effective soundproofing provides a peaceful ambiance which only rolling and wind noise occasionally disturbs.

• Price/equipment: 40%
Base models are better placed to compete with the American models which are less spoiled in terms of gadgets but more effective in difficult terrain.

• Fuel consumption: 40%
It is never really economical, even in quiet use, and it can reach horrendous rates depending on power requirements.

CONCLUSION

• Overall rating: 57.0%
The Montero addresses itself more to suburbanites in quest of all weather safety than to trail drivers. 😐

SPECIFICATIONS & PRICES

Model	Version/Trim	Body/ Seats	Wheel-base in	Lgth x Wdthx Hght. inx inx in	Curb weight lb	Towing capacity std. lb	Susp. type ft/rr	Brake type ft/rr	Steering system	Turning diameter ft	Steer. turns nber	Fuel. tank gal.	Standard tire size	Standard powertrain	PRICES US. $ 1994
MITSUBISHI	**General warranty: 3 years / 36,000 miles; Power Train: 5 years / 60,000 miles; corrosion perforation: 7 years / 100,000 miles.**														
Montero		4dr.wgn. 5	107.3	185.2x66.7x71.9	4134	4000	i/r	dc/dc	pwr.bal.	38.7	3.28	24.3	235/75R15	V6/3.5/M5	-
Montero		4dr.wgn. 5	107.3	185.2x66.7x71.9	4123	4000	i/r	dc/dc	pwr.bal.	38.7	3.28	24.3	235/75R15	V6/3.5/A4	-
Montero	RS	4dr.wgn. 5	107.3	185.2x66.7x71.9	4132	4000	i/r	dc/dc	pwr.bal.	38.7	3.28	24.3	235/75R15	V6/3.5/M5	-
Montero	RS	4dr.wgn. 5	107.3	185.2x66.7x71.9	4145	4000	i/r	dc/dc	pwr.bal.	38.7	3.28	24.3	235/75R15	V6/3.5/A4	-
Montero	LS	4dr.wgn. 5	107.3	185.2x66.7x71.9	4145	4000	i/r	dc/dc	pwr.bal.	38.7	3.28	24.3	235/75R15	V6/3.5/A4	$25,050
Montero	SR	4dr.wgn. 5	107.3	186.6x70.3x74.0	4222	4000	i/r	dc/dc	pwr.bal.	38.7	3.28	24.3	235/75R15	V6/3.5/A4	$31,775

See page 393 for complete 1995 Price List

MITSUBISHI 3000 GT
DODGE Stealth

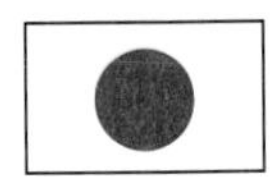

Poor man's Ferrari...

It would be a lot of honor for a Ferrari F355 to be compared to the Dodge Stealth or to the Mitsubishi 3000 GT. In terms of technical sophistication, the commoners are doing better than the princess, which saw daylight late, in putting pleasure into every day driving (Is the geisha girl more desirable than a princess?)

MITSUBISHI 3000GT VR-4

The Stealth R/T and its Mitsubishi 3000GT twin are without a doubt the most representative sports coupes of the 90's and address themselves to a broad group of fans and buyers. The turbo all wheel drive is as fast and sophisticated as an old 348 Ferrari, while the base model has more room, is more practical and less costly than a Nissan 300 ZX or a Mazda RX7. At Dodge, the trims are Standard, R/T and R/T Turbo and at Mitsubishi standard, LS and VR-4

STRONG POINTS

• Technical: 100%

The Stealth 3000GT derives, just like the Talon, from the Mitsubishi Galant platform. Since 1991 this car, which did not hold high sales in North America, offered an advanced technology, 4-wheel drive traction and steering and a mass of sophisticated electronics. The steel unit body is very rigid which explains the high mass, while its aerodynamic finesse varies between 0.30 and 0.33. Updated, the revised body design gained in simplicity and the spoiler at the bottom of the back light is, to say the least, original. The independent suspension shares parts with the Talon and Gallant all-wheel drives. Adjustable damping is standard on the R/T Turbo and VR-4 and optional on the others. The "Touring" position reacts as a function of speed while the "Sport" maintains a firm response at all times. Toe-in at the rear wheels parallels that of the front wheels. Vented disc brakes at all 4 wheels is provided with an ABS which comes standard on the R/T Turbo, SL and VR-4 and is optional on the base models or R/T. The engine on the R/T Turbo and VR-4 has a double overhead cam with a turbo for each bank of cylinders, and develops a max of 320 hp. The V6 on the R/T and SL is identical but naturally aspirated and delivers 220 hp while the standard engine has a single overhead cam and is rated at 164 hp. The R/T Turbo and VR-4 have a full time all wheel drive, while the rest are

DATA

Category: front-wheel drive or AWD sports coupe.
Class : GT

HISTORIC

Introduced in: 1991
Modified in: 1994: aesthetic and retractable roof (Mit.)
Made in: Nagoya, Japan by Mitsubishi.

RATINGS

Safety: 90 %
Satisfaction: 90 %
Depreciation: 60 % RT/Tbo : 55 %
Insurance: 4.5 % ($1,000)
Cost per mile: $0.47 RT/Tbo: $0.80

NUMBER OF DEALERS

In U.S.A.: 500 Mitsubishi, 1,887 Dodge.

SALES ON THE AMERICAN

Model	1992	1993	Result
Stealth	16,926	14,556	- 10.7 %
3000 GT	11,710	13,006	+10.0 %

MAIN COMPETITORS

CHEVROLET Corvette, MAZDA RX-7, NISSAN 300 ZX, PORSCHE 928, TOYOTA Supra.

EQUIPMENT

DODGE Stealth	base	R/T	R/T Tbo			
MITSUBISHI 3000 GT				base	SL	VR-4
Automatic transmission:	O	O	-	O	O	-
Cruise control:	S	S	S	S	S	S
Power steering:	S	S	S	S	S	S
ABS brake:	O	O	S	O	O	S
Air conditioning:	S	S	S	S	S	S
Air bag (2):	S	S	S	S	S	S
Leather trim:	-	O	O	-	O	S
AM/FM/ Radio-cassette:	S	S	S	S	S	S
Power door locks:	S	S	S	S	S	S
Power windows:	S	S	S	S	S	S
Tilt steering column:	S	S	S	S	S	S
Dual power mirrors:	S	S	S	S	S	S
Intermittent wipers:	S	S	S	S	S	S
Light alloy wheels:	S	S	S	S	S	S
Sunroof:	O	O	O	O	O	O
Anti-theft system:	-	O	S	-	O	S

S : standard; O : optional; - : not available

COLORS AVAILABLE

Exterior: Silver, Black, Blue, Red, White, Yellow.
Interior: Dark Gray.

MAINTENANCE

First revision: 5,000 miles
Frequency: 6 months
Diagnostic plug: Yes

WHAT'S NEW IN 1995 ?

- Convertible with a retractable rigid top on the Mitsubishi 3000 GT.
- Chromed aluminum rims and Yokohama 245/40 ZA 18 available as an option.
- Enhanced standard equipment.

Model/ versions *: standard	ENGINE Type / timing valve / fuel system	Displacement cu/in	Power bhp @ rmn	Torque lb.ft @ rpm	Compres. ratio	TRANSMISSION Driving wheels / transmission	Final ratio	PERFORMANCE Acceler. 0-62 mph s	Stand. 1/4 mile s	Stand. 5/8 mile s	Braking 62-0 mph ft	Top speed mph	Lateral acceler. G	Noise level dBA	Fuel economy mpg city	hwy	Gasoline type / octane
base	V6* 3.0 SOHC-12-MPSFI	181	164 @ 5500	185 @ 4000	8.9 :1	front - M5	4.153	8.8	16.2	29.0	135	125	0.85	67	25.0	35.0	R 87
						front - A4	3.958	9.4	17.5	30.5	141	119	0.85	67	24.0	34.0	R 87
R/T	V6* 3.0 DOHC-24-MPSFI	181	222 @ 6000	205 @ 4500	10.0 :1	front - M5*	4.153	7.8	15.0	27.7	138	144	0.87	68	23.0	32.0	S 91
						front - A4	3.958	8.3	15.6	28.4	135	137	0.87	68	22.0	31.0	S 91
R/T Tbo	V6*T3.0 DOHC-24-MPSFI	181	320 @ 6000	315 @ 2500	8.0 :1	all- M6*	3.870	5.5	13.9	26.0	118	155	0.90	68	21.0	32.0	S 91

satisfied with a front wheel drive.

• **Safety: 90%**
Extremely rigid, their structure is highly impact-resistant; also, the front occupants are protected by ? standard air bags.

• **Satisfaction : 90%**
Few owners of exotic cars can boast of a reliability and ease of maintenance equivalent to that of the Stealth 3000 GT, which allows daily problem-free use. Now it matters little that spare parts are rare and costly.

• **Seats: 90%**
They offer excellent comfort at the front with efficient lateral and lumbar support. At the rear, however, only children will be at ease.

• **Quality/fit/finish: 90%**
Precise assembly, meticulous finish, and a quality of the materials, on the over all, make these cars the model of their kind.

• **Handling: 80%**
The reactions of the suspension are never brutal, compared to others. All is smoothness and the nuances, as well as the handling in curves, go from neutral on the all wheel drive to a predictable understeer on the front wheel drive whose weight and power transfer are different.

• **Conveniences: 80%**
Numerous and practical storage spots include a glove box, door pockets, and a holder in the center console.

• **Steering: 80%**
Precise, well-assisted, and with a modulation that is a function of speed, it is agreeable in all circumstances, but a somewhat large steering diameter penalizes maneuverability.

• **Suspension: 70%**
In "Touring" the comfort is practically sedan-like. It becomes more Spartan on the "Sport" whose firmness, indispensable for high performance, is more marked.

• **Driver's compartment: 70%**
Visibility is limited, because the driver sits low, which makes the dashboard look massive. Also, the high belt line and the bulky spoiler of the Turbo do not help seeing. The newest steering wheel, does not interfere as much with reading the numerous and visible gauges, while the controls are typical of Mitsubishi. Firm in urban driving, the clutch pedal has a stroke which is a little long for sports use.

• **Performance: 70%**
The weight/power ratio is favorable with a Turbo engine, whose takeoff and acceleration are more impressive, than the naturally aspired ones where you have to shift more often to overcome inertia and high weight.

MITSUBISHI 3000GT convertible

DODGE Stealth R/T Turbo

• **Access: 60%**
It's not difficult to take place, thanks to long doors which open fairly wide. However, large people should avoid attempting to enter the rear under penalty of intense frustration.

• **Insurance: 60%**
The premium is almost reasonable, compared to more common cars or to exotics with equivalent performance.

• **Fuel consumption: 50%**
It is in the average of vehicles with equivalent displacement because it rarely goes under 22 mpg in reasonable driving!

• **Depreciation: 50%**
The value of its resale holds above average, even though potential buyers do not push and shove each other in recession times.

• **Braking: 50%**
Powerful and well-balanced, it resists fading, but during panic stops, you notice the beginning of many wheel lock ups which are more disagreeable than dangerous.

WEAK POINTS

• **Interior space: 30%**
In spite of the respectable bulk of these sports cars, the interior space is limited and only accommodates two adults up front and two young children at the rear.

• **Price/equipment: 40%**
Even though competitive, considering the potential and the advanced technical content of these models, the Dodge equipment is less generous than that of the Mitsubishi.

• **Trunk: 40%**
It doesn't contain much if the rear seats are occupied because its reduced height and tall sill do not enhance baggage handling.

• **Noise level: 40%**
A sports fan will not complain that the body soundproofing is too efficient or that it dampens too much, those engines' sounds are part of the cars driving pleasure.

CONCLUSION

• **Overall rating: 66.5%**
Those that look for the appearance or sensations of an exotic car for the price of a large family sedan will have no choice but to opt for these two cars. ☺

SPECIFICATIONS & PRICES

Model	Version/Trim	Body/ Seats	Interior volume cu.ft.	Trunk volume cu ft.	Drag coef. Cd	Wheel-base in	Lgth x Wdthx Hght. inx inx in	Curb weight lb	Towing capacity std. lb	Susp. type ft/rr	Brake type ft/rr	Steering system	Turning diameter ft	Steer. turns nber	Fuel. tank gal.	Standard tire size	Standard powertrain	PRICES US. $ 1994
DODGE		**General warranty: 3 years / 36,000 miles; Power Train 6 years / 60,000 miles; surface corrosion: 3 years / unlimited.**																
Stealth	base	3dr.cpe.2+2	82.3	11	0.33	97.2	180.5x72.4x49.1	3064	NR	i/i	dc/dc	pwr.r/p.	37.4	2.8	19.8	225/65HR15	V6/3.0/M5	**$21,145**
Stealth	R/T	3dr.cpe.2+2	82.3	11	0.33	97.2	180.5x72.4x49.1	3164	NR	i/i	dc/dc	pwr.r/p.	37.4	2.8	19.8	225/55VR15	V6/3.0/M5	**$23,931**
Stealth	R/T Turbo	3dr.cpe.2+2	82.3	11	0.33	97.2	179.7x72.4x49.3	3792	NR	i/i	dc/dc/ABS	r/p.4WS	37.4	2.8	19.8	245/45ZR17	V6T/3.0/M6	**$37,894**
MITSUBISHI		**General warranty: 3 years / 36,000 miles; Power Train: 6 years / 60,000 miles; corrosion perforation: 7 years / 100,000 miles.**																
3000GT	base	3dr.cpe.2+2	82.3	11	0.33	97.2	178.9x72.4x49.1	3219	NR	i/i	dc/dc	pwr.r/p.	37.4	2.8	19.8	225/65VR16	V6/3.0/M5	**$27,450**
3000GT	SL	3dr.cpe.2+2	82.3	11	0.33	97.2	178.9x72.4x49.1	3373	NR	i/i	dc/dc/ABS	pwr.r/p.	37.4	2.8	19.8	225/55VR16	V6/3.0/M5	**$31,975**
3000GT	VR-4	3dr.cpe.2+2	82.3	11	0.33	97.2	178.9x72.4x49.3	3803	NR	i/i	dc/dc/ABS	r/p.4WS	37.4	2.8	19.8	245/45ZR17	V6T/3.0/M6	**$41,275**

See page 393 for complete 1995 Price List

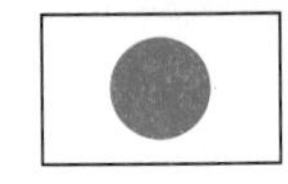

Nissan on target...

To reinforce its North American market position, Nissan counted on the Altima, its only representative in the very hot subcompact market. In striking a good balance, it batted a thousand, but will it continue to deprive it of the V6 for long?

The Altima begins its 3rd year on the market. It succeeded at the end of 92 to the Stanza of which it kept the platform and the main drive train. More American than Japanese, it was designed in San Diego at NDI, the line begun in Michigan, and is manufactured in Tennessee. Its local content is 70% because the drive train comes from Mexico. It is a 3-volume 4-door sedan offered in 4 trims, XE, GXE, SE and GLE.

STRONG POINTS

• Safety: 80%
In spite of the lateral panels, installed in one piece to give it maximum stiffness, the structural resistance could still be improved and only one air bag is available standard.

• Satisfaction: 80%
The reliability does not pose a problem because 83% of the owners claim full satisfaction.

• Quality/fit/finish: 80%
The exterior and interior presentations are well-detailed, the materials have a good appearance and assembly, like the finishing, is strict. Although it's not showy the faux appliqué livens up the proceedings.

• Driver's compartment: 80%
It is simple and fast to find the best driving position, thanks to different seat and steering column adjustments. Visibility is excellent toward the front and the side but reduced for 3/4 rear by the thick "C" pillar. The rounded, simple and functional dashboard has easily read instruments and conventional controls, except for the switches laid out below the steering, which are practically invisible.

• Insurance: 80%
If the Altima is on an equal footing with a Mazda 626, only the Camry premium is lower.

• Suspension: 70%
Its smoothness is as appreciable on the freeway as on small rugged roads and its wheel travel is sufficient, so only major road defects provoke the temper of the rear axle.

• Handling: 70%
The Altima is controllable in driving and stable in most instances. However, low suspension rates lead to a roll which accelerates the appearance of understeer. Body motions are better controlled on the SE and especially on the GLE, which travels more rapidly through a closed turn.

• Technical: 70%
The Altima is an ecological car and its conception was the object of particular car concerning environmental problems. The plastics from which it is built are recyclable, the paints are water-based and its air conditioning works without CFC. All trims receive a 2.4 L 4-cylinder which develops 150 hp with either a manual or automatic transmission. The SE is equipped with a front wheel drive, while the rear uses a variable toe-in to improve its handling. Without setting any special

DATA

Category: front-wheel drive compact sedans.
Class : 4

HISTORIC

Introduced in: 1993
Modified in: -
Made in: Smyrna, TN, U.S.A.

RATINGS

Safety: 90 %
Satisfaction: 83 %
Depreciation: 33 % (2 years)
Insurance: 5.0 % ($585)
Cost per mile: $0.40

NUMBER OF DEALERS

In U.S.A.: 1,100

SALES ON THE AMERICAN

Model	1992	1993	Result
Altima	NA		

MAIN COMPETITORS

BUICK Skylark, CHEVROLET Cavalier-Corsica, CHRYSLER Cirrus, DODGE-Stratus, FORD Contour, HONDA Accord, MAZDA 626, MERCURY Mystique, OLDSMOBILE Achieva, PONTIAC Grand Am, SUBARU Legacy, TOYOTA Camry, VOLKSWAGEN Passat.

EQUIPMENT

NISSAN Altima	XE	GXE	SE	GLE
Automatic transmission:	O	O	O	S
Cruise control:	S	S	S	S
Power steering:	S	S	S	S
ABS brake:	-	O	S	S
Air conditioning:	O	S	S	S
Air bag (2):	S	S	S	S
Leather trim:	-	-	O	S
AM/FM/ Radio-cassette:	S	S	S	S
Power door locks:	-	S	S	S
Power windows:	-	S	S	S
Tilt steering column:	S	S	S	S
Dual power mirrors:	S	S	S	S
Intermittent wipers:	S	S	S	S
Light alloy wheels:	O	O	O	S
Sunroof:	-	O	O	S
Anti-theft system:	-			

S : standard; O : optional; - : not available

COLORS AVAILABLE

Exterior: White, Ruby, Beige, Blue, Platinum, Charcoal,Teal.
Interior: Fabric: Brown, Neutral, Black, Blue-Gray.
Leather: Charbon, Teal, Blue-Gray.

MAINTENANCE

First revision: 7,500 miles
Frequency: 7,500 miles
Diagnostic plug: Yes

WHAT'S NEW IN 1995 ?

- Slight cosmetic changes including a new grill and redesigned taillights.

Model/ versions *: standard	ENGINE Type / timing valve / fuel system	Displacement cu/in	Power bhp @ rmn	Torque lb.ft @ rpm	Compres. ratio	TRANSMISSION Driving wheels / transmission	Final ratio	PERFORMANCE Acceler. 0-62 mph s	Stand. 1/4 mile s	Stand. 5/8 mile s	Braking 62-0 mph ft	Top speed mph	Lateral acceler. G	Noise level dBA	Fuel economy mpg city	hwy	Gasoline type / octane
base	L4* 2.4 DOHC-16-SFI	146	150 @ 5600	154 @ 4400	9.2 :1	front - M5*	3.650	8.9	16.6	29.7	128	124	0.82	65	29.0	39.0	R 87
						front - A4	3.619	9.5	17.0	30.8	131	118	0.82	65	25.0	37.0	R 87
SE	L4* 2.4 DOHC-16-ISFI	146	150 @ 5600	154 @ 4400	9.2 :1	front - M5	3.895	8.5	16.0	29.2	131	118	0.85	65	29.0	39.0	R 87

records for originality, its lines are harmonious but their aerodynamic efficiency is on the average, with a Cd of 0.34.

• Performance: 70%
The engine gives the impression of going faster than the chronometer that measures it because the weight is high and the weight-to-power ratio of 20 lbs/hp is average. This explains why fuel consumption is a little higher than average. Takeoffs are laborious but passings are a little hardier.

• Steering: 70%
Direct and well-modulated, its precision, just as its ratio, is sufficient. A turning diameter which is a little too large limits maneuverability.

• Seats: 70%
Well-formed and well-padded even though, the seat bench, which includes false headrests (no adjustment or support), lacks height.

• Conveniences: 70%
The AC is effective and easy-to-adjust, and its outlets, for air exchange, heat and cold, are well-spread out through the interior. Storage space includes a large glove box, small door pockets, a box and a tub in the center console.

• Interior space: 60%
It is only average because rear legroom is limited and only 4 people can fit comfortably. However, the height under the roof will accommodate large folks.

• Noise level: 60%
It is low at cruise speed, but on any full-powered takeoff, the engine howls in some inharmonious ways.

• Trunk: 60%
It is not proportioned to the total bulk, because it lacks depth and cannot be transformed, however, there is a passage behind the center armrest to install skis or long objects.

• Price/equipment: 55%
Slightly lower than those of the competition, the prices are justified by the equipment level, which is satisfactory even on the base model. For those harder to please, it is the GXE which offers the best compromise.

• Fuel consumption: 55%
The Altima consumes a gallon of gas every 24 miles in normal driving but can easily get down to 19 in a sports mode.

• Depreciation: 50%
The Altima keeps better resale value than the Stanza because of the reliability and excellent Nissan warranty, which takes good care of its buyers. It is a courageous action which has already paid off and shows a new way.

• Braking: 50%
Its effectiveness is better cold than hot and the ABS should be delivered standard, because panic stops are not linear when the Altima is not equipped with it. With harsh use, the distances extend rapidly and the efficiency goes up in smoke.

WEAK POINTS

• Access: 40%
While it's easy to get in up front, it is more difficult in the back where the rounded door shape obliges head bowing, a normal custom of Japanese culture.

CONCLUSION

• Overall rating: 63%
The Altima offers nothing exceptional, but is the result of some compromises, which make it easy on the eye and pleasant to drive. Also the operating budget is quiet reasonable. A V6 engine would improve its position and attraction. :-|

OWNER'S SUGGESTIONS

-V6 engine
-More width

SPECIFICATIONS & PRICES

Model	Version/Trim	Body/ Seats	Interior volume cu.ft.	Trunk volume cu ft.	Drag coef. Cd	Wheel-base in	Lgth x Wdthx Hght. inx inx in	Curb weight lb	Towing capacity std. lb	Susp. type ft/rr	Brake type ft/rr	Steering system	Turning diameter ft	Steer. turns nber	Fuel. tank gal.	Standard tire size	Standard powertrain	PRICES US. $ 1994
NISSAN		General warranty: 3 years / 50,000 miles; Power train: 6 years / 60,000 miles; perforation corrosion & antipollution: 6 years / unlimited.																
Altima	XE	4dr.sdn.5	93	14	0.35	103.1	180.5x67.1x55.9	2828	1000	i/i	dc/dr	pwr.r/p.	37.4	2.8	15.8	205/60R15	L4/2.4/M5	$13,999
Altima	GXE	4dr.sdn.5	93	14	0.35	103.1	180.5x67.1x55.9	2939	1000	i/i	dc/dr	pwr.r/p.	37.4	2.8	15.8	205/60R15	L4/2.4/M5	$15,154
Altima	SE	4dr.sdn.5	93	14	0.34	103.1	180.5x67.1x55.9	3027	1000	i/i	dc/dc/ABS	pwr.r/p.	37.4	2.8	15.8	205/60R15	L4/2.4/M5	$18,179
Altima	GLE	4dr.sdn.5	93	14	0.34	103.1	180.5x67.1x55.9	3093	1000	i/i	dc/dc/ABS	pwr.r/p.	37.4	2.8	15.8	205/60R15	L4/2.4/A4	$19,179

See page 393 for complete 1995 Price List

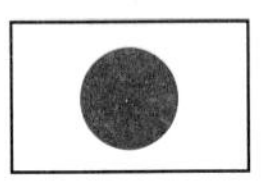

Too far ahead?

After the Multi and the Colt station wagon, the Axxess has been a pioneer in a search for intelligent compromise between a classic utility and a station wagon. The failure in the United States on the arrival of new models seem to comfort the Axxess at the moment when it will disappear.

The Axxess is positioned between a mini-pickup and a classic station wagon, where the roof height is just average and the sliding side doors are its major originality. It is only offered in 2 wheel drive, XE and SE, the all wheel drive being retired. The engine remains a 2.4 L with 12 valves which is also used on the Altima, the 240 SX and the standard Hardbody pickup.

STRONG POINTS

• Quality/fit/finish: 80%
The appearance is pleasant and well-detailed, and the quality of the material and of the finish brings to these station wagons an original and luxurious touch, that is part of their charm.

• Satisfaction: 75%
These station wagons give few problems and benefit from a good warranty.

• Interior space: 75%
Since the drive train takes much space up front, the length of the interior is limited toward the back. The occupants of the center bench seat dispose of enough legroom, which is more than you can say of the second bench seat, sold as an option on the 7-passenger version; that one can only be used by children.

• Driver's compartment: 70%
Seated high, the driver enjoys excellent visibility and the controls are organized in traditional fashion. The gauges are easy to read and well laid out, but you cannot find the automatic shifter position among the main indicators.

• Technical: 70%
The simple rounded access forms have no wrinkles but only develop a modest aerodynamic coefficient of 0.36. The unit body is built of Durasteel, includes a 4 wheel independent suspension and mixed braking which has never heard of ABS.

• Safety: 70%
Even though it is not equipped with ABS or an air bag, the Axxess has reinforcements in the door and roof, better resist lateral impacts and roll over. Open headrests, mounted standard in the rear seats, give it some advantage over most of its domestic competition of which this point remains neglected.

• Seats: 70%
Their proportions and padding are adequate but the front seats support better than the bench, which is not very contoured.

• Suspension: 70%
Soft and smooth it provides more comfort on the freeway than on little sinuous back roads in poor shape where it exhibits considerable roll and acts harshly to pavement defects..

• Storage: 60%
In existence in the 7-seat version, its capacity is limited by the size of the shock absorber towers on the 5-passenger version. Also, the

DATA

Category: front-wheel drive and AWD compact wagons.
Class : 4

HISTORIC

Introduced in: 1988
Modified in: -
Made in: Kyushiu, Japan.

RATINGS

Safety: 60 %
Satisfaction: 75 %
Depreciation: 62 %
Insurance: 5.6 % ($585)
Cost per mile: $0.42

NUMBER OF DEALERS

In U.S.A.: 1,100

SALES ON THE AMERICAN

Model	1992	1993	Result
Axxess	Not sold in U.S.A.		

MAIN COMPETITORS

DODGE-PLYMOUTH Caravan-Voyager (2.5L), DODGE-PLYMOUTH Colt Wagon, EAGLE Summit Wagon, MAZDA MPV (2.6L), TOYOTA Previa.

EQUIPMENT

NISSAN Axxess 4x2	XE	SE
Automatic transmission:	O	O
Cruise control:	S	S
Power steering:	O	O
ABS brake:	-	-
Air conditioning:	O	S
Air bag:	-	-
Leather trim:	-	-
AM/FM/ Radio-cassette:	S	S
Power door locks:	S	S
Power windows:	S	S
Tilt steering column:	S	S
Dual power mirrors:	S	S
Intermittent wipers:	S	S
Light alloy wheels:	-	S
Sunroof:	O	O
Anti-theft system:	O	O

S : standard; O : optional; - : not available

COLORS AVAILABLE

Exterior: Silver, Royal-Blue, Chrystal-Blue, Garnet, Brown, Drak Gray, White.
Interior: Gray, Blue.

MAINTENANCE

First revision: 7,500 miles
Frequency: 7,500 miles
Diagnostic plug: No

WHAT'S NEW IN 1995 ?

- All wheel drive retired from the catalog.
- Air conditioning without CFC.
- New sliding sunroof.
- Light alloy standard on the SE.

Model/ versions *: standard	ENGINE Type / timing valve / fuel system	Displacement cu/in	Power bhp @ rmn	Torque lb.ft @ rpm	Compres. ratio	TRANSMISSION Driving wheels / transmission	Final ratio	PERFORMANCE Acceler. 0-62 mph s	Stand. 1/4 mile s	Stand. 5/8 mile s	Braking 62-0 mph ft	Top speed mph	Lateral acceler. G	Noise level dBA	Fuel economy mpg city	hwy	Gasoline type / octane
base	L4* 2.4 SOHC-12-MPSFI	146	138 @ 5600	148 @ 4400	9.1 :1	front - M5*	3.895	9.0	16.8	29.5	154	106	0.72	66	25.0	34.0	R 87
						front - A4	3.876	10.0	17.1	31.0	151	103	0.72	66	22.0	29.0	R 87

center bench folds but cannot be removed, which still limits the cargo volume.

• Steering: 60%
Even though precise, it has too much reduction, which penalizes nimble urban driving, and its power assist is too high, which makes it sensitive on the road, especially in the cross winds.

• Performance: 60%
They are more comparable to that of a car than to a utility, but the engine is always asked for more power, particular with an automatic transmission, which makes one regret the absence of a V6.

• Price/equipment: 60%
The Axxess seems to have a yen for high cost, taking into account it is not a multipurpose machine (not truly off road). Also the sticker prices of its competition invite comparison. This is explained by the strength of the Yen and by very complete equipment, even on a base model.

• Insurance: 60%
The premium of the Axxess is equal to that of a compact sedan and, for instance, higher than that of the Caravan-Voyager minivans.

• Access: 60%
Very practical in tight parking, the sliding rear doors do not supply enough opening to allow large bodied people to easily board and the tight space between the seats interdicts any passage toward the rear of the interior.

• Noise level: 50%
Effective sound insulation provides a low noise level at cruise speed but the engine strongly signals its presence during take-off and acceleration and both wind and rolling noises increase with speed.

• Fuel consumption: 50%
Urged to by the weight and the bulk, the engine is more voracious than average, but the fuel tank is sizable and cruise range can reach 300 miles.

• Handling: 50%

Very soft, the suspension rolls considerably, and the body motions which result are more spectacular than dangerous. Understeering remains easy to control.

WEAK POINTS

• Braking: 40%
Easy to modulate in normal usage, it lacks as much efficiency as fade-resistance. During panic stops the distances are long, the trajectory is uncertain and without ABS, the front wheels lock very fast.

• Depreciation: 40%
Due to the strong competition in this market segment, the resale value of the Axxess has again lost points.

CONCLUSION

• Overall rating: 58.5%
After being the first and only representative of the hybrid formula, the Axxess will take its retirement at the moment when others, like the Honda Odyssey arrive with the same objective of filling the middle ground between station wagons and minivans. 😐

OWNER'S SUGGESTIONS

-Air bag and ABS standard
-Removable bench seat
-Firmer suspension (2WD)
-Longer rear floor pan
-More storage
-A longer version with conventional doors as on the Mitsubishi Expo
-A more powerful V6 engine
-A 4 wheel drive assisted by a viscous coupling instead of a transfer case

SPECIFICATIONS & PRICES

Model	Version/Trim	Body/ Seats	Interior volume cu.ft.	Trunk volume cu ft.	Drag coef. Cd	Wheel-base in	Lgth x Wdthx Hght. inx inx in	Curb weight lb	Towing capacity std. lb	Susp. type ft/rr	Brake type ft/rr	Steering system	Turning diameter ft	Steer. turns nber	Fuel. tank gal.	Standard tire size	Standard powertrain	PRICES US. $ 1994
NISSAN		**General warranty: 3 years / 50,000 miles; Power Train: 6 years / 60,000 miles; perforation corrosion & antipollution: 6 years / unlimited.**																
Axxess	XE	5dr.wgn.5	115	35.5	0.36	102.8	171.9x66.5x64.6	2937	2000	i/i	dc/dr	pwr.r/p.	34.8	3.26	17.2	195/70R14	L4/2.4/M5	-
Axxess	SE	5dr.wgn.5	115	35.5	0.36	102.8	171.9x66.5x64.6	2912	2000	i/i	dc/dr	pwr.r/p.	34.8	3.26	17.2	195/70R14	L4/2.4/M5	-

See page 393 for complete 1995 Price List

NISSAN Hardbody

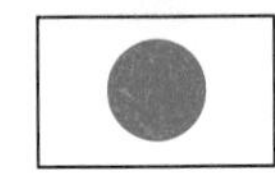

Erosion...

With time, the pick-up market has become more competitive, the Americans more pertinent and the buyers more difficult, just like the business. During this time, Nissan went to sleep and little by little it lost its leadership role, which it had in this market in the beginning of the 80's.

At one time the Hardbody trucks had a great popularity because they were the first to offer an extended cab (King Cabs) and a V6 engine (1986). They are sold with a standard box and cab, or extended (King Cab) with 2WD and 4WD in the following trim: base, XE, and XE V6. The 2.4 L 4-cylinder, one of the most powerful in its category, is standard on the base model, while the 3.0 L V6 is fitted to the XE-V6. The standard transmission is a manual 5-speed, and the option is a 4-speed automatic. The 4x4 comes with shields under the powertrain and fuel tank.

STRONG POINTS

• Satisfaction: 90%
Its rating remains high because the reliability remains good and the guarantee one of the best in the market.

• Seats: 70%
Those in the XE, identical to those in the Pathfinder, are fully equivalent to the ones you would find in a sports car, while the bench seat, on the base models, is an excellent excuse not to go to work.

• Quality/fit/finish: 70%
The assembly and finish in these utility machines are of a quality equal to that of the cars of the marque. The cab appearance has been improved by a new dashboard design, and some of the trim color combinations do not have the utility look.

• Safety: 70%
The structural rigidity and the passenger protection are among the best for this type of vehicle. However, the driver's protection suffers from the absence of an air bag and ABS as standard equipment.

• Driver's compartment: 70%
The driver is better seated in versions equipped with individual bucket

DATA

Category: rear-wheel drive or all-wheel drive pick-ups.
Class : utilitaires

HISTORIC

Introduced in: 1965
Modified in: 1981, 1986.
Made in: Smyrna, Tenessee, U.S.A.

RATINGS

Safety: 60 %
Satisfaction: 90 %
Depreciation: 58 %
Insurance: 6.5 % ($585)
Cost per mile: $0.40

NUMBER OF DEALERS

In U.S.A.: 1,100

SALES ON THE AMERICAN

Model	1992	1993	Result
Hardbody	121,194	110,001	- 9.3 %

MAIN COMPETITORS

DODGE Dakota, FORD Ranger, CHEVROLET S-10, GMC Sonoma, ISUZU, MAZDA B, MITSUBISHI Mighty Max, TOYOTA.

EQUIPMENT

NISSAN Hardbody 4x2 & 4X4	std	DLX XE	DLX XE	K.Cab SE-V6	K.Cab
Automatic transmission:	-	O	O	O	
Cruise control:	-	-	-	S	
Power steering:	-	O	O	O	
ABS brake:	S	S	S	S	
Air conditioning:	O	O	O	O	
Air bag:	-	S	S	S	
Leather trim:	-	-	-	-	
AM/FM/ Radio-cassette:	-	O	O	S	
Power door locks:	-	-	-	O	
Power windows:	O	O	O	O	
Tilt steering column:	O	O	O	S	
Dual power mirrors:	-	O	O	S	
Intermittent wipers:	S	S	S	S	
Light alloy wheels:	-	-	-	O	
Sunroof:	-	-	-	O	
Anti-theft system:	-	-	-	O	

S : standard; O : optional; - : not available

COLORS AVAILABLE

Exterior: White, Red, Black, Garnet, Silver, Blue.
Interior: Gray, Brown.

MAINTENANCE

First revision: 7,500 miles
Frequency: 7,500 miles
Diagnostic plug: No

WHAT'S NEW IN 1995 ?

- No major change except for the names of the models and the rearrangement of the range.

ENGINE						TRANSMISSION		PERFORMANCE										
Model/ versions *: standard	Type / timing valve / fuel system	Displacement cu/in	Power bhp @ rmn	Torque lb.ft @ rpm	Compres. ratio	Driving wheels / transmission	Final ratio	Acceler. 0-62 mph s	Stand. 1/4 mile s	Stand. 5/8 mile s	Braking 62-0 mph ft	Top speed mph	Lateral acceler. G	Noise level dBA	Fuel economy mpg city	hwy	Gasoline type / octane
base- XE	L4* 2.4 SOHC-8-MPSFI	146	134 @ 5200	154 @ 3600	8.6 :1	rear - M5*	3.545	13.5	19.0	37.5	180	97	NA	69	28.0	35.0	R 87
						rear - A4	3.700	14.0	19.5	38.0	190	100	NA	69	25.0	34.0	R 87
base- XE 4x4	L4* 2.4 SOHC-8-MPSFI	146	134 @ 5200	154 @ 3600	8.6 :1	all - M5*	4.375	14.0	19.7	37.8	187	97	NA	69	22.0	29.0	R 87
V6 & SE	V6* 3.0 SOHC-12-MPSFI	181	153 @ 4800	180 @ 4000	9.0 :1	rear - M5*	3.700	12.0	18.5	35.0	184	103	NA	68	23.0	31.0	R 87
						rear - A4	3.900	12.8	18.8	35.7	190	100	NA	68	22.0	31.0	R 87
V6 & SE 4x4	V6* 3.0 SOHC-12-MPSFI	181	153 @ 4800	180 @ 4000	9.0 :1	all - M5*	4.375	12.0	18.6	35.4	180	103	NA	68	19.0	25.0	R 87
						all - A4	4.625	12.8	19.0	36.2	177	100	NA	68	19.0	25.0	R 87

seats with reclinable backs that gain added lateral and lumbar support than the standard bench. The rudimentary instrumentation expands with the lengthening of the bill to become as numerous as sports car gauges on the XE V6 where you find two odometers. Most of the controls are laid out in Japanese fashion and visibility is excellent, except laterally on the "King Cab" where the "B" pillar is thick and inconvenient.

• **Handling: 60%**
It is average because the base model suspension is too soft and that of the 4x4 are too harsh, associated with big tires, which rebound too much. Traveling empty on a wet road, you have to be careful.

• **Steering: 60%**
It is more precise, and with a better ratio with a power assist than on the manual of the base model where driving is more painful. However, the power assist is a little too high which makes it light at speed and calls for constant vigilance when the wind is blowing.

• **Technical: 60%**
The steel body of the Hardbody is mounted on a ladder frame with 5-cross members. The front suspension is independent, while at the rear, a rigid axle is tied in to leaf springs. The brakes have discs at the front and drums at the rear and only the 4x4's receive a standard ABS acting only on the rear wheels.

• **Access: 60%**
It is more difficult to climb into the 4x4 versions or into the back of the "King Cab" than into the 2-wheel drive version which has a lower ground clearance.

• **Price/equipment: 60%**
The prices have climbed steadily in the last few years, but the equipment of the base model is just as lean and it's only on the XE-V6 version that it is more complete and similar to that of a car.

• **Insurance: 50%**
There isn't an enormous difference between the premium of the least expensive one and that of the best equipped which costs as much as that of a compact car.

• **Suspension: 50%**
That of the 4x4 reacts harshly to the least road defects and those reactions are amplified by the enormous tires of the XE-V6. Off road, it is better not to go too fast unless you want to disjoint all the bones of the body, as in a rodeo.

• **Depreciation: 50%**
It is mostly the usage that will decide the resale value of these vehicles and you will get more out of a pickup that has served as a second car than from a work horse that has been pounded on all its life.

• **Interior space: 50%**
The normal cab can only accept two people, while the King Cab will handle 2 children on the rear fold down seats, space better suited for baggage storage.

• **Conveniences: 50%**
It consists of a glove box and a hollow in the dashboard of the base models, while the XE-V6 has, in addition, a small lock box in the console and pockets in the doors.

• **Performance: 50%**
In spite of acceptable takeoffs and pickups, one notices a lack of low rpm torque with these 2 engines, inconvenient when driving under load or off road with the V6.

WEAK POINTS

• **Storage: 30%**
Non-existent in the regular cab, it offers appreciable rear volume in a King Cab when the jump seats are raised, but it is not easily accessible.

• **Braking: 40%**
It hasn't improved much these past years because stopping distances are still very long, the front wheels lock up early and the brake fade-resistance doesn't stand up to many tests, standard ABS is essential.

• **Fuel consumption: 40%**
Neither of the two engines is really economical, especially when asked to put out enough power to cope with heavy loads or to travel through difficult terrain.

CONCLUSION

• **Overall rating: 56.0%**
Instead of aesthetic face lifts of modest usefulness, Nissan would have done better to bring its utilities up to level, with air bags and ABS. 😐

SPECIFICATIONS & PRICES

Model	Version/Trim	Body/ Seats	Wheel-base in	Lgth x Wdthx Hght. inx inx in	Curb weight lb	Towing capacity std. lb	Susp. type ft/rr	Brake type ft/rr	Steering system	Turning diameter ft	Steer. turns nber	Fuel. tank gal.	Standard tire size	Standard powertrain	PRICES US. $ 1994
NISSAN	**General warranty: 3 years / 50,000 miles; Power Train: 6 years / 60,000 miles; perforation corrosion & antipollution: 6 years / unlimited.**														
Hardbody 4x2															
	base XE	2dr. p-u.2	104.3	174.6x65.0x62.0	2824	2000	i/r	dc/dr	bal.	33.5	3.8	15.9	195/75R14	L4/2.4/M5	**$9,459**
	King Cab XE	2dr. p-u.2+2	116.1	190.0x65.0x62.0	2910	3500	i/r	dc/dr	pwr.bal.	36.8	3.8	15.9	195/75R14	L4/2.4/M5	**$11,979**
	King Cab XE-V6	2dr. p-u.2+2	116.1	190.0x65.0x62.0	3241	3500	i/r	dc/dr	pwr.bal.	36.8	3.0	15.9	215/75R14	V6/3.0/M5	**$14,679**
Hardbody 4x4															
	XE	2dr. p-u.2	104.3	174.6x65.0x62.0	3470	3500	i/r	dc/dr/ABS	pwr.bal.	35.4	3.0	15.9	235/75R15	L4/2.4/M5	**$14,069**
	King Cab XE-V6	2dr. p-u.2+2	116.1	190.0x66.5x67.1	3904	3500	i/r	dc/dr/ABS	pwr.bal.	38.7	3.0	21.1	235/75R15	V6/3.0/M5	**$16,379**

See page 393 for complete 1995 Price List

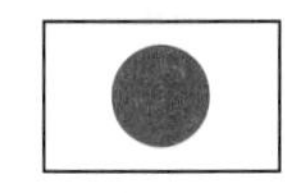

The best recruit of the year...

The latest Maxima vintage promises to be as great a year as those it replaces. In a different body, we have discovered a multi-faceted very homogenous sedan with heady performance and a racy behavior in a bouquet of personalized versions.

The Maxima range in the segment of luxury cars that it inaugurated fourteen years ago. This year it retains its championship title and also gains one of the best recruit of the year by coming out with the highest percentage points of the ninety five innovations. This 4-door sedan has radically changed lines and is much more collected, in a style inspired by the Altima. It continues to be offered in the popular GXE, the sports-like SE and the luxurious GLE. Those versions differ only in the level of their deluxe equipment, and their presentation. Besides this, their drive train is absolutely identical.

STRONG POINTS

• Satisfaction: 90%
The owners are unanimous: the Maxima cause few concerns. There is little reason for any of this to change with the new model. Ninety percent of the owners would again make the same purchase. However they mention that spare parts are hard to come by and rather expensive.

• Technical: 90%
The unit body of the latest Maxima has an independent suspension and disc brakes on all four wheels. Up front there is a modified McPherson suspension and at the rear Nissan introduces a multiarm system based on the Scott-Russell design. The unique 3.0 L V6 has a double overhead cam which delivers 190 hp. Last year it was available only on the SE and was derived from the 300ZX. The GXE and SE come standard with a manual 5-speed and the GLE has a 4-speed automatic which is optional on the first two. The lines are a little more anonymous but nonetheless very efficient, with a Cd of 0.32. Only the sports SE comes standard with a limited slip differential and a viscous coupling.

• Safety: 90%
The old model has already been recognized as one of the most rigid in the world and Nissan now announces that the new structure is 10% more resistant than the precious one. As it should be these days, the front seat occupants are protected by two air bags but the controversial ABS is standard only on the GLE.

• Performance: 85%
To accelerate from 0 to 62 mph in less than 8 seconds in a medium sedan with an automatic is not common place and the passing ability is muscular. The great advantage of this powertrain is to be just as much at ease in cruise or if you are in search of sports performance without any noise or untimely vibrations. The weight-to-power ratio is a favorable16 lbs/hp which explains both the interesting performance and the reasonable fuel consumption.

• Quality/fit/finish: 80%
Solidly constructed, painstakingly assembled and carefully finished, it gives a strong reassuring impression of neatness. If the interior of the GXE is not as flamboyant as that of the SE or GLE, its appearance still

DATA

Category: front-wheel drive luxury sedans.
Class : 7

HISTORIC

Introduced in: 1981 (L6 propulsion)
Modified in: 1984 (V6-T/A);1994: V6 190 hp;.1986,1991,1995: body.
Made in: Oppama, Japan.

RATINGS

Safety: 90 %
Satisfaction: 95 %
Depreciation: 50 %
Insurance: 5.0 % ($720 SE:$800)
Cost per mile: $0.45

NUMBER OF DEALERS

In U.S.A.: 1,100

SALES ON THE AMERICAN

Model	1992	1993	Result
Maxima	84,583	87,602	+ 3.5 %

MAIN COMPETITORS

ACURA Vigor, AUDI 90, BMW 3 Series, INFINITI J30, LEXUS ES300, MAZDA Millenia, MERCEDES BENZ C Class, SAAB 900, TOYOTA Avalon, VOLVO 850 GLT.

EQUIPMENT

NISSAN Maxima	GXE	SE	GLE
Automatic transmission:	O	O	S
Cruise control:	S	S	S
Power steering:	S	S	S
ABS brake:	O	O	S
Air conditioning:	S	S	S (Autom.)
Air bag (2):	O	S	S
Leather trim:	O	S	S
AM/FM/ Radio-cassette:	S	S	S
Power door locks:	S	S	S
Power windows:	S	S	S
Tilt steering column:	S	S	S
Dual power mirrors:	S	S	S
Intermittent wipers:	S	S	S
Light alloy wheels:	-	S	S
Sunroof:	O	O	O
Anti-theft system:	O	O	O

S : standard; O : optional; - : not available

COLORS AVAILABLE

Exterior: White, Beige, Pewter, Ruby, Blue, Ebony, Black.
Interior: Fabric & Leather: Gray-Green, Charcoal, Beige.

MAINTENANCE

First revision: 7,500 miles
Frequency: 7,500 miles
Diagnostic plug: Yes

WHAT'S NEW IN 1995 ?

- New, completely redesigned model.

	ENGINE					TRANSMISSION		PERFORMANCE										
Model/ versions *: standard	Type / timing valve / fuel system	Displacement cu/in	Power bhp @ rmn	Torque lb.ft @ rpm	Compres. ratio	Driving wheels / transmission	Final ratio	Acceler. 0-62 mph s	Stand. 1/4 mile s	Stand. 5/8 mile s	Braking 62-0 mph ft	Top speed mph	Lateral acceler. G	Noise level dBA	Fuel economy mpg city	hwy	Gasoline type / octane
1994 base	V6* 3.0 SOHC-12-SFI	182	160 @ 5200	182 @ 2800	9.0 :1	front - A4*	3.642	9.0	16.7	29.6	164	125	0.78	66	22.0	33.0	R 87
SE	V6* 3.0 DOHC-24-SFI	182	190 @ 5600	190 @ 4000	10.0 :1	front - M5*	3.823	7.5	14.8	27.5	157	138	0.81	66	25.0	33.0	S 91
						front - A4	3.619	8.3	16.1	29.0	167	131	0.81	66	24.0	34.0	S 91
1995	V6* 3.0 DOHC-24-MPSFI	182	190 @ 5600	205 @ 4000	10.0 :1	front - M5*	3.823	7.0	14.8	27.6	138	138	0.83	66	27.0	35.0	S 91
						front - A4	3.619	7.7	15.4	28.5	148	131	0.83	66	25.0	35.0	S 91

does not look inexpensive.

• Handling: 80%
Thanks to a new multiple arm rear suspension it has become more self-assured and remains neutral in most instances and practically infinite in normal use. Even a brutal change in pavement quality does not cause it to lose its self-possession. It could be put in any hands without fear because it forgives all errors. The multiple arm rear suspension optimizes stability in a straight line or in curves, offers less camber change in rapid lane changes and reduces rear suspension lift during panic stops.

• Steering: 80%
Rapid, precise, with good power assist modulation and ratio, it offers excellent maneuverability, thanks to a reduced steering diameter.

• Insurance: 80%
For a luxury car, the Maxima costs less premium money than a Millenia or a Lexus ES300. However, the premium is still higher than that of an Accord or a deluxe Camry

• Seats: 80%
In spite of their simple form, they are very comfortable, provide both good lateral and lumbar support, and the padding shape is inspired by German style.

• Driver's compartment: 80%
The simple and elegant dashboard presentation adds much class to the Maxima. While the different gauges and controls are well laid out, the center portion does not project far enough and is also not oriented toward the driver. The radio and AC controls are too far away. Driving position, just like visibility, is excellent but seat controls are placed too far back. It's hard to admit that in a car of this class there is no dashboard reminder of the shift level position of the automatic so that you have to take your eyes off the road to know the gear.

BEST ROOKIE OF THE YEAR

• Suspension: 80%
In spite of a generally firm response, it is never brutal and the shock absorber quality, as well as the wheel travel, absorbs all road defects.

• Conveniences: 80%
The storage spots are numerous and convenient in many well thoughtout details make its usage pleasant and rational. Front seat belts have a height adjustment.

• Interior space: 70%
It has improved, especially at the rear seats. It has ample head and legroom, but it is only a 4-seater, because a 5th occupant will find the space tight.

• Access: 70%
Even though the doors open wide, access to the rear seats is a little more limited because of the narrower rear doors.

• Noise level: 70%
Effective soundproofing provides a plush quiet at cruise. But wind and rolling noises fill the void left by the powertrain silence.

• Trunk: 60%
Its storage volume is proportional to the format of this model. It has lost its ability to be transformed since the bench seat is now fixed and only the trap door in the center armrest allows longer objects to be stored.

• Fuel consumption: 60%
Very reasonably, it stays around 21 mpg, thanks to excellent efficiency and the weight reduction sufficiently rare to be worth mentioning.

• Price/equipment: 50%
The price of the different Maxima versions is fully justified and should allow this model to loom large between the insipid high end Accord and the unaffordable Lexus ES300 which are nothing to write home about. Except for ABS which remains a philosophical choice, the standard GXE model has the same mechanical attributes as the sports SE or the luxury GLE and its general equipment is honorable.

• Depreciation: 50%
It maintains a favorable value because of the high satisfaction rate and good reliability. Good new cars make good used cars.

WEAK POINTS

• Brakes: 45%
Even though they give good initial bite and are easy to modulate in spite of a "hollow" pedal the distances are fairly long with ABS because they do not come down below 138 ft to slow from 62 mph. In intensive use the fade-resistance of the pads is good but the pedal gets a little harder.

CONCLUSION

• Overall rating: 73.0%
The latest Maxima offers a lot to a lot of people for a reasonable budget. The only question one can ask concerns the evolution of the aesthetics, but then you can't discuss taste and colors. ☺

SPECIFICATIONS & PRICES

Model	Version/Trim	Body/ Seats	Interior volume cu.ft.	Trunk volume cu ft.	Drag coef. Cd	Wheel-base in	Lgth x Wdthx Hght. inx inx in	Curb weight lb	Towing capacity std. lb	Susp. type ft/rr	Brake type ft/rr	Steering system	Turning diameter ft	Steer. turns nber	Fuel. tank gal.	Standard tire size	Standard powertrain	PRICES US. $ 1994
NISSAN		**General warranty: 3 years / 50,000 miles; Power Train: 6 years / 60,000 miles; perforation corrosion & antipollution: 6 years / unlimited.**																
1994																		
Maxima	GXE	4dr.sdn.5	96	14	0.32	104.3	187.6x69.3x55.1	3146	1000	i/i	dc/dc	pwr.r/p.	36.7	3.1	18.5	205/65R15	V6/3.0/A4	**$20,960**
Maxima	SE	4dr.sdn.5	96	11.2	0.32	104.3	187.6x69.3x55.1	3265	1000	i/i	dc/dc/ABS	pwr.r/p.	36.7	3.1	18.5	205/65R15	V6/3.0/M5	**$22,025**
Maxima	SE	4dr.sdn.5	96	11.2	0.32	104.3	187.6x69.3x55.1	3318	1000	i/i	dc/dc/ABS	pwr.r/p.	36.7	3.1	18.5	205/65R15	V6/3.0/A4	**$22,960**
Maxima	Brougham	4dr.sdn.5	96	11.2	0.32	104.3	187.6x69.3x55.1	3146	1000	i/i	dc/dc/ABS	pwr.r/p.	36.7	3.1	18.5	205/65R15	V6/3.0/A4	-
1995																		
Maxima	GXE	4dr.sdn.5	99.6	14.5	0.32	106.3	187.7x69.7x56.0	3018	1000	i/i	dc/dc	pwr.r/p.	34.8	2.9	18.5	205/65R15	V6/3.0/M5	**$22,429**
Maxima	SE	4dr.sdn.5	99.6	14.5	0.32	106.3	187.7x69.7x56.0	3051	1000	i/i	dc/dc	pwr.r/p.	34.8	2.9	18.5	215/65R15	V6/3.0/M5	**$23,529**
Maxima	SE	4dr.sdn.5	99.6	14.5	0.32	106.3	187.7x69.7x56.0	3113	1000	i/i	dc/dc	pwr.r/p.	34.8	2.9	18.5	215/65R15	V6/3.0/A4	**$24,464**
Maxima	GLE	4dr.sdn.5	99.6	14.5	0.32	106.3	187.7x69.7x56.0	3093	1000	i/i	dc/dc/ABS	pwr.r/p.	34.8	2.9	18.5	205/65R15	V6/3.0/A4	-

See page 393 for complete 1995 Price List

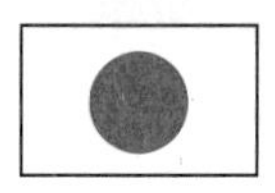

Count down...

The days of the Pathfinder are numbered and its replacement has already begun to make headlines in specialty magazines. The pioneer has gotten older and can no longer stand comparison with its usual competitors which are today more safety and performance-oriented.

It is Nissan which has opened the door to the sports utility vehicles as we know them today, because its Pathfinder was the first, derived from a 4x4 pickup, to offer a 3.0 L V6 and later a 4-door body which are quite popular by all the criteria of this category. Most of the competitors imitated it, and now it is the only one that has not been renewed. This year it is again offered as a 5-door in XE and SE versions with standard manual transmission or optional automatic or in the LE with the automatic as standard.

STRONG POINTS

• Satisfaction: 90%
Its ruggedness, excellent reliability, and complete Nissan warranty will reassure the owners who complain of the lack of availability of spare parts.

• Quality/fit/finish: 80%
The robustness of its build is typical of utilities, the material's quality and the finishing care are those of a car and the updating of the dashboard allows the pathfinder to go a little further but its replacement has already been announced.

• Storage: 80%
Just bringing the spare tire into the cab limits its capacity. It can be increased by using part or all of the bench seat to dispose of a vast flat floor.

• Driver's compartment: 70%
The driver sits high and disposes of a good visibility in spite of the "B" pillar thickness. By losing its square shape, the dashboard became a little more anonymous, but instrumentation is well-organized and legible. The main controls are well laid out except those of the cruise control and the radio.

• Steering: 70%
Its assistance and ratio are too high, making it sensitive and limiting

DATA

Category: rear-wheel drive or AWD sport utility.
Class : utility

HISTORIC

Introduced in: 1986
Modified in: 1990: 4-door & more powerful V6; 1994:interior touch-ups.
Made in: Kyushiu, Japan.

RATINGS

Safety: 60 %
Satisfaction: 75%
Depreciation: 46 %
Insurance: 5.0 % ($720-$800)
Cost per mile: $0.48

NUMBER OF DEALERS

In U.S.A.: 1,100

SALES ON THE AMERICAN

Model	1992	1993	Result
Pathfinder	36,704	48,770	+ 24.3 %

MAIN COMPETITORS

CHEVROLET Blazer, FORD Explorer, GMC Jimmy, ISUZU Rodeo & Trooper, JEEP Cherokee & Grand Cherokee, SUZUKI Sidekick, TOYOTA 4Runner.

EQUIPMENT

NISSAN Pathfinder	XE	SE	LE
Automatic transmission:	O	O	S
Cruise control:	S	S	S
Power steering:	S	S	S
ABS brake:	S	S	S
Air conditioning:	S	S	S
Air bag:	-	-	-
Leather trim:	-	O	O
AM/FM/ Radio-cassette:	S	S	S
Power door locks:	O	S	S
Power windows:	O	S	S
Tilt steering column:	S	S	S
Dual power mirrors:	O	S	S
Intermittent wipers:	S	S	S
Light alloy wheels:	-	S	S
Sunroof:	-	O	O
Anti-theft system:	-	-	-

S : standard; O : optional; - : not available

COLORS AVAILABLE

Exterior: Graphite, Garnet, Red, Black, Duck Blue, Astral Blue, Beige.
Interior: Gray, Blue, Red.

MAINTENANCE

First revision: 7,500 miles
Frequency: 7,500 miles
Diagnostic plug: No

WHAT'S NEW IN 1995 ?

- **No major change.**

ENGINE					TRANSMISSION		PERFORMANCE									
Model/ versions *: standard	Type / timing valve / fuel system	Displacement cu/in	Power bhp @ rmn	Torque lb.ft @ rpm	Compres. ratio	Driving wheels / transmission	Final ratio	Acceler. 0-62 mph s	Stand. 1/4 mile s	Stand. 5/8 mile s	Braking 62-0 mph ft	Top speed mph	Lateral acceler. G	Noise level dBA	Fuel economy mpg city	Fuel economy hwy	Gasoline type / octane
base	V6* 3.0 SOHC-12-SFI	181	153 @ 4800	180 @ 4000	9.0 :1	rear/4 - M5*	4.375	12.5	18.5	35.7	187	100	0.74	68	18.0	24.0	R 87
						rear/4 - A4	4.625	13.8	19.2	36.2	190	97	0.74	68	18.0	23.0	R 87

maneuverability in spite of the short turning diameter.

• Technical: 70%
Derived from the Hardbody pickup, from which it borrows most of its drive train, the Pathfinder has a unit body, joint or ladder type chassis, with 5 cross members. The front suspension is independent while at the rear, the rigid axle includes a Panhard rod and helical spring. The brakes are mixed on the XE and have 4 wheel discs on the SE and LE automatic with ABS and limited slip standard as well as automatic front hub locks.

• Conveniences: 70%
Storage includes a large size glove box, door pockets, and a well-designed center console.

• Seats: 70%
The bench seat offers less support than the better contoured front seats. Their padding is acceptable, but the optional leather trim is particularly slippery.

• Access: 60%
The tall ground clearance does not ease climbing in but front seats are more accessible than rear ones because of doors which are too narrow.

• Suspension: 60%
It admits its age by the lack of quality of its ground contact. While it is rather comfortable on the freeway, it becomes jumpy as soon as the pavement goes bad and actually brutal in passing bumps in off road.

• Insurance: 60%
The premium for sports models with four drive wheels is always higher than those of other vehicles because of the risks in leaving the beaten pads.

• Safety: 60%
From this point of view the Pathfinder has aged the most, in spite of the reinforcements installed in the doors last year. The structure could be more rigid and the absence of air bags deprives the occupants of effective protection.

• Interior space: 60%
Useful interior volume remains one of the most interesting because rear legroom is remarkable but it lacks a few centimeters in width to make everything perfect.

• Noise level: 50%
It is strongly dominated by the noise of the large tires on asphalt, the wind and the engine during strong acceleration. It is relatively comfortable at constant speeds.

• Depreciation: 50%
Contrary to some of its competitors, the Pathfinder keeps good resale value thanks to its reputation.

• Performance: 50%
The accelerations, just like the takeoffs are penalized by a weight that nears two tons, giving a rather unfavorable weight-to-power ratio (28 lbs/hp). At high speed on the road, the engine is silent and silky at high rpm, but is terribly lacking in torque under 2000 rpm; as soon as it is forced it becomes noisy and rough... The spacing of the stick shift ratios is not ideal because first gear pulls "too long" and there is also a giant gap between second and third. The shift pattern of the automatic leads to very high rpm in intermediate speeds which increases the fuel consumption.

WEAK POINTS

• Fuel consumption: 30%
The imposing Pathfinder's weight and its lack of aerodynamic finesse plus its engine power cause a high fuel consumption especially with a heavy load or in difficult terrain.

• Braking: 30%
Its efficiency, just as its resistance to fade is average. In intensive use, the friction material heats rapidly and while stopping distances are very long, they remain more or less straight, thanks to rear wheel ABS.

• Price/equipment: 40%
The Pathfinder is expensive if you take into account that its equipment is not complete, especially in terms of safety. Its capacity as a multi-purpose vehicle and its obstacle-crossing ability justify the investment, especially on the SE and LE.

• Handling: 40%
Average on the XE in which the suspension is too soft and has a marked role, it is more effective on the LE. The adjustable shocks of the LE are good at controlling body motion when set in a "Sport" mode. The height of the center of gravity limits cornering speed and the rebound of its large tires takes away from driving precision. Off road, the Pathfinder offers good obstacle crossing capacity, thanks to its short overhang. The traction quality is assisted by a limited slip.

CONCLUSION

• Overall rating: 59.5%
It's time for Nissan to think about renewing its Pathfinder which has lost its status as a reference in sports utility vehicles as well as some sales . ☹

SPECIFICATIONS & PRICES

Model	Version/Trim	Body/ Seats	Interior volume cu.ft.	Trunk volume cu ft.	Drag coef. Cd	Wheel-base in	Lgth x Wdthx Hght. inx inx in	Curb weight lb	Towing capacity std. lb	Susp. type ft/rr	Brake type ft/rr	Steering system	Turning diameter ft	Steer. turns nber	Fuel. tank gal.	Standard tire size	Standard powertrain	PRICES US. $ 1994
NISSAN		**General warranty: 3 years / 50,000 miles; Power Train: 6 years / 60,000 miles; perforation corrosion & antipollution: 6 years / unlimited.**																
Pathfinder XE		5dr.wgn.5	91.3	31.4	NA	104.3	171.9x66.5x65.7	4090	3500	i/i	dc/dr/ABS	re.pwr.bal.	35.4	3.0	21.1	235/75R15	V6/3.0/M5	**$19,429**
Pathfinder SE		5dr.wgn.5	91.3	31.4	NA	104.3	171.9x66.5x66.7	4246	3500	i/i	dc/dc/ABS	re.pwr.bal.	35.4	3.0	21.1	235/75R15	V6/3.0/M5	**$25,009**
Pathfinder LE		5dr.wgn.5	91.3	31.4	NA	104.3	171.9x66.5x66.1	4215	3500	i/i	dc/dc/ABS	re.pwr.bal.	35.4	3.0	21.1	235/75R15	V6/3.0/A4	**$28,999**

See page 393 for complete 1995 Price List

Less popular ...

The Sentra which is one of the most sold cars in the world, sees its sales decrease because of the success of the compacts, not much more expensive but offering more power and more comfort for a more similar MPG. Its future: get fatter or disappear.

Very simplified since 1990, the Sentra range consists of 2-door DLX and XE coupes, because the 4-door DLX, XE and GXE will be redesigned for the beginning of 1996. It is the bottom of the popular range of Nissan in North America. Since of the retreat of the Micra it is the most sold Nissan and accounts, by itself, for a third of the sales.

STRONG POINTS

• Price/equipment: 80%
Affordable in its standard version, the Sentra prices climb rapidly as soon as they have a little equipment and at equal prices there are other, more attractive, sub-compacts. It is bizarre that at Nissan, safety items such as the air bags and ABS are deluxe accessories.

• Technical: 80%
The two bodies are based on the same platform and use the same powertrain. The suspensions, front and rear, are independent, and the brakes are mixed but the ABS is optional only on the XE and GXE. The only engine available on these cars is a 1.6 L DOHC 4-cylinder which develops 110 hp and is derived from that of the Infinity G20. It has an aluminum block and receives a balance shaft with eight counter weights designed to eliminate vibrations. The steel unit body has a rather conservative line and an average 0.35 drag coefficient.

• Fuel consumption: 80%
It is one the major drawing cards of the Sentra because its little 4-cylinder rarely drinks more than a gallon per 31 miles even when it is whipped like a race horse.

• Satisfaction: 80%
The first buyers say they are satisfied with their experience with these models whose warranty is a major advantage. But one notes that cars built outside of Japan are unable to meet the same reliability rates.

DATA

Category: front-wheel drive sub-compact sedans.
Class : 3

HISTORIC

Introduced in: 1981
Modified in: 1986 & 1991
Made in: Aguascalientes, Mexique & Smyrna TE, U.S.A.

RATINGS

Safety: 75 %
Satisfaction: 83 %
Depreciation: 60 %
Insurance: 6.5 % ($450-$525)
Cost per mile: $0.32

NUMBER OF DEALERS

In U.S.A.: 1,100

SALES ON THE AMERICAN

Model	1992	1993	Result
Sentra	150,091	168,846	+ 11.1 %

MAIN COMPETITORS

DODGE-PLYMOUTH Colt, HONDA Civic Hbk, HYUNDAI Accent, MAZDA 323 Hbk, TOYOTA Tercel, VOLKSWAGEN Golf.

EQUIPMENT

	NISSAN Sentra 4-door DLX	XE	GXE	NISSAN Sentra 2-door DLX	XE
Automatic transmission:	O	O	O	O	O
Cruise control:	-	S	S	-	S
Power steering:	O	O	S	O	O
ABS brake:	-	O	S	-	-
Air conditioning:	S	S	S	-	-
Air bag:	-	O	O	-	-
Leather trim:	-	-	-	-	-
AM/FM/ Radio-cassette:	O	S	S	O	S
Power door locks:	-	-	S	-	-
Power windows:	O	O	S	-	-
Tilt steering column:	O	S	S	S	S
Dual power mirrors:	-	S	S	-	S
Intermittent wipers:	-	S	S	-	S
Light alloy wheels:	-	-	S	-	-
Sunroof:	-	O	O	-	O
Anti-theft system:	-	-	-	-	-

S : standard; O : optional; - : not available

COLORS AVAILABLE

Exterior: White, Red, Ruby, Green, Ebony, Beige, Gray, Blue, Silver.
Interior: Gray, Brown, Blue.

MAINTENANCE

First revision: 7,500 miles
Frequency: 7,500 miles
Diagnostic plug: Yes

WHAT'S NEW IN 1995 ?

- The new Sentra sedan was introduced in January of 1995 and the coupe will continue its career in 1995 under its present form.

ENGINE						TRANSMISSION		PERFORMANCE										
Model/ versions *: standard	Type / timing valve / fuel system	Displacement cu/in	Power bhp @ rmn	Torque lb.ft @ rpm	Compres. ratio	Driving wheels / transmission	Final ratio	Acceler. 0-62 mph s	Stand. 1/4 mile s	Stand. 5/8 mile s	Braking 62-0 mph ft	Top speed mph	Lateral acceler. G	Noise level dBA	Fuel economy mpg city	hwy	Gasoline type / octane
Sentra	L4* 1.6 DOHC-16-MPSFI	97	110 @ 6000	108 @ 4000	9.5 :1	front - M5*	3.895	9.5	17.0	31.0	151	112	0.78	68	35.0	50.0	R 87
						front - A4	3.827	10.5	17.6	31.8	157	109	0.78	68	31.0	46.0	R 87

• Driver's compartment: 80%
The driver rapidly finds the most comfortable seating position thank to seat and steering column adjustments and visibility remains good at all angles in spite of roof support thickness. Controls are laid down in standard fashion and most fall well under hand except for the cruise control switches mounted on the left. The gauges are easy to read and they are in sufficient amount. However we have noted that the headlights don't have much power and the second speed on the wind shield wiper is not fast enough.

• Safety: 70%
The structural resistance of the Sentra body shell is satisfactory, but only the sedans are entitled to an air bag, standard on the GXE and optional on the XE.

• Quality/fit/finish: 70%
The general impression is sober and without fantasy with careful assembly and finish and materials superior to that of the competition.

• Seats: 70%
The lateral support is superior at the front, the lumbar support is no more effective at the front than the rear but the padding is OK.

• Suspension: 70%
Ride softness is a main advantage of the Sentra. It never becomes disagreeable when the roads are bad because its damping is of good quality.

• Steering: 70%
Precise, well-modulated, and with a good ratio, its low turning diameter gives it an amazing maneuverability in town.

• Insurance: 60%
The premium is not as economical as the models themselves as it comes dangerously close to that of higher level units.

• Performance: 60%
There is no thunder and lightning there, especially with a 4-speed automatic whose pickups and accelerations are rather placid and the transmission ratios are too long.

• Access: 60%
It is easier to take a seat at the front than rear where narrow doors have an insufficient opening angle which will disturb large or tall people.

• Depreciation: 60%
It is normal for cars that are on the market in great numbers and Nissan's good warranty serves as an added attraction.

• Conveniences: 60%
In the "E" the storage is symbolic while in the XE and GXE you gets the rights to door pockets plus a glove box and hollows in the center console.

• Noise level: 50%
It remains in the comfort zone on cruise on freeways, but the engine remains noisy and vibrates (in spite of its balance shaft) during accelerations and wind plus rolling noises join the concert.

• Trunk: 50%
Even though it lacks length, its volume is sufficient and can be increased toward the interior by dropping the back of the bench seat and its access is eased by the cutout in its opening.

• Handling: 50%
In spite of the small 13 inch wheels, and the softness of the suspension, the roll is well under control and the understeer is easy to master. It is predictable and easy to handle with good quality tires.

WEAK POINTS

• Braking: 40%
It is not among the more powerful or best balanced because without ABS the front wheels lock rapidly in panic stops. The stopping distances are very long, and the trajectories are uncertain which calls for steering corrections. However, even when abused, the fade-resistance of the brake pads are satisfactory.

• Interior space: 40%
There is more space at the front seat than at the back, where the length and width are lacking. Even though the seat height is ample.

CONCLUSION

• Overall rating: 64%
At the eve of being renewed, the 94 sedan will still be available until January of 1995 and will remain a good buy so long as the price/equipment ratio is favorable. By using our price guide, which begins on page 393, you can get the best value. :-|

OWNER'S SUGGESTIONS

-More comfortable seats
-More attractive presentation
-More practical door handles
-More legroom in rear seats
-More effective brakes
-More complete equipment

SPECIFICATIONS & PRICES

Model	Version/Trim	Body/ Seats	Interior volume cu.ft.	Trunk volume cu ft.	Drag coef. Cd	Wheel-base in	Lgth x Wdthx Hght. inx inx in	Curb weight lb	Towing capacity std. lb	Susp. type ft/rr	Brake type ft/rr	Steering system	Turning diameter ft	Steer. turns nber	Fuel. tank gal.	Standard tire size	Standard powertrain	PRICES US. $ 1994
NISSAN		**General warranty: 3 years / 50,000 miles; Power Train: 6 years / 60,000 miles; perforation corrosion & antipollution: 6 years / unlimited.**																
Sentra	E	2dr.cpe.4	84	11.7	0.34	95.7	170.3x65.6x53.9	2346	1000	i/i	dc/dr	pwr.r/p.	30.2	3.07	13.2	175/70R13	L4/1.6/M5	**$10,199**
Sentra	XE	2dr.cpe.4	84	11.7	0.34	95.7	170.3x65.6x53.9	2346	1000	i/i	dc/dr	pwr.r/p.	30.2	3.07	13.2	175/70R13	L4/1.6/M5	**$12,549**
1994																		
Sentra	E	4dr.sdn.4	83	12	0.35	95.7	170.3x65.6x53.9	2346	1000	i/i	dc/dr	pwr.r/p.	30.2	3.07	13.2	155/80R13	L4/1.6/M5	**$11,049**
Sentra	XE	4dr.sdn.4	83	12	0.35	95.7	170.3x65.6x53.9	2359	1000	i/i	dc/dr	pwr.r/p.	30.2	3.07	13.2	175/70R13	L4/1.6/M5	**$12,749**
Sentra	GXE	4dr.sdn.4	83	12	0.35	95.7	170.3x65.6x53.9	2403	1000	i/i	dc/dr	pwr.r/p.	30.2	3.07	13.2	175/70R13	L4/1.6/M5	**$14,819**

See page 393 for complete 1995 Price List

Change of course...

There are so many models available in the sports coupe battle field that the buyer group constantly evolves and transforms itself. It's not just the young wolves in love with the absolute that buy the 240SX but rather old foxes looking for comfort and in search of young chicks.

If front wheel drive brings more safety, especially in winter driving, it doesn't generate much driving pleasure excitment and these days, rear wheel drives at popular prices are few. Next to Camaro/Firebird and Mustang of domestic manufacture, the Nissan 240SX is among the last representatives of this race in the process of extinction. After disappearing from the market during 1994, it is back as a unique 2-door, 3-volume coupe in version SE and LE. The engine is a 2.4 L with a 156 hp from a 16 valve DOHC. It comes with a standard stick shift or an optional automatic and a limited slip is standard.

STRONG POINTS

• Safety: **90%**
Thanks to the good structural rigidity, the extra reinforcements, and 2 air bags up front the 240X reaches the maximum rate.

• Satisfaction: **85%**
The reliability index of the 240X is high and the maintenance doesn't pose any problem to its owners, as it represents the best of both worlds.

• Technical : **80%**
The 240SX has been reworked. The platform of the previous version has been widened and lengthened to improve its handling. Its steel unit body has very conservative lines and shows an identically conservative aerodynamic coefficient of 0.32.
There is independent suspension, 4 wheel disc brakes and the front and rear suspension geometries include antidive and antilift. The bottom end of the engine is identical to that which equips the Altima and the Axxess but differs in its DOHC 16 valve head.

DATA

Category: rear-wheel drive sport coupes.
Class : 3S

HISTORIC

Introduced in: 1969 (200SX).
Modified in: 1977, 1979, 1983, 1989, 1995.
Made in: Kyushiu, Japan

RATINGS

Safety: 90 %
Satisfaction: 85 %
Depreciation: 58 %
Insurance: 5.5 % ($885)
Cost per mile: $0.47

NUMBER OF DEALERS

In U.S.A.: 1,100

SALES ON THE AMERICAN

Model	1992	1993	Result
240SX	-	23,047	

MAIN COMPETITORS

ACURA Integra, CHEVROLET Camaro, EAGLE Talon, FORD Mustang, FORD Probe, HONDA Prelude, MAZDA MX-6, MITSUBISHI Eclipse, PONTIAC Firebird, VW Corrado.

EQUIPMENT

NISSAN 240SX	SE	LE
Automatic transmission:	O	O
Cruise control:	S	S
Power steering:	S	S
ABS brake:	O	S
Air conditioning:	S	S
Air bag (2):	S	S
Leather trim:	-	S
AM/FM/ Radio-cassette:	S	S
Power door locks:	S	S
Power windows:	S	S
Tilt steering column:	S	S
Dual power mirrors:	S	S
Intermittent wipers:	S	S
Light alloy wheels:	S	S
Sunroof:	O	O
Anti-theft system:	-	-

S : standard; O : optional; - : not available

COLORS AVAILABLE

Exterior: -
Interior: -

MAINTENANCE

First revision: 7,500 miles
Frequency: 7,500 miles
Diagnostic plug: Yes

WHAT'S NEW IN 1995 ?

-New version of the former 2-door 240SX coupe.

Model/ versions *: standard	Type / timing valve / fuel system	ENGINE Displacement cu/in	Power bhp @ rmn	Torque lb.ft @ rpm	Compres. ratio	TRANSMISSION Driving wheels / transmission	Final ratio	PERFORMANCE Acceler. 0-62 mph s	Stand. 1/4 mile s	Stand. 5/8 mile s	Braking 62-0 mph ft	Top speed mph	Lateral acceler. G	Noise level dBA	Fuel economy mpg city	hwy	Gasoline type / octane
base	L4* 2.4 DOHC-16-SFI	146	155 @ 5600	160 @ 4400	9.5 :1	rear - M5*	4.083	8.3	15.8	28.5	125	125	0.82	68	27.0	35.0	M 89
						rear - A4	4.083	9.0	16.4	29.1	131	119	0.82	68	25.0	36.0	M 89

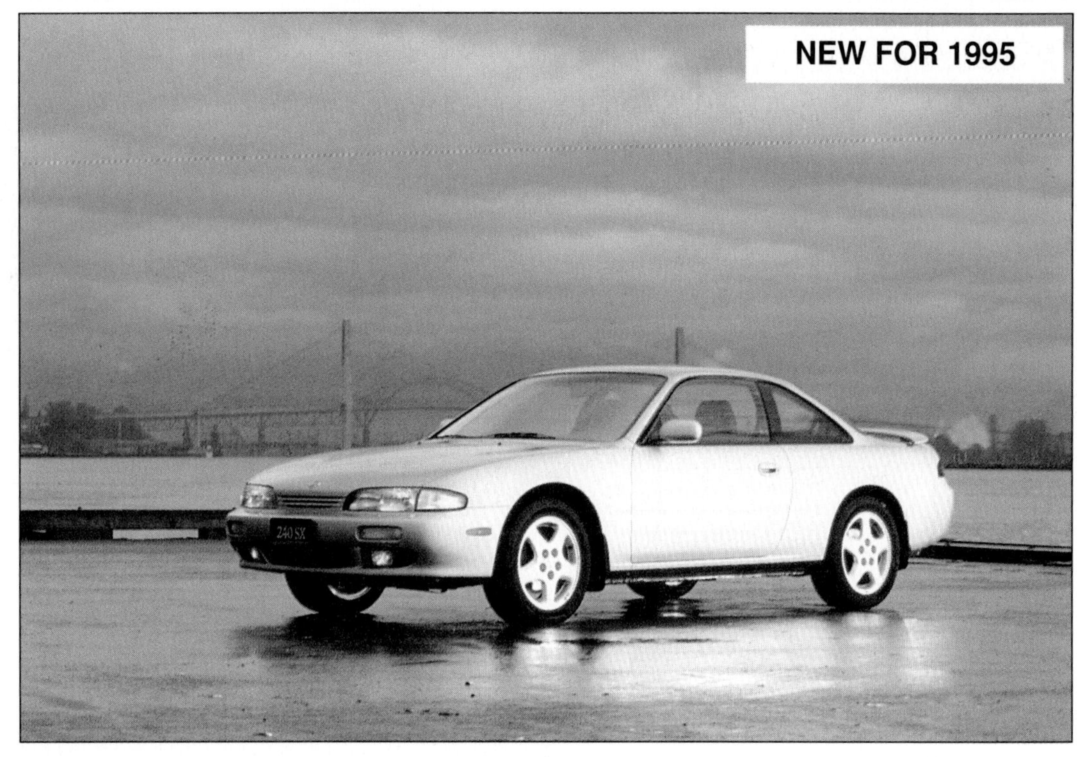

NEW FOR 1995

• Quality/fit/finish: 80%
Assembly and finish are careful, but the interior presentation is a disappointment, especially with leather trim.

• Driver's compartment: 80%
Discovering the best driver's position is somewhat laborious, but it is as satisfactory as the visibility. The instrumentation is clear and easy to read, day or night, thanks to an electro-luminescent device which alters the gauge-light intensity as a function of ambient light. The controls are conventional, but switches placed behind the steering wheel are difficult to reach and the center console does not jut out enough, so the controls placed there are not within hand's reach.

• Steering: 80%
Rapid, precise, and well-modulated, it provides better directional stability when speed increases than on the previous model. Maneuverability is excellent thanks to the turning diameter that is exceptionally short (less than 33 ft).

• Insurance: 75%
In relation to price, this model has a favorable premium rate, but costs practically double for a single, unmarried person under 25, which explains why the clientele is now quite different.

• Handling: 70%
A limited slip differential optimizes driveability when traction conditions are precarious. This does not mean you should not be cautious in winter driving. The excellent control of the drivetrains delivers sure-footed and fun handling. You can oversteer at the rear at will and imagine driving talents without taking any risks. Drifts are predictable and progressive, thanks to a good suspension balance.

• Performance: 70%
Even though honest, the figures deceive somewhat because the power reserve never appears sufficient and the transmission pulls too long in each gear. Driving is distinctly more fun with a well-staged stick shift but becomes more antiseptic with the automatic. The 3.0 L V6 engine of the 300ZX deflated to 150 hp would elicit more emotions.

• Seats: 70%
In spite of their contours, they could provide more lateral support at the hips.

• Suspension: 70%
Even though firm, it constitutes a good compromise, and pavement must be in sad shape before the ride becomes disagreeable.

• Braking: 60%
For once we have to admit the brakes are sufficiently powerful and easy to control, but the ABS should be standard on the SE because the rear wheels have a tendency to lock up too early.

• Conveniences: 60%
Storage spaces are well laid out and in sufficient amounts, but their volume is stingy. The AC is efficient and easy to set, but the rear windows are fixed.

• Noise level: 50%
It is average but rises quickly when the sunroof is opened.

• Price/equipment: 50%
Due to the strength of the yen, the price seems less competitive compared to that of some of its competitors. However, the equipment is adequate on the SE and more complete on the LE which benefits, among other things, from 16-inch wheels.

• Fuel consumption: 50%
Reasonable at speeds that are, it climbs like an arrow as soon as the driver tries to raise his adrenaline rate.

• Access: 50%
Large-sized people will have less problem seating themselves up front than in the back where, traditionally for this type of car, space is at a premium. However, young children will be comfortable.

WEAK POINTS

• Interior space: 20%
It is limited, especially in height. You sit elbow to elbow because the transmission tunnel takes up a fair amount of room. This, in turn, also limits legroom in width.

• Trunk: 30%
Readily accessible, it has seen its volume increase a bit but not enough to move your wardrobe.

• Depreciation: 40%
The resale value after three years falls right when the main warranty expires.

CONCLUSION

• Overall rating: 63.0%
The new buyers of the 240SX appreciate its classic and elegant appearance and its comfort/equipment aspect along with a considerable driving pleasure. Some extra vitamins would not hurt this engine which offers more pleasure in the European version. 😐

SPECIFICATIONS & PRICES

Model	Version/Trim	Body/ Seats	Interior volume cu.ft.	Trunk volume cu ft.	Drag coef. Cd	Wheel-base in	Lgth x Wdthx Hght. inx inx in	Curb weight lb	Towing capacity std. lb	Susp. type ft/rr	Brake type ft/rr	Steering system	Turning diameter ft	Steer. turns nber	Fuel. tank gal.	Standard tire size	Standard Power Train	PRICES US. $ 1994
NISSAN		General warranty: 3 years / 50,000 miles; Power Train: 6 years / 60,000 miles; perforation corrosion & antipollution: 6 years / unlimited.																
240 SX	SE	2dr.cpe.2+2	70.7	8.6	0.32	99.4	177.2x68.1x50.8	2813	1000	i/i	dc/dc	pwr.r/p.	31.5	3.1	17.2	195/60HR15	L4/2.4/M5	NA
240 SX	LE	2dr.cpe.2+2	70.7	8.6	0.32	99.4	177.2x68.1x50.8	2875	1000	i/i	dc/dc/ABS	pwr.r/p.	31.5	3.1	17.2	205/60VR16	L4/2.4/A4	NA

See page 393 for complete 1995 Price List

NISSAN 300 ZX

Yellow fever...

The Japanese have redefined the criteria of modern sports cars by making them sophisticated and accessible, and today they monopolize this segment. Thanks to its strong personality, the ZX coupe has played a big role in this conquest for the entire length of its career.

The 300ZXs, renewed in 1990, have marked their era by their aesthetics and technical achievements just like the first 240s did in their time. Last year, the convertible version joined the standard 2-seater coupe and the 2+2. They are equipped with a naturally aspirated engine and only the 2-seater coupe receives the twin turbo engine with its two intercoolers.

STRONG POINTS

• Technical: 90%
The steel unit body has a very successful line in both aesthetics and aerodynamics, with 0.32 drag coefficient. There is an independent suspension and disc brakes at all four wheels. The differential has a limited slip with a viscous coupling. Only the Bi-Turbo is provided standard with the HICAS system, which gives 4-wheel steering by an electric command. The power steering is modulated as a function of speed, and the damping can be switched into touring or sport modes.

• Safety: 90%
Two air bags are supplied standard for the front passengers, allowing a maximum rating, but some reinforcements are still needed to maximize structural stiffness.

• Steering: 90%
It nears perfection as much by its precision as by modulation and ratio, and is well-adapted to the handling. Too bad that the steering column is not adjustable in any direction.

• Satisfaction: 90%
The 300Z's do not give much concern to their owners, who are delighted with the excellent Nissan guaranty but mention that spare parts are rare and expensive.

• Quality/fit/finish: 80%
The 300ZX's are assembled with equal care inside and out. Finish is meticulous and the materials are of good quality.

DATA

Category: rear-wheel drive sports coupes.
Class: GT

HISTORIC

Introduced in: 1969
Modified in: 1970, 1975, 1978, 1983, 1990; 1994: convertible
Made in: Oppama, Japan

RATINGS

Safety: 90 %
Satisfaction: 88 %
Depreciation: 61 %
Insurance: 4.0 % ($1,065 -$1,200)
Cost per mile: $0.80

NUMBER OF DEALERS

In U.S.A.: 1,100

SALES ON THE AMERICAN MARKET

Model	1992	1993	Result
300 ZX	9,628	9,168	- 4.8 %

MAIN COMPETITORS

ACURA NSX, DODGE Stealth R/T & R/T Turbo, MAZDA RX-7, MITSUBISHI 3000GT, PORSCHE 968 & 911, TOYOTA Supra.

EQUIPMENT

NISSAN 300ZX	2 se	2+2	Turbo	Conv.
Automatic transmission:	O	O	O	-
Cruise control:	S	S	S	S
Power steering:	S	S	S	S
ABS brake:	S	S	S	S
Air conditioning:	S	S	S	S
Air bag:	S	S	S	S
Leather trim:	S	S	S	S
AM/FM/ Radio-cassette:	S	S	S	S
Power door locks:	S	S	S	S
Power windows:	S	S	S	S
Tilt steering column:	S	S	S	S
Dual power mirrors:	S	S	S	S
Intermittent wipers:	S	S	S	S
Light alloy wheels:	S	S	S	S
Sunroof:	S	S	S	S
Anti-theft system:	S	S	S	S

S : standard; O : optional; - : not available

COLORS AVAILABLE

Exterior: White, Garnet, Red, Bue, Black, Silver.
Interior: Charcoal, Tan.

MAINTENANCE

First revision: 7,500 miles
Frequency: 6 months/7,500 miles
Diagnostic plug: Yes

WHAT'S NEW IN 1995 ?

- No major change.

ENGINE						TRANSMISSION		PERFORMANCE										
Model/ versions *: standard	Type / timing valve / fuel system	Displacement cu/in	Power bhp @ rmn	Torque lb.ft @ rpm	Compres. ratio	Driving wheels / transmission	Final ratio	Acceler. 0-62 mph s	Stand. 1/4 mile s	Stand. 5/8 mile s	Braking 62-0 mph ft	Top speed mph	Lateral acceler. G	Noise level dBA	Fuel economy mpg city	Fuel economy mpg hwy	Gasoline type / octane
base	V6* 3.0 DOHC-24-MPSFI	180	222 @ 6400	198 @ 4800	10.5 :1	rear - M5*	4.083	7.0	15.3	27.7	125	140	0.86	70	22.0	31.0	S 91
						rear - A4	4.083	8.1	16.0	28.5	131	137	0.86	70	22.0	30.0	S 91
2+2	V6* 3.0 DOHC-24-MPSFI	180	222 @ 6400	198 @ 4800	10.5 :1	rear - M5*	4.083	7.5	15.7	28.2	134	137	0.84	70	22.0	31.0	S 91
						rear - A4	4.083	8.8	16.5	28.9	138	131	0.84	70	22.0	30.0	S 91
Turbo	V6T* 3.0 DOHC-24-MPSFI	180	300 @ 6400	283 @ 3600	8.5 :1	rear - M5*	3 692	5.8	14.0	27.0	115	153	0.88	70	21.0	31.0	S 91
			280 @ 6400			rear - A4	3.692	6.5	14.7	27.6	125	150	0.88	70	21.0	31.0	S 91

• Driver's compartment: 80%
The driver's seat has good lateral and lumbar support, in spite of simplistic adjustments which do not allow changes in thigh support. It is inconceivable that the steering column is not adjustable in any direction, which would help tall people find a more comfortable driving position. Visibility is not fabulous from 3/4 rear, where it is obstructed by a large "B" pillar, or at the rear where the rear window is narrow and distorting. The instrumentation and the main controls are well laid out but those disposed on both sides of the instrument cluster need to be gotten used to. The standard shift is firm and long clutch-pedal stroke soon become tiresome in urban driving.

• Performance: 80%
Our test instruments show that power to the wheels of the naturally aspirated version does not exceed 200 HP and that of the BI-Turbo 260 hp, according to DIN standards. If the numbers are lower than those given by the builder, there is still enough power that this should be the engine of choice for rapidly losing a few points or the entire license. The ease with which high speeds are reached is impressive because the engines are refined and smooth, even in urban driving. The automatic transmission marvelously complements the naturally aspirated engine, while the manual transmission and the turbo allow exotic takeoffs and accelerations without being truly extraordinary.

• Handling: 80%
In spite of the viscous coupler in the 300ZX differential, they don't demonstrate as much aplomb on a wet road as a 4-wheel drive. Some test rides will be necessary to learn how to have fun without danger. On a dry road, it easily gets into the turns, and its maneuverability allows you to pilot with finesse from crisp reactions from steering and the accelerator. You can heel and toe between the accelerator and the brake pedal. On the wet, you need caution because the rear swings out easily and getting back on the gas should be done with considerable discernment, especially with the Bi-Turbo.

• Insurance: 80%
For this type of car the rate is very modest since those who have the means to purchase this fantasy are settled and reasonable, at least in theory.

• Braking: 70%
It is at the level of the performance, powerful, balanced and easy to modulate; and the fade-resistance is reassuring.

• Seats: 70%
Well-designed, they hold you laterally and support your back in an ideal fashion. Also, their padding is not too firm. Fabrics used in a base model are less slippery than leather trim standard in other versions.

• Suspension: 60%
Even though firm, it offers an excellent compromise between handling and comfort, superior to most of its competitors except for the Stealth/3000GT.

• Conveniences: 50%
Neither numerous nor practical, storage spaces boil down to one glove box and one pocket in the center console, both minuscule.

• Fuel consumption: 50%
Taking into account the potential of this engine, which doesn't show a greater thirst than other quieter ones, it's reasonable.

WEAK POINTS

• Interior space: 20%
Lowering the dashboard provides an impression of space. The available interior volume falls between the Dodge Stealth and the Mazda RX-7. The bench of the 2+2 would only accommodate very young children. Their dimensions are generously sized to accept 50% of a normal adult.

• Trunk: 30%
It contains little, due to lack of height, but the panel disposed between the rear-wheel wells allows trinket storage.

• Noise level: 30%
Noises coming from the drive and power trains are well-filtered and those of the wind appear only at high speed, but the glass roof panels allow a high noise entry from the exterior, and it would be good to put in a double lid.

• Access: 40%
The lower roof height complicates access to the front seats for tall people, but it is worse at the rear of the 2+2, where everything is symbolic.

• Price/equipment: 40%
More expensive than a Stealth, the 300ZX is better compared to the RX-7, Porsche 968, and Supra. The 300ZX comes with complete equipment that includes a T sunroof mounted standard on its coupes.

• Depreciation: 40%
Resale value falls rapidly after 3 years, in the absence of a warranty and the prospect of exotic bills.

CONCLUSION

• Overall rating: 63.0%
In spite of the fall in sales at the announcement of a new model, fans will find the 300ZX the essence of a great sports coupe. However, its character necessitates that the pilot should know how to drive.

SPECIFICATIONS & PRICES

Model	Version/Trim	Body/ Seats	Interior volume cu.ft.	Trunk volume cu ft.	Drag coef. Cd	Wheel-base in	Lgth x Wdthx Hght. inx inx in	Curb weight lb	Towing capacity std. lb	Susp. type ft/rr	Brake type ft/rr	Steering system	Turning diameter ft	Steer. turns nber	Fuel. tank gal.	Standard tire size	Standard Power Train	PRICES US. $ 1994
NISSAN		**General warranty: 3 years / 50,000 miles; Power Train: 6 years / 60,000 miles; perforation corrosion & antipollution: 6 years / unlimited.**																
300ZX	base	3dr.cpe.2	70	10	0.32	96.5	169.5x70.5x48.4	3362	NR	i/i	dc/dc/ABS	pwr.r/p.	34.1	2.7	19	225/50VR16	V6/3.0/M5	**$33,699**
300ZX	2+2	3dr.cpe.2+2	75	12	0.32	101.2	178.0x71.0x48.1	3430	NR	i/i	dc/dc/ABS	pwr.r/p.	35.4	2.7	19	225/50VR16	V6/3.0/M5	**$36,489**
300ZX	Turbo	3dr.cpe.2	70	10	0.32	96.5	169.5x70.5x48.4	3532	NR	i/i	dc/dc/ABS	pwr.r/p.	34.1	2.4	19	225/50VR16 245/45ZR16 (re)	V6T/3.0/M5	**$40,099**
300ZX	convertible	2dr. con.2	70	5.8	0.35	96.5	169.5x70.5x49.5	3455	NR	i/i	dc/dc/ABS	pwr.r/p.	34.1	2.7	18.7	225/50VR16	V6/3.0/M5	**$40,879**

See page 393 for complete 1995 Price List

Cult object...

The 911 remains the quintessential, Porsche and it isn't for nothing that the firm hangs onto it like a life preserver on the high seas. For more than 30 years it evolved, blow by blow, to maintain itself among the *Who's Who* of sports cars. But in spite of its technicality, its reputation is overdone.

Those who are waiting for a revolution have been disappointed because the last 911 strongly resembles the preceding one. Only cosmetic touches and improvements have come about to try maintaining the technical value of these models at the height of their reputation. The Carrera in 2- and 4-wheel drives is available in coupe and convertible and as of this writing, the future of the RS America and the Turbo seems uncertain. They are equipped with the same flat opposed 3.6 L six cylinders and the choice of 2- or 4-wheel drive stick shifts. Tiptronic 4-speed automatic is available only on the Carrera.

STRONG POINTS

• **Performance: 95%**
The true Porsche fans search out performance before all and are not disappointed because the flat 6 is as powerful as it is civilized, which allows daily use. In spite of a fairly long response time, the Turbo provides the most impressive sensations. However, you have to be a skilled pilot to get the most out of the engine and still remain on the road. The manual transmission ratios are perfectly staged and the shift lever control is very precise, while the Tiptronic is very disconcerting by its smoothness and its efficiency, in spite of some gaps between the ratios.

• **Safety: 90%**
The structural rigidity has been further improved and the doors extensively modified to offer optimal lateral protection. Also, the presence of 2 standard air bags allows these models to reach the maximum rate.

• **Technical: 90%**
The Porsche is of unit body design, built of galvanized steel, and its originality stems from the location of its power train, cantilevered behind the rear wheels. The suspension has been modified, and the old axles have given way to a multiple-arm rear suspension that improved lateral accelerations to 1.0 G and has compelled the engineers to recalculate all of the suspension settings. On the Carrera 4, power goes to the front drivetrain by a shaft turning in a tube. Power is split between the drivetrains in a ratio of 61/39 percent, and each one of the two carriers has a limited slip differential. The Tiptronic automatic transmission consists of a ZF box controlled by sophisticated module which decides on gear selection as a function of the need of the moment. On one hand it can work as a conventional automatic and in another it can shift without working the clutch. In this mode the last 3 gear ratios are blocked between 1,200 and 2,000 rpm to avoid any slip. The aerodynamic finesse of their lines is good and based on the size of their tires and front openings, the drive coefficient varies from 0.31 to 0.37.

• **Handling: 85%**
The Carrera has lost nothing ofits understeering temperament, which switches easily when you work the accelerator or the brakes. On a wet road, considerable prudence is a good bet, especially with the Turbo.

DATA

Category: high performance rear-wheel drive sports coupes.
Class: GT

HISTORIC

Introduced in: 1963: 901; 1964: 911.
Modified in: 1973: Targa; 1976: Turbo; 1982: Convertible; 1992: AWD 1988: reviewed 1989: Tiptronic.
Made in: Zuffenhausen, Stuttgart, Germany.

RATINGS

Safety: 90 %
Satisfaction: 80 %
Depreciation: 50%
Insurance: 3.7 % ($1,860-$3,425)
Cost per mile: $1.42

NUMBER OF DEALERS

In U.S.A.: 210

SALES ON THE AMERICAN MARKET

Model	1992	1993	Result
911	2,873	2,412	- 16.1 %

MAIN COMPETITORS

ACURA NS-X, DODGE Stealth R/T Turbo, MAZDA RX-7, MERCEDES 300SL, NISSAN 300ZX, TOYOTA Supra.

EQUIPMENT

PORSCHE 911	Carrera	Carrera 4	Turbo
Automatic transmission:	O	-	-
Cruise control:	S	S	S
Power steering:	S	S	S
ABS brake:	S	S	S
Air conditioning:	S	S	S
Air bag:	S	S	S
Leather trim:	S	S	S
AM/FM/ Radio-cassette:	S	S	S
Power door locks:	S	S	S
Power windows:	S	S	S
Tilt steering column:	S	S	S
Dual power mirrors:	S	S	S
Intermittent wipers:	S	S	S
Light alloy wheels:	S	S	S
Sunroof:	S	S	S
Anti-theft system:	S	S	S

S : standard; O : optional; - : not available

COLORS AVAILABLE

Exterior: Red, White, Silver, Black, Green, Blue, Gray, Turquoise.
Interior: Gray, Red, Blue, Tan.

MAINTENANCE

First revision: 15,000 miles
Frequency: 15,000 miles
Diagnostic plug: Yes

WHAT'S NEW IN 1995 ?

- **Cosmetic body changes.**
- **More powerful base engine.**
- **New 6-speed manual transmission.**
- **Roomier baggage trunk.**
- **Automatic brake differential optional on a Carrera and standard on a Carrera 4.**
- **New double exhaust system.**
- **Sub-frame suspension and brakes.**

ENGINE						TRANSMISSION		PERFORMANCE										
Model/ versions *: standard	Type / timing valve / fuel system	Displacement cu/in	Power bhp @ rmn	Torque lb.ft @ rpm	Compres. ratio	Driving wheels / transmission	Final ratio	Acceler. 0-62 mph s	Stand. 1/4 mile s	Stand. 5/8 mile s	Braking 62-0 mph ft	Top speed mph	Lateral acceler. G	Noise level dBA	Fuel economy mpg city	Fuel economy mpg hwy	Gasoline type / octane
Carrera	H6* 3.6 SOHC-12-EFI	220	272 @ 6100	244 @ 5000	11.3 :1	rear-M6*	3.44	5.9	13.8	25.2	125	165	0.86	70	20.0	32.0	S 91
Tiptronic						rear-A4	3.56	6.8	14.3	26.0	131	162	0.85	70	20.0	32.0	S 91
Carrera 4	H6* 3.6 SOHC-12-EFI	220	272 @ 6100	244 @ 4800	11.3 :1	all-M6*	3.44	5.8	14.0	25.8	128	162	0.83	70	20.0	30.0	S 91
Turbo	H6* 3.6T SOHC-12-EFI	220	355 @ 5500	384 @ 4200	7.5 :1	rear-M5*	3.44	5.0	13.2	23.5	141	168	0.90	72	16.0	28.0	S 91

More antiseptic, the Carrera 4 pushes the limits further because its power transmission to the ground is hard to fault, no matter what the road condition. It is, however, heavier and less agile.

• **Quality/fit/finish:** 80%
Constructed in a craftsman like way, which allows optimum quality control, and using first-class materials, the Porsche has an exemplary presence but a style denuded of all fantasy.

• **Insurance:** 80%
The rate is one of the lowest on the market, but with prices as flamboyant, it is better to insure your investment only for the good season.

• **Seats:** 80%
Their wraparound shape provides lateral and lumbar support with considerable efficiency, but the padding is not soft.

• **Satisfaction:** 80%
Reliability is satisfactory but maintenance and parts are ruinous.

• **Driver's compartment:** 80%
The driver's position has been improved by an electrical height adjuster, but the pedal assembly was inherited from the Beetle, and most of the controls are confusing and require getting used to. In fact, they need serious revision.

• **Steering:** 70%
Precise and fast, it is still vague in the center and sensitive to road condition which it generously telegraphs. Finally, its maneuverability does not correspond to its format.

• **Braking:** 60%
While at the level of the performance, the length of the stops has nothing impressive, but the control is precise, the balance perfect and the fade-resistance exceptional.

• **Access:** 60%
The 911 offers only 2 seats for normal people and the rear seats are unusable, even children would find them claustrophobic.

• **Suspension:** 50%
The new Carrera suspension is more tolerable in terms of comfort and wheel travel is a little more generous.

• **Depreciation:** 50%
Porsches begin to lose a little more of their value, probably because of persistent rumors about the company's poor position.

WEAK POINTS

• **Trunk:** 10%
The new front-end design has increased it somewhat, but it doesn't contain much because of its tortured forms. However, lowering the rear seat bench forms a platform for holding any excess. Also, it is possible to have a 911 delivered without this bench seat as an option.

• **Interior space:** 20%
The seats suffer from the lack of width between the doors and are also not long enough.

• **Fuel consumption:** 25%
Considering the level of the engine and its power, it is no longer as economical as one would have you believe.

• **Noise level:** 25%
It spite of being encapsulated, the engine exults with each acceleration or takeoff, and the thundering of the tires is more melodious than that of the exhaust.

• **Price/equipment:** 30%
It is complete, not a fantastic achievement considering the price of the bill. For the total price requested there are other choices.

• **Conveniences:** 40%
Storage is limited to a small glove box and door pockets. A chauffeur-driven Mercedes will deliver the rest of your needs.

CONCLUSION

• **Overall rating:** 60.0%
We have more and more problems finding something interesting in these cars, frozen in time, which became the object of a cult that has lost many disciples.

SPECIFICATIONS & PRICES

Model	Version/Trim	Body/ Seats	Interior volume cu.ft.	Trunk volume cu ft.	Drag coef. Cd	Wheel-base in	Lgth x Wdthx Hght. inx inx in	Curb weight lb	Towing capacity std. lb	Susp. type ft/rr	Brake type ft/rr	Steering system	Turning diameter ft	Steer. turns nber	Fuel. tank gal.	Standard tire size	Standard Power Train	PRICES US. $ 1994
PORSCHE		**Warranty: 2 years / unlimited; surface corrosion: 3 years / unlimited; perforation: 10 years / unlimited; antipollution: 5 years / 50,000 miles.**																
911 Carrera		2dr.cpe.2+2	55	5.0	0.33	89.4	167.7x68.3x51.8	3064	NR	i/i	dc/dc/ABS	pwr.r/p.	38.5	2.47	19.4	205/55ZR16º	H6/3.6/M6	**$59,900**
911 Carrera		2dr.con.2+2	55	5.0	0.33	89.4	167.7x68.3x51.8	3119	NR	i/i	dc/dc/ABS	pwr.r/p.	38.5	2.47	19.4	245/45ZR16+	H6/3.6/M5	**$64,990**
911 Carrera 4		2dr.cpe.2+2	55	5.0	0.33	89.4	167.7x68.3x51.8	3175	NR	i/i	dc/dc/ABS	pwr.r/p.	38.5	2.47	19.4	205/50ZR17º	H6/3.6/M5	**$78,450**
911 Carrera 4		2dr.con.2+2	55	5.0	0.33	89.4	167.7x68.3x51.8	3175	NR	i/i	dc/dc/ABS	pwr.r/p.	38.5	2.47	19.4	255/40ZR17+	H6/3.6/M5	-
911 RS America		2dr.cpe.2+2	55	5.0	0.34	89.4	168.3x65.0x51.6	2954	NR	i/i	dc/dc/ABS	pwr.r/p.	41.0	3.02	20.3	255/40ZR17	H6/3.6/M5	**$54,800**
911 Turbo 3.6		2dr.cpe.2+2	55	5.0	0.35	89.4	167.7x69.9x51.7	3274	NR	i/i	dc/dc/ABS	pwr.r/p.	37.7	2.81	20.3	265/35ZR18	H6T/3.6/M5	**$99,000**

º: front +: rear

See page 393 for complete 1995 Price List

Whoever doesn't lead, retreats...

On paper, the 928 is the most international and the most logical of Porsches because its attributes place it decisively in the clan of the highest caliber GT's. However, its lines and engineering have not changed in a long time, and the power gain doesn't mean anything more these days.

Created for the American market, the 928 has known a certain success elsewhere in the world because its character is more universal than that of the 911 and a progressive increase of its power and performance has brought it to be considered as exotic. Its lines and aerodynamic efficiency were frozen for several years and are beginning to get seriously dated. The GTS is offered in the form of a 3-door 2-volume coupe 5.4 L V8 with a choice of manual or automatic.

STRONG POINTS

• **Performance:** 90%

With a weight-to-power ratio of 10.36 lbs/hp the 928 has the wherewithal to surprise the least exotic. With good torque and power of 350 hp its V8 is as civilized for domestic usage as it is aggressive on the freeways, where its takeoffs and accelerations are very powerful. As a comparison, 350 horse power is readily available from a small block Chevy with a modest rebuilt.

• **Safety:** 90%

Seriously stiffened up, this structure satisfies and surpasses even the 1997 American standards. It resists impacts and the protection of the occupants is ensured by 2 air bags. The 928 is one of the rare models fitted out with a device that permanently measures tire pressure.

• **Technical :** 90%

The layout of the 928 is classic with a front mounted engine and a transaxle at the rear. This gives an even balance front to rear and the connection between the 2 elements is affected by a shaft that turns inside a rigid tube (as on the unlamented 61 Pontiac Tempest with a "rope" drive shaft). The galvanized steel unit body supports a line that has not changed in 16 years but it remains fairly efficient from an aerodynamic point of view since its coefficient is at 0.35. It is also helped by a polyurethane spoiler which creates a down force on the rear wheels. Its finesse insures good stability in a car capable of reaching 171 mph. To lighten the assembly, the hood and doors are aluminum. There is four-wheel independent suspension and disc brakes controlled by a Bosch ABS, plus added traction from a limited slip differential.

• **Quality/fit/finish:** 90%

The general presentation is quiet and comfortable, the assembly is sturdy, the finish is carefully thought out, down to the least details, and the quality of the material used corresponds to the size of the billing. However, interior decoration often involves colors and designs that are, to say the least, not used in a car of this class. One also notes the poor body protection, very vulnerable in urban use, where some car owners sometimes lack civility or even civilization.

• **Handling:** 90%

The 928 impresses by its stability, the precision of its guidance and the self-assurance of its reactions regardless of road condition. However its imposing bulk limits maneuverability on a sinuous road course. The purist will complain of the lack of sports driving character derived from its high weight and the suspension obvious orientation toward comfort rather than performance.

DATA

Category: high performance rear-wheel drive sport coupes.
Class: GT

HISTORIC

Introduced in: 1977
Modified in: 1985: V8 5.0L 32v 219 hp;1985: 234 hp;1986: 310 hp 1989: 330 hp, 1994: 345 hp; 1995: 350 hp.
Made in: Zuffenhausen, Stuttgart, Germany.

RATINGS

Safety: 90 %
Satisfaction: 82 %
Depreciation: 34 % (GTS: 1 year)
Insurance: 3.5 % ($1,860)
Cost per mile: $1. 43

NUMBER OF DEALERS

In U.S.A.: 210

SALES ON THE AMERICAN MARKET

Model	1992	1993	Result
928	181	119	- 34.3 %

MAIN COMPETITORS

BMW 850i, CHEVROLET Corvette, LEXUS SC400, MERCEDES 500SL.

EQUIPMENT

PORSCHE 928	GTS
Automatic transmission:	No charge
Cruise control:	S
Power steering:	S
ABS brake:	S
Air conditioning:	S
Air bag:	S
Leather trim:	S
AM/FM/ Radio-cassette:	S
Power door locks:	S
Power windows:	S
Tilt steering column:	S
Dual power mirrors:	S
Intermittent wipers:	S
Light alloy wheels:	S
Sunroof:	S
Anti-theft system:	S

S : standard; O : optional; - : not available

COLORS AVAILABLE

Exterior: Red, White, Silver, Black, Green, Blue, Gray, Turquoise.
Interior: Gray, Red, Blue, Tan.

MAINTENANCE

First revision: 2,500 miles
Frequency: 6 months/ 7,500 miles
Diagnostic plug: Yes

WHAT'S NEW IN 1995 ?

- The engine is a little more powerful.

	ENGINE					TRANSMISSION		PERFORMANCE										
Model/ versions *: standard	Type / timing valve / fuel system	Displacement cu/in	Power bhp @ rmn	Torque lb.ft @ rpm	Compres. ratio	Driving wheels / transmission	Final ratio	Acceler. 0-62 mph s	Stand. 1/4 mile s	Stand. 5/8 mile s	Braking 62-0 mph ft	Top speed mph	Lateral acceler. G	Noise level dBA	Fuel economy mpg city	hwy	Gasoline type / octane
928 GTS	V8* 5.4 DOHC-32-EFI	329	350 @ 5700	370 @ 4250	10.4 :1	rear - M5*	2.73	5.9	13.9	25.6	128	171	0.88	66	17.0	29.0	S 91
						rear - A4*	2.54	6.0	14.5	26.0	125	162	0.86	65	18.0	25.0	S 91

• Driver's compartment: 80%
The adjustment of the seat and of the steering wheel and gauge panel will satisfy the most demanding pilots, and everything is organized to make driving efficient and safe. Still, the fact that you are sitting low limits the visibility, especially with a tall body belt line and many blind spots.
• Seats: 80%
Up front they provide everything that could be wished for in terms of lateral and lumbar support as well as padding. Rear occupants must be content with Spartan shells wedged in by the tunnel.
• Steering: 80%
Even though the ratio is a little slow and not sufficiently fast for a car of this type, it is smooth and precise and its modulation is well-balanced.
• Satisfaction: 80%
These cars are more reliable than others of their type, but the guaranty is less generous overall than models that cost distinctly less. The owners complain of the cost of maintenance and parts.
• Noise level: 80%
At cruise speed, the inside feel is velvety, the drivetrain is discreet and the soundproofing effective although it allows you to hear wind and rolling noises.
• Insurance: 80%
Even though it represents a round sum, its percentage is reasonable in relation to the capital insured.
• Access: 60%
It's more delicate to enter at the rear, where space is limited, than at the front, where the doors open wide.
• Suspension: 60%
It is rough when the road is not perfect, but sedan-like when cruising at constant speed on the freeway.
• Braking: 50%
Taking into account the large mass, the four ventilated disks and an improved ABS, the stops are easily modulated, balanced and relatively short with remarkable fade resistance.
• Depreciation: 50%
It is stronger than on a 911 because the 928 doesn't have the same reputation, After several years it is not hard to find one at a ridiculous price, proof that it lacks respect.

WEAK POINTS

• Fuel consumption: 10%
The 928 contents itself with 19 miles per gallon in normal use, but can easily pass this rate when you want to race the police cars.
• Interior space: 20%
Compared to other GTs the interior of the 928 looks vast, with unusual length and width. Two adults will be at ease up front and 2 young children at the back, where space is limited by lack of roof height and by the center tunnel. This in turn sets the tone for the seat shape.
• Trunk: 30%
It lacks height and you need to block up the rear seat to carry excess baggage.
• Conveniences: 40%
Since the implantation of air bags, the glove box has been replaced by a deep well in the center console and the door pockets hold little.
• Price/equipment: 40%
This Porsche addresses itself to clients more involved with social status and comfort than pure performance. While its price compares advantageously to that of Italian divas, its entertainment fees are just as costly.

CONCLUSION

• Overall rating: 64.5%
The 928 remains, in ours eyes, the most realistic Porsche because it has nothing mythical except the name. However, when a car begins to have more success used than new, it is a sign that its future is behind it. ☺

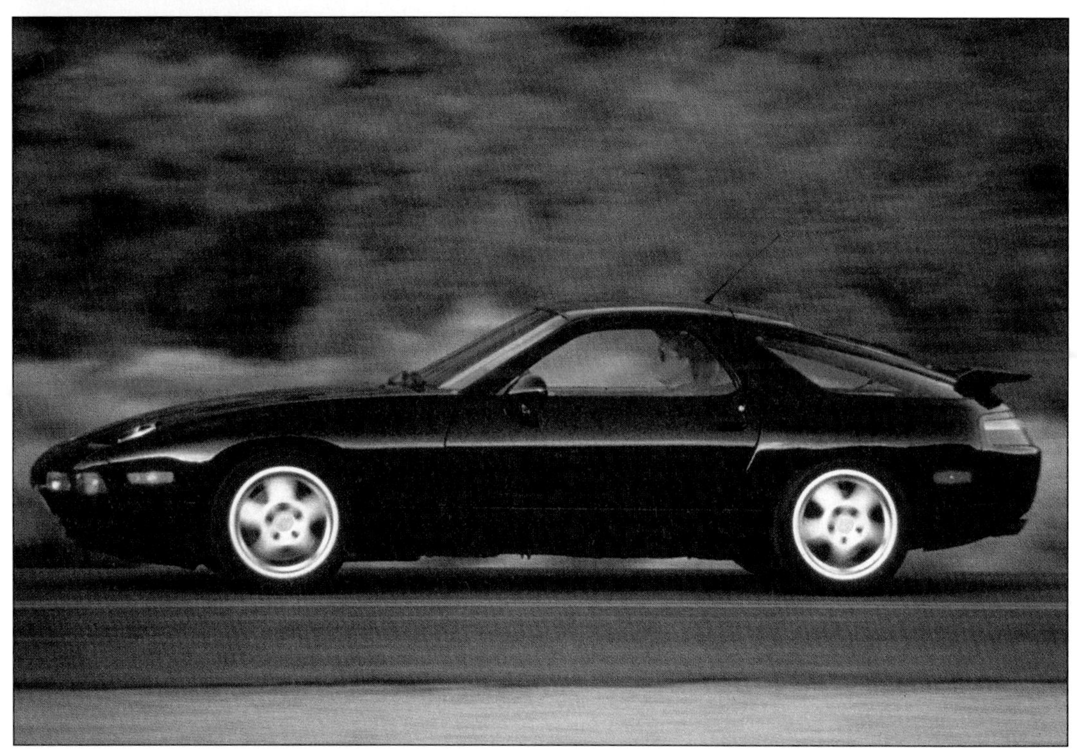

SPECIFICATIONS & PRICES

Model	Version/Trim	Body/ Seats	Interior volume cu.ft.	Trunk volume cu ft.	Drag coef. Cd	Wheel-base in	Lgth x Wdthx Hght. inx inx in	Curb weight lb	Towing capacity std. lb	Susp. type ft/rr	Brake type ft/rr	Steering system	Turning diameter ft	Steer. turns nber	Fuel. tank gal.	Standard tire size	Standard Power Train	PRICES US. $ 1994
PORSCHE		**Warranty: 2 years / unlimited; surface corrosion: 3 years / unlimited; perforation: 10 years / unlimited; antipollution: 5 years / 50,000 miles.**																
928	GTS man.	3dr.cpe.2+2	74	8.0	0.35	98.4	178.0x74.4x50.5	3593	NR	i/i	dc/dc/ABS	pwr.r/p.	38.4	3.0	22.8	225/45ZR17º 255/40ZR17+	V8/5.4/M5	**"$82,260**
928	GTS auto.	3dr.cpe.2+2	74	8.0	0.35	98.4	178.0x74.4x50.5	3638	NR	i/i	dc/dc/ABS	pwr.r/p.	38.4	3.0	22.8	225/45ZR17º 255/40ZR17+	V8/5.4/A4	**$82,260**

º: front +: rear

See page 393 for complete 1995 Price List

For masochists only...

The 968 constitutes the ultimate achievement of the 924, which was never considered a real Porsche by the fanatics of the make. It is mostly by its style and by its name, more so by action or driving pleasure, that it seduced buyers wanting to prove themselves.

After serving as the model for some Japanese competitors, the 968 is a 944 on which the ends were redesigned in the manner of a 928. Today it has been passed, especially when you consider the technical refinement and comfort of its competitors. It is sold in the form of a 3-door coupe or a 2-door convertible equipped with four cylinders with a 3.0 L. Standard transmission is a 6-speed manual and the Tiptronic is an extra-cost option.

STRONG POINTS

• Safety: 90%
The presence of 2 air bags and the stiffening of the structure allow it to offer maximum occupant protection.

• Technical: 90%
The changes that affected the 968 have been purely cosmetic because the main dimensions are identical to the 924 to the nearest inch, except for the weight which has taken on a few pounds. The galvanized steel body is equipped with an independent suspension and disc brakes with ABS on 4 wheels, while the transmission combined with the differential is installed between the rear wheels, as on a 928, to balance out the weight between the front and rear drivetrains. Also, the drive shaft need not carry added torque in lower gears, with the trans in the rear. Freshening up the lines improved aerodynamic efficiency because the drag coefficient has now reached 0.32. Its engine is the largest 4 cylinder on the market equipped with a Variocam: variable intact timing head to increase power and torque by varying valve timing through a Motronic module by changing the timing chain spacing between the cams. Two balanced shafts located on both sides of the crank cancel the vibrations normal for a four cylinder.

• Quality/fit/finish: 80%
The assembly is homogeneous, the fit and quality of the materials good, but the interior styling lacks warmth.

• Satisfaction: 80%
The Japanese charge less for service that is less frequent, less costly and less independent. Nothing quite matches the "gemutlich" greeting of the hauptman at the service desk.

• Performance: 80%
They impress less, in spite of an excellent weight-to-power ratio, than some Japanese that do distinctly better. In use, the engine is not very pleasant because it lacks low-end torque and stalls frequently when an inexperienced driver fails to give it enough rpm. It has, however, the advantage of offering more peak torque than its V6 competitors at lower rpm. The shift selection is not ideal because it is easy to confuse 1st and reverse, a problem uncured since early 924's. Also, the very hard clutch pedal makes you want to avoid urban driving and the 6th (Autobahn) gear has absolutely no usefulness in North America.

• Insurance: 80%
Its rate is reasonable, but the premium is as overvalued as the price of the model on which it is indexed.

• Handling: 80%
You need much time in a 968 to get well accustomed to it and wring

DATA

Category: rear-wheel drive sports coupes.
Class: GT

HISTORIC

Introduced in: 1975: 924; 1981: 944; 1992: 968.
Modified in: 1986: S & 2.7L; 1990: S2 & 3.0L.1992: body.
Made in: Zuffenhausen, Stuttgart, Germany.

RATINGS

Safety: 90 %
Satisfaction: 83 %
Depreciation: 58%
Insurance: 4.0 % ($1,560)
Cost per mile: $0.82

NUMBER OF DEALERS

In U.S.A.: 210

SALES ON THE AMERICAN MARKET

Model	1992	1993	Result
968	1,242	1,197	- 3.7 %

MAIN COMPETITORS

DODGE Stealth R/T Turbo, MAZDA RX-7, NISSAN 300 ZX, TOYOTA Supra.

EQUIPMENT

PORSCHE 968	coupe	convertible
Automatic transmission:	No charge	No charge
Cruise control:	S	S
Power steering:	S	S
ABS brake:	S	S
Air conditioning:	S	S
Air bag:	S	S
Leather trim:	O	O
AM/FM/ Radio-cassette:	S	S
Power door locks:	S	S
Power windows:	S	S
Tilt steering column:	S	S
Dual power mirrors:	S	S
Intermittent wipers:	S	S
Light alloy wheels:	S	S
Sunroof:	S	S
Anti-theft system:	S	S

S : standard; O : optional; - : not available

COLORS AVAILABLE

Exterior: Red, White, Silver, Black, Green, Blue, Gray, Turquoise.
Interior: Gray, Red, Blue, Tan.

MAINTENANCE

First revision: 2,500 miles
Frequency: 6 months/7,500 miles
Diagnostic plug: Yes

WHAT'S NEW IN 1995 ?

- Slightly more powerful engine (+5 hp).

Model/ versions *: standard	ENGINE Type / timing valve / fuel system	Displacement cu/in	Power bhp @ rmn	Torque lb.ft @ rpm	Compres. ratio	TRANSMISSION Driving wheels / transmission	Final ratio	PERFORMANCE Acceler. 0-62 mph s	Stand. 1/4 mile s	Stand. 5/8 mile s	Braking 62-0 mph ft	Top speed mph	Lateral acceler. G	Noise level dBA	Fuel economy mpg city	Fuel economy hwy	Gasoline type / octane
base	L4* 3.0 DOHC-16-DEFI	182	240 @ 6200	225 @ 4100	11.0 :1	rear-M6*	3.78	6.8	14.7	26.7	128	152	0.90	71	20.0	34.0	S 91
						rear-A4	3.25	8.2	15.0	27.2	134	149	0.90	70	18.0	32.0	S 91

the most out of it, even on a dry road where quality tires and suspension allow you to reach a high G force.

• Seats: 70%
Real bucket seats, they maintain well in every sense, but their padding is too hard and distance traveling is tiring.

• Steering: 70%
It is precise and its power assist firm to the point of transmitting all road defects. Strangely, maneuverability is only average because in spite of a short turning diameter, its speed is inadequate.

• Conveniences: 60%
A glove box of average size, door pockets and a small locker in the center console constitute the main storage spaces.

• Driver's compartment: 60%
Still not adjustable, the steering column does not provide a comfortable driving position, especially for tall drivers whose heads are so close to the top of the windshield that they must drive almost fully reclined. The bucket seat is not the problem because it remains perfect in all positions. Gauges are well laid out, but the switches seem to have been sown in a random pattern. The firmness of the shift selector and of the clutch pedal, as well as the low power steer assist, testify to a lapsed philosophy which cuts into driving pleasure. Visibility toward 3/4 rear is hampered by the roof shape and at the back by the narrowness of the rear window and the presence of a spoiler.

• Braking: 50%
Powerful, well-balanced, and easily modulated, it excels by its fade resistance rather than by its stopping distances, which offer nothing exceptional with an ABS.

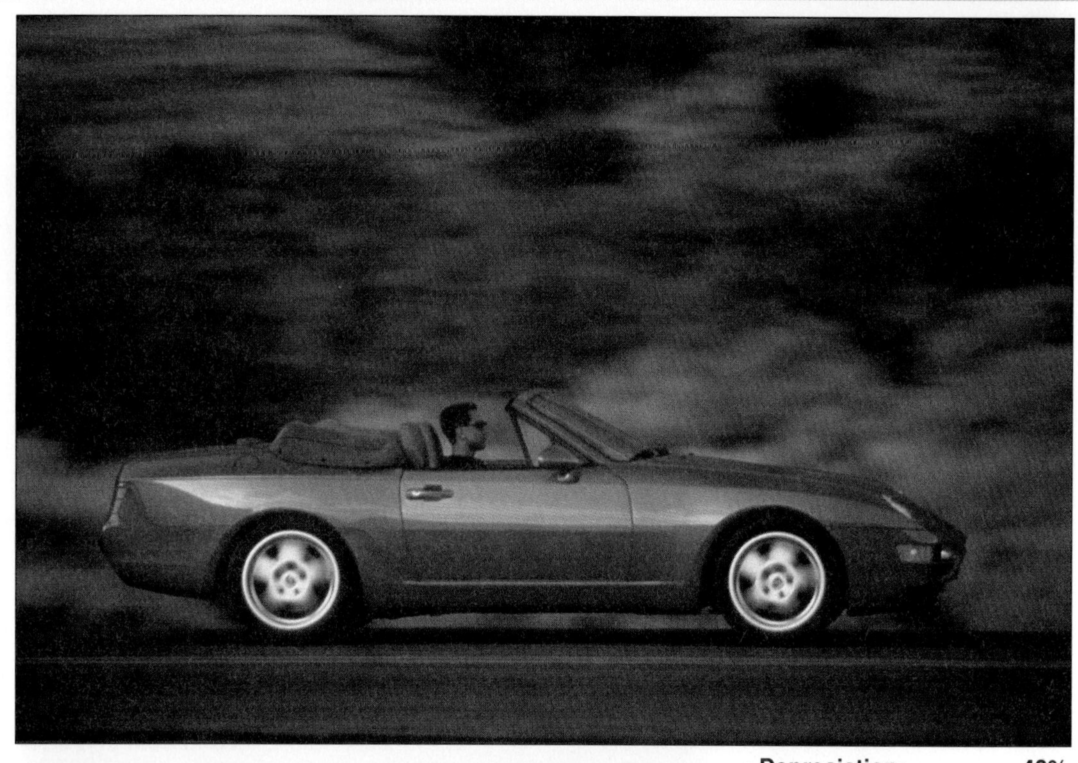

WEAK POINTS

• Noise level: 10%
In spite of serious soundproofing work it remains one of the highest we have registered on a car of this price.

• Interior space: 10%
The 968 must be considered a 2-seater because the very limited available space at the rear is more useful for baggage than to potential passengers. Narrowness of the interior will give large-body persons a disagreeable claustrophobic fear.

• Suspension: 30%
It is frankly uncomfortable on small roads in poor conditions, and it is better to pick freeways to arrive in good shape.

• Fuel consumption: 30%
It is often higher than that of V6's of equivalent displacement because it can drop to 19 mpg as soon as you want to play.

• Trunk: 40%
Even though large, it lacks height when you need to lower the rear seat back to increase of reasonable space.

• Access: 40%
It is complicated by the very enveloping wraparound bucket seats, the angle of the windshield and the low roof height. It is better to put on your helmet before boarding, just in case.

• Depreciation: 40%
A Porsche which loses as much as a vulgar automobile constitutes a statement of disapproval which this builder does not seem to sense, even when reading sales numbers.

• Price/equipment: 40%
The 968 has problems justifying its price/performance ratio because its Japanese competitors are light years ahead of it.

CONCLUSION

• Overall rating: 56.5%
The business of the Volkswagen/Audi/Porsche group is not scintillating because Dr. Piech obstinately directs it as he did 15 years ago. Today, builders listen to their buyers, but at Porsche there are fewer and fewer buyers to consult with. :-|

SPECIFICATIONS & PRICES

Model	Version/Trim	Body/ Seats	Interior volume cu.ft.	Trunk volume cu ft.	Drag coef. Cd	Wheel-base in	Lgth x Wdthx Hght. inx inx in	Curb weight lb	Towing capacity std. lb	Susp. type ft/rr	Brake type ft/rr	Steering system	Turning diameter ft	Steer. turns nber	Fuel. tank gal.	Standard tire size	Standard Power Train	PRICES US. $ 1994
PORSCHE		**Warranty: 2 years / unlimited; surface corrosion: 3 years / unlimited; perforation: 10 years / unlimited; antipollution: 5 years / 50,000 miles**																
968	coupe man.	3dr.cpe.2+2	63	12	0.34	94.5	170.9x68.3x50.2	3086	NR	i/i	dc/dcABS	pwr.r/p.	35.3	3.24	19.5	205/55ZR16º	L4/3.0/M6	**$39,950**
968	coupe autom.	3dr.cpe.2+2	63	12	0.34	94.5	170.9x68.3x50.2	3153	NR	i/i	dc/dcABS	pwr.r/p.	35.3	3.24	19.5	225/50ZR16+	L4/3.0/A4	**$43,100**
968	con. man.	2dr.con.2	-	-	0.36	94.5	170.9x68.3x50.2	3241	NR	i/i	dc/dcABS	pwr.r/p.	35.3	3.24	19.5	205/55ZR16º	L4/3.0/M6	**$51,900**
968	con. autom.	2dr.con.2	-	-	0.36	94.5	170.9x68.3x50.2	3306	NR	i/i	dc/dcABS	pwr.r/p.	35.3	3.24	19.5	225/50ZR16+	L4/3.0/A4	**$55,050**

º: front +: rear

See page 393 for complete 1995 Price List

Masterpieces in peril...

Bets are off in the car industry of her gracious Majesty's kingdom. After the Japanese, here are the Germans who succeeded in an English landing where they failed before. Today the Land Rover has BMW engines and Rolls-Royce, the supreme symbol of the monarchy is as the latter, besieged.

Bentleys and Rolls-Royce constitute the sum total of automobiliana. Not as much from the technical side where Mercedes hold the high road as from the luxury point of view. If the Rolls are the cars of kings of all sorts, the Bentley is generally considered the choice of people with taste. They are offered in convertibles, sedans with standard wheelbase or extended, and as limousines. They are all equipped with the new naturally aspirated 6.7 V8 or turbo charged, depending on the model, with a 4-speed automatic.

STRONG POINTS

• Noise level: 100%
Noise insulation is the most finely developed in the industry because on the road it stays at 62 dba while the average is 63. Keep in mind that there is a 10 times factor from one dba to the next. However, the weight of soundproofing material is "Top Secret".

• Quality/fit/finish: 100%
Reference supreme in the matter, these cars are built to last as long as the fortune of their owner. Fine leather, appliqué of walnut burl, and wool rugs decorate them like a high-society drawing room, in good taste and refinement.

• Safety: 90%
The mass and rigidity of the structure of these enormous cars play a major role in the protection of their occupants in case of collision or assassination attempts, and two air bags protect the front seat occupants. Bullet proofing is optional.

• Satisfaction: 90%
At this price level, the clients must be satisfied but registered in the past are several recall campaigns, confirming that nothing is perfect, not even a Rolls.

• Access: 90%
Climbing into your car loses all significance with these models whose

DATA

Category: rear-wheel drive luxury sedans, coupes and convertibles.
Class: 7

HISTORIC

Introduced in: 1980: Spirit, Spur, Brooklands; 1984:Limo. 1985: Continental-Corniche & Turbo R; 1995: Flying Spur.
Modified in: 1985: turbo engine.
Made in: Crewe: Spirit, Spur, Brooklands; at Muliner-Park-Ward, Continental & Corniche, and at Hooper, Limousine in London, England

RATINGS

Safety:	90 %
Satisfaction:	90 %
Depreciation:	50 %
Insurance:	3.2 % ($4,770)
Cost per mile:	$2.85

NUMBER OF DEALERS

In U.S.A.: NA

SALES ON THE AMERICAN MARKET

Model	1992	1993	Result
Bentley	Confidential		
Rolls-Royce	Confidential		

MAIN COMPETITORS

BMW 750iL, MERCEDES BENZ 600.

EQUIPMENT

Bentley/Rolls-Royce	all models
Automatic transmission:	S
Cruise control:	S
Power steering:	S
ABS brake:	S
Air conditioning:	S
Air bag:	S
Leather trim:	S
AM/FM/ Radio-cassette:	S
Power door locks:	S
Power windows:	S
Tilt steering column:	S
Dual power mirrors:	S
Intermittent wipers:	S
Light alloy wheels:	S
Sunroof:	O
Anti-theft system:	S

S : standard; O : optional; - : not available

COLORS AVAILABLE

Exterior: White, Blue, Silver, Red, Beige.
Interior: Tan, Black, White, Gray, Blue,Cream.

MAINTENANCE

First revision: NA
Frequency: NA
Diagnostic plug: Yes

WHAT'S NEW IN 1995 ?

- New: 6.7 L engine, the seats, the headlights and the air bags.
- The turbo engine of the Bentley is now installed in the Rolls inaugurating the name Flying Spur.

Model/ versions *: standard	ENGINE Type / timing valve / fuel system	Displacement cu/in	Power bhp @ rmn	Torque lb.ft @ rpm	Compres. ratio	TRANSMISSION Driving wheels / transmission	Final ratio	PERFORMANCE Acceler. 0-62 mph s	Stand. 1/4 mile s	Stand. 5/8 mile s	Braking 62-0 mph ft	Top speed mph	Lateral acceler. G	Noise level dBA	Fuel economy mpg city	hwy	Gasoline type / octane
1)	V8*6.7 OHV-16-EFI	412	245 @ 4000	370 @ 2000	8.0: 1	rear-A4	3.07	9.8	17.2	31.5	190	131	0.78	63	12.0	19.0	S 91
2)	V8T*6.7 OHV-16-EFI	412	360 @ 4200	554 @ 2000	8.1: 1	rear-A4	2.69	7.5	16.0	28.8	180	149	0.80	63	12.0	19.0	S 91

1) Brooklands, Continental Convertible, Silver Spirit III, Silver Spur III, Silver Spur III Limo., Corniche.
2) Turbo R, Continental R, Flying Spur.

ROLLS ROYCE Silver Spur III

BENTLEY Turbo R

doors are generously sized. The chauffeur has doffed his hat and is holding the door.

• Seats: 90%
They would not be out-of-place in a refined dining room, but again, this is a most unusual old chap. They are better shaped at the rear than at the front,there the cushion doesn't sufficiently support the chauffeur's hips (too bad for those who drive themselves).

• Suspension: 90%
It is of velvet in most instances and bad pavement is hardly perceived because it is usually approached at parade speed. Comfort constitutes the true great advantage of these cars.

• Conveniences: 90%
They are both numerous and practical. Glove boxes, door pockets, center console storage, rear armrests storage, tablet-desk at the back of the front seats, mirrors, hand holds, all are there.

• Technical: 80%
It shines more by the complexity of its servo-controls and its electrical circuits than by its revolutionary or avant-garde concepts. The latest engine is really the ultimate version of what one knew for a long time, the transmission comes from GM, and the front load leveler is a Citroen patent. The unit body is made up of galvanized steel and zinc-treated panels, while the doors, hood and trunk lids are of anodized aluminum alloy. Finally, the assembly receives a dozen coats of primer and paint. The suspension is independent and the 4-wheel disc brakes are controlled by a Bosch ABS.

• Trunk: 80%
Vast and well-padded, it has simple shapes to allow locating a collection of Louis Vulton suitcases and, if that is insufficient, tell the chauffeur to follow you in the other Rolls-Royce.

• Performance: 80%
It is surprising to see how fast these giant cars can really be. To move a mass of 2.5 tons from 0 to 62 mph in less than 10 sec with a naturally aspirated engine or in 7.5 l with a turbo is surprising, but most of the time we have driven in perfectly moderate and dignified fashion.

• Steering: 80%
There too, surprise, because in spite of a power assist that is a little too strong, the precision and speed are those of trimmer models, and their maneuverability is very honorable considering the dimensions.

• Interior space: 70%
Even though it is not proportional to the outside bulk, four people use up most space while a fifth will not be fully comfortable in the center rear, which is most shocking.

• Driver's compartment: 70%
It is more spectacular than ergonomic because most controls are far from the driver's hand, the steering wheel rim is thin, the pedals are minuscule and the varnished wood glares back a lot.

• Handling: 70%
It matches the respective performance of these models. It is not really sports-like with a turbo engine, but it surprises by its audacity in some risky maneuvers and sometimes to the detriment of the brakes and tires which are put to severe testing. A larger wheel diameter and tire size would help to get into the turns better.

• Insurance : 50%
An annual $5,000 premium is considerable but proportionally a little less than that of more popular models.

WEAK POINTS

• Price/equipment: 00%
Because they are so superbly equipped, these vehicles are offered at prices that insult human intelligence. Technically they are neither exceptional nor revolutionary.

• Fuel consumption: 00%
It is here that you pay for the luxury of weight and performance, and you have to figure on 10 mpg and twice as much in town or at high speeds.

• Braking: 20%
Here there is no miracle possible because inertia is enormous, and in spite of the pads' quality and the disk venting, panic stops are balanced but distances are longer than average and fade resistance is ephemeral.

• Depreciation: 40%
Depending on where you are in North America, a Rolls or a Bentley will see its value fluctuate from single to double. Selling it in L.A. or at the tail end of the Yukon will make all the difference.

CONCLUSION

• Overall rating: 69.0%
If giving a grade to the Queen of the automobiles is high treason, it does however allow you to locate it on a scale of true value. Let's say that the acquisition of this rolling living room is as much a matter of means as taste. ☺

SPECIFICATIONS & PRICES

Model	Version/Trim	Body/ Seats	Interior volume cu.ft.	Trunk volume cu ft.	Drag coef. Cd	Wheel-base in	Lgth x Wdthx Hght. inx inx in	Curb weight lb	Towing capacity std. lb	Susp. type ft/rr	Brake type ft/rr	Steering system	Turning diameter ft	Steer. turns nber	Fuel. tank gal.	Standard tire size	Standard Power Train	PRICES US. $ 1994
BENTLEY			General warranty : 4 years/ unlimited.															
Brooklands		4dr sdn.5	98	18	NA	120.5	207.7x74.2x58.5	5203	NR	i/i	dc/dc/ABS	pwr.r/p.	38.4	3.5	28.5	235/70VR15	V8/6.7/A4	**$147,000**
Turbo R		4dr sdn.5	98	18	NA	120.5	207.7x74.2x58.5	5401	NR	i/i	dc/dc/ABS	pwr.r/p.	38.4	3.5	28.5	255/60VR16	V8T/6.7/A4	-
Continental R		2dr.cpe.5	NA	12.4	NA	120.5	210.2x73.6x57.5	5401	NR	i/i	dc/dc/ABS	pwr.r/p.	43.0	3.5	28.5	255/55VR17	V8T/6.7/A4	-
Continental Convert.		2dr.con. 5	NA	13.4	NA	120.5	207.5x72.2x59.8	5556	NR	i/i	dc/dc/ABS	pwr.r/p.	38.0	3.5	28.5	235/70VR15	V8/6.7/A4	**$290,000**
ROLLS ROYCE			General warranty : 4 years/ unlimited.															
Silver Spirit III		4dr sdn.5	98	19.4	NA	120.5	207.7x74.2x58.5	5357	NR	i/i	dc/dc/ABS	pwr.r/p.	39.7	3.5	28.5	235/70VR15	V8/6.7/A4	**$175,000**
Silver Spur III		4dr sdn.5	98	19.4	NA	124.4	211.4x74.2x58.5	5445	NR	i/i	dc/dc/ABS	pwr.r/p.	41.0	3.5	28.5	235/70VR15	V8/6.7/A4	**$225,000**
Silver Spur III Limo.		4dr.limo.5	NA	13.4	NA	148.4	235.4x72.2x60.4	5820	NR	i/i	dc/dc/ABS	pwr.r/p.	44.3	3.5	28.5	235/70VR15	V8/6.7/A4	-
Flying Spur		4dr sdn.5	98	19.4	NA	120.5	207.7x74.2x58.5	5401	NR	i/i	dc/dc/ABS	pwr.r/p.	39.7	3.5	28.5	255/60VR16	V8T/6.7/A4	-
Corniche		2dr.con.5	NA	13.4	NA	120.5	207.1x72.2x59.8	5556	NR	i/i	dc/dc/ABS	pwr.r/p.	39.7	3.5	28.5	235/70VR15	V8/6.7/A4	**$325,000**

See page 393 for complete 1995 Price List

Viking counterattack...

Associated with GM, the small Swedish car builder goes on a two-pronged offensive. After the spectacular renewal of the 900 sedan at the end of 93, a coupe and a convertible have joined the standard range for the '94 year. Modern and well-thought out, they announce a new era.

After 15 years of existence, the SAAB 900 has retaken life as a 5-door sedan, joined by a 3-door coupe and a convertible. The trims are S and SE for all models. The S models are powered by 2.3 L 4 cylinders and the SE by a V6 2.5 L of Opel descent, with the option of a 2.0 L turbo on the coupe and the convertible. The standard transmission is a manual 5-speed and 4-speed automatic is optional.

STRONG POINTS

• Safety: 90%
The 900 includes numerous accessories that make it particularly safe to use. In addition to a much stiffened unit body, it has received 2 standard front air bags, plus headrests at all four seats.

• Technical: 80%
The new steel unit body has kept the traditional attributes of the old models such as a windshield with little inclination and a third small lateral window. A styling with much roundness in the grill is easy to identify. While the rear section is not entirely part of the symphony, one can get used to it. The coupe has picked up the torch of its glorious predecessors; the convertible is 70% more rigid than the previous ones and its electrically powered retractable soft top, designed and fabricated by ASG, is insulated. The aerodynamic efficiency is on the honor roll because both the sedan and the coupe have an air drag coefficient of 0.30, and the convertible registers a 0.36 with the top closed. The transverse engine reduces front overhang and offers a better load distribution (60/40%). There is a McPherson style front suspension and a semi-rigid rear axle is assisted by 2 torsion bars and gas shocks. There are vented discs at the front and solid ones at the rear, plus a standard ABS that also serves as a traction control. The automatic transmission (Aisin-Warner) functions in 3 modes: normal, sport, and winter. The lateral offers the advantage of starting in 3rd on slippery roads and works up to 50 mph.

• Quality/fit/finish: 80%
The presentation is without frills, somewhat severe, but the quality of the assembly and materials and the care devoted to fit are udeniable.

• Driver's compartment: 80%
The dashboard is traditional of the make and much inspired by aircraft. Everything is well laid out, easy to interpret or to use at night. You can shut off the instrument lighting and it will only come back on in case of a problem (doubtful gadget). The key is on the center tunnel to protect the driver's knee in case of an accident and complicates life for thieves by blocking the transmission. The driver sits at ease, but visibility in the sedan and coupe is only average toward the back, and there are major dead spots on the convertible when the top is closed.

• Access: 80%
It is equally easy to climb into the front or the rear because the doors open wide and height is sufficient, even on a 2-door.

• Satisfaction: 80%
Based on the experience with the older models, the maintenance cost, the high price of parts, and the modest network of dealers, the lack of experience about reliability of this latest model requires consideration.

DATA

Category: front-wheel drive luxury sedans, coupes and convertibles.
Class : 7

HISTORIC

Introduced in: 1969
Modified in: 1978: style; 1984: tbo; 1985 Convertible, 1993: renewed; 1994: coupes and convertibles.
Made in: Trollhattan, Sweden & Nystad, Finland.

RATINGS

Safety: 90 %
Satisfaction: 78 % (previous model)
Depreciation: 50 % Tbo 58 % (previous model)
Insurance: 5.0 % ($860 - $995)
Cost per mile: $0.62

NUMBER OF DEALERS

In U.S.A.: 365

SALES ON THE AMERICAN MARKET

Model	1992	1993	Result
900	16,011	10,974	- 31.5 %

MAIN COMPETITORS

ACURA Vigor, AUDI A4, BMW 3-Series, INFINITI G 20, LEXUS ES 300, MAZDA Millenia, NISSAN Maxima, VOLVO 850.

EQUIPMENT

SAAB 900 sedan	S	SE	
SAAB 900 cpe & con.	**S**	**SE**	**Tbo**
Automatic transmission:	O	O	O
Cruise control:	S	S	S
Power steering:	S	S	S
ABS brake:	S	S	S
Air conditioning:	S	S	S
Air bag:	S	S	S
Leather trim:	-	S	S
AM/FM/ Radio-cassette:	S	S	S
Power door locks:	S	S	S
Power windows:	S	S	S
Tilt steering column:	S	S	S
Dual power mirrors:	S	S	S
Intermittent wipers:	S	S	S
Light alloy wheels:	-	S	S
Sunroof:	O	O	O
Anti-theft system:	S	S	S

S : standard; O : optional; - : not available

COLORS AVAILABLE

Exterior: White, Black, Blue, Red, Gray, Green, Silver.
Interior: Gray, Beige, Black, Tan.

MAINTENANCE

First revision: 3,000 miles
Frequency: 6,000 miles
Diagnostic plug: Yes

WHAT'S NEW IN 1995 ?

- Coupes and convertibles made available during the 1994 model year.

Model/ versions *: standard	ENGINE Type / timing valve / fuel system	Displacement cu/in	Power bhp @ rmn	Torque lb.ft @ rpm	Compres. ratio	TRANSMISSION Driving wheels / transmission	Final ratio	PERFORMANCE Acceler. 0-62 mph s	Stand. 1/4 mile s	Stand. 5/8 mile s	Braking 62-0 mph ft	Top speed mph	Lateral acceler. G	Noise level dBA	Fuel economy mpg city	Fuel economy mpg hwy	Gasoline type / octane
Turbo	L4* 2.0 DOHC-16-EFI	121	185 @ 5500	195 @ 2100	9.2: 1	front - M5	4.05	NA									
						front - A4	2.65	NA									
S	L4* 2.3 DOHC-16-EFI	140	150 @ 5700	155 @ 4300	10.5 :1	front - M5*	4.05	10.2	17.8	31.3	131	128	0.80	68	22.0	35.0	R 87
						front - A3	2.65	11.0	19.0	31.9	128	125	0.80	68	22.0	34.0	R 87
SE	V6 2.5 DOHC-24-EFI	152	170 @ 5900	167 @ 4200	10.8 :1	front - M5*	4.45	8.5	16.2	29.0	125	136	0.82	67	22.0	33.0	R 87
						front - A3	2.86	9.2	17.0	29.4	134	130	0.82	67	22.0	33.0	R 87

manual transmission (whose selection is slow and delicate) while the smoother V6 goes with the automatic. The liveliness of the turbo provides driving pleasure above 2,500 rpm that is worthy of its predecessors.

• Braking: 50%
The stops are easy to control, shorter and better balanced than in the past. We have seen no reports on friction material wear, which was a problem with preceding models.

• Fuel consumption: 60%
There is little difference between the two engines which become gluttonous when push hard.

• Interior space: 50%
While the total length is shorter than in the preceding model, the other dimensions have progressed to allow four and sometimes five people to travel comfortably thanks to carefully designed dimensions.

• Seats: 80%
Redesigned, their seating is comfortable, but the cloth trim offers a more superior lateral hold than the leather whose seat back is larger.

• Handling: 70%
The 900 has gained a stability superior to that of the previous model. Even though it remains frankly understeering, it does not change its sports-like character. It is just as at ease in a straight line as it is fast on a sinuous track. Adherence and drive capability are not lacking, thanks to quality tires and a traction regulator. The convertible road holding is not quite as precise due to lack of body stiffness.

• Steering: 70%
It has a good ratio, fine modulation, and is precise, allowing excellent control and sufficient maneuverability in spite of a long wheelbase.

• Suspension: 70%
Even though response is firm, it absorbs all road defects without harshness or bottoming.

• Conveniences: 70%
The storage includes a large glove box and door pockets, front and rear.

• Insurance: 70%
Proportionally, the 900 costs more to insure than the 9000.

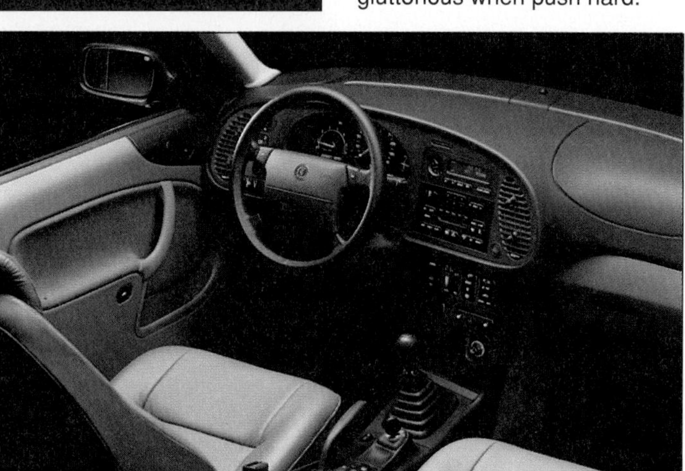

• Noise level: 60%
The rigidity and soundproofing have made the interior smooth as suede, but the engines are still noisier than average and the airstreams around the windshield let themselves be heard as soon as speed increases.

• Trunk: 60%
Its capacity is greater on the coupe then convertible and it can be increased toward the interior by lower the 60/40 split rear seat back.

• Performances: 60%
Each of the engines is interesting in its own way. The four cylinder goes well with the standard

WEAK POINTS

• Price/equipment: 40%
Even though they are not given away, their price is more competitive than in the past.

• Depreciation: 40%
The Saab's have always lost more value than other models in their category, and it is too soon to know if the new models will follow the same tendency.

CONCLUSION

• Overall rating: 67.0%
Well-renewed, the new family of cars presents Saab criteria in a more realistic way from all points of view. ☺

SPECIFICATIONS & PRICES

Model	Version/Trim	Body/ Seats	Interior volume cu.ft.	Trunk volume cu ft.	Drag coef. Cd	Wheel-base in	Lgth x Wdthx Hght. inx inx in	Curb weight lb	Towing capacity std. lb	Susp. type ft/rr	Brake type ft/rr	Steering system	Turning diameter ft	Steer. turns nber	Fuel. tank gal.	Standard tire size	Standard Power Train	PRICES US. $ 1994
SAAB		**Warranty: 3 years / 36,000 miles; Power Train: 6 years / 75,000 miles; corrosion perforation: 6 years / 100,000 miles.**																
900	S	2dr.con.4	82	10	0.36	102.3	182.6x67.4x56.5	-	NR	i/sr	dc/dc/ABS	pwr.r/p.	35.4	3.7	18.0	195/60VR15	L4/2.3/M5	**$33,275**
900	SE	2dr.con.4	82	10	0.36	102.3	182.6x67.4x56.5	3190	NR	i/sr	dc/dc/ABS	pwr.r/p.	35.4	3.7	18.0	155/60VR15	V6/2.5/M5	-
900	Turbo	2dr.con.4	82	10	0.36	102.3	182.6x67.4x56.5	3139	NR	i/sr	dc/dc/ABS	pwr.r/p.	35.4	3.7	18.0	205/50ZR16	L4T/2.0/M5	**$38,415**
900	S	2dr.cpe.4	91	24	0.30	102.3	182.6x67.4x56.5	2989	3527	i/sr	dc/dc/ABS	pwr.r/p.	35.4	3.7	18.0	195/60VR15	L4/2.3/M5	-
900	SE	2dr.cpe.4	91	24	0.30	102.3	182.6x67.4x56.5	3029	3527	i/sr	dc/dc/ABS	pwr.r/p.	35.4	3.7	18.0	195/60VR15	V6/2.5/M5	-
900	Turbo	2dr.cpe.4	91	24	0.30	102.3	182.6x67.4x56.5	3009	NR	i/sr	dc/dc/ABS	pwr.r/p.	35.4	3.7	18.0	205/50ZR15	L4T/2.0/M5	-
900	base	5dr.sdn.5	91	16	0.30	102.3	182.6x67.4x56.5	2998	3527	i/sr	dc/dc/ABS	pwr.r/p.	36.4	3.0	18.0	195/60R15	L4/2.3/M5	**$20,995**
900	base	5dr.sdn.5	91	16	0.30	102.3	182.6x67.4x56.5	3053	3527	i/sr	dc/dc/ABS	pwr.r/p.	36.4	3.0	18.0	195/60R15	L4/2.3/A4	**$21,890**
900	SE	5dr.sdn.5	91	16	0.30	102.3	182.6x67.4x56.5	3131	3527	i/sr	dc/dc/ABS	pwr.r/p.	36.4	3.0	18.0	195/60R15	V6/2.5/M5	**$26,280**
900	SE	5dr.sdn.5	91	16	0.30	102.3	182.6x67.4x56.5	3296	3527	i/sr	dc/dc/ABS	pwr.r/p.	36.4	3.0	18.0	195/60R15	V6/2.5/A4	**$27,175**

See page 393 for complete 1995 Price List

Waiting for a refreshment...

After the renewal of the 900, one can expect that the 9000 also gets a new skin because it is now 10 years that it carries its current silhouette... The big Saab finally gets a V6 which will give it a new dimension...

The 9000 continues to be offered in 4- and 5-door sedans with a naturally aspirated 2.3 L engine and a Turbo or a V6 3.6 L offered as an option on all models with a standard traction regulator. The stick shift is standard and the 4-speed automatic optional. The different trims are CF, CSE, Areo, CDE.

STRONG POINTS

• Safety: 90%
The rigidity of the 9000 structure was improved by 25% during the design changes done on the CS. It stands up much better to impact tests, and its handling improved just as much. Adding 2 standard front air bags as well as 3-point belts maintains their high index rating.

• Driver's compartment: 90%
Ergonomics has been a particular concern at Saab, and the 9000 is no exception, with its telescopically adjusted steering post and a thick rim which allows a good grip. The instruments and main controls are well laid out, but some require getting used to. The driver is comfortable and visibility is better on the 4-door than on the hatchback whose inclined rear window is rapidly obscured when it rains.

• Technical: 80%
The steel unit body has been redesigned on the CDE, it carries an average drag coefficient and differs by the trunk layout which is separated (CD) or has a hatchback (SC). The front suspension is independent, while the rear axle is still rigid. The different settings have been changed to insure ride comfort. Braking includes 4 discs with a 3rd generation ABS. The 4 cylinder has 2 balance shafts designed to eliminate the vibrations. It has a high output DOHC 16-valve head and a direct ignition system, specific to Saab, with a turbocharged Garrett T3 and an air-to-air heat exchanger. A 3.0 L, V6 derived from the 2.5 L, which is installed in the 900 also comes from Opel and is offered as

DATA

Category: front-wheel drive luxury sport sedans.
Class: 7

HISTORIC

Introduced in: 1984 2.0L 5 dr.
Modified in: 1985: S; 1988: CD 4 dr. & SPG; 1993:CS.
Made in: Trollhattan, Sweden.

RATINGS

	9000	9000Tbo
Safety:	90 %	90 %
Satisfaction:	85 %	80 %
Depreciation:	50 %	52 %
Insurance:	4.8% ($925)	3.9 % ($1,065)
Cost per mile:	$0.62	

NUMBER OF DEALERS

In U.S.A.: 365

SALES ON THE AMERICAN MARKET

Model	1992	1993	Result
9000	10,352	7,714	- 25.5 %

MAIN COMPETITORS

ACURA Legend, ALFA ROMEO 164, AUDI A6, BMW 5-Series, MERCEDES 300, VOLVO 960.

EQUIPMENT

SAAB 9000	CS Tbo	CSE	Aero Tbo	CDE
Automatic transmission:	O	O	O	O
Cruise control:	O	S	S	S
Power steering:	S	S	S	S
ABS brake:	S	S	S	S
Air conditioning:	S	S	S	S
Air bag:	S	S	S	S
Leather trim:	O	S	S	S
AM/FM/ Radio-cassette:	S	S	S	S
Power door locks:	S	S	S	S
Power windows:	S	S	S	S
Tilt steering column:	S	S	S	S
Dual power mirrors:	S	S	S	S
Intermittent wipers:	S	S	S	S
Light alloy wheels:	S	S	S	S
Sunroof:	O	O	S	S
Anti-theft system:	S	S	S	S

S : standard; O : optional; - : not available

COLORS AVAILABLE

Exterior: White, Black, Blue, Red, Gray, Silver.
Interior: Gray, Beige, Black, Tan.

MAINTENANCE

First revision: 3,000 miles
Frequency: 6,000 miles
Diagnostic plug: Yes

WHAT'S NEW IN 1995 ?

- New 3.0 L V6 optional on all models and delivered standard with a traction regulator.
- Aesthetic improvement to the front and rear CD.
- Adjustments to make the suspension more comfortable, the cruise control and adjustable steering column, alarm and sound system.
- Body colors.

Model/ versions *: standard	ENGINE Type / timing valve / fuel system	Displacement cu/in	Power bhp @ rmn	Torque lb.ft @ rpm	Compres. ratio	TRANSMISSION Driving wheels / transmission	Final ratio	PERFORMANCE Acceler. 0-62 mph s	Stand. 1/4 mile s	Stand. 5/8 mile s	Braking 62-0 mph ft	Top speed mph	Lateral acceler. G	Noise level dBA	Fuel economy mpg city	hwy	Gasoline type / octane
base	L4* 2.3 DOHC-16-EFI	139	150 @ 5500	156 @ 3800	10.1:1	front-M5*	4.45	8.5	16.0	31.6	121	125	0.78	66	22.0	35.0	R 87
						front-A4	4.28	9.8	16.9	32.0	127	119	0.78	66	20.0	35.0	R 87
Turbo	L4* 2.3T DOHC-16-EFI	139	200 @ 5000	243 @ 2000	8.5:1	front-M5*	4.45	7.8	15.0	28.4	118	137	0.82	67	23.0	36.0	R 87
				221 @ 2000	8.5:1	front-A4	4.28	8.3	15.9	29.1	125	125	0.80	67	20.0	33.0	R 87
Aero	L4* 2.3 DOHC-16-EFI	139	225 @ 5500	246 @ 1800	9.25:1	front-M5	NA										
option	V6 3.0 OHV-16-EFI	181	210 @ 6200	200 @ 3300	10.8:1	front-A4	NA										

an option on all models. Among the outstanding features is a variable intake manifold (VIM). Depending on the rpm and the computer, a valve directs the air fuel mixture to one of three channels of different lengths so as to tune frequency to the rpm.

• Satisfaction: 80%
In spite of the cost and frequency of some operations on replaceable parts, the warranty and robustness of the Saabs satisfy most owners.

• Access: 80%
It is easy to slide into a Saab 9000, because the doors are well-dimensioned and the interior height is ample.

• Seats: 80%
Lateral and lumbar supports are perfect, but we prefer cloth trim with its softer padding to leather which is more slippery and firmer.

• Quality/fit/finish: 80%
The assembly of the Saab is careful and its finish meticulous. The leather trim is remarkable and the equipment very complete, ranking it among the Deluxe models, as confirmed by its pricing.

• Insurance: 80%
Because of their purchase price, the premium is higher than that of its direct competitors.

• Interior space: 80%
It is difficult to admit the 9000 are classed as full-sized cars, compared to the Chevrolet Caprice, Ford Crown Victoria and recently the Chrysler New Yorker. However, the ample interior dimensions allow them to accept five persons.

• Trunk: 80%
It is larger on the hatchback 2-volume model where it is extensive, than on the 3-volume where it cannot be transformed. However, the hatchback counter balance needs improved calibration because it is heavy and difficult to manipulate.

• Suspension: 70%
It is firm but not to the point of being uncomfortable, enjoys effective damping and has enough wheel travel to absorb road unevenness without bottoming.

• Performance: 70%
Those of the naturally aspired 2.3 L are more lively with a manual stick shift than with an automatic whose first 2 ratios are too short to take advantage of the torque. The turbocharger gives the 9000 wings and sharp accelerations. Here power and torque are in abundance. Too bad that the slowness of the manual transmission synchronization adds to the response time of the turbo and that second also grinds. We cannot comment on the performances of the new V6 as we were unable to test it before going to press.

• Handling: 70%
It is more sports-like on the Aero whose setting are different. Other 9000's with more pedestrian adjustments get clumsier on twisty roads and the side-to-side balance can be improved by the Bosch traction control.

• Conveniences: 70%
They include a large glove box and practical pockets in the center console.

• Steering: 70%
Precise and well-assisted, it lacks spirit because the steering ratio is too high and interferes with response time.

• Braking: 60%
It is powerful and its modulation is precise. The ABS is sensitive to road defects and traction loss increases the stopping distances.

• Noise level: 60%
Discreet at constant speed, the engine signals its presence at the slightest acceleration.

• Fuel consumption: 50%
It is at the measure of the performance and cylinder displacement, but rises rapidly with power demand.

WEAK POINTS

• Price/equipment: 20%
In spite of complete equipment, their high price is the main Saab handicap.

• Depreciation: 40%
A turbo is not worth very much after the expiration of the warranty because no one wants to replace one after 60,000 miles. While the other versions lose less, they are still not a good deal.

CONCLUSION

• Overall rating: 69.0%
The Saab 9000 is a good car but costly to maintain and keep rolling, and you need to really want to affirm your personality to fall under its spell. ☺

SPECIFICATIONS & PRICES

Model	Version/Trim	Body/ Seats	Interior volume cu.ft.	Trunk volume cu ft.	Drag coef. Cd	Wheel-base in	Lgth x Wdthx Hght. inx inx in	Curb weight lb	Towing capacity std. lb	Susp. type ft/rr	Brake type ft/rr	Steering system	Turning diameter ft	Steer. turns nber	Fuel. tank gal.	Standard tire size	Standard Power Train	PRICES US. $ 1994
SAAB		**General warranty: 3 years / 36,000 miles; Power Train 6 years / 75,000 miles; corrosion perforation: 6 years / 100,000 miles.**																
9000	CS	5dr sdn.5	103	18	0.36	105.2	187.4x69.5x55.9	3108	2000	i/r	dc/dc/ABS	pwr.r/p.	35.8	3.0	17.4	195/65TR15	L4/2.3/M5	**$27,745**
9000	CSE	5dr sdn.5	103	24	0.34	105.2	183.5x69.5x55.9	3219	2000	i/r	dc/dc/ABS	pwr.r/p.	35.8	3.0	17.4	205/60ZR15	L4T/2.3/M5	**$33,045**
9000	Aero	4dr sdn.5	103	18	0.36	105.2	188.2x69.5x55.9	3208	NR	i/r	dc/dc/ABS	pwr.r/p.	35.8	3.0	17.4	195/65VR15	L4T/2.3/M5	**$38,690**
9000	CDE	4dr sdn.5	103	24	0.34	105.2	188.2x69.5x55.9	3175	NR	i/r	dc/dc/ABS	pwr.r/p.	35.8	3.0	17.4	195/65VR15	L4T/2.3/M5	**$32,685**

See page 393 for complete 1995 Price List

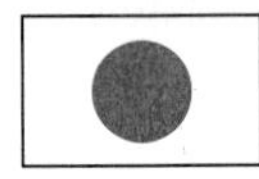

Back to square one...

At the moment of redefining its market position, Subaru has discovered that most of its buyers were especially attached to 4-wheel drive which no other builder offered in this battle corner. So, henceforth, the accent will be put on this transmission type rather than on front wheel drive.

This year the Impreza is offered in the form of a 2-door coupe, a sedan, and a 4-door station wagon in 2- or 4-wheel drive. Their common power train consists of a 1.8 L 4 cylinder with a manual 5-speed standard or a 3-speed automatic as an option.

STRONG POINTS

• Handling: 80%
In a distinct improvement, the Impreza is more at ease in a corner than was the Loyale. The roll is substantial and ends up with understeer with the front-wheel drive while the 4-wheel drive is neutral for longer and then begins to handle as a rear-wheel drive. Tires limit the evolution of these cars which would do much better if their sizes were better proportioned to the car. On a slippery road, the four-wheel drive shows a superior balance and driveability. Also its suspension is more rigid and benefits from the larger tire size.

• Steering: 80%
It is practically perfect; precise, easy-to-modulate with a good ratio and satisfactory maneuverability combine with a short turn diameter.

• Fuel consumption: 80%
Even though this type of engine has always been known as thirstier than average, it remains reasonable.

• Satisfaction: 80%
The first Imprezas' owners seem more satisfied than were those of the Loyale, especially in what concerns the quality of some electrical accessories.

• Quality/fit/finish: 70%
The assembly and finish are carefully done but the presentation as well as the appearance of some materials look money-saving, and some of the interior colors are, to say the least, bizarre.

DATA

Category: front-wheel drive or AWD sedans and wagons.
Class: 3

HISTORIC

Introduced in: 1993
Modified in: 1994: coupe.
Made in: Gunma & Yajima, Japan.

RATINGS

Safety: 80 %
Satisfaction: 80 %
Depreciation: 28 % (1 year)
Insurance: 7.0 % ($585)
Cost per mile: $0.32

NUMBER OF DEALERS

In U.S.A.: 727

SALES ON THE AMERICAN MARKET

Model	1992	1993	Result
Subaru	NA	23,577	

MAIN COMPETITORS

CHEVROLET Cavalier , DODGE-PLYMOUTH Neon, Colt, EAGLE Summit, FORD-MERCURY Escort-Tracer, FORD Contour, HONDA Civic 4-dr., HYUNDAI Elantra, MAZDA Protegé, MERCURY Mystique, SATURN, TOYOTA Corolla, VOLKSWAGEN Golf-Jetta.

EQUIPMENT

SUBARU Impreza	L	LS
Automatic transmission:	O	O
Cruise control:	-	O
Power steering:	O	S
ABS brake:	S	S
Air conditioning:	S	S
Air bag:	-	S
Leather trim:	-	-
AM/FM/ Radio-cassette:	S	S
Power door locks:	-	O
Power windows:	-	O
Tilt steering column:	-	O
Dual power mirrors:	S	S
Intermittent wipers:	S	S
Light alloy wheels:	O	S
Sunroof:	-	
Anti-theft system:	-	

S : standard; O : optional; - : not available

COLORS AVAILABLE

Exterior: Dark Gray, White, Ruby, Silver, Blue.
Interior: Gray, Blue.

MAINTENANCE

First revision: 3,000 miles
Frequency: 7,500 miles
Diagnostic plug: No

WHAT'S NEW IN 1995 ?

- 2-door coupe introduced at the end of 1994.

Model/ versions *: standard	ENGINE Type / timing valve / fuel system	Displacement cu/in	Power bhp @ rmn	Torque lb.ft @ rpm	Compres. ratio	TRANSMISSION Driving wheels / transmission	Final ratio	PERFORMANCE Acceler. 0-62 mph s	Stand. 1/4 mile s	Stand. 5/8 mile s	Braking 62-0 mph ft	Top speed mph	Lateral acceler. G	Noise level dBA	Fuel economy mpg city	Fuel economy mpg hwy	Gasoline type / octane
4x2	H4* 1.8 SOHC-8-MFI	111	110 @ 5600	110 @ 4400	9.5 :1	front - M5*	3.900	12.6	18.5	33.7	131	100	0.70	67	30.0	42.0	R 87
						front - A3	4.110	13.3	19.4	34.8	138	97	0.70	67	29.0	39.0	R 87
4x4	H4* 1.8 SOHC-8-MFI	111	110 @ 5600	110 @ 4400	9.5 :1	front - M5*	4.110	13.2	18.4	34.2	131	97	0.70	67	27.0	37.0	R 87
						front - A3	4.110	14.0	19.0	35.3	144	94	0.70	67	26.0	35.0	R 87

• Safety: 70%
The resistance of the unit body has been improved by adding reinforcements. The presence of an air bag on the driver's side and 3-point seat belts at all other places insures good occupant protection.

• Driver's compartment: 70%
The most convenient driver's position is readily found and visibility is good, even though one is seated low and the body belt line is high. The dashboard is not very ergonomical, because the console is recessed and the radio and AC controls are far from the hand. However, those grouped around the steering wheel fall right under the fingertips. The gauges are simple, in sufficient numbers, and easy to consult.

• Technical: 70%
The Impreza unit body is based on the larger Legacy platform. The suspension is independent, with a McPherson layout at all 4-wheels and braking is mixed on the standard model or 4-discs with ABS on the LS with 4-wheel drive. The full-time 4-wheel drive can only be used in rain or snow or on a gravel road. Otherwise, it must be disengaged to keep it from being damaged. The 4-cylinder pancake engine is unique to the Subaru. While the body lines are modern, the aerodynamic efficiency, with a drag coefficient of 0.35, is just average.

• Access: 70%
It is difficult to take a seat at the back, where the door cutout and available legroom are tighter than before.

• Seats: 70%
Even though firm, they are better shaped, at least at the front. The bench seat has no contouring and allows no good lateral hold nor lumbar support.

• Suspension: 70%
It is very successful in its capability of absorbing small and medium road defects, but reacts harshly to big bumps.

• Trunk: 60%
Its volume is sufficient and one can swing the bench seat forward to dispose of more space. Storage room is also readily accessible since the hatchback comes down even with the bumpers.

• Braking: 60%
Its modulation is very precise and the balance surprising, especially without ABS. Its fade resistance is amazing for a Japanese car, but the stopping distances are still long.

• Conveniences: 60%
Storage spaces are numerous, in sufficient volume, and the heating system is now effective at the rear seats.

• Price/equipment: 50%
If the LS versions are costly for what they offer, the L is better positioned, but there is not much in the base model.

WEAK POINTS

• Insurance: 30%
The Impreza is relatively costly to insure because its index reflects the insurers' mistrust about its repairability.

• Interior space: 30%
Four people can be seated, thanks to generous width and height, but there is not enough rear legroom. .

• Performance: 30%
The Impreza's performances are not anymore exciting than those of the Loyale. The engine offers good torque, but the power is very average and it is noisy on takeoff and vibrates at idle. Takeoffs and accelerations are longer than average, especially with an automatic, which makes the manual and its engine braking interesting.

• Noise level: 40%
At cruise speed, the power train is discreet and only rolling noise, dependent on the pavement, intrudes into the interior quietness.

• Depreciation: 40%
It is stronger than average, because Subaru is still considered a marginal builder and the quality of its products has left some buyers disatisfied.

CONCLUSION

• Overall rating: 60.5%
Impreza is a simple car, unique in its category by its 4-wheel drive and even more unique by offering it at a competitive price.

SPECIFICATIONS & PRICES

Model	Version/Trim	Body/ Seats	Interior volume cu.ft.	Trunk volume cu ft.	Drag coef. Cd	Wheel-base in	Lgth x Wdthx Hght. inx inx in	Curb weight lb	Towing capacity std. lb	Susp. type ft/rr	Brake type ft/rr	Steering system	Turning diameter ft	Steer. turns nber	Fuel. tank gal.	Standard tire size	Standard power train	PRICES US. $ 1994
SUBARU		Warranty: general: 3 years/ 36,000 miles; Power train: 5 years / 60,000 miles; corrosion & antipollution: 5 years / unlimited.																
Impreza	4x2	4dr.sdn.4	88	11	0.38	99.2	172.0x67.1x55.1	2326	1000	i/i	dc/dr	pwr.r&p.	33.5	3.2	13.2	165/80HR13	H4/1.8/M5	$10,999
Impreza	4x2	4dr.wgn.4	88	25.4	0.40	99.2	172.0x67.1x55.1	2405	1000	i/i	dc/dr	pwr.r&p.	33.5	3.2	13.2	165/80HR13	H4/1.8/M5	-
Impreza	4x4	4dr.sdn.4	88	11	0.38	99.2	172.0x67.1x55.1	2500	1000	i/i	dc/dr	pwr.r&p.	33.5	3.2	13.2	175/70HR14	H4/1.8/M5	-
Impreza	4x4	4dr.wgn.4	88	25.4	0.40	99.2	172.0x67.1x55.1	2579	1000	i/i	dc/dr	pwr.r&p.	33.5	3.2	13.2	175/70HR14	H4/1.8/M5	-
Impreza	4x2 L	4dr.sdn.4	88	11	0.38	99.2	172.0x67.1x55.1	2326	1000	i/i	dc/dr	pwr.r&p.	33.5	3.2	13.2	165/80HR13	H4/1.8/M5	$11,499
Impreza	4x2 L	4dr.wgn.4	88	25.4	0.40	99.2	172.0x67.1x55.1	2405	1000	i/i	dc/dr	pwr.r&p.	33.5	3.2	13.2	165/80HR13	H4/1.8/M5	$13,399
Impreza	4x4 L	4dr.sdn.4	88	11	0.38	99.2	172.0x67.1x55.1	2500	1000	i/i	dc/dr	pwr.r&p.	33.5	3.2	13.2	175/70HR14	H4/1.8/M5	$14,499
Impreza	4x4 L	4dr.wgn.4	88	25.4	0.40	99.2	172.0x67.1x55.1	2579	1000	i/i	dc/dr	pwr.r&p.	33.5	3.2	13.2	175/70HR14	H4/1.8/M5	$14,899
Impreza	4x2 LS	4dr.sdn.4	88	11	0.38	99.2	172.0x67.1x55.1	2326	1000	i/i	dc/dr	pwr.r&p.	33.5	3.2	13.2	165/80HR13	H4/1.8/M5	$15,699
Impreza	4x4 LS	4dr.sdn.4	88	11	0.38	99.2	172.0x67.1x55.1	2405	1000	i/i	dc/dr	pwr.r&p.	33.5	3.2	13.2	175/70HR14	H4/1.8/M5	$17,199
Impreza	4x4 LS	4dr.wgn.4	88	25.4	0.40	99.2	172.0x67.1x55.1	2579	1000	i/i	dc/dr	pwr.r&p.	33.5	3.2	13.2	175/70HR14	H4/1.8/M5	$16,899
Impreza	4x2	2dr.cpe.4	84.4	11.1	0.38	99.2	172.0x67.1x55.5	NA	1000	i/i	dc/dr	pwr.r&p.	33.5	3.2	13.2	175/70HR14	H4/1.8/M5	-
Impreza	4x4	2dr.cpe.4	84.4	11.1	0.38	99.2	172.0x67.1x55.5	NA	1000	i/i	dc/dr	pwr.r&p.	33.5	3.2	13.2	175/70HR14	H4/1.8/M5	-

See page 393 for complete 1995 Price List

Orphan...

The poor little Justy never knew success even when it was in on the program. Yet, it had the advantage of being the only mini-compact with a 4-wheel drive, which made it widely known in snow country.

The little Justy was the great unknown, which never received the success that it deserved, despite technical innovations: for one thing it was the only one to offer all-wheel drive in its category. It is no longer available except as a 5-door with a hatchback in the unique GL 4-wheel drive. The engine is a 1.2 L 3-cylinder matched up with a part-time 4-wheel drive that can be engaged on demand.

STRONG POINTS

• **Price/equipment:** **90%**
The equipment of the GL is fairly complete in this category: at this price level, you can't have everything...

• **Fuel consumption:** **90%**
It is a little higher than average for a 4-wheel drive but that is not a high price to pay for extra safety.

• **Technical:** **80%**
The Justy is a steel unit body with independent 4-wheel suspension and braking with discs and drums. Its line has not changed for years and does not offer the best aerodynamics but this is secondary on a car designed to circulate mostly in urban areas. The part-time 4-wheel can be engaged by a button just under the gear shift knob to provide traction at the rear. This device, which can remain locked out on dry pavement confers a serious advantage for winter driving when the road is slippery and makes it the smallest 4-wheel drive in North America. The 3-cylinder 1.2 L 9-valve does not date from yesterday but provides it with a favorable weight-to-power ratio of 28 lbs/hp. The only transmission available in 1995 is a 5-speed manual because the continuously variable automatic has fortunately disappeared.

• **Satisfaction:** **80%**
The Justy satisfies the vast majority of those who have selected it.

DATA

Category: AWD sub-compact sedans.
Class: 2

HISTORIC

Introduced in: 1983
Modified in: 1986: 1.2L; 1987: ECVT.
Made in: Gunma, Japan.

RATINGS

Safety:	60 %
Satisfaction:	80 %
Depreciation:	67 %
Insurance:	9.5 % ($500)
Cost per mile:	$0.30

NUMBER OF DEALERS

In U.S.A.: 727

SALES ON THE AMERICAN MARKET

Model	1992	1993	Result
Justy	2,978	2,860	-4.0 %

MAIN COMPETITORS

FORD Aspire, GEO Metro, HYUNDAI Accent, SUZUKI Swift.

EQUIPMENT

SUBARU Justy	GL
Automatic transmission:	-
Cruise control:	-
Power steering:	O
ABS brake:	-
Air conditioning:	O
Air bag:	-
Leather trim:	-
AM/FM/ Radio-cassette:	O
Power door locks:	-
Power windows:	-
Tilt steering column:	-
Dual power mirrors:	-
Intermittent wipers:	S
Light alloy wheels:	-
Sunroof:	-
Anti-theft system:	-

S : standard; O : optional; - : not available

COLORS AVAILABLE

Exterior: Silver, Red, White, Lavender.
Interior: Gray.

MAINTENANCE

First revision:	3,000 miles
Frequency:	7,500 miles
Diagnostic plug:	No

WHAT'S NEW IN 1995 ?

-The only version available for 95 is the GL 4RM 5-door with a 5-speed manual.

Model/ versions *: standard	ENGINE Type / timing valve / fuel system	Displacement cu/in	Power bhp @ rmn	Torque lb.ft @ rpm	Compres. ratio	TRANSMISSION Driving wheels / transmission	Final ratio	PERFORMANCE Acceler. 0-62 mph s	Stand. 1/4 mile s	Stand. 5/8 mile s	Braking 62-0 mph ft	Top speed mph	Lateral acceler. G	Noise level dBA	Fuel economy mpg city	Fuel economy mpg hwy	Gasoline type / octane
GL	L3* 1.2 SOHC-9-MPFI	73	73 @ 5600	71 @ 2800	9.1 :1	front/4-M5*	4.80	12.8	18.8	35.7	144	94	0.67	70	34.0	44.0	R 87

Apart from a few brake problems, owners feel that it is still too corrosion-sensitive.

• **Quality/fit/finish: 70%**
While the assembly is robust and the materials are of acceptable quality, the presentation and finish look mostly utilitarian, particularly the shiny dashboard plastics.

• **Driver's compartment: 70%**
The driver is comfortable behind the Justy steering wheel, visibility is good and the dashboard is simple and angular, but conveys most of the driving information needed. Most controls are laid out in conventional fashion. The shifts of the manual are precise, the clutch feels progressive and the steering wheel falls in the right place.

• **Safety: 60%**
As all small cars, the Justy is vulnerable to collisions and no air bag is offered; only 3-point seat belts. However, the four-wheel drive is a plus.

• **Conveniences: 60%**
Storage space includes a large size glove box, door pockets and a tablet built into the top of the dashboard.

• **Access: 50%**
It will be difficult for heavy duty people to get into the back where the doors are narrow and the space is tight.

• **Seats: 50%**
In relation to some of its competitors, the Justy offers seats which, while they lack being fully shaped, are at least amply padded.

• **Suspension: 50%**
In spite of small wheels, a short wheelbase and lack of wheel travel, the ride is not as uncomfortable as one would first think.

WEAK POINTS

• **Interior space: 20%**
The very compact format limits the usable volume, especially at the rear seats. Here head and legroom space is closely accounted for. However, four medium-sized people can sit there for short trips.

• **Performance: 30%**
They are adequate for driving around town but on the road, long trips are tedious. With a manual shift the drivetrain is livelier in the first gear ratios but the old low displacement engine has problems coping with the power requirements of 4-wheel drive.

• **Noise level: 30%**
At cruise speeds, the engine calms itself a little but wind and rolling noises take over to maintain the unharmonious musical atmosphere.

• **Depreciation: 30%**
It is preferable to use it up to the bitter end because it is not worth much after three years.

• **Trunk: 30%**
The trunk does not contain much when the bench seat is used, but added space is gained by lowering the 2-part seat back.

• **Insurance: 40%**
Proportionally, it is expensive to insure but its modest price gives you an affordable premium.

• **Handling: 40%**
Its square shape makes it sensitive to crosswinds. With its little tires and simplistic suspension, the Justy needs 4-wheel drive to keep out of trouble.

• **Braking: 40%**
Progressive, stable and fade-free are qualities that surprise in a Japanese car, but it does not have enough efficiency to ensure short stops.

• **Steering: 40%**
Vague in the center, too many turns lock-to-lock and not enough power assist: you have to wind and wind to park it in town. Its maneuverability is adequate and you can easily weave through town.

CONCLUSION

• **Overall rating: 52.5%**
At the price at which it is sold, the Subaru is far from a panacea. It will make a good little second car for urban use, for which it is sufficiently robust and reliable, providing you fit it with adequate tires so that you can benefit from 4-wheel drive in the winter. ☹

OWNERS' SUGGESTIONS

-More effective heating and ventilation
-A higher volume trunk
-Better soundproofing
-Remote control for the fuel tank door
-Improved waterproofing
-Fewer oil leaks

SPECIFICATIONS & PRICES

Model	Version/Trim	Body/ Seats	Interior volume cu.ft.	Trunk volume cu ft.	Drag coef. Cd	Wheel-base in	Lgth x Wdthx Hght. inx inx in	Curb weight lb	Towing capacity std. lb	Susp. type ft/rr	Brake type ft/rr	Steering system	Turning diameter ft	Steer. turns nber	Fuel. tank gal.	Standard tire size	Standard power train	PRICES US. $ 1994
SUBARU		**Warranty: general: 3 years/ 36,000 miles; Power train: 5 years / 60,000 miles; corrosion & antipollution: 5 years / unlimited.**																
Justy	GL4WD	5dr.sdn.4	79	10	0.39	90.0	145.5x60.4x55.9	2004	NR	i/i	dc/dr	r&p.	29.5	4.3	9.3	165/65R13	L3/1.2/M5	**$9,603**

See page 393 for complete 1995 Price List

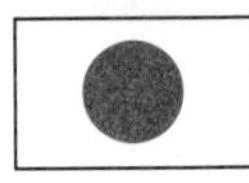

More of the same...

One could say that Subaru in its frenzied conservatism, has not yet taken the opportunity to renew its star model, the Legacy, to give it a little advance on competition which has already moved forward. Another missed rendez vous.

The Legacy is the model that sells best at Subaru, at the moment, endowing it with a new touch of youthfulness. Not to spook the customers that the Japanese make has so much trouble gathering, they have paddled softly and barely moved. The latest brood is offered in the form of sedans and station wagons in front- or 4-wheel drive, equipped with a 2.2 L 4-cylinder naturally aspirated since the Turbo is not available, at least not at this moment. The manual transmission is standard and the automatic is optional. The various versions are L, LS and LSi.

STRONG POINTS

• Satisfaction: 95%
Their general reliability is equal to other Japanese models and some owners advise of few problems, more annoying than serious.

• Safety: 90%
The structure has been substantially reinforced and two air bags protect the front seat occupants. However, the ABS brakes are considered a luxury and are only standard on the LS and LSi.

• Technical: 80%
One cannot say that the lines of the new Legacy are bold because it passes completely unnoticed and has only an average aerodynamic finesse. The platform derived from the preceding model has been redone. Its main dimensions have been lengthened, weight is lowered, and the 4-wheel independent suspension reworked to increase wheel travel. Brakes are 4-discs on all versions, but vented up front and with ABS on the LS and LSi. The engine remains a pancake VW style flat four, but with 4-valves per cylinder and a single overhead cam for each head. The power and torque have been improved, showing Subaru's attachment to this type of power plant. Four- wheel drives

DATA

Category: front-wheel drive or AWD compact sedans and wagons.
Class: 4

HISTORIC

Introduced in: 1989
Modified in: -
Made in: Gunma, Japan.

RATINGS

Safety: 90 %
Satisfaction: 94 %
Depreciation: 60 %
Insurance: 5.5 % ($725)
Cost per mile: $0.40

NUMBER OF DEALERS

In U.S.A.: 727

SALES ON THE AMERICAN MARKET

Model	1992	1993	Result
Legacy	64,981	60,209	- 7.4 %

MAIN COMPETITORS

BUICK Skylark, CHRYSLER Cirrus-Stratus, HONDA Accord, MAZDA 626, MITSUBISHI Galant, NISSAN Altima, OLDSMOBILE Achieva, Grand Am, TOYOTA Camry, VOLKSWAGEN Passat.

SUBARU Legacy Sedan Wagon	EQUIPMENT base	 L+ Brighton	 LS	 LSi
Automatic transmission:	O	O	O	S
Cruise control:	O	S	S	S
Power steering:	S	S	S	S
ABS brake:	O	O	S	S
Air conditioning:	O	S	S	S
Air bag (2):	S	S	S	S
Leather trim:	-	-	-	S
AM/FM/ Radio-cassette:	O	S	S	S
Power door locks:	O	S	S	S
Power windows:	O	S	S	S
Tilt steering column:	S	S	S	S
Dual power mirrors:	S	S	S	S
Intermittent wipers:	S	S	S	S
Light alloy wheels:	-	O	S	S
Sunroof:	-	O	S	S
Anti-theft system:	-	-	-	-

S : standard; O : optional; - : not available

COLORS AVAILABLE

Exterior: White, Silver, Sapphire, Ruby, Green, Taupe, Pearl Spruce.
Interior: Gray, Taupe.

MAINTENANCE

First revision: 3,000 miles
Frequency: 7,500 miles
Diagnostic plug: No

WHAT'S NEW IN 1995 ?

- All new model for 1995.
- Turbo engine eliminated.
- Front-wheel drive available with traction control.

ENGINE					TRANSMISSION			PERFORMANCE									
Model/ versions *: standard	Type / timing valve / fuel system	Displacement cu/in	Power bhp @ rmn	Torque lb.ft @ rpm	Compres. ratio	Driving wheels / transmission	Final ratio	Acceler. 0-62 mph s	Stand. 1/4 mile s	Stand. 5/8 mile s	Braking 62-0 mph ft	Top speed mph	Lateral acceler. G	Noise level dBA	Fuel economy mpg city	hwy	Gasoline type / octane
4x2	H4* 2.2 SOHC-16-MPSFI	135	135 @ 5400	140 @ 4400	9.5 :1	front - M5*	3.45	11.0	17.2	32.3	144	112	0.76	67	28.0	37.0	R 87
						front - A4	3.45	12.2	18.7	34.8	148	109	0.76	67	26.0	38.0	R 87
4x4	H4* 2.2 SOHC-16-MPFI	135	135 @ 5400	140 @ 4400	9.5 :1	front - M5*	3.90	11.0	17.2	32.3	144	112	0.76	67	25.0	35.0	R 87
						front - A4	3.90	12.2	18.7	34.8	148	109	0.76	67	25.0	35.0	R 87

can receive an electronic FWD/TCS traction control. This is a full-time 4WD with a manual shift and part-time with the automatic. Here, torque is controlled with electronic sensors connected to the ABS. The transmission has been revised, with softened shifts.

• Access: 80%
It is easy to get into these models whose length and door opening are sufficient, especially now that there is more legroom at the back.

• Quality/fit/finish: 75%
Although the overall look is anonymous, the construction seems solid, the materials' quality honest and it is carefully finished.

• Driver's compartment: 70%
Even freshened up, the dashboard does not express any more originality than the rest of the car. The instruments are easy to read and the controls conventional. Visibility is excellent and the most effective driver's position is easy to find since the standard steering column is adjustable.

• Suspension: 70%
Comfort is more evident with either 2- or 4-wheel drive, because the suspension has a lower spring rate, better damping and more wheel travel. A good combination to absorb pavement defects.

• Steering: 70%
Precise and well-modulated, its ratio is a little too high, even though this really does not interfere with maneuverability, thanks to a short steering diameter.

• Conveniences: 70%
Storage spaces include a glove box that is not enormous, recesses in the console and a sufficient number of door pockets.

• Insurance: 60%
The premium is realistic, thanks to the benefits of the 4-wheel drive during winter.

• Fuel consumption: 60%
Reasonable in normal travel with front-wheel drive, it climbs rapidly on the Turbo 4WD, if you maintain steady speeds.

• Noise level: 60%
Inspite of good body stiffness, wind, body and engine noises are all too present and call for more effective soundproofing.

• Seats: 60%
They are not ergonomic champions because their forms are evasive and their firm padding recalls the sinister memory of the older models of this make.

• Trunk: 60%
That of the sedans is roomy and the storage space of the station wagons is vast and practical, with a volume of 35 cu.ft. and they are both equally accessible.

• Handling: 60%
Even though the 2WD traction control is effective, it can't equal 4WD in a tight turn. Traction is optimal as soon as the system is engaged, regardless of road condition, for as long as the tires are adequate. There is more roll on the 2WD which does not interfere with getting into the turn as long as the speed is moderate. With 4WD the car stays neutral longer.

• Interior space: 60%
Larger base dimensions have benefited the interior volume, especially at the rear seats which have gained more length than width.

• Price/equipment: 50%
That of the front wheel drive is competitive with equipment similar to that of other manufacturers but the LSi 4WD gets paid well for this advantage. The standard equipment is interesting, because even the least expensive versions have standard conveniences, which are billed as an option elsewhere.

• Performance: 50%
One never has the impression of really disposing of 135 hp, because the engine is rough and always out of breath. However, takeoffs are better than accelerations and the first two transmission gears are shorter. Just in passing, gear selection is imprecise and hazardous - inconceivable in this day and age.

WEAK POINTS

• Brakes: 40%
In spite of the presence of 4-wheel disc brakes, the results are not impressive because during sudden stops the distances are long but straight, with or without ABS. However, the pedal remains difficult to modulate with any precision.

• Depreciation: 40%
Even though it is in the market average, it exceeds that of most models in this category, which are in demand.

CONCLUSION

• Overall rating: 65.0%
Improved from many points of view in relation to the preceding model, you can't see what would excite someone to purchase a Legacy, except for 4WD. ☺

SPECIFICATIONS & PRICES

Model	Version/Trim	Body/ Seats	Interior volume cu.ft.	Trunk volume cu ft.	Drag coef. Cd	Wheel-base in	Lgth x Wdthx Hght. inx inx in	Curb weight lb	Towing capacity std. lb	Susp. type ft/rr	Brake type ft/rr	Steering system	Turning diameter ft	Steer. turns nber	Fuel. tank gal.	Standard tire size	Standard power train	PRICES US. $ 1994
SUBARU		Warranty: general: 3 years/ 36,000 miles; Power train: 5 years / 60,000 miles; corrosion & antipollution: 5 years / unlimited.																
Legacy	base 4x2	4dr.sdn.4/5	-	12.6	-	103.5	180.9x67.5x55.3	2571	2000	i/i	dc/dc	pwr.r&p.	34.8	3.2	15.9	185/70HR14	H4/2.2/M5	$13,999
Legacy	Brit 4x2	4dr.wgn.4/5	-	36.1	-	103.5	183.9x67.5x57.1	2890	2000	i/i	dc/dc	pwr.r&p.	34.8	3.2	15.9	185/70HR14	H4/2.2/M5	$14,999
Legacy	L+ 4x4	4dr.sdn.4/5	-	12.6	-	103.5	180.9x67.5x55.3	2756	2000	i/i	dc/dc	pwr.r&p.	34.8	3.2	15.9	185/70HR14	H4/2.2/A4	-
Legacy	Brit 4x4	4dr.wgn.4/5	-	36.1	-	103.5	183.9x67.5x57.1	2851	2000	i/i	dc/dc	pwr.r&p.	34.8	3.2	15.9	185/70HR14	H4/2.2/A4	-
Legacy	LS 4x4	4dr.sdn.4/5	-	12.6	-	103.5	180.9x67.5x55.3	3029	2000	i/i	dc/dc	pwr.r&p.	34.8	3.2	15.9	195/60HR14	H4/2.2/M5	$19,700
Legacy	LS 4x4	4dr.wgn.4/5	-	36.1	-	103.5	183.9x67.5x57.1	3119	2000	i/i	dc/dc/ABS	pwr.r&p.	34.8	3.2	15.9	195/60HR14	H4/2.2/A4	$20,400
Legacy	LSi 4x4	4dr.sdn.4/5	-	12.6	-	103.5	180.9x67.5x55.3	3044	2000	i/i	dc/dc/ABS	pwr.r&p.	34.8	3.2	15.9	195/60HR14	H4/2.2/A4	$21,850
Legacy	LSi 4x4	4dr.wgn.4/5	-	36.1	-	103.5	183.9x67.5x57.1	3135	2000	i/i	dc/dc/ABS	pwr.r&p.	34.8	3.2	15.9	195/60HR14	H4/2.2/A4	$22,850

A daring bet...

One should not go by appearances alone because Subaru, known as a manufacturer who could not be more conservative, has suddenly caught the folly of grandeur and launched the SVX project on which it counts more from the point of view of image and reputation than pure sales.

Wanting no doubt to diversify its production into a sector with higher profits, the Subaru ventured into building an exotic sports coupe. Even though the first two years of sales have not been conclusive, the Japanese builder continues to tempt its fate in an effort to establish a new profit making niche, when the product is selling. The economic situation does not favor this type of model at the moment because they are more numerous on the market and most are better established than the SVX.

STRONG POINTS

• Technical: 90%

The body line, which does not lack personality, was started by Giugiaro and completed by Fuji stylists in Japan. It is, to say the least, original, especially in what concerns the side windows, and is inspired by competition cars. An efficient profile drops the drag coefficient to 0.29, lowest and best. Galvanized panels are used in the steel unit body. The 4-wheel independent suspension is supplemented with 4-wheel disc brakes and standard ABS. The drivetrain includes elements in which Subaru has invested many years of development and for which it is known, a 4WD and a pancake engine. The latter is a 3.3 L DOHC 6-cylinder horizontally opposed with 4-valves per cylinder. It develops 230 hp thanks to a turbo and an "Iris" injection system which improves fuel delivery and torque at low rpm. The full-time 4WD uses a differential with a viscous limited slip coupling between the rear wheels. The automatic transmission offers three operating modes: "normal", "power" and "manual" which allow automatic shifts with different shift patterns or manual shifts which hold in a particular gear.

• Safety: 90%

While the structure appears more rigid, two air bags protect the front seats and the horrible motorized seat belts are replaced by others with 3-point supports.

• Satisfaction: 80%

The number of units sold and the commentaries received are few, but enough to show that the number of very satisfied customers is at 82%.

• Quality/fit/finish: 80%

The interior presentation is as original as the body and distinguishes itself by the use of a simile deer to decorate the dashboard, a bizarre set of trim colors, a meticulous assembly and faux wood.

• Driver's compartment: 80%

It is easy to quickly find the best driving position, because the seat holds well and has an effective lumbar adjustment; also, the steering column is adjustable.The different gauges and controls are well-grouped and easy to use. The console is ergonomic and the driver's hand can reach the main accessories. Visibility is good, thanks to a large glazed area, but the fixed windows and the backlight gather dirt and bring on claustrophobia.

DATA

Category: AWD sport coupes.
Class: GT

HISTORIC

Introduced in: 1989
Modified in: marketed in 1991.
Made in: Gunma, Japan

RATINGS

Safety: 90 %
Satisfaction: 80 %
Depreciation: 50 %
Insurance: 6.5 % ($1,450)
Cost per mile: $0.80

NUMBER OF DEALERS

In U.S.A.: 727

SALES ON THE AMERICAN MARKET

Model	1992	1993	Result
SVX	3,667	3,859	+ 5.0 %

MAIN COMPETITORS

ACURA Legend coupe, CHEVROLET Corvette, DODGE Stealth R/T Turbo, MAZDA RX-7, NISSAN 300 ZX, TOYOTA Supra.

EQUIPMENT

SUBARU	SVX
Automatic transmission:	S
Cruise control:	S
Power steering:	S
ABS brake:	S
Air conditioning:	S
Air bag (2):	S
Leather trim:	S
AM/FM/ Radio-cassette:	S
Power door locks:	S
Power windows:	S
Tilt steering column:	S
Dual power mirrors:	S
Intermittent wipers:	S
Light alloy wheels:	S
Sunroof:	S
Anti-theft system:	-

S : standard; O : optional; - : not available

COLORS AVAILABLE

Exterior: White, Red, Green, Blue.
Interior: Gray,Beige.

MAINTENANCE

First revision: 3,000 miles
Frequency: 7,500 miles
Diagnostic plug: No

WHAT'S NEW IN 1995 ?

- **Two inflatable air bags standard.**
- **Heated front seats.**
- **Three-point seat belts replace the motorized system.**

Model/ versions *: standard	ENGINE Type / timing valve / fuel system	Displacement cu/in	Power bhp @ rmn	Torque lb.ft @ rpm	Compres. ratio	TRANSMISSION Driving wheels / transmission	Final ratio	PERFORMANCE Acceler. 0-62 mph s	Stand. 1/4 mile s	Stand. 5/8 mile s	Braking 62-0 mph ft	Top speed mph	Lateral acceler. G	Noise level dBA	Fuel economy mpg city	hwy	Gasoline type / octane
SVX	H6*3.3 DOHC-24-MPSFI	203	230 @ 4500	228 @ 4400	10.0 :1	all - A4	3.545	9.0	16.5	29.6	138	143	0.87	67	21.0	33.0	R 87

• Seats: 80%
Well-designed, they provide lateral hold and lumbar support even at the rear seats.

• Handling: 80%
Even though the North American version does not dispose as elsewhere of four-wheel steering, it has a good road handling thanks to 4WD and ample tires.
Once again we question the body stiffness, which emits strange noises at the door level when the road is not perfect and at the same time one notes that going into the turn is vague due to a poor front end geometry.

• Steering: 80%
Direct, precise, rapid and well-modulated, it also has a thick steering wheel rim which is easy to grip.
On a bumpy road the wheels often feed back shock directly into the steering wheel.

• Performance: 70%
They are disappointing, because assembly weight is high. Many coupes in this category accelerate from 0 to 62 mph in 5 to 7.0 seconds while the SVX does not come down below 9.0 seconds. Passing is to match, and lacks adequate punch to rate being called an exotic model.

• Access: 70%
It is easier to seat yourself up front than at the back even though the right front seat moves out of the way far enough to limit contortions.

• Suspension: 60%
It is not frankly uncomfortable but is firm and transmits all road imperfections.

• Insurance: 60%
The price and higher-than-average rates make for a hefty well-rounded premium, enough to make the undecided think twice.

• Conveniences: 60%
Storage spaces include a little glove box, a compartment in the center console and small door pockets. The radio is covered up with a protective plate to keep out greedy thieves.

• Noise level: 60%
At low cruising speeds on the parkway, the staccato engine noise is filtered out. The tires resonate on road joints.

• Depreciation: 50%
Resale value remains reasonable, in spite of a lack of enthusiasm for the car.

• Brakes: 50%
It is very ordinary in terms of effectiveness because stopping distances are fairly long, but it is easy to modulate, stable and fade-free.

WEAK POINTS

• Price/equipment: 10%
To invest over $30,000 in an SVX, you had to have fallen in love with it head over heels. The complete equipment justifies a good part of the asking price.

• Trunk: 30%
Deep rather than spacious, it lacks width because of the sizable suspension mounting points. However, the cutout in the opening allows easy access.

• Interior space: 40%
For a vehicle with a sports look, the interior is more spacious than average but the roof shape only allows it to be catalogued as a 2+2 rather than as a 4-seater since rear height and length are limited.

• Fuel consumption: 40%
It is fairly high, especially if a fast clip is maintained.

CONCLUSION

• Overall rating: 63.0%
With time, one gets used to the SVX and it would not be surprising if it came in vogue if its price came down by $6,000. Otherwise, the Dodge Stealth is a far better investment-pleasure. 😐

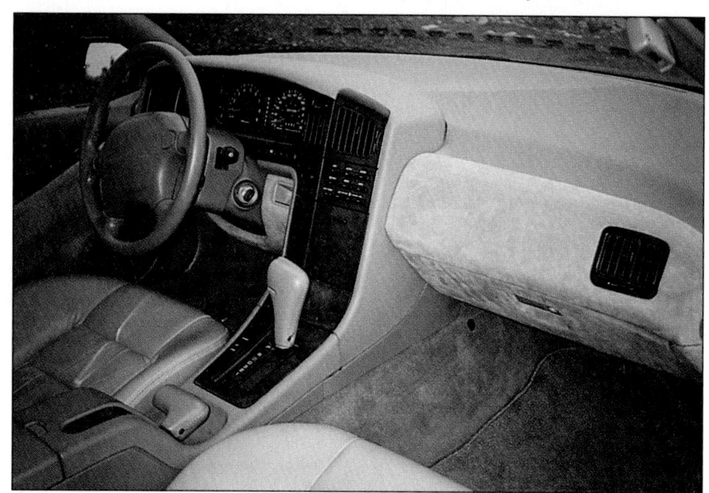

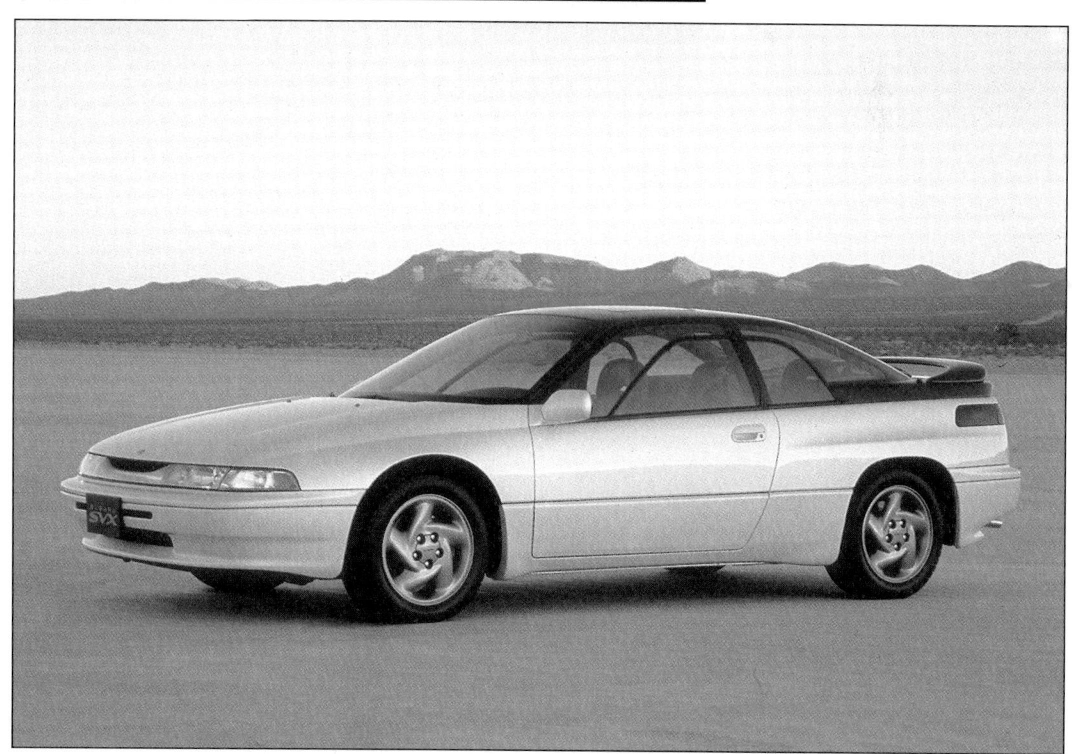

SPECIFICATIONS & PRICES

Model	Version/Trim	Body/ Seats	Interior volume cu.ft.	Trunk volume cu ft.	Drag coef. Cd	Wheel-base in	Lgth x Wdthx Hght. inx inx in	Curb weight lb	Towing capacity std. lb	Susp. type ft/rr	Brake type ft/rr	Steering system	Turning diameter ft	Steer. turns nber	Fuel. tank gal.	Standard tire size	Standard power train	PRICES US. $ 1994
SUBARU		**Warranty: general: 3 years/ 36,000 miles; Power train: 5 years / 60,000 miles; corrosion & antipollution: 5 years / unlimited.**																
SVX	LSi 4WD	2dr.cpe.2+2	85	8	0.29	102.7	183.1x69.7x1.1	3582	NR	i/i	dc/dc/ABS	pwr.r&p.	36.1	3.1	18.5	225/50VR16	H6/3.3/A4	**$33,850**

See page 393 for complete 1995 Price List

SUZUKI
GEO

Sidekick, Tracker

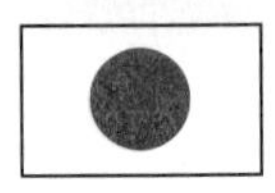

Flim flam show...

These little sports utilities have known mad success in the last years, more for their friendly show off ways than for their ability to find their way in the jungle. You will more often find them near center-city than roaming hills and dales.

The Suzuki Sidekick, Geo Tracker and Pontiac Sunrunner are identical to within a few equipment details, but only the Geo is offered as a 2-wheel drive while the Sidekicks are 4x4 and only the Sidekick is offered as a 4-door station wagon. The Suzuki is offered in a short version 2-door with a soft top in JA, JX and JLX or as a permanent hard top as JA and JX or an extended 4-door with a hard top in JX and JLX and the Geos are offered with a hard or soft top standard or LSi.

STRONG POINTS

• Driver's compartment: 80%

The driver is installed comfortably with a seat that provides lateral and lengthwise hold and the lumbar support is not bad. Visibility is best with the hard top while the soft top has major blind spots. The dashboard is well-organized for gauges and controls. However, the handles for the engine compartment and the gas tank flap are well-hidden.

• Technical: 80%

To gain a more favorable weight-to-power ratio, the unit body is built with thin gauge steel. The aerodynamic efficiency is not that of a small car but surpasses a conventional Jeep. All versions are now fitted with rear wheel ABS, but power assist is only available with 4-doors. Suspension has independent front wheels and the rigid axle with coil springs at the back. The engine on the 2-door is a 1.6 L 4-cylinder with 8-valves and a manual or automatic transmission, while the 4-door has a dual overhead cam and 16-valves that show a 15% power jump. However, the majority of the owners find this power insufficient and would prefer a 2.0 L which offers torque and power at a normal rpm and reduces noise and fuel consumption.

• Price/equipment: 80%

With time, these vehicles have become very expensive and comparing their price with a real Jeep or car, one will be very surprised, in as much as the equipment is proportional to the sticker price.

• Satisfaction: 80%

Few problems, except for a fragile hood and rust which gains rapidly once it shows up.

• Access: 70%

It is not easy to take a seat at the back even on the extended body whose doors are narrow and ground clearance is high.

• Seats: 70%

They hold well and have good contouring, but a somewhat softer padding would not hurt anyone.

• Quality/fit/finish: 70%

The interior appearance is pleasant even on the standard version assembly is accurate and the finish is meticulous. One would like some details, such as the inner wheel wells to be more substantial to reduce mud buildup at the junctions of the chassis and body. Mud holds salt and dampness and begins galvanic corrosion and rust.

DATA

Category: rear-wheel drive or AWD multipurpose vehicles.
Class: utility

HISTORIC

Introduced in: 1989
Modified in: 1991: 4dr version.;1992: 16V engine.
Made in: Ingersoll, Ontario, Canada.

RATINGS

		2 dr.	4 dr.	
Safety:		35 %	40 %	
Satisfaction:		80 %	85%	
Depreciation:		58 %	62 %	
Insurance:	($720)	8.5 %	7.0 %	($720)
Cost per mile:		$0.36	$0.36	

NUMBER OF DEALERS

In U.S.A.: NA, Suzuki, 4,466 Chevrolet-Geo

SALES ON THE AMERICAN MARKET

Model	1992	1993	Result
Sidekick	15,392	18,056	+ 14.8 %
Tracker	32,666	42,312	+ 22.8 %

MAIN COMPETITORS

JEEP Wrangler.

EQUIPMENT

SUZUKI Sidekick 4WD GEO Tracker	2WD	4WD	LSi	JA	JX	JLX
Automatic transmission:	O	O	O	O	O	O
Cruise control:	O	O	O	-	-	S
Power steering:	O	O	S	O	S	S
ABS brake:	O	S	S	S	S	S
Air conditioning:	-	O	O	-	O	S
Air bag:	-	-	-	-	-	-
Leather trim:	S	S	S	-	-	-
AM/FM/ Radio-cassette:	O	O	O	O	O	S
Power door locks:	-	-	S	-	-	S
Power windows:	-	-	S	-	-	S
Tilt steering column:	O	O	O	-	-	S
Dual power mirrors:	-	O	S	-	-	S
Intermittent wipers:	S	S	S	S	S	S
Light alloy wheels:	-	-	-	O	S	S
Sunroof:		-	-	-	-	-
Anti-theft system:	-			S	S	S

S : standard; O : optional; - : not available

COLORS AVAILABLE

Exterior: Blue, Magenta, White, Black, Red, Metallic Green.
Interior: Gray.

MAINTENANCE

First revision: 3,000 miles
Frequency: 6,000 miles
Diagnostic plug: No

WHAT'S NEW IN 1995 ?

- Soft top which can be installed or removed more easily (2-door).
- More powerful standard engine (2-door).
- New aluminum wheel on the JLX (4-door).

ENGINE						TRANSMISSION		PERFORMANCE										
Model/ versions *: standard	Type / timing valve / fuel system	Displacement cu/in	Power bhp @ rmn	Torque lb.ft @ rpm	Compres. ratio	Driving wheels / transmission	Final ratio	Acceler. 0-62 mph s	Stand. 1/4 mile s	Stand. 5/8 mile s	Braking 62-0 mph ft	Top speed mph	Lateral acceler. G	Noise level dBA	Fuel economy mpg city	Fuel economy mpg hwy	Gasoline type / octane
1) 2dr. JA	L4* 1.6 SOHC-8-IE	97	80 @ 5400	94 @ 3300	8.9 :1	rear/4 - M5*	5.12	13.5	18.8	34.6	151	87	0.77	72	28.0	34.0	R 87
						rear/4 - A3	4.62	15.5	20.0	38.8	157	84	0.76	72	27.0	34.0	R 87
2) 2dr.-4dr.	L4* 1.6 DOHC-16-IEMP	97	95 @ 5600	98 @ 4000	9.5 :1	rear/4 - M5*	5.12	13.3	18.0	34.2	168	90	0.78	70	28.0	36.0	R 87
						rear/4 - A4	4.30	15.5	19.5	38.4	184	87	0.77	70	27.0	32.0	R 87

1) 2-door 2) option 2-door & standard 4WD

Sidekick, Tracker

• **Fuel consumption:** 70%
It is not economical, even in normal use, and the comparison with competitors will make one think.

• **Steering:** 60%
Imprecise in the center, with too high a ratio (3.5 turns lock-to-lock) on the manual, but smooth and well-modulated when it is power-assisted.

• **Suspension** 50%
These little sports utilities are real shake machines especially when you frequent bad roads.

• **Conveniences:** 50%
Storage areas are less numerous on the economy models than on the deluxe, which receive door pockets.

• **Handling:** 50%
Not really dangerous, but you have to watch it because these vehicles are not automobiles and you have to get into tight turns carefully. They are also sensitive to crosswind and to expansion joints. Off road the ground clearance (7.8 inches) is not sufficient and it is easy to get hung up. The Tracker 4x2 is more secure than the 4x4 and handles more like a sports car. It has an uncanny way of getting around on a twisty road when in expert's hands.

• **Safety:** 50%
It remains a delicate item on this type of vehicle because its structure is vulnerable in spite of additional reinforcements in the doors and other places. Tests with dummies show head injuries; air bags are not offered as an option.

WEAK POINTS

• **Noise level:** 10%
The engine, tires and wind get together marvelously to provide a permanent concert.

• **Trunk:** 25%
Reduced to its simplest expression when the rear seats are occupied on short body versions; the 4-door storage is more spacious and modular, thanks to the divided back of the bench seat.

• **Braking:** 20%
It still lacks bite and stopping distances are long but rear wheel ABS stabilizes panic stops.

• **Performance:** 30%
In spite of more powerful engines, they are a drag on the 4x4, which is heavier and even on the 4-door. The torque due to a good transmission gearing is appreciable off road. It is different on the Tracker 4x2 which is snappier and more fun to drive.

• **Interior space:** 30%
Two average height adults are comfortably seated up front. In the back, only the station wagon offers two real seats of adequate length and height.

• **Insurance:** 45%
The 4X4s are always more expensive to insure, particularly since these vehicles address themselves mainly to the younger market.

• **Depreciation:** 40%
It has increased since the market is saturated with many of them.

CONCLUSION

• **Overall rating:** 53.0%
Interesting by their versatility, these machines combine automobiles, utilities and week-end leisure cars. ☹

SPECIFICATIONS & PRICES

Model	Version/Trim	Body/Seats		Wheel-base in	Lgth x Wdthx Hght. inx inx in	Curb weight lb	Towing capacity std. lb	Susp. type ft/rr	Brake type ft/rr	Steering system	Turning diameter ft	Steer. turns nber	Fuel. tank gal.	Standard tire size	Standard power train	PRICES US. $ 1994
SUZUKI		**General warranty: 3 years / 36,000 miles.**														
Sidekick	JS	short ST	3dr.con.4	86.6	142.5x64.2x65.1	2427	1000	i/r	dc/dr/ABS	w/r.	32.2	3.8	11.1	205/75R15	L4/1.6/M5	**$11,449**
Sidekick	JX	short ST	3dr.con.4	86.6	142.5x64.2x65.1	2489	1000	i/r	dc/dr/ABS	w/r.	32.2	3.8	11.1	205/75R15	L4/1.6/M5	**$12,849**
Sidekick	JLX	short ST	3dr.con.4	86.6	142.5x64.2x65.1	2518	1000	i/r	dc/dr/ABS	w/r.	32.2	3.8	11.1	205/75R15	L4/1.6/M5	-
Sidekick	JS	short HT	3dr.sdn.4	86.6	142.5x64.2x65.1	-	1000	i/r	dc/dr/ABS	w/r.	32.2	3.8	11.1	205/75R15	L4/1.6/M5	**$12,999**
Sidekick	JX	short HT	3dr.sdn.4	86.6	142.5x64.2x65.1	-	1000	i/r	dc/dr/ABS	w/r.	32.2	3.8	11.1	205/75R15	L4/1.6/M5	**$14,309**
Sidekick	JX	long HT	4dr.sdn.4	97.6	158.7x64.4x66.5	2679	1000	i/r	dc/dr/ABS	pwr.w/r.	35.4	3.45	14.5	205/75R15	L4/1.6/M5	-
Sidekick	JLX	long HT	4dr.sdn.4	97.6	158.7x64.4x66.5	2729	1000	i/r	dc/dr/ABS	pwr.w/r.	35.4	3.45	14.5	205/75R15	L4/1.6/M5	**$15,719**
Sidekick	JLX	long HT	4dr.sdn.4	97.6	158.7x64.4x66.5	2758	1000	i/r	dc/dr/ABS	pwr.w/r.	35.4	3.45	14.5	205/75R15	L4/1.6/A4	**$16,669**
GEO		**General warranty: 3 years / 36,000 miles.**														
Tracker		2WD short	3dr.con.2/4	86.6	142.5x64.2x64.3	2246	1000	i/r	dc/dr/ABS	w/r.	32.2	3.8	11.1	195/75R15	L4/1.6/M5	**$11,015**
Tracker		4WD short	3dr.con.2/4	86.6	142.5x64.2x65.1	2401	1000	i/r	dc/dr/ABS	w/r.	32.2	3.8	11.1	205/75R15	L4/1.6/M5	**$12,285**
Tracker	LSi	4WD short	3dr.sdn.4	86.6	142.5x64.2x65.1	2443	1000	i/r	dc/dr/ABS	pwr.w/r.	32.2	3.45	11.1	205/75R15	L4/1.6/M5	**$13,650**

See page 393 for complete 1995 Price List

Swift / Metro

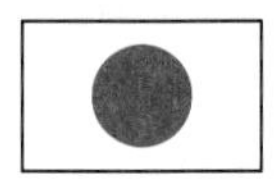

Mini-cars for mini-budgets...

Very popular among young drivers, these little cars have just been redesigned with a more modern appearance but on basics close to the preceding ones. To cosmetics is joined safety because they are stiffer and provided with air bags.

Created by Suzuki, which also produces them for GM, these vehicles are identical in their power train and also within a few equipment details. The Swift-Metro-Firefly change appearance in 1995 with a youthful and agreeable style. The Swift is offered in 3 and 4-doors(in Canada) with a standard 1.0 L, 3-cylinder (in Canada) and a DLX with a 1.3 L. In Canada, as well as in the United States, Geo markets the Standard or LSi Metro in sedans with 3 or 4-door. All of these models receive standard, a manual 5-speed or a 3-speed automatic.

STRONG POINTS

• Fuel consumption: 100%
They remain the most economic gasoline cars on the North American market with a consumption of45 to 55 mpg, depending on the engine and transmission.

• Price/equipment: 90%
These little cars are inexpensive and are often the first car purchased by a young person, so their equipment is often abbreviated, but offers the essentials. Four-door sedans are the best choice.

• Safety: 80%
In spite of work on stiffening, impact tests are not favorable to mini-cars whose structure is not strong and steel is thin. It is the price one pays for the benefits of economic operation, but at least this year, the models are equipped standard with 2 front seat air bags.

• Satisfaction: 70%
No big problems, but some owners are annoyed by lack of quality in various accessories and the rapid aging of these cars.

• Technical: 70%
Their steel unit body includes 4 wheel independent suspension of a

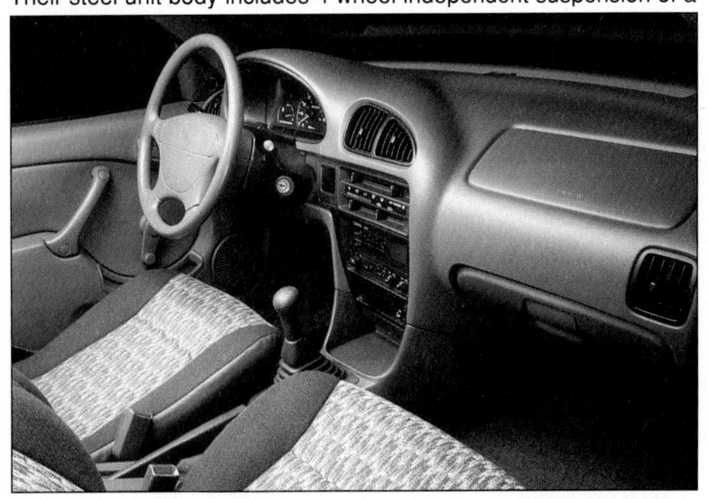

DATA

Category: front-wheel drive sub-compact sedans
Class: 2

HISTORIC

Introduced in: 1984: Forsa 3 dr. 3 cyl.
Modified in: 1987: Forsa 4 dr.; 1988: Swift 4 cyl.
Made in: Kosai, Japan & Ingersoll, Ontario, Canada.

RATINGS

	Swift	Metro
Safety:	40 %	40 %
Satisfaction:	90 %	85 %
Depreciation:	62 %	65 %
Insurance:	($445) 7 %	6.5 % ($445)
Cost per mile:	$0.29	$0.29

NUMBER OF DEALERS

In U.S.A.: NA, Suzuki,4,466 Chevrolet-Geo

SALES ON THE AMERICAN MARKET

Model	1992	1993	Result
Suzuki Swift	6,298	6,608	+ 4.7 %
Geo Metro	83,173	83,399	+ 0.3 %

MAIN COMPETITORS

FORD Aspire, Hyundai Accent 3dr, SUBARU Justy, .

EQUIPMENT

SUZUKI Swift / GEO Metro	Hbk	DLX	base	LSi
Automatic transmission:	O	O	O	O
Cruise control:	-	-	-	-
Power steering:	-	S	-	S
ABS brake:	-	S	O	O
Air conditioning:	O	O	O	O
Air bag (2):	S	S	S	S
Leather trim:	-	-	-	-
AM/FM/ Radio-cassette:	O	O	O	O
Power door locks:	-	-	-	-
Power windows:	-	-	-	-
Tilt steering column:	-	-	-	-
Dual power mirrors:	-	-	-	-
Intermittent wipers:	S	S	S	S
Light alloy wheels:	-	-	-	-
Sunroof:	-	-	-	-
Anti-theft system:	O	O	-	-

S : standard; O : optional; - : not available

COLORS AVAILABLE

Exterior: White, Red, Blue, Black, Gray Green, Magenta.
Interior: Gray, Black.

MAINTENANCE

First revision: 3,000 miles
Frequency: 6,000 miles
Diagnostic plug: No

WHAT'S NEW IN 1995 ?

- All new model in 1995, with 2 air bags in the front seats.
- Five-door sedan eliminated.

Model/ versions *: standard	Type / timing valve / fuel system	Displacement cu/in	Power bhp @ rmn	Torque lb.ft @ rpm	Compres. ratio	Driving wheels / transmission	Final ratio	Acceler. 0-62 mph s	Stand. 1/4 mile s	Stand. 5/8 mile s	Braking 62-0 mph ft	Top speed mph	Lateral acceler. G	Noise level dBA	Fuel economy mpg city	hwy	Gasoline type / octane
1)	L3* 1.0 SOHC-6-IE	60	55 @ 5700	58 @ 3300	9.5 :1	front - M5*	4.39	13.8	19.3	37.2	157	90	0.71	70	52.0	66.0	R 87
						front - A3	4.02	15.0	20.0	38.5	161	87	0.71	70	43.0	52.0	R 87
2)	L4* 1.3 SOHC-8-IE	81	70 @6000	74 @ 3500	9.5 :1	front - M5*	3.79	13.0	18.8	36.4	167	100	0.75	68	43.0	56.0	R 87
						front - A3	3.61	14.0	19.5	37.5	190	94	0.75	68	36.0	46.0	R 87

(ENGINE / TRANSMISSION / PERFORMANCE)

1) base 2) Swift Hbk & DLX SE, Metro LSi

Swift / Metro

McPherson type, as well as disc brakes and drums, but the ABS is only available as an option. Their new silhouettes are easy on the eye and have the merit of reconciling a modern look with an honest aerodynamic finesse and reduced bulk.

• Quality/fit/finish: 70%
The assembly is careful but some of the materials are fragile, the paint coats thin and the interior appearance lacks luster. The finish looks very utilitarian and there is a lot of rattles.

• Driver's compartment: 60%
The driver's position is low, but is satisfactory for visibility, and the instrumentation, simplified to the max, provides essential information. The controls are conventional and well laid out, but those on the console are far from the hand.

• Access: 60%
It remains easier to enter the front of these mini-cars than to get into from the back, either because the doors are narrow (4dr) or because height is missing (3dr).

• Seats: 60%
With little curvature, they don't provide much lateral hold, but their lumbar support is good up front, while at the rear, the bench seats are flat and their seats, just like the backs, are short.

• Suspension: 60%
It doesn't react too harshly to road defects, but it lacks amplitude and sharply marks tar strip crossing, especially under load.

• Steering: 60%
Less imprecise, it retains too high a ratio, which makes it hard to drive on a wet road . Even though the turning diameter is short, maneuverability suffers from the number of turns, lock-to-lock.

• Conveniences: 50%
On the luxury versions, door pockets are available, there are hollows on the console and a small glove box is common to all models.

WEAK POINTS

• Insurance: 15%
Premium is relatively high because the large index compensates for the low price of these cars.

• Braking: 30%
Acceptable in normal use, it lacks effectiveness during panic stops and trajectories become unpredictable when the wheels lock up too fast without ABS.

• Trunk: 30%
Very reduced on the hatchback when the bench seat is used, it can be increased by lowering the back. The sedans trunk also communicates with the interior, which allows storing bulky objects.

• Interior space: 35%
It is surprising, taking into account the bulk of these vehicles, particularly on the 4-doors, which truly are small sedans that have nothing to do with econo-boxes. However, rear seat occupants should not be too big.

• Depreciation: 35%
With compacts in vogue, these minis lose much of their value after 3 years.

• Handling: 40%
It remains fundamentally understeering, and while not dangerous, it requires care in a cross wind, to which these cars are sensitive, but less so than their predecessors. The increase in wheelbase and track brought better stability. The engine noise dominates those of roll and wind but soundproofing is not very highly developed.

NEW FOR 1995

CONCLUSION

• Overall rating: 53.0%
Brought back to the taste of the day, these baby cars will continue to attract fans of economy and mini-budgets. 😐

OWNERS' SUGGESTIONS

(previous model)

-Better general quality (Tires, paint, brake linings, finish ,accessories)
-More neutral handling
-More precise steering
-Improved soundproofing
-Better performance
-More comfortable rear bench seat
-More powerful braking

SPECIFICATIONS & PRICES

Model	Version/Trim	Body/ Seats	Interior volume cu.ft.	Trunk volume cu ft.	Drag coef. Cd	Wheel-base in	Lgth x Wdthx Hght. inx inx in	Curb weight lb	Towing capacity std. lb	Susp. type ft/rr	Brake type ft/rr	Steering system	Turning diameter ft	Steer. turns nber	Fuel. tank gal.	Standard tire size	Standard power train	PRICES US. $ 1994
SUZUKI		**General warranty: 3 years / 50,000 miles; corrosion perforation: 5 years / unlimited.**																
Swift	Hbk	3dr.sdn.4	80.8	8.4	-	93.1	149.4x62.6x54.7	1808	NR	i/i	dc/dr	r&p.	31.5	3.7	10.6	155/80R13	L4/1.3/M5	**$7,659**
Swift	DLX	3dr.sdn.4	80.8	8.4	-	93.1	149.4x62.6x54.7	1808	NR	i/i	dc/dr	r&p.	31.5	3.7	10.6	155/80R13	L4/1.3/M5	-
Swift	base	4dr.sdn.4	91.7	10.3	-	93.1	164.0x62.6x55.1	1938	NR	i/i	dc/dr	r&p.	31.5	3.6	10.6	155/80R13	L4/1.3/M5	**$8,699**
GEO		**General warranty: 3 years / 50,000 miles; corrosion perforation: 5 years / unlimited.**																
Metro	base	3dr.sdn.4	80.8	8.4	-	93.1	149.4x62.6x54.7	1808	NR	i/i	dc/dr	r&p.	31.5	3.7	10.6	155/80R13	L3/1.0/M5	**$7,295**
Metro	LSi	3dr.sdn.4	80.8	8.4	-	93.1	149.4x62.6x54.7	1808	NR	i/i	dc/dr	r&p.	31.5	3.7	10.6	155/80R13	L3/1.0/M5	-
Metro	base	4dr.sdn.4	91.7	10.3	-	93.1	164.0x62.6x55.1	1938	NR	i/i	dc/dr	r&p.	31.5	3.6	10.6	155/80R13	L4/1.3/M5	**$7,795**
Metro	LSi	4dr.sdn.4	91.7	10.3	-	93.1	164.0x62.6x55.1	1938	NR	i/i	dc/dr	r&p.	31.5	3.6	10.6	155/80R13	L4/1.3/M5	-

See page 393 for complete 1995 Price List

More American than Japanese...

For 1995 Toyota renews both ends of the range. At the bottom the Tercel, and at the top it has named a new replacement for the defunct Cressida. It's a great sedan whose main mission is to carry 6 people on 2 bench seats like the Americans of yesterday.

After the retirement of the Cressida in 1992, the Toyota range was without a prestige model. This was not too serious because the buyers could always visit Lexus to find a deluxe model. However, there is a step between Toyota and Lexus which some would not climb for obvious budget reasons. To create the high end of the range, Toyota had to design it within the North American market and to use known elements, because it had to be manufactured on the continent. They did not have to search far, for the Camry is built in Georgetown, Kentucky, the styling center is in California, and the engineering department in Michigan and Arizona. That is how the Avalon was born. It is offered in the unique form of a 5/6 sedan on the standard XL of the deluxe XLS.

STRONG POINTS

• Technical: **90%**
Beginning with the Camry platform, the engineers lengthened its wheelbase by 4 inches and its length by 2.3 inches to dispose of a base allowing them to build a larger sedan than a Camry, able to accept 6 people when a bench seat is used up front. The steel unit body Avalon includes galvanized sections. In the power train, you will find the 3.0 L V6 from the Camry, with slight changes for extra horsepower. The automatic is a 4-speed and the disc brakes are at all 4 wheels. As on the Camry, the front and rear drivetrains are supported by sub-frames which allow isolating the body from road shake.

• Safety: **90%**
It goes without saying that the structure is extremely rigid and that the interior is protected by a cage of steel stampings; 2 air bags protect the front seat occupants.

• Noise level: **90%**
It is as low as in luxury limousines, thanks to eliminating noise vibration and harshness.

• Access: **80%**
The doors open wide, not to interfere with large-sized people and you reach the inside without any problems.

• Quality/fit/finish: **80%**
Toyota has taken care of not going into Lexus level refinements. It's just that the grain of the dashboard truly resembles the one of the deluxe range... As always at Toyota, assembly is very strict, finish is careful, and the materials quality without reproach. It is just that the baggage compartment trim has as much class as the one of a Tercel.

• Driver's compartment: **80%**
The driver soon finds an adequate seating combination and visibility is good all around. The dashboard is shielded by a vast sun visor,

DATA

Category: front-wheel drive full-size sedans.
Class: 6

HISTORIC

Introduced in: 1995
Modified in: -
Made in: Georgetown, Kentuky, U.S.A.

RATINGS

Safety: 90 %
Satisfaction: NA
Depreciation: NA
Insurance: NA
Cost per mile: 0.52 $

NUMBER OF DEALERS

In U.S.A.: 1,233

SALES ON THE AMERICAN MARKET

Model	1992	1993	Result
Avalon	New model on the market.		

MAIN COMPETITORS

ACURA Legend, CHEVROLET Caprice, CHRYSLER LH & LHS, FORD Taurus & Crown Victoria, GM B & H-Series, MAZDA Millenia, NISSAN Maxima.

EQUIPMENT

TOYOTA Avalon	XL	XLS
Automatic transmission:	S	S
Cruise control:	S	S
Power steering:	S	S
ABS brake:	O	S
Air conditioning:	S	S
Air bag (2):	S	S
Leather trim:	-	O
AM/FM/ Radio-cassette:	S	S
Power door locks:	S	S
Power windows:	S	S
Tilt steering column:	S	S
Dual power mirrors:	S	S
Intermittent wipers:	S	S
Light alloy wheels:	O	S
Sunroof:	O	O
Anti-theft system:	-	S

S : standard; O : optional; - : not available

COLORS AVAILABLE

Exterior: White, Black, Ruby,Green, Plum, Red, Platinum, Beige.
Interior: Black, Quartz,

MAINTENANCE

First revision: 4,000 miles
Frequency: 4,000 miles
Diagnostic plug: Yes

WHAT'S NEW IN 1995 ?

-Brand new model on the market.

ENGINE					TRANSMISSION			PERFORMANCE									
Model/ versions *: standard	Type / timing valve / fuel system	Displacement cu/in	Power bhp @ rmn	Torque lb.ft @ rpm	Compres. ratio	Driving wheels / transmission	Final ratio	Acceler. 0-62 mph s	Stand. 1/4 mile s	Stand. 5/8 mile s	Braking 62-0 mph ft	Top speed mph	Lateral acceler. G	Noise level dBA	Fuel economy mpg city	Fuel economy hwy	Gasoline type / octane
base	V6 3.0 DOHC-24-MFI	183	192 @ 5200	210 @ 4400	10.5 :1	front - A4	3.625	8.7	16.8	29.0	125	125	0.80	64	25.0	35.0	M 89

under which are well laid out gauges; all commands are within hand's reach except for the foot-operated parking brake. It would be really funny to find a shift lever under the steering wheel of such a car.

• Seats: 80%
They support relatively well, but do not have much shape and for once, not customary, it seems that the rear bench seat is more sculpted and retains better.

• Suspension: 80%
It is very elaborate and the handling does not hurt the comfort. The car absorbs major road unevenness without jostling the occupants.

• Interior space: 70%
Even if three people cannot pretend to sit up front, the main dimensions of the interior are generous in all directions and one feels at ease with a lot of room for the head and legs in the back.

• Performance: 75%
With a weight-to-power ratio of 16.5 lbs/hp, it is not surprising that the Avalon accelerates from 0 to 62 mph in just under 8.5 seconds and covers the quarter mile in 16.8 seconds. The engine is prompt to react and you don't have to beg for it to rev up. The automatic 4-speed seconds it well, and the shifts are smooth; however, if you down shift manually, the engine braking is virtually absent and there is a disagreeable feel of free wheeling.

• Trunk: 70%
It has a large capacity, but as we said above, its interior trim is econo and the opening a little tight, even though the cutout comes down to bumper level.

• Handling: 70%
Very balanced, the Avalon never feels loose in a tight corner and its roll, lift, or dive seem very well controlled. It doesn't reach an excessive roll angle, doesn't mind being driven hard and its stability in a straight line or curve is great.

• Steering: 70%
It is without a doubt the part that we liked the least because, while precise and fast, it acts dead and doesn't relay vehicle reactions. Perhaps its power assist is a little high.

• Conveniences: 70%
Storage does not lack, because in addition to a large size glove box, there are pockets in the front doors, in the back of the rear front seats, and the retractable center armrest has a container for a cellular phone.

• Braking: 60%
For a car of its format, the Avalon brakes very well in a stable straight line fashion and the stops are short, under 135 ft, and their fade resistance is quite honorable.

• Fuel consumption: 60%
In normal driving it is around 24 mpg, which is nothing exaggerated for a car of this format, capable of remarkable performance.

WEAK POINTS

• Price/equipment: 30%
The XL is the least luxurious of the two Avalons. In relation to the XL it does not have the right to leather trim, the antitheft, and the optional ABS. Its price is between the most luxurious Camry and the Lexus ES300.

CONCLUSION

• Overall rating: 73.5%
From an aesthetic point of view, the Avalon doesn't say much but it is much more elegant in terms of luxury and performance. ☺

SPECIFICATIONS & PRICES

Model	Version/Trim	Body/ Seats	Interior volume cu.ft.	Trunk volume cu ft.	Drag coef. Cd	Wheel-base in	Lgth x Wdthx Hght. inx inx in	Curb weight lb	Towing capacity std. lb	Susp. type ft/rr	Brake type ft/rr	Steering system	Turning diameter ft	Steer. turns nber	Fuel. tank gal.	Standard tire size	Standard power train	PRICES US. $ 1995
TOYOTA		**General warranty: 3 years / 36,000 miles; Power train: 5 years / 60,000 miles;corrosion: 5 years / unlimited; with no deductible or transfert fees.**																
Avalon	XL	4dr.sdn.5/6	105.4	15.4	0.31	107.1	190.2x70.3x56.1	3263	2000	i/i	dc/dc	pwr.r/p.	37.6	NA	18.5	205/65R15	V6/3.0/A4	**$22,758**
Avalon	XLS	4dr.sdn.5/6	105.4	15.4	0.31	107.1	190.2x70.3x56.1	3285	2000	i/i	dc/dc/ABS	pwr.r/p.	37.6	NA	18.5	205/65R15	V6/3.0/A4	**$26,688**

See page 393 for complete 1995 Price List

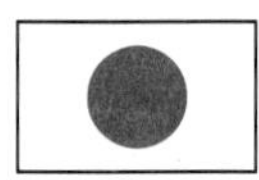

World standard...

Since its introduction, the Camry has become a world's standard in compact cars. To tell the truth, it shows up the intermediates when equipped with its V6 and it can play and win on two boards.

These two compact models know a phenomenal success, which is in the process of surpassing the one known by the compacts after the oil embargos of the '70s. In its 4-cylinder version, the Camry confronts the Honda Accord while the V6 is on the same playing field with the Ford and Mercury Taurus-Sable, the models which compete for the title of most-sold car in the United States. The Camry exists in 4-door sedans or station wagons and 2-door coupes in standard, LE, V6 and V6 LE.

STRONG POINTS

• Satisfaction: 90%
It reaches a high level, which makes the reputation of the marque.

• Safety: 90%
In spite of body reinforcements, which stiffen the structure, the presence of 2 air bags up front and 3-point seat belts, there is still a little work to do to reach the maximum rating.

• Technical: 80%
The Camry lines are not revolutionary, but their aerodynamics are effective with a drag coefficient that varies between 0.31 and 0.33. It has a steel unit body, and independent McPherson suspension at all 4 wheels, stabilizer bars, and gas shocks. The brakes are disc and drum with the 4-cylinder engine and 4 discs on the V6. For best noise and vibration insulation, the front and rear drivetrains are mounted on sub-frames, isolated from the body by rubber mounts. The body receives codings of different soundproofing material to cut down on resonance. The body, doors, hood and truck lid are ribbed to increase their stiffness. The base engine is a 2.2 L 4-cylinder which delivers 125 hp and the V6 is a new 188 hp 3.0 L. The 2 engines dispose of a choice of manual 5-speed standard, or an optional computer-controlled automatic.

• Quality/fit/finish: 80%
The assembly is very particular and the finish meticulous, but the plastics don't look rich, and the interior is fairly dull.

• Driver's compartment: 80%
Visibility is excellent with reduced blind spots, even on the station wagon whose backlight is more narrow. The driver is rather well installed, thanks to the combined seat and steering column adjustments. Gauges are easy to consult and controls are at your finger tips. However, some switches are hard to reach when the seat is moved forward.

• Access: 80%
The interior volume helps access because the doors are full sized and open wide.

• Suspension: 80%
It, together with the soundproofing, is the largest gain of this generation and it absorbs road defects well, thanks to its large wheel travel and well-calibrated shock absorbers.

DATA

Category: front-wheel drive compact (4cyl) and mid-size (6cyl) sedans.
Class: 4 & 5

HISTORIC

Introduced in: 1983
Modified in: 1986: V6 2.6L; 1988:all-wheel drive transmission.
Made in: Georgetown, KY, U.S.A. & Isustumi, Japan.

RATINGS

		L4	V6	
Safety:		92 %	92 %	
Satisfaction:		92 %	92 %	
Depreciation:		57.5 %	56 %	
Insurance:	($585)	4.6 %	4.2 %	($655)
Cost per mile:		$0.40	$0.45	

NUMBER OF DEALERS

In U.S.A.: 1,233

SALES ON THE AMERICAN MARKET

Model	1992	1993	Result
Camry	286,602	299,737	+ 4.4 %

MAIN COMPETITORS

BUICK Skylark & Regal, CHEVROLET Lumina, CHRYSLER Cirrus- Stratus & LH, FORD Taurus-Sable, Contour-Mystique, HONDA Accord, HYUNDAI Sonata, MAZDA 626, NISSAN Maxima & Altima, OLDSMOBILE Achieva & Cutlass Supreme, PONTIAC Grand Am & Grand Prix, SUBARU Legacy, VOLKSWAGEN Passat.

EQUIPMENT

TOYOTA Camry	DX	LE	SE	XLE
Automatic transmission:	O	S	O	S
Cruise control:	O	S	S	S
Power steering:	O	S	O	S
ABS brake:	O	O	O	S
Air conditioning:	O	S	S	S
Air bag (2):	S	S	S	S
Leather trim:	-	O	O	O
AM/FM/ Radio-cassette:	O	S	S	S
Power door locks:	O	S	S	S
Power windows:	O	S	S	S
Tilt steering column:	-	S	O	S
Dual power mirrors:	O	S	S	S
Intermittent wipers:	S	S	S	S
Light alloy wheels:	-	O	S	S
Sunroof:	-	O	O	O
Anti-theft system:	-	-	-	-

S : standard; O : optional; - : not available

COLORS AVAILABLE

Exterior: White, Silver, Black, Bordeaux, Red, Beige, Emerald,Taupe, Champagne.
Interior: Gray, Taupe, Oak.

MAINTENANCE

First revision: 4,000 miles
Frequency: 4,000 miles
Diagnostic plug: Yes

WHAT'S NEW IN 1995 ?

- **The 4-cylinder LE coupe is available with a manual 5-speed.**
- **Seat design on the LE versions.**
- **Hub caps for the LE sport.**
- **Seats with electrical adjustments, standard (SE, XLE)**

ENGINE						TRANSMISSION		PERFORMANCE										
Model/ versions *: standard	Type / timing valve / fuel system	Displacement cu/in	Power bhp @ rmn	Torque lb.ft @ rpm	Compres. ratio	Driving wheels / transmission	Final ratio	Acceler. 0-62 mph s	Stand. 1/4 mile s	Stand. 5/8 mile s	Braking 62-0 mph ft	Top speed mph	Lateral acceler. G	Noise level dBA	Fuel economy mpg city	hwy	Gasoline type / octane
base	L4 2.2 DOHC-16-MPFI	132	135 @ 5400	145 @ 4400	9.5:1	front- M5	3.94	11.0	18.2	32.3	128	109	0.78	65	27.0	38.0	R 87
						front-A4	3.94	11.8	18.6	32.8	135	106	0.78	65	25.0	35.0	R 87
V6	V6 3.0 DOHC-24-MPFI	152	188 @ 5200	203 @ 4400	9.6:1	front-A4	3.62	9.5	17.6	29.8	131	119	0.80	64	22.0	31.0	R 87

- **Insurance: 80%**
Even though expensive, it's in the middle of its category.
- **Conveniences: 80%**
Storage spots are numerous, practical and well laid out. The AC controls are equally well laid out and easy to adjust.
- **Handling: 70%**
In a straight line as in a wide curve, the Camry travels as if on rails, very insensitive to cross winds. With the firmer sport suspension, one is still more at ease in a wide curve than on a twisty path. The body rolls progressively on tight turns without excess which allows it to remain neutral for a long time.
- **Steering: 70%**
Smooth at low speeds, it is a little light when speed increases. Without losing anything of its precision or liveliness, its steering ratio allows good maneuverability. The power assist is too strong, tends to level out any road feel and requires steady attention.
- **Trunk: 60%**
It can be enlarged by lowering the backs of the bench seat. The storage area of the station wagon is vast and its regular forms allow it to accept much baggage.
- **Performance: 60%**
They have really not progressed. Since the new V6 is substantially more powerful, the weight is up. They allow good takeoff but passing is still a little placid. Response with a 4-cylinder and an automatic is anemic.
- **Seats: 60%**
They are not very comfortable, have little shape, and the poor padding does not insure adequate support on long trips.
- **Fuel consumption: 60%**
If the 4-cylinder is less thirsty in urban travel, on the freeways, at cruise speed, the V6 comes into its own.

- **Noise level: 60%**
It was one of the lowest levels measured on a high production car but it seems to be a little greater today. However, during strong takeoffs, the 4-cylinder reveals its presence and the V6 has an annoying roar at 2,500 rpm.
- **Price/equipment: 50%**
The good reputation of these models keeps the price high, which has not interfered with the Camry selling like hot cakes. The equipment is more complete on the LE versions than on the more stripped down base model.
- **Interior space: 50%**
The height under the sunroof is ample and the rear seats offer much legroom even when the front seats are pulled all the way back. In the sedan, you can get as an option, a retractable 3rd bench seat which provides 2 rear facing seats for children.

WEAK POINTS

- **Braking: 40%**
It remains perfectible in spite of its progress since stopping distances are long, with or without ABS, but in the second case, the wheels don't lock in a panic stop.
- **Depreciation: 40%**
Prices are falling because a great number of these cars are available on the used-car market.

CONCLUSION

- **Overall rating: 68.0%**
Camry has set the standard of the half-compact half-intermediate wagon and a number of competitors are trying to get inspired by it, without success. ☺

SPECIFICATIONS & PRICES

Model	Version/Trim	Body/ Seats	Interior volume cu.ft.	Trunk volume cu ft.	Drag coef. Cd	Wheel-base in	Lgth x Wdthx Hght. inx inx in	Curb weight lb	Towing capacity std. lb	Susp. type ft/rr	Brake type ft/rr	Steering system	Turning diameter ft	Steer. turns nber	Fuel. tank gal.	Standard tire size	Standard power train	PRICES US. $ 1994
TOYOTA		**General warranty: 3 years / 36,000 miles; Power train: 5 years / 60,000 miles;corrosion: 5 years / unlimited; with no deductible or transfert fees.**																
Camry	Dx	2dr.cpe.4	97	14.8	0.32	2620	187.8x69.7x52.0	2910	2000	i/i	dc/dr	pwr.r&p.	35.4	3.06	18.5	195/70R14	L4/2.2/M5	**$16,428**
Camry	DX	4dr.sdn.5	97	15	0.32	107.1	187.8x69.7x55.1	2932	2000	i/i	dc/dr	pwr.r/p.	35.4	3.06	18.5	195/70R14	L4/2.2/M5	**$16,718**
Camry	LE	2dr.cpe.4	97	14.8	0.32	2620	187.8x69.7x55.0	3064	2000	i/i	dc/dr	pwr.r&p.	35.4	3.06	18.5	195/70R14	L4/2.2/A4	**$19,628**
Camry	LE	4dr.sdn.5	97	15	0.32	107.1	187.8x69.7x55.1	3086	2000	i/i	dc/dr	pwr.r/p.	35.4	3.06	18.5	195/70R14	L4/2.2/A4	**$19,558**
Camry	LE	4dr.wgn.5	100	40.5	0.32	2620	189.4x69.7x56.3	3219	2000	i/i	dc/dr	pwr.r&p.	35.4	3.06	18.5	195/70R14	L4/2.2/A4	**$20,968**
Camry	SE	2dr.cpe.4	97	14.8	0.33	2620	187.8x69.7x55.0	3285	2000	i/i	dc/dc/ABS	pwr.r&p.	36.7	2.98	18.5	205/60VR15	V6/3.0/A4	**$22,618**
Camry	SE	4dr.sdn.5	97	15	0.33	107.1	187.8x69.7x55.1	3208	2000	i/i	dc/dc/ABS	pwr.r/p.	36.7	2.98	18.5	205/60VR15	V6/3.0/A4	**$22,908**
Camry	XLE	4dr.wgn.5	100	40.5	0.32	2620	189.4x69.7x56.3	3404	2000	i/i	dc/dc/ABS	pwr.r&p.	35.4	2.98	18.5	205/60VR15	V6/3.0/A4	**$25,208**

See page 393 for complete 1995 Price List

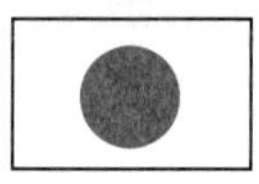

Laid back sport...

Renewed last year, the Celica coupe has again found a sympathetic and aggressive appearance which makes it recognizable at a glance. Targeted for an average buyer group, it is offered with a reasonable power train and economy that goes well with the social climate of our decade.

The Toyota Celica is in its 6th generation and by some aspects it is inspired by its big sister, the Supra. Its lines have returned to normal, after experimenting more or less successfully with some futuristic appearances in the generation before last. The lines of the 2 new bodies, a 2-door ST, and a 3-door GT are particularly successful. These cars differ very little because the rears are treated almost identically and you have to look carefully to differentiate the model with a hatchback from the one with a separate trunk.

STRONG POINTS

• Satisfaction: 95%
Celicas have always set reliability and durability records, even though this type of car is often manhandled by their sports-minded owners.

• Safety: 90%
The structure of the latest Celica has been stiffened to increase impact resistance and 2 air bags insure the protection of the front seat occupants.

• Technical: 80%
The steel unit body of the new Celicas is 10% larger than the model it replaces, which doesn't keep it from being 20% more rigid. Like the Camry and Supra, the Celica has 2 sub-frames which support the front suspension to isolate the body from noise and vibrations comlng directly from the drivetrain. The rear suspension has been reinforced and simplified it receives a wider track and the mounting points have been improved. A rotary valve mounted on the power assist system improves the steering feel as you turn the wheel. The brakes are mixed on the ST and have 4 discs on the GT but the ABS come only as an option. Cylinder displacement on the ST has gone from 1.6 to 1.8 while that of the GT has remained identical to the old model.

DATA

Category: front-wheel drive sport coupes.
Class: 3S

HISTORIC

Introduced in: 1971
Modified in: 1977, 1981, 1985 (front-wheel drive); 1989:Tbo 4x4; 1990.
Made in: Tahara, Japan.

RATINGS

Safety: 90 %
Satisfaction: 95 % (previous version)
Depreciation: 20 % (1 year)
Insurance: 6.4 % ($785)
Cost per mile: $0.43

NUMBER OF DEALERS

In U.S.A.: 1,233

SALES ON THE AMERICAN MARKET

Model	1992	1993	Result
Célica	41,750	29,237	- 30.0 %

MAIN COMPETITORS

ACURA Integra, EAGLE Talon, CHEVROLET Camaro, DODGE Stealth, FORD Mustang, FORD Probe, HONDA Prelude, MAZDA MX-3 & MX6, NISSAN 240SX, PONTIAC Firebird, TOYOTA MR2, VOLKSWAGEN Corrado.

EQUIPMENT

TOYOTA Celica	Liftback ST	Liftback GT
Automatic transmission:	O	O
Cruise control:	O	O
Power steering:	O	O
ABS brake:	O	O
Air conditioning:	O	O
Air bag (2):	S	S
Leather trim:	-	O
AM/FM/ Radio-cassette:	O	S
Power door locks:	O	S
Power windows:	O	S
Tilt steering column:	-	O
Dual power mirrors:	S	S
Intermittent wipers:	S	S
Light alloy wheels:	-	O
Sunroof:	-	-
Anti-theft system:	-	-

S : standard; O : optional; - : not available

COLORS AVAILABLE

Exterior: White, Black, Red, Crimson, Turquoise, Yellow, Blue, Silver.
Interior: Blue, Gray.

MAINTENANCE

First revision: 4,000 miles
Frequency: 4,000 miles
Diagnostic plug: Yes

WHAT'S NEW IN 1995 ?

- New convertible for sale in the US only.

Model/ versions *: standard	ENGINE Type / timing valve / fuel system	Displacement cu/in	Power bhp @ rmn	Torque lb.ft @ rpm	Compres. ratio	TRANSMISSION Driving wheels / transmission	Final ratio	PERFORMANCE Acceler. 0-62 mph s	Stand. 1/4 mile s	Stand. 5/8 mile s	Braking 62-0 mph ft	Top speed mph	Lateral acceler. G	Noise level dBA	Fuel economy mpg city	hwy	Gasoline type / octane
Liftback	L4* 1.8-DOHC-16-EFI	108	110 @ 5600	115 @ 2800	9.5 :1	front - M5*	4.058	9.5	17.5	32.8	138	112	0.87	68	32.0	44.0	R 87
						front - A4	2.821	10.6	18.2	33.4	144	109	0.87	68	31.0	42.0	R 87
GT	L4* 2.2-DOHC-16-EFI	132	135 @ 5400	145 @ 4400	9.5 :1	front - M5*	4.176	9.0	16.5	31.5	148	125	0.87	68	27.0	38.0	R 87
						front - A4	4.176	10.2	17.3	32.0	144	118	0.87	68	27.0	39.0	R 87

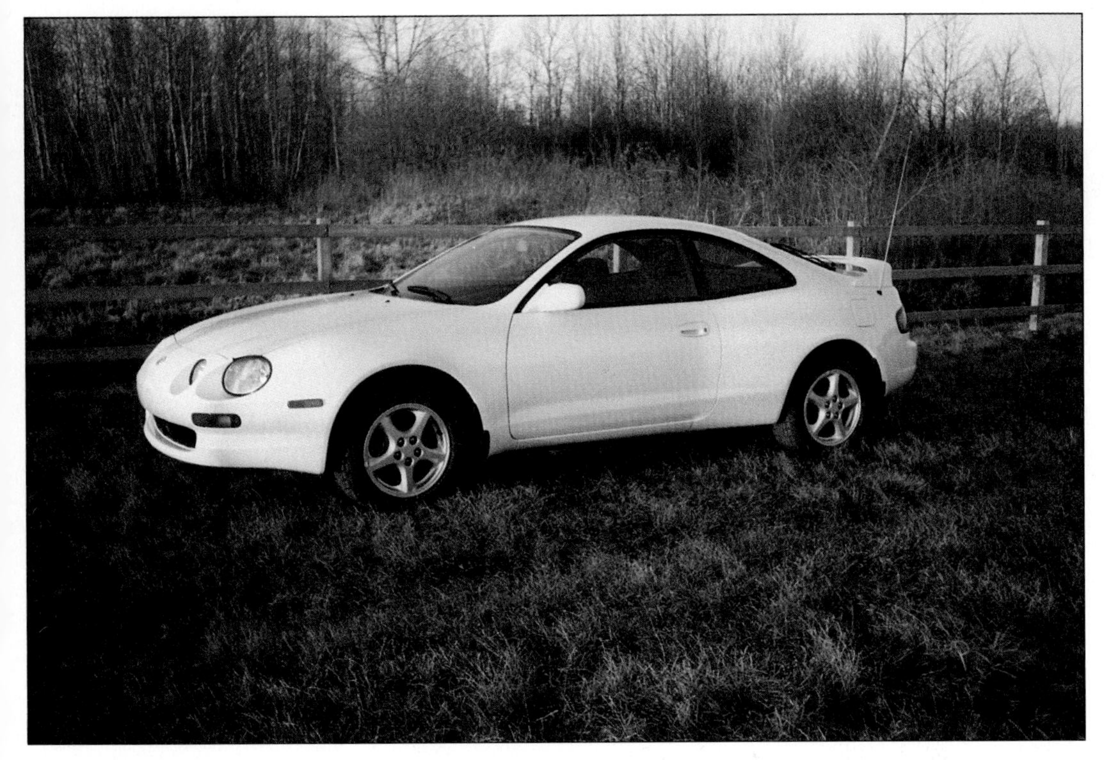

• **Fuel consumption: 80%**
It is very reasonable, even at high speeds. When respecting the legal limits it is possible to reach more than 35 mpg, remarkable for this type of car.
• **Quality/fit/finish: 80%**
In spite of the dull interior presentation, assembly is as strict as the finish but the quality of the materials makes it look very ordinary.
• **Driver's compartment: 80%**
Its well-organized dashboard is inspired by the Supra, but less rakish. The center portion of the console is oriented toward the driver, to put the controls within his hand's reach. Driving position is comfortable and visibility is satisfactory, even though the backlight is a little narrow and steeply inclined. The gauges have been smartly regrouped and controls are conventional.
• **Steering: 80%**
Well-modulated, it has an average steering ratio and is sufficiently incisive to provide a rapid control. Maneuverability is quite good and the steering diameter is reasonable; good for getting around in traffic.
• **Depreciation: 80%**
At the end of the first year the latest Celica has lost less than the average of its competitors. It must be said that it is the champion of reliability, which pays off.
• **Performance: 70%**
Even though the numbered results are slightly lower with the base engine, it is what gives the Celica the most character. Accelerations up to 60 mph are very close to that of the 2.0 L of the GT and there is very little difference in other circumstances, to the point that you wonder if it is really necessary to offer two engines which are this close. A 175 hp Turbo would make a major difference.
• **Handling: 70%**
The Celica is fun to drive because its reactions are crisp and clean. Pushed a little, it will reveal its understeering after a long period of neutrality. Equipped with a V6 it will provide fun close to that of the Corrado, which is the reference in its category.
• **Seats: 70%**
Even though its presentation is not very attractive, it offers good comfort even at the rear, where they are well-shaped for auxilliary seating.
• **Conveniences: 70%**
The practical side has not been neglected because in spite of the presence of a right side air bag, the glove box has ample volume and is backed up by a storage - armrest on the center console and two door pockets, plus trays and hollows here and there.
• **Access: 65%**
It is easy to take a seat in the back, except for the reduced roof height.
• **Suspension: 60%**
Firm but not excessively so, it is never disagreeable even on bad pavement.
• **Insurance: 60%**
The Celica premium is in the average of the sports coupes and can be considered normal under the circumstances.
• **Braking: 50%**
It is disappointing on both modes, and shows similar results by the lack of bite and the long panic stop distances. However, it brakes straight and has good fade resistance.
• **Noise level: 50%**
Lightening the body translates into a thinner, more resonant structure and the soundproofing materials are used in smaller quantities. The noise level is not forbidding but it is always present and persistent.
• **Price/equipment: 50%**
They are not given away but the Celicas do not lack for qualities to offset the financial effort they require. The reliability, driving pleasure, comfort and practical side will not make you hesitate long.

WEAK POINTS

• **Interior space: 40%**
It is not really a fault, because just like the trunk it derives from the particular shape of the vehicle. However, height and length are still more limited at the rear.
• **Trunk: 40%**
Its initial volume is more than adequate and it can be increased still further by removing the rear bench seat. However, its high sill does complicate handling heavy baggage.

CONCLUSION

• **Overall rating: 68.0%**
The Celica is back to being what it has always been; a popular sports car with good driving pleasure, a practical use and an economical operation. For once, it is the least expensive model that carries our vote. ☺

SPECIFICATIONS & PRICES

Model	Version/Trim	Body/ Seats	Interior volume cu.ft.	Trunk volume cu ft.	Drag coef. Cd	Wheel-base in	Lgth x Wdthx Hght. inx inx in	Curb weight lb	Towing capacity std. lb	Susp. type ft/rr	Brake type ft/rr	Steering system	Turning diameter ft	Steer. turns nber	Fuel. tank gal.	Standard tire size	Standard power train	PRICES US. $ 1994
TOYOTA		**General warranty: 3 years / 36,000 miles; power train: 5 years / 60,000 miles;corrosion:5 years / unlimited; without deductibles or transfert fees.**																
Celica Liftback ST		2dr.cpe.2+2	-	16.2	0.32	100.0	174.2x68.9x50.8	2414	1000	i/i	dc/dr/ABS	pwr.r&p.	34.1	2.9	15.9	185/70R14	L4/1.6/M5	**$16,168**
Celica Liftback GT		3dr.cpe.2+2	-	16.2	0.32	100.0	174.2x68.9x50.8	2579	1000	i/i	dc/dc/ABS	pwr.r&p.	34.1	2.9	15.9	205/55R15	L4/2.2/M5	**$18,428**
Celica convertible		2dr.con.2+2	-	16.2	0.32	100.0	174.2x68.9x50.8	2754	NR	i/i	dc/dc/ABS	pwr.r&p.	34.1	2.9	15.9	205/55R15	L4/2.2/M5	-

See page 393 for complete 1995 Price List

Corolla-Prizm

The world car...

The Corolla remains to this day the only world car sold in 54 countries, manufactured in several countries on five continents, it is the stereotype of the vehicle that solves the transportation problems of the greatest number of people on earth. One has a tendency to forget it...

If the Corolla is the most popular compact in the world, the Prizm offers the best perceptible quality in the USA if one believes in the studies of J.D. Power. These medium-sized cars are identical within a few equipment details. The Corolla, which is in its seventh generation, is one of the most-sold cars in the world. It is offered in the form of sedans and 4-door station wagons in base or LE, while the Prizm is only available as a sedan - base and LSi. The latter is marketed only in the US by the Geo Division of General Motors and is manufactured in Fremont, California. The base versions are powered by a 1.6 L 4-cylinder DOHC engine with 16 valves which delivers 105 hp, while the LE/LSi and station wagon with a 1.8 L provide 115 hp.

STRONG POINTS

• Safety: **90%**
The body has been reinforced at a number of locations, including the doors, for greater impact resistance. Two air bags at the front seats give it the maximum rating. Note that the 3-point seat belts are adjustable in height.

• Satisfaction: **90%**
The reliability and durability stems from the care given to design, assembly and finishing which attain a unique level.

• Quality/fit finish: **80%**
The new interior is simple, clean and functional. The dashboard plastics have a good appearance and so does the fabric trim of the seats and doors. The assembly and finish are adjusted, and the front part of the body is protected by a flexible coating to keep paint from chipping under the impact of road gravel.

• Driver's compartment: **80%**
When the steering column is adjustable, the driving position becomes

DATA

Category: front-wheel drive compact sedans and wagons.
Class: 3

HISTORIC

Introduced in: 1966
Modified in: 1970-74-79-83-89-93.
Made in: Cambridge, Ont., Canada Fremont, CA, U.S.A.

RATINGS

Safety: 90 %
Satisfaction: 90 %
Depreciation: 50 %
Insurance: 5.5 % ($520)
Cost per mile: $0.32

NUMBER OF DEALERS

In U.S.A.: 1,233 Toyota, 4,466 Chevrolet-Geo

SALES ON THE AMERICAN MARKET

Model	1992	1993	Result
Corolla	196,118	193,749	- 1.3 %
Prizm	74,346	74,053	- 0.4 %

MAIN COMPETITORS

CHEVROLET Cavalier, DODGE-PLYMOUTH Colt-Neon, EAGLE Summit, FORD Escort, HONDA Civic 4dr., HYUNDAI Elantra, MAZDA Protegé, NISSAN Sentra, PONTIAC Sunfire, SATURN SL1 & SL2, SUBARU Impreza, VOLKSWAGEN Jetta.

EQUIPMENT

TOYOTA Corolla sedan	**base**	**DX**	**LE**		
TOYOTA Corolla wagon		**DX**			
GEO Prizm				**base**	**LSi**
Automatic transmission:	O	O	O	O	O
Cruise control:	O	O	S	O	O
Power steering:	O	S	S	O	O
ABS brake:	O	O	O	S	S
Air conditioning:	O	O	O	S	O
Air bag (2):	S	S	S	S	S
Leather trim:	-	-	-	-	-
AM/FM/ Radio-cassette:	O	O	O	O	O
Power door locks:	-	-	S	O	O
Power windows:	O	O	O	O	O
Tilt steering column:	-	O	S	O	S
Dual power mirrors:	-	O	O	O	O
Intermittent wipers:	O	O	S	O	O
Light alloy wheels:	-	-	O	-	O
Sunroof:	-	O	O	-	O
Anti-theft system:	-	-	-	-	-

S : standard; O : optional; - : not available

COLORS AVAILABLE

Exterior: White, Red, Beige, Green, Taupe.
Interior: Corolla: Blue, Gray, Taupe. Prizm: Light & Dark Gray.

MAINTENANCE

First revision: 4,000 miles
Frequency: 4,000 miles
Diagnostic plug: Yes

WHAT'S NEW IN 1995 ?

- **Bumpers with matching colors.**
- **Better quality sound system in the station wagon**

Model/ versions *: standard	ENGINE Type / timing valve / fuel system	Displacement cu/in	Power bhp @ rmn	Torque lb.ft @ rpm	Compres. ratio	TRANSMISSION Driving wheels / transmission	Final ratio	PERFORMANCE Acceler. 0-62 mph s	Stand. 1/4 mile s	Stand. 5/8 mile s	Braking 62-0 mph ft	Top speed mph	Lateral acceler. G	Noise level dBA	Fuel economy mpg city	hwy	Gasoline type / octane
1)	L4* 1.6 DOHC-16-EFI	97	105 @ 5800	100 @ 4800	9.5 :1	front - M5*	3.722	11.0	17.6	32.2	134	109	0.76	65	33.0	45.0	R 87
						front - A3	3.526	11.9	18.2	33.0	131	106	0.76	65	31.0	38.0	R 87
2)	L4 1.8 DOHC-16-EFI	107	105 @ 5200	117 @ 2800	9.5 :1	front - M5*	3.722	9.8	16.5	30.4	148	112	0.78	67	34.0	45.0	R 87
						front - A4	2.821	11.0	18.3	32.7	141	109	0.78	67	32.0	44.0	R 87

1) base sedan Corolla/Prizm 2) Corolla sedan LE & wagon, Prizm LSi

adequate; the seats of the deluxe versions offer a superior hold and lumbar support compared to the base versions. Visibility is excellent from all angles and the ergonomics are carefully planned. The dashboard with its rounded forms is tied into a center console which is oriented toward the driver. This makes the radio and AC controls easily accessible. The main controls are gathered around the steering wheel, whose thick rim fits the hand. Based on the version, the instrumentation is grouped in 3- or 4-pods, kept well in sight.

• Fuel consumption: 80%
It is reasonable considering the plusher format of these models and rarely drops below 28 mpg.

• Price/equipment: 80%
These cars have become costly especially when you consider that their equipment is rather limited and that the long option list includes safety items that should be standard.

• Insurance: 75%
The index and the high prices combine into an expensive premium for a conservative car oriented toward the great public.

• Technical: 70%
The lines of the latest Corolla/Prizm edition are more aerodynamic, without breaking any records, with a drag coefficient of 0.33 for the sedans and 0.35 for the station wagon. The steel unit body has a McPherson independent 4-wheel suspension. Front end geometry is designed for the best stability and the more neutral feel in curves. Brakes are mixed and the ABS is only available as an option. Power steering is available only on the deluxe versions.

• Suspension: 70%
It is softer and, like the Camry, better at soaking up road faults, even though its drivetrains are not mounted on sub-frames.

• Access: 70%
There is easy access to either body, with ample door width and roof height, even at the rear seats.

• Seats: 70%
The LE/LSi offer ample hold and lumbar support, more effective than the base model, whose seats have less shape and whose padding is thinner and harder.

• Noise level: 70%
The Corolla/Prizms are now more silent thanks to body rigidity and effective soundproofing. In fact, the sound level is equal to that of more costly and sophisticated machines.

• Conveniences: 70%
There is an ample amount of storage areas, which include a glove box, (LE/LSi) and hollows on the center console.

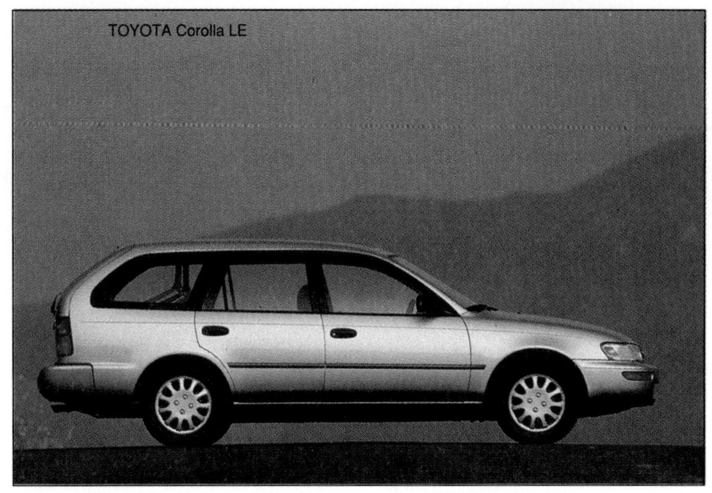
TOYOTA Corolla LE

• Steering: 70%
The manual is not comfortable compared to the power assist which is smooth and precise, but has too high a ratio which calls for windmilling the steering wheel during parking.

• Handling: 60%
It has improved and the Corolla holds the road better than in the past. New settings improve directional ability and cornering. The Corolla/Prizms are easier to get into a turn and to set up. It results in a neutral feel and if you push harder, the understeering appears slowly.

• Trunk: 50%
Their cargo volume is as easy to use on the sedan as on the wagon. Both have simple, easy-to-load shapes and can be enlarged by folding down one or both sides of the split rear seats.

• Interior space: 50%
These cars accept 4 adults in comfort, because the length and height are ample. However, a fifth person will not be comfortable for long in the middle of the bench seat.

• Depreciation: 50%
You can recover a good percentage of the original in a resale because the Corolla/Prizms are scarce and expensive in the used car market.

WEAK POINTS

• Performance: 40%
The two engines differ by the torque level, higher on the 1.8 L which helps low end acceleration. Otherwise, their performances are very close. The accelerations are not thunder-like because the weight-to-power ratio is 22 lbs/hp for the 1.6 L and 9.5 for the 1.8 L.

• Brakes: 45%
Considering that the front wheels block immediately during panic stops, one cannot understand that Toyota considers this ABS as an optional luxury. Even though the stopping distances are slightly shorter than before, they are still too long for a car that only weighs a ton.

CONCLUSION

• Overall rating: 68.0%
Agreeable to look at, to use and to resell, the Corolla/Prizms are part of a select group, the few best cars in the world. ☺

SPECIFICATIONS & PRICES

Model	Version/Trim	Body/ Seats	Interior volume cu.ft.	Trunk volume cu ft.	Drag coef. Cd	Wheel-base in	Lgth x Wdthx Hght. inx inx in	Curb weight lb	Towing capacity std. lb	Susp. type ft/rr	Brake type ft/rr	Steering system	Turning diameter ft	Steer. turns nber	Fuel. tank gal.	Standard tire size	Standard power train	PRICES US. $ 1994
TOYOTA		**General warranty: 3 years / 36,000 miles; power train: 5 years / 60,000 miles;corrosion: 5 years / unlimited; without deductibles or transfert fees**																
Corolla	base	4dr.sdn.4/5	89.3	12.7	0.33	97.0	172.0x66.3x53.5	2326	NR	i/i	dc/dr	r&p.	32.2	4.09	13.2	175/65R14	L4/1.6/M5	**$12,098**
Corolla	DX	4dr.sdn.4/5	89.3	12.7	0.33	97.0	172.0x66.3x53.5	2392	NR	i/i	dc/dr	pwr.r&p.	32.2	3.35	13.2	185/65R14	L4/1.8/M5	**$13,188**
Corolla	DX	4dr.wgn.4/5	91.1	31.4	0.36	97.0	172.0x66.3x55.3	2425	NR	i/i	dc/dr	pwr.r&p.	32.2	3.35	13.2	185/65R14	L4/1.8/M5	**$14,298**
GEO		**General warranty: 3 years / 36,000 miles.**																
Prizm	base	4dr.sdn.4/5	89.3	12.7	0.33	97.0	172.0x66.3x53.5	2348	NR	i/i	dc/dr	pwr.r&p.	32.2	3.35	13.2	175/65R14	L4/1.6/M5	**$11,070**
Prizm	LSi	4dr.sdn.4/5	89.3	12.7	0.33	97.0	172.0x66.3x52.4	2359	NR	i/i	dc/dr	pwr.r&p.	32.2	3.35	13.2	185/65R14	L4/1.8/M5	**$11,840**

See page 393 for complete 1995 Price List

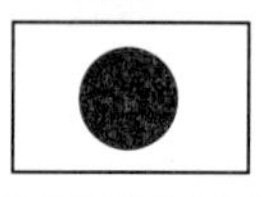

Globe-trotter...

Wherever you go in the world, especially if it is a far-away and hard-to-reach place, you will without fail find a Land Cruiser. In many countries, it has taken over the watch from the Land Rover, and is the ideal vehicle to plan a world tour or to go for a ride in the country.

In North America, the Land Cruiser is marketed only in the United States, where it has sold near 8,962 copies in 1993. Its size is impressive, compared to a 4Runner which looks like its baby... It is sold only as a 4-door wagon, whose hatchback opens in two parts and is offered in a unique trim to which can be grafted several optional assemblies

STRONG POINTS

• Satisfaction: 90%
According to its owners, the Land Cruiser works like a clock and gives no major problems.

• Interior space: 80%
You don't lack room aboard a Land Cruiser, thanks to its generous dimensions. Five people can normally be seated, 8 when the optional third seat is installed in the storage area. Height and width are largely sufficient but the lengthwise space between the seats is a bit short, especially if the front seats are pulled back.

• Storage: 80%
Even though its access is not the easiest considering the way the hatchback opens, its volume is remarkable and it can be increased further by folding the median seat bench. Some baggage space still remains, even if a third bench is installed.

• Quality/fit/finish: 80%
The general presentation is that of a utility and has nothing to do with the luxury of a Range Rover, or a Grand Cherokee, even with optional leather seats.

• Access: 80%
Even without a footrest, it is not hard to climb aboard, thanks to several handles available for this purpose but it is not easy at all to access the third bench.

• Seats: 70%
They are well-padded, do not provide enough lateral hold but are equipped with headrests even at the rear seats.

• Suspension: 70%
The presence of rigid axles does not make itself felt too much because there is ample wheel travel to absorb large road bumps and dips. At the most, you can complain of a shake coming from the front end, which disappears as though by magic when the vehicle is loaded.

• Driver's compartment: 70%
It resembles, at first glance, most compact pickups and the 4Runner, with its massive dashboard on which the gauges and commands are installed under a large shield. The driver is comfortable, even though the seat is too widened out to provide lateral support. The gauges are grouped into easy-to-read clusters and the main controls are set up in ergonomic fashion, except for the AC, under the steering. The radio, just like the gear shift and transfer box, is far away. The locks for the

DATA

Category: 4WD all-terrain multipurpose vehicle.
Class: utility

HISTORIC

Introduced in: 1951
Modified in: 1980, 1990: body.
Made in: Hino, Japan.

RATINGS

Safety: 75 %
Satisfaction: 88 %
Depreciation: 57 %
Insurance: 4.5 % ($785)
Cost per mile: $0.48

NUMBER OF DEALERS

In U.S.A.: 1,233

SALES ON THE AMERICAN MARKET

Model	1992	1993	Result
Land Cruiser	7,907	8,962	+ 11.8 %

MAIN COMPETITORS

GMC Suburban, ISUZU Trooper, JEEP Grand Cherokee, LAND ROVER Range Rover-Discovery, MITSUBISHI Montero.

EQUIPMENT

TOTYOTA Land Cruiser	base
Automatic transmission:	S
Cruise control:	O
Power steering:	S
ABS brake:	O
Air conditioning:	S
Air bag:	O
Leather trim:	O
AM/FM/ Radio-cassette:	S
Power door locks:	O
Power windows:	O
Tilt steering column:	S
Dual power mirrors:	O
Intermittent wipers:	S
Light alloy wheels:	O
Sunroof:	O
Anti-theft system:	S

S : standard; O : optional; - : not available

COLORS AVAILABLE

Exterior: White, Red, Green, Blue, Beige, Silver, Dark Gray.
Interior: Blue, Gray.

MAINTENANCE

First revision: 4,000 miles
Frequency: 4,000 miles
Diagnostic plug: Yes

WHAT'S NEW IN 1995 ?

- No major change.

	ENGINE				TRANSMISSION			PERFORMANCE									
Model/ versions *: standard	Type / timing valve / fuel system	Displacement cu/in	Power bhp @ rmn	Torque lb.ft @ rpm	Compres. ratio	Driving wheels / transmission	Final ratio	Acceler. 0-62 mph s	Stand. 1/4 mile s	Stand. 5/8 mile s	Braking 62-0 mph ft	Top speed mph	Lateral acceler. G	Noise level dBA	Fuel economy mpg city	Fuel economy mpg hwy	Gasoline type / octane
L-Cruiser	L6* 4.5 DOHC-24-EFI	273	212 @ 4600	275 @ 3200	9.0: 1	all - A4	4.100	11.8	18.4	34.8	148	106	0.72	67	12.0	15.0	R 87

central front or rear differentials are part of a sophisticated, uncommon and optional system. It is activated by a single switch on the dashboard.

• Technical: 70%
The Land Cruiser has a steel body mounted on a ladder frame with 5 cross members. The front and rear axles are rigid, retained by radius rods and carried by coil springs, 2 in the front and 4 in the back; and there are 2 stabilizer bars. The engine, an in-line 4.5 L 6 cylinder with 2 overhead cams and 24 valves developing 212 hp, is coupled to a 4-speed automatic. The transmission is a full-time 4-wheel drive and the front and rear differentials can be locked.

• Safety: 70%
The structure of this utility is very rigid, both in torsion and in beaming, but one cannot find any air bag and the safety of the occupants is insured by 3-point seat belts, set up to ease the use of children's seats. Reinforcements are included in the doors to reduce any lateral impact.

• Conveniences: 70%
The storage areas include a large glove box, door pockets and the driver gets a footrest.

• Steering: 60%
Its assistance is well-modulated and has good precision, allowing accurate steering. However, the steering ratio and turning diameter limit the maneuverability.

• Performance: 50%
In spite of its weight, this mastodon takes off quite decently but passing is more of an effort, especially with a load. The big 6-cylinder does not lack torque and is capable of pulling 5,000 lbs (draw bar) without batting an eye.

WEAK POINTS

• Fuel consumption: 00%
This ambulating loft does not run on water and can swallow 14 mpg in difficult terrain.

• Handling: 40%

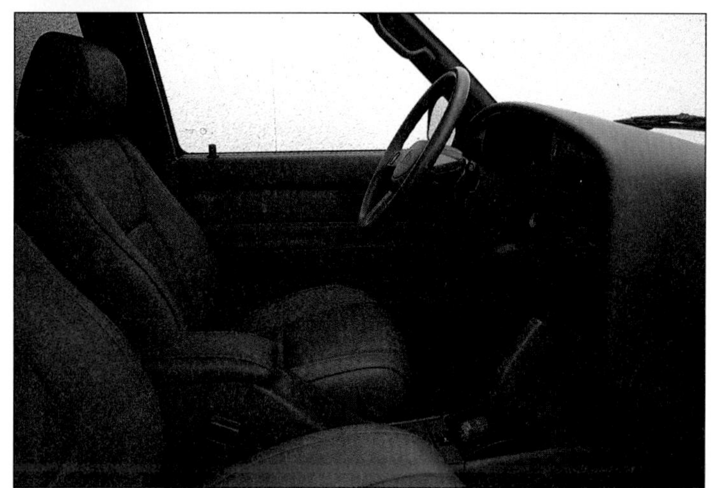

Be it on the road or on a trail, the big Toyota goes as if on rails with the assist of 4-wheel drive and a large suspension travel. Of course, in tight corners it is less agile and slalom is not its thing, but it surprises by its rapid response and its self-assured stability even on bumpy trails because side motions are small and not dangerous. Off road, its obstacle-crossing ability is limited by its extensive overhang, a spare tire under the rear floor and its bulk does not allow it to slip in just anywhere.

• Braking: 40%
The optional 4 discs and ABS insure distinctly shorter and more stable stops than the mixed brakes offered standard. Pedal modulation is progressive and its fade resistance average.

• Noise level: 40%
The noise of the big tires and of the airstreams that swirl around the windshield above the 60 mph is noticeable, but the power train is discreet and at cruise speed, the Land Cruise is not any noisier than a mini-van.

•Price/equipment: 40%
As all special vehicles, the Land Cruiser is costlier and its equipment is more complete, but without the flourish.

• Insurance: 40%
In spite of the size of the premium, you can consider it average for a vehicle of this nature and value.

• Depreciation: 40%
Depending on its state and a more or less intensive use, it retains good resale value.

CONCLUSION

• Overall rating: 59.0%
The Land Cruiser is, at the same time, the refuge of families in quest of safety and the preferred means of locomotion of globe-trotters the world over. 😐

SPECIFICATIONS & PRICES

Model	Version/Trim	Body/ Seats	Wheel-base in	Lgth x Wdthx Hght. inx inx in	Curb weight lb	Towing capacity std. lb	Susp. type ft/rr	Brake type ft/rr	Steering system	Turning diameter ft	Steer. turns nber	Fuel. tank gal.	Standard tire size	Standard power train	PRICES US. $ 1994
TOYOTA	General warranty: 3 years / 36,000 miles; power train: 5 years / 60,000 miles;corrosion:5 years / unlimited; without deductibles or transfert fees.														
Land Cruiser	base	5dr.wgn. 5/7	112.2	188.2x76.0x73.2	4760	5000	r/r	dc/dr	pwr.bal.	40.4	3.4	25.1	275/70R16	L6/4.5/A4	$34,268

See page 393 for complete 1995 Price List

Vital transplant...

Toyota tries to maintain the success of its Previa mini-van until its replacement, derived from the Camry, arrives next year. To remedy the lack of power, it has grafted to it a compressor. But the real problem comes from the basic concept, which is outdated.

The Previa is the Japanese mini-van which has known the most success, thanks to the originality of its futuristic lines and practical layout. Unfortunately, its sales dropped in relation to the price increase due to the strength of the yen. In 1995 it continues to be offered in the form of a single volume 4-door, equipped with a 2.4 L 4 cylinder with a 4-speed automatic with 2 or 4-wheel drive and in base or LE finish. However, for those who want extra power, the engine with a compressor provides 16 additional hp.

STRONG POINTS

• Interior space: 100%
It is not really modular but the interior is spacious and the seats are arranged in conventional fashion, but it isn't easy to move from the front to the rear of the interior.

• Access: 90%
The doors are amply-sized, but the sliding door and the hatchback are heavy and hard to operate.

• Safety: 90%
The presence of two air bags and retractable seat belts guarantees a satisfactory occupant protection, but the structure of this type of vehicle is more vulnerable to collisions than others, especially in the frontal area.

• Satisfaction: 90%
The legendary Toyota's reliability spreads happiness among the owners, who then justify the extra cost.

• Storage: 80%
The hatchback provides full access and you can add a sizable amount of bulky baggage without any need to fold to the side the two halves of the rear bench.

• Seats: 80%
They hold and support well, but the padding is firm. The standard bench seat backs are reclinable, so that converting them into a bed is fast and practical.

• Quality/fit/finish: 80%
The assembly looks robust and the finish is careful, but some plastics don't look rich. In spite of the size of the left wheel well, driver's position and visibility are good. Standard on the Previa is increased glazed surface and amply-dimensioned rear-view mirrors (but not adjustable from inside). The dashboard is ergonomic and practical, with an original design, gauges are legible and the controls well-grouped.

• Suspension: 80%
It absorbs the road defects and the wheel travel is ample.

• Insurance: 80%
Even though the unit body is harder to repair than others in case of impact, its index is equal to that of competitors.

• Technical: 70%
First created for the Japanese market, the mechanical drivetrain of the

DATA

Category: rear-wheel drive or AWD compact vans.
Class: utility

HISTORIC

Introduced in: 1990
Modified in: -
Made in: Kariya, Japan.

RATINGS

Safety: 95 %
Satisfaction: 89 %
Depreciation: 48 %
Insurance: 4.2 % ($655)
Cost per mile: $0.48

NUMBER OF DEALERS

In U.S.A.: 1,233

SALES ON THE AMERICAN MARKET

Model	1992	1993	Result
Previa	39,047	31,245	- 20.0 %

MAIN COMPETITORS

CHEVROLET Lumina, DODGE-PLYMOUTH Caravan-Voyager, FORD Aerostar-Windstar, PONTIAC Trans Sport, MAZDA MPV, MERCURY Villager, NISSAN Quest, VOLKSWAGEN Eurovan.

EQUIPMENT

TOYOTA Previa	Dx	2WD LE	Dx	4WD LE	4WD S/C
Automatic transmission:	S	S	S	S	S
Cruise control:	O	S	O	S	S
Power steering:	S	S	S	S	S
ABS brake:	O	O	O	O	O
Air conditioning:	S	S	S	S	S
Air bag (2):	S	S	S	S	S
Leather trim:	-	O	-	O	O
AM/FM/ Radio-cassette:	S	S	S	S	S
Power door locks:	O	S	O	S	S
Power windows:	O	S	O	S	S
Tilt steering column:	S	S	S	S	S
Dual power mirrors:	S	S	S	S	S
Intermittent wipers:	S	S	S	S	S
Light alloy wheels:	-	O	-	O	O
Sunroof:	-	O	-	-	O
Anti-theft system:	O	O	O	O	S

S : standard; O : optional; - : not available

COLORS AVAILABLE

Exterior: White, Burgundy, Red, Beige, Turquoise, Green.
Interior: Blue, Gray, Mocha.

MAINTENANCE

First revision: 4,000 miles
Frequency: 4,000 miles
Diagnostic plug: Yes

WHAT'S NEW IN 1995 ?

- New engine with compressor adding 16 hp.

Model/ versions *: standard	ENGINE Type / timing valve / fuel system	Displacement cu/in	Power bhp @ rmn	Torque lb.ft @ rpm	Compres. ratio	TRANSMISSION Driving wheels / transmission	Final ratio	PERFORMANCE Acceler. 0-62 mph s	Stand. 1/4 mile s	Stand. 5/8 mile s	Braking 62-0 mph ft	Top speed mph	Lateral acceler. G	Noise level dBA	Fuel economy mpg city	Fuel economy hwy	Gasoline type / octane
2x4	L4* 2.4 DOHC-16-EFI	149	138 @ 5000	161 @ 5000	9.3 :1	rear -A4	4.3	12.5	18.7	34.6	151	106	0.71	66	21.0	29.0	R 87
4x4	L4* 2.4 DOHC-16-EFI	149	138 @ 5000	161 @ 5000	9.3 :1	rear/4 -A4	4.3	13.2	19.5	36.2	141	103	0.73	66	20.0	27.0	R 87
4x4 S/C	L4* 2.4 DOHC-16-EFI	149	154 @ 4000	201 @ 3600	8.9 :1	rear/4 -A4	3.73	-	-	-	-	-	-	-	20.0	27.0	R 87

Previa is derived from the pickups of the marque. Installed in a central-forward position, the engine transmits power to an antique rigid axle. It's a 2.4 L 4-cylinder inclined at 75 degrees with a double overhead cam and 16 valves. It develops 138 hp naturally aspirated and 154 hp with a compressor placed directly under the front seats, which somewhat complicates its accessibility. The very efficient steel unit body gives it a drag coefficient of 0.35 (or 0.34 with a rear spoiler). The suspension is independent up front, with coil springs at all 4 corners and the 4-wheel drive has a center differential with a viscous coupling.

• Steering: 70%
It is precise, well-modulated, and provides good road feel, but its large gear reduction complicates road handling in a cross wind; maneuverability is only average.

• Noise level: 60%
Acceptable at cruise speed, it climbs during takeoffs, and is joined by wind noise from 50 mph.

• Conveniences: 60%
The storage areas are well-distributed and the glove box door serves as a small desk. The two AC systems are not too much to condition the environment of this traveling veranda on four wheels.

• Fuel consumption: 50%
It corresponds to that of a 3.0 L V6, with more noise and vibrations and less pleasurability.

• Depreciation: 50%
They are lower than others, because the demand remains high and so do the prices.

WEAK POINTS

• Handling: 30%
The oversteer is less strong on a 4-wheel drive than with just 2-rear wheels driven but is touchier to drive under less-than-idealconditions. Body movements are well-controlled but it is sensitive to a cross wind.

• Braking: 30%
While simple decelerations are progressive, panic stops bring on a rapid front wheel lockup and a strong nose dive. The 4 discs and ABS offered as an option are not a luxury because stopping distances exceed the average in most instances.

• Performance: 30%
The short ratio of the automatic transmission allows good takeoff and acceleration but the 4-cylinder engine, even with a compressor, is not at the level particularly with a 4-wheel drive with a higher weight. Most owners want a 3.0 L V6, but it will not be available until the Previa is redesigned. The next generation, forecast for 1996, will be based on the platform and drivetrain of the Camry.

• Price/equipment: 30%
Very expensive, especially with the 4-wheel drive the equipment is not very generous and you need to look into the options list for additional refinement.

CONCLUSION

• Overall rating: 67.0%
Previa has become too expensive to play a role in this category, because it lacks two vital elements: front or rear-wheel drive with a V6. ☺

OWNERS' SUGGESTIONS

-A more affordable price
-A V6 3.0 L engine
-An easier sliding door
-More effective windshield washers
-Quieter rear-view mirrors
-More powerful air conditioning

SPECIFICATIONS & PRICES

Model	Version/Trim		Body/ Seats	Wheel base in	Lgth x Wdthx Hght. inx inx in	Curb weight lb	Towing capacity std. lb	Susp. type ft/rr	Brake type ft/rr	Steering system	Turning diameter ft	Steer. turns nber	Fuel. tank gal.	Standard tire size	Standard power train	PRICES US. $ 1994
TOYOTA		**General warranty: 3 years / 36,000 miles; power train: 5 years / 60,000 miles;corrosion:5 years / unlimited; without deductibles or transfert fees.**														
Previa	DX	4x2	4dr.van. 5/7	112.8	187.4x70.9x68.7	3616	3500	i/r	dc/dr	pwr.r&p.	37.1	3.5	19.8	215/65R15	L4/2.4/A4	**$22,818**
Previa	LE	4x2	4dr.van. 5/7	112.8	187.4x70.9x68.7	3735	3500	i/r	dc/dc	pwr.r&p.	37.1	3.5	19.8	215/65R15	L4/2.4/A4	**$26,578**
Previa	DX	4x4	4dr.van. 5/7	112.8	187.4x70.9x69.1	3834	3500	i/r	dc/dr	pwr.r&p.	37.1	3.5	19.8	215/65R15	L4/2.4/A4	**$26,148**
Previa	LE	4x4	4dr.van. 5/7	112.8	187.4x70.9x69.1	3955	3500	i/r	dc/dc	pwr.r&p.	37.1	3.5	19.8	215/65R15	L4/2.4/A4	**$29,718**
Previa	LE S/C	4x4	4dr.van. 5/7	112.8	187.4x70.9x69.1	4094	3500	i/r	dc/dc	pwr.r&p.	37.1	3.5	19.8	215/65R15	L4C/2.4/A4	**$31,298**

See page 393 for complete 1995 Price List

Vanity...

On the board of directors of all car builders there is brainpower that thinks their make would do better if it would produce one of the fastest cars in the world. This very expensive image does not satisfy the public, which dreams of performance at more accessible prices.

The Supra was redesigned last year. Today it has become a real Japanese Gran Turismo offered standard with either a naturally aspirated or a Turbo in the USA or exclusively with a Turbo in Canada, with a standard manual transmission or an optional automatic.

STRONG POINTS

• Performance: 95%
They constitute the center of attraction of the Supra because the power claimed is readily available and comes on in progressive fashion. The numbers produce strong emotions because they are very close to those achieved at the wheel of more exotic cars, especially when they are obtained without overstressing the drivetrain. To accelerate from 0 to 62 mph in more or less five seconds, is not readily available on a large production car and no one will be able to experiment with a top speed of 175mph because the Supra is voluntarily reined to 155. A goal of technical simplicity guided the designers in rejecting 4-wheel drive or traction control and to be content instead with a Torsen differential to transmit the message.

• Satisfaction: 90%
The Supra rolls up miles and remains in good shape which satisfies a large number of owners.

• Technical: 90%
Pure performance is the only objective of the latest Supra. Its style, which is muscular and functional, is not of great originality and the presence of a spoiler patterned on a "Road Runner" does not achieve anything. However, in spite of the tires and the main air intakes, the body has a good aerodynamic efficiency and the drag coefficient varies from 0.31 to 0.32. The steel unit body (and its aluminum front hood) includes independent suspension with upper and lower control arms at all 4 wheels and includes antidive geometry at the front and

DATA

Category: rear-wheel drive GT coupes.
Class: GT

HISTORIC

Introduced in: 1978
Modified in: 1985 and 1993: reshaped.
Made in: Tahara, Japan.

RATINGS

Safety: 90 %
Satisfaction: 88 % (previous version)
Depreciation: 40 % (2 years)
Insurance: 3.6 % ($1,270)
Cost per mile: $0.80

NUMBER OF DEALERS

In U.S.A.: 1,233

SALES ON THE AMERICAN MARKET

Model	1992	1993	Result
Supra	1,193	2,901	+143 %

MAIN COMPETITORS

ACURA NSX, CHEVROLET Corvette, DODGE Stealth R/T Turbo, MAZDA RX-7, NISSAN 300 ZX, PORSCHE 911, 928.

TOYOTA Supra	base	Turbo
Automatic transmission:	O	O
Cruise control:	S	S
Power steering:	S	S
ABS brake:	S	S
Air conditioning:	S	S
Air bag (2):	S	S
Leather trim:	O	O
AM/FM/ Radio-cassette:	S	S
Power door locks:	S	S
Power windows:	S	S
Tilt steering column:	S	S
Dual power mirrors:	S	S
Intermittent wipers:	S	S
Light alloy wheels:	S	S
Sunroof:	O	S
Anti-theft system:	S	S

S : standard; O : optional; - : not available

COLORS AVAILABLE

Exterior: White, Silver, Black, Red, Blue.
Interior: Fabric: Black. Leather: Black or Ivory.

MAINTENANCE

First revision: 4,000 miles
Frequency: 4,000 miles
Diagnostic plug: Yes

WHAT'S NEW IN 1995 ?

- No major change.

Model/ versions *: standard	Type / timing valve / fuel system	Displacement cu/in	Power bhp @ rmn	Torque lb.ft @ rpm	Compres. ratio	Driving wheels / transmission	Final ratio	Acceler. 0-62 mph s	Stand. 1/4 mile s	Stand. 5/8 mile s	Braking 62-0 mph ft	Top speed mph	Lateral acceler. G	Noise level dBA	Fuel economy mpg city	hwy	Gasoline type / octane
ENGINE						TRANSMISSION		PERFORMANCE									
base	L6* 3.0 DOHC-24-IE	183	220 @ 5800	210 @ 4800	10.0 :1	rear - M5*	4.272	7.0	15.5	27.7	125	149	0.95	65	22.0	33.0	M 89
						rear - A4	4.272	7.8	16.2	28.6	131	146	0.95	65	22.0	33.0	M 89
Turbo	L6T* 3.0 DOHC-24-IE	183	320 @ 5600	315 @ 4000	8.5 :1	rear - M6	3.133	5.0	13.5	26.3	128	155+	0.96	66	21.0	31.0	S 91
						rear - A4	3.769	5.7	13.9	26.8	134	155+	0.96	66	22.0	31.0	S 91

+electronicaly governed speed

rear. The steering has a variable assist proportioned to vehicle speed and damping control. Large diameter vented discs are used with ABS; improved with a lateral acceleration center that reduces stopping distances in turns. The grade Z Michelin tires, especially devised for this application, are used with light alloy 17 inch rims.The heart of the new Supra is its in-line 3.0 L 6-cylinder with DOHC and 24 valves naturally aspirated or turbocharged with two sequential Turbos. The maximum power of this version is 320 hp as against 230 for naturally aspirated. The latter receives a manual 5-speed, while the Turbo uses a 6-speed Getrag. Both are optionally available with a 4-speed automatic and a Torsen limited slip.

• Safety: 90%
The unit body rigidity is the main concern of builders of these models, because it improves handling, the structure in case of impact and also the active safety. Two air bags and ABS are standard.

• Quality/fit/finish: 90%
The least you can say is that the presentation lacks imagination. However, assembly is meticulous and the finish carefully carried out. The appearance of the plastic distinctly lacks class on a car of this type.

• Insurance: 90%
Proportionally advantageous, the premium represents a fair sum.

• Handling: 90%
The great quality of the Supra stems from its general balance and the progressivity of its reactions, which never surprise the driver and have a tendency to go into oversteer. Road handling remains neutral for longer. Its reactions recall those of a 928 Porsche without the weight.

• Driver's compartment: 80%
The driver's area is very ergonomic and it is enough to move the hand forward to find, at the left as at the right, all that is necessary to control the vehicle. The vast center console is well-organized and the gauges grouped into three pods that fall naturally under the eyes. The most original control is the automatic shift whose form is very functional. The handbrake is well-located and the steering wheel offers a good grip but is amazingly ordinary looking.

• Steering: 80%
Even though the Supra is not really agile because of its weight and bulk, its guidance remains precise and the power assist well-modulated. Perhaps one would have wished for a faster steering ratio but reaction time and maneuverability are very good.

• Suspension; 70%
Contrary to other cars of the same type, that of the Supra has seemed to us as almost too smooth, too comfortable, too civilized, without detracting from its character.

• Noise level: 60%
The efficient soundproofing and the rigidity of the assembly give it the silence of a plush sedan at cruise speed.

• Braking: 60%
As powerful as it is easy to drive, it allows stopping the 3,300 pounds in less than 135 ft, a high safety level especially considering the excellent fade resistance of the brake pads.

• Depreciation: 60%
It is strong, considering the sum invested, and at that price, the buyers do not rush in.

• Access: 50%
The large doors allow more convenient seating front than rear, where room is more limited.

• Seats: 50%
In spite of their multiple adjustments, they do not provide a lateral hold as satisfactory as their lumbar support. It is incredible that precisely the most important item of the design was neglected.

WEAK POINTS

• Price/equipment: 00%
In spite of its qualities, the Supra is much too expensive, and its presentation is too average to justify such an investment.

• Interior space: 20%
Realistically, the interior only offers two seats because no normally constituted human being will ever fit the rear bench seat, except crossways.

• Trunk: 30%
As always on this type of vehicle, it does not contain much due to lack of height.

• Fuel consumption: 45%
Reasonable at legal speeds, it depends essentially on the weight of the driver's right foot.

CONCLUSION

• Overall rating: 62.0%
After the flop of the RX-7, the market needs not another unaffordable supercoupe. The Dodge Stealth and the Nissan 300ZX owe their success to their price, rather than their performances.

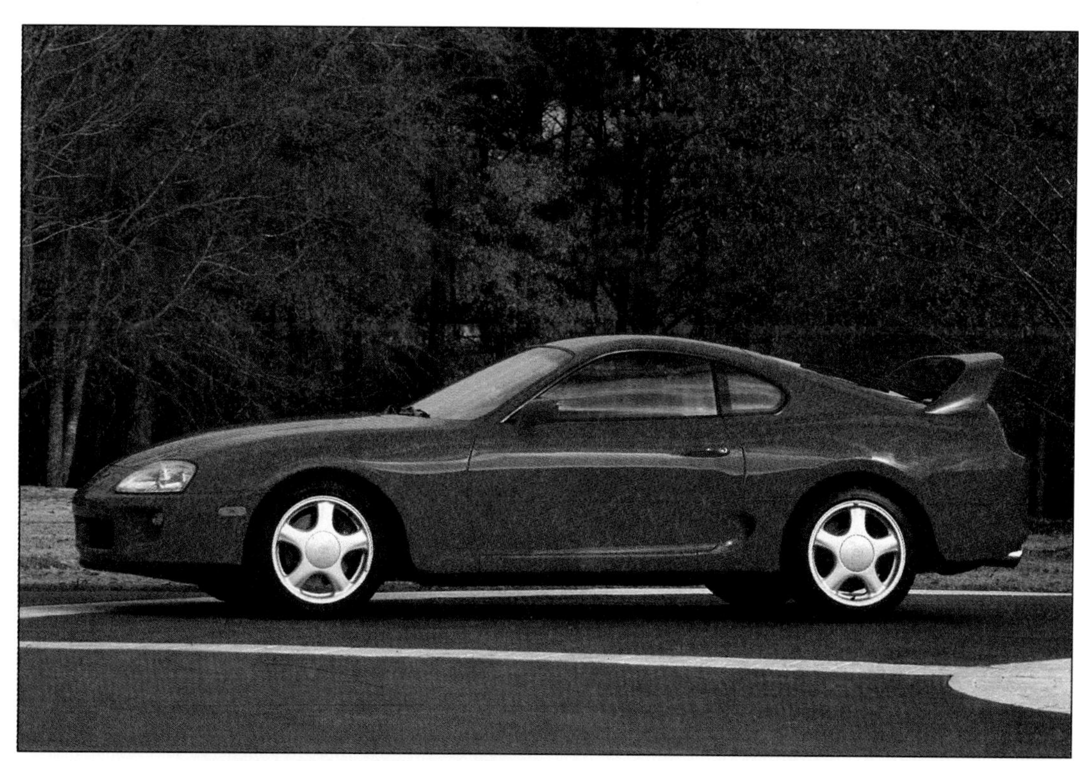

SPECIFICATIONS & PRICES

Model	Version/Trim	Body/ Seats	Interior volume cu.ft.	Trunk volume cu ft.	Drag coef. Cd	Wheel-base in	Lgth x Wdthx Hght. inx inx in	Curb weight lb	Towing capacity std. lb	Susp. type ft/rr	Brake type ft/rr	Steering system	Turning diameter ft	Steer. turns nber	Fuel. tank gal.	Standard tire size	Standard power train	PRICES US. $ 1994
TOYOTA		**General warranty: 3 years / 36,000 miles; power train: 5 years / 60,000 miles;corrosion:5 years / unlimited; without deductibles or transfert fees.**																
Supra	base	2dr.cpe.2+2	69.5	10	0.31	100.4	177.8x71.3x49.8	3210	NR	i/i	dc/dc/ABS	pwr.r&p.	35.8	3.0	18.5	225/50ZR16	L6/3.0/M5	**$35,800**
Supra	base	2dr.cpe.2+2	69.5	10	0.31	100.4	177.8x71.3x49.8	3265	NR	i/i	dc/dc/ABS	pwr.r&p.	35.8	3.0	18.5	225/50ZR16	L6/3.0/A4	**$36.700**
Supra	Turbo	2dr.cpe.2+2	69.5	10	0.32	100.4	177.8x71.3x49.8	3505	NR	i/i	dc/dc/ABS	pwr.r&p.	35.8	3.0	18.5	235/45ZR17 255/40ZR17 (re)	L6T/3.0/M5	**$42,800**
Supra	Turbo	2dr.cpe.2+2	69.5	10	0.32	100.4	177.8x71.3x49.8	3514	NR	i/i	dc/dc/ABS	pwr.r&p.	35.8	3.0	18.5	235/45ZR17 255/40ZR17 (re)	L6T/3.0/A4	**$42,800**

TOYOTA Tercel-Paseo

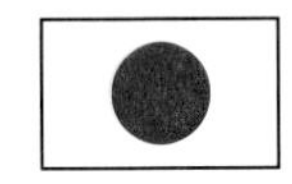

Growing pains...

Just like children, automobiles never cease to grow and the latest Tercel does not escape this phenomena which, one day or another, will force the builders to introduce a still smaller model to fill out the bottom end of the mini-compact range.

The Paseo coupe does not change for 1995, but the Tercel adopts a new body together with the more powerful engine already in the Paseo. Without wanting to pass judgment on the lines of the new arrival, it makes one think of the old Ford Escort rather than of a famous German.The Tercel is available in a 2- and 4-door, 3-volume sedan and a 3-volume, 2-door Paseo. The latter is sold at the level of a unique finish while the Tercel exists in versions S and DX.

STRONG POINTS

• Price/equipment: T: 90%; P: 80%
Even though they are always costlier than average, these cars justify their price by the excellent reputation for reliability and durability, more so than by the richness of their equipment, which is however in progress.

• Satisfaction: 85%
If the number of very satisfied owners is lower than for other models of the make, it does make for envy among the domestic builders.

• Safety: 80%
Two air bags and retractable belts compensate for the vulnerability of these small cars which however, have been made more rigid on the latest Tercels.

• Fuel consumption: 80%
Even though these models are not the most economical on the market, their fuel economy is good considering their more generous format and satisfactory performance.

• Quality/fit/finish: 70%
The interior has distinctly improved with plastics that have a richer appearance and are no longer malodorous. The assembly has remained strict and the finish carefully done; also the feel of a car that is economical at any cost, which characterized the preceding model, has disappeared.

• Driver's compartment: 70%
Even though the steering column is not adjustable, the driver readily gets comfortable. The dashboard is simple but well-organized and of good appearance. Instrumentation is legible and the controls conventional and well laid out. Visibility is better on the Tercel than on the Paseo whose tall tail and the presence of a spoiler reduce the view toward the rear.

• Suspension: 70%
The more consistent shock absorber control and longer wheel travel give these cars a driving comfort more characteristic of compacts than sub-compacts.

• Access: T: 70%; P: 50%
It has become easier to reach the rear Tercel 2-door bench seat than those of the Paseo which lacks height.

• Noise level: T: 60%; P: 50%
It has improved on the Tercel because even though the engine allows

DATA

Category: front-wheel drive sub-compact sedans and coupes.
Class: 3

HISTORIC

Introduced in: 1978
Modified in: 1982, 1987, 1991, 1995.
Made in: Takaoka, Japan.

RATINGS

Safety: 90 %
Satisfaction: 85 %
Depreciation: 53 %
Insurance: 7.0 % ($445)
Cost per mile: $0.30

NUMBER OF DEALERS

In U.S.A.: 1,233

SALES ON THE AMERICAN MARKET

Model	1992	1993	Result
Tercel	102,043	93,820	- 8.1 %
Paseo	22,838	24,466	+6.7 %

MAIN COMPETITORS

Tercel: DODGE Colt, HONDA Civic Hbk, GEO Metro (4dr.) HYUNDAI Accent, MAZDA 323 Hbk, NISSAN Sentra, VW Golf.
Paseo: FORD Escort GT, HONDA Civic SI & Del Sol, HYUNDAI Scoupe, MAZDA MX-3(1.6L) 323, SATURN SC1.

EQUIPMENT

	TOYOTA Tercel S	TOYOTA Tercel DX	TOYOTA Paseo base
Automatic transmission:	-	O	O
Cruise control:	-	O	O
Power steering:	-	O	O
ABS brake:	O	O	O
Air conditioning:	O	O	O
Air bag (2):	S	S	S
Leather trim:	-	-	-
AM/FM/ Radio-cassette:	O	O	O
Power door locks:	-	-	-
Power windows:	-	-	-
Tilt steering column:	-	-	-
Dual power mirrors:	-	S	S
Intermittent wipers:	S	S	S
Light alloy wheels:	-	-	O
Sunroof:	-	-	O
Anti-theft system:	-	-	-

S : standard; O : optional; - : not available

COLORS AVAILABLE

Exterior: Tercel: White, Silver, Red, Green, Ruby.
Paseo: White, Black, Red, Green, Ruby.
Interior: Black, Gray.

MAINTENANCE

First revision: 4,000 miles
Frequency: 4,000 miles
Diagnostic plug: Yes

WHAT'S NEW IN 1995 ?

- **Only two versions of the Tercel S and DX, the LS disappears.**
- **More powerful 1.5 L engine.**

Model/ versions *: standard	ENGINE Type / timing valve / fuel system	Displacement cu/in	Power bhp @ rmn	Torque lb.ft @ rpm	Compres. ratio	TRANSMISSION Driving wheels / transmission	Final ratio	PERFORMANCE Acceler. 0-62 mph s	Stand. 1/4 mile s	Stand. 5/8 mile s	Braking 62-0 mph ft	Top speed mph	Lateral acceler. G	Noise level dBA	Fuel economy mpg city	Fuel economy mpg hwy	Gasoline type / octane
Tercel	L4* 1.5 DOHC-16-EFI	91	93 @ 5400	100 @ 4400	9.4 :1	front - M4*	3.52	11.5	18.2	34.1	138	100	0.75	68	36.0	46.0	R 87
						front - M5*	3.72	11.0	18.0	33.7	144	103	0.75	68	35.0	45.0	R 87
						front - A3/A4	2.82	13.5	18.6	36.5	148	100	0.75	68	35.0	45.0	R 87
Paseo	L4* 1.5 SOHC-16-EFI	91	100 @ 6400	91 @ 3200	9.4 :1	front - M5*	3.94	10.5	17.3	33.0	144	106	0.77	68	36.0	44.0	R 87
						front - A4	2.82	10.8	17.7	33.5	151	103	0.77	68	31.0	43.0	R 87

you to hear takeoffs and accelerations, the aerodynamic and rolling noises are less pronounced than on the preceding model.

• Steering: 60%
The standard manual is heavy at low speeds and lightens up as speed increases. Its high steering ratio makes it vague in the center. The standard power assist of the Paseo is more direct, precise and better modulated, in brief, more pleasant.

• Conveniences: 60%
Few, the storage areas include a glove box, door pockets and hollows on the dash of the deluxe versions.

• Seats: 60%
They offer good hold and lumbar support, thanks to improved shape and padding.

• Technical: 60%
The steel unit body, of which some components are galvanized, uses a McPherson-type suspension up front and a torsional axle at the rear. The steering is manual and braking mixed on the standard and one is surprised not to find an ABS listed, even as an option. The simple lines show good aerodynamic efficiency with a Cd of 0.32 on a Tercel and 0.33 on the Paseo.

• Handling: T: 55%; P: 60%
The Tercel has become much more stable with less roll, more crisp and faster; it has almost become so much fun to drive that one wonders if the Paseo can add anything except for a sports feel.

• Insurance: 50%
The Tercel-Paseo do not cost any more to insure than any other sub-compact, which sets them off immediately as an interesting purchase for a young car owner, always penalized by the insurance companies.

• Performance: 50%
They are always more interesting with a stick shift than with an automatic, especially under load. A few extra horsepowers make the difference, but that still does not make them into dragsters. Accelerations are crisper with a manual transmission whose shifts are fast and precise than with the automatic which has satisfactory ratios and shifts.

NEW FOR 1995

TOYOTA Paseo

WEAK POINTS

• Trunk: T: 30%; P: 20%
Roomy considering its size, it is possible to increase its content by lowering the rear bench seat. However, the tight Paseo opening makes it difficult to handle baggage.

• Interior space: T: 35%; P: 30%
Four good-sized adults will be at ease in the Tercel whose leg and headroom have been increased. This is not the case with the Paseo, limited to two adults and two children.

• Braking: 40%
Its modulation and fade resistance are adequate and efficiency is progressing in spite of the fact that the front wheels block rapidly, which lengthens the stopping distance.

• Depreciation: 45%
It keeps a good resale value and bargains are rare and costly.

CONCLUSION

• Overall rating: T: 61.0%
P: 58.5%

More plush, the Tercel quietly mounts an attack on the Corolla because it is no longer a little car. However, its comfort and performances gain much together with its driving pleasure.

OWNERS' SUGGESTIONS
(last year's model)
-Front seats that slide back further with better support
-Improved rearward Paseo visibility to better evaluate its length
-Larger wheels

SPECIFICATIONS & PRICES

Model	Version/Trim	Body/ Seats	Interior volume cu.ft.	Trunk volume cu ft.	Drag coef. Cd	Wheel-base in	Lgth x Wdthx Hght. inx inx in	Curb weight lb	Towing capacity std. lb	Susp. type ft/rr	Brake type ft/rr	Steering system	Turning diameter ft	Steer. turns nber	Fuel. tank gal.	Standard tire size	Standard power train	PRICES US. $ 1994
TOYOTA		**General warranty: 3 years / 36,000 miles; power train: 5 years / 60,000 miles;corrosion:5 years / unlimited; without deductibles or transfert fees.**																
Tercel	S	2dr.sdn.4/5	80.4	9.3	0.32	93.7	162.2x65.4x53.1	1960	NR	i/si	dc/dr	r&p.	31.5	3.8	11.9	155/80R13	L4/1.5/M5	**$8,698**
Tercel	DX	2dr.sdn.4/5	80.4	9.3	0.32	93.7	162.2x65.4x53.1	1984	NR	i/si	dc/dr	r&p.	31.5	3.8	11.9	155/80R13	L4/1.5/M5	**$10,148**
Tercel	DX	4dr.sdn.4/5	81	9.3	0.32	93.7	162.2x65.4x53.1	2015	NR	i/si	dc/dr	r&p.	31.5	3.8	11.9	155/80R13	L4/1.5/M5	**$10,248**
Paseo	man.	2dr.cpe.2+2	77.2	7.7	0.33	93.7	163.2x65.2x50.2	2070	NR	i/si	dc/dr	pwr.r&p.	32.5	2.7	11.9	185/65R14	L4/1.5/M5	**$12,838**
Paseo	autom.	2dr.cpe.2+2	77.2	7.7	0.33	93.7	163.2x65.2x50.2	2160	NR	i/si	dc/dr	pwr.r&p.	32.5	2.7	11.9	185/65R14	L4/1.5/A4	**$13,638**

See page 393 for complete 1995 Price LIst

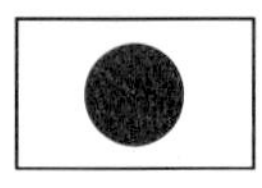

The most reliable...

The 4Runner can boast of being at the antipode of its North American competitors, because unlike them with their Sunday dress, over-equipment and eye-catchiness, it has more the image of a workaholic; effective but retiring. However, it has what others do not; an unfailing reliability.

After the mini-vans, the 4-door sport utilities are one of the most dynamic segments of the market and there the 4Runner held a good position, but the arrival of new competitors has caused it to lose many sales. It is available in only one 4-door body with an SR5 trim, a 4WD transmission and a 2.4 L 4-cylinder or a 3.0 L V6 with a manual 5-speed standard or an optional 4-speed or a 2WD with a V6 and only an automatic.

STRONG POINTS

- **Satisfaction:** 95%

It is, without a doubt, the most reliable sports utility on the market because owners do not tire of complimenting it on this subject.

- **Quality/fit/finish:** 80%

Interior presentation is more that of a utility than a car but assembly and finish show care and the quality of materials is satisfactory.

- **Driver's compartment:** 80%

Identical to that of pickups, the dashboard is well-organized, with legible instruments and well laid out main controls. The elevated driver's position gives him good visibility and the ideal driver position is easy to find.

- **Technical:** 70%

The 4Runner stems directly from the compact Toyota pickup, that is with a steel body mounted on the 5 cross member chassis, also all steel. The front suspension, whose pivot points are located above the axis of the chassis, is independent with torsion bars, while at the rear, the rigid axle connects through leaf springs. Braking is mixed and the ABS acts on the rear wheels, with the 4-cylinder engine. On a V6, it functions only in 2-wheel drive and pulls itself out of action when the FWD is engaged.

DATA

Category: 2 or 4WD all-terrain multipurpose vehicles.
Class: utility

HISTORIC

Introduced in: 1985
Modified in: 1988: V6 engine; 1989: new body.
Made in: Tahara, Japan.

RATINGS

Safety: 70 %
Satisfaction: 94 %
Depreciation: 48%
Insurance: 6.0 % ($785)
Cost per mile: $0.48

NUMBER OF DEALERS

In U.S.A.: 1,233

SALES ON THE AMERICAN MARKET

Model	1992	1993	Result
4 Runner	39,917	46,652	+ 14.5 %

MAIN COMPETITORS

CHEVROLET Blazer, FORD Explorer, ISUZU Rodeo & Trooper, GMC Jimmy, JEEP Cherokee-Grand Cherokee, NISSAN Pathfinder, SUZUKI Sidekick 4dr.

EQUIPMENT

TOYOTA 4Runner SR5	base	V6	V6 4RM	Limited
Automatic transmission:	O	O	O	O
Cruise control:	O	O	O	S
Power steering:	S	S	S	S
ABS brake:	S	S	S	S
Air conditioning:	O	O	O	S
Air bag:	-	-	-	-
Leather trim:	O	O	O	O
AM/FM/ Radio-cassette:	O	O	S	S
Power door locks:	O	O	O	S
Power windows:	O	O	O	S
Tilt steering column:	O	S	S	S
Dual power mirrors:	O	O	O	S
Intermittent wipers:	S	S	S	S
Light alloy wheels:	O	S	O	O
Sunroof:	-	O	O	O
Anti-theft system:	-	-	-	-

S : standard; O : optional; - : not available

COLORS AVAILABLE

Exterior: White, Beige, Pewter, Black, Garnet, Green, Blue.
Interior: Fabric: Blue, Gray. Leather: Oak.

MAINTENANCE

First revision: 4,000 miles
Frequency: 4,000 miles
Diagnostic plug: Yes

WHAT'S NEW IN 1995 ?

-"Limited" edition SR5 V6.

ENGINE						TRANSMISSION		PERFORMANCE										
Model/ versions *: standard		Type / timing valve / fuel system	Displacement cu/in	Power bhp @ rmn	Torque lb.ft @ rpm	Compres. ratio	Driving wheels / transmission	Final ratio	Acceler. 0-62 mph s	Stand. 1/4 mile s	Stand. 5/8 mile s	Braking 62-0 mph ft	Top speed mph	Lateral acceler. G	Noise level dBA	Fuel economy mpg city	hwy	Gasoline type / octane
base	4x4	L4* 2.4 SOHC-8-EFI	144	116 @ 4800	140 @ 2800	9.3 :1	rear/4 -M5*	4.100	13.0	19.7	35.8	151	100	0.69	67	23.0	28.0	R 87
V6	4x2	V6* 3.0 SOHC-12-EFI	180	150 @ 4800	180 @ 3400	9.0 :1	rear -A4	3.900	10.0	16.8	31.8	200	106	0.70	66	21.0	25.0	R 87
V6	4x4	V6* 3.0 SOHC-12-EFI	180	150 @ 4800	180 @ 3400	9.0 :1	rear/4 -M5*	4.300	10.5	17.5	32.6	168	103	0.72	66	17.0	23.0	R 87
							rear/4 -A4	4.556	11.7	18.8	34.0	180	100	0.72	66	16.0	21.0	R 87

• **Seats: 70%**
One is better accommodated up front, where the seats hold and support better than on the bench seat, which is too flat. However, the padding is consistent.

• **Interior space: 70%**
Four people will be at ease and a fifth can get accommodation in the middle of the rear bench seat, which lacks in legroom.

• **Storage: 70%**
The volume of the storage area is huge and relatively accessible in spite of a sizable ground clearance. The hatchback does not come down until the window has been fully lowered.

• **Safety: 70%**
Rather than offer leather seats, Toyota would do well to equip its 4Runner with air bags, because while its structure resists impact well (75%), the protection of the occupants is mediocre and that of the driver outright bad.

• **Insurance: 70%**
The high price and index give a spicy premium.

• **Steering: 70%**
It does not transmit any shock from the wheels and does not reveal the state of the road, because its power assist is better for off road than on the road where it is too sensitive. However, in spite of the high ratio, the reactions are fairly rapid and precise.

• **Conveniences: 60%**
The storage includes a large glove box, a hollow, door pockets and a center console box.

• **Suspension: 60%**
It surprises by its good manners in spite of the presence of a rigid axle and leaf springs. The dampers effectively absorb road defects.

• **Fuel consumption: 50%**
It is never economical with an average that ranges between 19 to 21 mpg with a V6.

• **Access: 50%**
Getting into a 4Runner is not easy because the doors are narrow, especially at the back where legroom is insufficient and the size of the ground clearance asks for a step, in addition to the handles already installed in strategic locations.

• **Depreciation: 50%**
It is lower than that of some of its competitors which allows recovering part of the investment.

• **Noise level: 50%**
Soundproofing is effective on this utility machine but tire and wind noises overshadow those of the power train.

WEAK POINTS

• **Braking: 30%**
In normal driving its modulation and precision are beyond criticism, but the panic stopping distances are very long and 4-discs with total ABS would improve this worrisome situation.

• **Handling: 30%**
In spite of the sizable ground clearance, handling in a curve is surprising and traction in the curve remains effective even on bumpy or wet pavement. In a straight line, it is imperturbable and cross-winds have little effect on the vehicle path. Thanks to its compact dimensions and reduced overhang, the 4Runner is very efficient off road, where its clearance exceeds that of many of its competitors in spite of a spare tire mounted under the rear floor board.

• **Performance: 40%**
The 4-cylinder engine with manual stick shift is livelier to drive in spite of its unfavorable weight-to-power ratio. Torque at the rear wheels is compensated for by the short final drive ratio. While the V6 is smoother, it lacks low end torque and its presence is not justified for off-road. The 2WD is lighter and more pleasant to drive.

• **Price/equipment: 40%**
A fad or winter traction is not sufficient excuses for acquiring this type of vehicle, whose price is indecent for the little equipment and gadgets that it includes in the standard package.

CONCLUSION

• **Overall rating: 60.0%**
The 4Runner remains the choice of those who want to drive off road with their mind at ease, regardless of the condition of the path or the size of the budget needed to roll in this go-anywhere machine. 😐

OWNERS' SUGGESTIONS

- Air bags
- More powerful brakes
- Footrests
- More affordable price
- More economical 6-cylinder with additional torque
- More practical hatchback
- More complete equipment
- More luxurious finish

SPECIFICATIONS & PRICES

Model	Version/Trim		Body/ Seats	Wheel-base in	Lgth x Wdthx Hght. inx inx in	Curb weight lb	Towing capacity std. lb	Susp. type ft/rr	Brake type ft/rr	Steering system	Turning diameter ft	Steer. turns nber	Fuel. tank gal.	Standard tire size	Standard power train	PRICES US. $ 1994
TOYOTA		**General warranty: 3 years / 36,000 miles; power train: 5 years / 60,000 miles;corrosion:5 years / unlimited; without deductibles or transfert fees.**														
4Runner	4x4	SR5	2dr.wgn. 4/5	103.3	176.8x66.5x67.3	3814	3500	i/r	dc/dr/ABS	re.pwr.bal.	37.4	3.9	17.2	225/75R15	L4/2.4/M5	**$19,998**
4Runner	4x2	V6-SR5	2dr.wgn. 4/5	103.3	176.8x66.5x67.3	-	3500	i/r	dc/dr/ABS	re.pwr.bal.	37.4	3.9	17.2	225/75R15	V6/3.0/M5	**$21,028**
4Runner	4x4	V6-SR5	2dr.wgn. 4/5	103.3	176.8x66.5x67.3	4125	3500	i/r	dc/dr/ABS	re.pwr.bal.	37.4	3.9	17.2	225/75R15	V6/3.0/A4	**$21,938**
4Runner	4x4	V6-SR5	2dr.wgn. 4/5	103.3	176.8x66.5x67.3	4180	3500	i/r	dc/dr/ABS	re.pwr.bal.	37.4	3.9	17.2	225/75R15	V6/3.0/A4	**$22,988**

See page 393 for complete 1995 Price List

Ignored...?

It seems that the Toyota pickups are less popular now than at one time. The exchange rate of the yen to local currency certainly has something to do with it, but in the case of the T100, sales have improved slightly, yet they were not starting with anything much...

The compact pickups and the T100 Toyota share more than appearances would let you think. They are offered in 2WD and 4WD with 4- or 6-cylinders and the transmissions are standard manual or optional automatic. Trim levels are standard or SR5. If the compact offers the choice of standard body, no body, single or extended cab (Xtracab), the T100 has one cab and a unique pickup bed.

STRONG POINTS

• Satisfaction: 85%
Few complaints, except for minor problems often stemming from owners' abuse.

• Quality/fit/finish: 80%
These pickups enjoy a good reputation for robustness and durability, owing to their strict build and finish.

• Technical: 80%
These utilities are based on a steel ladder bar chassis with 5-cross members on which the cab is mounted. The front suspension uses control arms and torsion bars while the rear suspension has a rigid axle with leaf springs. The 4WD versions benefit from the Hi Trac independent front suspension with torsion bars that anchor above the chassis to protect them when the vehicle is driven off road. For the same reason, the engine, transfer case and fuel tank are protected by skid plates. The brakes are mixed, the optional ABS on the base versions is unusual, with hydraulic rather than electronic controls. The engine equipping the base models is a 116 hp, 2.4 L, 4-cylinder (2.7 L and 150 hp for the T100), while that of the SR5 is a 3.0 L V6 which delivers 150 hp.

• Safety: 75%
The pickups don't do any better than their competitors with the

TOYOTA T100

DATA

Category: rear-wheel drive or AWD compact pick-ups.
Class: utility

HISTORIC

Introduced in: 1977, 1993:T100
Modified in: 1983, 1989: touch-ups.
Made in: Tahara & Hino, Japan.

RATINGS

Safety: 75 %
Satisfaction: 85 %
Depreciation: 52 %
Insurance: 4x2: 6.3 % ($585) 4x4 & T100: ($660 to $725)
Cost per mile: $0.36

NUMBER OF DEALERS

In U.S.A.: 1,233

SALES ON THE AMERICAN MARKET

Model	1992	1993	Result
Pick-up	175,150	182,498	+ 4.1 %
T100	1,461	22,028	+ 1,507 %

MAIN COMPETITORS

DODGE Dakota, FORD Ranger, CHEVROLET S-10, GMC Sonoma, ISUZU, MAZDA B, NISSAN Hardbody.

EQUIPMENT

	TOYOTA Pick-up base	TOYOTA Pick-up SR5	TOYOTA T100 base	TOYOTA T100 V6
Automatic transmission:	O	O	O	O
Cruise control:	O	O	O	O
Power steering:	O	O	O	O
ABS brake:	O	S (V6)	O	O
Air conditioning:	O	S (V6)	O	O
Air bag (left):	O	S	S	S
Leather trim:	-	-	-	-
AM/FM/ Radio-cassette:	O	O	O	O
Power door locks:	-	O	-	-
Power windows:	-	O	-	-
Tilt steering column:	-	S (V6)	-	O
Dual power mirrors:	O	O	O	O
Intermittent wipers:	-	O	-	O
Light alloy wheels:	-	O	-	O
Sunroof:	-	O (V6)	-	-
Anti-theft system:	-	-	-	-

S : standard; O : optional; - : not available

COLORS AVAILABLE

Exterior: White, Silver, Black, Garnet, Green, Blue, Beige, Red.
Interior: Gray, Blue,Oak.

MAINTENANCE

First revision: 4,000 miles
Frequency: 4,000 miles
Diagnostic plug: Yes

WHAT'S NEW IN 1995 ?

- 3.4 L V6 (T100).
- Driver's side air bag (T100).
- ABS brakes available for 2-or 4WD.
- The 4X4 pickups are offered only with an extended bed and 6-cylinder.

Model/ versions *: standard	ENGINE Type / timing valve / fuel system	Displacement cu/in	Power bhp @ rmn	Torque lb.ft @ rpm	Compres. ratio	TRANSMISSION Driving wheels / transmission	Final ratio	PERFORMANCE Acceler. 0-62 mph s	Stand. 1/4 mile s	Stand. 5/8 mile s	Braking 62-0 mph ft	Top speed mph	Lateral acceler. G	Noise level dBA	Fuel economy mpg city	hwy	Gasoline type / octane
base	L4* 2.4 SOHC-8-EFI	144	116 @ 4800	140 @ 2800	9.3 :1	rear/4 - M5*	3.583	12.5	19.5	35.5	151	94	0.69	69	27.0	36.0	R 87
						rear/4 - A4	3.727	13.0	19.2	36.6	167	90	0.69	69	25.0	30.0	R 87
SR5	V6* 3.0 SOHC-12-EFI	181	150 @ 4800	180 @ 3400	9.0 :1	rear/4 - M5*	3.417	9.5	17.3	30.5	157	103	0.70	67	19.0	26.0	R 87
						rear/4 - A4	3.417	11.0	18.0	31.8	164	100	0.70	67	21.0	29.0	R 87
T-100 base	L4* 2.7 DOHC-16-EFI	164	150 @ 4800	177 @ 4000	9.5 :1	rear/4 - M5	3.615	NA							20.0	25.0	R 87
	V6* 3.4 DOHC-24-EFI	206	190 @ 4800	220 @ 3400	9.6 :1	rear/4 - M5	3.769	11.8	18.5	34.7	148	106	0.75	67	20.0	26.0	R 87
						rear/4 - A4	3.769	13.0	19.5	35.8	157	103	0.75	67	20.0	26.0	R 87

structure and passenger suffering only average damage in the event of a collision, but at least the driver is protected by an air bag.

• **Steering: 70%**
The off-road advantage of a large steering ratio deserts the pickup on the road, where it feels light and sensitive at cruise speed.

• **Driver's compartment: 70%**
The benches of these models do not offer a lateral or lumbar support as effective as individual bucket seats. Also the adjustments between the steering wheels and the pedals are less precise, since the seat backs are not reclinable. That of the T100 SR5 is distinctly more comfortable and contoured. Visibility is satisfactory, instrumentation easy to read and the controls set up conventionally. Note that the handbrake is more practical than the footbrake. Finally, the manual shift is crisp and precise but the clutch pedal stroke is too long.

• **Access: 70%**
It is less difficult to get into the front seats even on the 4X4 with a high ground clearance than into the back of the Xtracab.

• **Price/equipment: 70%**
Reputation of these pickups for durability is costly. For the little equipment and comfort they give, they have problems facing the competition on this point.

• **Seats: 70%**
The bench seat is less comfortable than the individual bucket seats or those of the T100 deluxe, whose comfort is surprising.

• **Suspension: 60%**
Smooth on a good road with a 4X2 or when the box is loaded, the rear axle becomes untenable empty, when the pavement is weavy or has potholes.

• **Noise level: 50%**
Base versions are noisier than the SR5 whose soundproofing is as carefully done as on a car and whose V6 engine is silky.

• **Fuel consumption: 50%**
It depends on the load and on the speed but is not more economical than any others...

• **Performance: 50%**
Nervous and well-matched to the manual with short ratios, the 4-cylinders are not as smooth as the V6, whose higher torque agrees better with the automatic. However, performance depends on load and the choice of 2 or 4WD because the torque of these two engines is sometimes a little limited. Toyota should return to a high torque in-line 6-cylinder or a V6 built for torque rather than a peaky V6 for a car.

TOYOTA T100

WEAK POINTS

• **Interior space: 30%**
While the base cab offers enough room for 2 people on the compact and 3 on the T100, the Xtracab is more spacious, but less so than the Kingcab from Nissan or Spacecab from Isuzu and the rear volume would be better used for baggage storage than for passengers, because the jump seats can only be used by young kids.

• **Handling: 30%**
It is surprising that these utilities are as stable in a straight line as in a turn, from the moment that the pavement is in good condition and the body has a light load. Otherwise, on potholes or wet, the rear axle jumps non stop, hurting the traction in curves or in the straight. At the same time, it also detracts from a straight path. The big 4X4 tires are more at ease in the mud than on the road, where they eliminate all guidance or braking and bounce extensively on road defects, making the trajectory hard to forecast.

• **Storage: 30%**
On the normal cab it cannot contain any baggage and on the Xtracab it corresponds to the trunk of a mini-compact car.

• **Braking: 30%**
Even though easy to modulate, it lacks bite and stopping distances are long. ABS does not stabilize the trajectory because as soon as the back wheels lock, all steering control is lost.

• **Conveniences: 40%**
Only the SR5 can pretend to offer practical storage. While on other models, it comes down to a simple glove box.

• **Insurance: 40%**
They are as expensive to insure as they are to buy.

• **Depreciation: 40%**
The used prices depend much on the degree of utilization, just like a utility.

CONCLUSION

• **Overall rating: 56.0%**
If compact pickups have a better chance against their North American rivals, the T100 is part of the figuring, because it does not have the versatility, power or practical side of its domestic competitors. Of course, Toyota has one up their sleeve, a powerful 3.4 V6; the last word is yet to come.

SPECIFICATIONS & PRICES

Model	Version/Trim		Body/ Seats	Wheel-base in	Lgth x Wdthx Hght. inx inx in	Curb weight lb	Towing capacity std. lb	Susp. type ft/rr	Brake type ft/rr	Steering system	Turning diameter ft	Steer. turns nber	Fuel. tank gal.	Standard tire size	Standard power train	PRICES US. $ 1994
TOYOTA		**General warranty: 3 years / 36,000 miles; power train: 5 years / 60,000 miles;corrosion: 5 years / unlimited; without deductibles or transfert fees**														
Pick-up	2x4	short deck	2dr.p-u.2	103.0	174.6x66.5x60.6	2690	3500	i/r	dc/dr	bal.	35.0	4.2	17.2	195/75R14	L4/2.4/M5	**$9,818**
Pick-up	2x4	XCab.	2dr.p-u.4	121.9	193.1x66.5x61.0	3093	3500	i/r	dc/dr	pwr.bal.	41.3	3.6	19.3	205/75R14	V6/3.0/A4	**$14,568**
Pick-up	4x4	XCab.	2dr.p-u.4	121.9	193.1x66.5x67.3	3774	3500	i/r	dc/dr	pwr.bal.	43.0	3.9	19.3	225/75R15	V6/3.0/A4	**$15,988**
Pick-up	4x4	XCab.SR5V6	2dr.p-u.4	121.9	193.1x66.5x68.7	3814	3500	i/r	dc/dr	pwr.bal.	43.0	3.9	19.3	225/75R15	V6/3.0/A4	**$19,148**
T100	2x4	base	2dr.p-u.3	121.8	209.1x75.2x66.7	3318	4000	i/r	dc/dr	pwr.r&p.	37.7	3.84	24	215/75R15	L4/2.7/M5	**$14,698**
T100	2x4	SR5	2dr.p-u.3	121.8	209.1x75.2x66.7	3351	4000	i/r	dc/dr	pwr.r&p.	37.7	3.84	24	215/75R15	L4/2.7/M5	**$16,768**
T100	2x4	base	2dr.p-u.3	121.8	209.1x75.2x70.1	3428	5200	i/r	dc/dr	pwr.r&p.	43.3	3.28	24	235/75R15	V6/3.0/M5	**$15,438**
T100	2x4	SR5	2dr.p-u.3	121.8	209.1x75.2x70.1	3459	5200	i/r	dc/dr	pwr.r&p.	43.3	3.28	24	235/75R15	V6/3.0/A4	**$16,338**
T100	4x4	base	2dr.p-u.3	121.8	209.1x75.2x70.1	3873	5000	i/r	dc/dr	pwr.r&p.	43.3	3.28	24	235/75R15	V6/3.0/M5	**$18,168**
T100	4x4	SR5	2dr.p-u.3	121.8	209.1x75.2x70.1	3944	5000	i/r	dc/dr	pwr.bal.	43.3	3.28	24	235/75R15	V6/3.0/A4	**$21,078**

See page 393 for complete 1995 Price List

Always ahead...

With the Corrado VR6, Volkswwagen has the best mid-range sports coupe. If it was mischievous with 4-cylinders, it has acquired full maturity with the V6 which gives it a driving pleasure without par and gains the envy of many rivals, not the least of which is the 968 Porsche...

VW's Corrado coupe does not sell as well as most chroniclers of the automobile would wish. It must be said that the economic context does not really favor deluxe sports cars and that is what explains its disgrace. The Corrado is a 3-door, 2-volume coupe with a unique VR6 trim, equipped with a 2,8 L V6 with standard manual transmission or an optional automatic.

STRONG POINTS

• Safety: 90%
Very rigid, the structure stands up well in impact tests, receives 3-point seat belts. Two air bags installed at the front seats complete the occupant protection.

• Insurance: 90%
Relatively affordable for owners over 25 years of age, the premium will discourage those who have not yet gotten there.

• Performance: 90%
The V6 engine shows much liveliness with power and torque at practically all rpm and only asks to turn faster until the control folds its wings around 6,500 rpm. The excellent weight-to-power ratio explains how it can go from 0 to 62 mph in less than 7 seconds or to reach (on track) a maximum speed of 140 mph. The temperament, sharp and precise, makes this little rocket infinitely more fun to drive than a Porsche 968!

• Technical: 80%
It was designed beginning with a Golf/Jetta from which it keeps the McPherson independent front suspension and a semi-independent torsional axle at the back. The steel unit body has a very effective shape with drag coefficient of 0.32. To reduce the lift at the rear, a mobile spoiler extends automatically beginning with 50 mph and returns to its initial position at less than 12 mph, not to create visibility problems toward the rear. The 6 cylinders in the V are more narrow since the angle between them is only 15 degrees, allowing the use of a single cylinder head for the two cylinder banks and reaching minimal bulk size. This allowed it to be readily installed in place of the original 4-cylinder.The manual transmission is cable-operated to eliminate linkage problems and to elegantly get around the engine.

DATA

Category: front-wheel drive compact sport coupes.
Class: 3 S

HISTORIC

Introduced in: 1988
Modified in: 1990: imported in North America.
Made in: Osnabrück (at Karmann), Germany.

RATINGS

Safety: 90 %
Satisfaction: 85 %
Depreciation: 40 % (2 years)
Insurance: 5.1 % ($855)
Cost per mile: $0.47

NUMBER OF DEALERS

In U.S.A.: 648

SALES ON THE AMERICAN MARKET

Model	1992	1993	Result
Corrado	3,436	2,111	-38.6 %

MAIN COMPETITORS

EAGLE Talon TSi, DODGE Stealth, FORD Mustang, FORD Probe, HONDA Prelude, MAZDA MX-3 & MX-6, TOYOTA Celica.

EQUIPMENT

VOLKSWAGEN Corrado	VR6
Automatic transmission:	O
Cruise control:	S
Power steering:	S
ABS brake:	S
Air conditioning:	S
Air bag (2):	S
Leather trim:	O
AM/FM/ Radio-cassette:	S
Power door locks:	S
Power windows:	S
Tilt steering column:	S
Dual power mirrors:	S
Intermittent wipers:	S
Light alloy wheels:	S
Sunroof:	-
Anti-theft system:	-

S : standard; O : optional; - : not available

COLORS AVAILABLE

Exterior: White, Red, Black, Yellow, Green, Burgundy.
Interior: Fabric: Gray. Leather: Black, Black-Beige.

MAINTENANCE

First revision: 3,000 miles
Frequency: 6 months/6,000 miles
Diagnostic plug: Yes

WHAT'S NEW IN 1995 ?

- Two air bags and an antitheft system standard.

ENGINE						TRANSMISSION		PERFORMANCE									
Model/ versions *: standard	Type / timing valve / fuel system	Displacement cu/in	Power bhp @ rmn	Torque lb.ft @ rpm	Compres. ratio	Driving wheels / transmission	Final ratio	Acceler. 0-62 mph s	Stand. 1/4 mile s	Stand. 5/8 mile s	Braking 62-0 mph ft	Top speed mph	Lateral acceler. G	Noise level dBA	Fuel economy mpg city	Fuel economy mpg hwy	Gasoline type / octane
VR6	V6* 2.8 DOHC-12-EFI	170	178 @5800	177 @ 4200	10.0 :1	front - M5*	3.64	7.0	15.2	26.8	128	140	0.86	68	22.0	32.0	R 87
						front - A4	4.15	8.0	16.0	27.2	131	137	0.85	68	21.0	32.0	R 87

• **Quality/fit/finish: 80%**
Always manufactured in Germany, the Corrado is solid and well finished, and if its interior presentation is lugubrious, the materials are of good quality.

• **Driver's compartment: 80%**
It is easy to find the best driving position and visibility is good toward the front and sides, while at the rear the narrow backlight and the spoiler, (when extended) reduce the field of vision. The dashboard is typically VW with easy to read analog gauges and the controls are simple and robust. Regrettably, the foot rest is among the missing.

• **Handling: 80%**
Unit body rigidity, a sports suspension and a low center of gravity endow the Corrado a superb balance under all conditions. Cornering, its neutral feel is improved by bushings with controlled deflections that make the rear wheels slightly self steering. It nimbly gets into the turn in an incisive, fashion without getting out of shape.

• **Steering: 80%**
Practically ideal for this type of vehicle, it is fast and precise and its modulation well calibrated to permanently inform you about road conditions.

• **Conveniences: 80%**
Storage includes a small glove box, a large console recess, door pockets and two tablets under the dash.

• **Seats: 80%**
As always at VW, they are traditionally firm but the side wings hold well and there is effective lumbar support.

• **Satisfaction: 80%**
Vehicles still manufactured in Germany, like the Corrado, have a better reputatiopn than those assembled in Mexico.

• **Suspension: 70%**
Firm without being as stiff as one would fear in a German sports coupe, it does not allow you to ignore any road defect.

• **Insurance: 70%**
Relatively affordable for owners over 25 years of age, the premium discourages those who have not reached this age.

• **Access: 70%**
It is not as easy to reach the bench seat than to climb in front, because of the meager headroom.

• **Trunk: 60%**
Little, when the rear seats are occupied it is not really flat because the spare tire takes up room. One can open it up by folding the bench seat back, but its high sill comlicates baggage handling.

• **Braking: 50%**
In spite of the presence of four disc brakes, its major weight and the ABS mounted standard keep the Corrado from breaking as hard as the first version tested. Panic stops are particularly straight and fade resistance of the pads is excellent.

• **Fuel consumption: 50%**
It is not exaggerated, even at high speeds and stays around 25 miles per gallon.

• **Depreciation: 50%**
It is lower than that of other VW models which have declined in popularity due to their Mexican quality.

• **Noise level: 50%**
The soundproofing is too effective because one has a problem hearing the suggestive roar of the engine during accelerations.

WEAK POINTS

• **Interior space: 30%**
If you take into account the compact format of this coupe, its useful interior volume allows 4-people to be seated with enough head and leg room in the back.

• **Price/equipment: 40%**
Even though the Corrado is expensive, it is a better buy than the infamous Porsche 968 and in spite of its FWD, it provides a pleasure which justifies the investment.

CONCLUSION

• **Overall rating: 67.5%**
Nothing goes at VW any more because they are not even capable of selling what is good. The Corrado is the most inspired of the sports coupes actually available and it is not manufactured in Mexico. What a chance! ☺

OWNERS' SUGGESTIONS

- **More powerful headlights.**
- **Less independent service.**
- **Effective winterizing.**
- **Larger tires.**
- **Modify the rear spoiler.**

SPECIFICATIONS & PRICES

Model	Version/Trim	Body/ Seats	Interior volume cu.ft.	Trunk volume cu ft.	Drag coef. Cd	Wheel-base in	Lgth x Wdthx Hght. inx inx in	Curb weight lb	Towing capacity std. lb	Susp. type ft/rr	Brake type ft/rr	Steering system	Turning diameter ft	Steer. turns nber	Fuel. tank gal.	Standard tire size	Standard powertrain	PRICES US. $ 1994
VOLKSWAGEN		**Model sold only in Canada.**																
Corrado	SLC	3dr.cpe.4	80	15	0.32	97.2	159.3x66.5x51.6	2808	1000	i/i	dc/dc/ABS	pwr.r&p.	34.4	3.17	18.5	205/50R15	V6/2.8/M5	**$26,150**
Corrado	SLC	3dr.cpe.4	80	15	0.32	97.2	159.3x66.5x51.6	2853	1000	i/i	dc/dc/ABS	pwr.r&p.	34.4	3.17	18.5	205/50R15	V6/2.8/A4	**$27,025**

See page 393 for complete 1995 Price List

Engine failure...

The Eurovan would enjoy a great success if it was offered with an adequate engine. This problem is not new because it was also that of all preceding models. However, Volkswagen was so late in converting to a V6 that it does not manufacture enough to equip the Eurovan.

The Eurovan is marketed in a short wagon of 5- or 7-seats in finishes CL and GL, or extended to 10-seat in GL only. There are also commercial versions with or without glass, trucks with a single or a double cam and a camping van unique in the range of a car builder. The Eurovans are provided standard with a 5-cylinder 2.5L engine with a 5-speed manual or an optional 4-speed automatic. The 110 hp Diesel is an option on all versions but only with a manual transmission.

STRONG POINTS

• Interior space: 100%
The interior is sufficiently large to allow seven people to spend time in it and to circulate with ease between the benches; the 10-seater is a real bus.

• Access: 100%
It is easy to climb on board because the openings are ample and so is the inside height.

• Storage: 90%
You can pile a considerable amount of baggage into it without having to fold down the rear seat.

• Quality/fit/finish: 80%
The assembly and finish are in the VW tradition the quality of the materials is excellent, but the general presentation is that of a utility rather than a minivan.

• Satisfaction: 80%
The real German quality still has its followers, even at a stout price, as long as the product is reliable and durable.

• Technical: 80%
In spite of its careful design, the steel unit body still has a utility look,

DATA

Category: front-wheel drive compact vans.
Class : utility

HISTORIC

Introduced in: 1990
Modified in: 1992: imported in North America;1993: diesel engine.
Made in: Germany.

RATINGS

Safety: 60 %
Satisfaction: 80 %
Depreciation: 57 %
Insurance: 4.8 % ($655)
Cost per mile: $0.42

NUMBER OF DEALERS

In U.S.A.: 648

SALES ON THE AMERICAN MARKET

Model	1992	1993	Result
Eurovan	2,679	5,634	+ 110 %

MAIN COMPETITORS

CHEVROLET Lumina, CHEVROLET Astro, CHRYSLER Town & Country, DODGE Caravan, FORD Aerostar, Windstar, MAZDA MPV, MERCURY Villager, NISSAN Quest, PLYMOUTH Voyager, PONTIAC Trans Sport, TOYOTA Previa.

EQUIPMENT

VOLKSWAGEN Eurovan	CV	GL	Transporter
Automatic transmission:	O	O	O
Cruise control:	O	S	O
Power steering:	S	S	S
ABS brake:	O	O	O
Air conditioning:	O	O	O
Air bag:	O	O	-
Leather trim:	O	O	O
AM/FM/ Radio-cassette:	S	S	O
Power door locks:	S	S	S
Power windows:	S	S	-
Tilt steering column:	-	O	O
Dual power mirrors:	S	S	S
Intermittent wipers:	S	S	S
Light alloy wheels:	O	O	-
Sunroof:	-	-	-
Anti-theft system:	-	-	-

S : standard; O : optional; - : not available

COLORS AVAILABLE

Exterior: White, Blue, Gray, Red, Green, Brown, Bordeau, Beige, Silver.
Interior: Gray, Blue, Beige.

MAINTENANCE

First revision: 3,000 miles
Frequency: 6 months/6,000 miles
Diagnostic plug: No

WHAT'S NEW IN 1995 ?

- **Two air bags optional.**
- **Electric windows standard (CV).**
- **Entry lights (CV).**

Model/ versions *: standard	ENGINE Type / timing valve / fuel system	Displacement cu/in	Power bhp @ rmn	Torque lb.ft @ rpm	Compres. ratio	TRANSMISSION Driving wheels / transmission	Final ratio	PERFORMANCE Acceler. 0-62 mph s	Stand. 1/4 mile s	Stand. 5/8 mile s	Braking 62-0 mph ft	Top speed mph	Lateral acceler. G	Noise level dBA	Fuel economy mpg city	hwy	Gasoline type / octane
base	L5* 2.5 SOHC-8-IE	150	110 @ 4500	140 @ 2200	8.5 :1	front - M5*	4.62	14.5	18.3	36.6	157	100	0.73	68	20.0	28.0	R 87
						front - A4	4.56	15.5	19.8	37.8	167	97	0.73	68	19.0	24.0	R 87
Diesel	L5* 2.4 SOHC-8-I	147	77 @ 3700	154 @ 1800	22.5: 1	front - M5*	-	NA									

(Diesel engine only available in Canada)

which does not keep it from being aerodynamically very effective since its drag coefficient is 0.36. The powertrain is disposed transversely, above the rear wheels, which allows a roomier cab and improved handling, thanks to better weight distribution. The 4-wheel independent suspension is very sophisticated, with a system of double forks and fore and aft torsion bars at the front. The mixed disc and drum brakes are standard and 4-wheel discs with ABS are optional.

• Insurance: 80%
The Diesel version is less expensive to insure but the "gasoline" is in the middle of its category.

• Storage space: 80%
The GL Eurovan is the world champion of pocket emptiers of which there is a profusion, allowing children to hide their treasures.

• Driver's compartment: 70%
Seated as in a bus, the driver sees well from all angles. The dashboard is simple, functional and the controls easy to access. The total is ergonomic but armrests are needed for support.

• Seats: 70%
Their contours provide satisfactory hold and lumbar support and the benches have individual armrests but their padding has nothing soft about it and they are often disagreeable on a bad road.

• Suspension: 70%
Its smoothness allows it to absorb road irregularities but body movements are well-controlled. Tar strips and road joints cause very horse-like bucking.

• Steering 60%
It is very fuzzy in the center, takes many turns lock-to-lock, lacks returnability and often runs out of power-assist.

• Safety: 50%
In spite of numerous reinforcements, the vans are more vulnerable than cars in the event of a collision. The occupants' protection includes 3-point seat belts and only optional air bags. However, there are headrests at all seats.

WEAK POINTS.

• Performance: 10%
The 100 horsepower of the 5-cylinder engine has a hard time driving the respectable weight of the vehicle, which can exceed two tons and has a weight/power ration of 42 lbs/hp. Same remark applies to the diesel which has slightly more torque but whose take-offs and acceleration are very long. Passing requires much anticipation, not to speak of anxiety. Once launched, the Eurovan maintains a high cruising speed but its fuel consumption strongly resents it.

• Braking: 30%
The pedal is spongy but powerful, stable and fade-free.

• Price/equipment: 30%
The standard version offers very limited equipment while that of the GL is more complete and refined. Included are a heater in the windshield washer and heated rear-view mirrors, very useful in winter driving.

• Handling: 40%
It is reassuring, in spite of the body height and the softness of the suspension (independent and particularly well-adapted). Roll is well-controlled, getting into the turn is easy in either wide curves or tight turns. In a straight line, it is sensitive to crosswinds but nothing like its predecessors. In the winter time, the balance on a slippery road is impressive.

• Noise level: 40%
The cab resonates to many powertrain, rolling and wind noises. As speed increases, so does the body's amplifier action.

• Fuel consumption: 40%
It is lower with the automatic but equivalent to that of a 3.0 L V6, without the performance. The diesel is the queen of economy but it demands a certain philosophy and a slow drawing pipe to get used to this type of driving.

• Depreciation: 40%
Volkswagen's reputation is in a dip and depreciation rates have risen to those of its rivals.

CONCLUSION

• Overall rating: 62.0%
Eurovan is meant for families rather than to single folks seeking a versatile vehicle. Until it is offered with a 4.3 L V6 engine, it will be better to look elsewhere than to accept "time lag".

SPECIFICATIONS & PRICES

Model	Version/Trim	Body/ Seats	Drag coef. Cd	Wheel-base in	Lgth x Wdthx Hght. inx inx in	Curb weight lb	Towing capacity std. lb	Susp. type ft/rr	Brake type ft/rr	Steering system	Turning diameter ft	Steer. turns nber	Fuel. tank gal.	Standard tire size	Standard powertrain	PRICES US. $ 1994
VOLKSWAGEN	**Model sold only in Canada.**															
Eurovan	GL	4dr.van.5/7	0.36	115.0	186.6x72.4x75.6	3814	2000	i/i	dc/dr	pwr.r&p.	38.4	3.5	21.1	205/65R15	L5/2.5/M5	**$20,420**
Eurovan	GL Diesel	4dr.van.5/7	0.36	115.0	186.6x72.4x75.6	3858	2000	i/i	dc/dr	pwr.r&p.	38.4	3.5	21.1	205/65R15	L5D/2.4/M5	-
Eurovan	Camper	4dr.van.4	0.36	115.0	186.6x72.4x77.6	4718	2000	i/i	dc/dr	pwr.r&p.	38.4	3.5	21.1	205/65R15	L5/2.5/M5	**$21,850**
Eurovan	CV Diesel	4dr.van.5/7	0.36	115.0	186.6x72.4x77.6	4718	2000	i/i	dc/dr	pwr.r&p.	38.4	3.5	21.1	205/65R15	L5D/2.4/M5	-

(Diesel engine only available in Canada)

See page 393 for complete 1995 Price List

Bottom of the barrel...

One cannot say that things are going very well for Volkswagen these days. The Golf and Jetta constitute the best example of an excellent product whose reliability was zapped by greed. Their quality and credibility improved but the damage is done.

Renewed two years ago, the Golf and Jetta have not retrieved the public enthusiasm of the 80's. The Golf are 3- and 5-door (hatchback) sedans or 2-door convertibles offered in CL, GL and convertible. The Jetta is a 4-door available in CL, GL and GLX. The CL is equipped with a 1.8 L engine while the GL and convertible receive the 2.0 L and the GLX the 2.8 L V6. The diesel remains offered as an option on the GL since VW realizes that a good portion of its Canadian sales stem from this type of engine with low emission rates.

STRONG POINTS

• Handling: 80%
The Golf-Jetta surprise by their stability because their wheels seem welded to the road and they do not transmit any bump, while still communicating the pavement condition. They are neutral in tight corners since the rear axle offers a good grip and roll is controlled. One can be less emphatic about the GLX which is too powerful for its potential and cannot be handed to just anyone, because of its touchiness, especially on a wet road. Too soft, the suspension provokes body motions that are not compatible with balance.

• Steering: 80%
Both smooth and precise, it has lost some of its speed, but its assistance is well-modulated and the maneuverabily is excellent, except on the GLX where the feel is too light and it also suffers from a large torque steer which can make maneuvers tricky.

• Driver's compartment: 80%
The dashboard is familiar, but its forms are rounder and its compartmenting less strict. The gauge cluster is compact, well-organized and easy to read. The main controls are standardized and fall at your fingertips; the steering wheel offers a good grip. The driver soon finds his best position, thanks to simple seat adjustments and the optional steering column setting. Visibility is excellent in spite of the thick posts and the rear-view mirrors are well-located.

• Technical: 80%
The latest Golf-Jetta resemble their predecessors, except for a more bulbous front, more rounded forms and oval headlights. The steel unit body provides better aerodynamics and the reduced drag with a Cd that went from 0.34 to 0.32. The front end geometry includes McPherson struts and automatic stabilization with a negative offset (It launches and stops straighter). At the rear, VW has perfected a torsional axle whose track changes automatically to compensate for understeering by an induced geometry. The brakes are mixed on the CL/GL and 4-discs with ABS on the GLX.

• Satisfaction: 80%
The reliability has not returned to the level of the Japanese and a number of fans have been disappointed.

• Trunk: 80%

DATA

Category: front-wheel drive sub-compact sedans and convertibles. front-wheel drive compact sedans. (Jetta).
Class : 3S & 3

HISTORIC

Introduced in: 1974: Rabbit; 1975: GTi & Jetta; 1985: Golf.
Modified in: 1985 & 1993 : body.
Made in: Mexico:Golf/Jetta;convertible: Osnabrück, Germany.

RATINGS

	Golf	Jetta
Safety:	65 %	70 %
Satisfaction:	79 %	77 %
Depreciation:	43.5 %	45 % (2 years)
Insurance:	($655) 9.0 %	7.6 % ($785)
Cost per mile:	$0.30	$0.32

NUMBER OF DEALERS

In U.S.A.: 648

SALES ON THE AMERICAN MARKET

Model	1992	1993	Result
Golf	7,359	4,693	- 36.2 %
Jetta	29,907	14,583	-51.3 %

MAIN COMPETITORS

Golf: DODGE Colt, HONDA Civic Hbk, HYUNDAI Accent, MAZDA 323, NISSAN Sentra, TOYOTA Tercel.
Jetta: ACURA Integra 4p.,CHEVROLET Cavalier, CHRYSLER Neon, FORD Escort-Contour, HONDA Civic 4dr., HYUNDAI Elantra, MAZDA Protegé, MERCURY Mystique, PONTIAC Sunfire, SATURN SL1 &SL2, SUBARU Impreza, TOYOTA Corolla.

EQUIPMENT

VOLKSWAGEN Golf	CL	GL		Diesel	Con.
VOLKSWAGEN Jetta	**CL**	**GL**	**GLX**		
Automatic transmission:	O	O	O	-	O
Cruise control:	O	O	S	O	O
Power steering:	S	S	S	S	S
ABS brake:	-	-	S	O	O
Air conditioning:	O	O	S	O	O
Air bag (2):	-	O	S	O	S
Leather trim:	-	-	O	-	-
AM/FM/ Radio-cassette:	O	S	S	S	S
Power door locks:	S	S	S	S	S
Power windows:	O	O	S	O	O
Tilt steering column:	-	O	S	O	S
Dual power mirrors:	S	S	S	S	S
Intermittent wipers:	S	S	S	S	S
Light alloy wheels:	-	S	S	O	S
Sunroof:	O	O	S	O	-
Anti-theft system:	S	S	S	S	S

S : standard; O : optional; - : not available

COLORS AVAILABLE

Exterior: White, Gray, Blue, Red, Silver, Black, Green, Suede, Purple.
Interior: Gray.

MAINTENANCE

First revision: 3,000 miles
Frequency: 6 months/6,000 miles
Diagnostic plug: Yes

WHAT'S NEW IN 1995 ?

- New versions Jetta GLX and Golf GTi with a V6 engine.
- New Golf convertible.

Model/ versions *: standard	ENGINE Type / timing valve / fuel system	Displacement cu/in	Power bhp @ rmn	Torque lb.ft @ rpm	Compres. ratio	TRANSMISSION Driving wheels / transmission	Final ratio	PERFORMANCE Acceler. 0-62 mph s	Stand. 1/4 mile s	Stand. 5/8 mile s	Braking 62-0 mph ft	Top speed mph	Lateral acceler. G	Noise level dBA	Fuel economy mpg city	hwy	Gasoline type / octane
CL	L4* 1.8 SOHC-8-EFI	109	90 @ 5250	106 @ 2500	9.0 :1	front - M5*	3.67	12.6	18.5	33.5	128	109	0.80	67	30.0	43.0	R 87
Diesel	L4*T 1.9 SOHC-8-MI	116	75 @ 4200	107 @ 2400	22.5 :1	front - M5*	3.67	15.5	20.7	37.0	131	103	0.79	68	39.0	49.0	D #2
GL	L4* 2.0 SOHC-8-EFI	121	115 @ 5400	121 @ 3200	10.4 :1	front - M5*	3.67	10.5	16.2	32.5	121	121	0.81	67	28.0	41.0	R 87
& cabrio						front - A4	4.22	12.0	17.0	33.3	134	118	0.81	67	24.0	35.0	R 87
GLX	V6* 2.8 DOHC-24-MPSFI	170	172 @ 5800	173 @ 4200	10.0 :1	front - M5*	3.39	7.7	-	-	-	137	-	-	22.0	25.0	R 87
(Diesel engine only available in Canada)						front - A4	3.63	NA									

With a higher capacity and more accessibility than before, you can gain extra room by folding down the 60/40 rear bench backs. The trunk of the Jetta is so deep that it is dangerous to lean into it.

• Safety: 70%
The unit body has been substantially stiffened by reinforcements built into strategic areas for better impact protection of the occupants. Air bags will only be available as standard on the GLX convertible.

• Quality/fit/finish: 70%
The Golf-Jetta sold in North America are not being manufactured in Germany as is the convertible, but in Mexico and the quality problems they have known have not helped VW's credibility.

• Seats: 70%
They maintain better laterally, but the lumbar support is more effective on the GL and the GLX than on the CL. The rear headrests add to safety and comfort.

• Suspension: 70%
Relatively soft, it smoothly absorbs road defects, thanks to ample wheel travel.

• Fuel consumption: 70%
Normal for their displacement, only the Diesel version earns the right to the "economy" label.

• Access: 60%
It is not easy to get into the back seat of the 2-door Golf. Its seat back tilts, but the cushion does not slide far enough forward to liberate entry room.

• Performance: 60%
Thanks to good gear ratio staging, the performance of the standard Diesel and V6 are more impressive than those of the 2.0 L which equips the GL and is without fun. Even though the V6 would allow an inspired drive (by risking tickets), we have not found the sensations experienced in Europe at the wheel of a Golf VR6.

• Conveniences: 60%
Large door pockets, a recess in the center console, shelves under the dashboard and a locking cassette storage cabinet are well-located and form the main storage.

• Price/equipment: 60%
The Golf/Jetta remain more costly than their competitors with equal equipment, that is with very few things, since the main safety elements are stickered as an option.

• Depreciation: 50%
It is stronger than forecast, more supply than demand...

• Braking: 50%
Panic stops call for a stopping distance of 130 ft which seems to be the norm. The ABS functions well even on a bumpy road.

• Noise level: 50%
Even though the body is rigid and streamlined, mechanical noises remains high.

• Interior space: 50%
The volume has not progressed much with respect to the previous model but the main dimensions provide for adequate leg and headroom.

WEAK POINTS

• Insurance: 40 %
Many thefts have caused the rise in the premium of these vehicles above those of their competitors.

CONCLUSION

• Overall rating: 65.5%
Only quality and reliability will bring back the confidence that these models should never have lost. ☺

SPECIFICATIONS & PRICES

Model	Version/Trim	Body/ Seats	Interior volume cu.ft.	Trunk volume cu ft.	Drag coef. Cd	Wheel-base in	Lgth x Wdthx Hght. inx inx in	Curb weight lb	Towing capacity std. lb	Susp. type ft/rr	Brake type ft/rr	Steering system	Turning diameter ft	Steer. turns nber	Fuel. tank gal.	Standard tire size	Standard powertrain	PRICES US. $ 1994
VOLKSWAGEN		General warranty: 3 years / 36,000 miles; Power Train: 5 years / 60,000 miles; antipollution: 5 years / 50,000 miles; corrosion perforation: 6 years.																
Golf	CL	3dr.sdn.5	88	17.5	0.32	97.4	158.3x66.7x56.1	2463	1000	i/si	dc/dr	pwr.r&p.	32.6	3.2	14.5	175/70R13	L4/1.8/M5	-
Golf	CL	5dr.sdn.5	88	17.5	0.32	97.4	158.3x66.7x56.1	2526	1000	i/si	dc/dr	pwr.r&p.	32.6	3.2	14.5	175/70R13	L4/1.8/M5	-
Golf	GL	3dr.sdn.5	88	17.5	0.32	97.4	158.3x66.7x56.1	2586	1000	i/si	dc/dr	pwr.r&p.	32.6	3.2	14.5	185/60R14	L4/2.0/M5	-
Golf	GL	5dr.sdn.5	88	17.5	0.32	97.4	158.3x66.7x56.1	2667	1000	i/si	dc/dr	pwr.r&p.	32.6	3.2	14.5	185/60R14	L4/2.0/M5	$12,525
Golf	GL Diesel	3dr.sdn.5	88	17.5	0.32	97.4	158.3x66.7x56.1	2216	1000	i/si	dc/dr	pwr.r&p.	32.6	3.2	14.5	185/60R14	L4D/1.9/M5	
Golf	GL Diesel	5dr.sdn.5	88	17.5	0.32	97.4	158.3x66.7x56.1	2271	1000	i/si	dc/dr	pwr.r&p.	32.6	3.2	14.5	185/60R14	L4D/1.9/M5	
Golf	Cabriolet	2dr.con.5	75.6	9.5	0.38	97.4	160.6x66.1x55.1	2943	NR	i/si	dc/dr	pwr.r&p.	32.6	3.2	14.5	195/60HR14	L4/2.0/M5	NA
Jetta	CL	4dr.sdn.5	88	19.4	0.32	97.4	173.4x66.7x56.1	2579	1000	i/si	dc/dc	pwr.r&p.	32.6	3.2	14.5	185/60HR14	L4/1.8/M5	-
Jetta	GL	4dr.sdn.5	88	19.4	0.32	97.4	173.4x66.7x56.1	2683	1000	i/si	dc/dc	pwr.r&p.	32.6	3.2	14.5	185/60HR14	L4/2.0/M5	$13,750
Jetta	GL Diesel	4dr.sdn.5	88	19.4	0.32	97.4	173.4x66.7x56.1	2765	1000	i/si	dc/dc	pwr.r&p.	32.6	3.2	14.5	185/60HR14	L4D/1.9/M5	$15,700
Jetta	GLX	4dr.sdn.5	88	19.4	0.34	97.3	173.4x66.7x56.1	2932	1000	i/si	dc/dc/ABS	pwr.r&p.	34.2	3.17	14.5	205/50HR15	V6/2.8/M5	$20,850

(Diesel engine only available in Canada)

See page 393 for complete 1995 Price List

Fountain of youth...

Passat has just received an updating that was due. Most of its major defects have been corrected and today it is in its best form. The message still has to be transmitted, because having good products is not enough, one should know how to sell them. This is just a question of attitude.

For more than twenty years, 6.2 million Passats have sold throughout the world. The fourth generation, which was recently introduced, is available as a sedan or a 4-door station wagon in two trims: GLS base with a Diesel and GLX with a V6. The standard GLS transmission is manual and that of the GLX is automatic, offered as an option on the GLS. The body not only received an aesthetic face-lift along the front but also a practical one where the volume of the interior and trunk were rearranged.

STRONG POINTS

• Safety: 90%
The Passat is now fully safe, thanks to the addition of many reinforcements, including in the doors, and two air bags protecting the front occupants.

• Technical: 80%
The steel unit body has good aerodynamic efficiency since its drag coefficient varies from 0.31 to 0.33 depending on the model. Its unique line distinguishes it clearly from stereotyped Japanese products by the fact that it does not have a grill. The front suspension is a McPherson type and the steering wheels have a negative offset to reduce the effect of any difference in traction in takeoff and braking and to provide better self-alignment. The GLS engine is a 1.9 L Diesel which produces 75 hp and that of the GLX is the same V6 which already equips the Corrado and the Jetta GLX.

• Handling: 80%
The Passat posts an excellent road holding, thanks to the intelligent trade offs in the suspension, which in spite of its softness generates very little roll and allows a good lateral acceleration coefficient (0.83).

• Quality/fit/finish: 80%
As most cars built in Germany, the Passat projects a strong image of sturdiness and uniformity. The interior design is less austere than before and the materials are of good quality.

• Driver's compartment: 80%
The dashboard is compact and ergonomic and its controls well laid out. The manual transmission selector is cable-controlled, of great smoothness and the precision is acceptable. The driving position is most effective and easy to find. Visibility is good, except toward the rear of the sedan where the backlight takes a peculiar shape.

• Steering: 80%
It allows precise guidance, a good power assist, a correct ratio and maneuverability is normal.

• Access: 80%
It is easy to take a seat since the doors are well-proportioned and the roof height sufficient.

• Seats: 80%
Well-formed, they provide ample hold and lumbar support and their

DATA

Category: front-wheel drive compact sedans and wagons.
Class : 4

HISTORIC

Introduced in: 1973
Modified in: 1980,1988: renewed.
Made in: Emden, Germany.

RATINGS

Safety: 90 %
Satisfaction: 73 %
Depreciation: 57.5 %
Insurance: 4.8 % ($865)
Cost per mile: $0. 40

NUMBER OF DEALERS

In U.S.A.: 648

SALES ON THE AMERICAN MARKET

Model	1992	1993	Result
Passat	12,578	11,970	- 4.9 %

MAIN COMPETITORS

BUICK Skylark, CHRYSLER Cirrus-Stratus HONDA Accord, MAZDA 626, NISSAN Altima, OLDSMOBILE Achieva, PONTIAC Grand Am, SUBARU Legacy, TOYOTA Camry.

EQUIPMENT

VOLKSWAGEN Passat	GLS	GLX
Automatic transmission:	-	O
Cruise control:	S	S
Power steering:	S	S
ABS brake:	S	S
Air conditioning:	S	S
Air bag (2):	S	S
Leather trim:	O	O
AM/FM/ Radio-cassette:	S	S
Power door locks:	S	S
Power windows:	S	S
Tilt steering column:	S	S
Dual power mirrors:	S	S
Intermittent wipers:	S	S
Light alloy wheels:	S	S
Sunroof:	O	O
Anti-theft system:	S	S

S : standard; O : optional; - : not available

COLORS AVAILABLE

Exterior: White, Red, Blue, Green, Gray, Black, Purple.
Interior: Gray, Black, Blue, Beige.

MAINTENANCE

First revision: 3,000 miles
Frequency: 6 months/6,000 miles
Diagnostic plug: Yes

WHAT'S NEW IN 1995 ?

- **Aesthetic body redesign.**
- **Improvement in the interior volume and trunk.**
- **Increased safety level.**
- **Elimination of gasoline 4-cylinder engines.**

ENGINE						TRANSMISSION		PERFORMANCE										
Model/ versions *: standard	Type / timing valve / fuel system	Displacement cu/in	Power bhp @ rmn	Torque lb.ft @ rpm	Compres. ratio	Driving wheels / transmission	Final ratio	Acceler. 0-62 mph s	Stand. 1/4 mile s	Stand. 5/8 mile s	Braking 62-0 mph ft	Top speed mph	Lateral acceler. G	Noise level dBA	Fuel economy mpg city	Fuel economy mpg hwy	Gasoline type / octane
GLS diesel	L4* 1.9 SOHC-16-EFI	116	75 @ 4400	100 @ 2400	22.5 :1	front - M5*	3.94	17.8	23.1	37.6	131	103	0.80	68	39.0	49.0	D
GLX VR6	V6* 2.8 DOHC-12-SFI	170	172 @ 5800	177 @ 4200	10.0 :1	front - M5*	3.39	8.5	15.8	29.0	135	140	0.83	66	22.0	32.0	R 87
						front - A4	3.70	10.3	16.7	30.0	131	137	0.83	66	20.0	33.0	M 89

(Diesel engine only available in Canada)

padding is softer than in the past.

• Insurance: 75%
The index is average, but the rise in price gives it a premium to match.

• Satisfaction: 75%
Even though manufactured in Germany by a responsible company, the problems encountered with the fuel injection electronics and the automatic transmission controls have caused a drop in the popularity of these models.

• Interior space: 70%
Even though spacious, the interior will comfortably accept only four adults but a fifth can take a seat in the rear if essential.

• Trunk: 70%
As spacious on the sedans as on the station wagon, its opening has been improved to ease baggage handling.

• Conveniences: 70%
Conveniences are as numerous as they are practical and well-spread throughout the interior.

• Suspension: 70%
Smoother than before, its wheel travel allows it to effectively absorb road unevenness. However, the tires still tramp against the tar strips and road joints.

• Fuel consumption: 70%
The V6s are infinitely thirstier than the diesels considering the performance and non-negligible weight of the car.

• Noise level: 60%
The soundproofing has been improved in many ways, to the point where the diesel is no longer much noisier than the gasoline engine.

• Performance 60%
The Diesel engine seems full of spirit, but the tach denies this impression because it takes a little under 20 seconds to get from 0 to 62 mph: a time double that of a slow car. However, it offers an interesting torque which favors accelerations even underload. The choice is now easier, because economy fans will now opt for the Diesel and the sports fan for the V6 which brings a different dimension to the Passat. Smooth and full of torque, it provides standing starts and accelerations but not as much as the lighter Corrado .

• Braking: 50%
It shows up as efficient and easy to modulate and the ABS insures a smooth balance. Less harsh in its feel, the pedal is more progressive and easy to modulate.

WEAK POINTS

• Price/equipment: 40%
The price of the Passat is less competitive than when it was first introduced. The increase in standard equipment explains in part the price rise which risks lowering sales.

• Depreciation: 40%
Those who have had the third generation Passat, have paid full price for their confidence in VW, because the numerous problems which they experienced dropped the price.

CONCLUSION

• Overall rating: 70.0%
Volkswagen has an urgent need to regain the trust that its numerous fanatics had in its products. It is not an accident that the Passat and the Golf-Jetta have known so many reliability problems. It results from a lowering of quality to improve profitability or the neglect of addressing them rapidly to avoid their propagation. One already knows of the distress climate at Audi a few years ago and guess what, it was under the regime of Dr. Piech who today directs Volkswagen. ☺

OWNERS' SUGGESTIONS

-Improve the reliability
-15 inch wheels on the GLS
-Manual transmission ratios too long
-Better rear visibility
-More luxurious appearance

SPECIFICATIONS & PRICES

Model	Version/Trim	Body/ Seats	Interior volume cu.ft.	Trunk volume cu ft.	Drag coef. Cd	Wheel-base in	Lgth x Wdthx Hght. inx inx in	Curb weight lb	Towing capacity std. lb	Susp. type ft/rr	Brake type ft/rr	Steering system	Turning diameter ft	Steer. turns nber	Fuel. tank gal.	Standard tire size	Standard powertrain	PRICES US. $ 1994
VOLKSWAGEN	**General warranty: 3 years / 36,000 miles; Power Train: 5 years / 60,000 miles; antipollution: 5 years / 50,000 miles; corrosion perforation: 6 years.**																	
Passat	GLS Diesel	4dr.sdn.5	95.5	17.5	0.32	103.3	181.3x67.7x56.3	2976	2000	i/si	dc/dc/ABS	pwr.r&p.	34.1	3.33	18.5	195/60R14	L4D/1.9/M5	-
Passat	GLS Diesel	5dr.wgn.5	99.3	16.4	0.36	103.3	180.9x67.7x58.5	3020	2000	i/si	dc/dc/ABS	pwr.r&p.	34.1	3.33	18.5	195/60R14	L4D/1.9/M5	-
Passat	GLX VR6	4dr.sdn.5	95.5	17.5	0.32	103.3	181.3x67.7x56.3	3190	2000	i/si	dc/dc/ABS	pwr.r/p.	34.1	3.08	18.5	215/50R15	V6/2.8/M5	**$23,075**
Passat	GLX VR6	5dr.wgn.5	99.3	16.4	0.36	103.3	180.9x67.7x58.5	3256	2000	i/si	dc/dc/ABS	pwr.r&p.	34.1	3.08	18.5	215/50R15	V6/2.8/M5	**$23,500**

(Diesel engine only available in Canada)

See page 393 for complete 1995 Price List

Swedish Cinderella...

The front wheel drive that Volvo did not want is in the process of saving its life. After waiting long enough to surrender to the evidence, Volvo finally took the plunge and it is now a success story. Only dummies do not change their mind...

The 850 continues to prove that Volvo had reason to switch its gun to the other shoulder about front wheel drive since sales are staying up and they are really selling, even if Volvo does not know how many they are really selling. They continue to be offered in sedans or station wagons with 4-doors in three trims with a 2.4 L 5-cylinder. In 1995 the GLE, which constitutes the bottom of the range, will come with a 10-valve engine and the naturally aspirated 20-valve in a 2.3 L turbo. The transmissions are: the standard manual or an optional automatic, except for the Turbo, which is only available with an automatic.

STRONG POINTS

• Safety: 90%
The 850 structure is characterized by the implementation up front of two superimposed beams, a real cage which protects against front, side and rear impacts as well as roll overs. A cross member stiffens the rear platform under the bench seat and a full box placed between the front seats absorbs lateral impact so that all parts of the body help handle it. The doors are fitted with a rail and with cushioning to reduce lateral impacts to the occupants. There are two standard air bags and the belts are tensioned.

• Satisfaction: 90%
The owners appreciate the effort Volvo made to improve product and service but that is expensive.

• Technical: 80%
The body of the 850 does not break with tradition and its appearance retains all the attributes of the marque which have made the success of the 240 and the 700. It is a steel unit body where the weight is distributed 60/40. The front suspension resembles a McPherson while the rear uses a "Delta Link". It is a semi-independent which includes

DATA

Category: front-wheel drive luxury sedans ans wagons.
Class : 7

HISTORIC

Introduced in: 1992
Modified in: Imported in 1993.1994: wagons and turbo.
Made in: Gand, Belgium & Torslanda, Sweden

RATINGS

Safety: 90 %
Satisfaction: 90 %
Depreciation: 33% (2 years)
Insurance: ($785) base 3.8 % 3.7 % ($855) turbo
Cost per mile: 0.62 $

NUMBER OF DEALERS

In U.S.A.: 405

SALES ON THE AMERICAN MARKET

Model	1992	1993	Result
Volvo	5,202	28,367	+ 545 %

MAIN COMPETITORS

ACURA Vigor, AUDI A4, INFINITI J30, LEXUS ES300, MERCEDES BENZ C-Class, NISSAN Maxima, SAAB 900.

EQUIPMENT

VOLVO 850	GLE	GLT	Turbo
Automatic transmission:	O	O	S
Cruise control:	O	S	S
Power steering:	S	S	S (autom.)
ABS brake:	S	S	S
Air conditioning:	O	S	S
Air bag (2):	S	S	S
Leather trim:	O	O	S
AM/FM/ Radio-cassette:	S	S	S
Power door locks:	S	S	S
Power windows:	S	S	S
Tilt steering column:	S	S	S
Dual power mirrors:	S	S	S
Intermittent wipers:	S	S	S
Light alloy wheels:	S	S	S
Sunroof:	O	S	S
Anti-theft system:	S	S	S

S : standard; O : optional; - : not available

COLORS AVAILABLE

Exterior: Black,White,Red,Blue,Green,Silver,Graphite,Gray,Sand,Teal.
Interior: Blue, Taupe, Gray.

MAINTENANCE

First revision: 10,000 miles; Turbo : 5,000 miles
Frequency: 10,000 miles; Turbo : 5,000 miles
Diagnostic plug: Yes

WHAT'S NEW IN 1995 ?

- **Side air bags.**
- **New standard 10-valve 138 hp engine.**
- **Bumpers, door handles and lower body in body color.**
- **Cup carriers for front and rear passengers.**

ENGINE					TRANSMISSION			PERFORMANCE									
Model/ versions *: standard	Type / timing valve / fuel system	Displacement cu/in	Power bhp @ rmn	Torque lb.ft @ rpm	Compres. ratio	Driving wheels / transmission	Final ratio	Acceler. 0-62 mph s	Stand. 1/4 mile s	Stand. 5/8 mile s	Braking 62-0 mph ft	Top speed mph	Lateral acceler. G	Noise level dBA	Fuel economy mpg city	Fuel economy mpg hwy	Gasoline type / octane
GLE	L5* 2.4 DOHC-10-MPFI	149	138 @ 5400	152 @3600	10.0 :1	front - M5	4.00	NA									
						front - A4	2.74	NA									
GLT	L5* 2.4 DOHC-20-MPFI	149	168 @ 6200	162 @3300	10.5 :1	front - M5	4.00	8.9	16.5	30.0	121	128	0.78	66	24.0	37.0	R 87
						front - A4	2.74	9.6	17.0	31.5	128	125	0.78	66	24.0	36.0	R 87
Turbo	L5T* 2.3 DOHC-20-MPFI	142	222 @ 5200	221 @2100	8.5 :1	front - M5	2.54	7.6	15.6	28.7	131	146	0.80	68	23.0	34.0	M 89

torsion bars, control amplitude and suspension rate and also includes a stabilizer bar. Disc brakes are used on all 4 wheels, together with ABS and the valve controls the balance between the front and rear brakes. The 5-cylinder engine is modular between the 4- and the 6- with which it shares some elements. It is mounted transversely and benefits from an intake control developed by Volvo under the name VVIS.

• Quality/fit/finish: 80%
Numerous details show excellent quality control. The interior presentation is severe but the choice of colors is interesting; quality and finish are first class.

• Driver's compartment: 80%
The driving position is ideal and easy to find, thanks to multiple adjustments of the seat and steering wheel in both axis. Visibility is excellent from all angles and the lateral rear-view mirrors are well-sized. The driver finds all necessary information on a typically Volvo dashboard. The instruments are new, clear and the controls well laid out, with the exception of the cruise control and the window and rear-view mirror controls. These are placed on the center console. It would be helpful to invert the position of the radio, which one uses more often than those of the air conditioning.

• Access: 80%
It is not complicated to get on board, since the doors open wide and are well-sized.

• Seats: 80%
They provide comfortable lumbar support but one would appreciate an improved lateral hold, especially with leather trim.

• Suspension: 80%
It is soft and isolates the car from road defects, even though you can sense some pavements through the front drive train.

• Steering: 80%
Rapid and precise, its assistance is a little strong under some circumstances, but maneuverability is favored by a short turning diameter.

• Insurance: 80%
The premium is realistic, considering the price and the index based on a conservative clientele.

• Performance: 75%
The weight-to-power ratio is more favorable on the GLT with a Turbo than with the 10-valve engine, which lacks power and prefers the manual transmission. It presents at least the advantage of being more economical, but you can't have everything. Take-offs are more rapid than passings with the 20-valve but it is the Turbo which reminds more of the 740 by its ability to accelerate and its high torque. It is a real pleasure to step on the gas, because the response time is reduced to a minimum.

• Handling: 65%
It is very clean, no matter what the type of pavement, and cross-wind has very little influence on the trajectory, but the low rate and high amplitude of the suspension result in a large roll. This accelerates a considerable understeer which depends on the speed.

• Interior space: 65%
The interior accepts 4 adults and a young child at the center of the rear bench seat where a special seat has been incorporated into the center tunnel. The dimensions are ample and there is enough head and legroom.

• Conveniences: 65%
Storage space includes a glove box and a storage box located within the front center armrest. The front door pockets are not very practical.

• Noise level: 60%
The power train, front suspension and steering are mounted on the separate sub-frame which isolates the interior from vibrations. The well-cushioned interior is only disturbed in a minor way by the noise of tires crossing transverse tar strips or of the engine during strong accelerations.

• Trunk: 60%
Shorter and taller than usual, it offers ample volume even when the rear seats are occupied but can be increased by lowering the 60/40 bench seat back and there is a ski trap fitted within the center armrest.

• Braking: 60%
It is very effective in terms of distance and easy to modulate in normal use or during panic stops where the ABS works perfectly.

• Fuel consumption: 60%
The fuel consumption limits the performance, which is that of a V6 or equivalent.

• Depreciation: 60%
In the midst of a honeymoon the resale value stays above average.

WEAK POINTS

• Price/equipment: 30%
Affordable in the standard version, the 850 has paid well for its sports temperament.

CONCLUSION

• Overall rating: 71.0%
With the 850, Volvo holds the chicken that lays the golden eggs, for as long as it lasts. ☺

SPECIFICATIONS & PRICES

Model	Version/Trim	Body/ Seats	Interior volume cu.ft.	Trunk volume cu ft.	Drag coef. Cd	Wheel-base in	Lgth x Wdthx Hght. inx inx in	Curb weight lb	Towing capacity std. lb	Susp. type ft/rr	Brake type ft/rr	Steering system	Turning diameter ft	Steer. turns nber	Fuel. tank gal.	Standard tire size	Standard powertrain	PRICES US. $ 1994
VOLVO		**General warranty: 4 years / 50,000 miles; corrosion: 8 years / unlimited; antipollution: 5 years / 50,000 miles.**																
850	GLE	4dr.sdn. 5	97	15	0.32	104.9	183.5x69.3x55.7	3232	1000	i/si	dc/dc/ABS	pwr.r&p.	33.5	3.2	19.3	195/60VR15	L5/2.4/M5	**$24,680**
850	GLE	5dr.wgn.5	97	37	0.34	104.9	185.4x69.3x56.9	3342	1000	i/si	dc/dc/ABS	pwr.r&p.	33.5	3.2	19.3	195/60VR15	L5/2.4/M5	-
850	GLT	4dr.sdn. 5	97	15	0.32	104.9	183.5x69.3x55.7	3232	1000	i/si	dc/dc/ABS	pwr.r&p.	33.5	3.2	19.3	195/60VR15	L5/2.4/M5	**$27,110**
850	GLT	5dr.wgn.5	97	37	0.34	104.9	185.4x69.3x56.9	3342	1000	i/si	dc/dc/ABS	pwr.r&p.	33.5	3.2	19.3	195/60VR15	L5/2.4/M5	**$28,410**
850	Turbo	4dr.sdn. 5	97	15	0.32	104.9	183.5x69.3x55.7	3278	1000	i/si	dc/dc/ABS	pwr.r&p.	34.5	3.2	19.3	205/50ZR16	L5T/2.3/A4	**$30,435**
850	Turbo	5dr.wgn.5	97	37	0.34	104.9	185.4x69.3x56.9	3386	1000	i/si	dc/dc/ABS	pwr.r&p.	34.5	3.2	19.3	205/50ZR16	L5T/2.3/A4	**$31,735**

See page 393 for complete 1995 Price List

Jet Set...

Volvo has just revamped its 960 so that it can join the *Who's Who* of world automobiles. Its body shares the same format with more modest models, but its content is up to par and it now has all the attributes of a luxury car.

The 900-Series is the high end of the range of the Swedish car builder's range. It consists of three 4-door and three 5-door station wagons in three versions, depending on the powertrain that animates them. The 944 and 945 are equipped with a 2.3 L naturally aspirated 4-cylinder with SOHC while the 944 and 945 turbos have picked up the torch from the 744 and 745. The 964 and 965 received considerable changes and are equipped with a new 3.0 L, in-line 6-cylinder, 24-valve engine. All of these models come standard with a 4-speed automatic. In 1995 the 960 receives a new rear drive train with multiple control arms and major body changes.

STRONG POINTS

• Safety: 90%

The big Volvos do a good job at resisting frontal and side impacts, since their structure is one of the most rigid and the doors, as well as the floor and roof, include substantial reinforcements. Two air bags, retractable seat belts and the child's seat integrated into the center armrest equip all of the 900s.

• Technical: 80%

The steel unit body has independent suspension at the front and rigid axle at the back, except for the 960 sedan on which the four wheels are independent. The engine is a 4-cylinder, an old acquaintance inherited from the 700-Series while the in-line 6-cylinder, more modern, is in its second generation. With this engine, Volvo joins the BMW and Mercedes club which have always favored this power plant for its simplicity, its balance and high rpm torque, thanks to effectiveness of its manifolding and of the Bosch Motronic computer, which manages ignition and fuel delivery. It also owes part of its power and economy to a higher compression ratio. An electronically controlled automatic

DATA

Category: rear-wheel drive luxury sedans and wagons.
Class : 7

HISTORIC

Introduced in: 1990: 900 series 4 cylinder.
Modified in: 1992: L6 engine.
Made in: Torslanda, Sweden & Halifax, Canada.

RATINGS

Safety: 90 %
Satisfaction: 90 %
Depreciation: 48 %
Insurance: 940 ($725) 3.7 % 3.7 % ($925) 960
Cost per mile: $0.62

NUMBER OF DEALERS

In U.S.A.: 405

SALES ON THE AMERICAN MARKET

Model	1992	1993	Result
Volvo	27,047	32,495	+ 16.3 %

MAIN COMPETITORS

ACURA Legend, ALFA ROMEO 164, AUDI A6, BMW 525, LEXUS ES300, INFINITI J30, MAZDA Millenia, MERCEDES C-Class, NISSAN Maxima, SAAB 9000.

EQUIPMENT

VOLVO	940	940 Tbo	960
Automatic transmission:	S	S	S
Cruise control:	S	S	S
Power steering:	S	S	S (autom.)
ABS brake:	S	S	S
Air conditioning:	S	S	S
Air bag (2):	S	S	S
Leather trim:	O	O	S
AM/FM/ Radio-cassette:	S	S	S
Power door locks:	S	S	S
Power windows:	S	S	S
Tilt steering column:	O	O	S
Dual power mirrors:	S	S	S
Intermittent wipers:	S	S	S
Light alloy wheels:	O	O	S
Sunroof:	O	O	S
Anti-theft system:	-	-	S

S : standard; O : optional; - : not available

COLORS AVAILABLE

Exterior: Black,White,Red,Blue,Green,Silver,Graphite,Gray,Sand,Teal.
Interior: Blue, Taupe, Gray.

MAINTENANCE

First revision: 10,000 miles
Frequency: 10,000 miles
Diagnostic plug: Yes

WHAT'S NEW IN 1995 ?

- **New model 960 with different appearance, engine and rear drive train.**
- **Lateral air bags.**
- **Bumpers, door handles and lower body in body color.**
- **Cup holders for front and rear passengers.**

Model/ versions *: standard	ENGINE Type / timing valve / fuel system	Displacement cu/in	Power bhp @ rmn	Torque lb.ft @ rpm	Compres. ratio	TRANSMISSION Driving wheels / transmission	Final ratio	PERFORMANCE Acceler. 0-62 mph s	Stand. 1/4 mile s	Stand. 5/8 mile s	Braking 62-0 mph ft	Top speed mph	Lateral acceler. G	Noise level dBA	Fuel economy mpg city	hwy	Gasoline type / octane
940	L4* 2.3 SOHC-8-EFI	141	114 @ 5400	136 @ 2150	9.8 :1	rear - A4*	4.10	12.0	18.5	35.2	125	112	0.74	67	23.0	36.0	R 87
940 Turbo	L4T* 2.3 SOHC-8-EFI	141	162 @ 4800	195 @ 3450	8.7 :1	rear - A4*	4.10	8.8	16.8	29.6	131	125	0.76	67	23.0	32.0	R 87
960	L6* 3.0 DOHC-24-EFI	178	181 @ 5200	199 @ 4100	10.7 :1	rear - A4*	3.73	9.0	17.2	30.0	122	125	0.78	66	20.0	33.0	R 87

functions in three modes: E for economy, S for sport and W for winter. While the first two differ by the length of their shift times, the latter annuls the automatic shifts and keeps the transmission in gear like a manual. 4-wheel disc brakes with ABS are used on all models. The body derives from the 700 which has been stiffened and cosmetically retouched to make it more modern. That of the 960 has a more refined front, "a la 850".

• Seats: 80%
They provide ample lateral and lumbar support, both front and rear and their Nordically-inspired padding is satisfactory, though fairly stiff.

• Suspension: 80%
It does not underline the road defects as harshly as the last 960 and is appreciably smoother.

• Quality/fit/finish: 80%
The Volvos maintain their reputation of a solid constitution and a robust assembly plus the traditional materials quality and finish of the marque. However, only the 960 can pretend to the status of a deluxe car.

• Driver's compartment: 80%
The most comfortable driving position is easy to find and is accompanied by excellent visibility in spite of the rear seat headrests. The dashboard is classic Volvo because it resembles that of the old 700. Massive, well-organized, with easy-to-read gauges, it also uses familiar controls .

• Satisfaction: 80%
In a distinct rise in relation to that of the 700-Series, its reliability is considerably improved in a number of areas.

• Performance: 70%
They are very conservative, with the base engine and remind one of the ancient 240, while the Turbo has a fiery temperament and at the limit it is too live for the chassis and its antique rear end. The

6-cylinder is velvety and the accelerations put a wide smile on the one pushing on the accelerator.

• Trunk: 70%
More spacious, it is also more accessible, because of the cut-out of its opening that comes down to the level of the bumpers.

• Steering: 70%
It has too high a ratio to be fast but is soft and precise. A short turning diameter results in good maneuverability.

• Conveniences: 70%
The interior includes a good-sized glove box, door pockets, a recess and a box on the center console.

• Access: 70%
It is easy to get into these intermediate-sized cars whose doors are sufficiently long and the roof shape is well-designed at the rear.

• Interior space: 65%
It remains at the 700 level, but the inside height is greater, thanks to the new roof design.

• Noise level: 60%
Main noise is generated from the power train and wheels; airstreams could be more discreet on the 940, while the environment is more cushiony on the 960.

• Insurance: 60%
Because the clientele is conservative, just like the cars, the rates are reasonable.

• Braking: 55%
Powerful, progressive and well-balanced, its fade-free operation is excellent even when abused.

• Fuel consumption: 50%
Considering the weight and performance of these cars, the fuel expense is realistic and rarely goes below 22 mpg.

• Depreciation: 50%
Lower than average since its reliability has improved but maintenance and parts costs are high once the warranty has ended and the dealer network is limited.

• Handling: 50%
It remains tied to an antique rear drive train with a rigid axle on those versions that are still equipped that way. While the new independent 960 powertrain with multiple arms is much more stable under all conditions and its traction in corners is superior to that of the 940. Be it with the turbo or the 960, giving full gas on a slippery road requires a little tactfulness.

WEAK POINTS

• Price/equipment: 30%
The 940s have the most difficulty justifying their price because it is hard to consider them as the deluxe car that the 960 became. The base 940 is ordinary, rustic and tractable in spite of some qualities and many of its Japanese competitors are more evolved and efficient. These models are however, relatively well-equipped, particularly the 960.

CONCLUSION

• Overall rating: 67.0%
While the only pleasure of the standard 940 is an affordable price, the 940 Turbo and the 960 are much better at justifying their status as a luxurious sports car. However, the 960 should show a larger format. ☺

SPECIFICATIONS & PRICES

Model	Version/Trim	Body/ Seats	Interior volume cu.ft.	Trunk volume cu ft.	Drag coef. Cd	Wheel-base in	Lgth x Wdthx Hght. inx inx in	Curb weight lb	Towing capacity std. lb	Susp. type ft/rr	Brake type ft/rr	Steering system	Turning diameter ft	Steer. turns nber	Fuel. tank gal.	Standard tire size	Standard powertrain	PRICES US. $ 1994
VOLVO		General warranty: 4 years / 50,000 miles; corrosion: 8 years / unlimited; antipollution: 5 years / 50,000 miles.																
944	GLE	4dr.sdn.5	94	17	0.35	109.1	191.7x69.3x55.5	3208	2000	i/i	dc/dc/ABS	pwr.r&p.	32.2	3.5	19.8	185/65R15	L4/2.3/A4	**$22,900**
945	GLE	5dr.wgn.5	95	39	0.35	109.1	189.3x69.3x56.5	3238	2000	i/r	dc/dc/ABS	pwr.r&p.	32.2	3.5	19.8	185/65R15	L4/2.3/A4	**$24,200**
944	Turbo	4dr.sdn.5	94	17	0.35	109.1	191.7x69.3x55.5	-	2000	i/i	dc/dc/ABS	pwr.r&p.	32.2	3.5	19.8	185/65R15	L4T/2.3/A4	**$23,900**
945	Turbo	5dr.wgn.5	95	39	0.35	109.1	189.3x69.3x56.5	-	2000	i/r	dc/dc/ABS	pwr.r&p.	32.2	3.5	19.8	185/65R15	L4T/2.3/A4	**$25,200**
964	SE	4dr.sdn.5	94	17	0.35	109.1	191.8x68.9x56.6	3521	2000	i/i	dc/dc/ABS	pwr.r&p.	31.8	3.5	21.1	205/55VR16	L6/3.0/A4	**$33,950**
965	SE	5dr.wgn.5	95	39	0.35	109.1	191.8x68.9x57.6	3611	2000	i/i	dc/dc/ABS	pwr.r&p.	31.8	3.5	21.1	195/65HR15	L6/3.0/A4	**$35,250**

See page 393 for complete 1995 Price List

HELP US FURTHER IMPROVE ROAD REPORT
BY GIVING US YOUR COMMENTS

-We are calling for your comments and suggestions! They will help continuously improve the quality of our information, the way it is presented, and point to new sections of interest for you! Your thoughts and road testing experience are what helps us to stay the best in our field.

- Please list in sequence what you liked best in ROAD REPORT ?

1) ____________________

2) ____________________

3) ____________________

4) ____________________

5) ____________________

- What do you feel should be improved or is missing in ROAD REPORT?

1) ____________________

2) ____________________

3) ____________________

4) ____________________

5) ____________________

- What would you like to find on the cover of ROAD REPORT ?

1) ____________________

2) ____________________

3) ____________________

4) ____________________

5) ____________________

- Other suggestions:

- ____________________

- ____________________

- ____________________

Please fill in, cut off and send this page to:
ROAD REPORT: Fax: (514) 441 20-76

Name: ____________________

Address: ____________________ Apt. ________

City: ____________________ State: ____________

Zip code: ____________________

Tel: (   ) ____________ Fax: (   ) ____________

BECOME ONE OF OUR TESTERS AND PARTICIPATE IN EDITING ROAD REPORT

Fill in this sheet with your comments and send it back to us so we ca note YOUR EXPERIENCE in our comments.
MAKE: ....Model:..............................
Year:..............................New:................Used:..............
Trim:............................Options:..
Mileage:..
Engine:............................Transmission:..
Fuel consumption in town: ....
Highway:..................
Purchase cost?.....................................
- **For what reason did you pick this model** ?
....
....
- **What did you like best about your car** ?
1)..
2)..
3)..
4)..
5)..
- **What did you like least** ?
1)..
2)..
3)..
4)..
5)..
- **What improvements would you like to see** ?
1)..
2)..
3)..
4)..
5)..
- **How do you rate it for driving** ?
Performance..
Steering...
Braking...
Commamnds and controls...............................
- **How is the comfort** ?
Interior space...
Seats...
Suspension...
Noise level...
Air conditioning..
- **What do you think of the assembly quality/fit/finish**?
-Excellent............................
-Good.....................................
-Poor..............................
-Bad..............................

- **Did you have mechanical problems with your car**?
-No......................................
-Yes......................................
-If yes, what kind ?
....
....
....
....
....
- **Have you been satisfied with your dealer's work** ?
-Yes.......................................
-No......................................
- **What do you think of the manufacturer's warranty** ?
-Excellent................................
-Good...................................
-Poor................................
-Bad..............................
- **What do you think of the manufacturer's service?**
-Excellent...............................
-Good....................................
-Poor.................................
-Bad................................
- **Do you own a second car** ?
-No................Yes................
-If yes, make:..............................Model............................
- **Starting over, would you buy the same model** ?
....
- **Will your next car be of the same make**?
....
- **To what age group do you belong** ?
-16-30 :...............................
-31-40 :...............................
-41-50 ....
-51 and up:...............................

Please fill in, cut out and send this report to
ROAD REPORT

Fax: (514) 441 20-76

Name:______________________________
Address:______________________Apt.______
City:________________State:____________
Zip code:______________________
Tel: (   )______________
Fax:(   )______________

Thanks for your collaboration,
The "ROAD REPORT" team.

COMPARATIVE EVALUATION BY CATEGORY

1995 BEST BUYS

As in previous years, the Road Report team tested more than 50 different models and carefully recorded the parameters making up the value of each vehicle. The various test drivers have collected their impressions and analyzed the results measured by their instruments. The vehicles are classified into exact categories, corresponding to their global size and power. Their performance has been recorded according to a system we have perfected over the years. Ninety percent of these ratings are based on irrefutable mathematical formulas, giving each vehicle an equal chance. The compilation of these ratings gives a percentage result which can be used to evaluate and compare the different models. The system is self compensating. What a van gains in interior space and accessibility, it loses in performance or comfort, and conversely for a sport coupe. **No system is perfect, but ours is particularly severe. Don't be surprised if the subjective commentary of the tester is different from the rating which reflects the more mathematical reality. The driving impressions of a model can be more or less influenced by a number of elements, while the rating is an analysis which coldly reflects the reality.** It's easy to measure the dimensions or volume of a cabin, but it's more difficult to quantify the quality of the material of the seats or the softness of the suspension. Also, there is no clearly defined criteria to evaluate the quality of assembly or finish.

See page 8 for the rating table and explanations.

(Canadian Price Range)

Class 2-Mini-compacts-$9,000 and less (12 500$ and less)

1st FORD Aspire 57.5%

2nd GEO Metro 53.0%

3rd SUBARU Justy 52.5%

Models	10 20 30 40 50 60 70 80 90 100%
FORD Aspire	57.5%
GEO Metro & SUZUKI Swift	53.0%
SUBARU Justy	52.5%

1995 BEST BUYS

Class 3S - Sub-compacts - $9,000 to $12,000 (12 500$ to 15 000$)

1st MAZDA 323 64.0 %

2nd HONDA Civic 63.4%

3rd TOYOTA Tercel 61.0 %

Models	10 20 30 40 50 60 70 80 90 100%
MAZDA 323	64.0%
HONDA Civic	63.4%
TOYOTA Tercel	61.0%
DODGE Colt & MITSUBISHI Mirage	60.5%
HYUNDAI Accent	59.0%

Class 3 - Compacts - $12,000 to $15,000 (15 000$ to 17 500$)

1st TOYOTA Corolla 68.0 %

2nd VW Golf 65.5 %

3rd MAZDA Protegé 64.0%
NISSAN Sentra 64.0%
SATURN SL 64.0%

Models	10 20 30 40 50 60 70 80 90 100%
TOYOTA Corolla	68.0%
VW Golf	65.5%
MAZDA Protegé	64.0%
NISSAN Sentra	64.0%
SATURN SL	64.0%
HONDA Civic 4dr.	63.4%
DODGE Neon	61.0%
SUBARU Impreza	60.5%
CHEVROLET Cavalier & PONTIAC Sunfire	60.5%
FORD Escort	60.0%
HYUNDAI Elantra	60.0%

Class 4 - Compacts - $15,000 to $19,000 (17 500$ to 25 000$)

1st CHRYSLER Cirrus 70.0 %

1st VW Passat 70.0 %

3rd HONDA Accord 68.5 %

Models	Score (10–100%)
CHRYSLER Cirrus	70.0%
VW Passat	70.0%
HONDA Accord	68.5%
INFINITI G20	68.5%
TOYOTA Camry 4 cyl.	68.0%
MITSUBISHI Galant	67.5%
SAAB 900	67.0%
FORD Contour / MERCURY Mystique	66.5%
ACURA Integra 4dr.	66.0%
MAZDA 626	66.0%
SUBARU Legacy	65.0%
NISSAN Altima	63.0%
HYUNDAI Sonata	63.0%
BUICK Skylark / PONTIAC Grand Am / CHEVROLET Corsica/OLDSMOBILE Achieva	58.0%
DODGE Spirit / PLYMOUTH Acclaim	54.0%

Class 5 - Mid-size - $19,000 to $23,000 (25 000$ to 27 500$)

1st CHRYSLER LH 70.0 %

2nd TOYOTA Camry 68.0%

3rd FORD Taurus 62.7 %

Models	Score (10–100%)
CHRYSLER Concorde / DODGE Intrepid & EAGLE Vision	70.0%
TOYOTA Camry	68.0%
FORD Taurus MERCURY Sable	62.7%
BUICK Regal & CHEVROLET Lumina & OLDSMOBILE Cutlass Supreme & PONTIAC Grand Prix	60.0%
BUICK Century OLDSMOBILE Cutlass Ciera	54.0%

Class 6 - Full-size - $19,000 to $25,000 (27 500$ to 30 000$)

1st CHRYSLER New Yorker/LHS 69.0 %

2nd GM H-Series 66.5 %

3rd GM B-Series 62.0 %

Models	10 20 30 40 50 60 70 80 90 100%
CHRYSLER New Yorker	69.0%
BUICK Le Sabre & OLDSMOBILE 88 & PONTIAC Bonneville	66.5%
BUICK Roadmaster & CHEVROLET Caprice	62.0%
FORD Crown Victoria & MERCURY Grand Marquis	59.0%

Class 7 - Deluxe sedans - $25,000 to $35,000 (30 000$ to 50 000$)

1st NISSAN Maxima 73.0 %

2nd LEXUS ES300 72.5 %

3rd VOLVO 850 71.0 %

Models	10 20 30 40 50 60 70 80 90 100%
NISSAN Maxima	73.0%
LEXUS ES300	72.5%
VOLVO 850	71.0%
OLDSMOBILE Aurora	69.0%
SAAB 9000	69.0%
MAZDA Millenia	68.0%
MERCEDES C-Class	68.0%
ACURA Vigor	66.5%
ALFA ROMEO 164	66.0%
AUDI A4	65.5%
CADILLAC De Ville	65.5%
INFINITI J30	65.0%
MITSUBISHI Diamante	64.5%
BMW 3-Series	64.2%
LINCOLN Continental	64.0%

Non classée: TOYOTA Avalon.

1995 BEST BUYS

Class 7 - Deluxe sedans - $35,000 to $50,000 (50 000$ to 75 000$)

1st LEXUS LS400 74.0 %

2nd MERCEDES E 70.0 %

3rd INFINITI Q45 69.0 %

Scale: 10 20 30 40 50 60 70 80 90 100%

Models	Score
LEXUS LS400	74.0%
MERCEDES E-Class	70.0%
INFINITI Q45	69.0%
ACURA Legend	68.5%
BMW 5-Series	68.0%
CADILLAC Seville	66.5%
AUDI A6	66.0%
JAGUAR XJ6	64.0%

Class 3S - Sport coupes - $11,000 to $17,000 (15 000$ to 25 000$)

1st MAZDA MX3 65.0 %

2nd HONDA del Sol 64.5 %

3rd SATURN SC 64.0 %

Scale: 10 20 30 40 50 60 70 80 90 100%

Models	Score
MAZDA MX3	65.0%
HONDA del sol	64.5%
SATURN SC	64.0%
TOYOTA Paseo	58.5%
MAZDA MX-5	56.5%
HYUNDAI Scoupe	56.0%

1995 BEST BUYS

Class 5 - Sport coupes - $17,000 to $25,000 (25 000$ to 35 000$)

1st TOYOTA Celica 68.0 %

2nd VW Corrado 67.5 %

3rd ACURA Integra 66.0 %

Models	Rating (10–100%)
TOYOTA Celica	68.0%
VW Corrado	67.5%
ACURA Integra	66.0%
MAZDA MX-6	66.0%
MITSUBISHI Eclipse &EAGLE Talon	65.0%
FORD Mustang	65.0%
FORD Probe	64.0%
BUICK Regal, CHEVROLET Monte Carlo OLDSMOBILE Cutlass, PONTIAC Grand Prix	63.0%
NISSAN 240SX	63.0%
HONDA Prelude	60.5%
FORD T-Bird, MERCURY Cougar	60.0%
CHEVROLET Camaro & PONTIAC Firebird	56.0%

GT Sport coupes - $25,000 to $70,000 (35 000$ to 80 000$)

1st LEXUS SC400 70.0 %

2nd DODGE Stealth 66.5 %

3rd NISSAN 300ZX 63.0 %

Models	Rating (10–100%)
LEXUS SC400	70.0%
DODGE Stealth & MITSUBISHI 3000 GT	66.5%
NISSAN 300ZX	63.0%
TOYOTA Supra	62.0%
ACURA NSX	61.5%
MAZDA RX-7	61.5%
CHEVROLET Corvette	56.5%

1995 BEST BUYS

Compact pick-ups - $10,000 to $17,000 (15 000$ to 35 000 $)

1st CHEVROLET S-10 58.5 %

2nd NISSAN Pick-up 56.0%
& TOYOTA & T100 56.0 %

3rd FORD Ranger 53.0%
& MAZDA B 53.0 %

Models	10 20 30 40 50 60 70 80 90 100%
CHEVROLET S-10	58.5%
TOYOTA Pick-up & T100	56.0%
NISSAN Pick-up	56.0%
FORD Ranger MAZDA B-Series	53.0%
DODGE Dakota	52.0%

All-terrain - $12,000 to $30,000 (15 000$ to 35 000$)

1st JEEP Gd Cherokee 64.0 %

2nd ISUZU Trooper 61.0 %

3rd TOYOTA 4Runner 60.0 %

Models	10 20 30 40 50 60 70 80 90 100%
JEEP Grand Cherokee	64.0%
ISUZU Trooper	61.0%
TOYOTA 4Runner	60.0%
NISSAN Pathfinder	59.5%
TOYOTA Land Cruiser	59.0%
CHEVROLET Blazer	58.5%
MITSUBISHI Montero	57.0%
FORD Explorer &	56.5%
ISUZU Rodeo MAZDA Navajo	55.0%
GEO Tracker& SUZUKI Sidekick	53.0%
JEEP Cherokee	52.5%
JEEP YJ-Wrangler	44.5%

1995 BEST BUYS

Van category - $12,500 to $35,000 (12 500$ to 35 000$)

1st
CHRYSLER Town & Country
DODGE Caravan
PLYMOUTH Voyager
67.5 %

2nd TOYOTA Previa 67.0 %

3rd VW Eurovan 62.0 %

Models	10 20 30 40 50 60 70 80 90 100%
CHRYSLER Town & Country / DODGE Caravan / PLYMOUTH Voyager	67.5%
TOYOTA Previa	67.0%
FORD Windstar	62.0%
VW Eurovan	62.0%
MAZDA MPV	59.0%
CHEVROLET Lumina / PONTIAC Trans Sport & OLDSMOBILE Silhouette	61.0%
MERCURY Villager & NISSAN Quest	60.0%
FORD Aerostar	56.7%
CHEVROLET Astro	56.0%

1995 ROOKIE OF THE YEAR

NISSAN Maxima

Our choice this year went to the Nissan Maxima rather than the Toyota Avalon because we did have the chance to extensively test all the classified models which is not the case with the non-classified ones. The Maxima offers an excellent price-pleasure-efficiency ratio even with the base model.
For more details see the analysis page on 88 and the road test report on pages 314-315.

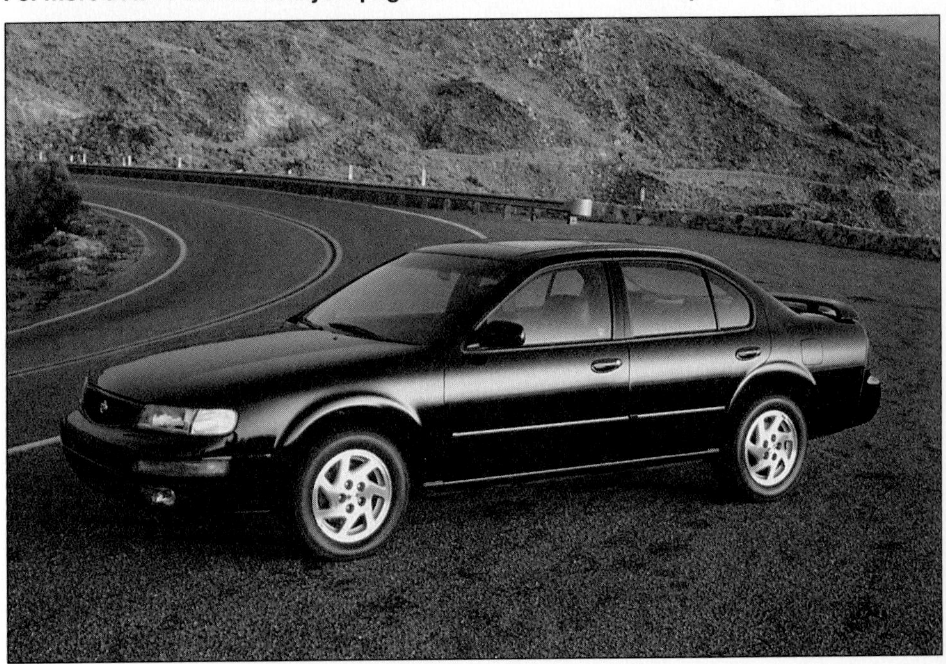

1995 NEW MODELS CHART

NISSAN Maxima	**73.0%**
CHRYSLER Cirrus	70.0%
OLDSMOBILE Aurora	69.0%
VOLVO 960	67.0%
FORD Contour	66.5%
SUBARU Legacy	65.0%
EAGLE Talon	65.0%
MAZDA 323/Protegé	64.0%
CHEVROLET Monte Carlo	63.0%
NISSAN 240SX	63.0%
BUICK Riviera	62.5%
FORD Windstar	62.5%
DODGE Neon	61.0%
CHEVROLET Lumina	60.0%
HYUNDAI Accent	59.0%
CHEVROLET Blazer	58.5%
FORD Explorer	56.5%
CHEVROLET Cavalier & PONTIAC Sunfire	53.0%
GEO Metro	53.0%
Not classified:	
TOYOTA Avalon	73.5%
JAGUAR XJ6/12	64.0%
TOYOTA Tercel	61.0%

1995 AMERICAN PRICES

ATTENTION: Prices shown are given as a guideline and are interpreted as follows:

Mini price: the lowest negotiable price.
Avrg price: average sale price for the model.
Maxi price: manufacturer's suggested retail price.

ACURA

Model	Trim	Type	Engine	Price U.S. $ mini	avrg	maxi
Integra	RS	3dr.sdn.	L4/1.8/M5	$13,630	$14,545	$15,460
	LS	3dr.sdn.	L4/1.8/M5	$15,985	$17,065	$18,140
	RS	4dr.sdn.	L4/1.8/M5	$14,295	$15,258	$16,220
	LS	4dr.sdn.	L4/1.8/M5	$16,510	$17,725	$18,940
	GS-R	3dr.cpe.	L4/1.8/M5	$17,920	$19,135	$20,350

• **Destination & handling charge: $420**

Model	Trim	Type	Engine	mini	avrg	maxi
Legend	L	4dr.sdn.	V6/3.2/A4	$30,635	$33,067	$35,500
	LS	4dr.sdn.	V6/3.2/A4	$34,127	$36,913	$39,700
	LS	2dr.cpe.	V6/3.2/M6	$37,948	$40,574	$43,200

• **Destination & handling charge: $420**

Model	Trim	Type	Engine	mini	avrg	maxi
NSX		2dr.cpe.	V6/3.0/M5	$63,208	$67,855	$72,500

• **Destination & handling charge: $420**

Model	Trim	Type	Engine	mini	avrg	maxi
Vigor	GS	4dr.sdn.	L5/2.5/M5	$24,535	$26,440	$28,350

• **Destination & handling charge: $420**

AUDI

Model	Trim	Type	Engine	mini	avrg	maxi
A4 (90)	S	4dr.sdn.	V6/2.8/M5	$21,890	$23,780	$25,670
Quattro	CS	4dr.sdn.	V6/2.8/M5	$21,519	$23,795	$26,070
	Cabrio.	2dr.sdn.	V6/2.8/A4	$30,600	$33,250	$35,900

• **Destination & handling charge: $445**

Model	Trim	Type	Engine	mini	avrg	maxi
A6 (100)	*CS	4dr.sdn.	V6/2.8/A4	$24,720	$27,660	$30,600
*Quat.	(S4)S6	4dr.sdn.	L5T/2.2/M5	$37,915	$41,595	$45,270
	Quat.CS	4dr.wgn.	V6/2.8/A4	$27,570	$30,370	$33,170

• **Destination & handling charge: $445**

BMW

Model	Trim	Type	Engine	mini	avrg	maxi
318	i	4dr.sdn.	L4/1.8/M5	$21,400	$23,190	$24,975
	i	2dr.con.	L4/1.8/M5	$29,070	$31,560	$34,050
	is	2dr.cpe.	L4/1.8/M5	$22,950	$24,810	$26,675
	ti	3dr.sdn.	L4/1.8/M5			NA
325	i	4dr.sdn.	L6/2.5/M5	$26,985	$29,220	$31,450
	is	2dr.cpe.	L6/2.5/M5	$28,085	$30,418	$32,750
	i	2dr.con.	L6/2.5/M5	$33,885	$36,745	$39,600

Destination & handling charge: $470

Model	Trim	Type	Engine	mini	avrg	maxi
525	i	4dr.sdn.	L6/2.5/M5	$29,655	$32,480	$35,300
530	i	4dr.sdn.	V8/3.0/M5	$36,730	$39,740	$42,750
540i		4dr.sdn.	V8/4.0/A5	$41,360	$44,655	$47,950

• **Destination & handling charge: $470**

Model	Trim	Type	Engine	mini	avrg	maxi
740	i	4dr.sdn.	V8/4.0/A5	$49,600	$53,750	$57,900

• **Destination & handling charge: $470**

Model	Trim	Type	Engine	mini	avrg	maxi
840	Ci	2dr.cpe.	V8/4.0/M5	$61,230	$65,565	$69,900

• **Destination & handling charge: $470**

BUICK

Model	Trim	Type	Engine	mini	avrg	maxi
Century	Custom	4dr.sdn.	V6/3.0/A4	$16,155	$17,060	$17,965
	Special	4dr.sdn.	V6/3.0/A4			$16,160
	Special	5dr.wgn.	V6/3.0/A4			$17,160

• **Destination charge: $525 (non-negotiable prices)**

Model	Trim	Type	Engine	mini	avrg	maxi
Le Sabre	Custom	4dr.sdn.	V6/3.8/A4			$20,410
	Ltd	4dr.sdn.	V6/3.8/A4			$24,110

• **Destination charge: $575 (non-negotiable prices)**

Model	Trim	Type	Engine	mini	avrg	maxi
Park Avenue	Ultra	4dr.sdn.	V6/3.8/A4	$29,150	$31,115	$33,084

• **Destination charge: $625**

Model	Trim	Type	Engine	mini	avrg	maxi
Regal	Cust.	2dr.cpe.	V6/3.1/A4			$17,960
	Cust.	4dr.sdn.	V6/3.1/A4			$18,660
	Ltd	4dr.sdn.	V6/3.8/A4			$21,235
	GSport	2dr.cpe.	V6/3.8/A4			$19,460
	GSport	4dr.sdn.	V6/3.8/A4	$20,320	$21,095	$21,870

• **Destination charge: $525 (non-negotiable prices)**

Model	Trim	Type	Engine	mini	avrg	maxi
Riviera		2dr.cpe.	V6/3.8/A4	$24,940	$26,290	$27,632

• **Destination charge: $625**

Model	Trim	Type	Engine	mini	avrg	maxi
Roadmaster	Base	4dr.sdn.	V8/5.7/A4	$22,690	$23,976	$25,265
	Estate	5dr.wgn.	V8/5.7/A4	$24,325	$25,695	$27,070
	Ltd	4dr.sdn.	V8/5.7/A4	$24,715	$26,135	$27,555
	Ltd Estate	5dr.wgn.	V8/5.7/A4	$26,700	$28,100	$29,465

• **Destination charge: $575**

Model	Trim	Type	Engine	mini	avrg	maxi
Skylark	Cust.	2dr.cpe.	V6/3.1/A4			$13,700
	Cust.	4dr.sdn.	V6/3.1/A4			$13,700
	Ltd	2dr.cpe.	L4/2.3/A4			$14,700
	Ltd	4dr.sdn.	V6/3.1/A4			$14,700
	GS	2dr.cpe.	V6/3.1/A4			$16,400
	GS	4dr.sdn.	V6/3.1/A4			$16,400

• **Destination charge: $485 (non-negotiable prices)**

CADILLAC

Model	Trim	Type	Engine	mini	avrg	maxi
De Ville	base	2dr.sdn.	V8/4.9/A4	$32,740	$33,820	$34,900
	Concours	4dr.sdn.	V8/4.9/A4	$36,895	$38,150	$39,400

• **Destination charge: $625**

Model	Trim	Type	Engine	mini	avrg	maxi
Fleetwood		4dr.sdn.	V8/5.7/A4	$33,360	$34,475	$35,595

• **Destination charge: $625**

Model	Trim	Type	Engine	mini	avrg	maxi
Eldorado	coupe	2dr.cpe.	V8/4.6/A4	$33,795	$36,010	$38,220
	Touring	2dr.sdn.	V8/4.6/A4	$37,441	$39,490	$41,535

• **Destination charge: $625**

Model	Trim	Type	Engine	mini	avrg	maxi
Seville	SLS	4dr.sdn.	V8/4.6/A4	$37,795	$39,865	$41,935
	STS	4dr.sdn.	V8/4.6/A4	$41,405	$43,670	$45,935

• **Destination charge: $625**

1995 AMERICAN PRICES

*1994 Price

CHEVROLET

Model	Trim	Type	Engine	Price U.S.$ mini	avrg	maxi
Beretta	Base,	2dr.cpe.	L4/2.2/M5	$12,035	$12,515	$12,995
	Z26	2dr.cpe.	V6/3.1/M5	$15,135	$15,715	$16,295
• Destination charge: $495						
Camaro	RS	3dr.cpe.	V6/3.4/M5	$13,365	$13,805	$14,250
	RS	2dr.con.	V8/5.7/M5	$18,260	$18,875	$19,495
	Z28	2dr.con.	V8/5.7/A4	$16,799	$17,357	$17,915
	Z28	2dr.cpe.	V8/5.7/M6	$21,645	$22,369	$23,095
• Destination charge: $500						
Caprice	Classic	4dr.sdn.	V8/4.3/A4	$18,275	$19,290	$20,310
	Classic	5dr.wgn.	V8/5.7/A4	$20,576	$21,710	$22,840
• Destination charge: $585						
Cavalier	Base	2dr.cpe.	L4/2.2/M5	$9,758	$9,910	$10,060
	Base	4dr.sdn.	L4/2.2/M5	$9,960	$10,110	$10,265
	LS	4dr.sdn.	L4/2.2/M5	$11,955	$12,200	$12,465
• Destination charge: $485						
Corsica	Base	4dr.sdn.	L4/2.2/A3	$12,875	$13,385	$13,890
• Destination charge: $495						
Corvette	LT1	3dr.cpe.	V8/5.7/M6	$32,725	$35,035	$37,345
	LT1	2dr.con.	V8/5.7/M6	$38,740	$41,480	$44,225
• Destination charge: $560						
Impala	SS	4dr.sdn.	V8/5.7/A4	$21,305	$22,105	$22,910
• Destination charge: $585						
Lumina	base	4dr.sdn.	V6/3.1/A4	$14,365	$15,180	$15,995
	LS sed.	2dr.sdn.	V6/3.1/A4	$15,695	$16,595	$17,495
• Destination charge: included						
Lumina APV Ut.						
	base	5dr.wgn.	V6/3.1/A3	$16,300	$16,950	$17,595
• Destination charge: $540						
Monte Carlo						
	LS	2dr.cpe.	V6/3.1/A4	$13,845	$14,657	$15,470
	Z34	2dr.cpe.	V6/3.4/A4	$16,980	$17,975	$18,970
• Destination charge: $535						

CHRYSLER

Model	Trim	Type	Engine	Price U.S.$ mini	avrg	maxi
Cirrus	LX	4dr.sdn.	V6/2.5/A4	$16,085	$17,030	$17,970
	LXi	4dr.sdn.	V6/2.5/A4	$17,775	$18,840	$19,900
• Destination charge: included						
Concorde		4dr.sdn.	V6/3.3/A4	$18,445	$19,495	$20,550
• Destination & handling charge: $535						
Le Baron	GTC	2dr.con.	V6/3.0/A4	$16,735	$17,105	$17,469
• Destination & handling charge: $530						
New-Yorker		4dr.sdn.	V6/3.5/A4	$22,883	$24,240	$25,596
LHS		4dr.sdn.	V6/3.5/A4	$26,320	$27,960	$29,595
• Destination & handling charge: $595						
Town & Country						
	4x2	4dr.wgn.	V6/3.3/A4	$25,575	$26,625	$27,680
	4x4	4dr.wgn.	V6/3.3/A4	$27,455	$28,615	$29,775
• Destination & handling charge: $560						

DODGE

Model	Trim	Type	Engine	Price U.S.$ mini	avrg	maxi
Caravan	base	4dr.wgn.	L4/2.5/M5	$15,095	$15,625	$16,160
	SE	4dr.wgn.	L4/2.5/A3	$17,521	$18,188	$18,855
	LE	4dr.wgn.	V6/3.3/A3	$21,665	$22,524	$23,380
	ES	4dr.wgn.	V6/3.3/A3	$22,125	$23,010	$23,890
• Destination & handling charge: $560						
Intrepid		4dr.sdn.	V6/3.3/A4	$16,137	$17,055	$17,974
	ES	4dr.sdn.	V6/3.3/A4	$18,770	$19,805	$20,844
• Destination & handling charge: $535						
Neon	Highline	2dr.cpe.	L4/2.0/M5	$10,670	$10,955	$11,240
	Sport	2dr.cpe.	L4/2.0/M5	$10,970	$13,115	$13,567
	base	4dr.sdn.	L4/2.0/M5	$9,035	$9,267	$9,500
	Highline	4dr.sdn.	L4/2.0/M5	$10,670	$10,955	$11,240
	Sport	4dr.sdn.	L4/2.0/M5	$12,363	$12,815	$13,267
• Destination & handling charge: $500						
Grand Caravan						
	base	4dr.wgn.	V6/3.0/A4	$17,287	$17,945	$18,605
	SE	4dr.wgn.	V6/3.3/A4	$18,160	$18,875	$19,595
	SE 4x4	4dr.wgn.	V6/3.3/A4	$20,560	$21,415	$22,270
	LE	4dr.wgn.	V6/3.3/A4	$21,888	$22,784	$23,680
	LE 4x4	4dr.wgn.	V6/3.3/A4	$23,065	$24,715	$25,755
	ES	4dr.wgn.	V6/3.3/A4	$22,346	$23,270	$24,190
	Es 4x4	4dr.wgn.	V6/3.3/A4	$24,075	$25,175	$26,265
• Destination & handling charge: $560						
Spirit	base	4dr.sdn.	L4/2.5/M5	$13,455	$13,889	$14,323
• Destination & handling charge: $505						
Stealth	R/T	3dr.cpe.	V6/3.0/M5	$22,105	$23,020	$23,931
	R/TTbo	3dr.cpe.	V6T/3.0/M5	$34,641	$36,267	$37,894
• Destination & handling charge: $430 (1994 prices)						
Viper		2dr.con.	V10/8.0/M6	$50,900	$53,455	$56,000
• Destination & handling charge: $700						

DODGE (Imported)

Model	Trim	Type	Engine	Price U.S.$ mini	avrg	maxi
Colt	base	2dr.cpe.	L4/1.5/M5	$9,567	$9,683	$9,799
	ES	2dr.cpe.	L4/1.5/M5	$10,374	$10,525	$10,675
• Destination & handling charge: $430						

DODGE (Utility)

Model	Trim	Type	Engine	Price U.S.$ mini	avrg	maxi
Dakota 4x2						
	regular	2dr.p-u.	L4/2.5/M5	$9,771	$10,025	$10,286
	extd	2dr.p-u.	L4/2.5/M5	$11,020	$11,310	$11,602
	ClubCab	2dr.p-u.	L4/2.5/M5	$14,925	$15,318	$15,711
Dakota 4x4						
	regular	2dr.p-u.	V6/3.9/M5	$14,555	$14,940	$15,325
	extd	2dr.p-u.	V6/3.9/M5	$15,230	$15,630	$16,034
	Club Cab	2dr.p-u.	V6/3.9/M5	$18,100	$18,590	$19,050
• Destination & handling charge: $495						
Ram1500 4x2						
WSregular cab		2dr.p-u.	V6/3.9/M5	$12,290	$12,585	$12,938
LT regular cab		2dr.p-u.	V6/3.9/M5	$13,950	$14,315	$14,683
ST reg cab		2dr.p-u.	V8/5.2/M5	$15,700	$16,085	$16,527
ST reg crew cab		2dr.p-u.	V8/5.2/M5	$16,830	$17,275	$17,716
Laramie SLT extd crew cab			V8/5.9/M5	$17,050	$17,500	$17,997
Power Ram 4x4 1500						
base regular		2dr.p-u.	V8/5.2/M5	$16,880	$17,325	$17,769
extd		2dr.p-u.	V8/5.2/M5	$17,220	$17,665	$18,109
ST extd		2dr.p-u.	V8/5.2/M5	$19,185	$19,690	$20,194
Laramie SLT		2dr.p-u.	V8/5.9/M5	$19,500	$20,000	$20,525

*1994 Price

1995 AMERICAN PRICES

Model	Trim	Type	Engine	Price U.S.$ mini	avrg	maxi
2500						
ST extd crew cab		2dr.p-u.	V8/5.9/M5	$18,780	$19,290	$19,791
Laramie SLT extd crew cab			V8/5.9/M5	$21,225	$21,780	$22,340
3500						
base		2dr.p-u.	V8/5.9/M5	$20,675	$21,230	$21,789
ST		2dr.p-u.	V8/5.9/M5	$20,250	$20,810	$21,360
SLT		2dr.p-u.	V8/5.9/M5	$22,760	$23,360	$23,960

• **Destination & handling charge: $600**

EAGLE

Model	Trim	Type	Engine	mini	avrg	maxi
Summit	DL	2dr.cpe.	L4/1.5/M5	$9,567	$9,683	$9,799
	ES	2dr.cpe.	L4/1.5/M5	$10,374	$10,525	$10,675
	DL	4dr.wgn.	L4/1.8/M5	$12,400	$12,755	$13,114
	LX	4dr.wgn.	L4/2.4/M5	$13,525	$13,930	$14,340
	T.I.	4dr.wgn.	L4/2.4/M5	$14,150	$14,585	$15,018

• **Destination & handling charge: $430 (1994 prices wgn.)**

Model	Trim	Type	Engine	mini	avrg	maxi
Talon	ESI	3dr.cpe.	L4/2.0/M5	$13,600	$13,980	$14,362
	TSI 4WD	3dr.cpe.	L4T/2.0/M5	$18,450	$18,950	$19,448

• **Destination & handling charge: $430**

Model	Trim	Type	Engine	mini	avrg	maxi
Vision	ESi	4dr.sdn.	V6/3.3/A4	$17,595	$18,645	$19,697
	TSi	4dr.sdn.	V6/3.5/A4	$20,335	$21,605	$22,871

• **Destination & handling charge: $525**

FORD

Model	Trim	Type	Engine	mini	avrg	maxi
Aerostar 4x2						
	XLTregular	4dr.van.	V6/3.0/A4	$15,3451	$16,350	$17,355
	XLT extd	4dr.van.	V6/3.0/A4	$18,650	$19,655	$20,655
Aerostar 4x4						
	XLT extd	4dr.van.	V6/4.0/A4	$20,785	$21,885	$22,980

• **Destination & handling charge: $540**

Model	Trim	Type	Engine	mini	avrg	maxi
Aspire	base	3dr.sdn.	L4/1.3/M5	$7,935	$8,185	$8,440
Aspire		5dr.sdn.	L4/1.3/M5	$8,505	$8,780	$9,055
	SE	3dr.sdn.	L4/1.3/M5	$8,870	$9,140	$9,415

• **Destination & handling charge: $295**

Model	Trim	Type	Engine	mini	avrg	maxi
Escort	LX	3dr.cpe.	L4/1.9/M5	$9,828	$10,130	$10,435
	LX	5dr.sdn.	L4/1.9/M5	$10,232	$10,550	$10,870
	LX	5dr.wgn.	L4/1.9/M5	$10,750	$11,085	$11,425
	GT	3dr.sdn.	L4/1.8/M5	$11,940	$12,330	$12,720

• **Destination & handling charge: $375**

Model	Trim	Type	Engine	mini	avrg	maxi
Contour	GL	4dr.sdn.	L4/2.3/M5	$13,025	$13,440	$13,855
	LX	4dr.sdn.	L4/2.3/M5	$14,315	$14,775	$15,230
	SE	4dr.sdn.	V6/2.5/M5	$14,315	$14,775	$15,231

• **Destination & handling charge: NA**

Model	Trim	Type	Engine	mini	avrg	maxi
Crown Victoria						
	base	4dr.sdn.	V8/4.6/A4	$18,970	$19,565	$20,160
	LX	4dr.sdn.	V8/4.6/A4	$20,750	$21,360	$21,970

• **Destination & handling charge: $575**

Model	Trim	Type	Engine	mini	avrg	maxi
Mustang	Base	2dr.cpe.	V6/3.8/M5	$13,275	$13,800	$14,330
	Base	2dr.con.	V6/3.8/M5	$19,105	$19,950	$20,795
	GT	3dr.cpe.	V8/5.0/M5	$16,480	$17,195	$17,905
	GT	3dr.con.	V8/5.0/A4	$20,730	$21,655	$22,585

• **Destination & handling charge: $475**

Model	Trim	Type	Engine	mini	avrg	maxi
Probe		3dr.cpe.	L4/2.2/M5	$13,070	$13,625	$14,180
	GT	3dr.cpe.	V6/2.5/M5	$15,220	$15,880	$16,545

• **Destination & handling charge: $360**

Model	Trim	Type	Engine	mini	avrg	maxi
Taurus	GL	4dr.sdn.	V6/3.0/A4	$16,270	$16,930	$17,585
	GL	5dr.wgn.	V6/3.8/A4	$17,365	$17,925	$18,680
	LX	4dr.sdn.	V6/3.0/A4	$18,085	$18,745	$19,400
	LX	5dr.wgn.	V6/3.8/A4	$19,288	$20,150	$21,010
	SHO	4dr.sdn.	V6/3.0/M5	$23,015	$24,080	$25,140

• **Destination & handling charge: $535**

Model	Trim	Type	Engine	mini	avrg	maxi
T-bird	LX	2dr.cpe.	V6/3.8/A4	$15,995	$16,700	$17,400
	SC	2dr.cpe.	V8/4.6/M5	$21,010	$21,960	$22,910

• **Destination & handling charge: $495**

Model	Trim	Type	Engine	mini	avrg	maxi
Windstar	Cargo	4dr.van.	V6/3.0/A4	$16,325	$17,035	$17,745
	GL	4dr.van.	V6/3.0/A4	$18,100	$18,845	$19,590
	LX	4dr.van.	V6/3.8/A4	$21,860	$22,810	$23,760

• **Destination & handling charge: NA**

FORD (Utility)

Model	Trim	Type	Engine	mini	avrg	maxi
Bronco	XL	2dr.wgn.	L6/4.9/M5	$20,050	$20,910	$21,785
	XLT	2dr.wgn.	L6/4.9/M5	$22,575	$23,560	$24,540
	E.B.	2dr.wgn.	L6/4.9/M5	$25,400	$26,505	$27,610

• **Destination & handling charge: $540**

Model	Trim	Type	Engine	mini	avrg	maxi
Explorer 4x2						
	*XL	2dr.wgn.	V6/4.0/M5	$15,860	$16,650	$17,240
Explorer 4x4						
	*XLT	4dr.wgn.	V6/4.0/M5	$20,615	$21,510	$22,410
	*Ltd	4dr.wgn.	V6/4.0/M5	$26,250	$27,390	$28,535

• **Destination & handling charge: 700 $**

Model	Trim	Type	Engine	mini	avrg	maxi
Pick-ups 4x2						
F 250 XL reg. cab.		2dr.p-u.	V8/5.0/M5	$14,415	$15,040	$15,670
F 250 XLT crew cab		2dr.p-u.	V8/5.0/M5	$16,900	$17,640	$18,373
Pick-ups 4x4						
F 250 XL reg. cab.		2dr.p-u.	V8/5.0/M5	$17,615	$18,380	$19,146
F 250 XL crew cab		2dr.p-u.	V8/5.0/M5	$19,160	$19,995	$20,826

• **Destination & handling charge: $600**

Model	Trim	Type	Engine	mini	avrg	maxi
Ranger 4x2						
	*XL r.ca	2dr.p-u.	L4/2.3/M5	$8,650	$9,020	$9,389

• **Destination & handling charge: $460**

GE0

Model	Trim	Type	Engine	mini	avrg	maxi
Metro	Base	3dr.sdn.	L3/1.0/M5	$7,740	$7,910	$8,085
	LSI	3dr.sdn.	L3/1.0/M5	$8,040	$8,210	$8,385
	Base	4dr.sdn.	L4/1.3/M5	$8,740	$8,910	$9,085
	LSI	4dr.sdn.	L4/1.3/M5	$9,140	$9,310	$9,485

• **Destination charge: $310**

Model	Trim	Type	Engine	mini	avrg	maxi
Tracker 4x2						
	Soft top	2dr.con.	L4/1.6/M5	$11,360	$11,515	$11,670
Tracker 4x4						
	Soft top	2dr.con.	L4/1.6/M5	$12,625	$12,780	$12,935
	Soft top	LSI		$13,995	$14,150	$14,305
	Hard top	2dr.sdn.	L4/1.6/M5	$12,705	$12,860	$13,015
	Hard top	LSI		$14,175	$14,330	$14,485

• **Destination charge: $310**

Chevrolet/GMC (Utility)

Model	Trim	Type	Engine	mini	avrg	maxi
Astro 4x2	Cargo	4dr.van.	V6/4.3/A4	$16,540	$17,160	$17,785
	pass.	4dr.van.	V6/4.3/A4	$16,250	$16,867	$17,492
Astro	4x4	4dr.van.	V6/4.3/A4	$18,450	$19,065	$19,692

• **Destination charge: $590**

Model	Trim	Type	Engine	mini	avrg	maxi
Jimmy	4x2					
	SL	3dr.wgn.	V6/4.3/M5	$16,995	$17,635	$18,274
	SL	5dr.wgn.	V6/4.3/M5	$18,880	$19,520	$20,157
Jimmy	4x4					
	SL	3dr.wgn.	V6/4.3/M5	$18,700	$19,345	$19,980
	SLE	5dr.wgn.	V6/4.3/M5	$20,925	$21,565	$22,205

• **Destination charge: $590**

1995 AMERICAN PRICES

*1994 Price

Model	Trim	Type	Engine	Price U.S.$ mini	avrg	maxi
Pick-ups S-10 4x2						
reg. cab.	base	2dr.p-u.	L4/2.2/M5	$9,665	$10,030	$10,395
crew cab	LS	2dr.p-u.	L4/2.2/M5	$11,780	$12,140	$12,510
Pick-ups T-10 4x4						
cab. reg	base	2dr.p-u.	V6/4.3/M5	$14,035	$14,400	$14,765
crew cab	LS	2dr.p-u.	V6/4.3/M5	$15,689	$16,280	$16,870

Destination charge: $480

Model	Trim	Type	Engine	mini	avrg	maxi
Blazer	4x2	2dr.wgn.	V6/4.3/M5	$16,875	$17,500	$18,145
		4dr.wgn.	V6/4.3/M5	$18,581	$19,205	$19,851
Blazer	4x4	2dr.wgn.	V6/4.3/M5	$18,635	$19,260	$19,905
		4dr.wgn.	V6/4.3/M5	$20,683	$21,318	$21,953

• **Destination charge: $485**

Pick-ups *All models are available in those trim: Cheyenne, Sierra & Silverado.

Model	Trim	Type	Engine	mini	avrg	maxi
4x2						
C-1500	reg crew cab	2dr.p-u.	V6/4.3/M5	$14,020	$14,550	$15,077
C-1500	extd crew cab	2dr.p-u.	V8/5.0/M5	$16,643	$17,170	$17,700
C-1500	regular cab	2dr.p-u.	V8/5.0/M5	$12,330	$12,860	$13,387
C-2500	reg crew cab	2dr.p-u.	V8/5.0/M5	$15,800	$16,340	$16,871
C-2500	extd crew cab	2dr.p-u.	V8/5.0/M5	$17,770	$18,300	$18,826
4x4						
K-1500	regular cab	.2dr.p-u.	V6/4.3/M5	$16,300	$16,860	$17,389
K-1500	extd cab	2dr.p-u.	V8/5.0/M5	$18,125	$18,650	$19,182
K-1500	reg crew cab	2dr.p-u.	V8/5.7/M5	$16,600	$17,150	$17,688
K-1500	extd crew cab	2dr.p-u.	V8/5.7/M5	$18,900	$19,425	$19,956
K-2500	extd cab	2dr.p-u.	V8/5.7/M5	$18,175	$18,700	$19,233
K-2500	reg crew cab	2dr.p-u.	V8/5.7/M5	$20,130	$20,890	$21,648

• **Destination charge: $610**

Model	Trim	Type	Engine	mini	avrg	maxi
Tahoe	Base	3dr.wgn.	V8/5.7/A4	$20,300	$21,065	$21,830

• **Destination charge: $610**

Model	Trim	Type	Engine	mini	avrg	maxi
Safari	4X2					
Base		4dr.van.	V6/4.3/A4	$17,075	$17,716	$18,358
Safari	4X4					
Base		4dr.van.	V6/4.3/A4	$19,275	$19,900	$20,558

• **Destination charge: $610**

Model	Trim	Type	Engine	mini	avrg	maxi
Suburban						
C1500	2WD SL	5dr.wgn.	V8/5.7/A4	$20,075	$20,800	$21,587
C2500	2WD SL	5dr.wgn.	V8/5.7/A4	$21,300	$22,065	$22,819
K1500	4WD SL	5dr.wgn.	V8/5.7/A4	$22,375	$23,130	$23,893
K2500	4WD SL	5dr.wgn.	V8/5.7/A4	$23,520	$24,280	$25,038

*Cheyenne versions are similar in price as the SL versions.

• **Destination charge: $650**

HONDA

Model	Trim	Type	Engine	mini	avrg	maxi
Accord						
Coupe	LX	3dr.sdn.	L4 2.2 M5	$15,590	$16,570	$17,550
	LX	3dr.sdn.	L4/2.2/A4	$16,340	$17,320	$18,300
	EX	3dr.sdn.	L4/2.2/M5	$17,300	$18,700	$20,110
	EX	3dr.sdn.	L4/2.2/A4	$18,050	$19,450	$20,860
Sedan	DX	4dr.sdn.	L4 2.2 M5	$13,150	$13,975	$14,800
	DX	4dr.sdn.	L4/2.2/A4	$13,800	$14,725	$15,550
	LX	4dr.sdn.	L4/2.2/M5	$16,000	$16,925	$17,750
	LX	4dr.sdn.	L4/2.2/A4	$16,750	$17,675	$18,500
	LX	4dr.sdn.	V6 2.7 A4	$19,800	$21,050	$22,300
	LX	5dr.wgn.	L4/2.2/A4	$17,700	$18,630	$19,460
	EX	4dr.sdn.	L4/2.2/M5	$18,550	$19,430	$20,310
	EX	4dr.sdn.	L4/2.2/A4	$19,300	$20,180	$21,060
	EX	4dr.sdn.	V6 2.7 A4	$22,155	$23,550	$24,950
	EX	5dr.wgn.	L4/2.2/A4	$20,330	$21,210	$22,090

• **Destination & handling charge: $350**

Model	Trim	Type	Engine	mini	avrg	maxi
Civic	DX	2dr.cpe.	L4/1.5/M5	$10,346	$10,970	$11,590
	DX	2dr.cpe.	L4/1.5/A4	$11,145	$11,770	$12,570
	EX	2dr.cpe.	L4/1.6/M5	$12,600	$13,320	$14,030
	EX	2dr.cpe.	L4/1.6/A4	$13,350	$14,074	$14,780
	CX	3dr.sdn.	L4/1.5/M5	$9,160	$9,450	$9,750
	DX	3dr.sdn.	L4/1.5/M5	$10,500	$10,800	$11,100
	DX	3dr.sdn.	L4/1.5/A4	$11,480	$11,780	$12,080
	VX	3dr.sdn.	L4/1.5/M5	$11,200	$11,500	$11,800
	Si	3dr.sdn.	L4/1.6/M5	$12,940	$13,240	$13,540
	DX	4dr.sdn.	L4/1.5/M5	$11,380	$11,680	$11,980
	DX	4dr.sdn.	L4/1.5/A4	$12,130	$12,430	$12,730
	LX	4dr.sdn.	L4/1.5/M5	$12,720	$13,020	$13,320
	LX	4dr.sdn.	L4/1.5/A4	$13,470	$13,770	$14,070
	EX	4dr.sdn.	L4/1.6/M5	$14,633	$15,415	$16,200
	EX	4dr.sdn.	L4/1.6/A4	$15,383	$16,165	$16,950

• **Destination & handling charge: $350**

Model	Trim	Type	Engine	mini	avrg	maxi
Del Sol	Si	2dr.cpe.	L4/1.6/M5	$15,270	$16,110	$16,950
	Si	2dr.cpe.	L4/1.6/A4	$16,020	$16,860	$17,700
	vtec	2dr.cpe.	L4 1.6 M5	$17,390	$18,295	$19,200

• **Destination & handling charge: $350**

Model	Trim	Type	Engine	mini	avrg	maxi
Passport	4x2					
	DX	4dr.wgn	L4/2.6/M5	$14,430	$15,330	$16,230
	LX	4dr.wgn.	V6/3.2/M5	$17,460	$18,550	$19,635
	LX	4dr.wgn.	V6/3.2/A4	$18,380	$19,470	$20,555
	4x4					
	LX	4dr.wgn.	V6/3.2/M5	$19,850	$21,074	$22,300
	LX	4dr.wgn.	V6/3.2/A4	$21,000	$22,214	$23,450
	EX	4dr.wgn.	V6/3.2/M5	$22,665	$24,030	$25,400
	EX	4dr.wgn.	V6/3.2/A4	$23,800	$25,180	$26,550

• **Destination & handling charge: $375**

Model	Trim	Type	Engine	mini	avrg	maxi
Prelude	SR	2dr.cpe.	L4/2.3/M5	$16,500	$17,550	$18,600
	SR	2dr.cpe.	L4/2.3/A4	$17,250	$18,250	$19,350
	SR-V	2dr.cpe.	L4/2.2/M5	$22,175	$23,585	$25,000

• **Destination & handling charge: $350**

HYUNDAI

Model	Trim	Type	Engine	mini	avrg	maxi
Accent	L	3dr.sdn.	L4/1.5/M5			-
	L	3dr.sdn.	L4/1.5/A4			-
	GL	3dr.sdn.	L4/1.5/M5			-
	GL	3dr.sdn.	L4/1.5/A4			-
	L	4dr.sdn.	L4/1.5/M5			-
	L	4dr.sdn.	L4/1.5/A4			-
	GL	4dr.sdn.	L4/1.5/M5			-
	GL	4dr.sdn.	L4/1.5/A4			-

• **Destination & handling charge: $405**

Model	Trim	Type	Engine	mini	avrg	maxi
Elantra	GL	4dr.sdn.	L4/1.6/M5	$9,481	$9,840	$10,199
	GL	4dr.sdn.	L4/1.8/A4	$10,881	$11,140	$11,499
	GLS	4dr.sdn.	L4/1.8/M5	$10,981	$11,240	$11,599
	GLS	4dr.sdn.	L4/1.8/A4	$11,705	$12,015	$12,324

• **Destination & handling charge: $405**

Model	Trim	Type	Engine	mini	avrg	maxi
Scoupe	base	2dr.cpe.	L4/1.5/M5	$9,226	$9,610	$9,995

Model	Trim	Type	Engine	Price U.S.$ mini	avrg	maxi
	base	2dr.cpe.	L4/1.5/A4	$9,900	$10,285	$10,670
	LS	2dr.cpe.	L4/1.5/M5	$10,330	$10,883	$11,435
	LS	2dr.cpe.	L4/1.5/A4	$11,005	$11,560	$12,110
	Turbo	2dr.cpe.	L4T/1.5/M5	$11,070	$11,705	$12,255

• Destination & handling charge: $405

Model	Trim	Type	Engine	mini	avrg	maxi
Sonata						
	base	4dr.sdn.	L4/2.0/M5	$12,155	$12,775	$13,399
	base	4dr.sdn.	L4/2.0/A4	$12,965	$13,565	$14,209
	base	4dr.sdn.	V6/3.0/A4	$14,705	$15,310	$15,919
	GLS	4dr.sdn.	V6/3.0/A4	$15,770	$16,585	$17,399

Destination & handling charge: $405

INFINITI

Model	Trim	Type	Engine	mini	avrg	maxi
G20	base	4dr.sdn.	L4/2.0/M5	$19,235	$21,055	$22,875
	"t"	4dr.sdn.	L4/2.0/M5	$22,335	$24,155	$25,975

• Destination & handling charge: $450

Model	Trim	Type	Engine	mini	avrg	maxi
J-30		4dr.sdn.	V6/3.0/A4	$32,180	$35,365	$38,550
	"t"	4dr.sdn.	V6/3.0/A4	$34,180	$37,365	$40,550

• Destination & handling charge: $450

Model	Trim	Type	Engine	mini	avrg	maxi
Q45	base	4dr.sdn.	V8/4.5/A4	$44,725	$48,560	$52,400
Q45	"t"	4dr.sdn.	V8/4.5/A4	$48,125	$51,960	$55,800
Q45	F.A.S.	4dr.sdn.	V8/4.5/A4	$51,675	$55,510	$59,350

• Destination & handling charge: $450

ISUZU

Model	Trim	Type	Engine	mini	avrg	maxi
Rodeo (1994 prices)						
	S 4x4	5dr.wgn.	L4/2.6/M5	$17,190	$18,280	$19,369
	LS 4x4	5dr.wgn.	V6/3.1/M5	$21,228	$22,575	$23,919
Trooper						
	S	5dr.wgn.	V6/3.2/M5	$18,925	$20,185	$21,450
	LS	5dr.wgn.	V6/3.2/A4	$23,455	$25,250	$27,050

• Destination & handling charge: $400

JAGUAR

Model	Trim	Type	Engine	mini	avrg	maxi
XJ6	Sovereign	4dr.sdn.	L6/4.0/A4	$45,685	$49,565	$53,450
	Vanden Plas	4dr.sdn.	L6/4.0/A4	$53,320	$57,760	$62,200
XJR super charged		4dr.sdn.	L6C/4.0/A4	$56,100	$60,550	$65,000
XJ12	base	4dr.sdn.	V12/6.0/A4	$66,342	$71,795	$77,250

• Destination & handling charge: $580

Model	Trim	Type	Engine	mini	avrg	maxi
XJS	4.0L	2dr.cpe.	L6/4.0/M5	$45,680	$49,560	$53,400
	6.0L	2dr.cpe.	V12/6.0/A4	$61,860	$67,104	$72,350
XJS	4.0L	2dr.con.	L6/4.0/A4	$52,900	$57,245	$61,550
	6.0L	2dr.con.	V12/6.0/A4	$70,550	$76,550	$82,550

• Destination & handling charge: $580

JEEP

Model	Trim	Type	Engine	mini	avrg	maxi
Cherokee 4x2						
	base	2dr.wgn.	L4/2.5/M5	$12,684	$13,160	$13,639
	base	4dr.wgn.	L4/2.5/M5	$13,700	$14,200	$14,677
	Sport	2dr.wgn.	L6/4.0/M5	$15,100	$15,580	$16,060
	Sport	4dr.wgn.	L6/4.0/M5	$16,125	$16,600	$17,094
	RHD	4dr.wgn.	L6/4.0/M5	$17,450	$17,950	$18,355
Cherokee 4x4						
	SE	2dr.wgn.	L6/4.0/M5	$14,125	$14,675	$15,154
	SE	4dr.wgn.	L6/4.0/M5	$15,130	$15,700	$16,188
	Sport	2dr.wgn.	L6/4.0/M5	$16,505	$17,100	$17,572
	Sport	4dr.wgn.	L6/4.0/M5	$17,500	$18,125	$18,606
	Country	4dr.wgn.	L6/4.0/A4	$19,000	$19,630	$20,170

• Destination & handling charge: $495

Model	Trim	Type	Engine	mini	avrg	maxi
Grand Cherokee 4x2						
	SE	4dr.wgn.	L6/4.0/M5	$21,500	$22,100	$22,643
	Limited	4dr.wgn.	L6/4.0/A4	$27,100	$27,550	$28,260
Grand Cherokee 4x4						
	SE	4dr.wgn.	L6/4.0/M5	$23,400	$23,995	$24,580
	Limited	4dr.wgn.	L6/4.0/A4	$29,000	$29,795	$30,687

• Destination & handling charge: $495

Model	Trim	Type	Engine	mini	avrg	maxi
Wrangler	S	2dr.con.	L4/2.5/M5	$10,990	$11,405	$11,818
	SE	2dr.con.	L4/2.5/M5	$14,425	$14,931	$15,437

• Destination & handling charge: $495

LEXUS

Model	Trim	Type	Engine	mini	avrg	maxi
ES300	base	4dr.sdn.	V6/3.0/A4	$26,400	$28,950	$31,500

• Destination & handling charge: $470

Model	Trim	Type	Engine	mini	avrg	maxi
GS300	base	4dr.sdn.	L6/3.0/A4	$34,990	$38,845	$42,700

• Destination & handling charge: $470

Model	Trim	Type	Engine	mini	avrg	maxi
LS400	base	4dr.sdn.	V8/4.0/A4	$42,600	$46,900	$51,200
SC300	base	2dr.cpe.	L6/3.0/M5	$34,750	$37,820	$40,900
SC400	base	2dr.cpe.	V8/4.0/A4	$40,170	$44,285	$48,400

• Destination & handling charge: $470

LINCOLN

Model	Trim	Type	Engine	mini	avrg	maxi
Continental (1994 price)						
	Executive	4dr.sdn.	V6/3.8/A4	$29,500	$31,725	$33,750
	Signature	4dr.sdn.	V6/3.8/A4	$31,300	$33,500	$35,600

• Destination & handling charge: $625

Model	Trim	Type	Engine	mini	avrg	maxi
Mark VIII		2dr.sdn.	V8/4.6/A4	$34,500	$36,650	$38,800

• Destination & handling charge: $625

Model	Trim	Type	Engine	mini	avrg	maxi
Town Car						
	Executive	4dr.sdn.	V8/4.9/A4	$32,555	$34,475	$36,400
	Signature	4dr.sdn.	V8/4.9/A4	$34,655	$36,575	$38,500
	Cartier	4dr.sdn.	V8/4.9/A4	$37,350	$39,350	$41,200

• Destination & handling charge: $625

MAZDA

Model	Trim	Type	Engine	mini	avrg	maxi
Protegé	DX	4dr.sdn.	L4/1.8/M5	$11,195	$11,595	$11,995
	LX	4dr.sdn.	L4/1.8/M5	$12,395	$12,895	$13,395
	ES	4dr.sdn.	L4/1.8/M5	$15,145	$15,645	$16,145

• Destination & handling charge: $440

Model	Trim	Type	Engine	mini	avrg	maxi
626						
	DX	4dr.sdn.	L4/2.0/M5	$13,845	$14,270	$14,695
	DX	4dr.sdn.	L4/2.0/A4	$14,645	$15,070	$15,495
	LX	4dr.sdn.	L4/2.0/M5	$15,700	$16,450	$17,195
	LX	4dr.sdn.	L4/2.0/A4	$16,500	$17,250	$17,995
	LX	4dr.sdn.	V6/2.5/M5	$17,510	$18,450	$19,395
	LX	4dr.sdn.	V6/2.5/A4	$18,310	$19,250	$20,195
	ES	4dr.sdn.	V6/2.5/M5	$20,000	$21,200	$22,395
	ES	4dr.sdn.	V6/2.5/A4	$20,800	$22,000	$23,195

• Destination & handling charge: $440

Model	Trim	Type	Engine	mini	avrg	maxi
MX-6						
	RS	2dr.cpe.	L4/2.0/M5	$16,820	$17,695	$18,573
	RS	2dr.cpe.	L4/2.0/A4	$17,620	$18,495	$19,373
	LS	2dr.cpe.	V6/2.5/M5	$19,250	$20,450	$21,648
	LS	2dr.cpe.	V6/2.5/A4	$20,050	$21,250	$22,448

• Destination & handling charge: $440

Model	Trim	Type	Engine	mini	avrg	maxi
Miata	MX-5	2dr.con.	L4/1.8/M5	$16,100	$16,800	$17,500
	MX-5	2dr.con.	L4/1.8/A4	$16,950	$17,650	$18,350

• Destination & handling charge: $440

1995 AMERICAN PRICES

*1994 Price

Model	Trim	Type	Engine	Price U.S.$ mini	avrg	maxi
Millenia	base	4dr.sdn.	V6/2.5/A4	$23,355	$24,675	$25,995
	S (Miller)	4dr.sdn.	V6/2.3/A4	$28,075	$30,035	$31,995

• **Destination & handling charge: $440**

Model	Trim	Type	Engine	mini	avrg	maxi
MX-3	RS	3dr.cpe.	L4/1.6/M5	$13,355	$13,900	$14,440
	RS	3dr.cpe.	L4/1.6/A4	$14,155	$14,700	$15,240

• **Destination & handling charge: $440**

Model	Trim	Type	Engine	mini	avrg	maxi
929	base	4dr.sdn.	V6/3.0/A4	$31,890	$33,840	$35,795

• **Destination & handling charge: $440**

Model	Trim	Type	Engine	mini	avrg	maxi
RX-7	base	2dr.cpe.	R2/1.3/M5	$31,740	$34,120	$36,500
	Touring	2dr.cpe.	R2/1.3/M5	$35,940	$38,320	$40,700

• **Destination & handling charge: $440**

Model	Trim	Type	Engine	mini	avrg	maxi
Pick-ups 4x2						
B-2300						
base short deck		2dr.p-u.	L4/2.3/M5	$8,800	$9,130	$9,460
Se short deck.		2dr.p-u.	L4/2.3/M5	$10,700	$11,035	$11,385
base Cab plus		2dr.p-u.	L4/2.3/M5	$11,500	$11,835	$12,195
B-3000						
base short deck		2dr.p-u.	V6/3.0/M5	$11,250	$11,600	$11,955
SE short deck		2dr.p-u.	V6/3.0/M5	$12,660	$12,950	$13,270
SE extd deck		2dr.p-u.	V6/3.0/M5	$11,900	$13,230	$12,600
base Cab plus		2dr.p-u.	V6/3.0/M5	$12,065	$13,395	$12,765
SE Cab plus		2dr.p-u.	L4/2.3/M5	$12,790	$13,125	$13,445
B-4000						
base extd deck		2dr.p-u.	V6/4.0/M5	$12,075	$12,425	$12,775
LE Cab plus		2dr.p-u.	V6/4.0/M5	$14,875	$15,250	$15,630

• **Destination & handling charge: $495 (1994 prices)**

Model	Trim	Type	Engine	mini	avrg	maxi
Pick-ups 4x4						
B-3000						
SE short deck		2dr.p-u.	V6/3.0/M5	$14,075	$14,425	$14,775
base Cab plus		2dr.p-u.	V6/3.0/M5	$15,125	$15,475	$15,835
SE crew cab		2dr.p-u.	V6/3.0/M5	$15,900	$16,305	$16,655
B 4000						
base short deck		2dr.p-u.	V6/4.0/M5	$15,925	$16,330	$16,675
SE Cab plus		2dr.p-u.	V6/3.0/M5	$16,895	$17,310	$17,635
LE Cab plus		2dr.p-u.	V6/3.0/M5	$18,995	$19,200	$19,840

• **Destination & handling charge: $495 (1994 prices)**

Model	Trim	Type	Engine	mini	avrg	maxi
MPV 4x2						
Base	5 p.	4dr.wgn.	L4/2.6/A4	$19,535	$20,535	$21,135
LX	7 p.	4dr.wgn.	V6/3.0/A4	$20,285	$21,285	$21,985
LXE	8 p.	4dr.wgn.	V6/3.0/A4	$22,485	$23,475	$24,375
MPV 4x4						
LX	7 p.	4dr.wgn.	V6/3.0/A4	$23,480	$24,480	$25,380
LTE	7p.	4dr.wgn.	V6/3.0/A4	$25,780	$26,780	$27,670

• **Destination & handling charge: $470**

Model	Trim	Type	Engine	mini	avrg	maxi
Navajo 4x2						
	DX	2dr.wgn.	V6/4.0/M5	$17,000	$17,890	$18,785
	LX	2dr.wgn.	V6/4.0/M5	$18,215	$19,010	$20,005
Navajo 4x4						
	DX	2dr.wgn.	V6/4.0/M5	$18,785	$19,580	$20,575
	LX	2dr.wgn.	V6/4.0/M5	$20,005	$20,800	$21,795

• **Destination & handling charge: $505**

MERCEDES-BENZ

Model	Trim	Type	Engine	mini	avrg	maxi
C-Class	220 SE	4dr.sdn.	L4/2.2/M5	$27,020	$28,985	$30,950
	280	4dr.sdn.	L6/2.8/M5	$32,370	$34,335	$36,300

• **Destination & handling charge: $475**

Model	Trim	Type	Engine	mini	avrg	maxi
E-Class						
	300 D	4dr.sdn.	L6D/3.0/A4	$35,730	$38,365	$41,000
	320 E	4dr.sdn.	L6/3.2/A4	$38,230	$40,865	$43,500
	320	5dr.wgn.	L6/3.2/A4	$42,235	$44,865	$47,500
	320 E	2dr.con.	L6/3.2/A4	$68,495	$73,750	$79,000
	320 CE	2dr.cpe.	L6/3.2/A4	$54,450	$58,725	$63,000
	420	4dr.sdn.	V8/4.2/A4	$46,685	$49,595	$52,500

• **Destination & handling charge: $475**

Model	Trim	Type	Engine	mini	avrg	maxi
S-Class						
	320	4dr.sdn.	L6/3.2/A5	$56,050	$60,975	$65,900
	350 D	4dr.sdn.	L6D/3.5/A4	$56,050	$60,975	$65,900
	420 S	4dr.sdn.	V8/4.2/A4	$62,805	$68,350	$73,900
	500 S	4dr.sdn.	V8/5.0/A4	$74,150	$80,825	$87,500
	500 S	2dr.cpe.	V8/5.0/A4	$77,303	$84,600	$91,900
	600 S	4dr.sdn.	V12/6.0/A4	112,475	121,385	130,300
	600 S	2dr.cpe.	V12/6.0/A4	115,065	124,180	133,300

• **Destination & handling charge: $475**

Model	Trim	Type	Engine	mini	avrg	maxi
SL-Class						
	320 SL	2dr.con.	L6/3.2/A5	$66,375	$72,335	$78,300
	500 SL	2dr.con.	V8/5.0/A4	$75,900	$82,900	$89,900
	600 SL	2dr.con.	V12/6.0/A4	103,665	111,885	120,100

• **Destination & handling charge: $475**

MERCURY

Model	Trim	Type	Engine	mini	avrg	maxi
Cougar	XR7	2dr.cpe.	V8/5.0/A4	$15,000	$15,930	$16,860

• **Destination & handling charge: $495**

Model	Trim	Type	Engine	mini	avrg	maxi
Mystique	GS	4dr.sdn.	L4/2.3/M5	$13,025	$13,440	$13,855
	LS	4dr.sdn.	V6/2.5/M5	$14,315	$14,775	$15,230

• **Destination & handling charge: NA**

Model	Trim	Type	Engine	mini	avrg	maxi
Sable	GS	4dr.sdn.	V6/3.0/A4	$16,735	$17,470	$18,210
	GS	5dr.wgn.	V6/3.0/A4	$17,885	$18,580	$19,360
	LS	4dr.sdn.	V6/3.0/A4	$18,995	$19,690	$20,470
	LS	4dr.wgn.	V6/3.0/A4	$20,095	$20,800	$21,570

• **Destination & handling charge: $535**

Model	Trim	Type	Engine	mini	avrg	maxi
G-Marquis	GS	4dr.sdn.	V8/4.6/A4	$20,025	$20,645	$21,270
	LS	4dr.sdn.	V8/4.6/A4	$21,455	$22,074	$22,690

• **Destination & handling charge: $575**

Model	Trim	Type	Engine	mini	avrg	maxi
Tracer						
	sedan	4dr.sdn.	L4/1.9/M5	$10,655	$10,970	$11,280
	wagon	4dr.wgn.	L4/1.9/M5	$11,175	$11,490	$11,800
	LTS	4dr.sdn.	L4/1.9/M5	$12,515	$12,825	$13,140

• **Destination & handling charge: $375**

Model	Trim	Type	Engine	mini	avrg	maxi
Villager	GS	4dr.van.	V6/3.0/A4	$16,680	$17,530	$18,375
	LS	4dr.van.	V6/3.0/A4	$20,975	$22,064	$23,155
	Nautica	4dr.van.	V6/3.0/A4	$22,300	$23,470	$24,635

• **Destination & handling charge: $540 (1994 prices)**

MITSUBISHI

Model	Trim	Type	Engine	mini	avrg	maxi
Diamante	LS	4dr.sdn.	V6/3.0/A4	$27,260	$31,255	$35,250

• **Destination & handling charge: $470**

Model	Trim	Type	Engine	mini	avrg	maxi
Eclipse	RS	3dr.cpe.	L4/2.0/M5	$13,100	$13,730	$14,359
	RS	3dr.cpe.	L4/2.0/A4	$13,800	$14,425	$15,059
	GS	3dr.cpe.	L4/2.0/M5	$14,771	$15,550	$16,329

Model	Trim	Type	Engine	Price U.S.$ mini	avrg	maxi
	GS	3dr.cpe.	L4/2.0/A4	$15,451	$16,230	$17,019
	GS-T	3dr.cpe.	L4T/2.0/M5	$17,924	$18,960	$19,999
	GS-T	3dr.cpe.	L4T/2.0/A4	$18,700	$19,750	$20,829
	GSX	3dr.cpe.	L4T/2.0/M5	$20,545	$21,740	$22,929
	GSX	3dr.cpe.	L4T/2.0/A4	$21,535	$22,540	$23,739
• Destination & handling charge: $420						
Galant	S	4dr.sdn.	L4/2.4/M5	$13,110	$13,730	$14,349
	S	4dr.sdn.	L4/2.4/A4	$14,010	$14,630	$15,249
	ES	4dr.sdn.	L4/2.4/A4	$16,308	$17,490	$18,669
	LS	4dr.sdn.	L4/2.4/A4	$17,700	$19,000	$20,269
• Destination & handling charge: $420						
Mighty-Max		2dr.p-u.	L4/2.4/M5	$9,625	$9,990	$10,359
		2dr.p-u.	L4/2.4/A4	$10,575	$10,840	$11,299
• Destination & handling charge: $420						
Mirage	base	2dr.cpe.	L4/1.5/M5	$9,567	$9,683	$9,799
	LS	2dr.cpe	L4/1.8/M5	$11,583	$12,076	$12,569
	LS	2dr.cpe	L4/1.8/A4	$12,243	$12,735	$13,229
• Destination & handling charge: $420						
Montero	LS	4dr.wgn.	V6/3.0/M5	$24,585	$26,105	$27,625
	LS	4dr.wgn.	V6/3.0/A4	$25,435	$26,955	$28,475
	SR	4dr.wgn.	V6/3.5/A4	$29,975	$32,300	$34,625
• Destination & handling charge: $445						
3000 GT	base	3dr.cpe.	V6/3.0/M5	$23,968	$26,210	$28,450
	base	3dr.cpe.	V6/3.0/A4	$24,840	$27,050	$29,325
	SL	3dr.cpe.	V6/3.0/M5	$28,545	$31,145	$33,750
	SL	3dr.cpe.	V6/3.0/A4	$29,410	$31,995	$34,625
	VR-4	3dr.cpe.	V6/3.0/M6	$36,340	$39,695	$43,050
• Destination & handling charge: $470						

NISSAN

Model	Trim	Type	Engine	mini	avrg	maxi
Altima	XE	4dr.sdn.	L4/2.4/M5	$13,415	$14,100	$14,799
	GXE	4dr.sdn.	L4/2.4/M5	$14,415	$15,100	$15,799
	SE	4dr.sdn.	L4/2.4/M5	$17,485	$18,170	$18,869
	GLE	4dr.sdn.	L4 2.4 A4	$18,500	$19,190	$19,889
• Destination & handling charge: $380						
Sentra	Limited	2dr.cpe.	L4/1.6/M5	$11,490	$12,120	$12,759
	Limited	4dr.sdn.	L4/1.6/M5	$11,710	$12,340	$12,979
• Destination & handling charge: $380						
Maxima	GXE	4dr.sdn.	V6/3.0/M5	$17,355	$18,675	$19,999
	SE	4dr.sdn.	V6/3.0/M5	$18,955	$20,250	$21,599
	GLE	4dr.sdn.	V6 3.0 A4	$22,375	$23,470	$24,819
• Destination & handling charge: $380						
240 SX	base		L4/2.4M5	$15,660	$16,580	$17,499
	coupe SE		L4/2.4M5	$18,990	$20,105	$21,219
	con.		L4/2.4A4	$21,445	$22,700	$23,969
• Destination & handling charge: $380						
300 ZX	base	3dr.cpe.	V6/3.0/M5	$30,785	$32,900	$35,009
	Turbo	3dr.cpe.	V6/3.0/M5	$37,185	$39,300	$41,409
	2+2	3dr.cpe.	V6/3.0/M5	$33,575	$35,690	$37,799
	con.	3dr.con.	V6/3.0/M5	$37,965	$40,075	$42,189
• Destination & handling charge: $380						
Quest	XE	5dr.wgn.	V6/3.0/A4	$17,645	$17,743	$19,839
	GXE	5dr.wgn.	V6/3.0/A4	$21,860	$23,235	$24,609
• Destination & handling charge: $380						

Model	Trim	Type	Engine	Price U.S.$ mini	avrg	maxi
Pathfinder	XE 4x2	4dr.wgn.	V6/3.0/M5	$18,535	$19,560	$20,589
	XE 4x4	4dr.wgn.	V6/3.0/M5	$20,055	$21,145	$22,229
	SE 4x4	4dr.wgn.	V6/3.0/M5	$23,905	$25,220	$26,539
	LE 4x4	4dr.wgn.	V6/3.0/A4	$27,304	$28,830	$30,359
• Destination & handling charge: $380						
Pick-ups Hardbody 2WD						
	Base	2dr.p-u.	L4/2.4/M5	$9,532	$9,730	$9,929
	XE	2dr.p-u.	L4/2.4/M5	$10,458	$10,730	$11,009
K.cab	XE	2dr.p-u.	L4/2.4/M5	$11,690	$12,190	$12,689
K.cab	SE-V6	2dr.p-u.	V6/3.0/M5	$13,930	$14,620	$15,309
Hardbody 4x4						
Short deck	XE	2dr.p-u.	L4/2.4/M5	$13,960	$14,545	$15,129
K.cab	XE V6	2dr.p-u.	V6/3.0/M5	$16,190	$16,960	$17,729
K.cab	SE V6	2dr.p-u.	V6/3.0/A4	$17,767	$18,540	$19,309
• Destination & handling charge: $380						

OLDSMOBILE

Model	Trim	Type	Engine	mini	avrg	maxi
Cutlass Ciera						
	S	4dr.sdn.	V6/3.1/A4			$14,460
• Destination charge: $535 (non-negotiable prices)						
Cutlass Cruiser						
	S	5dr.wgn.	V6/3.1/A4			$17,595
• Destination charge: $535 (non-negotiable prices)						
Cutlass Supreme						
	Base	2dr.cpe.	V6/3.1/A4			$17,460
	Base	4dr.sdn.	V6/3.1/A4			$17,460
	SL	2dr.con.	V6/3.4/A4			$25,460
• Destination charge: $535 (non-negotiable prices)						
98-Regency						
	base	4dr.sdn.	V6/3.8/A4			$26,060
	Elite	4dr.sdn.	V6/3.8/A4			$27,160
• Destination charge: $635 (non-negotiable prices)						
88-Royale	base	4dr.sdn.	V6/3.8/A4			$20,410
	LS	4dr.sdn.	V6/3.8/A4			$22,710
	LSS	4dr.sdn.	V6/3.8/A4			$24,010
• Destination charge: $585 (non-negotiable prices)						
Achieva	S	2dr.sdn.	L4/2.3/M5			$13,500
	SC	2dr.sdn.	L4/2.3/M5			$13,500
	S	4dr.sdn.	L4/2.3/M5			$15,200
	SL	4dr.sdn.	L4/2.3/A4			$15,200
• Destination charge: $495 (non-negotiable prices)						
Aurora	sedan	4dr.sdn.	V8/4.0/A4			$31,370
• Destination charge: $625 (non-negotiable prices)						
Silhouette	base	4dr.van	V6/3.8/A4			$20,255
• Destination charge: $540 (non-negotiable prices)						

PLYMOUTH

Model	Trim	Type	Engine	mini	avrg	maxi
Acclaim		4dr.sdn.	L4/2.5/M5	$13,455	$13,889	$14,323
• Destination & handling charge: $505						
Colt	base	2dr.cpe.	L4/1.5/M5	$9,567	$9,683	$9,799
	GL	2dr.cpe.	L4/1.5/M5	$10,374	$10,525	$10,675
• Destination & handling charge: $500						
Neon	Highline	2dr.cpe.	L4/2.0/M5	$10,670	$10,955	$11,240

1995 AMERICAN PRICES

*1994 Price

Model	Trim	Type	Engine	Price U.S.$ mini	avrg	maxi
	Sport	2dr.cpe.	L4/2.0/M5	$10,970	$13,115	$13,567
	base	4dr.sdn.	L4/2.0/M5	$9,035	$9,267	$9,500
	Highline	4dr.sdn.	L4/2.0/M5	$10,670	$10,955	$11,240
	Sport	4dr.sdn.	L4/2.0/M5	$12,363	$12,815	$13,267

• **Destination & handling charge: $500**

Model	Trim	Type	Engine	mini	avrg	maxi
Voyager	base	4dr.wgn.	L4/2.5/M5	$15,095	$15,625	$16,160
	SE	4dr.wgn.	L4/2.5/A3	$17,521	$18,188	$18,855
	LE	4dr.wgn.	V6/3.3/A3	$21,665	$22,524	$23,380

• **Destination & handling charge: $560**

Model	Trim	Type	Engine	mini	avrg	maxi
Grand Voyager						
	base	4dr.wgn.	V6/3.0/A4	$17,287	$17,945	$18,605
	SE	4dr.wgn.	V6/3.3/A4	$18,160	$18,875	$19,595
	SE 4x4	4dr.wgn.	V6/3.3/A4	$20,560	$21,415	$22,270
	LE	4dr.wgn.	V6/3.3/A4	$21,888	$22,784	$23,680
	LE 4x4	4dr.wgn.	V6/3.3/A4	$23,065	$24,715	$25,755

• **Destination & handling charge: $560**

PONTIAC

Model	Trim	Type	Engine	mini	avrg	maxi
Bonneville	SE	4dr.sdn.	V6/3.8/A4	$19,420	$20,400	$21,389
	SSE	4dr.sdn.	V6/3.8/A4	$24,420	$25,400	$26,389

• **Destination charge: $585**

Model	Trim	Type	Engine	mini	avrg	maxi
Firebird	base	3dr.cpe.	V6/3.4/M5	$14,445	$14,900	$15,359
	base	3dr.con.	V6/3.4/M5	$21,525	$21,980	$22,439
	Formula	3dr.cpe.	V8/5.7/M6	$18,405	$19,000	$19,599
	Formula	3dr.con.	V8/5.7/A4	$24,435	$25,032	$25,629
Trans Am	GTA	2dr.cpe.	V8/5.7/M6	$20,135	$20,851	$21,569
	GTA	3dr.con.	V8/5.7/A4	$26,200	$26,925	$27,639

• **Destination charge: $500**

Model	Trim	Type	Engine	mini	avrg	maxi
Grand Am	SE	2dr.cpe.	L4/2.3/M5	$12,580	$12,990	$13,399
	SE	4dr.sdn.	L4/2.3/M5	$12,680	$13,090	$13,499
	GT	2dr.cpe.	L4/2.3/M5	$14,355	$14,850	$15,349
	GT	4dr.sdn.	L4/2.3/M5	$14,455	$14,950	$15,449

• **Destination charge: $495**

Model	Trim	Type	Engine	mini	avrg	maxi
Grand-Prix	SE	2dr.cpe.	V6/3.1/A3	$16,825	$17,370	$17,919
	SE	4dr.sdn.	V6/3.1/A3	$15,940	$16,555	$17,169
	GT	4dr.sdn.	V6/3.4/M5	$17,765	$18,380	$18,994
	GTP	2dr.cpe.	V6/3.4/M5	$18,945	$19,560	$20,175

• **Destination charge: $535**

Model	Trim	Type	Engine	mini	avrg	maxi
Sunfire		2dr.cpe.	L4/2.0/M5	$11,040	$11,300	$11,559
		4dr.sdn.	L4/2.0/M5	$11,190	$11,450	$11,709
		2dr.con.	L4/2.0/M5	NA	NA	NA

• **Destination charge: $495**

Model	Trim	Type	Engine	mini	avrg	maxi
Trans Sport	SE	5dr.wgn.	V6/3.1/A3	$17,105	$17,765	$18,429

• **Destination charge: $540**

PORSCHE

Model	Trim	Type	Engine	mini	avrg	maxi
911		2dr.cpe.	H6/3.6/M6	$52,150	$56,025	$59,900
Carrera 4		2dr.cpe.	H6/3.6/M6	$58,150	$62,025	$65,900
Carrera 2		2dr.con.	H6/3.6/M6	$59,325	$63,760	$68,200
Carrera 4		2dr.con.	H6/3.6/M6	$65,325	$69,760	$74,200

• **Destination & handling charge: $725**

Model	Trim	Type	Engine	mini	avrg	maxi
928	GTS	2dr.cpe.	V8/5.4/M5	$70,750	$76,500	$82,260

• **Destination & handling charge: $725**

Model	Trim	Type	Engine	mini	avrg	maxi
968		2dr.cpe.	L4/3.0M6	$33,400	$36,680	$39,950
		2dr.con.	L4/3.0/M6	$44,255	$48,075	$51,900

• **Destination & handling charge: $725**

SAAB

Model	Trim	Type	Engine	mini	avrg	maxi
900 S		3dr.sdn.	L4/2.3 /M5	$21,855	$22,615	$23,375
		5dr.sdn.	L4/2.3 /M5	$22,175	$22,935	$23,695
		3dr.con.	L4/2.0 /M5	$29,040	$31,015	$32,995
900 SE		3dr.sdn.	L4T/2.0 /M5	$25,045	$27,020	$28,990
		5dr.sdn.	V6/2.5/M5	$24,745	$26,710	$28,680
		3dr.con.	L4T/2.0 /M5	$35,565	$37,540	$39,520
		3dr.con.	V6/2.5/M5	$36,115	$37,990	$40,070

• **Destination & handling charge: $470**

Model	Trim	Type	Engine	mini	avrg	maxi
9000	Aero	4dr.sdn.	L4T/2.3/M5	$35,970	$38,635	$41,300
	CDE	4dr.sdn.	V6/3.0/M5	$33,665	$36,335	$38,995
	CS	4dr.sdn.	L4T/2.3/M5	$26,750	$28,285	$29,845
	CSE	4dr.sdn.	L4T/2.3/A4	$31,170	$33,840	$36,510
	CSE	4dr.sdn.	V6/3.0/A4	$33,320	$35,990	$38,650

• **Destination & handling charge: $470**

SATURN

Model	Trim	Type	Engine	mini	avrg	maxi
Saturn	SL	4dr.sdn.	L4/1.9/M5	$9,175	$9,585	$9,995
	SL1	4dr.sdn.	L4/1.9/M5	$10,115	$10,555	$10,995
	SL2	4dr.sdn.	L4/1.9/M5	$11,115	$11,555	$11,995
	SC 1	2dr.cpe.	L4/1.9/M5	$11,015	$11,455	$11,895
	SC 2	2dr.cpe.	L4/1.9/M5	$12,115	$12,155	$12,995
	SW1	4dr.wgn.	L4/1.9/M5	$10,815	$11,255	$11,695
	SW2	4dr.wgn.	L4/1.9/M5	$11,815	$12,555	$12,695

• **Destination & handling charge: $360**

SUBARU

Model	Trim	Type	Engine	mini	avrg	maxi
Impreza	4x2					
	L	2dr.cpe.	H4/1.8/M5	$10,950	$11,400	$11,850
	L	2dr.cpe.	H4/1.8/A4	$12,850	$13,300	$13,750
Impreza 4x4						
	base	2dr.cpe.	H4/1.8/M5	$11,850	$12,300	$12,750
	L	2dr.cpe.	H4/1.8/A4	$13,850	$14,300	$14,750
	L	5dr.wgn.	H4/1.8/M5	$14,250	$14,700	$15,150
	LX	5dr.wgn.	H4/1.8/A4	$16,495	$16,945	$17,395
	L	4dr.sdn.	H4/1.8/M5	$13,850	$14,300	$14,750
	LX	4dr.sdn.	H4/1.8/A4	$16,095	$16,545	$16,995

• **Destination & handling charge: $475**

Model	Trim	Type	Engine	mini	avrg	maxi
Justy 4x4(1994 prices)						
	GL	5dr.sdn.	L3/1.2 /M5	$8780	$9,190	$9,603

• **Destination & handling charge: $475**

Model	Trim	Type	Engine	mini	avrg	maxi
Legacy 4x2						
	L	4dr.sdn.	H4/2.2/M5	$15,620	$16,120	$16,620
	L	5dr.wgn.	H4/2.2/A4	$16,320	$16,820	$17,320
Legacy 4x4						
	L	4dr.sdn.	H4/2.2/M5	$16,620	$17,120	$17,620
	LS	4dr.sdn.	H4/2.2/M5	$20,120	$20,620	$21,120
	LSi	4dr.sdn.	H4/2.2/M5	$22,620	$23,120	$23,620
	Brighton	5dr.wgn.	H4/2.2/M5	$14,999	$15,499	$15,999
	LS	5dr.wgn.	H4/2.2/M5	$20,820	$21,320	$21,820
	LSi	5dr.wgn.	H4/2.2/M5	$23,320	$23,820	$24,320

• **Destination & handling charge: $475**

Model	Trim	Type	Engine	mini	avrg	maxi
SVX(1994 prices)						
	LSi	3dr.cpe.	H6/3.3/A4	$30,150	$32,000	$33,850

• **Destination & handling charge: $475**

SUZUKI

Model	Trim	Type	Engine	mini	avrg	maxi
Swift	Hbk	3dr.sdn.	L4/1.3/M5	$8,247	$8,475	$8,699
	DLX	3dr.sdn.	L4/1.3/M5	$9,095	$9,325	$9,549

• **Destination & handling charge: $315**

1995 AMERICAN PRICES

Model	Trim	Type	Engine	Price U.S.$ mini	avrg	maxi
Sidekick	Soft top					
	JS	2dr.con.	L4/1.6/M5	$11,240	$11,468	$11,999
	JX	2dr.con.	L4/1.6/M5	$12,575	$13,035	$13,499
Sidekick 4 door						
	JS	4dr.wgn.	L4/1.6/M5	$12,575	$13,035	$13,499
	JX	4dr.wgn.	L4/1.6/M5	$13,195	$13,850	$14,509
	JLX	4dr.wgn.	L4/1.6/M5	$14,880	$15,600	$16,319

• **Destination & handling charge: $330**

TOYOTA

Model	Trim	Type	Engine	Price U.S.$ mini	avrg	maxi
Pick-ups 4x2						
	T-100	2dr.p-u.	L4/2.7/M5	$11,840	$12,675	$13,518
	T-100	2dr.p-u.	V6/3.4/M5	$12,820	$20,785	$14,498
Pick-ups 4x4						
	T-100	2dr.p-u.	V6/3.4/M5	$16,955	$18,065	$19,178
• **Destination & handling charge: $400**						
Pick-ups 4x2						
	reg	2dr.p-u.	L4/2.4/M5	$9,555	$9,950	$10,348
	Xtra cab	2dr.p-u.	L4/2.4/M5	$11,465	$12,160	$12,858
	reg	2dr.p-u.	V6/3.0/M5	$12,660	$13,355	$14,048
	Xtra cab	2dr.p-u.	V6/3.0/M5	$14,650	$15,355	$16,058
Pick-ups 4x4						
	reg	2dr.p-u.	L4/2.4/M5	$13,550	$14,320	$15,088
	Xtra cab	2dr.p-u.	L4/2.4/M5	$14,760	$15,735	$16,708
	Xtra cab	2dr.p-u.	V6/3.0/M5	$17,135	$18,175	$19,208
	SR5 V6	2dr.p-u.	V6/3.0/M5	$17,925	$18,965	$19,998
• **Destination & handling charge: $400**						
Avalon	XL	4dr.sdn.	V6/3.0/A4	$19,685	$21,220	$22,758
	XLS	4dr.sdn.	V6/3.0/A4	$23,615	$25,150	$26,688
• **Destination & handling charge: $400**						
Camry	DX	2dr.cpe.	L4/2.2/M5	$13,850	$14,990	$16,128
	LE	2dr.cpe.	L4/2.2/A4	$16,605	$17,935	$19,268
	LE V6	2dr.cpe.	V6/3.0/A4	$18,600	$20,095	$21,588
	DX	4dr.sdn.	L4/2.2/M5	$14,140	$15,280	$16,418
	LE	4dr.wgn.	L4/2.2/A4	$18,305	$19,635	$20,968
	LE	4dr.sdn.	L4/2.2/M5	$16,895	$18,225	$19,558
	LE V6	4dr.sdn.	V6/3.0/A4	$19,195	$20,525	$21,878
	XLE V6	4dr.sdn.	V6/3.0/A4	$21,715	$23,055	$24,398
	LE V6	4dr.wgn.	V6/3.0/A4	$26,900	$21,965	$23,308
• **Destination & handling charge: $400**						
Celica	ST	2dr.cpe.	L4/1.6/M5	$14,835	$15,860	$16,888
	GT	3dr.cpe.	L4/2.2/M5	$16,855	$18,075	$19,288
	GT	2dr.con.	L4/2.2/M5	$21,565	$22,780	$23,998
• **Destination & handling charge: $400**						
Corolla	Std	4dr.sdn.	L4/1.6/M5	$11,270	$11,820	$12,378
	DX	4dr.sdn.	L4/1.6/M5	$11,875	$12,680	$13,488
	LE	4dr.sdn.	L4/1.6/A4	$14,865	$15,670	$16,678
	base	5dr.wgn.	L4/1.6/M5	$12,805	$13,610	$14,618
• **Destination & handling charge: $400**						
Paseo		2dr.cpe.	L4/1.5/M5	$11,390	$12,115	$12,838
		2dr.cpe.	L4/1.5/A4	$12,1901	$12,915	$13,638
• **Destination & handling charge: $400**						
4Runner	4x2	4dr.wgn.	V6/3.0/M5	$18,860	$20,300	$21,748
4Runner	SR5	4dr.wgn.	L4/2.4/M5	$18,210	$19,650	$21,098
4Runner	SR5 V6	4dr.wgn.	V6/3.0/M5	$20,350	$21,800	$23,148
• **Destination & handling charge: $400**						
Previa 4X2						
	DX	4dr.wgn.	L4/2.4/A4	$19,390	$20,850	$22,318
	DX	4dr.wgn.	L4C/2.4/A4	$19,890	$21,650	$22,818
	LE	4dr.wgn.	L4/2.4/A4	$23,040	$24,810	$26,578
	LE	4dr.wgn.	L4C/2.4/A4	$23,540	$25,310	$27,078
Previa 4X4						
	DX	4dr.wgn.	L4/2.4/A4	$22,160	$23,900	$25,648
	DX	4dr.wgn.	L4C/2.4/A4	$22,660	$24,400	$26,148
	LE	4dr.wgn.	L4/2.4/A4	$25,760	$27,740	$29,718
	LE	4dr.wgn.	L4C/2.4/A4	$26,260	$28,240	$30,218
Destination & handling charge: $400						
Supra		2dr.cpe.	V6/3.0/M5	$31,065	$33,980	$36,900
		2dr.cpe.	V6/3.0/A4	$31,965	$34,880	$37,800
		2dr.cpe.	V6T/3.0/M6	$39,675	$43,135	$46,600
		2dr.cpe.	V6T/3.0/A4	$39,675	$43,135	$46,600
• **Destination & handling charge: $400**						
Tercel	Std	2dr.sdn.	L4/1.5/M4	$9,443	$9,720	$9,998
	Dx	2dr.sdn.	L4/1.5/M5	$10,215	$10,550	$11,028
	Dx	4dr.sdn.	L4/1.5/M5	$10,515	$10,850	$11,328

• **Destination & handling charge: $400**

VOLKSWAGEN

Model	Trim	Type	Engine	Price U.S.$ mini	avrg	maxi
Corrado	VR-6	2dr.cpe.	V6/2.8/M5	$23,060	$24,600	$26,150
• **Destination & handling charge: $390**						
Golf	CL	3dr.sdn.	L4/1.8/M5	$11,780	$12,150	$12,500
	GL	5dr.sdn.	L4/2.0/M5	$12,450	$13,850	$14,200
	GTI	3dr.sdn.	L4/2.0/M5	$13,500	$14,900	$15,250
	Cabrio.	2dr.sdn.	L4/2.0/M5	$18,225	$19,100	$19,975
• **Destination & handling charge: $390**						
Jetta	CL	4dr.sdn.	L4/1.8/M5	$12,445	$12,960	$13,475
	GL	4dr.sdn.	L4/2.0/M5	$14,645	$15,160	$15,675
	GLS	4dr.sdn.	L4/2.0/M5	$15,895	$16,460	$17,025
	GLX	4dr.sdn.	V6/2.8/M5	$18,410	$19,190	$19,975
• **Destination & handling charge: $390**						
Passat	GLX	4dr.sdn.	V6/2.8/M5	$18,940	$19,910	$20,890
	GLX	5dr.wgn.	V6/2.8/M5	$19,370	$20,340	$21,320

• **Destination & handling charge: $390**

VOLVO

Model	Trim	Type	Engine	Price U.S.$ mini	avrg	maxi
850 sdn	base	4dr.sdn.	L5/2.4/M5	$22,930	$23,800	$24,680
	base	4dr.sdn.	L5/2.4/A4	$23,830	$24,700	$25,580
	GLT	4dr.sdn.	L5/2.4/M5	$24,390	$25,750	$27,110
	GLT	4dr.sdn.	L5/2.4/A4	$25,290	$26,650	$28,010
	T	4dr.sdn.	L5/2.3/A4	$27,380	$29,210	$31,045
850 wgn	base	5dr.wgn.	L5/2.4/M5	$24,230	$25,100	$25,980
	base	5dr.wgn.	L5/2.4/A4	$25,130	$26,000	$26,880
	GLT	5dr.wgn.	L5/2.4/M5	$25,715	$27,065	$28,410
	GLT	5dr.wgn.	L5/2.4/A4	$26,615	$27,965	$29,310
	T	5dr.wgn.	L5/2.3/A4	$28,580	$30,410	$32,345
• **Destination & handling charge: $460**						
940 sedan		4dr.sdn.	L4/2.3/A4	$22,600	$22,980	$23,360
	TS	4dr.sdn.	L4T/2.3/A4	$23,600	$23,980	$24,360
940 wagon		5dr.wgn.	L4/2.3/A4	$23,925	$24,295	$24,660
	TS	5dr.wgn.	L4T/2.3/A4	$24,925	$25,295	$25,660
• **Destination & handling charge: $460**						
960 sedan		4dr.sdn.	L6/3.0/A4	$28,250	$29,050	$29,900
960 wagon		5dr.wgn.	L6/3.0/A4	$27,590	$29,395	$31,200

• **Destination & handling charge: $460**

1995 CANADIAN PRICES

ATTENTION: Prices shown are given as a guideline and are interpreted as follows:

Mini price: the lowest negotiable price.
Avrg price: average sale price for the model.
Maxi price: manufacturer's suggested retail price.

ACURA

Model	Trim	Type	Engine	Price canadian $ mini	avrg	maxi
Integra	RS	3dr.sdn.	L4/1.8/M5	16 430	17 210	17 995
	LS	3dr.sdn.	L4/1.8/M5	21 175	22 185	23 195
	RS	4dr.sdn.	L4/1.8/M5	17 750	18 520	19 295
	LS	4dr.sdn.	L4/1.8/M5	21 235	22 415	23 595
	GS-R	3dr.cpe.	L4/1.8/M5	23 665	24 980	26 295

• **Destination & handling charge: 595 $**

Model	Trim	Type	Engine	mini	avrg	maxi
Legend	L	4dr.sdn.	V6/3.2/A4	43 135	45 265	47 400
	LS	4dr.sdn.	V6/3.2/A4	46 775	49 085	51 400
	LS	2dr.cpe.	V6/3.2/M6	47 320	49 660	52 000

• **Destination & handling charge: 795 $**

Model	Trim	Type	Engine	mini	avrg	maxi
NSX		2dr.cpe.	V6/3.0/M5	75 590	80 755	85 900

• **Destination & handling charge: 795 $**

Model	Trim	Type	Engine	mini	avrg	maxi
Vigor	GS	4dr.sdn.	L5/2.5/M5	29 970	31 635	33 300

• **Destination & handling charge: 595 $**

AUDI

Model	Trim	Type	Engine	mini	avrg	maxi
A4 (90)	Sport	4dr.sdn.	V6/2.8/A4	31 590*	33 345*	35 100*
Quattro	Sport	4dr.sdn.	V6/2.8/M5	34 875*	36 810*	38 750*
	Cabrio.	2dr.sdn.	V6/2.8/A4	49 410*	52 155*	54 900*

• **Destination & handling charge: 1 050 $**

Model	Trim	Type	Engine	mini	avrg	maxi
A6 (100)	*CS	4dr.sdn.	V6/2.8/A4	43 425*	45 835*	48 250*
*Quat.	(S4)S6	4dr.sdn.	L5T/2.2/M5	51 835*	55 367*	58 900*
	Quat.CS	4dr.wgn.	V6/2.8/A4	52 715*	56 305*	59 900*

• **Destination & handling charge: 1 050 $**

BMW

Model	Trim	Type	Engine	mini	avrg	maxi
318	i	4dr.sdn.	L4/1.8/M5	26 445	27 670	28 900
	i	2dr.con.	L4/1.8/M5	35 990	38 445	40 900
	is	2dr.cpe.	L4/1.8/M5	27 360	28 630	29 900
	ti	3dr.sdn.	L4/1.8/M5	22 875	23 935	25 000
320	i	4dr.sdn.	L4/2.0/M5	30 710	32 805	34 900
325	i	4dr.sdn.	L6/2.5/M5	36 870	39 385	41 900
	is	2dr.cpe.	L6/2.5/M5	38 632	41 266	43 900
	i	2dr.con.	L6/2.5/M5	47 430	50 665	53 900

• **Destination & handling charge: 490 $**

Model	Trim	Type	Engine	mini	avrg	maxi
525	i	4dr.sdn.	L6/2.5/M5	47 425	50 160	52 900
530	i	4dr.sdn.	V8/3.0/M5	54 566	57 567	60 900
540i		4dr.sdn.	V8/4.0/A5	59 975	63 440	66 900

• **Destination & handling charge: 590 $**

Model	Trim	Type	Engine	mini	avrg	maxi
740	i	4dr.sdn.	V8/4.0/A5	70 310	75 105	79 900

• **Destination & handling charge: 590 $**

Model	Trim	Type	Engine	mini	avrg	maxi
840	Ci	2dr.cpe.	V8/4.0/M5	86 245	91 225	96 200

• **Destination & handling charge: 590 $**

BUICK

Model	Trim	Type	Engine	Price canadian $ mini	avrg	maxi
Century	Custom	4dr.sdn.	V6/3.0/A4	20 540	22 115	23 690
	Special	4dr.sdn.	V6/3.0/A4	19 200	21 420	21 635
	Custom	5dr.wgn.	V6/3.0/A4	20 310	21 600	22 885

• **Destination charge: 760 $**

Model	Trim	Type	Engine	mini	avrg	maxi
Le Sabre	base	4dr.sdn.	V6/3.8/A4	23 525	25 330	27 135
	Ltd	4dr.sdn.	V6/3.8/A4	26 170	28 180	30 185

• **Destination charge: 870 $**

Model	Trim	Type	Engine	mini	avrg	maxi
Park Avenue		4dr.sdn.	V6/3.8/A4	30 515	33 065	35 615
	Ultra	4dr.sdn.	V6/3.8/A4	34 740	37 570	40 400

• **Destination charge: 870 $**

Model	Trim	Type	Engine	mini	avrg	maxi
Regal	Cust.	2dr.cpe.	V6/3.1/A4	20 230	21 780	23 335
	Cust.	4dr.sdn.	V6/3.1/A4	20 580	22 160	23 735
	Ltd	4dr.sdn.	V6/3.8/A4	22 165	23 865	25 565
	GSport	2dr.cpe.	V6/3.8/A4	23 005	24 770	26 535
	GSport	4dr.sdn.	V6/3.8/A4	23 450	25 250	27 050

• **Destination charge: 760 $**

Model	Trim	Type	Engine	mini	avrg	maxi
Riviera		2dr.cpe.	V6/3.8/A4	32 580	35 300	38 025

• **Destination charge: 870 $**

Model	Trim	Type	Engine	mini	avrg	maxi
Roadmaster						
	Base	4dr.sdn.	V8/5.7/A4	27 250	29 340	31 430
	Estate	5dr.wgn.	V8/5.7/A4	29 060	31 290	33 515
	Ltd	4dr.sdn.	V8/5.7/A4	28 480	30 660	32 845
	Ltd Estate	5dr.wgn.	V8/5.7/A4	29 625	31 895	34 170

• **Destination charge: 870 $**

Model	Trim	Type	Engine	mini	avrg	maxi
Skylark	Cust.	2dr.cpe.	V6/3.1/A4	17 055	18 045	19 035
	Cust.	4dr.sdn.	V6/3.1/A4	17 055	18 045	19 035
	Ltd	2dr.cpe.	L4/2.3/A4	18 700	19 890	21 010
	Ltd	4dr.sdn.	V6/3.1/A4	18 765	19 925	21 085
	GS	2dr.cpe.	V6/3.1/A4	19 730	20 950	22 170
	GS	4dr.sdn.	V6/3.1/A4	19 800	21 020	22 245

• **Destination charge: 595 $**

CADILLAC

Model	Trim	Type	Engine	mini	avrg	maxi
De Ville	base	2dr.sdn.	V8/4.9/A4	39 120	42 250	45 385
	Concours	4dr.sdn.	V8/4.9/A4	46 085	49 260	52 440

• **Destination charge: 870 $**

Model	Trim	Type	Engine	mini	avrg	maxi
Fleetwood	Base	4dr.sdn.	V8/5.7/A4	38 705	41 805	44 910
	Brougham	4dr.sdn.	V8/5.7/A4	40 520	43 865	47 010

• **Destination charge: 870 $**

Model	Trim	Type	Engine	mini	avrg	maxi
Eldorado	coupe	2dr.cpe.	V8/4.6/A4	43 415	46 410	49 405
	Touring	2dr.sdn.	V8/4.6/A4	45 200	47 190	51 435

• **Destination charge: 870 $**

Model	Trim	Type	Engine	mini	avrg	maxi
Seville	SLS	4dr.sdn.	V8/4.6/A4	46 855	50 085	53 315
	STS	4dr.sdn.	V8/4.6/A4	48 015	51 325	54 635

• **Destination charge: 870 $**

1995 CANADIAN PRICES

*1994 Price

CHEVROLET

Model	Trim	Type	Engine	Price canadian $ mini	avrg	maxi
Beretta	Base,	2dr.cpe.	L4/2.2/M5	14 560	15 315	16 075
	Z26	2dr.cpe.	V6/3.1/M5	17 860	18 900	19 940
• Destination charge: 595 $						
Camaro	RS	3dr.cpe.	V6/3.4/M5	16 480	17 420	18 360
	RS	2dr.con.	V8/5.7/M5	22 780	24 080	25 375
	Z28	2dr.con.	V8/5.7/A4	20 240	21 395	22 550
	Z28	2dr.cpe.	V8/5.7/M6	25 880	27 355	28 830
• Destination charge: 655 $						
Caprice	Classic	4dr.sdn.	V8/4.3/A4	21 170	22790	24 415
	Classic	5dr.wgn.	V8/5.7/A4	24 345	26 210	28 080
• Destination charge: 870 $						
Cavalier	Base	2dr.cpe.	L4/2.2/M5	10 740	11 170	11 595
	Base	4dr.sdn.	L4/2.2/M5	11 200	11 650	12 095
	LS	4dr.sdn.	L4/2.2/M5	13 580	14 290	14 995
• Destination charge: 595 $						
Corsica	Base	4dr.sdn.	L4/2.2/A3	14 930	15 710	16 485
• Destination charge: 595 $						
Corvette	LT1	3dr.cpe.	V8/5.7/M6	38 980	42 510	46 040
	LT1	2dr.con.	V8/5.7/M6	46 165	49 820	53 480
	ZR1	2dr.cpe.	V8/5.7/M6	73 565	79 395	85 225
• Destination charge: 795 $						
Impala	SS	4dr.sdn.	V8/5.7/A4	24 330	26 145	27 965
• Destination charge: 870 $						
Lumina	base	4dr.sdn.	V6/3.1/A4	16 830	18 090	19 348
	LS sed.	2dr.sdn.	V6/3.1/A4	17 875	19 210	20 548
• Destination charge: 760 $						
Lumina APV Ut.						
	base	5dr.wgn.	V6/3.1/A3	19 615	20 730	21 850
	base	cargo .	V6/3.1/A4	17 260	18 245	19 230
• Destination charge: 745 $						
Monte Carlo						
	LS	2dr.cpe.	V6/3.1/A4	18 640	20 070	21 498
	Z34	2dr.cpe.	V6/3.4/A4	20 460	22 030	23 598
• Destination charge: 760 $						

CHRYSLER

Model	Trim	Type	Engine	mini	avrg	maxi
Cirrus	LX	4dr.sdn.	V6/2.5/A4	19 465	20 730	21 995
	LXi	4dr.sdn.	V6/2.5/A4	22 120	23 555	24 995
• Destination charge: 695 $						
Concorde		4dr.sdn.	V6/3.3/A4	21 175	22 545	23 920
• Destination & handling charge: 760 $						
Intrepid		4dr.sdn.	V6/3.3/A4	18 205	19 355	20 510
	ES	4dr.sdn.	V6/3.3/A4	21 390	22 785	24 180
• Destination & handling charge: 760 $						
Le Baron	GTC	2dr.con.	V6/3.0/A4	20 960	22 290	23 615
• Destination & handling charge: 625 $						
New-Yorker		4dr.sdn.	V6/3.5/A4	26 255	28 470	30 680
LHS		4dr.sdn.	V6/3.5/A4	30 910	33 570	36 225
• Destination & handling charge: 760 $						
Town & Country						
	4x2	4dr.wgn.	V6/3.3/A4	31 240	33 335	35 430
	4x4	4dr.wgn.	V6/3.3/A4	32 950	35 180	37 405
• Destination & handling charge: 810 $						

DODGE

Model	Trim	Type	Engine	mini	avrg	maxi
Caravan	base	4dr.wgn.	L4/2.5/M5	16 055	17 030	18 010
	SE	4dr.wgn.	L4/2.5/A3	17 700	18 805	19 910
	LE	4dr.wgn.	V6/3.3/A3	22 250	23 695	25 140
	ES	4dr.wgn.	V6/3.3/A3	23 785	25 345	26 905
• Destination & handling charge: 810 $						
Neon	Highline	2dr.cpe.	L4/2.0/M5	12 245	12 700	13 150
	Sport	2dr.cpe.	L4/2.0/M5	14 705	15 375	16 045
	base	4dr.sdn.	L4/2.0/M5	10 875	11 255	11 640
	Highline	4dr.sdn.	L4/2.0/M5	12 245	12 700	13 150
	Sport	4dr.sdn.	L4/2.0/M5	14 505	15 160	15 820
• Destination & handling charge: 600 $						
Grand Caravan						
	base	4dr.wgn.	V6/3.0/A4	18 895	20 090	21 290
	SE	4dr.wgn.	V6/3.3/A4	19 370	20 600	21 835
	SE 4x4	4dr.wgn.	V6/3.3/A4	23 780	25 340	26 900
	LE	4dr.wgn.	V6/3.3/A4	23 340	24 865	26 390
	LE 4x4	4dr.wgn.	V6/3.3/A4	25 880	27 600	29 320
	ES	4dr.wgn.	V6/3.3/A4	24 870	26 515	28 160
	Es 4x4	4dr.wgn.	V6/3.3/A4	27 280	29 105	30 935
• Destination & handling charge: 810 $						
Spirit	base	4dr.sdn.	L4/2.5/M5	15 080	15 750	16 425
• Destination & handling charge: 600 $						
Stealth	R/T	3dr.cpe.	V6/3.0/M5	30 175	32 320	34 465
	R/TTbo	3dr.cpe.	V6T/3.0/M5	41 705	44 860	48 015
• Destination & handling charge: 600 $						
Viper		2dr.con.	V10/8.0/M6	67 630	71 765	77 900
• Destination & handling charge: 800 $						

DODGE (Imported)

Model	Trim	Type	Engine	mini	avrg	maxi
Colt	base	2dr.cpe.	L4/1.5/M5	11 225	11 610	11 995
	GL	2dr.cpe.	L4/1.5/M5	11 970	12 425	12 885
• Destination & handling charge: 415 $						

DODGE (Utility)

Model	Trim	Type	Engine	mini	avrg	maxi
Dakota 4x2						
	regular	2dr.p-u.	L4/2.5/M5	13 085	13 180	13 275
	extd	2dr.p-u.	L4/2.5/M5	14 260	14 755	15 255
	ClubCab	2dr.p-u.	L4/2.5/M5	15 850	16 470	17 090
Dakota 4x4						
	regular	2dr.p-u.	V6/3.9/M5	16 405	17 035	17 670
	extd	2dr.p-u.	V6/3.9/M5	16 785	17 445	18 110
	Club Cab	2dr.p-u.	V6/3.9/M5	19 034	19 865	20 700
• Destination & handling charge: 660 $						
Ram1500 4x2						
WSregular cab		2dr.p-u.	V6/3.9/M5	14 130	14 960	15 790
LT regular cab		2dr.p-u.	V6/3.9/M5	15 510	16 765	18 025
ST reg cab		2dr.p-u.	V8/5.2/M5	16 085	17 400	18 715
ST reg crew cab		2dr.p-u.	V8/5.2/M5	18 170	19 675	21 185
Laramie SLT extd crew cab			V8/5.9/M5	21 275	23 085	24 900
Power Ram 4x4 1500						
base regular		2dr.p-u.	V8/5.2/M5	18 715	19 790	20 865
extd		2dr.p-u.	V8/5.2/M5	19 230	20 350	21 475

1995 CANADIAN PRICES

Model	Trim	Type	Engine	Price canadian $ mini	avrg	maxi
ST extd		2dr.p-u.	V8/5.2/M5	19 805	20 985	22 165
Laramie SLT		2dr.p-u.	V8/5.9/M5	22 655	24 115	25 575
2500						
ST extd crew cab		2dr.p-u.	V8/5.9/M5	22 665	24 580	26 500
Laramie SLT extd crew cab			V8/5.9/M5	25 200	27 365	29 535
3500						
base		2dr.p-u.	V8/5.9/M5	23 020	24 390	25 765
ST		2dr.p-u.	V8/5.9/M5	24 015	25 480	26 950
SLT		2dr.p-u.	V8/5.9/M5	26 240	27 925	29 615

• **Destination & handling charge: 870 $**

EAGLE

Model	Trim	Type	Engine	mini	avrg	maxi
Summit	DL	2dr.cpe.	L4/1.5/M5	11 225	11 610	11 995
	ES	2dr.cpe.	L4/1.5/M5	11 970	12 425	12 885
	DL	4dr.wgn.	L4/1.8/M5	16 900	17 725	18 555
	LX	4dr.wgn.	L4/2.4/M5	19 000	19 875	20 750
	T.I.	4dr.wgn.	L4/2.4/M5	20 440	21 415	22 395

• **Destination & handling charge: 415 $**

Model	Trim	Type	Engine	mini	avrg	maxi
Talon						
	ESI	3dr.cpe.	L4/2.0/M5	18 915	19 170	19 425
	TSI 4WD	3dr.cpe.	L4T/2.0/M5	26 420	27 390	28 360

• **Destination & handling charge: 650 $**

Model	Trim	Type	Engine	mini	avrg	maxi
Vision	ESi	4dr.sdn.	V6/3.3/A4	21 125	21 765	22 410
	TSi	4dr.sdn.	V6/3.5/A4	24 760	25 680	26 605

• **Destination & handling charge: 760 $**

FORD

Model	Trim	Type	Engine	mini	avrg	maxi
Aerostar 4x2						
	XLTregular	4dr.van.	V6/3.0/A4	17 875	19 035	20 195
	XLT extd	4dr.van.	V6/3.0/A4	20 350	21 670	22 995
Aerostar 4x4						
	XLT extd	4dr.van.	V6/4.0/A4	21 235	22 615	23 995

• **Destination & handling charge: 845 $**

Model	Trim	Type	Engine	mini	avrg	maxi
Aspire	base	3dr.sdn.	L4/1.3/M5	9 845	10 170	10 495
Aspire		5dr.sdn.	L4/1.3/M5	10 335	10 665	10 995
	SE	3dr.sdn.	L4/1.3/M5	10 675	11 035	11 395

• **Destination & handling charge: 660 $**

Model	Trim	Type	Engine	mini	avrg	maxi
Escort	LX	3dr.cpe.	L4/1.9/M5	11 920	12 455	12 995
	LX	5dr.sdn.	L4/1.9/M5	12 325	12 890	13 455
	LX	5dr.wgn.	L4/1.9/M5	12 325	12 890	13 455
	GT	3dr.sdn.	L4/1.8/M5	13 380	13 785	14 295

• **Destination & handling charge: 660 $**

Model	Trim	Type	Engine	mini	avrg	maxi
Contour	GL	4dr.sdn.	L4/2.3/M5	15 530	16 210	16 895
	LX	4dr.sdn.	L4/2.3/M5	16 340	17 065	17 795
	SE	4dr.sdn.	V6/2.5/M5	18 135	18 965	19 795

• **Destination & handling charge: 720 $**

Model	Trim	Type	Engine	mini	avrg	maxi
Crown Victoria						
	S	4dr.sdn.	V8/4.6/A4	21 300	22 245	23 195
	base	4dr.sdn.	V8/4.6/A4	22 015	23 005	23 995
	LX	4dr.sdn.	V8/4.6/A4	23 545	24 620	25 695

• **Destination & handling charge: 900 $**

Model	Trim	Type	Engine	mini	avrg	maxi
Mustang	Base	2dr.cpe.	V6/3.8/M5	15 485	16 140	16 795
	Base	2dr.con.	V6/3.8/M5	25 130	26 310	27 495
	GT	3dr.cpe.	V8/5.0/M5	18 775	19 935	21 095
	GT	3dr.con.	V8/5.0/A4	24 425	26 060	27 695

• **Destination & handling charge: 680 $**

Model	Trim	Type	Engine	mini	avrg	maxi
Probe		3dr.cpe.	L4/2.2/M5	15 910	16 600	17 295
	SE	3dr.cpe.	L4/2.2/M5	18 720	19 555	20 395
	GT	3dr.cpe.	V6/2.5/M5	19 275	20 135	20 995

• **Destination & handling charge: 720 $**

Model	Trim	Type	Engine	mini	avrg	maxi
Taurus	GL	4dr.sdn.	V6/3.0/A4	17 960	19 125	20 295
	GL	5dr.wgn.	V6/3.8/A4	17 960	19 125	20 295
	LX	4dr.sdn.	V6/3.0/A4	20 735	22 115	23 495
	LX	5dr.wgn.	V6/3.8/A4	20 735	22 115	23 495
	SE	4dr.sdn.	V6/3.0/A4	17 960	19 125	20 295
	SHO	4dr.sdn.	V6/3.0/M5	26 980	28 835	30 695

• **Destination & handling charge: 785 $**

Model	Trim	Type	Engine	mini	avrg	maxi
T-bird	LX	2dr.cpe.	V6/3.8/A4	20 160	21 325	22 495
	SC	2dr.cpe.	V6C/3.8/M5	25 160	26 675	28 195

• **Destination & handling charge: 800 $**

Model	Trim	Type	Engine	mini	avrg	maxi
Windstar	base	4dr.van.	V6/3.0/A4	18 315	19 455	20 595
	GL	4dr.van.	V6/3.0/A4	19 885	21 140	22 395
	LX	4dr.van.	V6/3.8/A4	22 565	24 030	25 495

• **Destination & handling charge: 850 $**

FORD (Utility)

Model	Trim	Type	Engine	mini	avrg	maxi
Bronco	XL	2dr.wgn.	L6/4.9/M5	22 185	23 140	24 095
	XLT	2dr.wgn.	L6/4.9/M5	24 070	25 130	26 195
	E.B.	2dr.wgn.	L6/4.9/M5	27 210	28 450	29 695

• **Destination & handling charge: 900 $**

Model	Trim	Type	Engine	mini	avrg	maxi
Explorer 4x2						
	*XL	2dr.wgn.	V6/4.0/M5	18 950	20 120	21 295
Explorer 4x4						
	*XLT	4dr.wgn.	V6/4.0/M5	24 365	25 980	27 595
	*Ltd	4dr.wgn.	V6/4.0/M5	31 235	33 365	35 495

• **Destination & handling charge: 700 $**

Model	Trim	Type	Engine	mini	avrg	maxi
Pick-ups 4x2						
F 250 XL reg. cab.		2dr.p-u.	V8/5.0/M5	16 565	17 930	19 295
F 250 XLT cab.all.		2dr.p-u.	V8/5.0/M5	19 995	21 695	23 695
Pick-ups 4x4						
F 250 XL reg. cab.		2dr.p-u.	V8/5.0/M5	20 330	22 060	23 795
F 250 XL crew cab		2dr.p-u.	V8/5.0/M5	21 920	23 805	25 695

• **Destination & handling charge: 850 $**

Model	Trim	Type	Engine	mini	avrg	maxi
Ranger 4x2						
	XL R.C	2dr.p-u.	L4/2.3/M5	11 310	11 800	12 295

• **Destination & handling charge: 850 $**

GEO

Model	Trim	Type	Engine	mini	avrg	maxi
Metro	Base	3dr.sdn.	L3/1.0/M5	9 260	9 630	9 995
	LSI	3dr.sdn.	L3/1.0/M5	9 490	9 670	10 245
	Base	4dr.sdn.	L4/1.3/M5	10 185	10 820	10 995
	LSI	4dr.sdn.	L4/1.3/M5	10 415	10 830	11 245

• **Destination charge: 500 $**

Model	Trim	Type	Engine	mini	avrg	maxi
Tracker 4x2						
	Soft top	2dr.con.	L4/1.6/M5	12 085	12 315	13 020
Tracker 4x4						
	Soft top	2dr.con.	L4/1.6/M5	13 880	14 420	14 955
	Soft top	LSI		15 770	16 380	16 990
	Hard top	2dr.sdn.	L4/1.6/M5	13 780	14 315	14 845
	Hard top	LSI		15 670	16 275	16 880

• **Destination charge: 500 $**

Chevrolet/GMC (Utility)

Model	Trim	Type	Engine	mini	avrg	maxi
Astro 4x2						
	Cargo	4dr.van.	V6/4.3/A4	19 540	20 655	21 770
	pass.	4dr.van.	V6/4.3/A4	20 130	21 280	22 425
Astro	4x4	4dr.van.	V6/4.3/A4	22 240	23 510	24 775

• **Destination charge: 820 $**

1995 CANADIAN PRICES

*1994 Price

Model	Trim	Type	Engine	Price canadian $ mini	avrg	maxi
Jimmy	4x2					
	SL	3dr.wgn.	V6/4.3/M5	20 410	21 570	22 740
	SL	5dr.wgn.	V6/4.3/M5	21 200	22 410	25 465
	SLS	3dr.wgn.	V6/4.3/M5	22 860	24 160	25 465
	SLS	5dr.wgn.	V6/4.3/M5	24 540	25 940	27 340
	SLT	5dr.wgn.	V6/4.3/M5	26 850	28 380	29 915
Jimmy	4x4					
	SL	3dr.wgn.	V6/4.3/M5	22 255	23 525	24 795
	SLE	5dr.wgn.	V6/4.3/M5	23 250	24 575	25 905
	SLS	3dr.wgn.	V6/4.3/M5	25 050	26 480	27 910
	SLS	5dr.wgn.	V6/4.3/M5	26 585	28 100	29 620
	SLT	5dr.wgn.	V6/4.3/M5	28 780	30 415	32 050

• **Destination charge**: 670 $

Model	Trim	Type	Engine	mini	avrg	maxi
Pick-ups S-10 4x2						
reg. cab. base		2dr.p-u.	L4/2.2/M5	12 560	12 980	13 400
crew cab	LS	2dr.p-u.	L4/2.2/M5	14 235	16 065	15 895
reg. cab. SS		2dr.p-u.	V6/4.3/A4	16 505	17 465	18 430
Pick-ups T-10 4x4						
cab. reg base		2dr.p-u.	V6/4.3/M5	17 290	17 875	18 465
crew cab LS		2dr.p-u.	V6/4.3/M5	18 530	19 610	20 690

• **Destination charge**: 660 $

Model	Trim	Type	Engine	mini	avrg	maxi
Blazer	4x2	2dr.wgn.	V6/4.3/M5	20 345	21 505	22 665
		4dr.wgn.	V6/4.3/M5	21 135	22 340	23 545
Blazer	4x4	2dr.wgn.	V6/4.3/M5	22 050	22 310	24 565
		4dr.wgn.	V6/4.3/M5	23 045	24 360	25 675

• **Destination charge**: 670 $

Pick-ups *All models are available in those trim: Cheyenne, Sierra & Silverado.

Model	Type	Engine	mini	avrg	maxi
4x2					
C-1500reg crew cab	2dr.p-u.	V6/4.3/M5	17 230	18 515	19 805
C-1500extd crew cab	2dr.p-u.	V8/5.0/M5	18 100	19 450	20 805
C-1500regular cab	2dr.p-u.	V8/5.0/M5	14 835	15 650	16 470
C-2500 reg cab.all.	2dr.p-u.	V8/5.0/M5	19 105	20 530	21 960
C-2500 extd crew cab	2dr.p-u.	V8/5.0/M5	20 165	21 670	23 175
4x4					
K-1500 regular cab .	2dr.p-u.	V6/4.3/M5	17 925	19 260	20 600
K-1500 extd cab	2dr.p-u.	V8/5.0/M5	18 185	19 540	20 900
K-1500 reg crew cab	2dr.p-u.	V8/5.7/M5	19 580	21 040	22 505
K-1500 extd crew cab	2dr.p-u.	V8/5.7/M5	20 450	21 975	23 505
K-2500 extd cab	2dr.p-u.	V8/5.7/M5	18 495	19 875	21 260
K-2500 reg crew cab	2dr.p-u.	V8/5.7/M5	20 935	22 495	24 060

• **Destination charge**: 870 $

Model	Trim	Type	Engine	mini	avrg	maxi
Tahoe	Base	3dr.wgn.	V8/5.7/A4	22 650	24 340	26 035
	LS	3dr.wgn.	V8/5.7/A4	26 835	28 840	30 840

• **Destination charge**: 670 $

Model	Trim	Type	Engine	mini	avrg	maxi
Safari	4X2					
Base cargo		4dr.van.	V6/4.3/A4	19 540	20 655	21770
SLX		4dr.van.	V6/4.3/A4	20 130	21 280	22 425
SLE		4dr.van.	V6/4.3/A4	22 310	23 580	24 855
SLT		4dr.van.	V6/4.3/A4	24 740	26 150	27 560
Safari	4X4					
SLX		4dr.van.	V6/4.3/A4	22 240	23 510	24 775
SLE		4dr.van.	V6/4.3/A4	24 420	25 810	27 205
SLT		4dr.van.	V6/4.3/A4	26 940	28 475	30 010

• **Destination charge**: 820 $

Model	Trim	Type	Engine	Price canadian $ mini	avrg	maxi
Suburban						
C1500	2WD SL	5dr.wgn.	V8/5.7/A4	22 190	23 845	25 500
C1500	2WD SLE	5dr.wgn.	V8/5.7/A4	27 875	29 955	32 040
C2500	2WD SL	5dr.wgn.	V8/5.7/A4	23 145	24 870	26 600
C2500	2WD SLE	5dr.wgn.	V8/5.7/A4	28 820	30 970	33 125
K1500	4WD SL	5dr.wgn.	V8/5.7/A4	24 595	26 430	28 270
K1500	4WD SLE	5dr.wgn.	V8/5.7/A4	30 285	32 545	34 810
K2500	4WD SL	5dr.wgn.	V8/5.7/A4	25 490	27 395	29 300
K2500	4WD SLE	5dr.wgn.	V8/5.7/A4	31 170	33 495	35 825

*Cheyenne versions are similar in price as the SL versions and the Silverado are identical to the SLE version.

• **Destination charge**: 870 $

HONDA

Model	Trim	Type	Engine	mini	avrg	maxi
Accord						
Coupe	LX	3dr.sdn.	L4 2.2 M5	17 530	18 460	19 395
	LX	3dr.sdn.	L4/2.2/A4	18 650	19 520	20 395
	EX-R	3dr.sdn.	L4/2.2/M5	23 405	24 550	25 695
	EX-R	3dr.sdn.	L4/2.2/A4	24 305	25 500	26 695
Sedan	LX	4dr.sdn.	L4 2.2 M5	18 290	19 140	19 995
	LX	4dr.sdn.	L4/2.2/A4	19 190	20 090	20 995
	EX	4dr.sdn.	L4/2.2/M5	20 355	21 325	22 295
	EX	4dr.sdn.	L4/2.2/A4	21 255	22 275	23 295
	*EX	4dr.sdn.	V6 2.7 A4	n.a.	n.a.	n.a.
	EX	5dr.wgn.	L4/2.2/A4	22 690	23 790	24 895
	EX-R	4dr.sdn.	L4/2.2/M5	24 125	25 310	26 495
	EX-R	4dr.sdn.	L4/2.2/A4	25 025	26 260	27 495
	*EX-R	4dr.sdn.	V6 2.7 A4	n.a.	n.a.	n.a.

• **Destination & handling charge**: 695 $

Model	Trim	Type	Engine	mini	avrg	maxi
Civic	CX	3dr.sdn.	L4/1.5/M5	10 315	10 655	10 995
	CX	3dr.sdn.	L4/1.5/A4	11 405	11 780	12 155
	SI	3dr.sdn.	L4/1.6/M5	15 515	16 255	16 995
	SI	3dr.sdn.	L4/1.6/A4	16 425	17 210	17 995
	LX	4dr.sdn.	L4/1.5/M5	12 775	13 385	13 995
	LX	4dr.sdn.	L4/1.5/A4	13 690	14 340	14 995
	EX	4dr.sdn.	L4/1.5/M5	15 240	15 965	16 695
	EX	4dr.sdn.	L4/1.5/A4	16 155	16 925	17 695

• **Destination & handling charge**: 695 $

Model	Trim	Type	Engine	mini	avrg	maxi
Del Sol	Si	2dr.cpe.	L4/1.6/M5	17 980	18 840	19 695
	Si	2dr.cpe.	L4/1.6/A4	18 890	19 790	20 695
	vtec	2dr.cpe.	L4 1.6 M5	20 720	21 705	22 695

• **Destination & handling charge**: 695 $

Model	Trim	Type	Engine	mini	avrg	maxi
Prelude	SR	2dr.cpe.	L4/2.3/M5	23 515	24 855	26 195
	SR	2dr.cpe.	L4/2.3/A4	24 410	25 800	27 195
	SR-V	2dr.cpe.	L4/2.2/M5	25 935	27 415	28 895

• **Destination & handling charge**: 695 $

HYUNDAI

Model	Trim	Type	Engine	mini	avrg	maxi
Accent	L	3dr.sdn.	L4/1.5/M5	8 455	8 725	8 995
	L	3dr.sdn.	L4/1.5/A4	9 115	9 405	9 695
	GL	3dr.sdn.	L4/1.5/M5	9 585	9 890	10 195
	GL	3dr.sdn.	L4/1.5/A4	10 240	10 565	10 895
	L	4dr.sdn.	L4/1.5/M5	9 395	9 695	9 995
	L	4dr.sdn.	L4/1.5/A4	10 055	10 375	10 695
	GL	4dr.sdn.	L4/1.5/M5	10 055	10 375	10 695
	GL	4dr.sdn.	L4/1.5/A4	10 710	11 050	11 395

• **Destination & handling charge**: 795 $

*1994 Price

1995 CANADIAN PRICES

Model	Trim	Type	Engine	Price canadian $ mini	avrg	maxi
Elantra	GL	4dr.sdn.	L4/1.6/M5	11 220	11 705	12 195
	GL	4dr.sdn.	L4/1.8/A4	12 300	12 835	13 370
	GLS	4dr.sdn.	L4/1.8/M5	13 060	13 625	14 195
	GLS	4dr.sdn.	L4/1.8/A4	13 840	14 440	15 045

• Destination & handling charge: 795 $

Model	Trim	Type	Engine	mini	avrg	maxi
Scoupe	base	2dr.cpe.	L4/1.5/M5	11 125	11 610	12 095
	base	2dr.cpe.	L4/1.5/A4	11 770	12 280	12 795
	LS	2dr.cpe.	L4/1.5/M5	12 415	12 955	13 495
	LS	2dr.cpe.	L4/1.5/A4	13 060	13 625	14 195

• Destination & handling charge: 795 $

Model	Trim	Type	Engine	mini	avrg	maxi
Sonata						
	GL	4dr.sdn.	L4/2.0/M5	13 745	14 520	15 295
	GL	4dr.sdn.	L4/2.0/A4	14 580	15 405	16 230
	GL	4dr.sdn.	V6/3.0/A4	15 540	16 415	17 295
	GLS	4dr.sdn.	L4/2.0/A4	16 985	17 840	18 695
	GLS	4dr.sdn.	V6/3.0/A4	17 625	18 660	19 695

• Destination & handling charge: 795 $

INFINITI

Model	Trim	Type	Engine	mini	avrg	maxi
G20	base	4dr.sdn.	L4/2.0/M5	23 930*	25 185*	26 440*
	G-20T	4dr.sdn.	L4/2.0/M5	24 835*	26 135*	27 440*

• Destination & handling charge: sans frais

Model	Trim	Type	Engine	mini	avrg	maxi
J-30		4dr.sdn.	V6/3.0/A4	40 275*	42 635*	45 000*
	J-30t	4dr.sdn.	V6/3.0/A4	42 960*	45 480*	48 000*

• Destination & handling charge: sans frais

Model	Trim	Type	Engine	mini	avrg	maxi
Q45	base	4dr.sdn.	V8/4.5/A4	63 360*	67 680*	72000*

• Destination & handling charge: sans frais

ISUZU

Model	Trim	Type	Engine	mini	avrg	maxi
Rodeo						
	S 4x4	5dr.wgn.	L4/2.6/M5	20 435*	21 325*	22 215*
	LS 4x4	5dr.wgn.	V6/3.1/M5	23 470*	24 560*	25 560*
Trooper						
	S	5dr.wgn.	V6/3.2/M5	23 630*	24 725*	25 825 *
	LS	5dr.wgn.	V6/3.2/A4	30 250*	31 835*	33 425 *

• Destination & handling charge: 525 $

JAGUAR

Model	Trim	Type	Engine	mini	avrg	maxi
XJ6	Sovereign	4dr.sdn.	L6/4.0/A4	66 000	70 500	75 000
	Vanden Plas	4dr.sdn.	L6/4.0/A4	73 040	78 020	83 000
XJR super charged		4dr.sdn.	L6C/4.0/A4	74 800	79 900	85 000
XJ12	base	4dr.sdn.	V12/6.0/A4	84 480	90 240	96 000

• Destination & handling charge: sans frais

Model	Trim	Type	Engine	mini	avrg	maxi
XJS	4.0L	2dr.cpe.	L6/4.0/M5	65 030	73 630	73 900
	6.0L	2dr.cpe.	V12/6.0/A4	80 870	86 385	91 900
XJS	4.0L	2dr.con.	L6/4.0/A4	74 360	79 430	84 500
	6.0L	2dr.con.	V12/6.0/A4	92 310	98 605	104 900

• Destination & handling charge: sans frais

JEEP

Model	Trim	Type	Engine	mini	avrg	maxi
Cherokee 4x2						
	base	2dr.wgn.	L4/2.5/M5	16 515	17 030	17 555
	base	4dr.wgn.	L4/2.5/M5	18 290	18 940	19 590
	Sport	2dr.wgn.	L6/4.0/M5	17 920	18 570	19 225
	Sport	4dr.wgn.	L6/4.0/M5	19 685	20 470	21 260
	Country	2dr.wgn.	L6/4.0/M5	20 830	21 730	22 630
	Country	4dr.wgn.	L6/4.0/M5	22 785	23 875	24 965
Cherokee 4x4						
	SE	2dr.wgn.	L6/4.0/M5	18 450	19 110	19 775
	SE	4dr.wgn.	L6/4.0/M5	19 575	20 265	20 955
	Sport	2dr.wgn.	L6/4.0/M5	19 950	20 695	21 445
	Sport	4dr.wgn.	L6/4.0/M5	20 970	21 795	22 625
	Country	2dr.wgn.	L6/4.0/A4	22 115	23 055	23 995
	Country	4dr.wgn.	L6/4.0/A4	24 070	25 200	26 330

• Destination & handling charge: 670 $

Model	Trim	Type	Engine	mini	avrg	maxi
Grand Cherokee 4x2						
	SE	4dr.wgn.	L6/4.0/M5	26 425	27 390	28 355
	Laredo	4dr.wgn.	L6/4.0/A4	27 645	27 585	29 530
	Limited	4dr.wgn.	L6/4.0/A4	32 340	33 640	34 940
Grand Cherokee 4x4						
	SE	4dr.wgn.	L6/4.0/M5	27 410	28 330	29 255
	Laredo	4dr.wgn.	L6/4.0/A4	28 425	29 425	30 430
	Limited	4dr.wgn.	L6/4.0/A4	33 600	34 940	36 280

• Destination & handling charge: 670

Model	Trim	Type	Engine	mini	avrg	maxi
YJ	SE	2dr.con.	L4/2.5/M5	16 480	17 080	17 680
	Sahara	2dr.con.	L4/2.5/M5	19 260	20 070	20 880

• Destination & handling charge: 670

LEXUS

Model	Trim	Type	Engine	mini	avrg	maxi
ES300	base	4dr.sdn.	V6/3.0/A4	39 335	42 015	44 700

• Destination & handling charge: 795

Model	Trim	Type	Engine	mini	avrg	maxi
GS300	base	4dr.sdn.	L6/3.0/A4	54 315	59 105	63 900

• Destination & handling charge: 795

Model	Trim	Type	Engine	mini	avrg	maxi
LS400	base	4dr.sdn.	V8/4.0/A4	64 515	70 205	75 900
SC400	base	2dr.cpe.	V8/4.0/A4	61 285	66 690	72 100

• Destination & handling charge: 795

LINCOLN

Model	Trim	Type	Engine	mini	avrg	maxi
Continental (1994 price)						
	Executive	4dr.sdn.	V6/3.8/A4	35 635	38 065	40 495
	SE	4dr.sdn.	V6/3.8/A4	34 050	36 370	38 695
	Signature	4dr.sdn.	V6/3.8/A4	36 690	39 190	41 695

• Destination & handling charge: 895

Model	Trim	Type	Engine	mini	avrg	maxi
Mark VIII		2dr.sdn.	V8/4.6/A4	44 885	47 390	49 895

• Destination & handling charge: 895

Model	Trim	Type	Engine	mini	avrg	maxi
Town Car						
	Executive	4dr.sdn.	V8/4.9/A4	38 270	40 780	43 295
	Signature	4dr.sdn.	V8/4.9/A4	39 950	42 570	45 195
	Cartier	4dr.sdn.	V8/4.9/A4	42 440	44 815	47 195

• Destination & handling charge: 895

MAZDA

Model	Trim	Type	Engine	mini	avrg	maxi
323	S	3dr.sdn.	L4/1.6/M5	10 000	10 310	10 620
	GS	3dr.sdn.	L4/1.5/M5	13 230	13 770	14 310
	LS	3dr.sdn.	L4/1.8/M5	13 515	14 520	15 530

• Destination & handling charge: 500

Model	Trim	Type	Engine	mini	avrg	maxi
Protegé	S	4dr.sdn.	L4/1.8/M5	12 820	13 395	13 970
	LX	4dr.sdn.	L4/1.8/M5	16 000	16 555	17 110
	ES	4dr.sdn.	L4/1.8/M5	17 130	17 775	18 420

• Destination & handling charge: 500

Model	Trim	Type	Engine	mini	avrg	maxi
626 Cronos						
	DX	4dr.sdn.	L4/2.0/M5	17 330*	18 260*	19 195*
	DX	4dr.sdn.	L4/2.0/A4	18 175*	19 160*	20 145*
	LX	4dr.sdn.	L4/2.0/M5	19 905*	20 975*	22 050*
	LX	4dr.sdn.	L4/2.0/A4	20 750*	21 875*	23 000*
	LX	4dr.sdn.	V6/2.5/M5	21 825*	23 015*	24 205*
	LX	4dr.sdn.	V6/2.5/A4	22 670*	23 910*	25 155*
	ES	4dr.sdn.	V6/2.5/M5	25 685*	27 065*	28 450*

1995 CANADIAN PRICES

*1994 Price

Model	Trim	Type	Engine	Price canadian $ mini	avrg	maxi
	ES	4dr.sdn.	V6/2.5/A4	26 530*	27 965*	29 400*
• Destination & handling charge: 500 $						
MX-6 Mystère						
	RS	2dr.cpe.	L4/2.0/M5	19 150*	20 180*	21 210*
	RS	2dr.cpe.	L4/2.0/A4	20 040*	21 125*	22 210*
	LS	2dr.cpe.	V6/2.5/M5	22 665*	23 885*	25 105*
	LS	2dr.cpe.	V6/2.5/A4	23 555*	24 830*	26 105*
• Destination & handling charge: 500 $						
Miata						
	MX-5	2dr.con.	L4/1.8/M5	19 325*	20 195*	21 070*
	MX-5	2dr.con.	L4/1.8/A4	21 160*	22 180*	23 205*
• Destination & handling charge: 500 $						
Millenia						
	base	4dr.sdn.	V6/2.5/A4	32 120	34 310	36 500
	S (Miller)	4dr.sdn.	V6/2.3/A4	37 310	39 855	42 400
• Destination & handling charge: 500 $						
MX-3 -Precidia						
	RS	3dr.cpe.	L4/1.6/M5	14 635*	15 290*	15 945*
	RS	3dr.cpe.	L4/1.6/A4	15 450*	16 160*	16 870*
	GS	3dr.cpe.	V6/1.8/M5	16 870*	17 625*	18 380*
	GS	3dr.cpe.	V6/1.8/A4	17 685*	18 495*	19 305*
• Destination & handling charge: 500 $						
929-Serenia						
	base	4dr.sdn.	V6/3.0/A4	35 480*	38 390*	41 300 *
• Destination & handling charge: 500 $						
RX-7						
	base	2dr.cpe.	R2/1.3/M5	40 175*	42 725*	45 280*
	Touring	2dr.cpe.	R2/1.3/M5	42 800*	45 090*	47 380 *
	Premium	2dr.cpe.	R2/1.3/M5	44 750*	47 175*	49 605*
• Destination & handling charge: 500 $						
Pick-ups 4x2						
B-2300						
base short deck		2dr.p-u.	L4/2.3/M5	11 090*	11 430*	11 775*
Se short deck.		2dr.p-u.	L4/2.3/M5	12 200*	12 890*	13 585*
base extd deck		2dr.p-u.	L4/2.3/M5	11 285*	11 915*	12 550*
Se extd deck		2dr.p-u.	L4/2.3/M5	12 465*	13 170*	13 875*
base crew cab		2dr.p-u.	L4/2.3/M5	13 015*	13 765*	14 520*
B-3000						
base short deck		2dr.p-u.	V6/3.0/M5	11 945	12 335	12 730
SE short deck		2dr.p-u.	V6/3.0/M5	13 575	14 345	15 120
base extd deck		2dr.p-u.	V6/3.0/M5	12 195	12 880	13 570
SE extd deck		2dr.p-u.	V6/3.0/M5	13 945	14 735	15 525
base crew cab		2dr.p-u.	V6/3.0/M5	13 900	14 705	15 510
SE crew cab		2dr.p-u.	L4/2.3/M5	15 245	16 135	17 030
B-4000						
base short deck		2dr.p-u.	V6/4.0/M5	12 300	12 725	13 150
SE short deck		2dr.p-u.	V6/4.0/M5	13 935	14 735	15 540
base extd deck		2dr.p-u.	V6/4.0/M5	12 555	13 270	13 990
Se extd deck		2dr.p-u.	V6/4.0/M5	14 305	15 125	15 945
base crew cab		2dr.p-u.	V6/4.0/M5	14 255	15 090	15 930
Se crew cab		2dr.p-u.	V6/4.0/M5	15 825	16 765	17 705
• Destination & handling charge: 900 $						
Pick-ups 4x4						
B-2300						

Model	Trim	Type	Engine	Price canadian $ mini	avrg	maxi
base short deck		2dr.p-u.	L4/2.3/M5	15 630	16 165	16 705
B-3000						
SE short deck		2dr.p-u.	V6/3.0/M5	17 710	18 765	19 825
base crew cab		2dr.p-u.	V6/3.0/M5	17 450	18 495	19 545
SE crew cab		2dr.p-u.	V6/3.0/M5	19 375	20 555	21 735
B 4000						
base short deck		2dr.p-u.	V6/4.0/M5	16 365	16 960	17 553
SE short deck		2dr.p-u.	V6/4.0/M5	18 035	19 120	20 210
base crew cab		2dr.p-u.	V6/3.0/M5	17 820	18 900	19 980
Se crew cab		2dr.p-u.	V6/3.0/M5	19 610	20 815	22 020
• Destination & handling charge: 900 $						
MPV 4x2						
Base	5 p.	4dr.wgn.	L4/2.6/A4	19 910	20 730	21 555
LX	7 p.	4dr.wgn.	V6/3.0/A4	23 025	24 745	26 490
LX	8 p.	4dr.wgn.	V6/3.0/A4	24 700	26 005	27 310
LTD	7 p.	4dr.wgn.	V6/3.0/A4	30 420	32 025	33 635
MPV 4x4						
Base	7 p.	4dr.wgn.	V6/3.0/A4	25 665	27 055	28 445
LX	7 p.	4dr.wgn.	V6/3.0/A4	27 090	28 575	30 060
LTD	7p.	4dr.wgn.	V6/3.0/A4	33 525	35 365	37 205
• Destination & handling charge: 500 $						

MERCEDES BENZ

Model	Trim	Type	Engine	mini	avrg	maxi
C-Class	220 SE	4dr.sdn.	L4/2.2/M5	31 495	33 245	34 995
	220	4dr.sdn.	L4/2.2/M5	37 075	39 135	41 195
	280	4dr.sdn.	L6/2.8/M5	43 875	46 310	48 750
• Destination & handling charge: 750 $						
E-Class						
	300 D	4dr.sdn.	L6D/3.0/A4	50 395	53 195	55 995
	320 E	4dr.sdn.	L6/3.2/A4	54 855	57 900	60 950
	320	5dr.wgn.	L6/3.2/A4	60 795	64 172	67 550
	320 E	2dr.con.	L6/3.2/A4	82 410	88 030	93 650
	320 CE	2dr.cpe.	L6/3.2/A4	71 150	76 000	80 850
	420	4dr.sdn.	V8/4.2/A4	64 195	68 570	72 950
• Destination & handling charge: 750 $						
S-Class						
	320	4dr.sdn.	L6/3.2/A5	78 320	83 660	89 000
	350 SD	4dr.sdn.	L6TD/3.5/A4	78 320	83 660	89 000
	420 S	4dr.sdn.	V8/4.2/A4	87 910	93 905	99 900
	500 S	4dr.sdn.	V8/5.0/A4	109 910	117 405	124 900
	500 S	2dr.cpe.	V8/5.0/A4	114 310	122 105	129 900
	600 S	4dr.sdn.	V12/6.0/A4	145 645	158 505	171 350
	600 S	2dr.cpe.	V12/6.0/A4	149 725	162 935	176 150
• Destination & handling charge: 750 $						
SL-Class						
	320 SL	2dr.con.	L6/3.2/A5	87 560	93 530	99 500
	500 SL	2dr.con.	V8/5.0/A4	105 695	114 295	122 900
	600 SL	déc 2p.	V12/6.0/A4	130 290	140 895	151 500
• Destination & handling charge: 750 $						

MERCURY

Model	Trim	Type	Engine	mini	avrg	maxi
Cougar	XR7	2dr.cpe.	V8/5.0/A4	19 810	20 950	22 095
• Destination & handling charge: 800 $						
Mystique	GS	4dr.sdn.	L4/2.0/M5	15 910	16 600	17 295
	LS	4dr.sdn.	V6/2.5/M5	19 000	19 845	20 695
• Destination & handling charge: 720 $						
Sable	GS	4dr.sdn.	V6/3.0/A4	18 775	19 985	21 195
	GS	5dr.wgn.	V6/3.0/A4	18 775	19 985	21 195
	LS	4dr.sdn.	V6/3.0/A4	21 375	22 785	24 195
	LS	4dr.wgn.	V6/3.0/A4	21 375	22 785	24 195

1995 CANADIAN PRICES

Model	Trim	Type	Engine	Price canadian $ mini	avrg	maxi
	LTS	4dr.sdn.	V6/3.8/A4	21 375	22 785	24 195

• **Destination & handling charge: 785 $**

Model	Trim	Type	Engine	mini	avrg	maxi
G-Marquis	GS	4dr.sdn.	V8/4.6/A4	23 390	24 440	25 495
	LS	4dr.sdn.	V8/4.6/A4	24 400	25 495	26 595

• **Destination & handling charge: 900 $**

Model	Trim	Type	Engine	mini	avrg	maxi
Villager	GS	4dr.van.	V6/3.0/A4	19 875	21 135	22 395
	LS	4dr.van.	V6/3.0/A4	23 690	25 240	26 795
	Nautica	4dr.van.	V6/3.0/A4	25 250	26 920	28 595

• **Destination & handling charge: 810 $**

NISSAN

Model	Trim	Type	Engine	mini	avrg	maxi
Altima	XE	4dr.sdn.	L4/2.4/M5	17 315	18 050	18 790
	XE	4dr.sdn.	L4/2.4/A4	18 200	19 000	19 790
	GXE	4dr.sdn.	L4/2.4/M5	19 375	20 380	21 390
	GXE	4dr.sdn.	L4/2.4/A4	20 265	21 325	22 390
	SE	4dr.sdn.	L4/2.4/M5	21 520	22 705	23 890
	SE	4dr.sdn.	L4/2.4/A4	22 410	23 650	24 890
	GLE	4dr.sdn.	L4 2.4 A4	24 670	26 030	27 390

• **Destination & handling charge: 895 $**

Model	Trim	Type	Engine	mini	avrg	maxi
Sentra	DLX	2dr.cpe.	L4/1.6/M5	11 545	11 765	11 990
	XE	2dr.cpe.	L4/1.6/M5	12 710	13 100	13 490
	DLX	4dr.sdn.	L4/1.6/M5	11 950	12 320	12 690
	DLX	4dr.sdn.	L4/1.6/A4	13 070	13 515	13 960
	XE	4dr.sdn.	L4/1.6/M5	13 410	14 900	14 390
	XE	4dr.sdn.	L4/1.6/A4	14 240	14 790	15 340
	GXE	4dr.sdn.	L4/1.6/M5	16 115	16 800	17 490

• **Destination & handling charge: 895 $**

Model	Trim	Type	Engine	mini	avrg	maxi
Maxima	GXE	4dr.sdn.	V6/3.0/M5	23 090	24 290	25 490
	GXE	4dr.sdn.	V6 3.0 A4	24 055	25 320	26 590
	SE	4dr.sdn.	V6/3.0/M5	25 425	26 910	28 390
	SE	4dr.sdn.	V6/3.0/A4	26 390	27 940	29 490
	GLE	4dr.sdn.	V6 3.0 A4	27 305	28 900	30 490

• **Destination & handling charge: 895 $**

Model	Trim	Type	Engine	mini	avrg	maxi
240 SX	Coupe SE		L4/2.4M5	21 400	22 445	23 490
			L4/2.4A4	22 280	23 385	24 490
	Coupe LE		L4/2.4M5	23 995	25 240	26 490
			L4/2.4A4	24 875	26 180	27 490

• **Destination & handling charge: 895 $**

Model	Trim	Type	Engine	mini	avrg	maxi
300 ZX	base	3dr.cpe.	V6/3.0/M5	46 265	47 180	48 090
	base	3dr.cpe.	V6/3.0/A4	47 160	48 130	49 090
	Turbo	3dr.cpe.	V6/3.0/M5	53 110	54 100	55 090
	Turbo	3dr.cpe.	V6/3.0/A4	54 005	55 050	56 090
	2+2	3dr.cpe.	V6/3.0/M5	48 775	49 435	50 590
	2+2	3dr.cpe.	V6/3.0/A4	49 670	50 630	51 590
	déc	3dr.cpe.	V6/3.0/M5	53 110	54 100	55 090

• **Destination & handling charge: 895 $**

Model	Trim	Type	Engine	mini	avrg	maxi
Axxess	XE	5dr.wgn.	L4/2.4/M5	17 510	18 150	18 790
	XE	5dr.wgn.	L4/2.4/A4	18 395	19 095	19 790
	SE	5dr.wgn.	L4/2.4/M5	18 695	19 490	20 290
	SE	5dr.wgn.	L4/2.4/A4	19 580	20 435	21 290

• **Destination & handling charge: 895 $**

Model	Trim	Type	Engine	mini	avrg	maxi
Quest	XE	5dr.wgn.	V6/3.0/A4	23 345	23 670	23 990
	GXE	5dr.wgn.	V6/3.0/A4	26 110	27 550	28 990

• **Destination & handling charge: 895 $**

Model	Trim	Type	Engine	mini	avrg	maxi
Pathfinder	XE	2dr.wgn.	V6/3.0/M5	23 090	24 290	25 490
	XE	2dr.wgn.	V6/3.0/A4	24 155	25 575	26 990
	SE	5dr.wgn.	V6/3.0/M5	28 000	29 545	31 090
	SE	5dr.wgn.	V6/3.0/A4	29 065	30 680	32 290
	LE	5dr.wgn.	V6/3.0/A4	26 920	28 400	29 890

• **Destination & handling charge: 895 $**

Model	Trim	Type	Engine	mini	avrg	maxi
Pick-ups Hardbody 2WD						
	Base	2dr.p-u.	L4/2.4/M5	11 545	11 770	11 990
	XE	2dr.p-u.	L4/2.4/M5	12 050	12 420	12 790
K.cab	XE	2dr.p-u.	L4/2.4/M5	13 965	14 475	14 990
K.cab	XE	2dr.p-u.	L4/2.4/A4	14 865	15 425	15 990
K.cab	SE-V6	2dr.p-u.	V6/3.0/M5	17 035	17 760	18 490
K.cab	SE-V6	2dr.p-u.	V6/3.0/A4	17 935	18 710	19 490
Hardbody 4x4						
Short deck	XE	2dr.p-u.	L4/2.4/M5	16 575	17 280	17 990
K.cab	XE V6	2dr.p-u.	V6/3.0/M5	19 760	20 725	21 690
K.cab	SE V6	2dr.p-u.	V6/3.0/A4	20 835	21 860	22 890

• **Destination & handling charge: 895 $**

OLDSMOBILE

Model	Trim	Type	Engine	mini	avrg	maxi
Cutlass Ciera						
	S	4dr.sdn.	V6/3.1/A4	19 490	20 725	21 960

• **Destination charge: 760 $**

Model	Trim	Type	Engine	mini	avrg	maxi
Cutlass Cruiser						
	S	fam 5p.	V6/3.1/A4	20 320	21 610	22 895

• **Destination charge: 760 $**

Model	Trim	Type	Engine	mini	avrg	maxi
Cutlass Supreme						
	Base	2dr.cpe.	V6/3.1/A4	20 575	22 150	23 730
	Base	4dr.sdn.	V6/3.1/A4	20 860	22 460	24 060
	Base	2dr.con.	V6/3.1/A4	27 635	29 755	31 875
	SL	2dr.con.	V6/3.4/A4	27 945	30 085	32 230

• **Destination charge: 760 $**

Model	Trim	Type	Engine	mini	avrg	maxi
98-Regency						
	base	4dr.sdn.	V6/3.8/A4			
	Elite	4dr.sdn.	V6/3.8/A4	29 605	31 815	34 025

• **Destination charge: 870 $**

Model	Trim	Type	Engine	mini	avrg	maxi
88-Royale	base	4dr.sdn.	V6/3.8/A4	23 600	25 410	27 220
	LS	4dr.sdn.	V6/3.8/A4	26 450	28 480	30 510
	LSS	4dr.sdn.	V6/3.8/A4	27 565	29 680	31 795

• **Destination charge: 870 $**

Model	Trim	Type	Engine	mini	avrg	maxi
Achieva	S	2dr.sdn.	L4/2.3/M5	16 025	16 960	17 895
	SC	2dr.sdn.	L4/2.3/M5	18 680	19 770	20 860
	S	4dr.sdn.	L4/2.3/M5	16 005	16 940	17 870
	SL	4dr.sdn.	L4/2.3/A4	18 495	19 570	20 650

• **Destination charge: 595 $**

Model	Trim	Type	Engine	mini	avrg	maxi
Aurora	sedan	4dr.sdn.	V8/4.0/A4	35 915	38 715	41 520

• **Destination charge: 870 $**

PLYMOUTH

Model	Trim	Type	Engine	mini	avrg	maxi
Acclaim		4dr.sdn.	L4/2.5/M5	15 080	15 750	16 425

• **Destination & handling charge: 625 $**

Model	Trim	Type	Engine	mini	avrg	maxi
Colt	base	2dr.cpe.	L4/1.5/M5	11 225	11 610	11 995
	GL	2dr.cpe.	L4/1.5/M5	11 970	12 425	12 885

• **Destination & handling charge: 415 $**

*1994 Price

Model	Trim	Type	Engine	Price canadian $ mini	avrg	maxi
Neon	Highline	2dr.cpe.	L4/2.0/M5	12 245	12 700	13 150
	Sport	2dr.cpe.	L4/2.0/M5	14 705	15 375	16 045
	Base	4dr.sdn.	L4/2.0/M5	10 875	11 255	11 640
	Highline	4dr.sdn.	L4/2.0/M5	12 245	12 700	13 150
	Sport	4dr.sdn.	L4/2.0/M5	14 405	15 160	15 280

• **Destination & handling charge: 600 $**

Model	Trim	Type	Engine	mini	avrg	maxi
Voyager	base	4dr.wgn.	L4/2.5/M5	16 055	17 030	18 010
	SE	4dr.wgn.	L4/2.5/M5	17 700	18 805	19 910
	LE	4dr.wgn.	V6/3.0/A3	22 250	23 695	25 140

• **Destination & handling charge: 810 $**

Model	Trim	Type	Engine	mini	avrg	maxi
Grand Voyager						
	base	4dr.wgn.	V6/3.3/A4	18 895	20 090	21 290
	SE	4dr.wgn.	V6/3.3/A4	19 370	20 600	21 835
	SE 4x4	4dr.wgn.	V6/3.3/A4	23 780	25 340	26 900
	LE	4dr.wgn.	V6/3.3/A4	23 340	24 865	26 390
	LE 4x4	4dr.wgn.	V6/3.3/A4	25880	27 600	29 320

• **Destination & handling charge: 810 $**

PONTIAC

Model	Trim	Type	Engine	mini	avrg	maxi
Bonneville	SE	4dr.sdn.	V6/3.8/A4	23 325	25 115	26 905
	SSE	4dr.sdn.	V6/3.8/A4	27 735	29 860	32 160
	SSEI	4dr.sdn.	V6C/3.8/A4	29 005	32 230	33 455

• **Destination charge: 870 $**

Model	Trim	Type	Engine	mini	avrg	maxi
Firebird	base	3dr.cpe.	V6/3.4/M5	17 105	18 080	19 056
	base	3dr.con.	V6/3.4/M5	24 800	26 210	27 625
	Formula	3dr.cpe.	V8/5.7/M6	21 705	22 940	24 180
	Formula	3dr.con.	V8/5.7/A4	28 365	29 980	31 600
Trans am	GTA	2dr.cpe.	V8/5.7/M6	23 330	24 660	25 990
	GTA	3dr.con.	V8/5.7/A4	30 275	32 000	33 730

• **Destination charge: 655 $**

Model	Trim	Type	Engine	mini	avrg	maxi
Firefly	base	3dr.cpe.	L3/1.0/M5	9 255	9 625	9 995
	SE	3dr.cpe.	L3/1.0/M5	10 185	10 590	10 995
	base	4dr.sdn.	L4/1.3/M5	9 490	9 870	10 245
	SE	4dr.sdn.	L4/1.3/M5	10 415	10 830	11 245

• **Destination charge: 500 $**

Model	Trim	Type	Engine	mini	avrg	maxi
Grand Am	SE	2dr.cpe.	L4/2.3/M5	15 025	15 900	16 775
	SE	4dr.sdn.	L4/2.3/M5	15 115	15 995	16 875
	GT	2dr.cpe.	L4/2.3/M5	17 425	18 440	19 460
	GT	4dr.sdn.	L4/2.3/M5	17 520	18 540	19 565

• **Destination charge: 595 $**

Model	Trim	Type	Engine	mini	avrg	maxi
Grand-Prix	SE	2dr.cpe.	V6/3.1/A3	20 375	21 940	23 500
	SE	4dr.sdn.	V6/3.1/A3	19 485	20 980	22 475
	GT	4dr.sdn.	V6/3.4/M5	21 870	23 545	25 225
	GTP	2dr.cpe.	V6/3.4/M5	22 320	24 030	25 745

• **Destination charge: 760 $**

Model	Trim	Type	Engine	mini	avrg	maxi
Sunfire		2dr.cpe.	L4/2.0/M5	11 385	11 840	12 295
		4dr.sdn.	L4/2.0/M5	11 850	12 320	12 795
		2dr.con.	L4/2.0/M5	ND	ND	ND

• **Destination charge: 595 $**

Model	Trim	Type	Engine	mini	avrg	maxi
Sunrunner						
2WD Soft top		2dr.con.	L4/1.6/M5	12 085	12 550	13 020
4WD GT hard top		2dr.con.	L4/1.6/M5	15 670	16 275	16 880
4WD Soft top		2dr.con.	L4/1.6/M5	13 880	14 415	14 955
4WD Hard top		2dr.con.	L4/1.6/M5	13 780	14 310	14 845
4WD GT soft top		2dr.con.	L4/1.6/M5	15 770	16 380	16 990

• **Destination charge: 500 $**

Model	Trim	Type	Engine	mini	avrg	maxi
Trans Sport	SE	5dr.wgn.	V6/3.1/A3	19 590	21 090	22 595

• **Destination charge: 745 $**

PORSCHE

Model	Trim	Type	Engine	mini	avrg	maxi
911 Carrera 2						
		2dr.cpe.	H6/3.0/M5	84 390	90 145	95 900
		2dr.cpe.	H6/3.6/M6	78 850	84 225	89 600
		2dr.con.	H6/3.6/M6	89 320	95 410	101 500
RS america		2dr.cpe.	H6/3.6/M5	69 435	74 165	78 900
	Turbo	2dr.cpe.	H6/3.6/M5	128 655	137 425	146 200

• **Destination & handling charge: 1050 $**

Model	Trim	Type	Engine	mini	avrg	maxi
928	GTS	2dr.cpe.	V8/5.4/M5	101 815*	108755*	115 700*

• **Destination & handling charge: 1050 $**

Model	Trim	Type	Engine	mini	avrg	maxi
968		2dr.cpe.	L4/3.0M6	53 910*	56 905*	59 900*
		2dr.con.	L4/3.0/M6	68 940*	73 670*	76 600*

• **Destination & handling charge: 1050 $**

SAAB

Model	Trim	Type	Engine	mini	avrg	maxi
900 S		3dr.sdn.	L4/2.3 /M5	25 055	26 175	27 295
		5dr.sdn.	L4/2.3 /M5	24 780	25 890	26 995
		3dr.con.	L4T/2.0 /M5	39 715	41 855	43 995
900 SE		3dr.sdn.	L4T/2.0 /M5	31 770	33 380	34 995
		5dr.sdn.	V6/2.5/M5	31 315	32 905	34 495
		3dr.con.	L4T/2.0 /M5	46 935	48 965	50 995

• **Destination & handling charge: 845 $**

Model	Trim	Type	Engine	mini	avrg	maxi
9000	Aero	4dr.sdn.	L4T/2.3/M5	42 925	45 460	47 995
	CDE	4dr.sdn.	V6/3.0/M5	40 345	42 380	45 995
	CS	4dr.sdn.	L4T/2.3/M5	31 575	33 785	35 995
	CSE	4dr.sdn.	L4T/2.3/A4	40 565	43 405	46 245

• **Destination & handling charge: 845 $**

SATURN

Model	Trim	Type	Engine	mini	avrg	maxi
Saturn	SL	4dr.sdn.	L4/1.9/M5	10 950	11 470	11 995
	SL1	4dr.sdn.	L4/1.9/M5	12 230	12 810	13 395
	SL2	4dr.sdn.	L4/1.9/M5	13 780	14 435	15 095
	SC 1	2dr.cpe.	L4/1.9/M5	13 180	13 805	14 435
	SC 2	2dr.cpe.	L4/1.9/M5	15 680	16 425	17 175
	SW1	4dr.wgn.	L4/1.9/M5	12 190	13 040	13 895
	SW2	4dr.wgn.	L4/1.9/M5	14 365	15 050	15 735

• **Destination & handling charge: 400 $**

SUBARU

Model	Trim	Type	Engine	mini	avrg	maxi
Impreza	4x2					
	L	2dr.cpe.	H4/1.8/M5	16 555	17 275	17 995
	L	2dr.cpe.	H4/1.8/A4	17 385	18 140	18 895
Impreza	4x4					
	L	2dr.cpe.	H4/1.8/M5	18 120	18 905	19 695
	L	2dr.cpe.	H4/1.8/A4	18 845	19 720	20 595
	L	5dr.wgn.	H4/1.8/M5	14 991*	15 643*	16 295*
	LS	5dr.wgn.	H4/1.8/A4	19 684*	20 540*	21 395*
	L	4dr.sdn.	H4/1.8/M5	14 439*	15 067*	15 695*
	LS	4dr.sdn.	H4/1.8/A4	19 135*	19 965*	20 795*

• **Destination & handling charge: 695 $**

Model	Trim	Type	Engine	mini	avrg	maxi
Justy 4x4						
	GL	5dr.sdn.	L3/1.2 /M5	11 495	11 995	12 495

• **Destination & handling charge: 695 $**

*1994 Price

1995 CANADIAN PRICES

Model	Trim	Type	Engine	Price canadian $ mini	avrg	maxi
Legacy 4x2						
		4dr.sdn.	H4/2.2/M5	16 555	17 275	17 995
	L	4dr.sdn.	H4/2.2/M5	20 695	21 845	22 995
	L	5dr.wgn.	H4/2.2/A4	21 415	22 605	23 795
Legacy 4x4						
	L	4dr.sdn.	H4/2.2/M5	22 225	23 460	24 695
	LS	4dr.sdn.	H4/2.2/M5	25 375	26 785	28 195
	LSi	4dr.sdn.	H4/2.2/M5	26 395	28 195	29 995
	brighton	5dr.wgn.	H4/2.2/M5	18 395	19 195	19 995
	LS	5dr.wgn.	H4/2.2/M5	25 515	27 255	28 995
	LSi	5dr.wgn.	H4/2.2/M5	27 100	28 945	30 795

• **Destination & handling charge**: **695 $**

Model	Trim	Type	Engine	mini	avrg	maxi
SVX						
		3dr.cpe.	H6/3.3/A4	39 145	42 070	44 995

• **Destination & handling charge**: **695 $**

SUZUKI

Model	Trim	Type	Engine	mini	avrg	maxi
Swift	Hbk	3dr.sdn.	L4/1.3/M5	9 730	10 110	10 495
	DLX	3dr.sdn.	L4/1.3/M5	10 375	10 785	11 195

• **Destination & handling charge**: **300 $**

Model	Trim	Type	Engine	mini	avrg	maxi
Sidekick	Soft top					
	JA	2dr.con.	L4/1.6/M5	13 460	13 975	14 495
	JX	2dr.con.	L4/1.6/M5	15 780	16 390	16 995
	JLX	2dr.con.	L4/1.6/M5	17 635	18 315	18 995
Sidekick 4 door						
	JX	4dr.wgn.	L4/1.6/M5	16 245	16 870	17 495
	JLX	4dr.wgn.	L4/1.6/M5	19 400	20 145	20 895

• **Destination & handling charge**: **400 $**

TOYOTA

Model	Trim	Type	Engine	mini	avrg	maxi
Pick-ups 4x2						
	T-100	2dr.p-u.	V6/3.0/M5	19 070	19 800	20 528
Pick-ups 4x4						
	T-100	2dr.p-u.	V6/3.0/M5	20 370	21 150	21 928

• **Destination & handling charge**: **695 $**

Model	Trim	Type	Engine	mini	avrg	maxi
Pick-ups 4x2						
	reg	2dr.p-u.	L4/2.4/M5	12 360	12 715	13 068
	Xtra cab	2dr.p-u.	L4/2.4/M5	14 915	15 305	16 278
Pick-ups 4x4						
	reg	2dr.p-u.	L4/2.4/M5	19 505	20 525	21 548
	Xtra cab	2dr.p-u.	V6/3.0/M5	20 380	21 450	22 518
	SR5 V6	2dr.p-u.	V6/3.0/M5	23 250	24 470	25 688
Avalon	XL	4dr.sdn.	V6/3.0/A4	28 110	29 584	31 058
	XLS	4dr.sdn.	V6/3.0/A4	31 190	32 825	34 458

• **Destination & handling charge**: **695 $**

Model	Trim	Type	Engine	mini	avrg	maxi
Camry	base	2dr.cpe.	L4/2.2/M5	18 225	19 010	19 798
	LE	2dr.cpe.	L4/2.2/A4	21 635	22 770	23 908
	LE V6	2dr.cpe.	V6/3.0/A4	25 415	26 765	28 118
	base	4dr.sdn.	L4/2.2/M5	18 815	19 625	20 438
	LE	4dr.wgn.	L4/2.2/A4	25 030	26 345	27 658
	LE	4dr.sdn.	L4/2.2/M5	22 150	23 315	24 478
	V6	4dr.sdn.	V6/3.0/A4	22 105	23 095	24 088
	LE V6	4dr.sdn.	V6/3.0/A4	25 975	27 360	28 748
	LE V6	4dr.wgn.	V6/3.0/A4	27 990	29 485	30 978

• **Destination & handling charge**: **695 $**

Model	Trim	Type	Engine	mini	avrg	maxi
Celica	base	2dr.cpe.	L4/1.6/M5	21 020	21 990	22 958
	GT	3dr.cpe.	L4/2.2/M5	21 865	23 010	24 158
	GT-S	3dr.cpe.	L4/2.2/M5	26 315	27 930	29 548

• **Destination & handling charge**: **695 $**

Model	Trim	Type	Engine	mini	avrg	maxi
Corolla	base	4dr.sdn.	L4/1.6/M5	12 240	12 615	12 988
	DX	4dr.sdn.	L4/1.6/M5	14 475	14 975	15 478
	LE	4dr.sdn.	L4/1.6/M5	15 700	16 335	16 968
	base	5dr.wgn.	L4/1.6/M5	17 045	17 835	18 628

• **Destination & handling charge**: **695 $**

Model	Trim	Type	Engine	mini	avrg	maxi
Paseo		2dr.cpe.	L4/1.5/M5	14 940	15 585	16 228
		2dr.cpe.	L4/1.5/A4	15 815	16 495	17 178

• **Destination & handling charge**: **695 $**

Model	Trim	Type	Engine	mini	avrg	maxi
4Runner	SR5	4dr.wgn.	L4/2.4/M5	22 765	23 745	24 728
4Runner	SR5 V6	4dr.wgn.	V6/3.0/M5	26 025	27 305	28 588

• **Destination & handling charge**: **695 $**

Model	Trim	Type	Engine	mini	avrg	maxi
Previa 4X2						
	base	4dr.wgn.	L4/2.4/A4	25 070	26 275	27 478
	LE	4dr.wgn.	L4/2.4/A4	29 005	30 395	31 788
Previa 4X4						
	LE	4dr.wgn.	L4C/2.4/A4	35 145	36 830	38 518

• **Destination & handling charge**: **695 $**

Model	Trim	Type	Engine	mini	avrg	maxi
Supra		2dr.cpe.	V6T/3.0/M6	61 520	64 875	68 228
		2dr.cpe.	V6T/3.0/A4	60 260	63 545	66 828

• **Destination & handling charge**: **695 $**

Model	Trim	Type	Engine	mini	avrg	maxi
Tercel	S	2dr.sdn.	L4/1.5/M4	10 470	10 735	10 998
	Dx	2dr.sdn.	L4/1.5/M5	11 275	11 590	11 908
	Dx	4dr.sdn.	L4/1.5/M5	11 470	11 905	12 118

• **Destination & handling charge**: **695 $**

VOLKSWAGEN

Model	Trim	Type	Engine	mini	avrg	maxi
Corrado	VR-6	2dr.cpe.	V6/2.8/M5	28 515	29 755	30 995

• **Destination & handling charge**: **675 $**

Model	Trim	Type	Engine	mini	avrg	maxi
Eurovan	Cargo	4dr.wgn.	L5/2.4D/M5	19 525	20 260	20 995
	Cargo	4dr.wgn.	L5/2.5/A4	20 455	21 225	21 995
	GLS	4dr.wgn.	L5/2.5/M5	26 500	27 495	28 495
	GLS	4dr.wgn.	L5/2.4D/M5	26 500	27 495	28 495
	Camper	4dr.wgn.	L5/2.4D/M5	32 195	33 595	34 995
	Camper	4dr.wgn.	L5/2.5/M5	34 035	35 515	36 995

• **Destination & handling charge**: **675 $**

Model	Trim	Type	Engine	mini	avrg	maxi
Golf	CL	3dr.sdn.	L4/1.8/M5	12 820	13 155	13 495
	CL	5dr.sdn.	L4/1.9TD/M5	15 035	15 515	15 995
	GL	5dr.sdn.	L4/2.0/M5	15 035	15 515	15 995
	GTI	3dr.sdn.	L4/2.0/M5	17 200	17 845	18 495
	GTI	3dr.sdn.	V6/2.8/M5	22 875	24 435	25 995
	Cabrio.	2dr.sdn.	L4/2.0/M5	23 050	24 770	6 495

• **Destination & handling charge**: **675 $**

Model	Trim	Type	Engine	mini	avrg	maxi
Jetta	GL	4dr.sdn.	L4/2.0/M5	17 475	18 235	18 995
	GL	4dr.sdn.	L4/1.9TD/M5	17 475	18 235	18 995
	GLX	4dr.sdn.	V6/2.8/M5	24 355	26 175	27 995

• **Destination & handling charge**: **675 $**

Model	Trim	Type	Engine	mini	avrg	maxi
Passat	GLS	4dr.sdn.	L4/1.9TD/M5	22 970	24 730	26 495
	GLS	5dr.wgn.	L4/1.9TD/M5	23 840	25 665	27 495
	GLX	4dr.sdn.	V6/2.8/M5	25 570	27 530	29 495
	GLX	5dr.wgn.	V6/2.8/M5	26 440	28 715	30 495

• **Destination & handling charge**: **675 $**

VOLVO

Model	Trim	Type	Engine	Price canadian $ mini	avrg	maxi
850 sdn.	O	4dr.sdn.	L5/2.4/M5	26 185	27 140	29 095
	A	4dr.sdn.	L5/2.4/A4	26 995	28 495	29 995
	OS	4dr.sdn.	L5/2.4/M5	26 730	28 715	29 700
	AS	4dr.sdn.	L5/2.4/A4	27 235	28 915	30 600
	GTOS	4dr.sdn.	L5/2.4/M5	31 060	33 175	35 295
	GTAS	4dr.sdn.	L5/2.4/A4	31 850	34 025	36 195
	To	4dr.sdn.	L5/2.3/M5	34 405	36 750	39 095
	T	4dr.sdn.	L5/2.3/A4	35 195	37 195	39 995
850 wgn.	O	5dr.wgn.	L5/2.4/M5	26 785	28 440	30 095
	A	5dr.wgn.	L5/2.4/A4	27 585	29 290	30 995
	OS	5dr.wgn.	L5/2.4/M5	27 325	29 010	30 700
	AS	5dr.wgn.	L5/2.4/A4	28 125	29 860	31 600
	GTOS	5dr.wgn.	L5/2.3/M5	31 940	34 115	36 295
	GTAS	5dr.wgn.	L5/2.3/A4	32 730	34 960	37 195

• **Destination & handling charge: 695 $**

Model	Trim	Type	Engine	mini	avrg	maxi
940 sedan		4dr.sdn.	L4/2.3/A4	24 835	24 970	27 595
	TS	4dr.sdn.	L4T/2.3/A4	27 365	29 055	30 750
940 wagon		5dr.wgn.	L4/2.3/A4	25 735	27 165	28 595
	TS	5dr.wgn.	L4T/2.3/A4	28 255	30 000	31 750

• **Destination & handling charge: 695 $**

Model	Trim	Type	Engine	mini	avrg	maxi
960 sedan		4dr.sdn.	L6/3.0/A4	35 665	38 330	40 995
960 wagon		5dr.wgn.	L6/3.0/A4	37 405	40 200	42 995

• **Destination & handling charge: 695 $**